Springer Natural Hazards

AF890674

The Springer Natural Hazards series seeks to publish a broad portfolio of scientific books, aiming at researchers, students, and everyone interested in Natural Hazard research. The series includes peer-reviewed monographs, edited volumes, textbooks, and conference proceedings. It covers all categories of hazards such as atmospheric/climatological/oceanographic hazards, storms, tsunamis, floods, avalanches, landslides, erosion, earthquakes, volcanoes, and welcomes book proposals on topics like risk assessment, risk management, and mitigation of hazards, and related subjects.

Sandipan Ghosh
Editor

The Himalaya Dilemma

Navigating Risk, Vulnerability, and Resilience in Geohazard-Prone Regions

Springer

Editor
Sandipan Ghosh
Department of Geography
Chandrapur College
Bardhaman, West Bengal, India

ISSN 2365-0656 ISSN 2365-0664 (electronic)
Springer Natural Hazards
ISBN 978-3-031-95082-7 ISBN 978-3-031-95083-4 (eBook)
https://doi.org/10.1007/978-3-031-95083-4

© The Editor(s) (if applicable) and The Author(s), under exclusive license to Springer Nature Switzerland AG 2025

This work is subject to copyright. All rights are solely and exclusively licensed by the Publisher, whether the whole or part of the material is concerned, specifically the rights of translation, reprinting, reuse of illustrations, recitation, broadcasting, reproduction on microfilms or in any other physical way, and transmission or information storage and retrieval, electronic adaptation, computer software, or by similar or dissimilar methodology now known or hereafter developed.
The use of general descriptive names, registered names, trademarks, service marks, etc. in this publication does not imply, even in the absence of a specific statement, that such names are exempt from the relevant protective laws and regulations and therefore free for general use.
The publisher, the authors and the editors are safe to assume that the advice and information in this book are believed to be true and accurate at the date of publication. Neither the publisher nor the authors or the editors give a warranty, expressed or implied, with respect to the material contained herein or for any errors or omissions that may have been made. The publisher remains neutral with regard to jurisdictional claims in published maps and institutional affiliations.

This Springer imprint is published by the registered company Springer Nature Switzerland AG
The registered company address is: Gewerbestrasse 11, 6330 Cham, Switzerland

If disposing of this product, please recycle the paper.

Lord Buddha Statue
Langza, Spiti,
Himachal Pradesh
Anupam Chakraborty

This book, "The Himalaya Dilemma," is dedicated the resilient people of the majestic Himalaya, who have lived with and adapted to the region's geohazards for centuries, and continue to inspire us with their courage, determination, and resourcefulness. This book honours the late Khadg Singh Valdiya (1937–2020), a well-known Indian geologist and an International Honorary Fellow of the Geological Society of America, who devoted his academic career to studying the environment and geology of the Himalaya. The book is also dedicated to the geoscientists, researchers, and practitioners who have devoted their careers to comprehending and reducing the effects of Himalayan geohazards and disasters. Their work has influenced and moulded the book's content.

May this book contribute to a deeper understanding of the complexities and challenges of the Himalaya, and inspire collective action to build a more resilient and sustainable future for this remarkable region.

Foreword

The Himalaya, a majestic and awe-inspiring mountain range, have been the cradle of civilization, culture, and biodiversity for centuries. As the highest mountain range on the Earth, they dominate the physical landscape and hold deep significance in the lives of millions of people who call this region home. Yet, beneath their towering peaks and serene beauty, the Himalaya is also deeply vulnerable to natural hazards. From landslides and avalanches to earthquakes and floods, the very forces that shape the Himalaya also pose risks that threaten the lives and livelihoods of those living in their shadow. In *The Himalaya Dilemma: Navigating Risk, Vulnerability, and Resilience in Geohazard-Prone Regions*, the authors delve into the complex interplay between the environment, human activity, and the myriad geohazards that define this stunning yet perilous region through 27 valuable chapters.

The sustainability of the Himalaya for future generations will depend on present investments; thus, many geoscientists and environmentalists are warning that urgent action against the impact of climate change is essential. This book is an important and timely addition to our awareness of the difficulties facing the Himalayan region, which is one of the planet's most geologically active and most impacted by global warming. Over the years, scientists, policymakers, and local communities have made significant strides in researching and addressing the risks associated with these natural disasters. However, despite ongoing efforts, the region continues to experience devastating events that wreak havoc on both infrastructure and human lives. The need for a comprehensive approach to managing the risks posed by these geohazards has never been more pressing.

In this context, *The Himalaya Dilemma* provides a much-needed framework for navigating the interconnected issues of risk, vulnerability, and resilience. The authors explore the multifaceted nature of these challenges, drawing on cutting-edge research from multiple disciplines, including geology, climatology, geography, sociology, and urban planning. Through an interdisciplinary lens, the book offers valuable insights into the mechanisms of geohazards in the region. It highlights the importance of understanding local contexts to develop effective risk mitigation strategies. Additionally, *The Himalaya Dilemma* makes a significant contribution by considering the human and cultural dimensions of disaster risk management. Natural hazards are not just scientific phenomena; they are events that have profound social, economic, and cultural consequences.

This collective approach is not only necessary for disaster response but also for long-term planning and sustainable development. In many parts of the Himalaya, development and risk reduction must go hand in hand. Infrastructure, urban planning, and environmental conservation must be carefully balanced to ensure that growth does not exacerbate the risks faced by vulnerable populations. *The Himalaya Dilemma* provides practical recommendations for policymakers, urban planners, and disaster managers on integrating geohazard risk reduction into development strategies.

Ultimately, *The Himalaya Dilemma* serves as a crucial call to action for all those concerned with the Himalayan's future. The book provides a blueprint for understanding the dynamics of geohazards in this unique and vital part of the world and offers actionable insights for navigating the challenges of risk, vulnerability, and resilience. It encourages readers to think critically about how we can safeguard the region's diverse populations and ecosystems in the face of a rapidly changing environment.

In reading this book, we are reminded that the Himalayan range is not just a distant and awe-inspiring mountain range; they are a vital, living system that affects the lives of millions. As we move forward in an era of unprecedented global change in the Anthropocene, it is our responsibility to protect and preserve these regions for future generations. *The Himalaya Dilemma* provides a powerful and insightful framework for doing just that.

Vineet Kumar Gahalaut, FASc, FNA, FNASc
Director, Wadia Institute of Himalayan Geology (WIHG)
Dehradun, Uttarakhand, India
https://www.wihg.res.in/

Preface

The Himalaya Mountain range, which forms a massive rampart on the northwest, north, and northeast sides of the Indian subcontinent, safely separates India from the rest of Eurasia. The main Himalaya is an arcuate region that trends from north-west to south-east and is 2400 km long and 300–400 km wide. The mountain ranges contain four lithostructurally distinct and physiographically contrasted terranes: the Tethys Himalaya, the Himachal (Lesser Himalaya), the Himadri (Great Himalaya), and the Siwalik. Nine of ten of the world's tallest peaks, including Mount Everest, are found in the Himalaya, the highest mountain range in the world. These mountains—known as the 'Third Pole'—are the source of some of Asia's largest rivers and have a role in controlling the temperature of our planet. Travellers and trekkers often describe the Himalaya as a breathtaking, challenging, and spiritual landscape, characterized by towering peaks, diverse ecosystems, and remote, often harsh, environments, offering a unique and transformative experience.

The Himalaya is portrayed as a holy, immovable, and unbeatable mountain range in ancient Indian texts such as the Vedas and Puranas. Mountain is frequently depicted as the home of gods, sages, and other celestial entities, and it is associated with spiritual meaning and life-changing stories. In the Rigveda, the Himalaya is called "*Himavanta*" (meaning "snow-abode"). The Himalaya is mentioned in the Puranas as the location of several austerities and transforming stories, as well as the residence of gods like *Shiva* and *Parvati*. In his *Kumarasambhava*, Kalidasa portrayed the Himalaya as a mountain with a deity-like spirit and spurs that plunge into the Eastern and Western oceans, serving as the earth's measuring rod.

The Himalaya is always a prime attention to the UNDP (United Nations Development Programme). The SECURE Himalaya project is a part of "Global Partnership on Wildlife Conservation and Crime Prevention for Sustainable Development" (Global Wildlife Program) funded by the Global Environment Facility (GEF). This project of UNDP promotes sustainable management of alpine pastures and forests in the high range Himalayan ecosystems to secure conservation of globally significant wildlife and ensuring livelihood of local indigenous human communities. This initiative supports the following SDGs (Suatnable Development Gaols): Goal 15—Life on Land, Goal 8—Decent Work and Economic Growth; and Goal 13—Climate Action. The Indian Himalayan Region (IHR) and the SDGs are related, as several of the SDGs specifically address the possibilities and difficulties that the region faces. Given its susceptibility to climate change, water shortage, and biodiversity loss, as well as its potential for sustainable development through resource management and responsible tourism, the IHR is a crucial region for the achievement of the SDGs.

The Himalaya, often referred to as the "roof of the world," stand as one of the most majestic yet geologically dynamic mountain ranges on Earth. While these towering peaks inspire awe and reverence, they also conceal an intricate web of environmental challenges, geological hazards, biodiversity loss, and socio-economic vulnerabilities. The delicate interplay between natural forces and human intervention in this fragile landscape has made disaster risk management and resilience-building crucial topics for scientific inquiry and policy formulation. This book, *The Himalaya Dilemma: Navigating Risk, Vulnerability, and Resilience in Geohazard-Prone Regions*, is a response to this crisis. The book is an endeavour to explore and understand the multifaceted risks that the Himalaya faces due to geohazards such as landslides, earthquakes, glacial lake outbursts, flash floods and climate change-induced instability. It brings together the collective knowledge and expertise of researchers, scientists, and practitioners from across India to provide a comprehensive understanding of the complex issues facing the Himalaya. The edited volume is a culmination of years of research, analysis, and reflection on the part of the authors, and is intended to contribute to the growing body of knowledge on the Himalaya and its challenges. This book is the result of extensive research, fieldwork, and interactions with geologists, climate scientists, geomorphologists, geographers, disaster management experts, policymakers, and, most importantly, the people who call the Himalaya their home. Their experiences, resilience, and wisdom have significantly enriched the discussions presented in this volume.

In recent decades, rapid urbanization, infrastructural expansion, and climate variability have intensified the vulnerabilities of the Himalayan region. While scientific advancements have improved hazard assessment and risk modelling, the region continues to grapple with catastrophic events that disrupt lives and livelihoods. The challenge lies not just in understanding these hazards but in integrating scientific knowledge with local wisdom to create effective mitigation strategies. This book is motivated by the urgent need to bridge the gap between research, policy, and community action. By bringing together perspectives from geology, environmental science, disaster management, and sustainable development, it provides a comprehensive overview of the risk landscape of the Himalayas. It also highlights success stories of resilience and adaptation, offering valuable lessons for researchers, decision-makers, and those living in geohazard-prone regions.

The Himalaya is a unique and fragile ecosystem, with a rich cultural and biological diversity. The region is home to over 10,000 plant species, 800 bird species, and 300 mammal species, many of which are found nowhere else in the world. However, the

Himalayan belts are also one of the most hazard-prone regions in the world, with a high incidence of earthquakes, landslides, floods, and forest fires. The region's geology, topography, and climate make it inherently vulnerable to these hazards, which are being exacerbated by climate change.

The impact of geohazards on the Himalaya is devastating. Landslides and floods have become a regular occurrence, causing widespread damage to infrastructure, agriculture, and human settlements. The 2013 Kedarnath floods in Uttarakhand, India, which killed over 5000 people, and the 2015 Nepal earthquake, which killed over 8000 people, are stark reminders of the region's vulnerability to geohazards. The economic costs of these disasters are also significant, with the 2015 Nepal earthquake estimated to have caused over $10 billion in damages.

Notwithstanding these difficulties, the various Himalayan areas exhibit remarkable adaptability and immense resilience. The people of the Himalaya have lived with and adapted to the region's geohazards for centuries, developing unique coping mechanisms and strategies to mitigate the impacts of these hazards. However, the scale and frequency of geohazards in the Himalayas are increasing, and the region's ability to adapt to these changes is being severely tested.

This book is divided into six parts (Part I–VI), each of which focuses on a specific aspect of the Himalaya dilemma. As an introductory note, the first chapter looks at how the region's ecosystems, infrastructure, and human populations are affected by seismic risk, climate change, flood hazard, changes in land use and land cover, and infrastructural development. The first part provides an overview of the Himalaya and its challenges, including the region's geology, active tectonics and seismic risk. The second part examines the factors and impacts of landslides in the different regions of Himalaya, including risk assessment and advanced modelling of susceptibility. The third part explores the climatic dimensions of geohazards, including the impacts on the Himalayan glaciers, vegetation, agriculture, and infrastructure. The fourth part discusses the various aspects of hydro-climatic hazards, like glacial lake outburst floods and urban floods. The fifth part provides a synthesis on anthropogenic perspective of hazards, emphasising the role of human activity in modifying the natural landscapes. The last part incorporates significant environmental challenges faced by the different regions of the Himalaya, including crisis and sustainability in the management of water, soil and forest resources.

The book is intended for a wide range of audiences, including researchers, scientists, policymakers, practitioners, and students. It is hoped that the book will contribute to a deeper understanding of the complex issues facing the Himalaya, and inspire collective action to build a more resilient and sustainable future for this remarkable region. By synthesizing complex scientific concepts into accessible narratives, the book aims to engage both technical and non-technical readers in meaningful discussions about the risks and opportunities that define the Himalaya.

As we stand at the crossroads of environmental change, population pressures, and evolving geohazards, the need for interdisciplinary research and proactive risk management has never been greater. The Himalaya will continue to be a landscape of both challenge and opportunity, demanding constant adaptation and resilience-building.

It is my hope that *The Himalaya Dilemma* serves as a valuable resource for those seeking to understand and navigate the uncertainties of this region. More importantly, I hope it sparks dialogue, innovation, and collective action toward a safer, more sustainable future for the communities living in the lap of the Himalaya.

Bardhaman, West Bengal, India

Dr. Sandipan Ghosh

Acknowledgments

I am deeply grateful to the individuals of Springer Nature that have supported to the creation of this edited book. Their support, guidance, and expertise have been invaluable in helping us to bring this project to fruition. I am indebted to Prof. Simon Shah-Rohlfs (Editor in Earth Science, Geography and Environment—Springer Spektrum and Springer Nature) for this book project.

I am deeply grateful to the esteemed authors who have contributed their expertise and knowledge to this book through their chapters. Their contributions have enriched our understanding of the Himalayan hazards and its complex challenges. We appreciate the time, effort, and expertise that each of these authors has brought to their chapters, and I am grateful for their contributions to this book.

First and foremost, I would like to express our sincere gratitude to the people of the Himalaya, who have shared their stories, wisdom, and experiences with me. Your resilience, adaptability, and generosity of spirit are an inspiration to all.

I am deeply grateful to the talented photographers of India who have contributed their stunning images to this book. Their photographs have brought the beauty, majesty, and diversity of the Himalaya to life, and have greatly enhanced the reader's experience. I would like to extend our sincere thanks to the following photographers: (a) Anupam Chakraborty: For his breathtaking image of the Langza Budha Statue of Ladakh Himalayan landscape, and culture; (b) Nillakkhya Dutta: For his captivating images of the Himalayan mountains, valleys, glaciers and villages. (c) Puskar Basu: For his stunning images of the Himalayan wildlife, mainly snow leopard, Himalayan brown bear, wild yak and Himalayan wolf; and (d) Kallol Mukherjee: for his exceptional photograph of Kangchenjunga and red panda.

Finally, I would like to acknowledge the beauty, majesty, and power of the Himalaya, which have inspired and awed us throughout this project. May this book contribute to a deeper understanding and appreciation of this incredible region and its people.

Contents

Landslide Susceptibility and Hazard Assessment

Climate Change and Fragile Environment

Editor and Contributors

About the Editor

Dr. Sandipan Ghosh is an applied geomorphologist with a postgraduate M.Phil. and Ph.D. in Geography from the University of Burdwan (West Bengal, India). His main research interests include the Quaternary environment, tectonic geomorphology, flood hydrology, and fluvial geomorphology. More than forty book chapters and national and international research articles in numerous esteemed geography and geosciences publications have been published by him. He is the author of two books: (1) *Laterites of the Bengal Basin: Characterization, Geochronology, and Evolution* (ISBN 978-3-030-22937-5) and (2) *Key Aspects of Himalayan Geology* (978-3-031-81186-9). He is also one of the editors of six books: (1) *Neo-Thinking on Ganges-Brahmaputra Basin Geomorphology* (ISBN 978-3-319-26442-4), (2) *Quaternary Geomorphology in India: Case Studies from the Lower Ganga Basin* (ISBN 978-3-319-90426-9), (3) *Anthropogeomorphology of Bhagirathi-Hooghly River System in India* (ISBN 978-0-367-55727-0), (4) *Drainage Basin Dynamics* (ISBN 978-3-030-79634-1), (5) *Fluvial Systems in the Anthropocene* (ISBN 978-3-031-11180-8), and (6) *Floods in the Ganga–Brahmaputra–Meghna Delta* (ISBN 978-3-031-21085-3). He has also served as an editor for the Journal of Geography and Cartography and the Asian Journal of Spatial Science.

Contributors

Shukla Acharjee Centre for Studies in Geography, Dibrugarh University, Dibrugarh, Assam, India

Moumita Akuria Department of Earth Sciences, Indian Institute of Technology Kanpur, Kanpur, India

Bikram Singh Bali Department of Earth Sciences, University of Kashmir, Srinagar, Jammu and Kashmir, India

Shreya Basak Department of Geography, University of Calcutta, Kolkata, West Bengal, India

Naseer Ahmad Bhat Department of Earth Sciences, University of Kashmir, Srinagar, Jammu and Kashmir, India

Gouri Sankar Bhunia Department of Geography and Applied Geography, University of North Bengal, Darjiling, India

Pankaj Chakraborty 21st Century GIS Academy, Kolkata, West Bengal, India

Rachita Chakraborty Department of Geography, Vidyasagar College, Kolkata, India

Monisha Chetia Department of Applied Geology, Dibrugarh University, Dibrugarh, Assam, India;
Department of Physics, Nanda Nath Saikia College, Titabar, Assam, India

Anup Das North Eastern Space Applications Centre (NESAC), Umiam, Meghalaya, India

Balai Chandra Das Department of Geography, Barasat Government College, Kolkata, West Bengal, India

Sunil Kumar De Department of Geography, North-Eastern Hill University (NEHU), Shillong, Meghalaya, India

Sreedeep Dey Department of Chemical Engineering, University of Calcutta, Kolkata, India

Kuldeep Dutta Centre for Earth Sciences and Himalayan Studies, Itanagar, Arunachal Pradesh, India;
Department of Geology, Physical Science Block, Sikkim University, Gangtok, Sikkim, India

Abhishek Ghosh Department of Geography, North-Eastern Hill University (NEHU), Shillong, Meghalaya, India

Arnab Ghosh 21st Century GIS Academy, Kolkata, West Bengal, India

Sandipan Ghosh Department of Geography, Chandrapur College, Chandrapur, India

Manisha Giri North Eastern Space Applications Centre (NESAC), Umiam, Meghalaya, India

Olivia Halder Department of Geography, University of Calcutta, Calcutta, West Bengal, India

Nabajit Hazarika Department of Environmental Science, Nagaland University, Nagaland, India;
Department of Environmental Biology and Wildlife Sciences, Cotton University, Guwahati, Assam, India

Mohammad Irfan Department of Earth Sciences, University of Kashmir, Srinagar, Jammu and Kashmir, India

Kabirul Islam Karajgram High School, Purba Bardhaman, West Bengal, India

Latonglila Jamir Department of Environmental Science, Nagaland University, Nagaland, India

Amit Kumar Jana Department of Geography and Applied Geography, University of North Bengal, Darjiling, India

Kshitiz Kandel University of Chinese Academy of Sciences, Beijing, China;
Research Insight-Nepal, Bharatpur, Chitwan, Nepal;
Kathmandu University, Dhulikhel, Nepal

Kaikho Khusulio Lovely Professional University, Phagwara, Punjab, India

Atul Kumar CEN Center for Earth System Research and Sustainability, Institute of Geography, University of Hamburg, Hamburg, Germany

Rohan Kumar Center for Space Research, Lovely Professional University, Jalandhar, Punjab, India

Prabir Kumar Kundu Department of Geography, Siliguri Mahila Mahavidyalaya, Siliguri, Darjiling, West Bengal, India

Surajit Kundu Department of Geography, Trivenidevi Bhalotia College, Raniganj, West Bengal, India

Siddhartha Kumar Lahiri Department of Applied Geology, Dibrugarh University, Dibrugarh, Assam, India

Sonam Lama Department of Geography, Darjiling Government College, Darjiling, West Bengal, India

A. K. Mahajan Department of Geology, Central University of Himachal Pradesh, Shahpur, Himachal Pradesh, India

Suraj Mal Centre for the Study of Regional Development, School of Social Sciences, Jawaharlal Nehru University, New Delhi, India

Kanchan Mishra Department of Geosciences, University of Tübingen, Tübingen, Germany

Shreeansh Mishra Department of Tourism and Hospitality Management, Central University of Tamil Nadu, Thiruvarur, India

Anil Kumar Misra Department of Geology, Physical Science Block, Sikkim University, Gangtok, Sikkim, India

Suranjita Mohapatra Amity Institute of Geoinformatics and Remote Sensing, Amity University, Noida, Uttar Pradesh, India

Malay Mukul Continental Deformation Laboratory, Department of Earth Sciences, IIT Bombay, Mumbai, India

Manas Mukul School of Computer Applications, Kalinga Institute of Industrial Technology, Bhubaneshwar, India

Chinmaya Mund Amity Institute of Geoinformatics and Remote Sensing, Amity University, Noida, Uttar Pradesh, India

Biswajit Nath Lab of Geoinformatics and Earth Observation Research (LGEOR), Department of Geography and Environmental Studies, Faculty of Biological Sciences, University of Chittagong, Chittagong, Bangladesh

Sohini Neogy Department of Geography, Banwarilal Bhalotia College, Asansol, India

Aastha Pandey University of Chinese Academy of Sciences, Beijing, China; Kathmandu University, Dhulikhel, Nepal

Rukumesh Paudyal University of Chinese Academy of Sciences, Beijing, China

Sarada Prasad Pradhan Department of Earth Sciences, Indian Institute of Technology Roorkee, Roorkee, India

Bakhat Rawat University of Chinese Academy of Sciences, Beijing, China;
Research Insight-Nepal, Bharatpur, Chitwan, Nepal;
Kathmandu University, Dhulikhel, Nepal

Sonam Rinzin Newcastle University, Newcastle Upon Tyne, UK

Vimha Ritse Department of Environmental Science, Nagaland University, Nagaland, India

Bikramjit Roy Department of Geography, Vidyasagar College, Kolkata, India

Arpita Saha Department of Geography, Chanchal College, Malda, West Bengal, India

Arup Kr. Saha Chandrapur College, Chandrapur, West Bengal, India

Ujwal Deep Saha Department of Geoenvironmental Research, Institute of Geography and Spatial Organization, Polish Academy of Sciences, Krakow, Poland

Arindam Sarkar Department of Geography, Purash Kanpur Haridas Nandi College (University of Calcutta), Kolkata, West Bengal, India

Dipro Sarkar Department of Earth Sciences, Indian Institute of Technology Kanpur, Kanpur, India

Taposhi Sarmah Department of Environmental Biology and Wildlife Sciences, Cotton University, Guwahati, Assam, India

Udo Schikchoff CEN Center for Earth System Research and Sustainability, Institute of Geography, University of Hamburg, Hamburg, Germany

Chhatra Mani Sharma Central Department of Environmental Science, Tribhuvan University, Kirtipur, Kathmandu, Nepal

Subodh Sharma Kathmandu University, Dhulikhel, Nepal

Swati Sharma Department of TransDisciplinary Research, Division of Research and Development, Lovely Professional University, Jalandhar, Punjab, India

Tariq Siddique Department of Geology, Aligarh Muslim University, Aligarh, India

Rajiv Sinha Department of Earth Sciences, Indian Institute of Technology Kanpur, Kanpur, India

Subhajit Sinha Department of Geology, University of Calcutta, Kolkata, India

Vinee Srivastava Department of Earth and Environmental Sciences, Indian Institute of Science Education and Research, Bhopal, India

Mandil Thapa Department of Humanities and Social Sciences, Tribhuvan University, Kirtipur, Nepal

Narayan Thapa Department of Geomatics Engineering, Kathmandu University, Dhulikhel, Nepal

Pawan Thapa Department of Geography and the Environment, University of Alabama, Tuscaloosa, AL, USA

Rutger Willem Vervoort The University of Sydney, Sydney, NSW, Australia

Ahsan Afzal Wani Department of Earth Sciences, University of Kashmir, Srinagar, Jammu and Kashmir, India

Nishchal Wanjari Department of Geology, Physical Science Block, Sikkim University, Gangtok, Sikkim, India

Living on the Edge of Fragile Majesty: An Introductory Note on Emerging Risks, Hazards and Disasters in the Himalaya

Sandipan Ghosh

Abstract The Himalaya, one of the most iconic and ecologically fragile mountain ranges in the world, is facing unprecedented threats from emerging risks, hazards, and disasters. This comprehensive synthesis reviews the current state of knowledge on the complex interplay between geological, hydrological, and atmospheric processes that underpin the region's vulnerability to natural hazards. This chapter examines the impacts of seismic risk, climate change, land use/land cover changes, and infrastructure development on the region's ecosystems, infrastructure, and human populations. The analysis reveals that the Himalaya is experiencing an increase in extreme weather events, such as cloudbursts, landslides, and floods, which are exacerbated by climate change. Several studies and researches also identify emerging risks, including earthquakes, glacial lake outburst floods, rockfalls, landslides and ecological risk, which pose significant threats to human settlements, infrastructure, and ecosystems. This synthesis highlights the urgent need for integrated risk management strategies, policy interventions, and community-based initiatives to mitigate the impacts of emerging risks, hazards, and disasters in the Himalaya. The findings have significant implications for sustainable development, disaster risk reduction, and environmental conservation in the region. This chapter serves as a precede to the investigation of the 'Himalaya Dilemma' in relation to risk, danger, and disaster by summarizing the next chapters.

1 Unwise to Neglect Himalayan Crisis

Known as the 'Abode of Snow,' the Himalaya is a breathtaking example of the majesty of the natural world. This amazing mountain range, which stretches over the whole of Asia, is a cultural and spiritual symbol in addition to being a natural wonder. We explore the Himalaya's beauty, significance, and many aspects in this article. Learn about the real wonders of the magnificent Himalaya. These mountains

S. Ghosh (✉)
Department of Geography, Chandrapur College, Bardhaman 713145, West Bengal, India
e-mail: sandipanghosh19@gmail.com

© The Author(s), under exclusive license to Springer Nature Switzerland AG 2025

S. Ghosh (ed.), *The Himalaya Dilemma*, Springer Natural Hazards,
https://doi.org/10.1007/978-3-031-95083-4_1

are more than simply peaks; they are lifelines, influencing weather patterns (regulating and stabilizing the climate), providing food and water security for millions, and serving as a cultural and spiritual symbol (Valdiya, 1998). The entire 2400 km length of the Himalaya is home to the main structural divisions, the lesser Himalaya fold-and-thrust belt, the Tethyan Himalaya sedimentary units, the Greater Himalaya Sequence (GHS) metamorphic rocks, the Indus–(Yarlung Tsangpo) suture zone, and the Sub-Himalaya Siwalik molasse basin (Searle & Treloar, 2019; Valdiya, 2016). The Himalaya extends across five countries—India, Nepal, Bhutan, China, and Pakistan. They are home to a diverse range of ecosystems and human populations that are increasingly vulnerable to natural hazards. The 'Third Pole' is currently very concerned about climate change. The Hindu Kush-Karakoram-Himalayan system (HKKH), referred to as the 'Third Pole', is a mountainous area that lies to the west and south of the Tibetan Plateau (Gao et al., 2019). This region is abundant in natural resources and includes all or parts of the four hotspots for biodiversity in the world. The region's 220 million residents rely on the mountain resources for a variety of ecosystem services, including drinking water, food production, and livelihoods. In addition, the 1.5 billion people who live in the downstream river basins, or one-sixth of the world's population, indirectly benefit from these resources. Data from the EM-DAT international disaster database of 2023 (https://www.emdat.be/) indicates that the 2,500-km-long Himalayan states and union territories—Ladakh, Jammu and Kashmir, Himachal Pradesh, Uttarakhand, Arunachal Pradesh, Manipur, Meghalaya, Mizoram, Nagaland, Sikkim, Tripura, and Assam—account for 18% of the nation's total land area, but have been the site of 35% of the major disasters over the past 110 years. At 68, the most recent decade (2013–2022) has the most catastrophes, accounting for 44% of all disasters recorded in India. Reminding these facts, this introductory chapter reviews key hazards affecting the Himalayan region, their causes, impacts, and mitigation measures.

Rising temperatures in the Himalayas contribute to glacier retreat, erratic monsoons, and shifting ecosystems. These changes intensify existing hazards and increase water scarcity, threatening millions who rely on Himalayan rivers. Integrated climate adaptation strategies are necessary. Changes in precipitation patterns have led to more extreme weather events, increasing the frequency of flash floods and droughts. Cross-border cooperation among Himalayan nations is critical for climate resilience. Alongside, the Himalayan orogenic belt is a seismically active region due to the ongoing collision between the Indian and Eurasian tectonic plates. Major earthquakes, such as the 2015 Gorkha earthquake in Nepal (M_w 7.8), have caused significant destruction. Seismic hazards are exacerbated by rapid urbanization and inadequate infrastructure. Historical earthquakes, including the 1934 Nepal-Bihar earthquake (M_w 8.1), highlight the recurring nature of seismic risks. Enhanced geotechnical studies and structural reinforcement are necessary to reduce vulnerability. The Himalaya faces multiple interconnected hazards that threaten billions of human lives and rich ecosystems. A multidisciplinary approach involving scientific research, policy action, and community participation is essential for reducing vulnerability, enhancing resilience and managing the dilemma of environmental risk and concern. Governments, researchers, and local communities must collaborate to implement sustainable mitigation strategies and adaptive policies that address the evolving risks in the Himalayan region.

The last few decades have recorded rapid and magnified natural catastrophes across the world, with scientists linking such extreme events with global warming. The Himalayas have recorded such calamities like earthquakes, avalanches, cloudbursts, intense rainfall landslide lake outburst floods (LLOFs are breaching of lakes formed by obstructions due to landslides), and glacial lake outburst floods (GLOFs are a sudden release of a significant amount of water retained in a glacial lake) in their most furious forms; ground studies indicate a warming climate as the cause. The warming climate has the maximum impact in the areas intermediate between glaciated and non-glaciated terrain—these are the areas that receive snowfall as well as rainfall. It is pertinent to mention that most of the disastrous catastrophes, including the Kedarnath floods of 2013, the Chamoli flash floods of 2021 and the subsidence in Joshimath, have occurred in such intermediary climatic zones. Such intermediary climatic zones, with picturesque glacial landscape and easy access offers a natural choice for tourism and subsequent developments. With the sudden change in topographic altitudes and availability of perennial sources of water, such places also offer the best suitable sites for the development of river valley projects. But while the entire Himalayan region is fragile due to its immature topography and tectonic settings under continued active erosional and weathering processes, the intermediary zones may be regarded as the most fragile and unstable belt on account of the interplay of glacial and fluvial (dominated by rivers and streams) processes.

2 Many-Splendoured Mountain—Himalaya

Before going into the main sections, it is needed to understand the natural beauty and uniqueness of the mighty Himalaya which is very dynamic and great diversity also. Kashmir, also referred to as "Paradise on Earth," is a well-liked travel destination because of its breathtaking scenery, which includes snow-capped mountains, verdant valleys, and glistening lakes (Fig. 1a). Home to rare creatures and a rich tapestry of flora and fauna, the Himalayas are a natural wonder with their towering snow-capped peaks, diversified ecosystems, and spectacular vistas (Fig. 1b). They provide a unique combination of raw beauty and quiet serenity (like Kanchenjunga national park). The tranquil beauty of the area is enhanced by the high-altitude lakes that mirror the nearby peaks. Renowned for its breathtaking natural beauty (Fig. 1c), Kashmir is home to a number of notable lakes, including as the well-known Dal Lake, Nigeen Lake, and the sizable freshwater Wular Lake, each of which offers distinctive scenery and experiences. The isolated Himalayan area of Zanskar Valley in Ladakh (Fig. 1d) is well known for its untamed natural beauty, dramatic, rocky terrain, and rich cultural legacy, all of which make it an alluring place for exploration and adventure.

Fig. 1 Breathtaking Himalaya a mesmerizing gift to India: **a** because of its magnificent unspoiled beauty, the Kashmir Himalaya is sometimes referred to as "Paradise on Earth"—abundance of verdant valleys, snow-capped mountains, stunning lakes, and idyllic scenery that conjures up images of heaven (photographed by Kallol Mukherjee); **b** the photograph shows the third-highest peak in the world, the Kanchenjunga, also known as "The Five Treasures of Snow," which is a component of the Great Himalaya Range and is in the eastern Himalayas on the boundary between Sikkim (India) and Nepal (photographed by Kallol Mukherjee); **c** trekkers on the Aharbal to Kousarnag trek frequently stop at Mahinag Lake, a charming campground tucked away in the Kashmir Valley that offers peaceful, pristine waters and breathtaking views of the neighbouring Pir Panjal ranges (photographed by Nilakkhya Dutta); and **d** Rangdum, a scenic landscape in the Suru Valley in the Tethys Himalayan (union territory of Ladakh), is surrounded by rocky mountains and glaciers, particularly Drang-drung, on one side and colourful hills on the other (photographed by Nilakkhya Dutta)

2.1 Restless Giant

Dwarfing other mountains of the world in seer height, the Himalaya is truly the *'Nagadhiraj'* (King of Mountains)—a colossus among the great mountains of the world (Valdiya, 1998). The very-snowy resplendent summits of the Great Himalaya, a broken by precipitous scarps, evoke admiration not unmixed with awe. Very young in age (~40–50 million years), the rugged Himalaya continues to be seized repeatedly with spells of tectonic restlessness or crustal disturbances. These disturbances indicate accumulating inner tectonic stress in the crust. The giant body of the mountain quivers in some segments or twitches spasmodically and breaks up violently in other places (Valdiya, 1998). These are the manifestations of the tectonic turmoil of growing, for the Himalaya is still growing. Fourteen peaks (Fig. 2) above 8000 m are recognized in the Himalaya—Karakoram range (Table 1). It is the titanic force of

Fig. 2 The pictorial illustration depicts Mount Everest (Sagarmatha or Qomolangma) in a holistic view: In October 2022, the International Union of Geological Sciences (IUGS) published a listing of 100 geological heritage sites worldwide, including the Ordovician Rocks of Mount Everest, which are recognized as the "highest rocks on the planet" due to their fossiliferous, marine limestone composition (collected from Siraj & Siraj, 2024)

compression resulting from the coming together and eventual collision of the continents of India and Asia that transformed the once enormous pile of layered rocks in to the Himalaya of immense splendour and magnificence. The ongoing convergence of these plates, with India moving northward into Asia, causes the Himalayas to rise by approximately 1 cm per year. This constant pressure and movement lead to the formation of active fault zones and frequent earthquakes.

2.2 *Controller of Climate*

Guarding the northern border like a mighty sentinel, the Himalaya has not only witnessed the unfolding history of evolution of the Indian civilization, but has also moulded destiny and lifestyles of its people (Valdiya, 1998). The 2400 km long and 250–300 km wide mountain barrier, which rises 500–8000 m above sea level, vitally affects the atmospheric circulation of the Asian continent including India and controls the climatic conditions here. The Himalaya is responsible for the peculiar

Table 1 Important peaks above 8000 m in the Himalaya

Sl. No	Name of peak	Height (m) from mean sea level	Mountain range
1	Mount Everest (Sagarmatha)	8,848.86	Mahalangur Himal range
2	K2 (Chogori)	8,611	Karakoram range
3	Kangchenjunga	8,586	Great Himalayan range
4	Lhotse	8,516	Mahalangur Himal range
5	Makalu	8,485	Mahalangur Himal range
6	Cho Oyu	8,201	Mahalangur Himal range
7	Dhaulagiri	8,167	Dhaulagiri Himal range
8	Manaslu	8,163	Mansiri Himal range
9	Nanga Parbat	8,126	Western Himalayan Range of Gilgit
10	Annapurna I	8,091	Nepal Himalya
11	Gasherbrum I (hidden peak)	8,068	Karakoram range
12	Broad peak	8,051	Karakoram range
13	Gasherbrum II	8,035	Karakoram range
14	Shishapangma	8,012	Central Himalayan range of Tibet

Source Ramanan and Ramanan (2023)

monsoon climate of India and its unique succession of six seasons. The Himalayas lead to distinct variations in rainfall, with the highest annual rainfall occurring on the southern slopes (e.g., Cherrapunji) and drier conditions in the rain shadow areas of the Tibetan Plateau. The Himalayas' uplift and the Tibetan Plateau's high elevation contribute to the strength of the Indian monsoon, creating a boundary between warm, moist air from India and cold, dry air from Tibet. The Himalayas act as a barrier, preventing cold continental air masses from Central Asia from reaching India in winter, moderating temperatures.

2.3 Bounty of the Himalaya

The Himalaya is often referred to as the 'Water Tower of Asia' because they are the source of major rivers like the Ganges, Indus, and Brahmaputra. Himalaya is the provider of life-giving water to the Indian sub-continent. Nearly 1,20,00,000 million m3 of water flowing down the Himalaya rivers annually sustain and nourish the teeming millions of the Indo-Gangetic Plains in which nearly 60% of India's population live in dense concentration (around 520 people per square kilometre). In addition, these waters have potential of generating 51,700–128,700 megawatts by the Ganga, around 60,000 megawatts by the Indus, and 40,500 megawatts by the Brahmaputra. Nearly, 525,020 million m^3 of water from the rivers of Ganga basin can be utilized for irrigation, whereas 24 billion m^3 of water of the Brahmaputra has irrigation potential for the people. The bounty of the Himalaya is also locked in the 1400 km^2 of snow and ice, spread over nearly 33,200 km^2 area in higher altitudes above the 4300–5800 m snowline (Valdiya, 1998). An ISRO (Indian Space Research Organization) study showed that there are 34,909 glaciers spread over 75,779 km2 in the Indus, Ganga and Brahmaputra basins, covering Great Himalaya, Trans-Himalaya and Karakoram region. The Himalaya is a part of 'Third Pole' which includes the Tibetan Plateau and its environs, which include the Himalayas in the south, the Pamir-Hindu Kush Mountain ranges in the west, the Hengduan mountains in the east, and the Tienshan and Qilian mountains in the north. It is socioeconomically significant, has the greatest ice mass outside of the polar regions, supplies freshwater resources to almost two billion people, controls the climate, and preserves wildlife. Third pole has major and largest Himalayan glaciers of the world: (a) Siachen (eastern Karakoram range), (b) Gangotri (Garhwal Himalaya), (c) Zemu (Eastern Himalaya), (d) Baltaro (Karakoram range), (e) Biafo Gyang (Karakoram range), (f) Machoi (Drass, Ladakh), (g) Pindari (Kumaon Himalaya), (h) Shafat (Ladakh), (i) Hispar (Karakoram range) and (j) Nubra (Karakoram range).

3 Climate Change and Deglaciation

The largest areas of land covered by glaciers and permafrost outside of the polar regions are found in the greater Himalayan region, also known as "the roof of the world" (Ericksson et al., 2009). It also has the most expansive and rocky high-altitude regions on Earth. The water resources from this area drain through ten of the largest rivers in Asia (Table 2), in the basins of which more than 1.3 billion people find their livelihoods. The area and its water resources are crucial for biodiversity, rainfed and irrigated agriculture, hydropower, global atmospheric circulation, and the manufacturing of goods that are sold in marketplaces across the world. Numerous factors are currently endangering this region's water resources. Since snow and glacier melting contribute up to 50% of the average annual flows in rivers, global warming is having a significant effect on the amount of snow and ice, which has significant short- and long-term effects on the availability of water downstream (Ericksson et al., 2009). Compared to the global average of 0.74 °C over the past century, the warming in the greater Himalayas has been significantly higher, with Nepal Himalaya experiencing 0.6 °C warming every ten years (Ericksson et al., 2009).

Particularly at high elevations (above 3000 m asl), a significant amount of the yearly precipitation falls as snow, which feeds the Himalayan glaciers. Outside of the polar regions, the highest concentrations of glaciers are found in the inner Asian and high Himalayan mountains (Ericksson et al., 2009). Over 112,000 km^2 of the broader Himalayan region are covered by glaciated areas (Table 3). With a total ice volume of about 3,420 km^3, the Himalayan range alone (a sub-region) has roughly

Table 2 Principal rivers of the Himalayan range—basin statistics

Sl. No.	River	River		River basin			
		Annual mean discharge (m^3/sec)	% of glacier melt in river flow	Basin area (km^2)	Population density (persons/ km^2)	Population × 1000	Water availability (m^3/ persons/ year)
1	Brahmaputra	21,261	~12	651,335	182	118,543	5656
2	Ganges	12,037	~9	1,016,124	401	407,466	932
3	Indus	5533	~50	1,081,718	145	178,483	978
4	Mekong	9001	~7	805,604	71	57,198	4963
5	Yangtze	28,811	~18	1,722,193	214	368,549	2465

Source Eriksson et al., (2009)

33,000 km^2 of glaciers, or 17% of the mountain area (compared to 2.2% in the Swiss Alps). These glaciers provide significant short- and long-term water storage facilities. In the west, glaciers experience winter buildup and summer ablation, while in the east, they primarily experience simultaneous summer accumulation and melt. The primary melting happens during the height of summer, but if it falls during the monsoon, it might not be as detrimental to the water supply as if it happened during the spring and fall shoulder seasons. The meltwater from snow and ice may prevent or mitigate severe drought when the monsoon is weak, delayed, or fails.

The Fourth Assessment Report of the IPCC (IPCC, 2007) states that there is a greater than 90% chance that human activity-induced greenhouse gas emissions are to blame for the warming that has been seen since the 1950s. Temperature projections for the twenty-first century (Fig. 3) indicate that warming has accelerated significantly compared to the twentieth century (Ruosteenoja et al., 2003; Ericksson et al., 2009; Shrestha et al., 2012). Every region of Asia is probably going to warm this century. Warming is strongest throughout South and Eastern Asia, slowest in Southeast Asia (like the global mean warming), and highest in the central, western, and northern regions of Asia's continent. Arid Asian regions and the Himalayan highlands, particularly the Tibetan Plateau, will experience substantial warming (Gao et al., 2003; Yao et al., 2006; Ericksson et al., 2009). Temperatures in the Indian subcontinent are expected to increase by 3.5–5.5 °C by 2100, whereas those on the Tibetan Plateau are expected to rise by 2.5 °C by 2050 and 5 °C by 2100, according

Table 3 Glaciated areas in Himalaya range

Sl. No.	River basin	No. of glaciers	Total area (km^2)	Total ice reserve (km^3)
1	Brahmaputra	4366	16,677	1971
2	Ganges	6694	6579	600
3	Indus	5057	8926	850

Source Eriksson et al., (2009)

to regional climate models (Rupa Kumar et al., 2006). Numerous studies indicate that during the past century, the Himalayas have warmed significantly more than the global average of 0.74 °C (IPCC, 2007). Nepal and the Tibetan Plateau have experienced warming that has increased steadily with height, indicating that this pattern of warming that increases with altitude is common throughout the wider Himalayan region. The average annual mean temperature over the Asian landmass, including the Himalaya (Fig. 2), is expected to rise by around 3 °C by the 2050s and by roughly 5 °C by the 2080s, according to Shrestha et al. (2012). By 2080, the region's average yearly precipitation will also rise by 10–30%.

Compared to the global average, the Himalayan glaciers (Fig. 3) are currently retreating more quickly (Dyurgerov & Meier, 2005). 82% of the glaciers in western China have receded during the final half of the twentieth century (Liu et al., 2006). A higher pace of retreat is shown by the glacial area on the Tibetan Plateau declining by 7% over the past 40 years and 4.5% over the past 20 years (Ren et al., 2003). In the Himalayas, "a decrease in precipitation combined with an increase in temperature" causes glacier retreat. If the warming and drying of the climate persist, the glaciers will melt more quickly (Ren et al., 2003). A 2 °C increase by 2050 is predicted to cause 35% of the current glaciers to melt and runoff to rise, reaching its peak between 2030 and 2050 (IPCC, 2007). In Ladakh Himalayan region (Fig. 4), snow cover fluctuates significantly with elevation, peaking in March and dipping to its lowest in August, with an average coverage of 68%. Although the Zanskar valley has not seen a substantial decline in snow cover, notable changes are observed between 4,500 and 5,500 m above sea level. Rising temperatures and decreasing solid precipitation are pronounced, likely driving glacier retreat in the region. Glacier retreat can cause surrounding slopes to become unstable, which can result in devastating landslides that can dam streams and occasionally cause floods (Ballantyne & Benn, 1994; Dadson & Church, 2005). Often combined with liquid precipitation, excessive meltwaters can cause debris flows or flash floods. Catastrophic rockslides in the Karakoram have been shown to significantly impact glaciers and may have caused glacial surges (Hewitt, 2005). The Indus, Ganges, and Brahmaputra are among the perennial rivers that provide water to hundreds of millions of people in South Asia. These rivers are nourished by the special water storage created by the 16,000 Himalayan glaciers. According to the current patterns of glacial melt, climate change will likely result in a significant reduction in low flow (Ericksson et al., 2009). Economic growth and food production, for instance, are expected to suffer as a result. In the coming decades, the condition might seem typical in the area, even with more water available to meet the demands of the dry season (Ericksson et al., 2009). Another significant development challenge is flooding. It is anticipated that more downstream floods will result from more unpredictable and direct rainfall runoff due to deglaciation (Fig. 5).

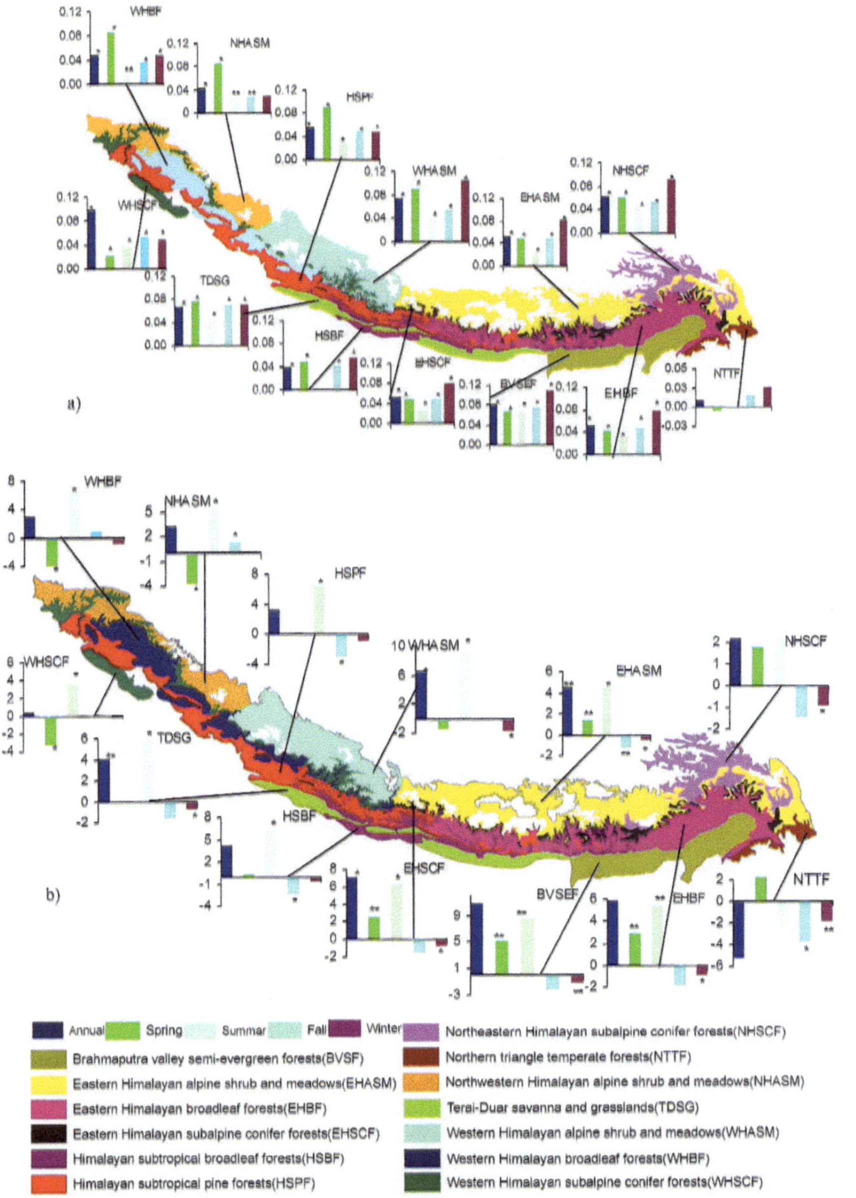

Fig. 3 Thematic maps representing current temperature and precipitation changes in the Himalaya—**a** annual and season trends of temperature (^{o}C yr^{-1}, shown in bar) in different ecoregions, and **b** annual and season trends of precipitation (mm yr^{-1}, shown in bar) in different ecoregions (*P $\leq$ 0.05, **P $\leq$ 0.10) (collected from Shrestha et al., 2012)

Fig. 4 **a** At the Kargil-Zanskar Road, close to Pensila Mountin Pass, is the Drang Drung, the biggest glacier (between 1971 and 2017, the glacier retreated by 925 m) in the Ladakh Himalaya—with an average elevation of 4779 m and a maximum length of 23 kms, the Drang Drung is perhaps the biggest glacier in Ladakh aside from Siachen Glacier in the Karakoram Range (photographed by Nilakkhya Dutta); and **b** the coalescence of valley glaciers in the Great Himalaya (photographed by Nilakkhya Dutta)

4 Flash Floods and Glacial Lake Outburst Floods

With water levels rising minutes to hours after heavy rainfall, a flash flood is characterized as a flood that happens quickly after an extreme storm event. These floods typically affect tiny basins that are less than 500 square kilometres in size. Short, intense rainstorms are usually linked to flash floods. Because climate change affects intricate atmospheric and meteorological systems, it is affecting the frequency of heavy downpours during storms and unexpected outbursts. As a result, they are distinguished by their rapid reaction times and possess the capability to significantly affect and harm communities across various climatic conditions worldwide. Despite their significance in science and society, little is known about the basic mechanisms that cause a flash flood. Climate change is causing more flash floods by increasing the amount of water in the atmosphere and leading to more extreme precipitation patterns. Alongside, urbanization and land use changes have exacerbated the impact of flash floods. The Himalaya, one of the most seismically active and ecologically fragile mountain ranges in the world, are prone to a wide range of natural hazards, including flash floods. Flash floods, which are rapid and extreme flows of water that

Fig. 5 Sacred Himalayan Landscapes during winter in Ladakh: **a** Gonbo Rangjon peak (5,520 m asl) is a stand-alone towering rocky cliff located south of Kargyak hamlet in the Lungnak valley, Zanskar (Ladakh Himalaya); **b** spectacular rugged landscape crafted by Zanskar River at Zangla (3,931 m asl) which is a village in Zanskar tehsil of Kargil district; **c** an important wetlands area called Ramsar Site, Tso Moriri is a lake with a closed drainage basin (4,522 m asl) in the Leh district of the union territory of Ladakh—it is the biggest high-altitude lake in India and the only one in Ladakh in this Trans-Himalayan biogeographic region; and **d** situated at an altitude of approximately 4998 m, this pristine freshwater lake, Mirpal Tso, remains relatively untouched and unspoiled, offering a rare opportunity to experience nature in its most pure and unadulterated form—tucked away in the rugged Changthang plateau of Ladakh, India, lies the breathtakingly beautiful lake of the Ladkah Himalaya [photographed by Nilakkhya Dutta]

can occur with little to no warning, pose a significant threat to the region's ecosystems, infrastructure, and human populations. Another serious hazard of Himalayan terrain is the glacial lake outburst flood (GLOF) which is the word used to describe the catastrophic release of a water reservoir that has formed on the surface, beneath, inside, or in front of a glacier. Morainic debris, bedrock, or glacial ice may make up the majority of the dam constructions that enclose the water reservoir. The Last Glacial Maximum and, more recently, the Little Ice Age and its aftermath saw the formation of numerous fragile lakes in high mountain locations. When the glaciers gradually recede, the ice dam's support is lost, and the lake may either drain disastrously or continue to be trapped behind the old glaciers' lateral moraines. This section provides an overview of the current state of knowledge on flash floods in the Himalaya, highlighting the key drivers, impacts, and future research directions.

June 14–17, 2013 had a lot of rain in Kedarnath and the surrounding areas. Significant flooding and landslides occurred downstream of the Mandakini River as a result of torrential rainfall and excessive ice melt from the Chorabari glacier, which boosted the river's volume. This severe flood episode was one of the deadliest natural disasters to hit India, with a significant loss of life and property. Chorabari Lake was breached due to the accumulation of rainwater from several occurrences of high rainfall spells that occurred between June 14 and June 17, 2013, causing severe flash floods throughout the area. It was reported that two distinct occurrences had an impact on Kedarnath in June 2013. A relatively minor flash flood occurred from the Saraswati catchment at 17:15 (India Standard Time) on June 16; second, the Chorabari lake overflowed and burst at 06:45 on June 17, causing a flood-cum-debris flow that drained its water in 5–10 min. In addition to Kedarnath, the subsequent incident caused unheard-of destruction several kilometres downstream throughout the Mandakini valley. Uncountable homes and buildings were damaged by landslides brought on by the floods, which also killed others who were stranded. Significant landslides and flash floods were caused by the intense rainfall. In certain areas, more than 70,000 people were stranded due to blocked or damaged roadways. There were more than three days of stranding in other significant places, such as the Valley of Flowers, Roopkund, and the Sikh holy site Hemkund. Data from the Uttarakhand government indicates that as of July 16, 2013, about 5,700 persons were 'presumed dead.' Included in this number were 934 locals. Subsequently, 6,054 people were reported dead.

The worst-case GLOF scenario of Sikkim, according to Sattar et al. (2019), is already disclosed during a moraine overtopping failure (South Lhonak Lake), resulting in a peak flood of 6064.6 m^3 s^{-1} and discharging a total water volume of 25.7×106 m^3. Sattar et al. (2019) predicted that a peak flood of 3828.08 m^3 s^{-1} could be achieved 4 h after the original dam breach event at Chungthang town of North Sikkim, which is only 62.35 km from the Lake. The flood wave has the potential to submerge villages along the flow channel's bank. When 14.7 million m^3 of frozen lateral moraine collapsed into South Lhonak Lake (SLL) on October 3, 2023, it created a 20-m tsunami-like impact wave, broke through the moraine, and drained about 50 million m^3 of water, setting off a multi-hazard cascade in the Sikkim Himalaya, India (Sattar et al., 2025). Hydropower plants along the Teesta River were overtaken by the 270 million m^3 of silt that was destroyed by the subsequent GLOF (Sattar et al., 2025). It is imperative to consider how human activity and climate change contribute to the escalation of natural disasters given the event's magnitude, human effect, and economic implications. One of Sikkim's biggest, fastest-growing, and most dangerous lakes, it has the capacity to seriously harm downstream communities in the case of a GLOF (Sattar et al., 2025). A catastrophic water cascade (Fig. 6) caused by an outburst at SLL on October 3, 2023, killed 55 persons, left 74 missing, and demolished the 1200-megawatt (MW) Teesta-III hydropower dam. Sikkim and West Bengal were affected by the flood cascade that occurred from October 3–4, 2021 while Bangladesh was affected across the border (Sattar et al., 2025). The GLOF cascade caused a total of 45 secondary landslides to be seen in the Darjiling Sikkim Himalaya. About 25,900 buildings, 59% of which were constructed within the last ten years, were destroyed by the flood cascade (Sattar et al., 2025). The majority of the impacted structures are found 200–385 kms downstream of Chungthang (north

Sikkim), with the most severely inundated area in Bangladesh occurring between 290 and 385 km downstream (Sattar et al., 2025). Due to the ongoing melting of snow and ice caused by climate change, many lakes in the Himalayas are becoming larger, raising the possibility of devastating floods in towns downstream. In the Himalayan states of Himachal Pradesh, Uttarakhand, Sikkim, and Arunachal Pradesh, the Indian government has initiated a $18 million (150 crore) endeavour to investigate the issue and potential remedies. In valleys downstream of the lakes, 47 dams have been classified by national disaster planning experts as being vulnerable to outburst floods.

5 Earthquakes and Active Tectonics

The Himalayan region is one of the most seismically active zones in the world. Due to the ongoing collision between the Indian Plate and the Eurasian Plate, the region frequently experiences earthquakes of varying intensities. The study of earthquake intensity in the Himalaya is crucial for understanding seismic hazards and implementing appropriate disaster preparedness measures (Searle & Treloar, 2019). Understanding earthquake intensity, implementing proper mitigation strategies, and enhancing disaster preparedness can significantly reduce potential damage and loss of life. Continuous research and technological advancements in seismic monitoring are essential to improve resilience against future earthquakes in this geologically dynamic region. The Himalaya was formed due to the collision of the Indian Plate with the Eurasian Plate around 50 million years ago. This ongoing convergence results in significant tectonic stress accumulation, leading to frequent seismic activity (Gahalaut & Kundu, 2011). The major thrusts responsible for earthquakes in the region include: (a) Main Himalayan Thrust (MHT): The primary fault where the Indian Plate subducts beneath the Eurasian Plate; (b) Main Boundary Thrust (MBT): A major fault zone where tectonic movements contribute to earthquake generation; and (c) Main Central Thrust (MCT): Another fault line responsible for intense seismic activity (Fig. 7). The highest seismic risk zone, known as earthquake Zone 5, is the Himalayan belt. It designates regions that are vulnerable to very large earthquakes (MSK IX or higher) and mandates that structures be built to resist the most powerful ones. On March 28, 2025, Myanmar and Thailand were hit by a strong 7.7-magnitude earthquake that left over 1,000 people dead and around 2,400 injured. A 'Great Himalayan Earthquake' of magnitude 8 or higher might strike the Himalayan orogenic zone, per Roger Bilham's (American geophysicist) constant warnings, endangering millions of lives in heavily populated areas. The only region in the globe where a significant earthquake may occur on land is the Himalaya, according to Bilham (2004). GPS measurements showed that India and southern Tibet converged at rate of 20 ± 3 mm year^{-1}, accumulating 2 m of potential slip favourable for mega earthquakes (Bilham et al., 2001). Every century, India falls 2 m below Tibet's southern border. Miserably, its northern edge does not travel easily; instead, it is held up (by friction) for hundreds of years before catching up in a matter of minutes. The inevitable and inescapable result of this motion is the

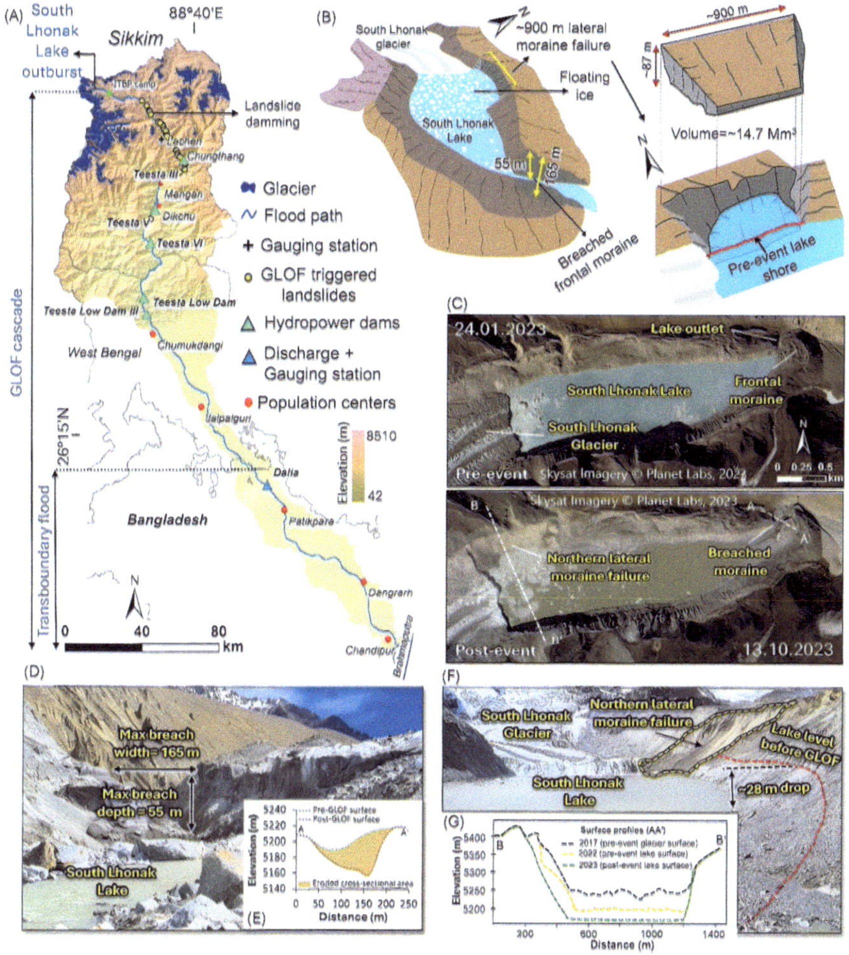

Fig. 6 Overview of the GLOF cascade (Teesta Basin) from the South Lhonak Lake (SLL) on 3rd and 4th October 2023, as developed by Sattar et al. (2025): **a** Teesta River flood-impacted length displaying SLL position, main flood processes (lake outburst, GLOF cascade), hydropower plants impacted, landslides caused by flooding, important population centres, discharge, and gauging stations; **b** conceptual diagram showing the SLL outburst incident; **c** high-resolution SkySat images of the lake before and after GLOF; **d**–**f** field photographs and diagram showing moraine failure and cross-sectional change; **g** the pre- and post-GLOF surface along cross section BB′ (collected from Sattar et al., 2025)

slip occurrences, which geophysicists refer to as earthquakes ($M_w > 8.0$). As stated by Wallace et al. (2015) and Gahalaut (2008), the Kangra rupture (following 1905 Kangra M_w7.8) may fail again today in a $M_w = 7.5$ earthquake (1.4 m of slip) or as a component of a considerably bigger earthquake ($Mw \approx 8.6$). With at least one major

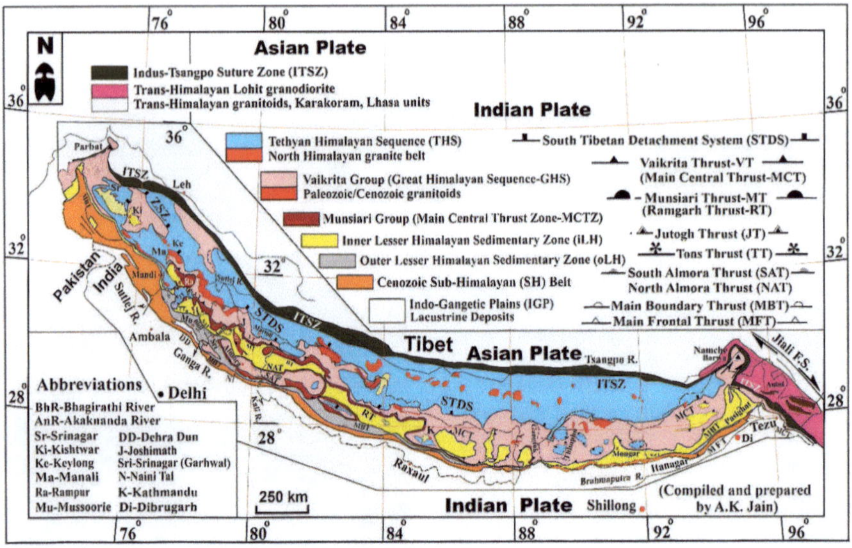

Fig. 7 An overview of geology in the Himalaya including main lithological units and tectonic settings (collected from Jain et al., 2020)

Himalayan earthquake "overdue in the region," seismologists are growing increasingly concerned. They have warned that both lowlands crowdy cities like New Delhi and mountain communities like Shimla are dreadfully unprepared for a megaquake.

The first known Himalayan earthquake occurred in Kathmandu in 1255, and a historical source states that "one third of the Kingdom of Nepal perished." An earthquake in June of 1505 destroyed a number of monasteries in southern Tibet, and another one struck Kabul, Afghanistan, a month later. An earthquake that struck Srinagar in 1555 and claimed hundreds of lives is the first known earthquake in Kashmir history. As the Indian Plate gradually underthrusts beneath the Tibetan Plate along the MHT, a belt of mountains spanning 2500 km is created (Arora & Malik, 2017). The plates are convergent at a pace of 4–5 cm/year, according to GPS measurements, with thrust mechanisms along the Himalayan arc allowing for 18 mm/year (Arora & Malik, 2017). Over the course of the last few centuries, the Himalaya has experienced a number of moderate to large earthquakes (Table 4 and Fig. 8) with magnitudes ranging from 7.8 to 8.4. Of these, five events are regarded as great earthquakes: the 1934 Bihar–Nepal earthquake (M_w 8.2), the 1950 Assam earthquake (M_w 8.4), the 1951 earthquake (Mw 8) following the Assam earthquake shock, the 1905 Kangra earthquakes (M_w 7.8), and the 2015 Nepal earthquake (M_w 7.8) (Arora & Malik, 2017). According to many studies, one of these earthquakes appears to have caused surface ruptures along the Himalayan Frontal Thrust (HFT) by rupturing the upper brittle section of the locked MHT. This region has seen a small number of large magnitude earthquakes ($6.6 \geq M_w \leq 7.8$), including the 1991 Uttarkashi earthquake (M_w 6.9), the 1999 Chamoli earthquake (M_w 6.6), the 1803

Central Himalaya earthquake (M_w 7.2), and the 1833 M_w 7.8 with an epicentre in the presumed rupture zone of the 1934 Bihar–Nepal earthquake (Bilham, 2019).

The study of how current tectonic processes, such as fault movements and earthquakes, alter the Earth's surface and geography, especially in areas that are still deforming, is known as active tectonics. It seeks to comprehend how faults and earthquakes accommodate plate movements as well as the rheologies and processes involved in continental deformation. The Main Central Thrust (MCT), Main Boundary Thrust (MBT), and Main Frontal Thrust (MFT) are three active fault segments that define the Himalayan region. These segments are essential to comprehending the seismicity and active tectonics of the area. At the moment, the Himalaya's frontal zone is at its most active. Along the Himalayan Front Fault (HFF), Main Boundary Fault (MBF), and related faults, active faulting has been occurring. Usually, these active faults create an imbricated thrust zone that dips north. On the MBF and HFF, dip-slip faulting is mostly seen along east–west trending faults, while active faulting in the Himalayan Range is characterized by right-lateral displacement with northward downthrow, particularly along northwest-southeast trending faults. Ghosh (2025) recently observes crustal deformation, active tectonics and high seismic strain between the MBT and MFT, developing topographic anomaly in the foreland parts of the Darjiling Sikkim Himalaya (Fig. 9). The GPS-derived total shortening rate is

Table 4 Major earthquakes in the Himalayan terrain since 1255

Sl. No.	Date of occurrence	Moment magnitude	Location	Fatalities
1	07-07-1255	7.8	Kathmandu, Nepal	2200
2	06-06-1505	8.2–8.8	Saldang, Nepal	6000
3	?-09-1555	7.6–8.0	Kashmir valley	600–60,000
4	01-09-1803	7.8	Garhwal, Uttarakhand	300
5	10-01-1869	7.4	Cachar, Assam	2
6	04-04-1905	7.8	Kangra, Himachal Pradesh	> 20,000
7	02-07-1930	7.1	Dhubri, Assam	~
8	15-01-1934	8	Nepal	10,700–12,000
9	15-08-1950	8.7	Assam	1500–3300
10	21-08-1988	6.9	Udayapur, Nepal	709–1450
11	29-03-1999	6.8	Chamoli, Uttarakhand	103
12	08-10-2005	7.6	Kashmir valley	86,000–87,351
13	25-04-2015	7.8	Nepal	8964
14	12-05-2015	7.3	Nepal	218
15	18-09-2011	6.9	Sikkim	111
16	07-01-2025	7.1	Tibet, Nepal	134–400

Source Rajendran and Rajendran (2022)

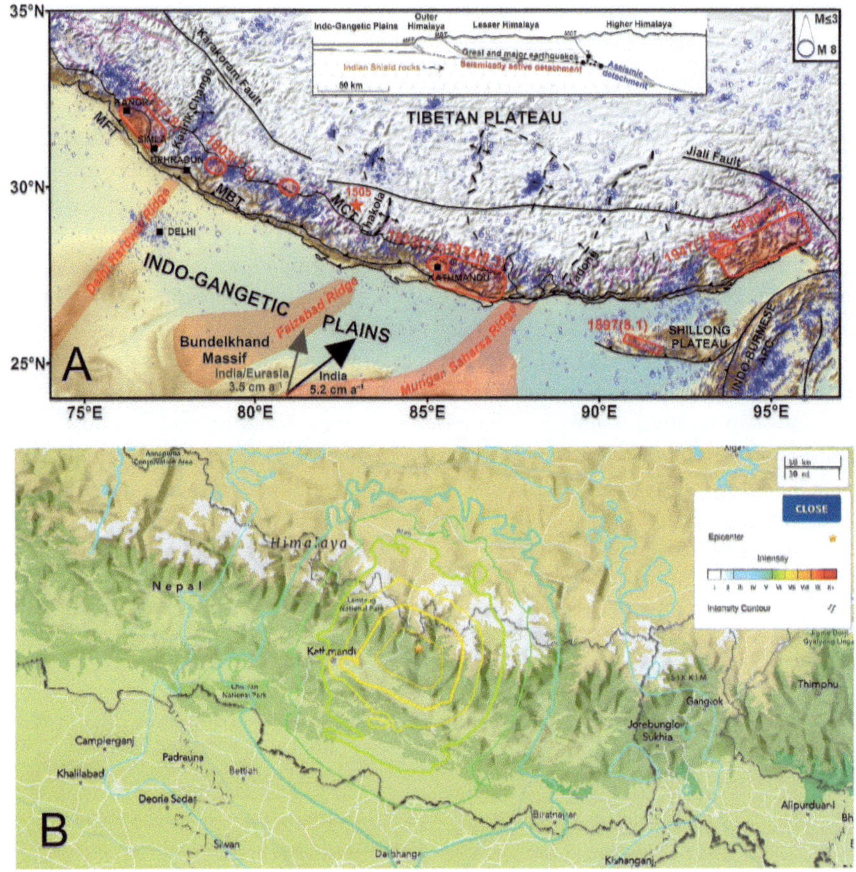

Fig. 8 Mega-earthquakes are characterized by their large magnitude and the potential for widespread damage along the zones of MFT, MBT and MCT: **a** key tectonic characteristics and the Himalayan arc's seismic activity. The Himalayan arc's $M_W > 7.2$ earthquake ruptures over the last 200 years are shown by red ellipses and rounded rectangles. Red circles also indicate earthquakes of magnitude $M_W > 7 > 7.2$. Inset diagram shows the seismically active detachment between two plates, triggering major earthquakes (collected from Gupta & Gahalaut, 2014). **b** USGS derived Gorkha earthquake intensity contour (M_W 7.8). The USGS claimed that the earthquake was triggered by a quick, sharp thrust (25th April 2015) that caught many people off guard and killed them only because of this push. The thrust occurred along the main fault line where the Indian plate, which carries India, is gradually descending into the Eurasian plate, which carries much of Europe and Asia

~11.1 ± 1.5 mm year^{-1} across a set of four east–west trending Quarterly faults, viz., Gorubathan, Matiali, Chalsa and Baradighi faults (Ghosh, 2025). The following are the likely dates of active fault scarps (24–78 m relief) that were formed by those faults (as surface splays of MBT and MFT): (a) Gorubathan scarps {33.8 ka, (b) Taljhora scarp ~ 37 ka, (c) Nagrakata scarp ~ 5.7 ka, (d) Matiali scarp ~ 33.8 ka, and (e) Chalsa scarp ~ 22.3 ka (Ghosh, 2025).

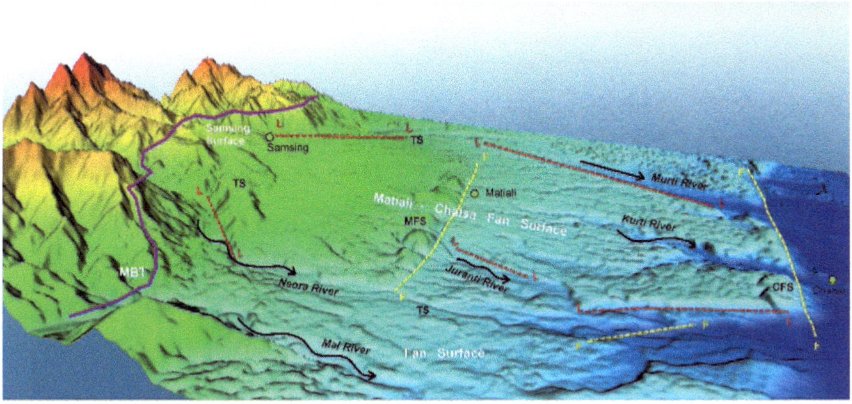

Fig. 9 An illustration of the topographical anomaly (fault scarps and triangle facets) in relation to the presence of Quaternary faults and lineaments in the piedmont fan area (Mataili-Chalsha Fan, Darjiling, Sikkim, Himalaya) in three dimensions [L—Lineament; F—Fault; MBT—Main Boundary Thrust; MFS—Matiali Fault Scarp; CFS—Chalsa Fault Scrap]

The previous seismotectonic studies (Ni, 1989; Nakata, 1989; Stevens and Avouac, 2015; Arora & Malik, 2017; Bilham, 2019; Rajendran and Rajendran, 2022) revealed the following active fault segments (demarcated as potential zones for future great earthquakes) of the Himalaya (Fig. 10):

1. Kashmir Segment—The Kashmir Basin, a 150-km-long Neogene-Quaternary intermontane basin with a complex faulting pattern surrounding the Pir Panjal range, is seismically very active. The Medlicott-Wadia Thrust, Kotli Thrust, Balapur Fault, and Himalayan Frontal Thrust are among the faults.
2. Kangra—In the meizoseismal zone of the 1905 Kanga earthquake in the Karakoram, a 60-km strike-slip fault that runs NNW to SSE has been found and given the name Kangra Valley Fault. A palaeo-seismic analysis along this fault showed that at least four earthquakes had previously occurred: Event I, which happened about 900 BC; Event II, which occurred between 890 BC and 720 AD; Event III, which occurred between 750 and 1290 AD; and Event IV, which occurred around 1620 AD.
3. Chandigarh—The Chandigarh Fault has displaced the Ghagar, Kalka, Pinjore, and Koshallia terraces (charcoal age of 1036–1401 AD), resulting in a sharp southwest-facing scarp (15–38 m). The Chandigarh Fault revealed 3.5 m of total displacement to develop the fault scarp during 1400–1500 AD. The Pinjore Garden Fault in Pinjore Dun revealed the occurrence of at least two major earthquake events with a total displacement of 2.85 m and a net displacement of 2 m. The surface expression of the active fault traces along HFT is discontinuous and has vertically displaced Late Pleistocene to Holocene alluvial fan surfaces.
4. Dehradun sector—Due to seismic activity along HFT, various fluvial terraces that were built by the two main rivers, the Ganga and Yamuna, passing by Dehradun were damaged. The displaced terrace surfaces' radiocarbon dates and seismic

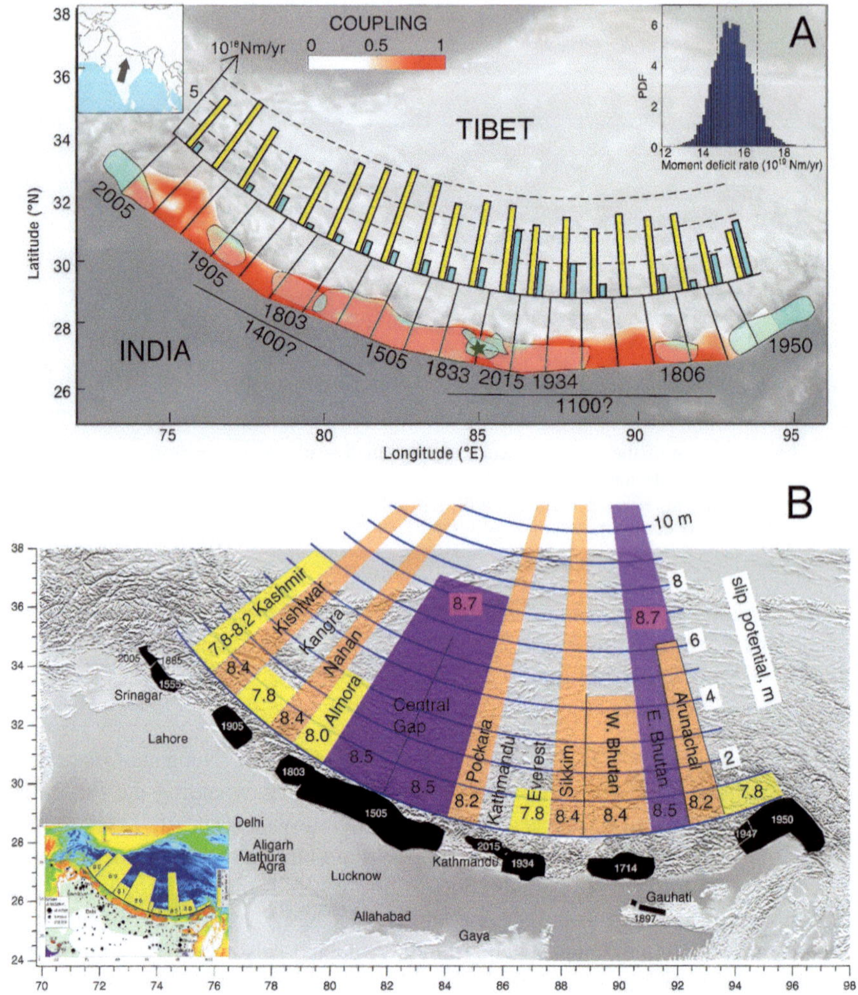

Fig. 10 Seismic coupling, the degree to which tectonic plates are locked and slip along a fault/thrust, and slip potential, the likelihood of future earthquakes, are crucial for understanding mega-earthquakes of the Himalayan belt: **a** light blue bars represent the approximate locations of earthquakes M_W 7.5–8.0 during the last 200 years and $\geq$ 8 over the last 500 years, which are used to calculate the rate of seismic moment release. The coupling pattern is displayed in red. The seismic moment accumulation distribution along the arc's strike is depicted by the yellow bars (collected from Stevens & Avouac, 2015). **b** Since the previous rupture in the designated parts, there have been five centuries of Himalayan rupture zones (black) and present slip potential (meter scale, right). The colours show the highest earthquake magnitude that might happen right now if a section fails all at once or partially slips. Mw > 8.7 earthquakes might occur in two of the violet-shaded zones. A M_W 8.4 earthquake might breach the six brown-shaded zones. Five regions, indicated in yellow, are now vulnerable to M_W > 7.7 earthquakes, which are comparable to the most recent Gorkha earthquake (Collected from Bilham, 2019)

reflection data indicated a slip rate of 13.8 ± 3.6 mm $year^{-1}$ on HFT. Between 1319 and 1380 and 1360 and 1442 AD, there were recent earthquakes.

5. Western Nepal—During an earthquake, seismic waves generate fluidization in the soft strata, resulting in sand blows and dikes and ball-pillow structures known as seismites in the area surrounding Kathmandu (along MFT). There have been three previous mega earthquakes: Event I (M_w 8.6) in 1100 AD, Event II ($M_w >$ 8.6) in 1200 AD, and Event III (M_w 8.1) in 1255 AD.
6. Central Nepal—In the Central Nepal region, a number of studies have been conducted to estimate surface ruptures. Both the 1818 and 1934 earthquakes, as well as the 2015 Gorkha earthquake, left the area stricken. With two slip maxima of 5.9 m and 6.32 m, the Mw 7.8 earthquake was caused by thrusting along the low-angle MHT, a decollement beneath the Himalaya that rises to the surface in the south as the HFT. Another Mw 5.5 earthquake was detected by the Government of India's National Centre for Seismology on February 28, 2016, near Chautara in Nepal's Sindhupal Chowk district along the MCT.
7. Eastern Himalaya—In order to fill in the gaps in the historical archives in the eastern and northeastern Himalaya and improve surface rupture data from previous earthquakes, many trenches have been dug across the irregular scarps along HFT. Researchers discovered that the Palaeoearthquake occurred in the Darjiling foothills in 1152, 1378, and 1255 AD. An earthquake of magnitude M_w 8.6 struck Tezu, Arunachal Pradesh, on August 15, 1950. On September 21, 2009, an earthquake of a magnitude of Mw 6.1 struck the eastern Himalayan region of Bhutan. Additionally, on May 4, 1714, Bhutan had a massive earthquake (M_w 8.1). On September 18, 2011, strike-slip faulting caused a M_w 6.9-mega earthquake near the Tista Lineament in Sikkim Himalya.

Latent strain acquired from ancestral incomplete ruptures is the driving force behind the large earthquakes that will occur, and the breadth of the Himalayan interseismic decoupling zone is essential in regulating seismic risks over the Himalayan arc (Bilham et al., 2017). Although the $M_w = 7.8$ Gorkha earthquake served as a model for interpreting various aspects of major Himalayan earthquakes, it was unable to close the knowledge gap between major earthquakes and the great $8 < M_w < 9$ earthquakes that are necessary to account for the majority of the convergence between the Indian Plate and southern Tibet (Bilham, 2019). Now, the slip-potential approach may be further enhanced by adding palaeoseismic data, better constraints to past rupture zones, and a spatially variable convergence rate (Fig. 8a). Focusing on the recurrence of $M_w > 7$ earthquakes that might serve as the precursors of future $M_w > 8$ earthquakes in the Himalayan belt is one potential strategy that could provide insight into the location of the next major earthquake, according to Bilham (2019). An assessment of the slip in the last 500 years of earthquakes reveals that there are seismicity gaps in areas with the potential for large earthquakes, primarily 15 segments (Fig. 8b), where the Main Himalayan Thrust (MHT) has not been ruptured by an earthquake. Seismic gaps and the location of future earthquakes can be summarized as follows (Bilham, 2019):

1. Due to $7.6 < M_w < 7.8$, the Kashmir section will break as a steep reversal fault, and the Kishtwar segment rupture can only occur when a M_w 8.4 earthquake triggers it.
2. An extremely uncommon $M_w < 8.4$ earthquake might shatter the Nahan segment, which has not experienced an earthquake since the Middle Ages.
3. A $M_w < 8.3$ earthquake might now cause the Almora portion to break since it has a slip potential of at least 3 m since the 1803 earthquake.
4. The Pockara section of Nepal may have been the location of the 1505 rupture's slippage, as it is situated between the end of the 1505 rupture and the 2015 Kathmandu segment. The area has a 10 m slip potential and can generate an earthquake with a magnitude of $8 < M_w < 8.2$.
5. Since there hasn't been a significant earthquake in the Sikkim region in the last 500 years, there is a present slip potential equal to $M_w = 8.4$.
6. A $M_w = 8.4$ earthquake might be caused by the 4.5–5 m slip potential seen in the Bhutan rupture zone of 1714.

6 Mass Movement and Erosion

The majority of human communities in the Himalaya, like Joshimath, are situated on sun-shadow slopes, or valley slopes that face north-to-easterly directional light. Consequently, these regions have experienced extended glacial activity, leading to a softer and extensively worn slope that is coated with thick fluvioglacial debris, which is the product of sediment erosion and deposition brought on by glacial meltwater. The debris is made up of pebble-sized rock fragments and huge boulders surrounded by clay. On such slopes, remnants of old landslides and some creep—slow, downslope movement of soil and earth materials—are typical, mostly because the ground is always wet. However, excessive ground saturation combined with unscientific surficial loading, slope modification, and erosion or toe removal causes the areas to become a "sinking zone" or debris slide. Many parts of Himachal Pradesh are situated on these hills, including Reckong Peo, Urni in the Kinnaur district, Prashar Lake Rari village near Bajaura in the Beas valley, and Kothi near Manali. The Himalaya's south-to-westerly slopes, on the other hand, which receive enough solar radiation, are steeper, divided, stony, and tainted with rockfall and rockslides. They are rarely chosen because they exhibit talus, or loose to semi-consolidated rock debris, at elevations that are acceptable for human habitation. The Himalayan region is facing an escalating crisis of landslides, exacerbated by climate change. As the frequency and intensity of heavy rainfall events, including cloudbursts, increase in the Himalayan foothills, the region's hilly slopes are becoming increasingly saturated and unstable. This has led to a surge in rainfall-triggered landslides across the Himalayas, with devastating consequences. In Himachal Pradesh alone, the 2023 monsoon season saw 509 deaths, 2,200 homes destroyed, and 10,000 damaged, with estimated losses of US$1.2 billion. The destruction was particularly severe during two brief periods in July and August 2023.

The Himalaya's most seismically susceptible Zone-V is where the Joshimath (1,875 m from mean sea level; an important Hindu pilgrim centre of west-central Himalaya, Uttarakhand) is located. Approximately 0.5 thick rubble paleo-landslides form the foundation of the settlement. High sheared, often folded, fractured, jointed, and faulted gneisses of the Higher Himalaya make up the bed rocks (Srivastava, 2023). Joshimath, a town in the Garhwal Himalayas, is at significant risk of landslides and land subsidence (Fig. 11) because of its location on unstable glacial moraines, as well as the effects of infrastructure development and climate change, which could cause building cracks and force residents to relocate (Srivastava, 2023). An impending ecological hazard was also warned about in 2010 by academics from the Disaster Mitigation Management Centre and Garhwal University. Because of the increasing anthropogenic pressure, Joshimath has started displaying symptoms of discomfort. According to geologists, the current issue developed as a result of inadequate planning and the government' disregard for these repeated warnings since 1978. More than 600 homes (up to January 2023) in Joshimath, a tiny village in Uttarakhand, India, have developed fractures as a result of unusually quick sinking, prompting authorities to evacuate thousands of residents. The Border Roads Organization (BRO) begins employing heavy gear to build the Helang bypass, which will cut the journey to Badrinath by around 30 km (Bera et al., 2023). This work may cause more landslides in the area close to Joshimath. The head race tunnel for the Tapovan Vishnugad Hydropower Project travels through the geologically fragile region just beneath the Joshimath (Bera et al., 2023). The head race tunnel was dug using a tunnel boring machine, but it also punctured a saturated, porous layer of aquifer that held and carried water. Land subsidence occurrences and hydrological instability are increased by such catastrophes. In order to ascertain how these hillslopes would react to different loading scenarios—such as gravity, rainfall, building load, domestic discharge, and seismic stress—a recent study, by Sundriyal et al (2023), employed a slope stability simulation based on continuum modelling. The displacement in these hillslopes may reach up to 20–25 m, according to the results, which will make matters worse (Sundriyal et al., 2023). Due to the towns' frequent extreme rainfall and the three major earthquakes that struck on September 1, 1803 (M_w 7.8), October 20, 1991 (M_w 6.8), and March 29, 1999 (M_w 6.6), a study with a hypocentral distance of less than 30 km is more feasible for decision-making (Sundriyal et al., 2023). According to Sreejith et al. (2024), the Joshimath landslide exhibits both seasonal and episodic kinematic changes, with severe rain episodes causing episodic accelerations. Extreme rainfall on October 19, 2021, set off the most recent landslide acceleration episode that led to the 2022 catastrophe. According to the investigation (Sreejith et al., 2024), there were instances of cascading movements brought on by intense rainstorms, which caused the landslide velocity to rise generally from −22 mm/yr in 2004–2010 to −325 mm/yr in 2022–2023. Another active landslide, 6 km southwest of Joshimath town, is detected by the InSAR analysis. It has been advancing since mid-2018 at a speed of around 75 mm per year (Sreejith et al., 2024). The people of Joshimath are reeling today. For what length of time will the earth engulf their houses? Subsidence can be a gradual process, after all. How many inches or feet the soil has been dropping over decades is a mystery to everyone. Studies on the town's potential sinking over time are non-existent.

The Darjiling Sikkim Himalaya has a history of frequent massive landslides, with notable events occurring in 1899, 1934, 1950, 1968, 1993, 2009, 2011, and 2015, often triggered by heavy rainfall or earthquakes. A case study (Ghosh, 2025) revealed that in 2020, 118 records f landslide events identified in the vicinity of the Darjiling Sikkim Himalaya (Fig. 12). 81 of the 118 landslides were classified as rock slides, which occur when a portion of the bedding plane of failure passes through compacted rock, causing the material to fall collectively rather than in separate chunks, and 37 as debris slides, which are defined as the rapid, chaotic movement of a mass of primarily unconsolidated and incoherent soil and rock pieces down a steep slope, often due to saturation or other factors (Ghosh, 2025). In this vulnerable region, North Sikkim highway, Gangtok—Nathula road and Rangpo—Gangtok NH31A are severely affected by landslides in each monsoon. The prime factor is

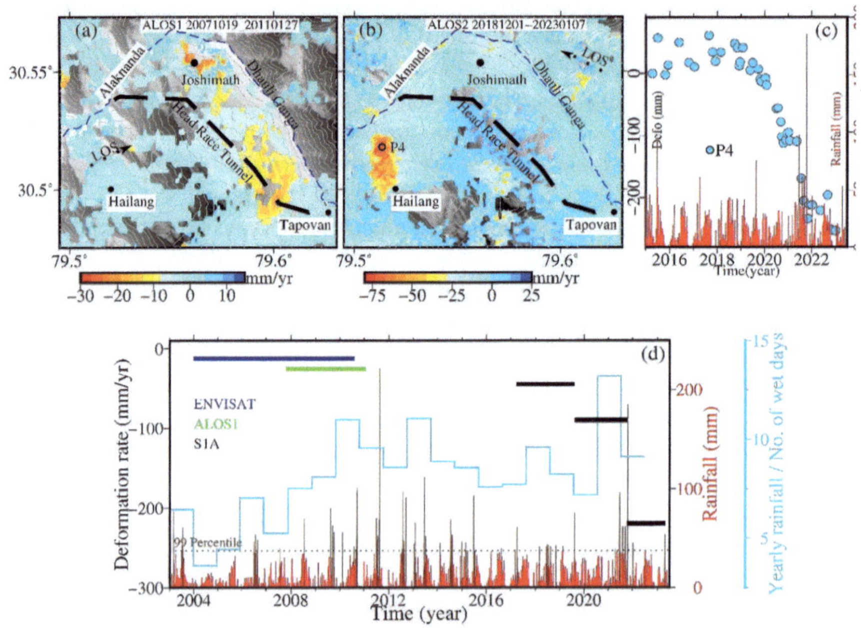

Fig. 11 Cascading landslide motion triggered by extreme rain events and surface deformation at Joshimath, NW Himalaya: **a** the mean Line of Sight (LOS) deformation rate was calculated by fitting the ALOS-1 Interferometric Synthetic Aperture Radar (InSAR) data series from October 19, 2007, to January 27, 2011 linearly. The mean LOS deformation rate was determined by fitting the ALOS-2 InSAR time series on December 1, 2018, January 7, 2023, and the time-series deformation at P4 in a linear fashion. **d** The average deformation rate derived from Sentinel-1 InSAR analysis, ALOS-1, and ENVISAT, as well as the yearly changes in rainfall intensity between 2004 and 2023 (light blue lines). Additionally, severe events (brown) and daily rainfall (red) are displayed (collected from Sreejith et al., 2024)

that Reyang formation, Kanchenjunga gneiss, Gorubathan and Chungthang formation across the faults and lineaments are more susceptible to severe mass movement in this region. area. Extensive deforestation erodes the soil and raises the possibility of landslides. Landslides are made more likely by unplanned and fast development, building projects, and insufficient drainage systems. Urbanization and unplanned agriculture are two examples of rapid changes in land use that can destabilize slopes. Among the regions known to have landslides are Kalimpong, Darjiling, Ghum, Mirik, Mangpoo, Gangtok and Kurseong.

One major factor contributing to land deterioration is soil erosion, which poses a serious environmental risk. In developing nations like India, where agriculture is the main industry, land degradation has become a serious environmental concern (Maurya et al., 2024). The projected amount of soil loss in India alone is 1,559 Mg km^{-2} $year^{-1}$, of which 550 million hectares are influenced by wind erosion and 1,100 million hectares by precipitation erosion (Maurya et al., 2024). Staple crop output is greatly impacted by this erosion, which has negative socioeconomic effects and reduces GDP (Gross Domestic Product) by 1.0–1.7% (Maurya et al., 2024). According to an estimation (Burbank et al., 2012), the rate of river erosion in the Himalayan watersheds is high—between 1- and 2.9-mm year, especially in the eastern regions—and is controlled by tectonic activity, climate, and runoff. Seventy-nine percent of the land degradation in the Himalaya region is caused by water erosion, which is mostly seen in river catchment regions. Significantly, soil erosion is a problem in the lower Himalaya area, which degrades the land and makes it unusable. Soil erosion affects over 54% of the land in the North Indian Himalayan state of Himachal Pradesh, with water-related processes responsible for 98% of this erosion (Maurya et al., 2024). Together with human activities, the complex and rocky landscape, geological forces, and many climatic factors have made the strain on natural resources—especially land resources in the lower Himalaya—even worse (Maurya et al., 2024).

The Brahmaputra and Ganga Basins comparison reveals that the eastern Himalaya erodes at a faster pace (2.9 mm/yr) than the western Himalaya (2.1 mm/yr) (Galy & France-Lanord, 2001). The Brahmaputra basin's increased runoff is probably the cause of this. The erosion rate within the range is first-orderly controlled by the monsoon's severity. According to a study by Burbank et al. (2012), the current rates of erosion in the dry, alpine Himalayan rainshadow are about 0.1 mm/year, while they are 2 mm/year in the Khudi catchment (Fig. 13) on the monsoon-dominated southern Himalayan flank (Nepal Himalaya). The top Marsyandi catchment of Nepal has the highest rates among the intervening catchments, which range from 0.2 to 0.8 mm/yr. The study by Maurya et al. (2024) has highlighted the substantial impact of land use on soil erosion in the Beas Valley, with an average soil loss of 19.9 t/ha/yr recorded for the year 2023. Soil erosion is a serious threat to the Beas River Basin in Kullu, Himachal Pradesh, affecting its environment and agricultural sustainability. The loss of 601 Mt of soil and 685.8, 99.8, 511.1, 22.6, 14.0, 57.1, and 43.0 thousand tons of nitrogen, phosphorus, potassium, manganese, zinc, calcium, and magnesium have been caused by water erosion in the northeastern Himalayan area alone (Mishra et al., 2022). Mishra et al. (2022) conducted a study in the Rani Khola watershed of the

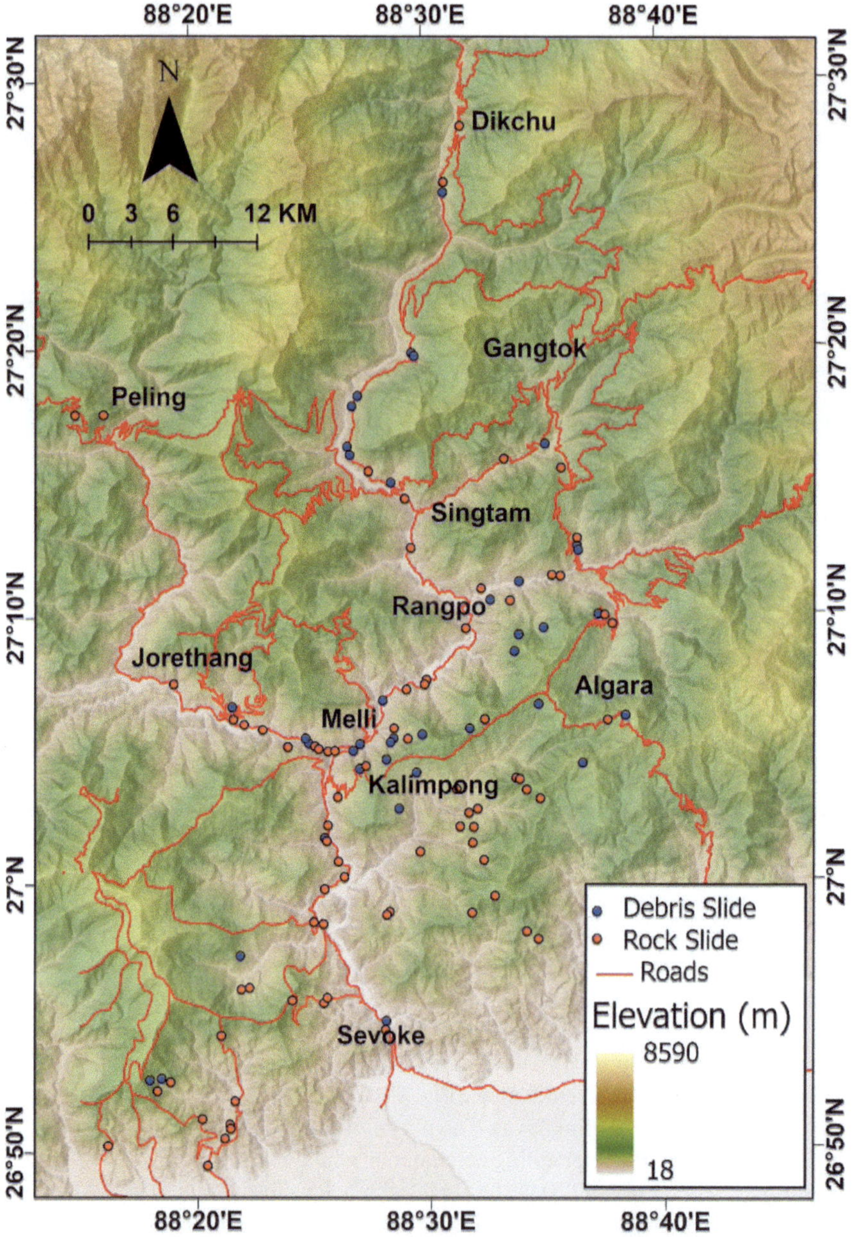

Fig. 12 Landslide vulnerable sites of the Darjiling Sikkim Himalaya—the region is highly susceptible to landslides due to its mountainous terrain, weak lithology, heavy rainfall, seismicity and human activities like deforestation, road constriction and urbanization, which exacerbate the problem

Darjiling Sikkim Himalaya and discovered that the cardamom-based agroforestry had the least amount of soil loss (3.23 Mg/ha/year), followed by terrace cultivation (3.23 Mg/ha/year). The study also found that the highest amount of soil loss was for barren land (7.73 Mg/ha/year), followed by mixed cropping (4.32 Mg/ha/year). In the Dehar watershed in the Himachal Himalaya, Prashanth et al. (2023) reported mean soil loss of 63.71, 60.99, and 66.71 t/ha/yr for 1999, 2010, and 2020, respectively. Given the high precipitation, high altitudinal location of the Dehar watershed, hilly terrain, steep to very steep slopes, forest fires, and unsustainable forest management practices, it is assumed that the Barren Land and Forest classes had a higher rate of soil erosion than other land use land cover classes (Prashanth et al., 2023). According to the RUSLE model (Joshi et al., 2023), the Aadhikhola watershed in the Nepal Himalaya experiences 41.4 tons of soil loss each year, with a high of 1302 tons and a minimum of 0.09 tons per year. In the Narayani basin, an erosion rate of 1.6 mm/year (average over 15 years) was calculated using cosmogenic nuclides (Joshi et al., 2023).

Key factors contributing to soil erosion are categorized as: (a) Steep Slopes and Active Tectonics: The Himalayan rugged terrain, steep convex slope (topographic threshold) and geological activity make them highly susceptible to erosion; (b) Deforestation and Land Use Changes: Reduced forest cover and unsustainable land management practices contribute to soil erosion; (c) Intense Rainfall: The region experiences heavy rainfall, which can lead to increased runoff and soil loss, and (d) Human Activities: Agricultural practices, road construction, and other development activities can further exacerbate soil erosion. The major consequences of soil erosion faced by the Himalaya are demarcated as:

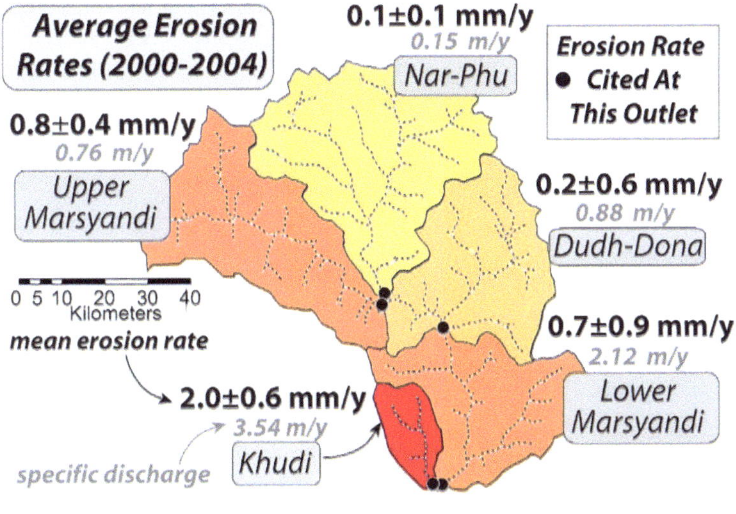

Fig. 13 Current rates of erosion in the Marsyandi watershed of Nepal Himalaya during the monsoon season. Rates calculated and averaged using data from stream gauging between 2000 and 2004 (Collected from Burbank et al., 2012)

- Land Degradation: Soil erosion leads to the loss of fertile topsoil, making land less productive.
- Reduced Agricultural Output: Loss of fertile land impacts crop yields and food security.
- Increased Runoff and Sedimentation: Erosion leads to increased runoff and sedimentation in rivers and water bodies, impacting water quality and infrastructure.
- Exacerbation of Landslides and Debris Flows: Soil erosion can destabilize slopes, increasing the risk of landslides and debris flows.
- Ecological Imbalances: Soil erosion can disrupt ecosystems and biodiversity, enhancing the fragility in ecologically sensitive zones.

7 Biodiversity at Risk

Like other regions of the world, the Himalayas are now seeing a rise in temperature. Three times faster than the global average, this region is rising by 0.6 °C per decade. Many animals are compelled by these warming temperatures to relocate to higher elevations where it is cooler. Nevertheless, this mobility may intensify competition for space and resources, especially at higher elevations, endangering biodiversity. With over 3,160 rare, endemic, and sensitive plant species that have unique therapeutic qualities, the Himalayas are one of 36 biodiversity hotspots. With trees holding over 62% of the total forest carbon, they represent one of the Himalayas' primary carbon sinks. Further carbon storage is made possible by the colder forest soils found in the northern biomes, such as the tundra and boreal forest, which contain undecomposed organic matter. A wide variety of species can be found in the Himalayan belts, including 300 animals, 979 birds, 10,000 vascular plants, and the Himalayan wolf, Himalayan tahr, red panda, Snow leopard, Himalayan brown bear, and Himalayan monal. The area is made up of a vast mountain range that spans 2,400 kms through China, Myanmar, Nepal, India, Bhutan, Pakistan, and Afghanistan. From tropical to alpine habitats with ice and rocks in the highest zone, it has a variety of temperature types and ecological zones. Within a small elevation range, all of these ecological zones are condensed. In addition to providing significant ecosystem services, the Himalayas and the associated Tibetan Plateau are the "third pole" and the source of the majority of Asia's main rivers, earning them the nickname "the world's water tower." The conservation and protection of these delicate ecosystems is critically important.

Every 190 m up or down, the average temperature in the central Himalayas fluctuates by almost one degree Celsius. The same amount of temperature change, in contrast, happens around every 150 kms in the northern hemisphere and every 197 kms in the southern hemisphere along a north–south line. Hiking in the mountains allows one to readily observe the noticeable variations in vegetation (in connection with variation with climate) within a relatively minor height shift. The transition from treeline to alpine grasslands is where the biodiversity changes most visibly.

Because of the vast amount of vegetation that absorbs carbon through photosynthesis, the Himalayan forests act as significant carbon sinks, meaning they absorb more carbon dioxide from the atmosphere than they release. As such, they are an essential part of the solution to climate change. The dense coniferous and broadleaf forests, which are found at different elevations in the mountain range, store a significant amount of carbon in their trees and soil. High-elevation coniferous forests frequently serve as important carbon sinks, although different forest types at different elevations contribute to varied carbon storage capabilities. In the principal forest types, the average carbon (C) store varied between 79 and 373 g of carbon per hectare (Gogoi et al., 2022). The distribution of ecosystem C stocks in the components of plants (24–55%), soils (43–75%), deadwood (1–4.8%), and litter (0.20–1.25%) differed significantly (Gogoi et al., 2022). The Himalaya's ability to absorb carbon dioxide can be greatly diminished by any extensive deforestation, which would raise atmospheric carbon dioxide levels. Himalayan forests' capacity to sequester carbon and maintain their health may be impacted by shifting weather patterns brought on by climate change.

Nearly one-third of the geographic area of the Himalaya province stretching from the Indus in the west to the Brahmaputra valley in the east, is covered with forests in various states of ecological health. The Himalaya, mainly eastern part, is recognized as 'biodiversity hotspot' which is a geographical area with a high level of biodiversity that is also threatened by extinction. Nearly 30% of the forest species found in the Himalaya are not found anywhere else in the world. Four main types of forests are recognized here: (a) tropical forest, (b) sub-tropical forest, (c) temperate forests intervening with grasslands, and (d) Alpine Forest. It is worthwhile to mention that the Indian Himalayan region is home to about 80 species, 10 subspecies and 14 varieties of rhododendrons. A significant portion of the species found in the region are endemic, meaning they are unique to that area. The Himalaya supports a vast array of plant species, with estimates ranging from 8,000 to 10,000 species of vascular plants. The region is also home to numerous plants with medicinal properties, used in traditional systems of herbal medicine.

Increasing human populations, deforestation, illegal taking of non-wood forest material, and disturbances by livestock and herders are the main causes of habitat destruction and fragmentation, which pose a threat to the red panda (*Ailurus fulgens*) which is a signature mammal of the Eastern Himalayan biome (Fig. 12a). Hunting is prohibited in all countries where the red panda is found; it is protected under CITES Appendix I. Since 2008, the IUCN Red List has classified it as 'Endangered' due to a declining population trend and an estimated 10,000 individuals worldwide. In the Eastern Himalaya (Nepal, Sikkim and Bhutan), red pandas occupy over half of their habitat. Red panda populations are declining over much of their habitat due to the clearing of their forest home, which is resulting in the loss of bamboo and breeding trees. Langtang National Park (an area of 1,710 km^2) in north-central Nepal and Khangchendzonga National Park (UNESCO world heritage site; an area of 850 km^2) in Sikkim, India are the important habitats for red pandas.

According to the IUCN Red List, the snow leopard (*Panthera uncia*) is classified as "Vulnerable" because its global population is estimated to be less than 10,000 mature individuals and is predicted to decline by roughly 10% by 2040. Poaching and habitat destruction resulting from Third Pole infrastructure developments are the primary threats to this species, which is found in alpine and subalpine zones at elevations between 3,000 and 4,500 m, spanning from eastern Afghanistan, the Himalaya, and the Tibetan Plateau to southern Siberia, Mongolia, and western China (Fig. 14). As masters of stealth and camouflage, snow leopards are uniquely suited to the chilly and untamed mountains of Asia, and their renowned ability to blend in with their environment has earned them the nickname "ghost of the mountains." The Himalayan treeline will probably change due to greenhouse gas emissions, and the alpine zone will diminish. This might result in a 30% reduction in snow leopard habitat. Human-wildlife conflict occurs when snow leopards feed on domestic cattle. The main causes of the steadily declining snow leopard population are poaching, livestock defence, and the lack of natural prey as a result of overgrazing by cattle. The loss and fragmentation of snow leopard habitat, especially in the southern regions of their range (Fig. 15), can result from warming temperatures, melting glaciers, and changing tree lines (Forrest et al., 2012).

With habitat degradation, changing hibernation periods, and possible changes in food availability, climate change poses a serious threat to the Himalayan brown bear (*Ursus arctos isabellinus*), which might result in a population reduction (Fig. 14b). The Himalayan brown bear inhabits the western Himalayas, extending from north-eastern Pakistan to the Himalayas in central Nepal via the Indian states of Jammu and Kashmir, Himachal Pradesh, and Uttarakhand. The tribal Chamba area and the Kugti and Tundah wildlife sanctuaries in Himachal Pradesh are their home. The favourite habitat of the Himalayan brown bear is the tree that bears buransh, the state flower of Himachal. The buransh tree is commercially cut because of its high value, thus destroying the habitat of brown bears. Under the worst climatic scenario, Tibetan and Himalayan brown bears were expected to lose more than 34% and 32% of their present occupied territory, respectively, inside the Hindu Kush Himalayan region by 2070 (Dai et al., 2021). Both Appendix II of CITES and Schedule I of the Indian Wildlife (Protection) Act of 1972 include the Himalayan brown bear. According to the IUCN Red List, this bear is considered 'Critically Endangered'. This species has the highest danger rating and is thought to be at extremely high risk of going extinct in the wild Himalaya.

Although there is still little information available on the ecology and population status of the Himalayan wolf (*Canis lupus chanco*), it is becoming more widely acknowledged that this species needs protection (Werhahn et al., 2019). Because it has evolved to a low-oxygen environment, the Himalayan wolf, which is typically found over 4,000 m in height, is located in Ladakh in the Himalaya, the Tibetan Plateau, and the Central Asian highlands. It may be found in northern India in the Ladakh Union Territory and the northeastern Himachal Pradesh area of Lahaul and Spiti. It was estimated in 2004 that there were 350 Himalayan wolves in India, covering an area of around 70,000 km^2. Threats to this wolf, which is classified as 'vulnerable', include habitat loss, changes in the natural environment, livestock depredation by humans, illegal hunting for fur and body parts, and hybridization with feral canines. Climate change impacts in the Himalayan biomes, such as altered

Fig. 14 Majestic mammals of the Himalaya are under threat of habitat loss and extinction: **a** Red Panda (Endangered) from the Singalila National Park (photographed by Kallol Mukerjee), **b** Himalayan brown bear (Critically Endangered) from Drass, Ladakh (photographed by Puskar Basu), **c** snow leopard (vulnerable) from Kibber, Himachal Pradesh (photographed by Puskar Basu), **d** wild yak (vulnerable) from Humla, Nepal (photographed by Puskar Basu), and **e** Tibetan wolf (vulnerable) from Hanle, Ladakh (photographed by Puskar Basu)

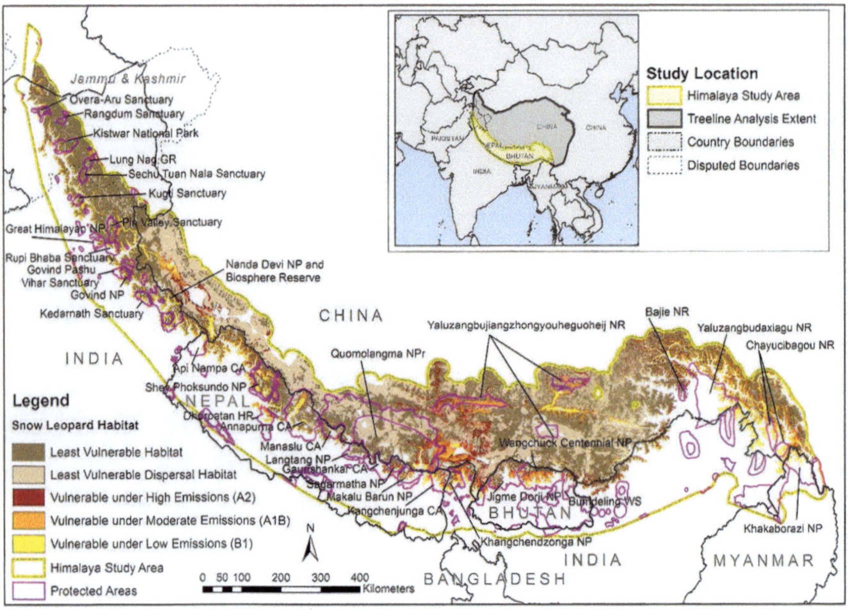

Fig. 15 The habitat of snow leopards in the Himalaya is susceptible to changes brought on by climate change. Country borders: ESRI ArcWorld; data sources: Protected Areas: WWF Bhutan, WWF Nepal. Boundaries between countries do not indicate formal approval or endorsement. NR stands for Nature Reserve, WS for Wildlife Sanctuary, NPr for Nature Preserve, GR for Game Reserve, NP for National Park, and CA for Conservation Area (collected from Forrest et al., 2012)

temperature and precipitation patterns, can affect prey abundance and distribution, impacting the Himalayan wolf's food source. The number of adult individuals in the wild Himachal Pradesh ranges from 2,275 to 3,792, which is surprising and alarming also.

8 Living with Risk and Hazards

The Himalaya, one of the most geologically active and environmentally fragile regions on Earth, presents unique challenges for those who live there. As mentioned earlier, the region is prone to multiple natural hazards, including earthquakes, landslides, glacial lake outburst floods (GLOFs), avalanches, and extreme weather events. Additionally, human activities such as deforestation, unplanned urbanization, and climate change exacerbate these risks. Understanding and mitigating hazards in the Himalaya is crucial for ensuring the safety and sustainability of communities living in this high-altitude terrain. People of the Himalaya are living unconditionally with high risk and hazards, and the important settlements, like Leh, Dehradun, Kedarnath,

Badrinath, Joshimath, Darjiling, Gangtok, Katmandu, Siliguri, New Delhi, Guwahati, etc., are on particularly shaky ground, having close proximity to the Himalayan seismic risk zone IV and V. The rough topography, seismic activity, heavy precipitation, and human activities of the Himalayan area make it very vulnerable to several natural dangers. When disrupted, these elements may cause devastating landslides, flash floods, cloud bursts, and other natural disasters. Ecosystems, infrastructure, human settlements, and the region's general environmental stability are all seriously threatened by these occurrences. Landslides in the Himalayas have claimed over 10,000 lives in the past ten years. Living with risk and hazard in the Himalaya is an ongoing challenge, but proactive measures can significantly reduce their impact. By combining traditional knowledge with modern science, investing in resilient infrastructure, and fostering community preparedness, Himalayan societies can navigate uncertainties and build a safer future. As climate change and human pressures continue to shape the region, adaptive and innovative risk management approaches will be essential for ensuring sustainable development in this fragile yet vital part of the world.

Risk in the Himalaya is the likelihood of adverse events occurring due to the region's unique geological, climatic, and socio-economic factors. It is influenced by multiple elements, including exposure, vulnerability, and coping capacity. The region's remoteness and challenging terrain further complicate risk management efforts. A hazard is any potential source of damage, harm, or adverse health effects on communities and ecosystems in the Himalaya. Some major hazards, affected most of the people, in the region include: (a) Seismic Hazards—The Himalaya is located in a seismically active zone due to the ongoing collision of the Indian and Eurasian plates, making earthquakes a constant threat; (b) Landslides—The steep slopes, loose soil, and heavy monsoonal rains trigger frequent landslides, disrupting transportation and endangering lives; (c) Glacial Lake Outburst Floods (GLOFs)—Melting glaciers have created unstable glacial lakes, which can suddenly burst and cause catastrophic flooding downstream; (d) Extreme Weather Events: The region experiences heavy snowfall, blizzards, and unpredictable weather patterns that impact agriculture, tourism, and livelihoods; (e) Avalanches: The steep and snow-covered terrain poses a high risk of avalanches, particularly in regions with human settlements and trekking routes; and (f) Water Scarcity: Despite having large water resources, many Himalayan communities struggle with seasonal water shortages due to changing precipitation patterns and glacial melt. In the Himalayas, human activity has made landslides more frequent. Indirect or direct causes of slope destabilization include uncontrolled urbanization, deforestation, and the construction of infrastructure including highways, dams, and hydropower projects. In order to build roads and highways in the Himalayas, it is common practice to cut into steep slopes, which can cause landslides and damage the slope's natural stability. There have been devastating and violent floods in the Himalaya. The eventual death toll may reach several thousand, according to fears. The number of communities devastated, properties ruined, highways swept away, and hydropower projects wrecked in Uttarakhand, a Himalayan state, is difficult to measure. The people who lived there are dead, injured, and battered, and the mountains are bleeding.

Loss of life, injuries, and displacement are common consequences of hazards in the Himalaya. Earthquakes, landslides, and floods have historically caused significant casualties. Limited access to healthcare and emergency services in remote areas exacerbates the situation. Hazards in the Himalaya severely affect economic stability. Damage to infrastructure, loss of agricultural land, and disruption of trade routes have long-term economic repercussions. The tourism industry, a major economic driver, is highly vulnerable to disasters such as avalanches and earthquakes. The fragile Himalayan ecosystem is highly susceptible to degradation due to hazards. Landslides lead to deforestation and soil erosion, while glacial melt contributes to changing river flows and biodiversity loss. Climate change intensifies these environmental challenges. To mitigate the loss and to prepare the roadmap of management the first step is Risk assessment in the Himalaya involves identifying potential hazards, analysing their likelihood, and evaluating their impact on local communities. Scientists and policymakers now use mathematical modelling, satellite imagery, remote sensing, and ground-based surveys to monitor glacial lakes, seismic activity, and weather patterns. Risk mitigation strategies now include: (a) Infrastructure Resilience: Building earthquake-resistant homes, retaining walls to prevent landslides, and improved road networks for emergency response; (b) Glacial Lake Management: Controlled drainage of unstable glacial lakes and early warning systems to mitigate GLOF risks; (c) Reforestation and Soil Conservation: Planting trees to stabilize slopes and prevent landslides, and (d) Sustainable Urban Planning: Designing settlements with proper drainage systems and avoiding construction in high-risk zones. Preparedness is key to reducing vulnerability in the Himalaya. Community-based disaster management programs, training on emergency response, and investment in climate-resilient agriculture help improve adaptation strategies. Traditional knowledge, combined with modern technology, plays a crucial role in enhancing disaster preparedness. Factors that contribute to resilience include: (a) Community Participation—Engaging local populations in disaster preparedness and response initiatives; (b) Diversified Livelihoods—Reducing dependence on single economic activities (such as tourism) to withstand economic shocks, and (c) Education and Awareness: Spreading knowledge on hazard risks, safety measures, and sustainable resource management.

9 Areas of Research Needs

The Himalayan region, a "data-deficient" area, requires focused research on climate change impacts, biodiversity, and sustainable development, with a need for international collaborations and improved data collection and analysis. Scientists argue that to effectively save the Himalaya, research must focus on understanding and addressing the impacts of climate change, promoting sustainable practices, and fostering regional cooperation, while also considering the socio-economic needs of local communities. To improve hazard management and risk mitigation in the

Himalaya, extensive research (mirco/meso/macro scale) is required in the several areas. Here's a breakdown of the research needs in the context of Himalayan hazards:

1. Since these cascading impacts increase the dangers to infrastructure, ecosystems, and human populations, research is required to understand how primary hazards (such as earthquakes, avalanches, and landslides) cause secondary hazards (such as debris flows, flooding, and glacial lake outburst floods).
2. To forecast future risks, it is necessary to investigate how climate change affects Himalayan hazards, such as glacier melt, altered precipitation patterns, and rising temperatures.
3. Research on the seismic hazard potential, including the frequency, size, and geographical distribution of earthquakes, as well as their effects on landslides and other hazards, is essential given the tectonic activity of the Himalayas.
4. In-depth study is necessary to evaluate the hazard potential and create early warning systems because of the growing quantity and size of glacier lakes in the Himalayas as well as the possibility of their outburst floods. In addition, flash floods brought on by brief periods of intense precipitation are a problem in the Himalayan region.
5. In order to more effectively target disaster planning and mitigation measures, studies are required to identify susceptible individuals and infrastructure. It is crucial to map regions vulnerable to various disasters, such as earthquakes, floods, and landslides, in order to assist infrastructure development and land-use planning.
6. Effective risk management requires the development of thorough and precise models for evaluating the risks associated with different hazards, taking into account the intricate relationships between numerous components.
7. Reducing damage and fatalities requires research on designing and creating infrastructure—such as buildings, bridges, and roadways—that is resistant to natural disasters. Building resilience in local communities requires providing them with the information and tools they need to anticipate and respond to natural catastrophes.
8. Research should concentrate on creating climate change adaptation plans, such as those for sustainable agriculture, water management, and ecosystem restoration.

10 About the Volume

10.1 Part I

The Himalaya, one of the most seismically active regions in the world, is facing an unprecedented tectonic dilemma. The region's unique geology, characterized by the collision of the Indian and Eurasian tectonic plates, makes it prone to powerful earthquakes (Valdiya, 1998). The Himalayan seismic zone is home to several major fault lines, including the Main Himalayan Thrust (MHT), which is capable of producing earthquakes of magnitude 8.0 or higher. The seismic risk in the Himalaya is exacerbated by several factors, including: (a) High population density: The Himalayan region is home to over 50 million people, many of whom live in urban areas with inadequate infrastructure and building codes; (b) Poor infrastructure: Many buildings, bridges, and roads in the Himalaya are not designed to withstand powerful earthquakes, making them vulnerable to damage and collapse; and (c) Lack of preparedness: Despite the high seismic risk, many communities in the Himalaya lack adequate emergency preparedness and response plans. This section comprises five different chapters focusing mainly of earthquake hazard (associated risks), level of active tectonics and tectonic geomorphology of the Himalaya. Chapter 1 shows that the Darjiling-Sikkim Himalaya serves as an example of hazard prone region, and the critical scenario is relevant to the entire Himalayan region, emphasizing the need for a comprehensive approach to mitigate these hazards. The Himalayas are prone to natural hazards like earthquakes, landslides, and glacial lake outburst floods (GLOFs). Earthquakes are triggered by mountain-building processes, while landslides are caused by monsoonal rains, topographic slopes, and brittle fault-related deformation. Chapter 2 explains that the Kashmir Basin's prolonged aseismic period suggests significant stress accumulation, increasing the likelihood of a future large-magnitude earthquake, highlighting the need for updated seismic hazard assessments. Chapter 3 deals with a study of tectonic geomorphology in the Baner Basin (Kangra), Himalayan foothills. It analysed geomorphic indicators and relative tectonics to understand topography and hazard potential. Morpho-analytical tools and statistical values created a relative tectonic map, informing regional planning and hazard mitigation initiatives. Chapter 4 includes a case study in the Neora River catchment which was examined ton know the impact of tectonic perturbations on fluvial processes using morphotectonic indices and channel analysis. Results showed tectonic deformities influenced channel profiles, with lithology dominating above Himalayan thrusts and surface deformations dominating on the Himalayan Frontal Belt. Finally, Chap. 5 depicts an overview of Himalayan active fault zones. Studies, using geologic, geomorphic, and GPS data, identify seismic gaps, including the Central Gap in Uttarakhand, with high potential for magnitude earthquakes, aiding vulnerability mapping and risk assessment.

10.2 Part II

The Himalaya is prone to landslides due to its steep terrain, fragile geology, and intense monsoon rainfall. Landslide susceptibility and hazard assessment are crucial for mitigating risks. Factors such as slope angle, lithology, land use, and rainfall intensity are used to assess landslide susceptibility (Valdiya, 2016). Hazard assessment involves evaluating the likelihood and potential impact of landslides. GIS-based mapping and modelling techniques are employed to identify high-risk areas and develop early warning systems to protect communities and infrastructure. Eight important chapters on the Himalaya's landslide-prone regions—the

Kashmir, Garhwal, Darjiling, Sikkim, and northeastern Himalayas—are included in this section. These chapters cover topics such as mapping landslide susceptibility, applying machine learning algorithms, and assessing landslide hazard geologically and geomorphologically. Chapter 6 encapsulate a detailed study in the Kashmir basin used frequency ratio, probabilistic likelihood, and adaptive neuro-fuzzy methods to identify landslide-susceptible zones. The Chap. 7 aimed to address escalating landslide occurrences, informing sustainable development strategies and targeted mitigation efforts to ensure regional sustainability and disaster preparedness in the Garhwal Himalaya. Chapter 8 includes a feasibility study which used LHEF/BIS and AHP methods for landslide susceptibility zonation in Manipur's Moa-Maran region. Results showed 70.9–72.63% accuracy of applied methods which are reasonably feasible for similar terrain conditions in other regions. Chapters 9–13 elaborates different studies on landslide susceptibility and hazard mapping the various parts of the Darjiling Sikkim Himalaya using machine learning algorithms, remote sensing and GIS tools. The methods of these chapters are crucial in landslide risk assessment because they enable more accurate and efficient prediction, mapping, and analysis of landslide-prone areas, ultimately improving disaster preparedness and mitigation efforts.

10.3 Part III

Climate change is posing serious ecological threats to the Himalayan area, a delicate environment and a crucial supply of water for billions of people. These threats include rapid warming, glacial melting, and changing precipitation patterns, which endanger water resources, biodiversity, and livelihoods. The average surface temperature in the Himalayas has risen by more than 1.6 °C, which is driving the glaciers to melt quickly and recede more quickly. The Himalaya's delicate natural areas are particularly susceptible to the ripple effects of climate change, such as melting glaciers, changed precipitation patterns, and an increase in extreme weather occurrences. The two chapters deals with various aspects: (a) The Himalaya is warming more quickly than the rest of the world, which causes temperatures to rise and snow and glaciers to melt; (b) glaciers, which are essential to the region's water supply, are melting more quickly, affecting the amount of water available for agriculture and human consumption; (c) changes in rainfall patterns, including a rise in the frequency and severity of extreme weather events like droughts and floods, are upsetting ecosystems and livelihoods; (d) habitat fragmentation and biodiversity loss due to rising temperatures and vegetation zone changes are affecting wild species; and (e) changes in temperature and water availability can impact tree growth and treeline dynamics. Chapter 14 deals with a study in Arunachal Pradesh (1988–2020) where the authors found declining snow cover, rising temperatures, and variable precipitation, highlighting the region's growing vulnerability to climate change. Chapter 15 assesses temperature-precipitation-vegetation interactions using satellite data in the Brahmaputra River Basin (2000–2019). Results showed strong correlations between skin surface temperature, precipitation, and vegetation changes. Findings support sustainable practices for terrestrial ecosystems, hydrology, and water resources in the region.

10.4 Part IV

Significant dangers to life, property, and infrastructure arise from the Himalayan region's extreme vulnerability to hydro-climatic hazards including floods, cloudbursts, and glacial lake outbursts, which are made worse by climate change and human activity. During the monsoon season, the Himalayan River basins are vulnerable to heavy rainfall, which can cause landslides and flash floods. Devastating floods and landslides can occur in the area as a result of abrupt, heavy rainfall occurrences known as cloudbursts. The probability of GLOFs, where glacial lakes can rupture and cause devastating floods downstream, rises as a result of glaciers melting as a result of climate change. The primary impact and risk can be summed up as follows: (a) the Himalayas' hydro-climatic hazards result in a high death toll and property damage; (b) landslides and floods can damage roads, bridges, and other infrastructure; (c) disruptions to agriculture, tourism, and other economic activities can result in significant financial losses; (d) climate change and dam construction can alter the availability of water, causing water scarcity and conflicts; and (e) hydropower projects in the Himalayas are also susceptible to hydro-climatic hazards, such as floods, landslides, and glacial lake outburst floods. Chapters 16 and 19 illustrate the causes, pattern and vulnerability of the GLOFs across the largest river basins of the Himalaya. The monitoring of glacial lakes, the mapping of their area, and the creation of inventory of potentially dangerous lakes are all made possible by geospatial technologies such as GIS and remote sensing. One of the primary findings is that the development and growth of glacial lakes are influenced by climate change, precipitation variability, and glacier retreat, all of which raise the danger of GLOFs. The deep learning model produced more precise and thorough risk projections, as explained in Chap. 20, and showed a markedly rising trend of urban floods in the Morang district of the Nepal Himalaya.

10.5 Part V

The believed geological era known as the 'Anthropocene', which is characterized by human impact on Earth, brings up important concerns including resource depletion, biodiversity loss, and climate change, all of which call for immediate action to address. According to an anthropogenic perspective, human activities like deforestation, uncontrolled urbanization, and infrastructure development have made the Himalayan area more vulnerable to natural disasters like floods and landslides by upsetting natural processes and destabilizing slopes. Tree removal weakens the soil, making landslides and soil erosion more likely. On steep slopes, intensive farming can cause soil erosion and increased runoff, which can exacerbate floods and landslides. In vulnerable regions, unchecked urban growth can cause landslides, floods, and excessive runoff. Infrastructure projects like building roads, dams, and other structures can change drainage patterns, destabilize slopes, and raise the risk of geohazards. Natural hazards can have a higher impact and cause more property and human casualties if there is inadequate infrastructure, waste management, and disaster planning. The impacts of the Anthropocene are often disproportionately felt by marginalized communities, exacerbating existing inequalities. The Himalayan river's sand mining hazard is covered in Chap. 21. This issue presents serious environmental and socioeconomic risks, such as habitat loss, channel erosion, deteriorating water quality, and threats to infrastructure and livelihoods, particularly in riverine area. Chapter 22 emphasizes tourism in the Himalaya, while offering economic benefits, but it presents several hazards, including environmental degradation, natural disasters, and unsustainable practices, requiring careful management for long-term sustainability.

10.6 Part VI

Rapid urbanization, deforestation, and unregulated tourism contribute to increased hazard risks in the Himalaya. Poor waste disposal, unsustainable hydropower projects, and encroachments into ecologically sensitive zones heighten vulnerabilities. Implementing environmentally sustainable development policies is necessary to minimize human-induced hazards. Himalayan hazards have severe implications for human settlements, agriculture, hydropower projects, and biodiversity. Loss of life, displacement, economic disruptions, and environmental degradation are common consequences. Sustainable development strategies must incorporate hazard mitigation. The tourism sector, a key economic driver, faces setbacks from frequent disasters. Biodiversity loss due to habitat destruction threatens endemic species and disrupts ecological balance. The Himalaya faces multiple interconnected hazards that threaten human lives and ecosystems. A multidisciplinary approach involving scientific research, policy action, and community participation is essential for reducing vulnerability and enhancing resilience. Governments, researchers, and local communities must collaborate to implement sustainable mitigation strategies and adaptive policies that address the evolving ecological risks in the Himalayan region. Although there is far less research on the water quality of the 'Asian Water Tower' than on its quantity, anticipated rises in upstream riverine chemical fluxes and lowland pollution discharge may worsen downstream water quality degradation, as highlighted in Chap. 23. Chapter 24 emphasizes the need for a multifaceted strategy that addresses climate change and promotes sustainable development in addition to reforestation, sustainable land management, community engagement, and the creation of protected areas in order to address the devastating effects of deforestation and advance sustainable conservation in the Himalayan eco-regions. Chapter 25 and 26 emphasize how climate change, overgrazing, and unsustainable land use threaten the Himalayan alpine meadows, which are essential ecosystems that affect local livelihoods, biodiversity, and water supplies. Community-based conservation, sustainable grazing methods, and mitigating the effects of climate change are all part of sustainability aims. Lastly, the authors of Chap. 27 mapped rill and gully erosion hotspots using geospatial technologies, determining that steep slopes, high rainfall intensity, and soil type are the main causes of erosion in the Darjiling Sikkim Himalaya's Lish and Gish Catchment.

References

Arora, S., & Malik, J. N. (2017). Overestimation of the earthquake hazard along the Himalaya: constraints in bracketing of medieval earthquakes from paleoseismic studies. *Geoscience Letters, 4*, 19. https://doi.org/10.1186/s40562-017-0083-6

Ballantyne, C. K., & Benn, D. I. (1994). Paraglacial slope adjustment and resedimentation following recent glacier retreat, Fabergstolsdalen, Norway. *Arctic and Alpine Research, 26*(3), 255–269.

Bera, B., Saha, S., & Bhattacharjee, S. (2023). Sinking and sleeping of Himalayan city Joshimath. *Quaternary Science Advances, 12*, 100100; https://doi.org/10.1016/j.qsa.2023.100100

Bilham, R. (2004). Earthquakes in India and the Himalaya: Tectonics, geodesy and history. *Annals of Geophysics, 47*, 839–858.

Bilham, R. (2019). Himalayan earthquakes: A review of historical seismicity and early 21st century slip potential. *Geological Society, London, Special Publications, 483*, 423–482. https://doi.org/10.1144/SP483.16

Bilham, R., Gaur, V. K., & Molnar, P. (2001). Himalayan seismic hazard. *Science, 293*(5534), 1442–1444.

Bilham, R., Mencin, D., Bendick, R., & Burgmann, R. (2017). Implication for elastic energy storage in the Himalaya from the Gorkha 2015 earthquake and other in complete ruptures of the Main Himalayan Thrust. *Quaternary International, 462*, 3–21.

Burbank, D. W., Bookhagen, B., Gabet, E. J., & Putkonen, J. (2012). Modern climate and erosion in the Himalaya. *Comptes Rendus Geoscience, 344,* 610–626: https://doi.org/10.1016/j.crte.2012.10.010

Dadson, S. J., & Church, M. (2005). Postglacial topographic evolution of glaciated valleys: A stochastic landscape evolution model. *Earth Surface Processes and Landforms, 30*(11), 1387–1403.

Dyurgerov, M. D., & Meier, M. F. (2005). *Glaciers and changing earth system: A 2004 snapshot, boulder (Colorado).* Institute of Arctic and Alpine Research, University of Colorado

Eriksson, M., Jianchu, X., Shrestha, A. B., Vaidya, R. A., Nepal, S., & Sandström, K. (2009). *The Changing Himalayas: Impact of Climate Change on Water Resources and Livelihoods in the Greater Himalayas* (p.23). International Centre for Integrated Mountain Development, Kathmandu.

Forrest, J. L., Wikramanayake, E., Shrestha, R., et al. (2012). Conservation and climate change: Assessing the vulnerability of snow leopard habitat to treeline shift in the Himalaya. *Biological Conservation, 150* (1), 129–135.

Gahalaut, V. K. (2008). Major and great earthquakes and seismic gaps in the Himalayan arc. *Memoir Geological Society of India* 1–21.

Gahalaut, V. K., & Kundu, B. (2011). Possible influence of subducting ridges on the Himalayan arc and on the ruptures of great and major Himalayan earthquakes. *Gondwana Research, 21*, 1080–1088.

Galy, A., & France-Lanord, C. (2001). Higher erosion rates in the Himalaya: Geochemical constraints on riverine fluxes. *Geology, 29*(1), 23–26.

Gao, J., Yao, T., Masson-Delmotte, V., Steen-Larsen, H. C., & Wang, W. (2019). Collapsing glaciers threaten Asia's water supplies. *Nature, 565*, 19–21. https://doi.org/10.1038/d41586-018-07838-4

Gao, X. J., Li, D. L., Zhao, Z. C., & Giorgi, F. (2003). Climate change due to greenhouse effects in Qinghai-Xizang Plateau and along the Qianghai-Tibet Railway. *Plateau Meteorology, 22*(5), 458–463.

Ghosh, S. (2025). *Key Aspects of Himalayan Geology* (pp. 117 and 141). Springer Nature. https://doi.org/10.1007/978-3-031-81186-9

Gogoi, A., Ahirwal, J., & Sahoo, U. K. (2022). Evaluation of ecosystem carbon storage in major forest types of Eastern Himalaya: Implications for carbon sink management. *Journal of Environmental Management, 302*(Part A), 113972. https://doi.org/10.1016/j.jenvman.2021.113972

Gupta, H., & Gahalaut, V. K. (2014). Seismotectonics and large earthquake generation in the Himalayan region. *Gondwana Research, 25*, 204–213.

Hewitt, K. (2005). The Karakoram anomaly? Glacier expansion and the 'elevation effects' Karakoram Himalaya'. *Mountain Research and Development, 25*(4), 332–340.

IPCC. (2007). Climate change 2007: Impacts, adaptation and vulnerability. In M. L. Parry, O. F. Canziani, J. P. Palutikof, P. J. van der Linden, C. E. Hanson (Eds.), *Contribution of Working Group II to the Fourth Assessment Report of the Intergovernmental Panel on Climate Change.* Cambridge University Press.

Jain, A. K., Mukherjee, P. K., & Singhal, S. (2020). Terrane characterization in the Himalaya since Paleoproterozoic. *Episodes, 43*(1), 346–357; https://doi.org/10.18814/epiiugs/2020/020021

Joshi, P., Adhikari, R., Bhandari, R., Shrestha, B., Shrestha, N., Chhetri, S., Sharma, S., & Routh, J. (2023). Himalayan watersheds in Nepal record high soil erosion rates estimated using the RUSLE model and experimental erosion plots. *Heliyon, 9*(5), e15800. https://doi.org/10.1016/j.heliyon.2023.e15800

Liu, S. Y., Ding, Y. J., Li, J., Shangguan, D. H., Zhang, Y. (2006). Glaciers in response to recent climate warming in Western China. *Quaternary Sciences, 26*(5), 762–771.

Maurya, S. K., Singh, V., Chand, K., & Mishra, P. K. (2024). Assessment of soil erosion in the Beas Valley, Kullu, Himachal Pradesh: A study of Western Himalayan landscape, Northern India. *Soil Science Annual, 75*(1), 1–12. https://doi.org/10.37501/soilsa/185558

Mishra, P. K., Rai, A., Abdelrahman, K., Rai, S. C., & Tiwari, A. (2022). Land degradation, overland flow, soil erosion, and nutrient loss in the Eastern Himalayas, India. *Land, 11*, 179. https://doi.org/10.3390/land11020179

Nakata, T. (1989). Active faults of the Himalaya of India and Nepal. *Geological Society of America Special Paper, 232*, 243–264.

Ni, F. (1989). Active tectonic of the Himalya. *Proceedings of the Indian Academy of Sciences (Earth and Planetary Science), 98*, 71–89.

Prashanth, M., Kumar, A., Dhar, S., Verma, O., Rai, S.K., & Kouser, B. (2023). Land use/land cover change and its implication on soil erosion in an ecologically sensitive Himachal Himalayan watershed, Northern India. *Frontiers in Forests and Global Change, 6*, 1124677. https://doi.org/10.3389/ffgc.2023.1124677

Rajendran, C. P., & Rajendran, K. (2022). *Earthquakes of the Indian Subcontinent* (p. 253). Springer Nature.

Ramanan, V., & Ramanan, J. (2023). *Mountains of our Destiny the Himalayas: An Expedition to the Indian Range*. The Hindu Press.

Ren, J. W., Qin, D. H., Kang, S. C., Hou, S. G., Pu, J. C., & Jin, Z. F. (2003). Glacier variations and climate warming and drying in the Central Himalayas. *Chinese Science Bulletin, 48*(23), 2478–2482.

Ruosteenoja, K., Carter, T. R., Jylhä, K., & Tuomenvirta, H. (2003). Future climate in world regions: an intercomparison of model-based projections for the new IPCC emissions scenarios. *The Finnish Environment, 644*, 83. Helsinki, Finnish Environment Institute.

Rupa Kumar, K., Sahai, A. K., Krishna Kumar, K., Patwardhan, S. K., Mishra, P. K., Revadkar, J. V., Kamala, K., Pant, G. B. (2006), High resolution climate change scenario for India for the 21st Century. *Current Science, 90*, 334–345.

Sattar. A., Cook, K. L., & Rai, S. K., et al. (2025). The Sikkim flood of October 2023: Drivers, causes, and impacts of a multihazard cascade. *Science 387*(6740). https://doi.org/10.1126/science.ads2659

Sattar, A., Goswami, A., & Kulkarni, A. V. (2019). Hydrodynamic moraine-breach modeling and outburst flood routing—A hazard assessment of the South Lhonak lake, Sikkim. *Science of the Total Environment, 668*, 362–378. https://doi.org/10.1016/j.scitotenv.2019.02.388

Searle, M. P., & Treloar, P. S. (2019). Introduction to Himalayan tectonics: A modern synthesis. *Geological Society London Special Publication, 483*, 1–17.

Shrestha, U. B., Gautam, S., & Bawa, K. S. (2012). Widespread climate change in the Himalayas and associated changes in local ecosystems. *PLoS ONE, 7*(5), e36741. https://doi.org/10.1371/journal.pone.0036741

Siraj, R., & Siraj, N. (2024). *Radhanath Sikdar and Mount Everest: A Himalayan mystery*. Empire Diaries, Kolkata. https://empirediaries.com/2024/05/17/radhanath-sikdar/

Sreejith, K. M., Jasir, M. C. M., Sunil, P. S., Rose, M. S., Saji, A. P., Agrawal, R., et al. (2024). Geodetic evidence for cascading landslide motion triggered by extreme rain events at Joshimath, NW Himalaya. *Geophysical Research Letters, 51*, e2023GL106427. https://doi.org/10.1029/2023GL106427

Srivastava, V. (2023). Sinking Himalayan town leaves thousands of homes at risk. https://doi.org/10.1038/d44151-023-00002-6

Stevens, V. L., Avouac, J. P. (2015). Interseismic coupling on the main Himalayan thrust. *Geophysical Research Letters, 42*(14), 5828–5837.

Sundriyal, Y., Kumar, V., Chauhan, N., Kaushik, S., Ranjan, R., & Punia, M. K. (2023). Brief communication: The northwest Himalaya towns slipping towards potential disaster. *Natural Hazards and Earth Systems Sciences, 23*, 1425–1431. https://doi.org/10.5194/nhess-23-1425-2023

Valdiya, K. S. (1998). *Dynamic Himalaya* (p. 178). Universities Press.

Valdiya, K. S. (2016). *The Making of India Geodynamic Evolution* (p. 924). Springer.

Wallace, K., Bilham, R., Blume, F., Gaur, V. K., & Gahalaut, V. (2015). Surface deformation in the region of the 1905 Kangra Mw = 7.8 earthquake in the period 1846 – 20001. *Geophysical Research Letters, 32*(15), 1–5.

Werhahn, G., Liu. Y., Meng, Y., et al. (2019). Himalayan wolf distribution and admixture based on multiple genetic markers. *Journal of Biogeography, 47*, 1272–1285.

Yao, T. D., Guo, X. J., Lonnie, T., Duan, K. Q., Wang, N. L., Pu, J. C., Xu, B. Q., Yang, X. X., & Sun, W. Z. (2006). δ^{18}O record and temperature change over the past 100 years in ice cores on the Tibetan Plateau. *Science in China: Series D Earth Science, 49*(1), 1–9.

Tectonic Hazard and Geomorphic Dilemma

Integrated Natural Hazards in the Himalaya: Insights from the Darjiling-Sikkim Himalaya

Malay Mukul, Vinee Srivastava, and Manas Mukul

Abstract Earthquakes, landslides, and glacial lake outburst floods (GLOFs) are commonly recognized as Himalayan natural hazards. Earthquakes originate from active mountain-building processes in the Himalaya, and as the Himalayan foreland is one of Earth's most densely populated regions, a large or great earthquake today will likely result in a significant loss of life and property. Prediction of earthquakes in time is an elusive goal at present. The only realistic option is strengthening existing commercial and residential infrastructure and ensuring that all future constructions can survive great earthquakes. The second significant Himalayan natural hazard is landslides. Monsoonal rains are the primary trigger for the Himalayan landslides. However, high topographic slopes and brittle fault-related deformation are critical factors contributing to Himalayan slope failure. Topographic slope growth results from neotectonics, and the internal architecture of brittle fault zones controls the strength of the fault rocks. The rocks in the brittle fault core and damage zone are fractured, and their grain size and strength are reduced. These regions are most susceptible to slope failure during and after the monsoon. Therefore, identifying and strengthening fault-affected and induced slopes is required to minimize the Himalayan landslide hazard. Changes in the Himalayan climate in response to Global warming have caused an increase in the melting and retreat of the Himalayan glaciers. When glaciers retreat, moraine-dammed glacier lakes develop and fill up. These lakes are the source of glacial lake outburst floods due to moraine dam failure. GLOFs result in catastrophic flooding downstream, with significant geomorphic and socioeconomic impacts. The worst-case natural hazard scenario in the Himalayas will involve large coeval earthquakes, landslides, and GLOF events. Himalayan natural hazard models must integrate seismic, landslides, and GLOF hazards for any realistic forecast of

M. Mukul (✉)
Continental Deformation Laboratory, Department of Earth Sciences, IIT Bombay, Mumbai, India
e-mail: malaymukul@iitb.ac.in

V. Srivastava
Department of Earth and Environmental Sciences, Indian Institute of Science Education and Research, Bhopal, India

M. Mukul
School of Computer Applications, Kalinga Institute of Industrial Technology, Bhubaneshwar, India

© The Author(s), under exclusive license to Springer Nature Switzerland AG 2025
S. Ghosh (ed.), *The Himalaya Dilemma*, Springer Natural Hazards,
https://doi.org/10.1007/978-3-031-95083-4_2

disasters and disaster preparedness. This chapter discusses the Himalayan integrated hazards scenario using the Darjiling-Sikkim Himalaya as an example and emphasizes that the scenario is relevant to the entire Himalayas.

Keywords Earthquakes · Seismic hazard · Landslides · Faults · GLOFs · Himalayan natural hazard · Darjiling-Sikkim Himalaya

1 Introduction

The Himalayan mountain chain is the product of the continent–continent collision of the Eurasian and Indian Plates and the destruction of the intervening Tethys Ocean over the last ~65 Ma. The Himalayan boundary is wedge-shaped in cross-section and arcuate in map view (Fig. 1). The Himalayan wedge is defined by a basal decollement named the Main Himalayan thrust (DeCelles et al., 1998; Nelson et al., 1996; Schelling & Arita, 1991; Srivastava & Mitra, 1994). The near-surface manifestation of this structure is recognized as the Main Frontal thrust (e.g., Nakata, 1989; Zhao et al., 1993; Lavé & Avouac, 2001; Bilham et al., 2001; Mukul, 2000; Mukul et al., 2007; Mukul, 2010; Srivastava et al., 2016, 2018). The Himalayan convergence continues today (Bilham, 2019; Bilham et al., 2001). This implies that the Himalayan deformation is manifested over different spatial and temporal scales. Deformation in the Himalaya typically ranges from plate scale (10^6 m) to even nanometer scale (10^{-9} m) (Mukul, 2005; Mamtani, 2025). Similarly, the Himalayan deformation can range from 10 s of millions of years to decadal time scales. This deformation can be quantified in terms of the total displacement vector consisting of translation, rotation, and pure strain (Mitra, 1994), both in space and time (e.g., McNaught & Mitra, 1996; Mukul, 1998, 1999; Mukul & Mitra, 1998; Bilham et al., 2001; Mukul et al., 2004, 2018; Srivastava et al., 2018; Bilham, 2019; Srivastava & Mukul, 2020). The active nature of the mountain belt is expressed in the seismicity observed in the belt that varies in magnitude, focal mechanism, and spatial distribution, resulting in variation in the Himalayan seismic hazard (Bilham, 2019; Ni & Barazangi, 1984). Decollement-related thrust earthquakes originating along the Main Himalayan thrust are the most destructive earthquakes possible in the Himalayan region. Nevertheless, large strike-slip and normal fault-related earthquakes have also been recorded in the Himalayan region. This chapter first discusses the significant Himalayan natural hazards and their integrated nature. Finally, we emphasize that Himalayan natural hazard models must integrate seismic, landslides, and GLOF hazards for any realistic forecast of disasters, disaster preparedness, and sustainable development.

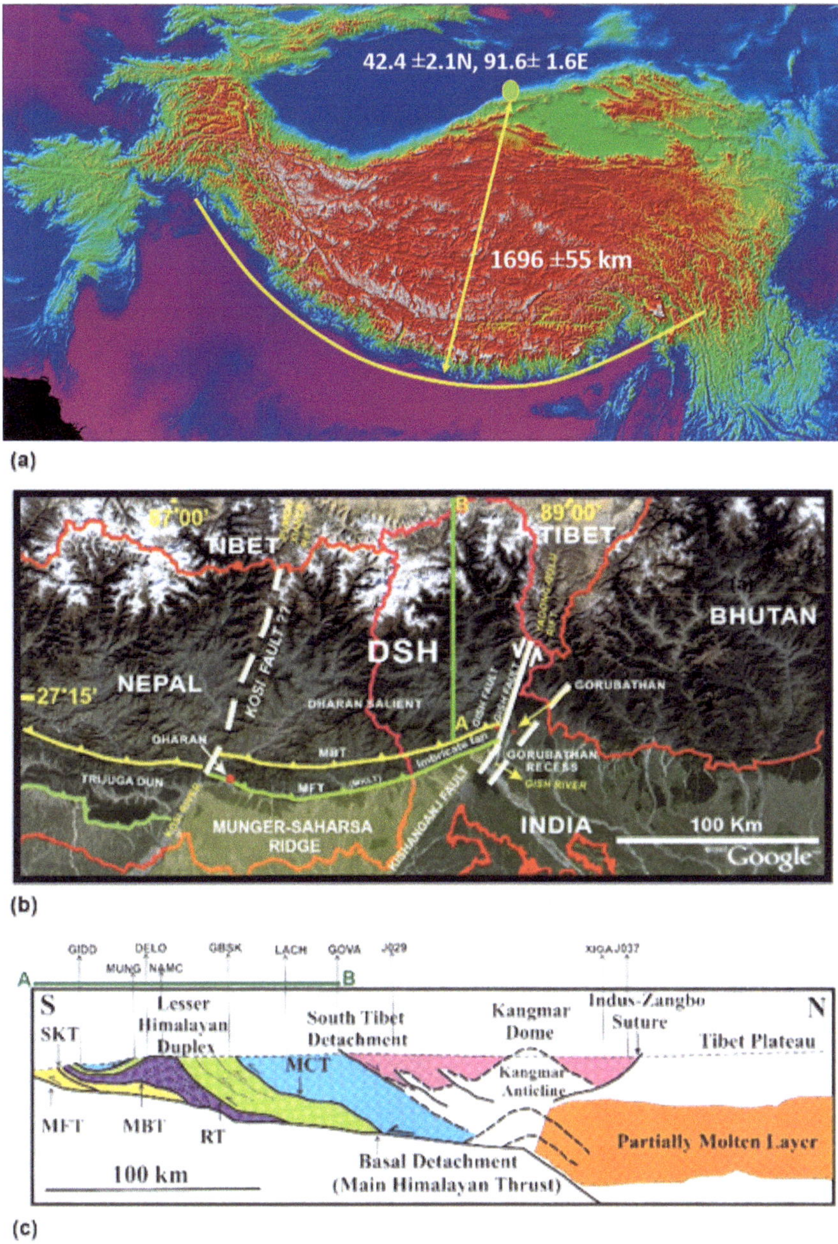

Fig. 1 The geometry of the Himalayan collision boundary. In the map view, the boundary is arcuate (**a**) (Bendick and Biham, 2001). The boundary, as seen along A-B in the Darjiling-Sikkim Himalaya (DSH) (**b**), is wedge-shaped (**c**), with the base of the wedge being defined by the Main Himalayan thrust (Mukul et al., 2024). The image in (**a**) was provided by and used with permission from Dr. Chris Duncan

2 The Himalayan Seismic Hazard

The Geological Survey of India (GSI) published India's first seismic zoning map in 1935. The Bureau of Indian Standards defined seven seismic zones (Zone 0 to Zone VI) in 1962 (IS 1893: 1962) (Fig. 2). The basis of the zonation is the frequency of significant to great earthquakes. Zone 0 was the lowest seismic hazard zone, and Zone VI was the highest. The zones were reduced to 5 in 1970 (Seismic Zones I to V), and Zones 0 and VI were eliminated. Eventually, Zone I was also eliminated in 2002. The present seismic zonation includes Zones II through V. The Himalaya falls in seismic Zones IV and V and is, therefore, one of the country's most vulnerable regions to seismic hazards (Fig. 2). Seismic Zone V is the most susceptible, typically characterized by the occurrence of great to large earthquakes in the past.

Decollement-related thrust earthquakes originating on the MHT are likely the most hazardous in the Himalaya. High-precision, Global Positioning System-based studies reveal that slip transfer along the MHT from the Himalayan hinterland in Tibet towards the foreland in the Himalayan wedge is the key to understanding the seismic hazard in the Himalaya (Bilham et al., 2001; Grandin et al., 2012; Mukul et al., 2018). The slip transfer is aseismic in the deeper, ductile part of the MHT. The aseismic slip is arrested at the brittle-ductile transition, which locks the frontal MHT south of the transition. During large and great earthquakes, the frontal MHT is unlocked, and slip transfer continues to the foreland during these seismic events (Bilham, 2019; Bilham et al., 2001; Mukul et al., 2007). In the extreme case, slip is transferred to the foreland and activates the Main Frontal thrust (MFT). Periodic reactivation of the MFT has been observed in paleoseismic and neotectonic studies in

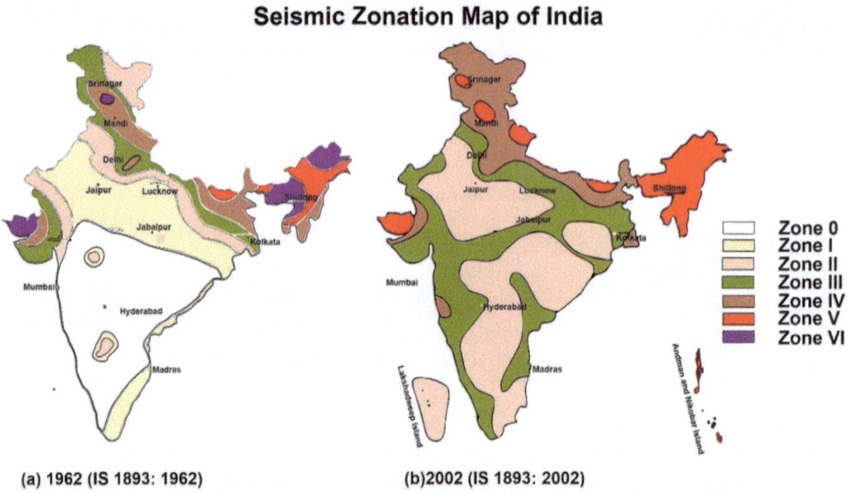

Fig. 2 The seismic zonation maps of India based on the Indian Bureau of Standards (IS 1893) as initially proposed in 1962 (**a**) with seven seismic zones from Zone 0 to Zone VI. The current 2002 zonation map (**b**) has dropped Zones 0, I, and VI

the Himalayan front (Wesnousky et al., 1999; Lavé and Avouac, 2001; Kumar et al., 2001, 2006, 2010; Burgess et al., 2012; Mugnier et al., 2013; Mukul & Singh, 2016; Srivastava et al., 2017, 2018; Rajendran et al., 2020; Brice et al., 2024). These studies provide two crucial insights. First, the brittle-ductile transition zone where the slip along the MHT changes from aseismic creep to seismic appears to nucleate all large earthquakes in the Himalaya. Second, the transfer of slip to the Himalayan foreland will only be possible when the Himalayan wedge is critical or super-critical, likely resulting in an in-sequence, great earthquake (Mukul et al., 2007, 2018). In the sub-critical Himalayan wedge, the slip transfer terminates north of the Himalayan front, resulting in out-of-sequence deformation. The April 25, 2015, magnitude 7.8 Gorkha earthquake was a recent example of a large out-of-sequence earthquake originating on the MHT. These studies conclude that the hypocentres of large and great Himalayan earthquakes will be located in the brittle-ductile transition zone along the MHT in the central Himalayan region near the surface trace of the Main Central thrust (MCT) in the Himalaya. The Lesser Himalayan Duplex (LHD) typically characterizes this zone. It is recognized as an active structure marked by a zone of low-to-intermediate interseismic seismicity in the Himalaya (e.g., Pandey et al., 1999; Kayal, 2001; Mukul et al., 2014, 2018; Mendoza et al., 2019).

Strike-slip earthquakes are also seen in the Himalaya (De & Kayal, 2004; Mukul, 2010; Mukul et al., 2009; Ni & Barazangi, 1984), apart from the thrust earthquakes. Strike-slip is the dominant earthquake mechanism in the Darjiling-Sikkim and the eastern Bhutan Himalaya. On September 18, 2011, the magnitude 6.9 Kanchenjunga earthquake was a recent example of a significant strike-slip earthquake in the Himalaya that occurred at ~50 km depth (Paul et al., 2015). From a hazard perspective, strike-slip earthquakes occur below the decollement and are unlikely to nucleate a large or great earthquake. They have been attributed to oblique deep-seated faults in the middle to lower crust and/or tear faults transverse to the Himalayan arc (Mukul, 2010, Mukul et al., 2014; Paul et al., 2015; Hubbard et al., 2021).

The Himalayan hinterland north of the South Tibet Detachment System (STDS) is characterized by normal fault-related earthquakes (e.g., Ni & Barazangi, 1984). The magnitude 7.1 quake that struck southern Tibet on January 7, 2025, at ~10 km along a N-S trending normal fault is an example of such an earthquake. The Himalayan hinterland shows an E-W extension along several N-S trending rifts (e.g., Yadong-Gulu Rift). In addition, N-S extension is also indicated by normal faulting along the E-W striking STDS. Like strike-slip faults, normal faults are also unlikely to nucleate a great earthquake, although they can host major earthquakes around magnitude 7.

2.1 Seismicity in the Darjiling-Sikkim Himalaya

Seismicity in the Darjiling-Sikkim Himalaya is very intriguing as earthquakes occur along thin-skinned style thrusts and deep-seated strike-slip faults, and moderate strike-slip seismicity dominates the recorded seismicity in the Darjiling-Sikkim Himalaya. Strike-slip hypocenters plot above and below the MHT at varying depths

(11–58 km) (de la Torre et al., 2007; Mukul et al., 2014; Paul et al., 2015) and are the dominant earthquake mechanism. Thrust earthquakes like the 14 February 2006 Sikkim earthquake of magnitude 5.3 (Mw) (Raju et al., 2007; Hazarika et al., 2010; Paul et al., 2015) have also occurred above and along the MHT, indicating near coeval strike-slip and thrust seismotectonics in the Darjiling-Sikkim and Bhutan Himalaya (e.g., Drukpa et al., 2006; Mukul et al., 2014 and references therein).

Some microseismicity has also been observed that exhibits strong epicentral concentration between the surface traces of the MBT and the MCT within the Lesser Himalayan Duplex (Baranowski et al., 1984; Ni & Barazangi, 1984; Ekström, 1987; Pandey et al., 1999; Kayal, 2001; De & Kayal, 2004; de la Torre et al., 2007; Mishra et al., 2010; Kumar et al., 2012; Mukul et al., 2014; Paul et al., 2015; Mukul et al., 2018).

2.1.1 Global Positioning System-Based Geodesy and Seismic Hazard in the Darjiling-Sikkim-Eastern Nepal Himalaya

The Indian plate is bounded by a divergent boundary in the Indian Ocean to the south and the Himalayan convergent collision boundary to the north. The composite ITRF GPS velocity measured at the southernmost station, Maldives, on the Indian Plate is 57.24 ± 0.44 mm/yr to the NE. This NE-directed velocity decreases to 55.06 ± 0.36 mm/yr at the center of the Indian Plate, Bangalore, and 51.33 ± 0.3 mm/yr at Hyderabad, farther to the north. The NE velocity remains around 53.17 ± 1.28 mm/yr at Durgapur in peninsular West Bengal south of the Darjiling-Sikkim Himalaya (Mukul et al., 2009, 2018) (Fig. 3).

GPS velocities were measured in the Dharan salient in the Darjiling-Sikkim-Eastern Nepal region between the Gish and the Kosi transverse structures (Figs. 1 and 3). In the foreland of the Dharan salient, the NE-directed velocity at Biratnagar, Nepal, was measured as 52.23 ± 3.16 mm/yr. The velocities measured in the frontal Dharan salient at Mirrik, Giddhapahar, and Mungpu in the Darjiling-Sikkim Himalaya remained NE-directed, between 51.5 and 53.6 mm/yr, and were statistically indistinguishable. This implied that the frontal Dharan salient was coupled with the Indian plate, and the MFT, MBT, and the frontal MHT were locked. Consequently, the India-fixed relative velocities for these stations were also statistically insignificant.

A reduction in velocity of ~2 mm/yr was observed across the Lesser Himalayan Duplex (LHD) in the Darjiling-Sikkim-Eastern Nepal Himalaya (Mukul et al., 2018; Adel et al., 2012). At the south end of the LHD, NE-directed velocities of ~51 mm/yr were observed at Labha, Delo, Soreng, and Namchi in the Dajiling-Sikkim Himalaya. At the north end of the LHD, NE-directed velocities of ~49 mm/yr were observed at Peling, Gangtok, and Kyongnosla in the Darjiling-Sikkim Himalaya, Khandbari, and Taplejung in the Eastern Nepal Himalaya (Table 1).

As the MHT in the frontal Himalaya appears locked and the frontal Himalaya is coupled with the Indian plate, the LHD is the region where slip along the MHT is transitioning from aseismic creep to the north to frictional sliding, getting arrested,

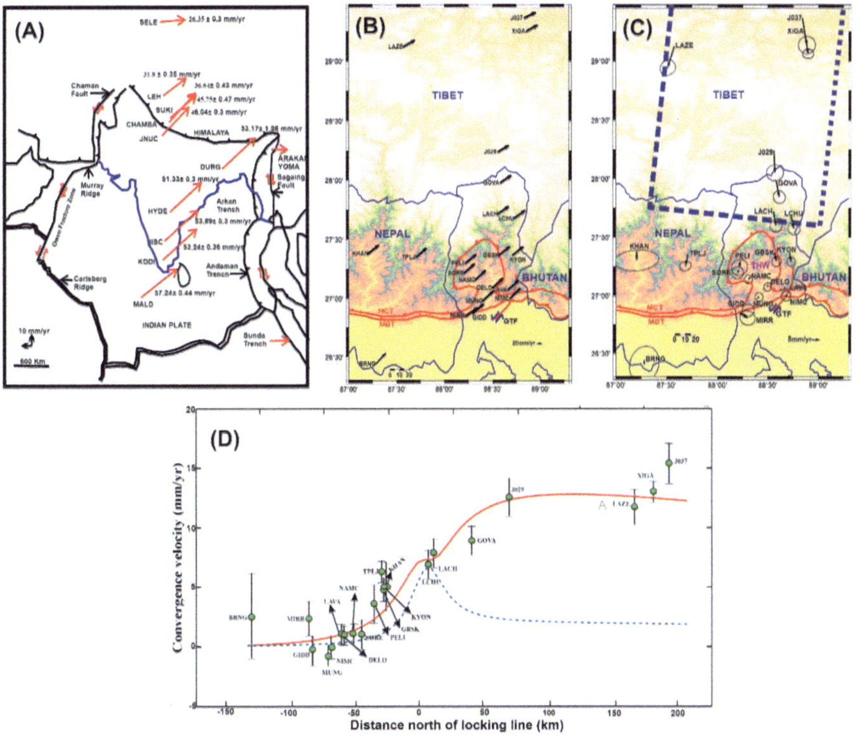

Fig. 3 International Terrestrial Reference Frame (ITRF) velocities in the (**a**) Indian Plate (after Mukul et al., 2009) and the (**b**) Darjiling-Sikkim-Nepal-Tibet area (Mukul et al., 2018). India-fixed velocities in the Darjiling-Sikkim-Nepal-Tibet Himalaya with the dislocation model to simulate these velocities (**c**) (Mukul et al., 2018). The frontal 150 km of the Main Himalayan Fault (MHT) is locked (**d**). The blue dashed line shows the modelled vertical uplift velocity. See Table 1 for station symbols and velocities

and converting to shortening. From a seismic hazard perspective, this zone will also be where future large-to-great thrust earthquakes will likely originate. This zone in the central part of the Darjiling-Sikkim-Eastern Nepal Himalaya is marked by micro- to moderate interseismic seismicity (e.g., Pandey et al., 1999; Kayal, 2001; De & Kayal, 2004; Mukul et al., 2014, 2018). Therefore, although earthquakes cannot be predicted in time, we know the Himalayan earthquakes will originate in the central part of the Himalaya. This insight can help save lives if the buildings and infrastructure in that part of the Himalaya are designed or reinforced to withstand large to great earthquakes.

The composite NE-directed IGS velocity decreased to ~48 mm/yr, as seen at Lachen, Gova, and Lachung in the Darjiling-Sikkim Higher Himalaya in the MCT thrust sheet and ~45–47 mm/yr in the Tethyan Himalaya in Tibet north of the South Tibet Detachment System at the Indo-Tibet border. The India-fixed, S-directed relative velocities for these stations increase from ~7 to 10 mm/yr in the Higher Himalaya

Table 1 The International Terrestrial Reference Frame (ITRF) 2005 and India-fixed velocities of GPS stations in Darjiling-Sikkim Himalaya and Tibet Plateau (modified from Mukul et al., 2018)

Station	Latitude (°N)	Longitude (°E)	Velocity-E (mm/yr) ITRF velocity	India-Fixed-E	Velocity-N (mm/yr) ITRF Velocity	India-Fixed-N	Composite ITRF Velocity (mm/yr)/E of N (°)
1. Peninsular India							
Maldives (MALD)	4.2	73.5	43.87 ± 0.99	-0.2 ± 1.11	34.26 ± 0.97	−0.17 ± 0.99	55.66 ± 0.98
							52.01 ± 1.00
Bangalore (IISC)	13.02	77.57	41.65 ± 0.36	− 0.05 ± 0.48	36.01 ± 0.36	3.27 ± 0.41	55.06 ± 0.36
							49.15 ± 0.37
Hyderabad (HYDE)	17.42	78.55	40.54 ± 0.55	− 0.42 ± 0.78	34.96 ± 0.55	0.01 ± 0.57	53.53 ± 0.55
							49.23 ± 0.59
Durgapur (DURG)	23.53	87.31	38.68 ± 1.30	− 1.50 ± 1.33	36.48 ± 1.29	3.03 ± 1.36	53.17 ± 1.29
							46.68 ± 1.39
2. Foreland of Darjiling-Sikkim-Eastern Nepal Himalaya							
Biratnagar (BRNG)	26.44	87.28	37.70 ± 2.83	− 1.46 ± 2.99	36.15 ± 3.49	2.70 ± 3.58	52.23 ± 3.16
							46.20 ± 3.5
Mirrik (MIRR)	26.84	88.22	43.62 ± 1.54	4.36 ± 1.58	31.09 ± 1.38	−2.37 ± 1.45	53.57 ± 1.49
							54.52 ± 1.54
Giddpahar (GIDD)	26.87	88.30	39.62 ± 1.19	0.35 ± 1.25	33.81 ± 1.16	0.34 ± 1.24	52.09 ± 1.18
							49.52 ± 1.29
Mungpu (MUNG)	26.98	88.40	39.97 ± 0.75	0.71 ± 0.84	34.21 ± 0.73	0.74 ± 0.86	52.61 ± 0.74

(continued)

Table 1 (continued)

Station	Latitude (°N)	Longitude (°E)	Velocity-E (mm/yr) ITRF velocity	India-Fixed-E	Velocity-N (mm/yr) ITRF Velocity	India-Fixed-N	Composite ITRF Velocity (mm/yr)/E of N (°)
							49.44 ± 0.8
Nim (NIMC)	27.00	88.69	39.11 ± 0.87	− 0.22 ± 0.95	33.48 ± 0.85	0.01 ± 0.97	51.48 ± 0.86
							49.43 ± 0.96
3. Lesser Himalaya-Dharan salient and Darjiling-Sikkim Himalayan wedge							
Labha (LAVA)	27.07	88.66	39.55 ± 0.75	0.26 ± 0.84	32.43 ± 0.73	−1.04 ± 0.86	51.15 ± 0.74
							50.65 ± 0.83
Delo (DELO)	27.09	88.50	39.01 ± 0.69	−0.23 ± 0.79	32.53 ± 0.67	−0.94 ± 0.81	50.79 ± 0.68
							50.18 ± 0.77
Namchi (NAMC)	27.16	88.32	38.26 ± 0.75	−0.91 ± 0.84	32.39 ± 0.73	−1.08 ± 0.86	50.13 ± 0.74
							49.75 ± 0.84
Soreng (SORE)	27.21	88.21	38.54 ± 1.14	−0.58 ± 1.20	32.42 ± 1.12	−1.04 ± 1.21	50.36 ± 1.13
							49.93 ± 1.28
Peling (PELI)	27.30	88.22	38.73 ± 1.60	−0.36 ± 1.65	29.85 ± 1.56	−3.61 ± 1.62	48.90 ± 1.59
							52.38 ± 1.84
Taplejung (TPLJ)	27.35	87.71	37.98 ± 0.55	−0.96 ± 1.17	31.24 ± 0.32	−2.21 ± 0.91	49.18 ± 0.47
							50.56 ± 0.5
Gangtok (GBSK)	27.37	88.57	40.24 ± 0.85	1.08 ± 0.94	28.69 ± 0.86	−4.78 ± 0.98	49.42 ± 0.85
							54.51 ± 0.99
Kyongnosla (KYON)	27.37	88.71	40.15 ± 0.78	0.95 ± 0.87	28.44 ± 0.75	−5.03 ± 0.88	49.20 ± 0.77

(continued)

Table 1 (continued)

Station	Latitude (°N)	Longitude (°E)	Velocity-E (mm/yr) ITRF velocity	India-Fixed-E	Velocity-N (mm/yr) ITRF Velocity	India-Fixed-N	Composite ITRF Velocity (mm/yr)/E of N (°)
							54.69 ± 0.89
Khandbari (KHAN)	27.38	87.21	38.97 ± 4.47	0.18 ± 4.47	29.91 ± 2.05	−3.53 ± 2.05	49.12 ± 3.75
							52.49 ± 3.70
4. Higher Himalaya- Dharan salient and Darjiling-Sikkim Himalayan wedge							
Lachung (LCHU)	27.68	88.73	40.76 ± 1.14	1.67 ± 1.21	26.55 ± 1.11	−6.92 ± 1.21	48.64 ± 1.13
							56.92 ± 1.32
Lachen (LACH)	27.72	88.56	40.38 ± 1.13	1.35 ± 1.20	25.58 ± 1.11	−7.89 ± 1.20	47.80 ± 1.12
							57.65 ± 1.34
Gova (GOVA)	27.98	88.59	40.59 ± 1.13	1.65 ± 1.20	24.54 ± 1.12	−8.93 ± 1.21	47.43 ± 1.13
							58.84 ± 1.35
5. Tethyan Himalaya and Tibet Plateau-Darjiling-Sikkim-Tibet Himalayan wedge							
Gamba (J029)	28.24	88.56	39.65 ± 1.71	0.81 ± 1.71	22.45 ± 1.61	−11.02 ± 1.61	45.56 ± 1.69
							60.48 ± 2.06
Lhatse (LAZE)	29.12	87.58	41.44 ± 1.39	3.22 ± 1.56	23.76 ± 1.11	−09.69 ± 1.48	47.77 ± 1.33
							60.17 ± 1.42
Xigatze (XIGA)	29.25	88.86	40.26 ± 1.15	1.72 ± 1.15	22.41 ± 0.89	−11.06 ± 0.89	46.08 ± 1.09
							60.90 ± 1.19
Zegong A (J037)	29.36	88.84	41.33 ± 1.71	2.84 ± 1.71	19.58 ± 1.71	−13.89 ± 1.71	45.73 ± 1.71
							64.65 ± 2.14

(continued)

Table 1 (continued)

Station	Latitude (°N)	Longitude (°E)	Velocity-E (mm/yr) ITRF velocity	India-Fixed-E	Velocity-N (mm/yr) ITRF Velocity	India-Fixed-N	Composite ITRF Velocity (mm/yr)/E of N (°)
6. India-fixed Composite Velocities – Higher and Tethyan Himalaya and Tibet Plateau							
Lachung (LCHU)	27.68	88.73		1.67 ± 1.21		−6.92 ± 1.21	7.12 ± 1.21
							-13.57 ± 9.0
Lachen (LACH)	27.72	88.56		1.35 ± 1.20		−7.89 ± 1.20	8.00 ± 1.2
							-9.74 ± 8.6
Gova (GOVA)	27.98	88.59		1.65 ± 1.20		−8.93 ± 1.21	9.1 ± 1.21
							-4.2 ± 8.59
Gamba (J029)	28.24	88.56		0.81 ± 1.71		−11.02 ± 1.61	11.05 ± 1.61
							-4.2 ± 8.86
Lhatse (LAZE)	29.12	87.58		3.22 ± 1.56		−09.69 ± 1.48	10.21 ± 1.5
							-18.38 ± 8.71
Xigatze (XIGA)	29.25	88.86		1.72 ± 1.15		−11.06 ± 0.89	11.19 ± 0.9
							-8.84 ± 6
Zegong A (J037)	29.36	88.84		2.84 ± 1.71		−13.89 ± 1.71	14.18 ± 1.71
							-11.56 ± 6.91

and ~11 to 14 mm/yr in the Tethyan Himalaya from south to north. The slip along the MHT that generates these surface velocities in the Higher and Tethyan Himalaya is attributed to aseismic creep (Mukul et al., 2018). These velocities also indicate that the slip along the MHT originates near the Indus-Zangbo suture (Fig. 3).

The active deformation in the Darjiling-Sikkim-Eastern Nepal Himalaya can be best explained by a geologically consistent dislocation model within the Dharan salient based on a reverse aseismic slip of ~18 mm/yr on an MHT equivalent single elastico-frictional dislocation dipping 06° → 006°, locked at a depth of ~16 km over a length of 160 km between 27.76° N, 87.36° E and 27.61° N, 88.99° E (Mukul et al., 2018). Seismotectonic and seismic hazard models must consider slip along the MHT as the causative mechanism for earthquakes in the Dharan salient in the Darjiling-Sikkim-Eastern Nepal Himalaya. Past seismic hazard models in the region are built around either the MBT (De, 2000) or the MCT (De & Kayal, 2004) being seismically active (Nath et al., 2005). These models are now redundant, and the only approach taking the MHT as the seismogenic fault (Bilham, 2019; Bilham et al., 2001) predicts > 10 m of potential slip in the Dharan salient and a possible M_w = 8.4 earthquake.

Therefore, we conclude that the seismic hazard in the Dharan salient is primarily related to decollement tectonics along the Main Himalayan thrust. The region is classified as Seismic Zone IV, which is not as hazardous as Seismic Zone V. However, GPS studies indicate that ~18 mm/yr of slip has been steadily accumulating in the LHD zone since the last great earthquake in the region. There are no reports of large-to-great thrust earthquakes in the region, so we do not know how much slip has accumulated along the decollement. Such a scenario is more hazardous than Seismic Zone V, where the strain was released during the last large-to-great earthquake, and strain accumulation has only occurred after the earthquake for a shorter time. This insight suggests that it might make more scientific sense to switch Seismic Zones IV and V. Therefore, the Darjiling-Sikkim-Eastern Nepal Himalaya seems overdue for a large-to-great earthquake whose hypocenter will be located between Gangtok and Lachen/Lachung. The next large earthquake in these areas will cause large-scale loss of life and property. The time to minimize or prevent this is, now instead of waiting for the event to happen and managing the disaster. This insight can help save lives if the buildings and infrastructure in the Darjiling-Sikkim-Eastern Nepal Himalaya are designed or reinforced to withstand great earthquakes.

3 The Himalayan Landslide Hazard

Landslides constitute a major global natural hazard that causes widespread destruction of infrastructure and loss of human lives and property (USGS 2013). Between 1998 and 2017, landslides affected an estimated 4.8 million people and caused more than 18,000 deaths (WHO). As per the World Bank Report (2020), on average, 3600 significant earthquake-triggered and 31,430 rainfall-triggered landslides (1980–2018) occur annually in India. Landslide hazard is a global issue, especially in tectonically active regions, e.g., the Himalaya, that already face the risk of earthquakes and are always on the brink of natural disasters that threaten millions of lives and infrastructure. These landslides are the consequence of natural causes such as rain and earthquakes. Still, they are also frequently aided by anthropogenic activities that make slopes further unstable and damage the affected regions.

Earthquakes and rainfall are the primary natural triggers for landslides. However, anthropogenic developmental activities such as roads, tunnels, and dams may also greatly aid these natural triggers, making a previously stable area more susceptible to landslides. The significance of human-induced triggers for landslides has become crucial as mountainous regions sustain millions of people and their livelihoods. The global rise in population in these areas has led to extensive infrastructure development, placing considerable stress on the fragile natural habitats of the mountains. Consequently, the impact of anthropogenic factors on landslides has intensified.

3.1 Landslide Disasters in Darjiling-Sikkim Himalaya

The Darjiling-Sikkim Himalaya has always been susceptible to landslides (Fig. 4), especially during the monsoons and occasionally during great earthquake events that culminate in significant loss of lives and property (Starkel, 1972; Naithani, 1999; Bhasin et al., 2002; Basu & De, 2003; Chakraborty et al., 2011 and references therein). As per the landslide inventory of India, a total of 1,569 landslides in Sikkim and 172 in Darjiling, West Bengal, between 2011 and 2017 were reported (The Landslide Atlas of India by NRSC-ISRO, 2023). The south, east, west, and north Sikkim rank from highest to lowest with 8th, 9th, 20th, and 51st, respectively, with decreasing landslide density regarding socio-economic parameters (NRSC-ISRO, 2023).

Various studies were done on the landslide occurrences, susceptibility, and identifying potential regions for reactivating landslide surfaces and their mitigation in the Darjiling-Sikkim Himalaya based on remote sensing, geophysical, and engineering geology techniques (Starkel, 1972; Naithani, 1999; Starkel et al., 2000; Bhasin et al., 2002; Sengupta et al., 2010; Mehrotra et al., 1996; Sengupta et al., 2010; Anbarasu et al., 2010; Anbalagan et al., 2015; Gupta et al., 2022; Joshi, 2022; Alam et al., 2024 and references therein). Over 125 years of historical landslide data indicate that the largest landslides were dominantly rainfall-induced, with some accompanied by cloud burst-driven flash floods in the Darjiling Himalaya (Fig. 5), causing numerous fatalities (Pal et al., 2016). During 1899, 1935, 1950, 1976, 1991, 1993, 2003–2006, 2009, 2011, and 2015 major landslides mainly confined to the Tista basin at Darjiling Town, Pulbazar, Kurseong, Rongtong, Tindharia, Kalimpong, Nimbong, and Mirik, which lies outside of the Tista Basin (cf, Figs. 2 and 4 of Pal et al., 2016 and references therein). In 1968, the great Tista flood due to the cloud burst caused a major landslide that devastated the Tista valley and impacted Darjiling town, Manipuri, Tista Bazar, Lebong, and Kalimpong, resulting in 667 fatalities and extensive destruction of several tea gardens (Pal et al., 2016).

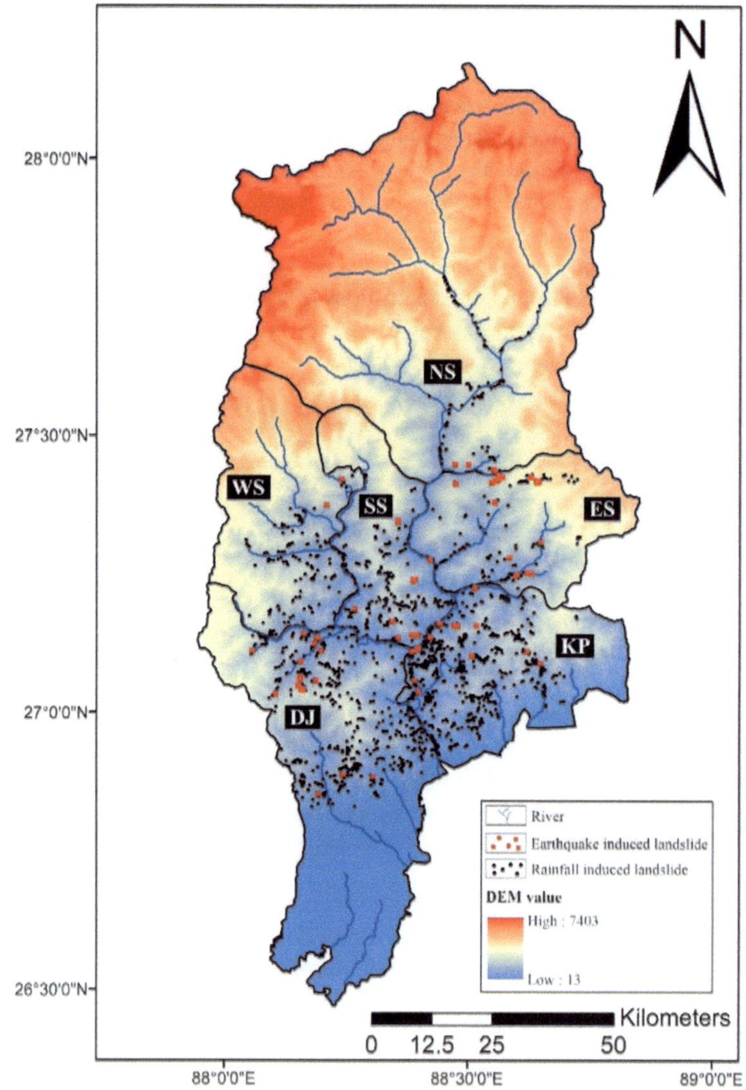

Fig. 4 The Darjiling-Kalimpong and Sikkim region's Landslide inventory map in the Darjiling-Sikkim Himalaya (DSH). The data was downloaded from the Bhukosh data portal and ranges from 2007 to 2018. (*Source* https://bhukosh.gsi.gov.in/). NS: North Sikkim, WS: West Sikkim, ES: East Sikkim, SS: South Sikkim, DJ: Darjiling, KP: Kalimpong

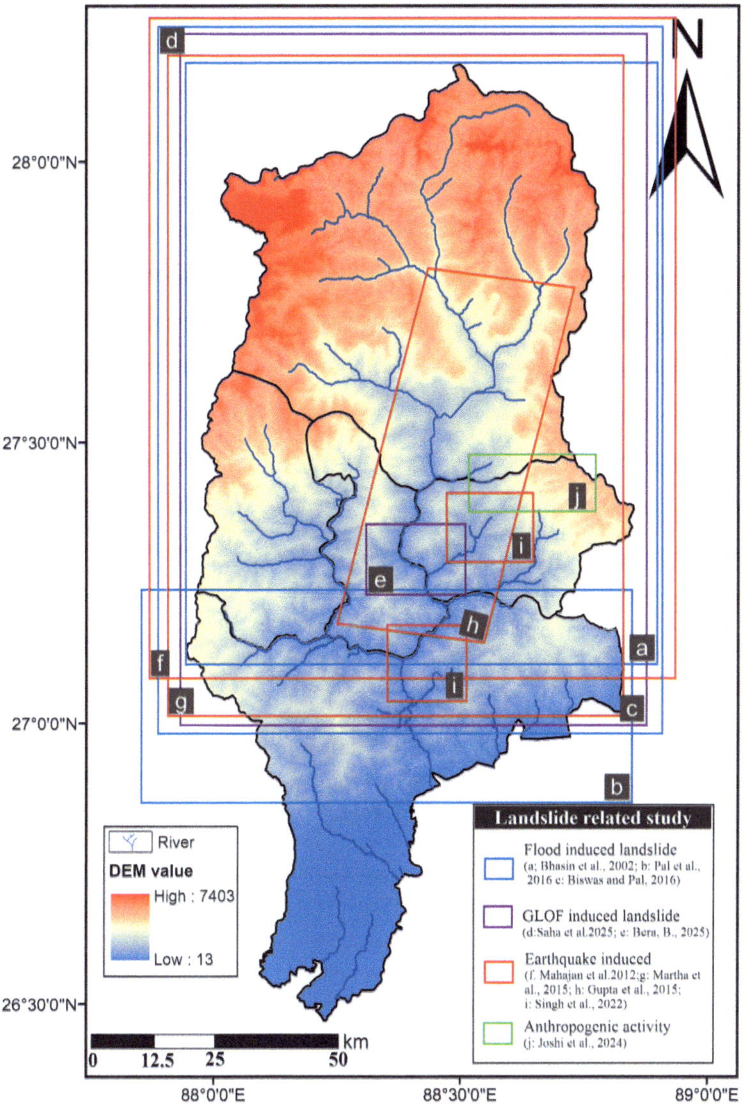

Fig. 5 Representative studies of Flood (Bhasin et al., 2002; 2012; Pal et al., 2016; Biswas and Pal, 2016), Earthquake (Mahajan et al., 2012; Martha et al., 2015; Gupta et al., 2015; Singh et al., 2022), GLOF (Saha et al., 2025; Bera, 2025), and Anthropogenic activity (Joshi et al., 2024) induced landslide occurrences in the Darjiling-Sikkim Himalaya

The Tista River flowed into the Ganga River till the 18th century; after a massive 1787 flood, it shifted its course and attained its present course into the Brahmaputra (Agarwal & Narain, 1991). The cloud burst-driven major flash flood events of 1968 and 2015 were accompanied by significant landslide occurrences in the Tista basin that resulted in extensive destruction of life and property (Agarwal & Narain, 1991; Bhattacharya, 2015). Besides monsoonal or cloudburst-induced extreme rainfall, earthquakes, and GLOF events also cause landslide occurrences. In 2011, the Sikkim earthquake triggered nearly 1,200 landslides across Sikkim, affecting the Mangan, Theng, Chungthang, Bichhu, and Singtham areas in North Sikkim. NH10 highway along the Tista River and its tributary streams suffered severe damage due to these landslides (Martha et al., 2015; Mahajan et al., 2012). In October 2023, flash floods caused by a glacial lake outburst flood (GLOF) due to moraine collapse in the South Lhonak Lake led to significant landslides in the lower catchments (Sattar et al., 2025; Saha et al., 2025, and references therein). Furthermore, on August 20, 2024, a landslide occurred at Dipu-Dara near Singtam in Gangtok District, covering an area of 0.02 km^2 and collapsing onto the powerhouse of the NHPC's 510 MW Tista V hydropower project along the Singtam-Dikchu road. Recent studies have examined landslide vulnerability in this region (Joshi et al., 2024; Poddar & Roy, 2024; Sonker et al., 2024).

All triggers respond to the rheological properties of the material that constitute the slopes that are susceptible to failure. This makes a region"s geology essential in understanding an area's landslide susceptibility. This becomes even more critical in tectonically active and structurally deformed regions like the Himalaya, where numerous faults, fractures, joints, and other deformation structures increase landslide susceptibility as they reduce the material strength along the slopes. Fault cores have an extreme grain-size reduction, forming powdery fault gouges with minimal strength. Therefore, the key issue in such deformed zones is to assess the landslide initiation susceptibility of a region and act before the initial landslide occurs. Various studies have been conducted in the Darjiling-Sikkim Himalaya to assess and determine the slope and for landslide susceptibility analysis, prediction, and mitigation (Pal et al., 2016; Sarkar, 2018; Nath et al., 2021; Mandal et al., 2022, and references therein). Deformation in the Himalayas also causes an increase in the topographic slope, especially in zones with maximum uplift, such as the Lesser Himalayan Duplex. Aseismic deformation in the Higher Himalaya will likely increase slopes even without earthquakes. These can occur over decadal (10 years) or neotectonic (1000 years) time scales. Therefore, identifying and strengthening fault-affected and induced slopes is crucial to recognizing and minimizing the Himalayan landslide hazard.

The Himalayan region is growing rapidly, and dams, tunnels, and roads are also being constructed to provide the community with a livelihood. Dams built in the Himalayas to meet the ever-increasing power demand constitute one such infrastructure. Studies show that the dams directly interact with landslides worldwide, implying that dams either have been built on pre-existing landslides or have been subjected to landslide activity during or after construction (e.g., Schuster, 2006). Studies conducted on landslides around the world focus mainly on assessing motion within pre-existing landslides using various conventional surveying tools/

geotechnical methods and modern techniques e.g. theodolite, Electronic Distance Measurement (EDM) levels, total station, inclinometers, wire extensometers (Bertachini et al., 2009; Calcaterra et al., 2012), GPS, terrestrial laser scanning, SAR interferometry, and photogrammetry (Gili et al., 2000; Malet et al., 2002; Yalçinkaya & Bayrak, 2002; Coe et al., 2003; Rott & Nagler, 2006; Tagliavini et al., 2007; Peyret et al., 2008; Bertachini et al., 2009; Calcaterra et al., 2012; Rawat et al., 2011; Wang 2012; Skilodimou et al., 2018). In addition, the anthropogenic impact of land use on landslides has also been studied (Schuster, 2006; Michoud et al., 2011). A few studies have also been conducted in India using episodic GNSS to assess the slope stability in pre-existing landslides in the eastern Sikkim Himalaya (Rawat et al., 2011) and to analyze the impact of haphazard road construction as the effect of anthropogenic activity on landslides in the Uttarakhand Himalayan region (Ramasamy et al., 2014).

Most studies, specifically in India, have focused on landslides affected by natural triggers with little or no focus on anthropogenically affected or induced landslides. Landslide occurrences have increased globally due to anthropogenic activities (Froude & Petley, 2018). Seven human-induced changes or actions potentially affecting slope stability have been identified. These are: 1. Slope re-profiling, 2. Groundwater flow perturbation and quick pore pressure change, 3. Surface water overland flow modifications, 4. Land-use changes and land degradation, 5. Inappropriate artificial structures, 6. Vibration and explosive use, and 7. Aging and degradation of infrastructure (Jaboyedoff et al., 2016). Land use in extensive cultivation, road, and dam constructions are the major anthropogenic activities causing landslides globally (Swanson & Dyrness, 1975; Schuster, 2006; Zhang et al., 2012; Skilodimou et al., 2018, and references therein). There are very few studies in India, especially in the Himalaya, that explore human impact on landslides (e.g., Haigh and Rawat 2011; Ramasamy et al., 2014). The impact of anthropogenic activity on the Himalayan landslides has been studied, focusing on road construction in the Uttarakhand Himalaya (Ramasamy et al., 2014). However, no studies are available on the slope stability around dams and its impact on the communities living around them in the Himalaya. The increasing population in the Himalayan Mountains and the subsequent construction of infrastructure to sustain them pose a threat through natural and anthropogenically aided disasters.

4 The Himalayan Glacial Lake Outburst Flood Hazard

Glacial Lake Outburst Floods (GLOFs) are catastrophic events resulting from the sudden release of water from a moraine-dammed lake formed by glacial melt. These lakes, typically situated in high-altitude regions, can be destabilized by various factors, including increased glacial melt, seismic activity, incessant rainfall, cloudbursts, and landslides. The rapid discharge of a massive volume of water from a glacial lake can cause significant downstream devastation, including loss of human life, disruption of livelihoods, damage to infrastructure, environmental degradation (such as erosion and landscape alteration), increased sediment loads in rivers, and substantial economic and psychological impacts on affected communities.

The earliest recorded GLOF event occurred in 1595 (Zryd, 2008) when the failure of a glacial lake dam on the Gietro Glacier in southwestern Switzerland resulted in a flood that claimed 140 lives. While GLOFs have historically occurred in various regions with glaciers, including the Andes (Emmer et al., 2022), the European Alps (Emmer et al., 2015), the Rocky Mountains in North America (Shaw et al., 2017), Iceland (Guan et al., 2015) Greenland (Carrivick et al., 2017; Grinsted et al., 2017), and Scandinavia (Xu et al., 2015), the high-mountain regions of Asia, encompassing the Hindu Kush (Aggarwal et al., 2017), Karakoram (Round et al., 2017), and Himalaya (Gurung et al., 2017; Watanabe et al., 1998; Zhang et al., 2024) exhibit exceptionally high vulnerability.

The Himalayas constitute a fragile ecosystem highly susceptible to natural hazards. The region's steep slopes and rugged terrain intensify this risk, making it highly prone to landslides, particularly during the monsoon season. Furthermore, the rapid development of infrastructure, including roads, railways, tunnels, and high-rise buildings, alongside the expansion of settlements within densely populated Himalayan towns, significantly increases the pressure on this fragile ecosystem. These anthropogenic activities are maximum in the river valleys, often disrupting natural drainage systems and increasing the likelihood of landslides and flash floods. The Himalayas are also home to approximately 8,000 glacial lakes (Ives et al., 2010), formed by the accumulation of meltwater from receding glaciers, mainly in the Higher and Tethyan Himalaya. Climate change, driven by global warming, has accelerated glacial retreat, leaving a large volume of water within these lakes. Glaciers continue to shrink alarmingly, and new glacial lakes are also emerging. The expanding volume of water in these glacial lakes significantly increases the risk of glacial lake outburst floods (GLOFs), posing a substantial threat to downstream ecosystems. Consequently, the impact of natural hazards in this region can be catastrophic, affecting millions of inhabitants. Moreover, disaster recovery efforts face significant challenges due to the rugged terrain, which hinders access to relief and assistance.

4.1 Distribution of Glacial Lakes in the Himalaya

The Himalayan region is widely acknowledged as one of the most susceptible to GLOFs globally (Carrivick & Tweed, 2016; Ives, 1986; Liu et al., 2014; Quincey et al., 2005; Worni et al., 2013; Yamada, 1993). Since 1990, the number and size of glacial lakes in the Himalaya have increased (Shugar et al., 2020). This phenomenon is likely attributed to the region's higher rate of warming than the global average (Shrestha et al., 2012). While many glacial lakes exist in the Himalaya, their distribution is uneven. Numerous studies indicate that glaciers in the eastern Himalaya are retreating at a faster rate than those in the western Himalaya (Kargel et al., 2005; Gardelle et al., 2011, 2013; Worni et al., 2013). Consequently, the most critical glacial lakes (Mohanty & Maiti, 2021), those with a high risk of outburst floods, are concentrated in the central and eastern Himalaya. Research interest in Himalayan glacial lakes has grown considerably, resulting in numerous published lake inventories. However, discrepancies in lake numbers often arise due to variations in the timeframes and methodologies employed by different studies. Nie et al. (2013) reported 4950 lakes across the Himalaya. Regional studies have further elucidated the distribution of these lakes. For instance, Zhang et al. (2015) identified 594 lakes in the Amu Darya basin, 1607 in the Indus basin, 364 in the Ganges basin, and 2247 in the Brahmaputra basin. In the Indian Himalaya, Bhambri et al. (2011) documented 1266 lakes in Uttarakhand, while 1104 glacial lakes were reported from Sikkim (Aggarwal et al., 2017). In neighboring Nepal, Jeelani et al. (2011) identified 1466 lakes. The escalating global temperatures and the rise in GLOFs, particularly in the Himalaya, have spurred increased research interest. A comprehensive and regularly updated database of all glacial lakes is crucial for effective monitoring, hazard assessment, and mitigation strategies. Such a database would be invaluable for researchers, policymakers, and disaster management agencies.

4.2 GLOF Events in the Himalaya

The International Centre for Integrated Mountain Development (ICIMOD, 2022) database, last updated in 2024, documents approximately 736 instances of GLOFs in the high mountain region of Asia. Of these, 493 events were reported from the Himalaya between 1533 and 2024, spanning Nepal, India, China, Bhutan, and Afghanistan (Fig. 6). China and Pakistan collectively accounted for approximately 70% of these Himalayan GLOFs, with 202 and 149 events recorded, respectively. India, Nepal, and Bhutan reported 60, 57, and 21 GLOF events, while Afghanistan recorded only 4. Out of the 60 recorded GLOFs in India, a significant portion, 40 events, originated from the Ladakh region. Sikkim followed with 10 reported events. Uttarakhand and Himachal Pradesh each experienced 2 GLOFs, while Arunachal Pradesh recorded a single event.

Glacial lakes can form through various mechanisms, including moraine damming, ice damming, supraglacial formation, and bedrock erosion. Among these, moraine-dammed lakes exhibit the highest susceptibility to GLOFs. These lakes are formed as retreating glaciers leave behind unstable piles of debris (moraines) that act as dams. According to the ICIMOD (2022), more than 50% of recorded GLOFs in the Himalaya originate from moraine-dammed Lakes. Ice-dammed lakes account for approximately 20% of GLOF occurrences. The remaining 30% are bedrock, supraglacial, and water-pocket glacial lakes. In the Ladakh region of India, approximately 63% of recorded GLOFs originated from ice-dammed lakes, while the remaining 27% were attributed to moraine-dammed lakes. In contrast, 9 out of the 10 recorded GLOF events in Sikkim stemmed from moraine-dammed lakes, with a single event originating from a supraglacial lake. The high proportion of GLOFs (90%) originating from moraine-dammed lakes in Sikkim is particularly concerning,

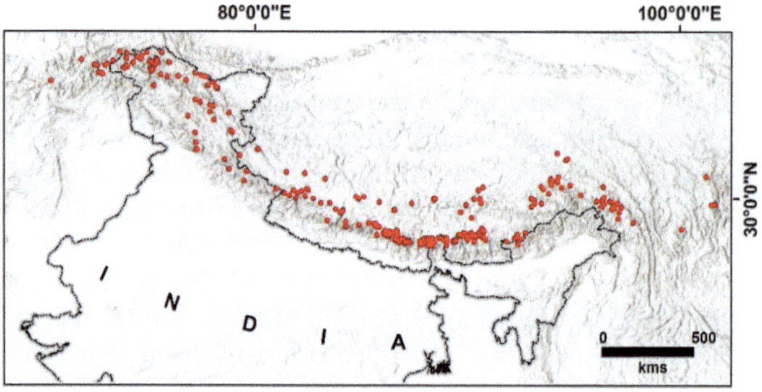

Fig. 6 Distribution of the GLOF events in the Himalaya between 1533 and 2024 (ICIMOD, 2022)

as these lakes are inherently susceptible to recurrent events due to the instability of the moraine dams. Between 2009 and 2024, nine GLOFs were recorded in India. Seven occurred in Ladakh, while one each was reported from Sikkim and Uttarakhand. The 2013 Uttarakhand GLOF, originating from the moraine-dammed Chorabari glacier in the Alakananda Basin, was catastrophic. This event resulted in the tragic loss of approximately 6,000 lives, the destruction of 138 residential structures, and damage to 56 houses (Allen et al., 2016). The most recent GLOF event in 2023 resulted from the collapse of the lateral moraine of the South Lhonak Lake in Sikkim, India. This catastrophic event resulted in the loss of 174 lives and widespread infrastructure damage. A total of 3 hydropower projects, 33 bridges, and 16 roads were destroyed. Furthermore, it triggered extensive landslides and severely impacted highways along the Tista River, with damage extending up to 139 km downstream (Sattar et al., 2025; Zhang et al., 2024).

4.3 Hazards of GLOF in Sikkim

Sikkim, a small Himalayan state encompassing 7,096 km^2 with a population exceeding 600,000, exhibits a significant portion of its northern and western districts within the cryosphere. Characterized by rugged terrain covering more than 43% of its area, the state experiences heavy rainfall and snowfall, with an annual precipitation of approximately 2,500 mm (GSI, 2012). Moreover, its location within the High Damage Earthquake Risk Zone (Zone IV) necessitates consideration of potential large and great earthquakes. Furthermore, Sikkim hosts over 1100 glacial lakes, collectively covering an area of 30.5 km^2 (Aggarwal et al., 2017). Global warming contributes to the expansion of these glacial lakes due to accelerated glacial retreat, increasing their susceptibility to GLOFs. This has consequently encouraged research interest (Aggarwal et al., 2017; Sattar et al., 2021; Tripathi et al., 2022; Hazra & Krishna, 2021; Chanda & Biswas, 2021; Remya et al., 2019; Rawat et al., 2022; Chowdhury et al., 2021) in investigating the GLOF susceptibility of these lakes.

Aggarwal et al. (2017) identified 12 glacial lakes in Sikkim as susceptible to GLOFs. Among these, South Lhonak and South Eastern Gurudongmar lakes, impounded by moraines, exhibited the highest susceptibility. These two lakes are situated within the Tista River basin, one of Sikkim's two major river basins, alongside the Rangeet River basin (Fig. 7).

The Gurudongmar Glacier Lake Complex comprises three large and one significantly smaller lake in the Upper Tista basin of the Northeast Sikkim Himalaya. West Upper and East Upper Gurudongmar Lakes lie to the south, overlooking Gurudongmar Lake. These Upper Gurudongmar Lakes are glacier-fed, moraine-dammed bodies of water. Water from the upper lakes subsequently flows into Gurudongmar Lake to the north, draining into the uppermost tributary of the Tista

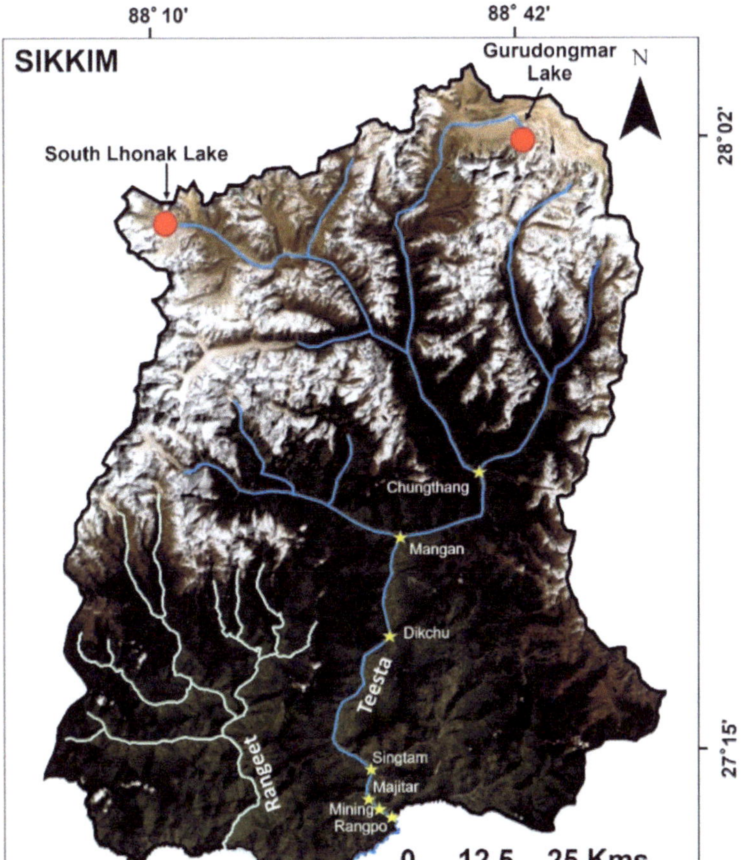

Fig. 7 Important towns located along the Tista basin vulnerable to GLOFs (Image from ArcGIS basemap database)

River (Fig. 8). A study by Chowdhury et al. (2021) quantified an increase of 1.5 km^2 in the combined area of the Gurudongmar Lakes between 1962 and 2018. Since 2010, the expansion rate has accelerated to 0.03 km^2 per year. Among the four lakes, East Upper Gurudongmar Lake exhibited the most rapid expansion, increasing its area by 1.20 km^2 at a rate of 0.22 km^2 per year. Overhanging the Gurudongmar Lake, the East Upper Gurudongmar Lake poses a significant risk due to its high susceptibility to a potential glacial lake outburst flood (GLOF).

Sattar et al. (2021) identified South Lhonak Lake (Fig. 8) as the largest proglacial lake in Sikkim, with a projected future volume of 114.8×10^6 m^3, rendering it highly susceptible to GLOFs. The study further highlighted the lake's expansion towards steep slopes. This proximity increases the risk of avalanches impacting the lake surface, potentially generating impulse waves and triggering a GLOF. Sharma et al., (2018) determined the average annual expansion rate of South Lhonak Lake to be

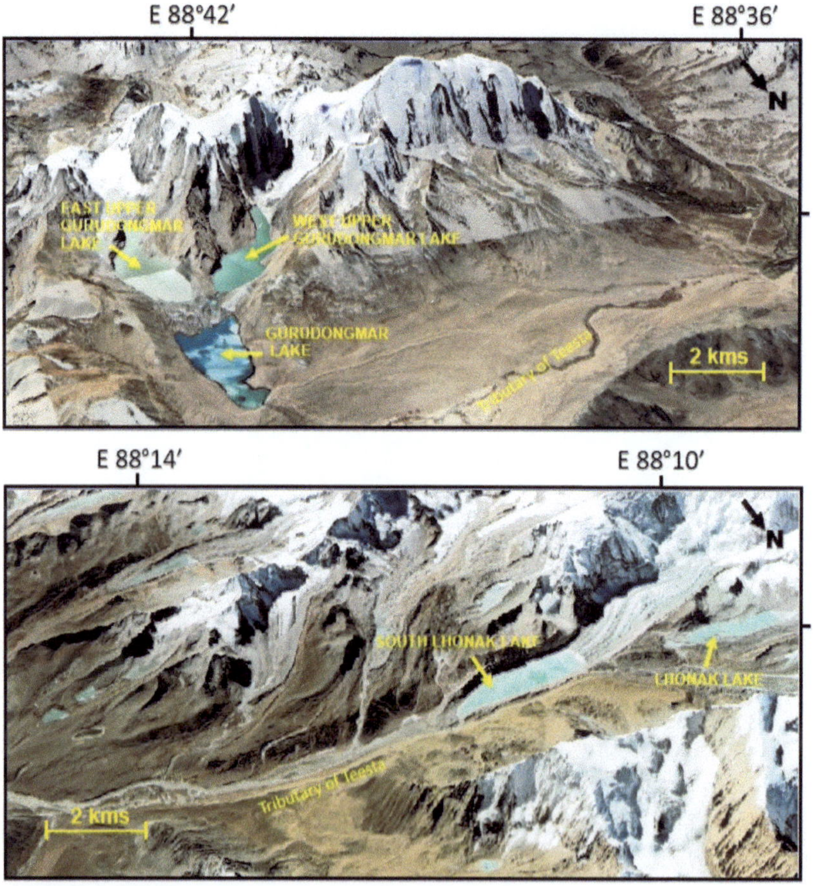

Fig. 8 Gurudongmar and South Lhonak Lake systems feeding the Tista River tributary (Image Source: Google Earth)

0.027 km^2 per year between 1976 and 2016. A significant acceleration in expansion was observed post-2008 compared to the pre-2000 period. The lake's surface area dramatically increased from 0.20 to 1.31 km^2, accompanied by a 1.6 km lengthening. Furthermore, the study employed bathymetric surveys to estimate the lake's water volume at 65.81×10^6 m^3.

The predicted GLOF susceptibility of South Lhonak Lake was confirmed on October 4, 2023, when the moraine-dammed lake experienced a sudden and substantial water discharge shortly after midnight. Initial media reports attributed the subsequent flooding to a cloudburst over South Lhonak Lake (https://www.bbc.com/news/world-asia-india-66991815), a claim echoed in Nath and Choudhury (2024). However, subsequent investigations established that the discharge was a GLOF triggered by a landslide involving an estimated 14–16 million m^3 of debris from a lateral moraine (Sattar et al., 2025; Zhang et al., 2024). This finding was corroborated by Yu et al. (2024), who suggested that regional heavy rainfall precipitated the landslide, affecting an area of 0.29 km^2 of the lateral moraine and consequently causing the GLOF. Analysis of pre-and post-GLOF satellite imagery by the Indian Space Research Organization (ISRO) in the quick aftermath of the event suggested (https://www.isro.gov.in/Satellite_studies_South_Lhonak_Lake.html) drainage of 105 hectares, approximating a 63% reduction in the total lake area. However, this estimation conflicts with findings presented by Zhang et al. (2024), who, based on an estimated release of 38.49 million m^3 of water, calculated a 15.38% decrease in lake area.

While the South Lhonak GLOF in the Sikkim Himalaya was comparatively small in scale, it is a stark reminder of the region's inherent vulnerability to such events. Both the South Lhonak Lake and the highly susceptible Gurudongmar Lakes in the area drain into the Tista River, posing a threat of future disasters of unprecedented magnitude. Several towns along the Tista, including Chungthang, Mangan, Dikchu, Singtam, Rangpo, Majitar, and Mining, have already experienced the impact of the South Lhonak GLOF (Fig. 7). The rapid and substantial discharge of water associated with these events results in a sudden increase in river discharge and flow velocity, leading to accelerated erosion of riverbeds and banks (Fig. 9).

This can cause significant damage to infrastructure located along the riverbanks. The eroded debris and sediment from the breached moraine dam and surrounding areas abrade the banks, further intensifying the erosional processes. Consequently, substantial sediment deposition occurs downstream. This sediment load, varying in size and composition, alters the river's sediment characteristics, impacting its hydraulic properties and potentially increasing the risk of future flooding. Furthermore, the deposited sediment can elevate the river channel, widen the river course, and diminish the river's storage capacity and associated reservoirs. These deposits also contribute to bank instability, making riverbanks more susceptible to landslides and collapse, perpetuating erosion of the surrounding terrain. Beyond the immediate destruction, the sediments deposited by a GLOF can continue influencing river systems for decades following the event.

The South Lhonak Lake GLOF event serves as a stark illustration of the multifaceted hazards associated with these glacial lake outbursts. The event began with a

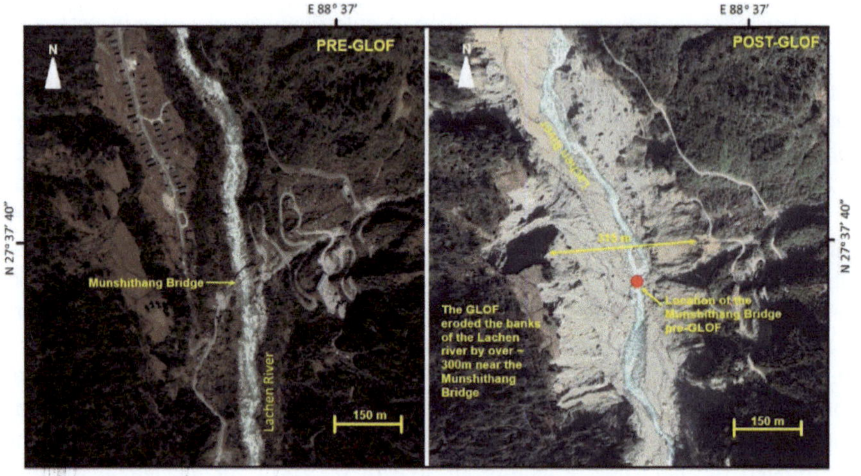

Fig. 9 Erosion of the banks of the Lachen River, north of the Chungthang town, due to the South Lhonak GLOF (Image source: Google Earth)

landslide impacting the lateral moraine, leading to a breach in the moraine-dammed lake. The ensuing rapid release of water surged downstream along the tributaries of the Tista River, causing extensive bank erosion and inundating the Chungthang town (Fig. 10). Critically, the floodwaters washed away the 1200 MW Sikkim Urja Stage III dam at Chungthang and numerous bridges, roads, and hamlets.

The substantial sediment load carried by the floodwaters was subsequently deposited in the lower Tista River, resulting in increased channel height and widening, which in turn led to inundation of the adjacent national highway and the Tista Bazaar,

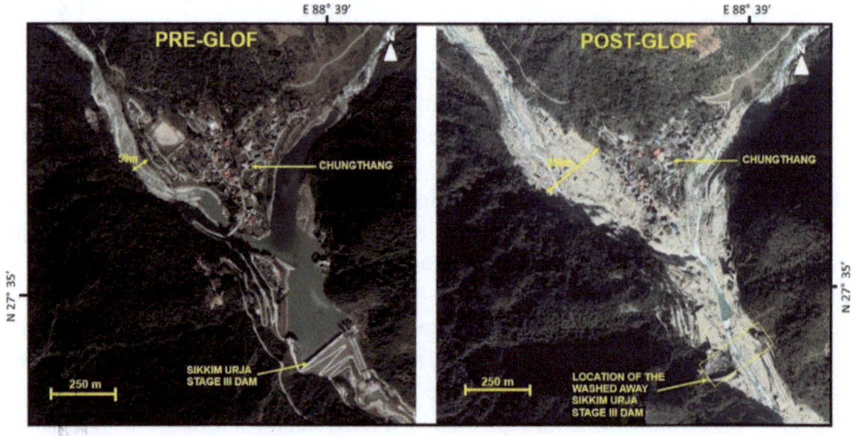

Fig. 10 Pre- and post-GLOF satellite imagery of the Chungthang town and Sikkim Urja Stage III dam (Image source: Google Earth)

a town near the Sikkim border in West Bengal. The destruction wrought by the South Lhonak GLOF, including the loss of property, homes, and infrastructure, offers a glimpse into the potential devastation that larger GLOF hazards could unleash in the Sikkim Himalaya.

5 Discussion: Integrated Natural Hazard in the Darjiling-Sikkim Himalaya

Earthquake, landslide, and GLOF hazards estimations indicate significant risks even independently in the Darjiling-Sikkim Himalaya. Therefore, the worst-case scenario would involve their simultaneous occurrence. The September 18, 2011, magnitude 6.9 Kanchenjunga earthquake triggered nearly 1,200 landslides across Sikkim (Martha et al., 2015; Mahajan et al., 2012). However, this did not cause a GLOF even though the epicenter was in the Kanchenjunga region (Paul et al., 2015). Nevertheless, suppose a great earthquake of magnitude 8 or greater does occur in the Darjiling-Sikkim Himalaya along the MHT in the Higher Himalaya between Lachen and Gangtok, as expected. In that case, the Higher and Tethyan Himalaya will be most affected. The glacial lakes with high GLOF hazards are also located in the Higher Himalayas, so they will unlikely stay unaffected, as they were during the 2011 earthquake. The Higher Himalaya also undergoes maximum vertical uplift and slope steepening. If the magnitude 6.9, 2011 earthquake triggered 1200 landslides, far more landslides would occur if an earthquake of magnitude 8 or greater occurred. If multiple GLOFs and landslides occur during a great earthquake, the destruction of roads and communication networks will be widespread. This will rule out all ground-based disaster management and aid. Movement of any heavy machinery to remove collapsed concrete buildings and slabs will also not be possible. This points to a grim scenario. Therefore, Himalayan natural hazard models must integrate seismic, landslides, and GLOF hazards for any realistic forecast of disasters and disaster preparedness. Our disaster preparation strategies should include the possibility of all three hazards occurring together in the Himalaya. If these red flags are ignored, the magnitude of future disasters will keep increasing as the population increases and more lives and infrastructure become vulnerable. We must plan to get out of harm's way when there is still time. We must embrace maintaining maximum sync and harmony with nature or be resigned to the consequences.

References

Agarwal, A., & Narain, S. (1991). *Floods, flood plain and environmental myths* (State of India's Environment: A Citizens' Report, New Delhi. Centre for Science and Environment.

Aggarwal, S., Rai, S. C., Thakur, P. K., & Emmer, A. (2017). Inventory and recently increasing GLOF susceptibility of glacial lakes in Sikkim, Eastern Himalaya. *Geomorphology, 295*, 39–54. https://doi.org/10.1016/j.geomorph.2017.06.014

Alam, S. A., Mandal, S., & Maiti, R. (2024). Geospatial intelligence for landslide susceptibility and risk analysis: Insights from NH31A and East Sikkim Himalaya settlements. *Natural Hazards Research, 5*, 187–208. https://doi.org/10.1016/j.nhres.2024.10.001

Allen, S. K., Rastner, P., Arora, M., Huggel, C., & Stoffel, M. (2016). Lake outburst and debris flow disaster at Kedarnath, June 2013: Hydrometeorological triggering and topographic predisposition. *Landslides, 13*, 1479–1491. https://doi.org/10.1007/s10346-015-0584-3Alam

Anbalagan, R., Kumar, R., Lakshmanan, K., Parida, S., & Neethu, S. (2015). Landslide hazard zonation mapping using frequency ratio and fuzzy logic approach, a case study of Lachung Valley, Sikkim. *Geoenvironmental Disasters, 2*, 1–17.

Anbarasu, K., Sengupta, A., Gupta, S., & Sharma, S. P. (2010). Mechanism of activation of the Lanta Khola landslide in Sikkim Himalayas. *Landslides, 7*, 135–147.

Baranowski, J., Armbruster, J., Seeber, L., & Molnar, P. (1984). Focal depths and fault plane solutions of earthquakes and active tectonics of the Himalaya. *Journal of Geophysical Research: Solid Earth, 89*(B8), 6918–6928.

Basu, S. R., & De, S. K. (2003). Causes and consequences of landslides in the Darjiling-Sikkim Himalayas, India. *Geographia Polonica, 76*(2), 37–52.

Bendick, R., & Bilham, R. (2001). How perfect is the Himalayan arc? *Geology, 29*(9), 791–794.

Bera, B. (2025). A massive landslide damaged NHPC's Teesta V hydropower station on 20th August 2024 at Balutar, Sikkim Himalayas. *Natural Hazards Research, 5*(2), 372–379.

Bertachini, E., Capitani, A., Capra, A., Castagnetti, C., Corsini, A., Dubbini, M., & Ronchetti, F. (2009). Integrated surveying system for landslide monitoring, Valoria landslide (Appennines of Modena, Italy). In *FIG working week*. Eilat, Israel.

Bhasin, R., Grimstad, E., Larsen, J. O., Dhawan, A. K., Singh R., Verma, S. K., & Venkatachalam, K. (2002). Landslide hazards and mitigation measures at Gangtok, Sikkim Himalaya. *Engineering Geology, 64*, 351–368.

Bhambri, R., Bolch, T., Chaujar, R. K., & Kulshreshtha, S. C. (2011). Glacier changes in the Garhwal Himalaya, India, from 1968 to 2006 based on remote sensing. *Journal of Glaciology, 57*(203), 543–556.

Bhattacharya, B. (2015). *Banvasi Tista, Bipanna Pray 20 Hazar (Flooded Tista, about 20 thousand affected)*. Kolkata. Anandabazar Patrika (A Bengali Newspaper).

Bilham, R. (2019). Himalayan earthquakes: A review of historical seismicity and early 21st century slip potential. *Geological Society, London, Special Publications, 483*(1), 423–482.

Bilham, R., Gaur, V. K., & Molnar, P. (2001). Himalayan seismic hazard. *Science, 293*(5534), 1442–1444.

Brice, A., Jayangondaperumal, R., Priyanka, R. S., Pandey, A., Mishra, R. L., Singh, I., ... & Dash, S. P. (2024). Paleoseismological evidence for segmentation of the Main Himalayan Thrust in the Darjeeling-Sikkim Himalaya. *Scientific Reports, 14*(1), 14537.

Biswas, S. S., & Pal, R. (2016). Causes of landslides in Darjeeling Himalayas during June-July, 2015. *Journal of Geography and Natural Disasters, 6*, 173. https://doi.org/10.4172/2167-0587.1000173

Burgess, W. P., Yin, A., Dubey, C. S., Shen, Z. K., & Kelty, T. K. (2012). Holocene shortening across the Main Frontal Thrust zone in the eastern Himalaya. *Earth and Planetary Science Letters, 357*, 152–167.

Calcaterra, S., Cesi C., Di Maio, C., Gambino, P., Merli, K., Vallario, M., & Vassallo, R. (2012). Surface displacements of two landslides evaluated by GPS and inclinometer systems: A case study in Southern Apennines, Italy. *Natural Hazards, 61*(1), 257–266.

Carrivick, J. L., Tweed, F. S., Ng, F., Quincey, D. J., Mallalieu, J., Ingeman-Nielsen, T., ... & Hubbard, A. (2017). Ice-dammed Lake drainage evolution at Russell Glacier, West Greenland. *Frontiers in Earth Science, 5*, 100. https://doi.org/10.3389/feart.2017.00100

Carrivick, J. L., & Tweed, F. S. (2016). A global assessment of the societal impacts of glacier outburst floods. *Global and Planetary Change, 144*, 1–16. https://doi.org/10.1016/j.gloplacha.2016.07.001

Chakraborty, I., Ghosh, S., Bhattacharya, D., & Bora, A. (2011). Earthquake induced landslides in the Sikkim-Darjeeling Himalayas—An aftermath of the 18th September 2011 Sikkim earthquake. *Report of Geological Survey of India, Kolkata, 8.*

Chanda, A., & Biswas, B. (2021). Glacial Lake Outflow Hazard and Risk Probability in Sikkim. *Recent technologies for disaster management and risk reduction: sustainable community resilience & responses* (pp. 3–26). Springer International Publishing.

Chowdhury, A., Kroczek, T., De, S. K., Vilímek, V., Sharma, M. C., & Debnath, M. (2021). Glacial lake evolution (1962–2018) and outburst susceptibility of gurudongmar lake complex in the Tista basin, Sikkim Himalaya (India). *Water, 13*(24), 3565.

Coe, J. A., Ellis, W. L., Godt, J. W., Savage, W. Z., Savage, J. E., Michael, J. A., Kibler, J. D., Powers, P. S., Lidke, D. J., & Debray, S. (2003). Seasonal movement of the Slumgullion landslide determined from global positioning system surveys and field instrumentation, July 1998–March 2002. *Engineering Geology, 68*, 67–101.

De, R. (2000). A microearthquake survey at the MBT zone: Sikkim Himalaya. *Journal of Geophysics (Hyderabad), 21*(2), 77–84.

De, R., & Kayal, J. R. (2004). Seismic activity at the MCT in Sikkim Himalaya. *Tectonophysics, 386*(3–4), 243–248.

De la Torre, T. L., Monsalve, G., Sheehan, A. F., Sapkota, S., & Wu, F. (2007). Earthquake processes of the Himalayan collision zone in eastern Nepal and the southern Tibetan Plateau. *Geophysical Journal International, 171*(2), 718–738.

DeCelles, P. G., Gehrels, G. E., Quade, J., & Ojha, T. P. (1998). Eocene-early Miocene foreland basin development and the history of Himalayan thrusting, western and central Nepal. *Tectonics, 17*(5), 741–765.

Drukpa, D., Velasco, A. A., & Doser, D. I. (2006). Seismicity in the Kingdom of Bhutan (1937–2003): Evidence for crustal transcurrent deformation. *Journal of Geophysical Research: Solid Earth, 111*(B6).

Ekström, G. A. (1987). *A broad band method of earthquake analysis* (Doctoral dissertation, Harvard University).

Emmer, A., Merkl, S., & Mergili, M. (2015). Spatiotemporal patterns of high-mountain lakes and related hazards in western Austria. *Geomorphology, 246*, 602–616. https://doi.org/10.1016/j.geomorph.2015.06.032

Emmer, A., Wood, J. L., Cook, S. J., Harrison, S., Wilson, R., Diaz-Moreno, A., ... & Poma, C. (2022). 160 glacial lake outburst floods (GLOFs) across the Tropical Andes since the Little Ice Age. *Global and Planetary Change, 208*, 103722.

Froude, M. J., Petley, D. N. (2018). Global fatal landslide occurrence from 2004 to 2016. *Natural Hazards Earth System Science, 18*, 2161–2181.

Gardelle, J., Arnaud, Y., & Berthier, E. (2011). Contrasted evolution of glacial lakes along the Hindu Kush Himalaya mountain range between 1990 and 2009. *Global and Planetary Change, 75*(1–2), 47–55.

Gardelle, J., Berthier, E., Arnaud, Y., & Kääb, A. (2013). Region-wide glacier mass balances over the Pamir-Karakoram-Himalaya during 1999–2011. *The Cryosphere, 7*(4), 1263–1286.

Gili, J., A., Corominas, J., & Rius, J. (2000). Using global positioning system techniques in landslide monitoring. *Engineering Geology, 55*(3), 167–192.

Grandin, R., Doin, M. P., Bollinger, L., Pinel-Puysségur, B., Ducret, G., Jolivet, R., & Sapkota, S. N. (2012). Long-term growth of the Himalaya inferred from interseismic InSAR measurement. *Geology, 40*(12), 1059–1062.

Grinsted, A., Hvidberg, C. S., Campos, N., & Dahl-Jensen, D. (2017). Periodic outburst floods from an ice-dammed lake in East Greenland. *Scientific Reports, 7*(1), 9966. https://doi.org/10.1038/s41598017-07960-9

GSI. (2012). *Geology and mineral resources of the states of India.* Geological Survey of India.

Guan, M., Wright, N. G., Sleigh, P. A., & Carrivick, J. L. (2015). Assessment of hydro-morphodynamic modelling and geomorphological impacts of a sediment-charged jökulhlaup, at Sólheimajökull, Iceland. *Journal of Hydrology, 530*, 336–349. https://doi.org/10.1016/j.jhydrol.2015.09.062

Gupta, V., Chauhan, N., Penna, I., Hermanns, R., Dehls, J., Sengupta, A., & Bhasin, R. K. (2022). Geomorphic evaluation of landslides along the Tista river valley, Sikkim Himalaya, India. *Geological Journal, 57*(2), 611–621. https://doi.org/10.1002/gj.4377

Gupta, V., Mahajan, A. K., & Thakur, V. C. (2015). A study on landslides triggered during Sikkim earthquake of September 18, 2011. *Himalayan Geology, 36*(1), 81–90.

Gurung, D. R., Khanal, N. R., Bajracharya, S. R., Tsering, K., Joshi, S., Tshering, P., ... & Penjor, T. (2017). Lemthang Tsho glacial Lake outburst flood (GLOF) in Bhutan: cause and impact. *Geoenvironmental Disasters, 4*, 1–13.

Hazarika, P., Kumar, M. R., Srijayanthi, G., Raju, P. S., Rao, N. P., & Srinagesh, D. (2010). Transverse tectonics in the Sikkim Himalaya: Evidence from seismicity and focal-mechanism data. *Bulletin of the Seismological Society of America, 100*(4), 1816–1822.

Hazra, P., & Krishna, A. P. (2021, July). AHP Based Assessment of Glof Susceptibility of South Lhonak Glacial Lake, Sikkim Himalaya, India. In *2021 IEEE International Geoscience and Remote Sensing Symposium IGARSS* (pp. 5489–5492). IEEE.

https://datacatalogfiles.worldbank.org/ddh-published/0037584/DR0045411/global-landslide-hazard-map-report.pdf?versionId=2023-01-18T20:43:00.6156776Z

Hubbard, M., Mukul, M., Gajurel, A. P., Ghosh, A., Srivastava, V., Giri, B., ... & Mendoza, M. M. (2021). Orogenic segmentation and its role in Himalayan Mountain building. *Frontiers in Earth Science, 9*, 641666.

ICIMOD. (2022). *GLOF database of High Mountain Asia*. ICIMOD. https://doi.org/10.26066/RDS.1973283

IS 1893: 1962. *Recommendations for earthquake resistant design of structures*. Bureau of Indian Standards, N Delhi, India.

Ives, J. D., Shrestha, R. B., & Mool, P. K. (2010). *Formation of glacial lakes in the Hindu Kush-Himalayas and GLOF risk assessment* (pp. 10–11). ICIMOD.

Ives, J. D. (1986). Glacier lake outburst floods and risk engineering in the Himalaya. In *ICIMOD occasional paper* (Vol. 5, p. 41). Kathmandu.

Jaboyedoff, M., Michoud, C., Derron, M. H., Voumard, J., Leibundgut, G., Sudmeier-Rieux, K., Michoud, C., Nadim, F., & Leroi, E. (2016). Human-induced landslides: Toward the analysis of anthropogenic changes of the slope environment. In *Landslides and engineered slopes. Experience, theory and practice* (pp. 217–232). https://doi.org/10.1201/b21520-20

Jeelani, G. H., Bhat, N. A., Shivanna, K., & Bhat, M. Y. (2011). Geochemical characterization of surface water and spring water in SE Kashmir Valley, western Himalaya: Implications to water–rock interaction. *Journal of Earth System Science, 120*, 921–932.

Joshi, M. (2022). Co-seismic landslides in the Sikkim Himalaya during the 2011 Sikkim earthquake: Lesson learned from the past and inference for the future. *Geological Journal, 57*(12), 5039–5060.

Joshi, M., Kumar, K., Rajeevan, K., & Barfal, S. S. (2024). Tragedies due to human folly: A learnt lesson from year 2023 Sikkim avalanche. *Quaternary Science Advances, 13*, 100162. https://doi.org/10.1016/j.qsa.2023.100162

Kargel, J. S., Abrams, M. J., Bishop, M. P., Bush, A., Hamilton, G., Jiskoot, H., Kääb, A., Kieffer, H. H., Lee, E. M., Paul, F., & Rau, F. (2005). Multispectral imaging contributions to global land ice measurements from space. *Remote Sensing of Environment, 99*(1–2), 187–219.

Kayal, J. R. (2001). Microearthquake activity in some parts of the Himalaya and the tectonic model. *Tectonophysics, 339*, 331–351

Kumar, M. R., Hazarika, P., Prasad, G. S., Singh, A., & Saha, S. (2012). Tectonic implications of the September 2011 Sikkim earthquake and its aftershocks. *Current Science* 788–792.

Kumar, S., Wesnousky, S. G., Rockwell, T. K., Ragona, D., Thakur, V. C., & Seitz, G. G. (2001). Earthquake recurrence and rupture dynamics of Himalayan Frontal Thrust, India. *Science, 294*(5550), 2328–2331.

Kumar, S., Wesnousky, S. G., Rockwell, T. K., Briggs, R. W., Thakur, V. C., & Jayangondaperumal, R. (2006). Paleoseismic evidence of great surface rupture earthquakes along the Indian Himalaya. *Journal of Geophysical Research: Solid Earth, 111*(B3).

Kumar, S., Wesnousky, S. G., Jayangondaperumal, R., Nakata, T., Kumahara, Y., & Singh, V. (2010). Paleoseismological evidence of surface faulting along the northeastern Himalayan front, India: Timing, size, and spatial extent of great earthquakes. *Journal of Geophysical Research: Solid Earth, 115*(B12). https://doi.org/10.1029/2009JB006789.

Lavé, J., & Avouac, J. P. (2001). Fluvial incision and tectonic uplift across the Himalayas of central Nepal. *Journal of Geophysical Research: Solid Earth, 106*(B11), 26561–26591.

Liu, J. J., Cheng, Z. L., & Su, P. C. (2014). The relationship between air temperature fluctuation and Glacial Lake Outburst Floods in Tibet, China. *Quaternary International, 321*, 78–87. https://doi.org/10.1016/j.quaint.2013.11.023

Mahajan, A. K., Gupta, V., & Thakur, V. C. (2012). Macroseismic field observations of 18 September 2011 Sikkim earthquake. *Natural Hazards, 63*, 589–603.

Malet, J-P., Maquaire, O., & Calais, E. (2002). The use of global positioning system techniques for the continuous monitoring of landslides: Application to the super-sauze earth flow (Alpes-de-Haute-Province, France). *Geomorphology, 43*, 33–54.

Mamtani, M. A. (2025). The future of structural geology in the 21st century-moving from mesoscale to nanoscale observations in tectonically deformed rocks. *Journal of the Geological Society of India, 101*(1), 10–23.

Mandal, K., Saha, S., & Mandal, S. (2022). Predicting the landslide susceptibility in Eastern Sikkim Himalayan region, India using boosted regression tree and REPTree machine learning techniques. In *Applied geomorphology and contemporary issues* (pp. 683–707). Springer International Publishing.

Martha, T. R., Govindharaj, K. B., & Kumar, K. V. (2015). Damage and geological assessment of the 18 September 2011 Mw 6.9 earthquake in Sikkim, India using very high-resolution satellite data. *Geoscience Frontiers, 6*(6), 793–805.

McNaught, M. A., & Mitra, G. (1996). The use of finite strain data in constructing a retrodeformable cross-section of the Meade thrust sheet, southeastern Idaho, USA. *Journal of Structural Geology, 18*(5), 573–583.

Mehrotra, G. S., Sarkar, S., Kanungo, D. P., & Mahadevaiah, K. (1996). Terrain analysis and spatial assessment of landslide hazards in parts of Sikkim Himalaya. *Journal Geological Society of India, 47*(4), 491–498.

Mendoza, M. M., Ghosh, A., Karplus, M. S., Klemperer, S. L., Sapkota, S. N., Adhikari, L. B., & Velasco, A. (2019). Duplex in the Main Himalayan Thrust illuminated by aftershocks of the 2015 M w 7.8 Gorkha earthquake. *Nature Geoscience, 12*(12), 1018–1022.

Michoud, C., Derron, M.-H., Jaboyedoff, M., Nadim, F., & Leroi, E. (2011). Classification of landslide-inducing anthropogenic activities. In *5th Canadian conference on geotechnique and natural hazards* (pp. 231–233).

Mishra, O. P., Chakrabortty, G. K., Singh, O. P., Ghosh, D. C., Mukherjee, K. K., & Das, P. C. (2010). A report on seismogenesis in the Sikkim–Darjiling Himalayas and assimilation of dynamic snap shots of the region: future vulnerability. *GSI Report (FSP: SEI/CHQ/CGD/2007/070)*, 57.

Mohanty, L., & Maiti, S. (2021). Probability of glacial lake outburst flooding in the Himalaya. *Resources, Environment and Sustainability, 5*, 100031.

Mugnier, J. L., Gajurel, A., Huyghe, P., Jayangondaperumal, R., Jouanne, F., & Upreti, B. (2013). Structural interpretation of the great earthquakes of the last millennium in the central Himalaya. *Earth-Science Reviews, 127*, 30–47.

Mukul, M. (1998). A spatial statistics approach to the quantification of finite strain variation in penetratively deformed thrust sheets: An example from the Sheeprock Thrust Sheet, Sevier Fold-and-Thrust belt, Utah. *Journal of Structural Geology, 20*(4), 371–384.

Mukul, M. (1999). Strain variation in fold-and-thrust belts: Implications for construction of retrodeformable models. *Proceedings of the Indian Academy of Sciences-Earth and Planetary Sciences, 108*, 207–221.

Mukul, M. (2000). The geometry and kinematics of the Main Boundary Thrust and related neotectonics in the Darjiling Himalayan fold-and-thrust belt, West Bengal, India. *Journal of Structural Geology, 22*(9), 1261–1283.

Mukul, M. (2005). Continental deformation and global positioning system-based geodesy. *Himalayan Geology, 26*(1), 193–198.

Mukul, M. (2010). First-order kinematics of wedge-scale active Himalayan deformation: Insights from Darjiling–Sikkim–Tibet (DaSiT) wedge. *Journal of Asian Earth Sciences, 39*(6), 645–657.

Mukul, M., & Mitra, G. (1998). Finite strain and strain variation analysis in the Sheeprock Thrust Sheet: An internal thrust sheet in the Provo salient of the Sevier Fold-and-Thrust belt, Central Utah. *Journal of Structural Geology, 20*(4), 385–405.

Mukul, M., Singh, V. (2016). Active tectonics and geomorphological studies in India During 2012–2016. *Proceedings of the Indian National Science Academy, 82(3),* 727–735.

Mukul, M., Jade, S., & Matin, A. (2009). Active deformation in the Darjiling-Sikkim Himalaya based on 2000–2004 geodetic global positioning system measurements. In P. Ghosh & S. Gangopadhyay (Eds.), *Indian statistical institute platinum jubilee volumes, numerical methods and models in earth science* (pp. 1–28). New India Publishing Agency.

Mukul, M., Jade, S., Ansari, K., & Matin, A. (2014). Seismotectonic implications of strike–slip earthquakes in the Darjiling–Sikkim Himalaya. *Current Science*, 198–210.

Mukul, M., Jade, S., Ansari, K., Matin, A., & Joshi, V. (2018). Structural insights from geodetic global positioning system measurements in the Darjiling-Sikkim Himalaya. *Journal of Structural Geology, 114*, 346–356. https://doi.org/10.1016/j.jsg.2018.03.007

Mukul, M., Jaiswal, M., & Singhvi, A. K. (2007). Timing of recent out-of-sequence active deformation in the frontal Himalayan wedge: Insights from the Darjiling sub-Himalaya, India. *Geology, 35*(11), 999–1002.

Mukul, M., Roy, D., Satpathy, S., & Kumar, V. A. (2004). Bootstrapped spatial statistics: A more robust approach to the analysis of finite strain data. *Journal of Structural Geology, 26*(3), 595–600.

Mukul, M., Srivastava, V., & Mukul, M. (2024). Structural control on the landscape evolution and avulsive behavior of rivers at mountain exits: The example of the Kosi River in eastern Nepal Himalaya. *Tectonophysics, 888*, 230442.

Nakata, T. (1989). Active Faults of the Himalaya of India and Nepal. *Geological Society of America, 232*(Special Paper), 243–264.

Nath, M., & Choudhury, D. P. (2024). Cloudburst in Sikkim. *Journal of the Geological Society of India, 100*(3): 456. https://doi.org/10.17491/jgsi/2024/173855

Nath, S. K., Sengupta, A., & Srivastava, A. (2021). Remote sensing GIS-based landslide susceptibility & risk modeling in Darjeeling–Sikkim Himalaya together with FEM-based slope stability analysis of the terrain. *Natural Hazards, 108*(3), 3271–3304.

Nath, S. K., Vyas, M., Pal, I., & Sengupta, P. (2005). A seismic hazard scenario in the Sikkim Himalaya from seismotectonics, spectral amplification, source parameterization, and spectral attenuation laws using strong motion seismometry. *Journal of Geophysical Research: Solid Earth, 110*(B1). https://doi.org/10.1029/2004JB003199.

Nelson, K. D., Zhao, W., Brown, L. D., Kuo, J., Che, J., Liu, X., ... & Edwards, M. (1996). Partially molten middle crust beneath southern Tibet: Synthesis of project INDEPTH results. *Science, 274*(5293), 1684–1688.

Ni, J., & Barazangi, M. (1984). Seismotectonics of the Himalayan collision zone: Geometry of the underthrusting Indian plate beneath the Himalaya. *Journal of Geophysical Research: Solid Earth, 89*(B2), 1147–1163.

Nie, Y., Liu, Q., & Liu, S. (2013). Glacial lake expansion in the Central Himalayas by Landsat images, 1990–2010. *PLoS ONE, 8*(12), e83973. https://doi.org/10.1371/journal.pone.0083973

Pal, R., Biswas, S. S., Mondal, B., & Pramanik, M. K. (2016). Landslides and Floods in the Tista Basin (Darjeeling and Jalpaiguri Districts): Historical Evidence, Causes and Consequences. J. Ind. Geophys. Union, 20(2), 66–72.

Pandey, M. R., Tandukar, R. P., Avouac, J. P., Vergne, J., & Heritier, T. (1999). Seismotectonics of the Nepal Himalaya from a local seismic network. *Journal of Asian Earth Sciences, 17*(5–6), 703–712.

Paul, H., Mitra, S., Bhattacharya, S. N., & Suresh, G. (2015). Active transverse faulting within underthrust Indian crust beneath the Sikkim Himalaya. *Geophysical Journal International, 201*(2), 1072–1083. https://doi.org/10.1093/gji/ggv058

Peyret, M., Djamour, Y., Rizza, M., Ritz, J. F., Hurtrez, J. E., Goudarzi, M. A., Nankali, H., Chéry, J., Le DortzK, & Uri, F. (2008). Monitoring of the large slow Kahrod landslide in Alborz mountain range (Iran) by GPS and SAR interferometry. *Engineering Geology, 100*(3–4), 131–141.

Poddar, I., & Roy, R. (2024). Application of GIS-based data-driven bivariate statistical models for landslide prediction: A case study of highly affected landslide prone areas of Teesta River basin. *Quaternary Science Advances, 13, 100150*. https://doi.org/10.1016/j.qsa.2023.100150

Quincey, D. J., Lucas, R. M., Richardson, S. D., Glasser, N. F., Hambrey, M. J., & Reynolds, J. M. (2005). Optical remote sensing techniques in high-mountain environments: Application to glacial hazards. *Progress in Physical Geography, 29*(4), 475–505. https://doi.org/10.1191/0309133305pp456ra

Rajendran, C. P., Singh, T. E. J. P. A. L., Mukul, M. A. L. A. Y., Thakkar, M. A. H. E. S. H., Kothyari, G. C. H., John, B., & Rajendran, K. (2020). Paleoseismological studies in India (2016–2020): Status and prospects. *Proceedings of the Indian National Science Academy, 86*(1), 585–607.

Raju, P. S., Rao, N. P., Singh, A., & Kumar, M. R. (2007). The 14 February 2006 Sikkim earthquake of magnitude 5.3. *Current Science, 93*, 848–850.

Ramasamy, S. M., Muthukumar, M., & Subagunasekar, M. (2014). Drainage and anthropogenic aberrations in Himalayan landslides, Uttarakhand, India. *Indian Landslides, 7*(1 & 2), 9–22.

Rawat, M. S., Joshi, V., Rawat, B. S., & Kumar, K., (2011). Landslide movement monitoring using GPS technology: A case study of Bakthang landslide, Gangtok, East Sikkim, India. *Journal of Development and Agricultural Economics, 3*(5), 194–200.

Rawat, M., Pandey, A., & Gupta, P. K. (2022). Geospatial analysis of glacial lake outburst flood (GLOF). *Geospatial technologies for resources planning and management* (pp. 141–160). Springer International Publishing.

Remya, S. N., Kulkarni, A. V., Pradeep, S., & Shrestha, D. G. (2019). Volume estimation of existing and potential glacier lakes, Sikkim Himalaya, India. *Current Science, 116*(4), 620–627.

Rott, H., & Nagler, T. (2006). The contribution of radar interferometry to the assessment of landslide hazards. *Advances in Space Research, 37*(4), 710–719.

Round, V., Leinss, S., Huss, M., Haemmig, C., & Hajnsek, I. (2017). Surge dynamics and lake outbursts of Kyagar Glacier, Karakoram. *The Cryosphere, 11*(2), 723–739. https://doi.org/10.5194/tc11-723-2017

Saha, S., & Bera, B. (2024). Rainfall threshold for prediction of shallow landslides in the Garhwal Himalaya. India. *Geosystem Geoenvironment, 3*(3), 100285. https://doi.org/10.1016/j.geogeo.2024.100285

Saha, S., Bera, B., Sengupta, D., Mukhopadhyay, U., Ghosh, D., Tamang, L., Bhattacharjee, S., & Sengupta, N. (2025). Multiple drivers of the recent South Lhonak glacial lake outburst flood in Sikkim Himalaya and its aftermath on the Teesta river valley. *Geosystems and Geoenvironment, 4*(2), 100375.

Sarkar, S. (2018). Landslide risk assessment in parts of the Darjeeling Himalayas, India. In *Integrating disaster science and management* (pp. 99–108). Elsevier.

Sattar, A., Cook, K. L., Rai, S. K., Berthier, E., Allen, S., Rinzin, S., ... & Bhat, S. Y. (2025). The Sikkim flood of October 2023: Drivers, causes and impacts of a multihazard cascade. *Science* eads2659.

Sattar, A., Goswami, A., Kulkarni, A. V., Emmer, A., Haritashya, U. K., Allen, S., ... & Huggel, C. (2021). Future glacial lake outburst flood (GLOF) hazard of the South Lhonak Lake, Sikkim Himalaya. *Geomorphology, 388*, 107783.

Schelling, D., & Arita, K. (1991). Thrust tectonics, crustal shortening, and the structure of the far-eastern Nepal Himalaya. *Tectonics, 10*(5), 851–862.

Schuster, R. L. (2006). Interaction of dams and landslides: Case studies and mitigation. *U. S. Geological Survey Professional Paper, 1723*, 107.

Sengupta, A., Gupta, S., & Anbarasu, K. (2010). Rainfall thresholds for the initiation of landslide at Lanta Khola in north Sikkim, India. *Natural Hazards, 52*, 31–42.

Sharma, R. K., Pradhan, P., Sharma, N. P., & Shrestha, D. G. (2018). Remote sensing and in situ-based assessment of rapidly growing South Lhonak glacial lake in eastern Himalaya, India. *Natural Hazards, 93*, 393–409.

Shaw, J., Stacey, C. D., Wu, Y., & Lintern, D. G. (2017). Anatomy of the Kitimat fiord system, British Columbia. *Geomorphology, 293*, 108–129. https://doi.org/10.1016/j.geomorph.2017.04.043

Shrestha, U. B., Gautam, S., & Bawa, K. S. (2012). Widespread climate change in the Himalayas and associated changes in local ecosystems. *PloS One, 7*(5), e36741.

Shugar, D. H., Burr, A., Haritashya, U. K., Kargel, J. S., Watson, C. S., Kennedy, M. C., Bevington, A. R., Betts, R. A., Harrison, S., & Strattman, K. (2020). Rapid worldwide growth of glacial lakes since 1990. *Nature climate change, 10*(10), 939–945.

Singh, R., Aryan, V., & Joshi, M. (2022). Understanding the flash flood event of 7th February 2021 in Rishi Ganga basin, Central Himalaya using remote sensing technique. *Remote Sensing Applications: Society and Environment, 26*, 100744.

Singh, S., Raju, A., & Banerjee, S. (2022). Detecting slow-moving landslides in parts of Darjeeling-Sikkim Himalaya, NE India: Quantitative constraints from PSInSAR and its relation to the structural discontinuities. *Landslides, 19*(10), 2347–2365. https://doi.org/10.1007/s10346-022-01900-z

Skilodimou, H. D., Bathrellos, G. D., Koskeridou, E., Soukis, K., & Rozos, D. (2018). Physical and anthropogenic factors related to landslide activity in the Northern Peloponnese, Greece. *Land, 7*, 85.

Sonker, I., Tripathi, J. N., Swarnim. (2024). Landslide susceptibility mapping using morphological and hydrological parameters in Sikkim Himalaya: Frequency ratio model and geospatial technologies. *Natural Hazards, 120*(7), 6797–6832. https://doi.org/10.1007/s11069-024-06491-7

Srivastava, P., & Mitra, G. (1994). Thrust geometries and deep structure of the outer and lesser Himalaya, Kumaon and Garhwal (India): Implications for evolution of the Himalayan fold-and-thrust belt. *Tectonics, 13*(1), 89–109.

Srivastava, V., & Mukul, M. (2020). Cataclastic strain from external thrust sheets in fold-thrust belts: Insights from the frontal Indian Himalaya. *Journal of Asian Earth Sciences, 188*, 104092.

Srivastava, V., Mukul, M., & Barnes, J. B. (2016). Main Frontal thrust deformation and topographic growth of the Mohand Range, northwest Himalaya. *Journal of Structural Geology, 93*, 131–148. https://doi.org/10.1016/j.jsg.2016.10.009

Srivastava, V., Mukul, M., Barnes, J. B., & Mukul, M. (2018). Geometry and kinematics of Main Frontal thrust-related fault propagation folding in the Mohand Range, northwest Himalaya. *Journal of Structural Geology, 115*, 1–18.

Srivastava, V., Mukul, M., & Mukul, M. (2017). Quaternary deformation in the Gorubathan recess: Insights on the structural and landscape evolution in the frontal Darjiling Himalaya. *Quaternary International, 462*, 138–161. https://doi.org/10.1016/j.quaint.2017.05.004

Starkel, L. (1972). The role of catastrophic rainfall in the shaping of the relief of the Lower Himalaya (Darjeeling Hills). *Geographia Polonica, 21*, 103–147.

Starkel, L., & Basu, S. (Eds.). (2000). *Rains, landslides, and floods in the Darjeeling Himalaya.* Indian National Science Academy.

Tagliavini, F., Mantovani, M., Marcato, G., Pasuto, A., & Silvano, S. (2007). Validation of landslide hazard assessment by means of GPS monitoring technique–a case study in the dolomites (Eastern Alps, Italy). *Natural Hazards Earth System Science, 7*, 185–193.

Tripathi, J. N., Sonker, I., Tripathi, S., & Singh, A. K. (2022). Climate change traces on Lhonak Glacier using geospatial tools. *Quaternary Science Advances, 8*, 100065.

Wang, G-Q. (2012). Kinematics of the Cerca del Cielo, Puerto Rico landslide derived from GPS observations. *Landslides, 9*(1), 117–130.

Watanabe, T., Khanal, N. R., & Gautam, M. P. (1998). The Nangama glacial lake outburst flood occurred on 23 June 1980 in the Kanchanjunga area, eastern Nepal. *Annals of the Hokkaido Geographical Society, 72*, 13–20.

Wesnousky, S. G., Kumar, S., Mohindra, R., & Thakur, V. C. (1999). Uplift and convergence along the Himalayan Frontal Thrust of India. *Tectonics, 18*(6), 967–976.

Worni, R., Huggel, C., & Stoffel, M. (2013). Glacial lakes in the Indian Himalayas—From an area-wide glacial lake inventory to on-site and modeling based risk assessment of critical glacial lakes. *Science of the Total Environment, 468*, S71–S84. https://doi.org/10.1016/j.scitotenv.2012.11.043

Xu, M., Bogen, J., Wang, Z., Bønsnes, T. E., & Gytri, S. (2015). Pro-glacial lake sedimentation from jökulhlaups (GLOF), Blåmannsisen, northern Norway. *Earth Surface Processes and Landforms, 40*(5), 654–665. https://doi.org/10.1002/esp.3664

Yalçinkaya, M., & Bayrak, T. (2002). GPS in landslides monitoring: A case study from North Eastern Turkey. In *International symposium on GIS, Istanbul, Turkey*.

Yamada, T. (1993). Glacier lakes and outburst floods in the Nepal Himalaya. *IAHS Publications, 218*, 319–330.

Yu, Y., Li, B., Li, Y., & Jiang, W. (2024). Retrospective analysis of glacial lake outburst flood (glof) using ai earth insar and optical images: A case study of south lhonak lake, sikkim. *Remote Sensing, 16*(13), 2307.

Zhang, G., Yao, T., Xie, H., Wang, W., & Yang, W. (2015). An inventory of glacial lakes in the Third Pole region and their changes in response to global warming. *Global and Planetary Change, 131*, 148–157. https://doi.org/10.1016/j.gloplacha.2015.05.013

Zhang, T., Wang, W., & An, B. (2024). A massive lateral moraine collapse triggered the 2023 South Lhonak Lake outburst flood, Sikkim Himalayas. *Landslides, 22*(2), 299–311.

Zhang, F., Liu, G., Chen, W., Liang, S., Chen, R., & Han, W. (2012). Human-induced landslide on a high cut slope: A case of repeated failures due to multi-excavation. *Journal of Rock Mechanics and Geotechnical Engineering, 4*, 367–374.

Zhao, W., Nelson, K. D., Che, J., Quo, J., Lu, D., Wu, C., & Liu, X. (1993). Deep seismic reflection evidence for continental underthrusting beneath southern Tibet. *Nature, 366*(6455), 557–559.

Zryd, A. (2008). *Les glaciers en mouvement: la population des Alpes face au changement climatique* (Vol. 47). Collection Savoir Suisse.

Quantitative Assessment of Relative Tectonic Activity and Its Implications for Seismic Hazard Analysis in the Pir Panjal and Zanskar Mountain Ranges, Northwest Himalaya

Ahsan Afzal Wani, Bikram Singh Bali, Mohammad Irfan, and Naseer Ahmad Bhat

Abstract The Kashmir Basin, located in India's seismically active Zone V, lies within a complex fault system, making it highly susceptible to earthquakes. Despite historical seismicity in Sidmatnagar, Kondabal, Nilnag, Baramulla, and Dodherhama, the region has not experienced a major earthquake in recent centuries, except for the 2005 Muzaffarabad earthquake (Mw 7.6), which had its epicenter outside the Kashmir Valley near Balakot. This prolonged aseismic period suggests significant stress accumulation, increasing the likelihood of a future large-magnitude event. To assess relative tectonic activity and seismic hazard potential, we conducted a quantitative analysis using geomorphic indices. The region was subdivided into 22 sub-basins, of which nine (five in the southern sector and four in the northern sector) were analyzed using K-means clustering of eight morphometric indices: hypsometric integral, asymmetry factor, mountain front sinuosity, basin shape, form factor, bifurcation ratio, and sinuosity index. These indices serve as proxies for active tectonics, allowing a systematic evaluation of landscape deformation and fault activity. The clustering approach facilitated the identification of spatial patterns in tectonic activity, offering insights into the structural controls shaping the region. The results categorize the region into three tectonic activity classes: low (Class 1), moderate (Class 2), and high (Class 3). The highest relative tectonic activity is concentrated along the Panjal Thrust and Balapur Fault, forming a structurally controlled 100 km corridor from Shopian to Baramulla. In contrast, the Zanskar Thrust exhibits the lowest activity

A. A. Wani · B. S. Bali · M. Irfan (✉) · N. A. Bhat
Department of Earth Sciences, University of Kashmir, Srinagar, Jammu and Kashmir, India
e-mail: irfanwani1506@gmail.com

A. A. Wani
e-mail: ahsanawani@gmail.com

B. S. Bali
e-mail: baligeol@gmail.com

N. A. Bhat
e-mail: bhatnaseerabn340@gmail.com

© The Author(s), under exclusive license to Springer Nature Switzerland AG 2025

S. Ghosh (ed.), *The Himalaya Dilemma*, Springer Natural Hazards,
https://doi.org/10.1007/978-3-031-95083-4_3

in the northern and northeastern segments. Regions classified as high-active zones display higher geomorphic deformation, characterized by straight mountain fronts, asymmetric drainage and elongated basin shapes. Seismic frequency-magnitude analysis suggests a higher probability of moderate-magnitude earthquakes in northern regions. This spatial variation in tectonic activity, with attenuations from south to north, emphasizing the urgent need for updated seismic hazard assessments, geodetic monitoring, and risk mitigation strategies in the Kashmir Valley.

Keywords Active tectonics · Morphometric indices · Relative tectonic activity · Pir Panjal Range · Seismic risk · Kashmir Valley

1 Introduction

Tectonic geomorphology investigates the complex interactions between active tectonic forces and surface processes that collectively shape landscapes in tectonically active regions (Delcaillau et al., 1998; Burbank & Anderson, 2012; Irfan et al., 2024a). This field integrates principles from geology, geomorphology and tectonics to analyze how structural forces such as faulting, folding and crustal uplift influences surface features and landscape evolution (Demoulin et al., 1998; Keller & Pinter, 2002). Landscapes hold valuable records of the rates and spatial patterns of tectonic deformation, preserving evidence of how tectonic forces have shaped and continue to shape the Earth's surface. To analyze and quantify these changes, tectono-geomorphic indices offer essential tools for interpreting landscape responses to tectonic forces (Keller & Pinter, 2002; Bali et al., 2016; Topal et al., 2025). Consequently, tectono-geomorphic investigations in tectonically active regions have becomes key approach for understanding both present and historical tectonic activity (Kirby & Whipple, 2001; Topal & Irfan, 2025). Several studies have demonstrated that bedrock channels, in particular, respond actively to the interactions between erosion and tectonic uplift, with tectono-geomorphic parameters offering a valuable means for assessing these responses (Keller & Pinter, 1996; Kirby & Whipple, 2012; Wani et al., 2020). By applying these parameters, researchers can characterize and quantify landscape evolution of any basin, providing deeper insights into the dynamic processes that shape continuously shape and transform tectonically active landscapes.

Numerous researchers have contributed to the classification of tectonic activity by developing methodologies based on diverse tectono-geomorphic parameters (Taloor et al., 2017; Dey et al., 2016; Wani & Bali, 2017; Faghih et al., 2015; Khabazi et al., 2012; Dehbozorgi et al., 2010; El-Hamdouni et al., 2008; Silva et al., 2003; Raj et al., 1999, 2003; Chamyal et al., 2002; Bull & McFadden, 1977). Some studies, such as those by Silva et al. (2003), focused on specific parameters within localized contexts to highlight unique tectonic signatures, thereby demonstrating the adaptability of these frameworks for regional tectonic analysis. The method introduced by El-Hamdouni et al. (2008) was particularly systematic, as they proposed a Relative

Tectonic Activity Index (IAT) that categorizes tectonic activity into four classes. This classification uses the summation and averaging of several index values to create a composite measure of tectonic influence, providing a systematic approach to assess tectonic activity levels. In this way, the IAT facilitates a more standardized approach to geomorphic assessments, enhancing comparisons across different tectonic settings.

Several intermontane basins of Himalayan region emerge during the Tertiary Period through combined tectonic processes, where both extensional and compressional forces shaped their structure and morphology (Burbank & Johnson, 1983; Irfan et al., 2025; Bali et al., 2025). Since the Last Glacial Maximum (LGM), these basins have progressively filled with sediments eroded from adjacent, uplifted mountain range, which were deposited by river systems, glacial processes and mass wasting events (Kuhle, 2005). The evolution of these basins has been shaped by the diverse tectonic processes, changing climatic conditions, and variations in erosion and deposition rates, all of which have played a significant role in their development (Nakata, 1989; Goswami & Pant, 2007; Menier et al., 2017). Studying these basins offers valuable insights into their formation mechanism, tectonic history and the geomorphological processes that have shaped their evolution. The study area features NW–SE trending faults associated with Main Central Thrust (MCT) systems, including the Panjal Thrust and Zanskar Thrust (Bhat, 1982; Bossart et al., 1988; Calkins et al., 1975; Searle & Rex, 1989). In the Southwestern region, subsidiary thrusts of the MCT system, particularly the Panjal Thrust and out-of-sequence Balapora Fault, traverse through the Quaternary deposits (Ahmed et al., 2013). The activity along these thrusts, combined with subsequent erosional processes, has resulted in prominent deformational features. These include beheaded and offset streams, uplifted terraces (Veshav, Sasara, Romushi and Rambiara rivers), triangular facets and fault scarps which reflect the tectonic impacts of the Panjal Thrust and Balapora Fault activity. The identification of these geomorphic markers is therefore useful for deciphering the ongoing tectonic activity in order to understand the regions landscape evolution (Irfan et al., 2024a).

The Himalaya stands as a striking example of an orogenic system formed by the continental-continental collision between the Indian and Eurasian tectonic plates (Dewey & Burke, 1973; Wani et al., 2021). This collision, one of the most significant tectonic events in Earth's history, has driven the uplift and deformation that characterize the Himalayan region (Bhat et al., 2024; Dewey & Bird, 1970; Irfan et al., 2024b). The resulting orogenic processes continue to shape the Himalayas, making them an ideal natural laboratory for studying mountain-building, tectonic deformation and the interplay between tectonic and surface processes. The Kashmir Basin, located north of the Himalayan Frontal Thrust (HFT) and south of the Indus Tsangpo Suture Zone (ITSZ), is of significant geological interest due to its prime location within the evolving Himalayan orogenic belt. It lies between the Pir Panjal Range (PPR) and the Zanskar Range (ZR), and is traversed by various compressional structures such as thrust faults and folds (Dar et al., 2024; Mir et al., 2023). This unique tectonic configuration provides an ideal context for investigating tectono-geomorphic parameters and analyzing drainage systems and patterns to understand relative tectonic activity and the role of tectonics in defining the geometry and structure of the Kashmir Basin.

2 Study Area

The Kashmir Basin, located in the northwestern part of the Indian subcontinent and within the Union Territory of Jammu and Kashmir, is a prominent intermontane basin in the northwest Himalaya (Fig. 1). Characterized by its unique geological setting, it is bordered by two major mountain ranges: the Zanskar Range to the northeast and the Pir Panjal Range to the southwest, both contributing to its distinctive geomorphology (Ahmad et al., 2014b; Wani et al., 2019; Bali & Wani, 2020; Bali & Wani, 2021; Bali et al., 2021; Bhat et al., 2024; Dar et al., 2024). The drainage system of the Kashmir Basin is primarily structured around the Jhelum River, which flows from southeast to northwest, traversing the valley and forming a complex network with its numerous tributaries. Originating from the Verinag spring in the southern part of the basin, the Jhelum River meanders through the valley before existing via the Baramulla gorge to join the Indus River system. This river is sustained by a network of 24 tributaries, dominantly sourced from snow and glaciers (Sarah et al., 2021, 2024), that drain the surrounding mountain ranges and valleys through surface and subsurface pathways (Sarah et al., 2025), creating an extensive drainage system that plays a crucial role in shaping the landscape and contributing to its unique geomorphology. The Kashmir Baisn spans approximately 135 km in length, stretching from the north-northwest to the south-southeast, and measures around 40 km in width (Fig. 1). The region features rugged terrain, marked by numerous high mountain ranges, including the towering peaks of the Greater Himalayas and the Pir Panjal Range (Wani et al., 2022; Irfan et al., 2025b). These mountains are not only visually impressive but also harbor several glaciers that play a crucial role in supplying water to the River Jhelum and its tributaries, nourishing the river systems that traverse the basin. The formation of the Kashmir Basin is closely linked to geological processes, particularly the desiccation of a Karewa lake, which emerged due to tectonic uplift in the Pir Panjal and Zanskar Ranges (Burbank and Reynolds, 1984). This tectonic activity has resulted in a dynamic landscape of profound geological significance. The basin's topography is further characterized by deep valleys, gorges, glacial valleys, and incised and meandering rivers, alluvial fans creating a complex interplay of features that reflects its rich geological history (Irfan et al., 2025c). This study is focused on nine sub-basins, four located within the Greater Himalayan Range and in the Pir Panjal Range.

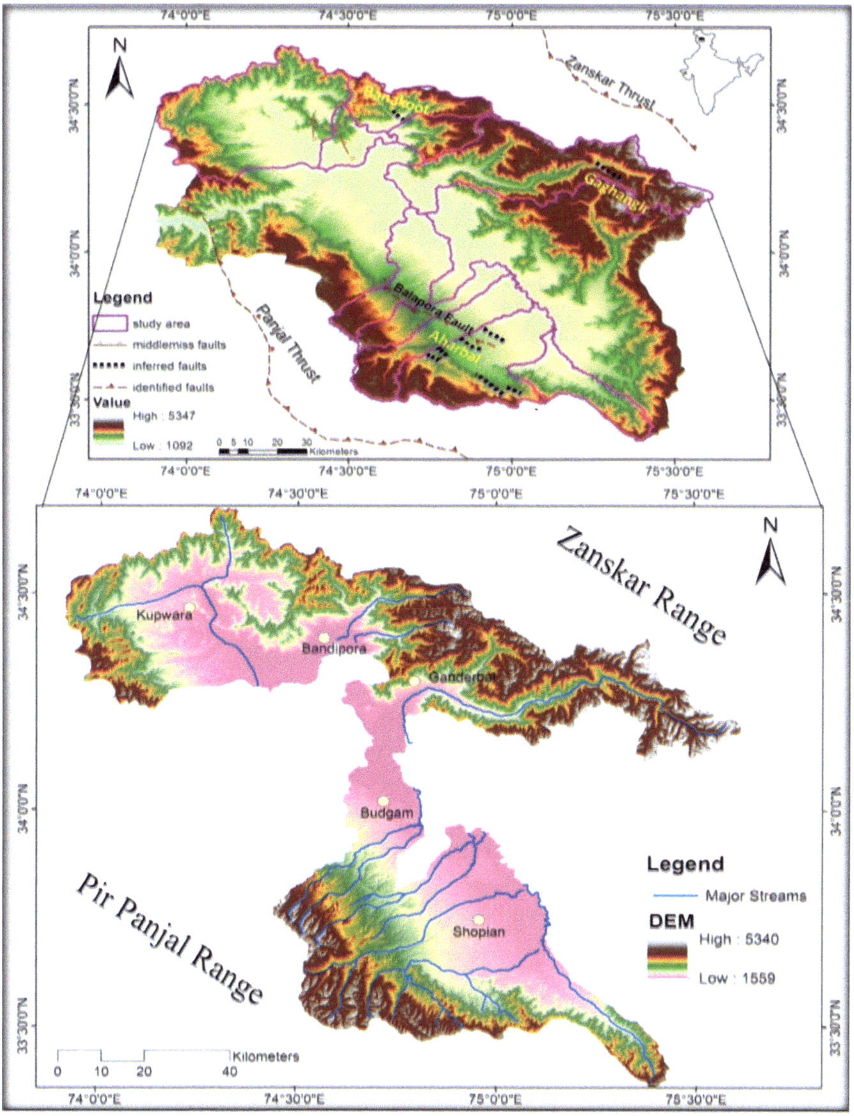

Fig. 1 Location map of the study area. Upper panel represents the Kashmir basin surrounded by two major thrusts. Lower panel represents the nine sub-basins of Jhelum basin five from Pir Panjal Range and four from Zanskar Range

3 Geological and Tectonic Setup

The Kashmir Basin, which evolved during the Neogene-Quaternary period, is an intermontane piggyback basin exhibiting prominent tectonostratigraphic characteristics that closely align with the regional orientation of the arcuate fold and thrust belt in the Northwest Himalaya (Ahmed et al., 2014a; Bali & Wani, 2021). This area is referred to as the Kashmir-Hazara Syntaxis (KHS), where the Himalayan fold-thrust belt curves in a hairpin bend in a north–south direction. The Kashmir Basin (KB), characterized by recurrent seismic activity, represents an ancient orogen-parallel compressional zone with a synclinal architecture composed of crustal-scale folds and thrust zones (Yousuf et al., 2021). It covers an area of about 13,605 km^2 and exhibits a discernible nappe-like structure, demarcated by the Zanskar Thrust (ZT) to the north and the Main Central Thrust (MCT) to the south (Fig. 2). These thrusts delineate the structural boundaries of the valley, representing zones of intense tectonic activity where rocks have been uplifted and deformed due to the ongoing collision between Indian and Eurasian plates. This region is marked by substantial altitude variations and a complex geological configuration, with rock formations ranging from the Precambrian to the Cenozoic (Thakur & Rawat, 1992; Bali et al., 2025). The basin contains well-formed successions of Paleozoic, Mesozoic, and Cenozoic rock formations distributed throughout the basin and its vicinity.

The basin is filled with a layer of unconsolidated sediments, approximately 1,300 m thick, composed of clays, sands, silts, gravels, and conglomerates, interspersed with occasional lignite beds referred to as Karewa. These sedimentary deposits provide significant evidence regarding the formation and evolution, where the sedimentation pattern has been influenced by sporadic and differential tectonic events (Agarwal & Agrawal, 2005; Burbank & Johnson, 1983; Burbank & Reynolds, 1984). The Balapur Fault is the sole confirmed structure in the region, extending approximately 95 km in a northwest-southeast direction (Ahmad et al., 2015). The Kashmir Valley exhibits a diverse and well-exposed geological record, ranging in age from the Precambrian to the Recent period. The Salkhala Formation represents the oldest rocks of the Kashmir Valley, dating back to the Precambrian era. These highly deformed metamorphic rocks, up to 6 km thick, consist of phyllites, slates, schists, and quartzites and are exposed around the fringes of the valley (Fig. 2).

The Dogra Slates, conformably overlying the Salkhala Formation, belong to the Paleozoic era and are characterized by finely cleaved slates with a thickness of up to 1,525 m. The Muth Quartzite, Devonian in age, forms the lower part of the Upper Paleozoic sequence and is prominently exposed in northwestern Kashmir. The Syringothyris Limestone, from the Permo-Carboniferous period, contains fossil evidence such as the characteristic brachiopod Syringothyris cuspidate. The Fenestella Shale, a Middle Carboniferous formation, follows the Syringothyris Limestone and contains a variety of marine fossils. The Agglomeratic Slate and Panjal Volcanics mark significant volcanic activity during the Upper Carboniferous to Lower Permian. These formations are overlain by the Gangamopteris Beds, rich in plant fossils, and the Zewan Formation, notable for its marine fossils. The Triassic sequence, comprising limestone and shale, is found in southern Kashmir, while the Karewa Formation, a vast sequence of glacio-fluvio-lacustrine sediments, covers much of the valley and plays a crucial role in its geological structure.

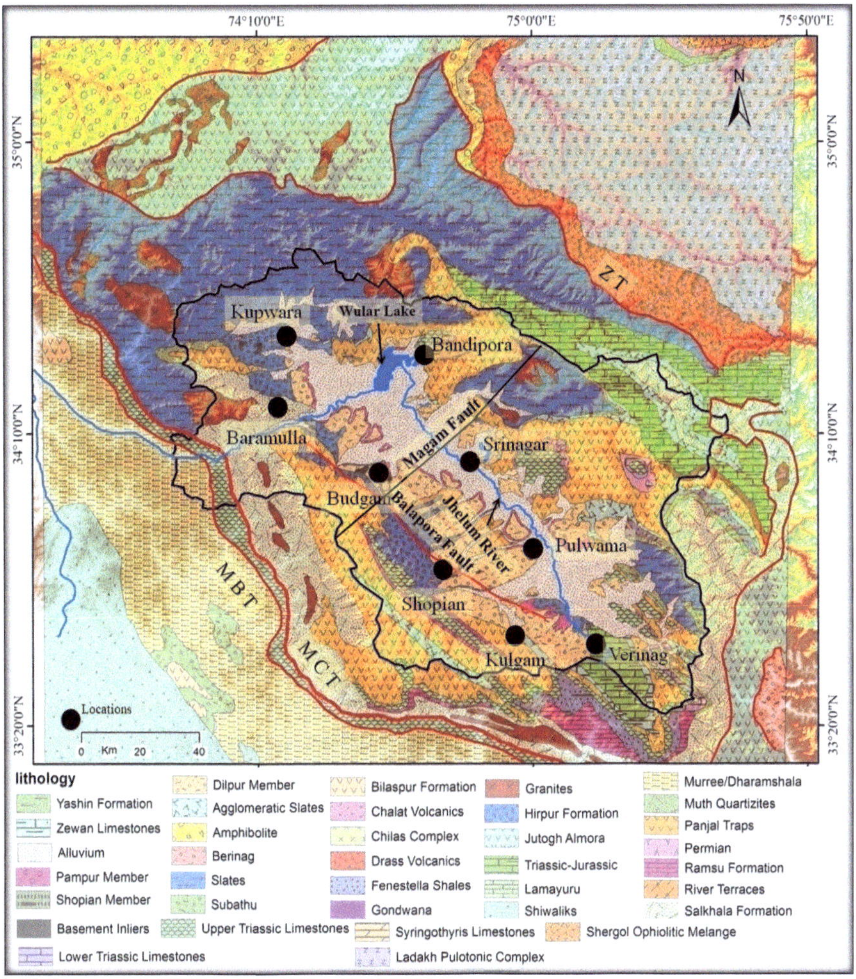

Fig. 2 Geological map of the Kashmir basin and surrounding areas—showing the traces of major faults that are present in and around Kashmir valley

4 Geomorphological Setting

The geomorphology of Kashmir basin presents a dynamic and diverse landscape shaped by an intricate interplay between tectonic forces, glacial, and fluvial processes. Situated in the north-western Himalayas, this basin is marked by rugged mountain ranges, deep valleys, and extensive alluvial plains. The Kashmir Basin, formed by the uplift of the Pir Panjal and Zanskar ranges, showcases both fluvial and glacial influences, which have shaped its distinctive landforms. In the higher elevations, glaciers have carved U-shaped valleys and left behind moraines, bearing witness to past glacial activity. Meanwhile, rivers like the Jhelum and its tributaries have sculpted V-shaped valleys and created fertile floodplains across the valley floor, supporting agricultural activity. The geomorphological map of the Kashmir Basin offers an insightful overview of the distribution of these varied landforms. It identifies major geomorphological units, including alluvial plains, plateaus, ridges, and areas of snow cover or glaciers (Fig. 3). The alluvial plains, typically aligned with river valleys, are fertile, flat regions formed by sediment deposition. Plateaus, as elevated flat areas, represent older and more stable surfaces, while ridges reflect linear elevated features shaped by active tectonic uplift or erosion. In the higher altitude regions, the map reveals areas covered by snow and glaciers, highlighting glacial contributions to the formation of U-shaped valleys. Much of the terrain is dominated by highly dissected hills and valleys, indicating ongoing, intense erosion in tectonically active zones, while more mature erosion patterns are evident in dissected hills. This geomorphological diversity underscores the active and evolving landscape of the Kashmir Basin, sculpted by the dynamic forces of tectonic uplift and surface processes that continually modify the landscape in this tectonically dynamic region.

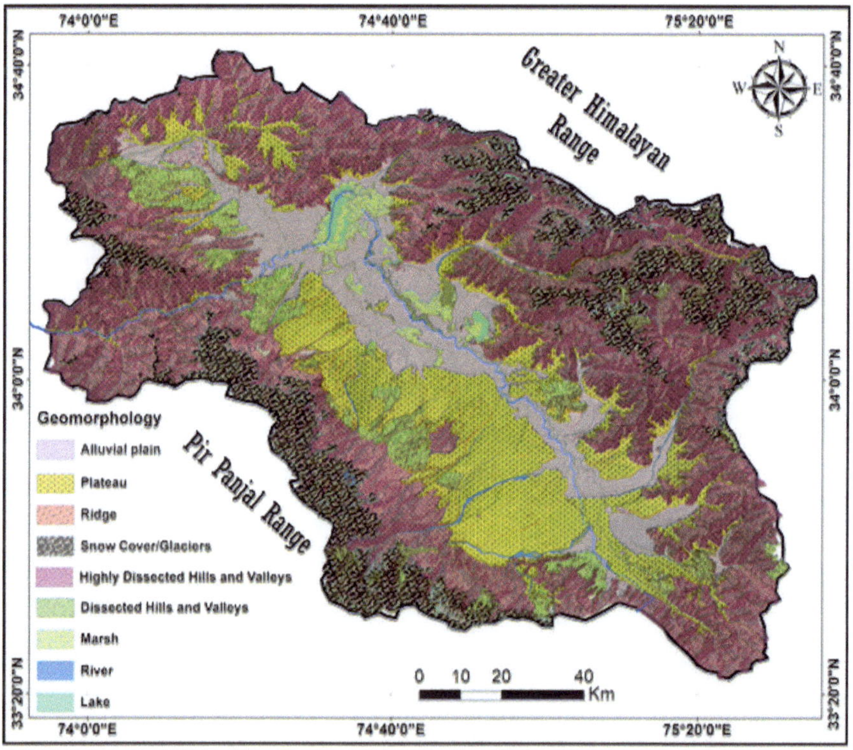

Fig. 3 Geomorphological map of the Kashmir Valley highlighting diverse landforms, including alluvial plains, plateaus, ridges, snow cover/glaciers, highly dissected hills and valleys, marshes, rivers, and lakes

5 Seismicity

Kashmir Basin falls under Zone V in India's seismic zonation map (1996). It is bordered by two well-known fault systems: the Panjal thrust to the south and the Zanskar Thrust to the north (Bhat, 1982). Additionally, the basin is dissected by several smaller faults and thrusts, such as Balapur Fault, Erin fault, Magam Fault, Laam fault and other faults identified by Middlemiss, making the area prone to earthquakes (Ahmed et al., 2018; Bali et al., 2025; Wani et al., 2025). The Balapur Fault, having shown no activity in the past, contributes to the regions high seismic risk (Ahmed et al., 2014a; Hough et al., 2009; Madden et al., 2011; Middlemiss, 1910, 1911). The neotectonic activity in this region is evidenced by numerous seismic events, including earthquakes, tectonic structures and liquefaction features found in the Plio-Pleistocene Karewa deposits. Historical and recorded seismic data (1972–2024, NEIC catalogue) reveal that the region has experienced numerous low-magnitude (<5 mb), high-frequency seismic events. Notable historical earthquakes that have periodically affected the Kashmir valley include the Baramulla earthquake of 1885, Nilnag Budgam in 1885, Kondabal earthquake of 1887 and the Wular/Sandimatnagar earthquake of 2082–2041 BC (Fig. 4; Ahmad et al., 2009, 2015; Ahmad & Shafi, 2014; Bilham et al., 2010; Bilham & Bali, 2013; Bali et al.,2025). The seismic pattern in the southern Kashmir valley indicates a high risk for future moderate earthquakes.

The 1500 m thick Karewa deposits above the valley floor exhibit a northeast tilt of 40°, suggesting disruption attributed to neotectonic activity or uplift due to the Panjal Thrust (Fig. 5a and b). Near Balapur village, the Rambiara River displays a braided and beheaded pattern, along with the development of unpaired river terraces (Ahmed et al, 2013). In the Southwestern part, small faults observed in the Pleistocene Karewa deposits further indicate neotectonic activity (Fig. 6a–d). These neotectonic features, along with the Balapur fault make the southern and southwestern part of the Kashmir Basin more susceptible to seismic hazards. Another fault identified near Raithan in Budgam shows preserved geomorphic features, such as oppositely dipping karewa beds, river braiding and meandering, and unpaired terraces (Fig. 7a–d). Similarly, near Tangmarg village, the Veshaw River features asymmetrical/unpaired river terraces on its sides (Fig. 5c and d).

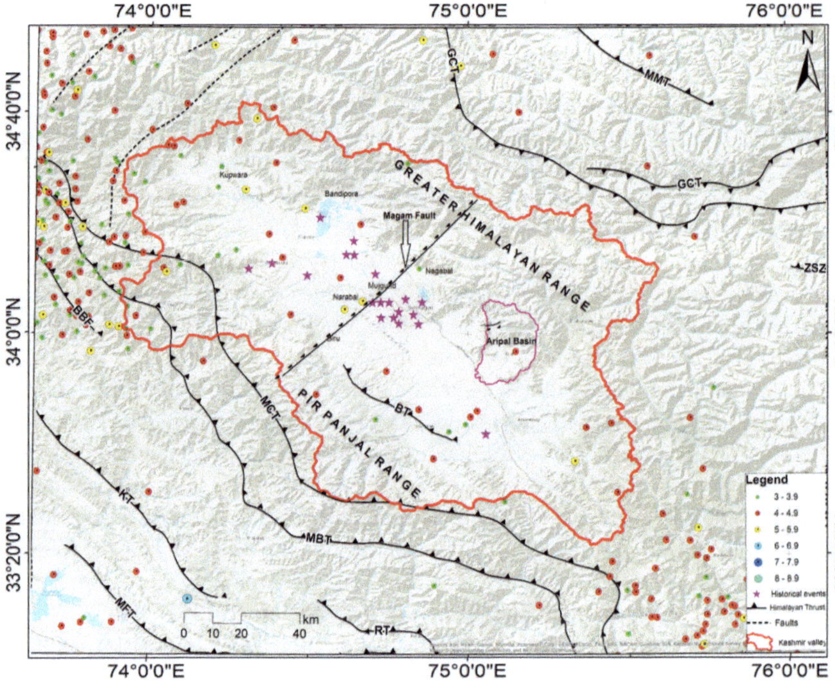

Fig. 4 Historical and instrumental records of seismic events of Kashmir valley and surrounding areas (after Bali et al., 2025; Ahmad et al., 2009)

6 Results and Discussion

6.1 Tectono-Geomorphic Analysis

Tectonic influences on basin evolution are effectively represented by structural, tectono-geomorphic and fluvial characteristics. Morphotectonic techniques, involving mathematical evaluation and measurement mechanism, provide a systematic approach for analyzing the geometry and structure of landscapes across regions. These techniques yield valuable insights into basin characteristics, illuminating on processes such as erosion, deposition, uplift rates, basin genesis and broader trends in landscape evolution (Keller & Pinter, 2002; Pike & Wilson, 1971; Strahler, 1952). The applications of Morph-tectonic methods enable researchers to quantify relative tectonic activity and assess how tectonic structures influence landscape dynamics, offering a detailed view of tectonic controls on geomorphological processes. This study, therefore, presents substantial evidence regarding the role of tectonic structures in the landscape evolution, highlighting the complex interplay between tectonics and surface processes.

Fig. 5 **a** Dipping Karewa beds near Kuil, Chandpora (**a** and **b**), Unpaired River terraces near Tangmarg (**c** and **d**) Where RD and LD are right bank and Left bank drainage divides respectively, Blue arrow indicates flow direction and red ellipse shows offset Veshaw river

6.2 Mountain Front Sinuosity (Smf)

Mountain front sinuosity (Smf) is a geomorphological index used to quantify the sinuosity or irregularity of the mountain front, which is the boundary where mountain ranges meet adjacent valleys (Bull & McFadden, 1977; Keller & Pinter, 1996). It measures how much the mountain front deviates from a straight line, providing insights into the tectonic activity and landscape evolution of a region. Younger, tectonically active mountain fronts are generally straight, whereas older mountain fronts, experiencing less tectonic disturbance, tend to be irregular and sinuous (Bull & McFadden, 1977). This parameter reflects the balance between geomorphic processes or erosion, which create an irregular and sinuous mountain front, and active tectonic deformation, which tend to form a more defined, straight front (Irfan et al., 2024a; Wani et al., 2019). A low sinuosity value close to 1 suggests a straight mountain front, commonly found in tectonically active regions where recent uplift has restricted significant erosional changes (Keller & Pinter, 1996). This index distinguishes between mountain fronts shaped predominantly by erosion, which tent to be sinuous, and those shaped by active tectonic deformation, which generally exhibit a straighter front (Wani et al., 2021; Irfan et al., 2024c). Mountain fronts with high tectonic activity typically exhibit Smf values between 1.0 and 1.6, reflecting more pronounced tectonic influence. In contrast, less active or inactive fronts tend to show

Fig. 6 **a** Normal fault near Chandpora with 20 feet offset of boulder bed **b** reverse fault with 2.5 feet offset **c** normal fault near Gundipora with 1.5 feet offset **d** inset of reverse fault shown on (B)

Fig. 7 **a** Northeast dipping karewa beds near Arigam **b** Southwest dipping Karewa beds near Raithan and offset stream indicated by green line **c** un paired river terraces and meandering of river **d** braided bar deposit above fault plane (Black arrow indicate flow direction)

higher Smf values, ranging from 1.6 to 3.0 for moderately active fronts and exceeding 3.0 for inactive ones (Bull & McFadden, 1977).

In this study, Smf values were quantified for nine sub-basins to gain insights into the geomorphological characteristics and tectonic activity within the region. Four sub-basins are located in the Greater Himalayan Range, while the remaining five are situated in the Pir Panjal Range. The average Smf values for each sub-basin were analyzed to determine the primary factory driving sub-basin evolution. The Smf values for the sub-basins, Sandran (1.13), Veshaw (1.09), Rambria (1.14), Romushi (1.15), Doodhganga (1.10), Pohru (1.30), Madhumati (1.25), Erin (1.20), and Sindh (1.29), range from 1.09 to 1.30, suggesting relatively straight mountain fronts likely influenced by active tectonic uplift. The overall average Smf value of 1.18 across these sub-basins further indicates a landscape shaped by tectonic forces, with straightened mountain fronts reflecting ongoing tectonic uplift processes. Quaternary to Recent sediment deposits along mountain fronts provide strong evidence of ongoing tectonic activity. This activity is marked by several geomorphological features, such as distorted alluvial fans, indicating shifts in sediment deposition patterns due to tectonic forces. Additionally, rivers in these areas show deep incision, characteristic of uplift regions where streams erode downwards in response to tectonic adjustments (Fig. 8a and c). A rejuvenated drainage pattern supports this evidence, as reorganization of drainage system often results from landscape shifts caused by tectonic events. We also observed triangular facets with long bases along these mountain fronts, pointing to active tectonic uplift (Fig. 8b). These facets, which form as steep, planar surfaces due to the faulting and uplift, are strong indicators of tectonic activity (Bahrami, 2013; Irfan et al., 2024a). Their presence suggests that tectonic forces are actively shape these fronts, with uplift processes outpacing erosion. This observation further emphasizes the role of tectonic stress in driving landscape evolution. The presence of incised V-shaped valleys, typically associated with river down cutting in tectonically active regions, and U-shapes valleys, often shaped by past glacial activity, highlight the complex interplay between tectonic forces and erosional processes (Fig. 8d). The interactions among these landforms indicate a moderate to high rate of tectonic activity in the Kashmir Basin. This tectonic influence, combined with glacial deposits and subsequent erosion, plays a significant role in shaping the mountain front geometry, underscoring the dynamic landscape evolution within the region.

6.3 Basin Asymmetry Factor (Af)

The asymmetry factor (Af) is a metric used to assess the asymmetry of river basins (Bull & McFadden, 1977). It helps to quantify the degree of imbalance between the left and right sides of drainage basins, which can provide insights into tectonic activity, erosional processes, or other factors influencing the evolution of the basin (Hare & Gardner, 1985). The Af values range from 1 to 100%, indicating how asymmetric the basin is along its main river course. When the value of Af is 50%,

Fig. 8 Geomorphic features present in the basins with higher Smf values (**a** and **c**) deeply incised V-valleys **b** Triangular facets **d** U-shaped valleys

the basin is perfectly symmetric. This means the areas on both sides of the river are equal. When the Af value is greater than 50%, it indicates that the left side of the basin is larger than the right side, suggesting that the river has been influenced by processes or factors that cause an imbalance, such as tectonic uplift, faulting or differential erosion. The higher or lower the Af value (approaching to 1 and 100%), the more significant the asymmetry.

In the present study, the Af was computed for all nine sub-basins, with values ranging from 22.20 for the Erin to 62.50 for the Sindh sub-basin. The quantified Af values for the sub-basins, which include Sandran (38.50), Veshaw (70.22), Rambria (40.20), Romushi (55.40), Doodhganga (50.30), Pohru (26.55), Madhumati (58.40), Erin (22.20), and Sindh (62.50), suggest varying degrees of asymmetry across the region, likely influenced by tectonic forces and geomorphological processes. Sub-basins like Veshaw (70.22) and Sindh (62.5) exhibit higher Af values, indicating substantial asymmetry. These values imply significant tectonic influence, potentially from faulting, uplift or tilting, which may have led to a more pronounced imbalance in drainage distribution. This high asymmetry could point to active or recent tectonic adjustments affecting the landscape on one side of these basins more than the other. Intermediate Af values, such as those for Romushi (55.40), Doodhganga (50.30) and Madhumati (58.40), suggest moderate asymmetry that could result from differential tectonic uplift or differential erosion. The results suggest that five basins including,

Veshaw, Madhumati, Romushi, Doodhganga and Sindh are tilted towards right side, while 4 basins such as Sandran, Rambria, Pohru and Erin are tilted towards left side.

The defining characteristic of asymmetrical basins is the presence of deformed and unpaired river terraces. These terraces are considered significant geomorphological indicators because they reflect the impact of external processes, such as tectonic forces, which contribute to the asymmetry of these basins. Field investigations into such basins have provided crucial evidence supporting the connection between tectonic activity and the formation of unpaired terraces. For example, the longer tributaries observed on the tilted side of the basin is typically the result of the rivers flow being influenced by the tectonic uplift. This differential uplift causes one side of the basin to experience more erosion, leading to the formation of longer tributaries on that side, while the opposite side remains less eroded, resulting in shorter tributaries. In basins with right-side tilt, the tributaries in the right side are generally shorter, while the tributaries on the left side are longer. This is because the right side of the basin has been uplifted more, causing greater erosion and reducing the length of the tributaries that drain into the river. On the other hand, the left side experiences less erosion and therefore, the tributaries are able to extend further into the basin. A similar pattern can be observed in lift-tilted basins, where the left side tributaries are longer than those on the right. Overall, the range in Af values reveals the differential impacts of tectonic and erosional processes across the sub-basins, indicating that some areas experience stronger tectonic influences than others, while some basins maintain more Geomorphologically balanced conditions.

6.4 Hypsometric Integral (Hi)

The Hypsometric Integral (Hi) is used to quantify the shape and elevation distribution of a landscape, particularly in relation to the overall topography of a region (Strahler, 1952). This index provides insights into the balance between erosion and uplift within a given terrain. High values close to 1, indicate a landscape with a more youthful, elevated terrain, typically meaning less erosional modifications and a steeper gradient. Low Hi values closer to 0, suggest a mature landscape that has been subject to more extensive erosion and is closer to a more even lower elevation profile.

In the present study, the Hi values for all sub-basins were calculated, ranging from 0.51 to 0.63. The range values of HI suggest a youthful landscape with ongoing tectonic uplift and erosion, indicating that the sub-basins are in a transitional stage of landscape development. These values reflect a balance between uplift and erosional processes, contributing to the evolving topography in the region. The quantified Hi values for the sub-basins, Sandran (0.58), Veshaw (0.56), Rambria (0.51), Romushi (0.49), Doodhganga (0.47), Pohru (0.52), Madhumati (0.60), Erin (0.57), and Sindh (0.63), suggest a range of geomorphic maturity stages across the region. Higher values, as observed in the Sindh (0.63) and Madhumati (0.60), imply relatively youthful landscapes with ongoing tectonic uplift and active erosion. In contrast, lower values, such as those in the Doodhganga (0.47) and Romushi (0.49), sub-basins, indicate more mature landscapes that may be experiencing less tectonic influence and higher levels of erosional balance. This variation in Hi values reflects the diverse tectonic and erosional dynamics influencing the geomorphology of each sub-basin.

6.5 Elongation Ratio (Eb)

The Elongation ratio (Eb) is used to assess the shape of a drainage basin and also know the influence of tectonics over a basin. It is defined as the ratio between of the diameter of a circle with the same area as the basin to the maximum basin length (Schumm, 1956). This parameter provides insights into basin hydrology and tectonic activity, as it reflects the degree to which a basins shape deviates from being perfectly circular. Chow (1964) classifies basin shapes based on varying Eb values: ratios ranging from 0.9 to 1.0 are considered circular, those between 0.8 and 0.9 are oval, ratios from 0.7 to 0.8 are categorized as elongated, and values below 0.7 indicate an elongated basin shape. Molin et al. (2004) suggest that young basins in areas of active uplift are typically characterized by a more elongated shape. Bull and McFadden (1977) used low values of basin elongation as a proxy to indicate recent tectonic activity within the basin.

In the present study, the Eb values for the Kashmir basin were quantified for nine sub-basins with the following results: Sandran (0.12), Veshaw (0.15), Rambria (0.11), Romushi (0.11), Doodhganga (0.13), Pohru (0.17), Madhumati (0.15), Erin (0.17), and Sindh (0.09). The Eb values for the sub-basins are shown in Table 1. The average Eb value across these sub-basins is 0.22, which indicates highly elongated basins, suggesting a strong tectonic influence, likely due to active faulting or structural controls that shape the basin morphology in the region. These low values of all sub-basins reflect the stretching of the landscape, potentially driven by regional tectonic forces associated with the ongoing convergence of the Indian and Eurasian plates. The sub-basins along the PPR display a highly elongated form, which is typically associated with significant tectonic influences. On the northern side of the Kashmir Basin, the Sindh sub-basin exhibits notably low value of 0.09, which indicate higher tectonic disturbance.

This interpretation is supported by the presence of various tectono-geomorphic indicators including triangular facets with long bases, wine-glass valleys with narrow outlets, alluvial fans, deeply incised channels and uplifted river terraces. These structural features run both parallel to and intersect the main river, highlighting the impact of tectonic forces that have actively shaped and deformed the landscape in this region. These values provide insights into the morphological characteristics of each sub-basin and contribute to understanding the tectonic and geomorphological processes shaping the region.

Table 1 Computed geomorphic indices for nine sub-basins and their statistical summary

Sl. No.	Smf	Hi	Af	Rb	Ka	Eb	Si	Ff
Sandran	1.13	0.40	48	4.50	0.30	0.12	1.14	0.20
Veshaw	1.09	0.49	70	3.60	0.38	0.15	1.40	0.34
Rambria	1.14	0.51	61.2	3.56	0.30	0.11	1.20	0.19
Romushi	1.15	0.38	63.3	3.08	0.24	0.11	1.32	0.28
Doodhganga	1.10	0.23	21.62	3.78	0.31	0.13	1.30	0.41
Pohru	1.30	0.23	47	4.25	0.57	0.17	1.36	0.70
Madhumati	1.25	0.49	68.52	4.60	0.41	0.15	1.12	0.33
Erin	1.20	0.46	32	5.10	0.46	0.17	1.12	0.34
Sindh	1.29	0.40	62.7	4.18	0.24	0.09	1.28	0.23

6.6 *Circularity Ratio (Ka)*

The circulatory ratio (Ka) is another geomorphic parameter used to define the shape of a basin, providing insights into its tectonic characteristics (Miller, 1953). The values of this index range from 0 to 1. A value closer to 1 indicates a more circular basin shape, while lower values suggest a more elongated basin. Higher values (closer to 1) typically indicate a basin with minimal tectonic deformation, suggesting that erosional processes are more dominant than tectonic forces. These basins are often associated with low to moderate relief. In contrast, low Ka values are typical of elongated basins, which are often influenced by tectonic activity or structural controlled, such as active faults or regional tectonic stresses. In this study, the Ka values were calculated for all nine sub-basins to assess their basin shapes and gain insights into the underlying geomorphological and tectonic processes. The obtained values are as follows: Sandran (0.30), Veshaw (0.38), Rambria (0.30), Romushi (0.24), Doodhganga (0.31), Pohru (0.57), Madhumati (0.41), Erin (0.46), and Sindh (0.24). The values range from 0.24 (Romushi and Sindh) to 0.57 (Pohru), with an average value of 0.35. This suggests that the basins generally exhibit an elongated shape, with some basins showing more circular characteristics, such as Pohru, while others are more elongated, indicating tectonic influences and structural controls that shape the morphology of these basins with time.

6.7 *Form Factor (Ff)*

The Form factor (Ff) is used to describe the shape of a river basin or watershed (Horton, 1945). This index provides insights into the shape and potential hydrological behavior of the basin. Higher Ff values (Close to 1) indicate more circular basins. These basins tend to have a more rapid runoff response, as water reaches the outlet faster. In contrast, lower Ff values indicate elongated basins, where runoff may take longer to reach the outlet, resulting in a more delayed response. In the present study, Ff values were computed for each sub-basin, with results ranging from 0.20 in the Sandran and Rambria sub-basins to 0.70 in the Pohru sub-basin. The specific Ff values for each sub-basin are as follows: Sandran (0.20), Veshaw (0.34), Rambria (0.20), Romushi (0.28), Doodhganga (0.32), Pohru (0.70), Madhumati (0.33), Erin (0.46), and Sindh (0.23). These values suggest significant variability in basin shapes, with lower Ff values indicating elongated basins and higher values pointing to more circular basins. This variation implies differences in runoff response, as elongated basins typically have slower response times compared to more circular basins, which may experience quicker runoff and higher peak flow.

6.8 Sinuosity Index (Si)

River patterns are significantly shaped by the underlying geology and active tectonic processes, as they influence river morphology, orientation and channel sinuosity of any region (Twidale, 2004). Tectonic activities, such as regional uplift or subsidence, alter the landscape, reshaping slopes and gradients that directly impact river behavior (Schumm, 1986). These changes in river courses and sinuosity patterns reflect how rivers adapt to structural alignments, adjusting to faults, folds and rock resistance within a tectonically active landscape (Irfan et al., 2024a). The Sinuosity Index (SI) is a measure used in geomorphology to quantify the degree of meandering or winding of a river or stream channel compared to the straight line distance between two points long the channel. The geometric structure of the drainage system and patterns can be influenced by subsurface geological structures, which can be determined using the SI index (Bali et al., 2025). This metric is an indicator of the degree of meandering in a river channel and serves as a tool for evaluating the influence of active tectonics on landscape evolution (Irfan et al., 2024a, 2024b; Wani et al., 2019). Mullar (1968) classified sinuosity index values to distinguish different channel patterns: a sinuosity index value of less than 1 indicates a straight river course, values between 1–1.5 signify a sinuous pattern, and values greater than 1.5 represent a meandering pattern. These classifications allow geomorphologists to interpret the tectonic and geological controls on river morphology, as well as to infer the level of landscape maturity and tectonic stability in a region.

In the Kashmir Basin, the SI values were calculated for each sub-basin to assess the meandering and winding characteristics of the river channels. The results ranged from 1.12 in the Madhumati and Erin sub-basins, indicating relatively sinuous pattern, to 1.42 in the Pohru sub-basin, suggesting a higher degree of sinuosity and even braided patterns. These variations reflect the complex interactions between the underlying geological structures, tectonic processes band the flow dynamics of the rivers. These variations in channel morphology across the basin, likely influenced by local geological structures and tectonic activity, which shape the flow dynamics and deposition processes within the river system. The specific SI values for each sub-basin are as follows: Sandran (1.14), Veshaw (1.40), Rambria (1.20), Romushi (1.32), Doodhganga (1.30), Pohru (1.42), Madhumati (1.12), Erin (1.12), and Sindh (1.28). The SI value for Madhumati and Erin exhibits low sinuosity, suggesting relatively straight rivers, which could be a result of steeper gradients, tectonic uplift processes that restricts lateral movement. On the other hand, the intermediate values (e.g., Sandran, Veshaw, Rambria) reflect high sinuosity, braided pattern and moderate meandering, indicating that the rivers are influenced by both tectonic processes, particularly by Balapur Fault and erosional processes that allow for some degree of sinuosity. Subbasins like Pohru exhibit higher sinuosity, possibly due to tectonic adjustments, and sediment deposition that facilitate lateral migration. These values further illustrate the diversity in the river morphologies across the basin, influenced by factors such as topography, tectonic forces and sediment transport.

6.9 Relative Tectonic Activity (IRTA)

The statistical summary of the geomorphic indices obtained from the analysis is presented in Tables 1 and 2. These indices provide valuable insights into the relative tectonic activity of the study area. The assessment is based on eight key geomorphic indices Smf, SI, Af, Hi, Rb, Ka, and Ff. the average of these indices was used to classify the region into three categories of relative tectonic activity: low (Class 1), Moderate (Class 2) and High (Class 3), following the methodology of El Hamdouni et al. (2008) and Wani and Bali (2017). The spatial distribution of these classes is depicted in Fig. 9. Regions classified under high relative tectonic activity exhibit distinct seismic and tectono-geomorphic characteristics. Notably, these areas experience frequent small-magnitude earthquakes, along with a historical record of large-magnitude seismic events (Fig. 4). The high tectonic activity is further corroborated by Class 3 values of the relative tectonic activity indicating significant ongoing deformation. Structurally, this zone is marked by the presence of high-angle thrust faults such as the Balapora Fault, which, despite being relatively silent in historical seismic records, contributes to regional tectonic adjustments. Additionally, key geomorphic indicators of active tectonics are observed in the high tectonic activity class. These include the presence of unpaired river terraces along the Veshaw (Fig. 5c and d), Rambaria and Sasara river basins. Such terraces suggest episodic uplift and river incision, typically associated with active tectonic settings. Furthermore, the region exhibits highlighy elongated river basins, convex up longitudinal river profiles, and a parallel drainage pattern with offset streams, particularly evident near Balapora, Tangmarg and Raithan. These characteristics indicate strong structural control on the drainage system, reinforcing the interpretation of high tectonic activity in this part of the Kashmir valley.

In addition to the above tectono-geomorphic features the braided streams, incised valleys, linear mountain fronts, elongate basins, low bifurcation and elongation ratio in the southern part of the area and presence of faults (Panjal thrust, Balapora fault and other small faults) which have formed important anticlines and fault scarp near

Table 2 The output tables of K-means clustering processes by IBM SPSS 21 for Smf index. **A** Initial cluster centers; **B** iteration history of the clustering; **C** final cluster centers; **D** analysis of variance; **E** number of cases in each cluster

(A) Initial cluster centers						
	Cluster					
	1		2		3	
Smf	1.30		1.20		1.09	
(B) Iteration history						
Iteration	Change in cluster centers					
	1		2		3	
1	0.000		0.025		0.005	
2	0.000		0.007		0.015	
3	0.000		0.000		0.000	
1	0.000		0.025		0.005	
(C) Final cluster centers						
	Cluster					
	1		2		3	
Smf	1.28		1.20		1.12	
(D) ANOVA						
	Cluster		Error		F	Sig.
	Mean square	df	Mean square	df		
Smf	0.024	2	0.001	6	34.647	0.001
(E) Number of cases in each cluster						
Cluster			1		1.000	
			2		3.000	
			3		5.000	
Valid					9.000	
Missing					0.000	

Shopian, beheaded and offset streams above Shopian town (Ahmed & Bhat, 2012; Ahmed et al., 2013) are responsible for high tectonic activity in these subasins. Also the Balapora fault has thrusted 9 m Pampur member over the Shopian member which is clearly visible near Balapora village (Ahmed et al., 2013). The uplift along the Panjal thrust has deformed the Karewa beds towards northeast at an angle of 40° (Fig. 5a and b). Balapora fault and other small faults in the Karewas with the development of tectonogeomorphic features can be a reason for the increase in relative tectonic activity towards the southern fringe of the study area. High tectonic activity based on Iat, already published literature and field observations suggest high tectonic activity is widespread towards the southern and southwestern part of the area. Based on Arian and Hashemi (2008), this area is under high seismic risk, with the chance of moderate earthquake in future. Because the area contains some smaller faults like

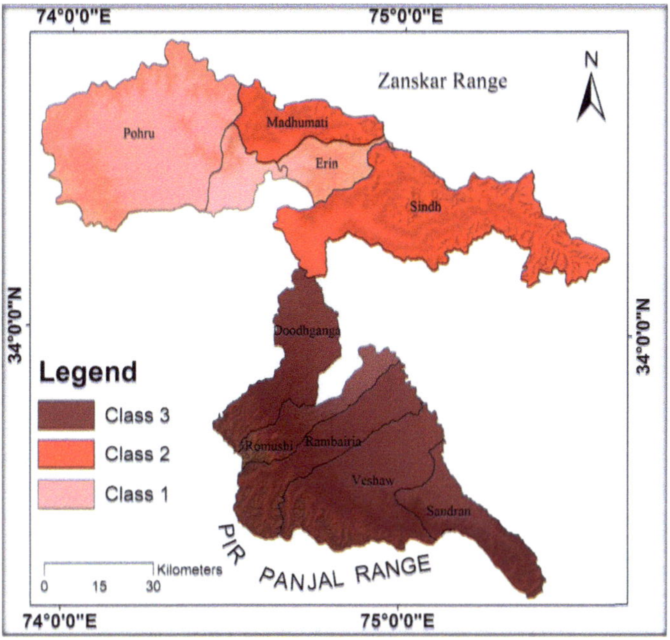

Fig. 9 Spatial distribution of relative tectonic activity classes of nine sub-basins in the study area

Tangmarg and Raithan faults which were identified in the field by preserving some geomorphic features like meandering, braiding, unpaired river terraces and offset stream near Tangmarg and opposite dipping karewa beds near Raithan. The beds dip towards the SW and NE on the right and left side of the fault plane facing south. However, in the northern part of the study area, the less elongate basins with moderate smf values, low bifurcation ratio values, high form factor and least frequency earthquakes and lesser number structures in the form of faults, folds, incised valleys and straight river course led to decrease in the relative tectonic activity to moderate and low classes. Another important aspect supports the reduction of relative tectonic activity is the least frequency of earthquakes recorded in this area. Seismic activity serves as a critical indicator of ongoing tectonic deformation, and the observed low seismicity suggests that the region is experiencing comparatively lower level of tectonic stress.

7 Conclusions

The computation and classification of geomorphic indices using K-means clustering demonstrate their effectiveness in evaluating the tectonic activity of the study area. Several geomorphic indices, including drainage basin asymmetry (Af), hypsometric integral (Hi), elongation ratio (Re), circulatory ratio (Rc), form factor (Ff), sinuosity index (Si), and mountain-front sinuosity (Smf), were analyzed for nine sub-basins within the Kashmir Valley—five located in the Pir Panjal Range and four in the Zanskar Range. To classify the tectonic activity, the geomorphic indices were categorized into three categories—high, moderate, and low—based on computed values, seismic activity, and field observations. A composite tectonic activity index (Iat) was derived by integrating eight geomorphic indices for each sub-basin, providing a comparative assessment of relative tectonic activity. The spatial distribution of tectonic activity reveals that Class 3 (high tectonic activity) accounts for the largest proportion of the area, covering approximately 3,394 km^2 (47%). Class 2 (moderate tectonic activity) spans 2,053 km^2 (28%), while Class 1 (low tectonic activity) occupies 2,070 km^2 (25%). Areas classified under Class 3 are predominantly located along major structural features such as the Panjal Thrust, Balapora Fault, and identified fault zones near Raithan and Tangmarg, as well as in the vicinity of significant anticlines. Class 2 zones are distributed in the northeaster segment of the Himalayan Range, whereas Class 1 zones are confined to the northernmost extent of the study area. Regions with high tectonic activity (Class 3) correspond to areas with frequent seismic events, particularly near the Panjal Thrust and Balapora Fault. The higher occurrence of low-magnitude, high-frequency earthquakes in these areas suggest the potential for moderate seismic events in the future. The finding of this study are consistent with previous research in the region, reinforcing that sub-basins categorized under Class 3 are highly susceptible to seismic hazards and should be prioritized for further geotechnical and seismic risk assessments.

References

Agarwal, K. K., & Agrawal, G. K. (2005). A genetic model of thrust bounded intermontane basinusing scaled sandbox analogue models: An example from the Karewa Basin, Kashmir Himalayas, India. *International Journal of Earth Science, 94*, 47–52.

Ahmad, B., & Shafi, M. (2014). Some more earthquakes from medival Kashmir. *Journal of Seismology* 1383–4649.

Ahmad, B., Bhat, M. I., & Bali, B. S. (2009). Historical record of earthquakes in the Kashmir Valley. *Himalayan Geology, 30*(1), 75–84.

Ahmad, B., Ahmed, S., Alam, A., Wang, S., & Bhat, M. S. (2015). Looking for missing links in Kashmir: An update on nineteenth-century seismicity. *Seismological Research Letters, 86*(4), 1–6.

Ahmad, B. Sana, H., & Alam, A. (2013). Macroseismic intensity assessmentof 1855 earthquake of northwest Kashmir Himalaya, using environmental seismic intensity scale (ESI2007). *Quarternary International, 355*, 1–6.

Ahmed, S., & Bhat, M. I. (2012). Tectonic geomorphology of Rambiara basin, SW Kashmir Valley reveals emergent out-of-sequence active fault system. *Himalayan Geology, 33*(2), 162–172.

Ahmed, S., Alam, A., Ahmad, B., Bhat, M. I., & Bhat, M. S. (2015). Geomorphic evidences of unrecognized Balapur fault segment in the southwest Kashmir basin of Northwest Himalaya. *Geomorphology* 159–172.

Ahmed, S., Alam, A., Ahmad, B., Wani, A. A., Bhat, M. I., Bhat, M. S., & Farooq, H. (2018). Tectono-geomorphic indices of Erin Basin, Northeast Kashmir valley, India. *Asian Earth Sciences, 151*, 16–30.

Ahmed, S., Bhat, M. I., Ahmad, B., & Alam, A. (2014a). Late quaternary landforms generated by the Balapur fault and its kinematic analysis on the northeastern flank of the Pir Panjal Range, Kashmir Valley, India. In *International Geographical Union (IGU). Conference Paper*, 4–5 June 2014.

Ahmad, S., Bhat, M. I., Madden, C. & Bali, B. S. (2014b). Geomorphic analysis reveals active tectonic deformation on the eastern flank of the Pir Panjal Range, Kashmir Valley, India. *Arabian Journal of Geosciences, 7*(6), 2225–2235.

Arian, M., & Hashemi, A. (2008). Seismotectonic zoning in the Zagros. *Journal of Sciences, 18*, 63–76.

Bahrami, S. (2013). Tectonic controls on the morphometry of alluvial fans around Danehkhoshk anticline, Zagros, Iran. *Geomorphology, 180*, 217–230.

Bali, B. S., & Wani, A. A. (2020). Earthquake hazard potential along the Panjal Thrust and the Zanskar Thrust, NW Himalaya. https://doi.org/10.21203/rs.3.rs-62862/v1

Bali, B. S., & Wani, A. A. (2021). Analysis of neotectonic structures in the piedmont region of PirPanjal Range NW Himalaya by integrating geomorphic indicators coupled with geophysical transects (GPR). *Natural Hazards, 2020*, 1–14. https://doi.org/10.1007/s11069-020-04428-4

Bali, B. S., Wani, A. A., & Bhat, G. R. (2021). GPR investigation of mining induced subsidence and its effects on surface structures: A case study of Srinagar City, J&K, India, NW Himalayas. *Journal of the Geological Society of India, 97*, 751–759. https://doi.org/10.1007/s12594-021-1756-5

Bali, B. S., Wani, A. A., Khan, R. A., & Ahmed, S. (2016). Morphotectonic analysis of the Madhumati watershed, Northeast Kashmir Valley. *Arabian Journal of Geosciences, 9*(5), 1–17.

Bali, B. S., Wani, A. A., Mir, S. A., Irfan, M., & Farooq, D. (2025). Rupture length as a proxy for quantifying palaeo-earthquake magnitude for future seismic micro-zonation in Kashmir, NW Himalaya, India. *Natural Hazards* 1–28. https://doi.org/10.1007/s11069-025-07189-0

Bhat, M. I. (1982). Thermal and tectonic evolution of Kashmir basin, vis-vis petroleum prospects. *Tectonophysics, 88*(1–2), 117–132.

Bhat, N. A., Bali, B. S., Mir, S. A., & Kumar, P. (2024). Seismotectonics and seismogenesis of Kashmir Valley, NW Himalaya, India from a local seismic network. *Journal of Earth System Science, 133*(2), 61.

Bilham, R., & Bali, B. S. (2013). A ninth century earthquake-induced landslide and flood in the Kashmir Valley, and earthquake damage to Kashmir's Medieval temples. *Bulletin of Earthquake Engineering*. https://doi.org/10.1007/s10518-013-9504-x

Bilham, R., Bali, B. S., Bhat, M. I., & Hough, S. (2010). *Historical earthquakes in Srinagar, Kashmir: clues from the Shiva temple at Pandrethan, GSA special paper 471 on ancient earthquakes* (M. Sintubin, I. S. Stewart, T. M. Niemi, & E. Altunel (Eds.)). ISBN 9780813724713.

Bossart, P., Dietrich, D., Greco, A., Ottiger, R., & Ramsay, J. G. (1988). The tectonic structure of the Hazara-Kashmir Syntaxis Southern Himalaya, Pakistan. *Tectonics, 7*, 273–297.

Bull, W. B., & McFadden, L. D. (1977). Tectonic geomorphology north and south of the Garlock Fault, California, in arid regions. *Proceedings Eighth Annual Geomorphic System* (pp. 115–138). State University.

Burbank, D. W., & Anderson, R. S. (2012). *Tectonic geomorphology* (2nd ed.). Wiley-Blackwell.

Burbank, D. W., & Johnson, G. D. (1983). Intermontane-basin development in the past 4 Myr in the north-west Himalaya. *Nature, 298*, 432–436.

Burbank, D. W., & Reynolds, R. G. (1984). Sequential late Cenozoic structural disruption of the northern Himalayan foredeep. *Nature, 311*(5982), 114–118.

Calkins, J. A., Oldfield, T. W., AMullah, S. K. M., & Ali, S. T. (1975). *Geology of the southern Himalayas in Hazara, Pakistan, and adjacent areas*. U.S. ~01. Survey of Professsional Paper 716-C, 29.

Chamyal, L. S., Maurya, D. M., Bhandari, S., & Raj, R. (2002). Late Quaternary geomorphic evolution of the lower Narmada valley, Western India: Implications for neotectonic activity along the Narmada-Son Fault. *Geomorphology, 46*(3–4), 177–202.

Chow, V. T. (1964). *Handbook of applied hydrology: A compendium of water resources technology*. McGraw Hill.

Dar, M. N., Bali, B. S., Mir, S. A., & Wani, A. A. (2024). Fault system dynamics of the Kashmir, NW Himalaya, India using continuous GPS observations and geomorphic evidences. *Journal of Geodynamics, 161*, 102044.

Dehbozorgi, M., Pourkermani, M., Arian, M., Matkan, A., Motamedi, H., & Hosseiniasl, A. (2010). Quantitative analysis of relative tectonic activity in the Sarvestan area, central Zagros, Iran. *Geomorphology, 121*, 329–341.

Delcaillau, B., Deffontaines, B., Floissac, L., Angelier, J., Deramond, J., Souquet, P., Chu, H. T., & Lee, J. (1998). Morphotectonic evidence from lateral propagation of an active frontal fold; Pakuashan anticline, foothills of Taiwan. *Geomorphology, 24*, 263–290.

Demoulin, A. (1998). Testing the tectonic significance of some parameters of longitudinal profiles: the case of the Ardenne (Belgium, NW Europe). *Geomorphology, 24*, 189–208.

Dewey, J. F., & Bird, J. M. (1970). Mountain belts and the new global tectonics. *Journal of Geophysical Research, 75*(14), 2625–2647.

Dewey, J. F., & Burke, K. C. (1973). Tibetan, Variscan, and Precambrian basement reactivation: Products of continental collision. *The Journal of Geology, 81*(6), 683–692.

Dey, S., Thiede, R. C., Schildgen, T. F., Wittmann, H., Bookhagen, B., Scherler, D., & Strecker, M. R. (2016). Holocene internal shortening within the northwest Sub-Himalaya: Out-of-sequence faulting of the Jwalamukhi Thrust, India. *Tectonics, 35*.

El Hamdouni, R., Irigaray, C., Fernández, T., Chacón, J., & Keller, E. (2008). Assessment of relative active tectonics, southwest border of the Sierra Nevada (southern Spain). *Geomorphology, 96*, 150–173.

Faghih, A., Nourbakhsh, A., & Kusky, T. M. (2015).GIS-based analysis of relative tectonic activity along the Kazerun Fault Zone, Zagros Mountains, Iran: Insights from data mining of geomorphic data. *Journal of Earth Science, 26*(5), 712–723.

Goswami, P. K., & Pant, C. C. (2007). Geomorphology and tectonics of Kota– Pawalgarh Duns, Central Kumaun Sub-Himalaya. *Current Science, 92*(5), 685–690.

Hare, P.W., & Gardner T.W. (1985). Geomorphic indicators of vertical neotectonism along converging plate margins, Nicoya Peninsula, Costa Rica. In M. Morisawa & J. T. Hack (Eds.), *Proceedings of Allen Unwin Boston Binghantom (15th Annual Tectonic Geomorphology Symposium)*, September 1984.

Horton, R. E. (1945). Erosional development of streams and their drainage basins; hydrophysical approach to quantitative morphology. *Geological Society of America Bulletin, 56*(3), 275–370.

Hough, S., Bilham, R., & Bhat, M. I. (2009). Kashmir Valley Megaearthquake, estimates of the magnitudes of the past seismic events foretell a very shaky future for this pastoral valley. *American Scientists, 97*, 42–49.

Irfan, M., Bali, B. S. & Afzal, A. (2024a). Late Quaternary tectono-geomorphological investigations of Zanskar Basin, NW Himalaya, India, using geospatial techniques. *Earth Surface Processes and Landforms, 49*(2), 684–705. https://doi.org/10.1002/esp.5730

Irfan, M., Bali, B. S., & Afzal, A. (2024b). Morphological characteristics and controlling factors of the piedmont fan systems in the Zanskar Basin, NW Himalaya, Ladakh. *Quaternary Science Advances* 100240. https://doi.org/10.1016/j.qsa.2024.100240

Irfan, M., Bali, B. S., Afzal, A., & Topal, S. (2025a). Morphology and controls of the alluvial fan systems along the Zanskar Shear Zone, Northwest Himalaya, India. *Geomorphology, 472*, 109607. https://doi.org/10.1016/j.geomorph.2025.109607

Irfan, M., Afzal, A., Bali, B. S., Topal, S., Naik, S., Ahmad, N., & Bashir, A. (2025b). Exploring the Geoheritage, cultural geomorphology, and geotourism potential of the Kashmir Region, Northwest Himalaya, India. *Geoheritage, 17*(2), 1–29. https://doi.org/10.1007/s12371-025-011 17-1

Irfan, M., Bali, B. S., & Afzal, A. (2025c). Corrigendum to "Morphological characteristics and controlling factors of the piedmont fan systems in the Zanskar region, Northwest Himalaya, India" [*Quat. Sci. Advan. 16*(2024) 100240]. *Quaternary Science Advances, 18*, 100249. https://doi.org/10.1016/j.qsa.2024.100249

Irfan, M., Mir, S. A., Bali, B. S., Afzal, A., Bhat, N. A., Bhat, N., Bashir, A., & Nazir, M. (2024c). Unveiling the geoheritage, cultural geomorphology and geotourism potential of Zanskar region, NW Himalaya, India. *Geomorphology, 462*, 109354. https://doi.org/10.1016/j.geomorph.2024.109354

Keller, E. A., & Pinter N. (1996). *Active tectonics: Earthquakes, uplift and landscapes* (Vol. 338). Prentice Hall.

Keller, E. A., & Pinter, N. (2002). *Active tectonics: Earthquakes, uplift, and landscape* (Second, pp. 121–147). Englewood Cliffs, Prentice Hall.

Khabazi, H., Moayeri, M., & Maryam, N. (2012).The effect of active tectonics on paleogeomophologic transformations of arid zones (case study: Dagh-e- Sorkh Basin Playa in Iran Country). *American Journal of Scientific Research, 63*, 14–25. ISSN 2301-2005.

Kirby, E., & Whipple, K. (2001). Quantifying rock uplift-rates via stream profile analysis. *Geology, 29*, 415–418.

Kirby, E., & Whipple, K. X. (2012). Expression of active tectonics in erosional landscapes. *Journal of Structural Geology, 44*, 54–75.

Kuhle, M. (2005). Glacial geomorphology and ice ages in Tibet and the surrounding mountains. *Island Arc, 14*(4), 346–367.

Madden, C., Trench, D., Meigs, A., Ahmad, S., Bhat, M. I., & Yule, J. D. (2010). Late quaternary shortening and earthquake chronology of an active fault in the Kashmir Basin, Northwest Himalaya. *Seismological Research Letters, 81*(2), 346.

Menier, D., Mathew, M., Pubellier, M., Sapin, F., Delcaillau, B., Siddiqui, N., Ramkumar, M., & Santosh, M. (2017). Landscape response to progressive tectonic and climatic forcing in NW Borneo: Implications for geological and geomorphic controls on flood hazard. *Scientific Reports, 7*(1), 457.

Middlemiss, C. S. (1910). Revision of Silurian-Trias sequence of Kashmir. *Records of the Geological Survey of India, 40*, 206–260.

Middlemiss, C. S. (1911). Sections in the Pir Panjal range and Sindh Valley, Kashmir. *Records of the Geological Survey of India, 41*, 115–144.

Miller, D. J. (1953). Late Cenozoic marine glacial sediments and marine terraces of Middleton Island, Alaska. *The Journal of Geology, 61*(1), 17–40.

Mir, S. A., Gahalaut, V. K., Wani, A. A., & Bali, B. S. (2023). Estimation of crustal strain in Kashmir Himalayan region of north India using continuous GPS measurements. *Geological Journal, 58*(7), 2668–2677.

Molin, P., Pazzaglia, F. J., & Dramis, F. (2004). Geomorphic expression of active tectonics in a rapidly-deforming forearc, Sila Massif, Calabria, Southern Italy. *American Journal of Science, 304*, 559–589.

Nakata, T. (1989). Active faults of the Himalayas of India and Nepal. *Special Papers Geological Society of America, 232*, 243–264.

Pike, R. J., & Wilson, S. E. (1971). Elevation relief ratio, hypsometric integral and geomorphic area-altitude analysis. *Geological Society of America Bulletin, 62*, 1079–1084.

Raj, R., Bhandari, S., Maurya, M., & Chamyal, L. S. (2003). Geomorphic indicators of active tectonics in the Karjan river basin, Lower Narmada Valley, western India. *Journal of Geological Society of India, 62*(6), 739–752.

Raj, R., Maurya, D. M., & Chamyal, L. S. (1999). Tectonic geomorphology of the Mahi River Basin, Western India. *Journal of Geological Society of India, 54*, 387–398.

Sarah, S., Shah, W., & Ahmed, S. (2021). Modeling and comparing streamflow simulations in two different montane watersheds of western Himalayas. *Groundwater for Sustainable Development, 15*, 100689.

Sarah, S., Shah, W., Somers, L. D., Deshpande, R. D., & Ahmed, S. (2024). Saturated hydraulic conductivity (Ksat) and topographic controls on baseflow contribution in high-altitude aquifers with complex geology. *Journal of Hydrology, 641*, 131763.

Sarah, S., Shah, W., Dar, F. A., Romshoo, S. A., & Arora, T. (2025). Hydrochemical composition and hydrodynamics as proxies to identify the stream water sources and pathways in high-altitude catchments. *Applied Geochemistry, 183*, 106344.

Schumm, S. A. (1956). Evolution of drainage systems and slope badlands at Perth Amboy, N. *Geological Society of America Bulletin, 67*, 597–646.

Schumm, S. A. (1986). *Alluvial river response to active tectonics* (pp. 80–94). Active tectonics.

Searle, M. P., & Rex, A. J. (1989). Thermal model for the Zanskar Himalaya. *Journal of Metamorphic Geology, 7*, 127–134.

Silva, P. G., Goy, J. L., Zazo, C., & Bardaji, T. (2003). Fault generated mountain fronts in Southeast Spain: Geomorphologic assessment of tectonic and earthquake activity. *Geomorphology, 50*, 203–226.

Strahler, A. N. (1952). Hypsometric (area-altitude) analysis of erosional topography. *Geological Society of America Bulletin, 63*, 1117–1142.

Taloor, A. K., Ray, P. K. C., Jasrotia, A. S., Kotlia, B. S., Alam, A., Kumar, S. G.,Kumar, R., Kumar, V., & Roy, S. (2017). Active tectonic deformation along reactivated faults in Binta basin in Kumaun Himalaya of north India: Inferences from tectono-geomorphic evaluation. *Zeitschrift für Geomorphologie, 61*(2):159–180. https://doi.org/10.1127/zfg/2017/0417

Thakur, V. C., & Rawat, B. S. (1992). Geologic Map of Western Himalaya, 1:1,000,000. Dehra Dun, India, Wadia Institute of Himalayan Geology.

Topal, S., & Irfan, M. (2025). Normal faulting and its role in the drainage divide migration in the Karıncalıdağ region, Menderes Massif, Western Türkiye. *Journal of Mountain Science* 1–12. https://doi.org/10.1007/s11629-024-9132-8

Topal, S., Akgün, E., Softa, M., & Irfan, M. (2025). Deformation assessments along the Amanos Mountains: Insights from morphometric analyses and tectonic implications for the Düziçi-İskenderun Fault Zone. Paper presented at the 77th Geological Congress of Türkiye with International Participation, April 14–18, Ankara, Türkiye.

Twidale, C. R. (2004). River patterns and their meaning. *Earth-Science Reviews, 67*(3–4), 159–218.

Wani, A. A., & Bali, B. S. (2017). Geomorphic analysis of relative tectonic activity in the Sindh Basin, Jammu and Kashmir Himalaya, northwest India. *Journal of Himalayan Geology, 38*(2), 171–183.

Wani, A. A., Bali, B. S., Ahmad, S., Nazir, U., & Meraj, G. (2022). Geospatial modeling in landslide hazard assessment: A case study along Bandipora-Srinagar Highway, NW Himalaya, J&K, India. In *Geospatial modeling for environmental management* (pp. 113–125). CRC Press.

Wani, A. A., Bali, B. S., Bhat,G. R., & Hussain, N. (2020). Impact of tectonics on drainage network evolution of Suru basin, Kargil N/w Himalaya, Jammu and Kashmir, India. *Environmental Earth Sciences, 79*(1).

Wani, A. A., Bali, B. S., & Mohammad, S. (2019). Drainage Characteristics of tectonically active area: An example from Mawar Basin, Jammu and Kashmir, India. *Journal of Geological Society of India, 93*(3), 313–320.

Wani, A. A., Bali, B. S., & Wani, M. I. (2021). Quantitative analysis and drainage dynamics of Mawar Basin, northwest Himalaya, J&K, India. *Journal of Research and Development, 21*, 73–87.

Wani, A. A., Bali, B. S., Mir, S. A., & Wani, M. I. (2025). Kinematics of the Kashmir Himalaya: Inferences from geological and geodetic data. *Geodesy and Geodynamics*. https://doi.org/10.1016/j.geog.2025.02.003

Yousuf, M., Bukhari, S. K., & Bhat, G. R. (2021). Using paleo-liquefaction features to determine the likely source, magnitude and ground accelerations of pre-historic earthquakes in the Kashmir Basin (Northwestern Himalaya), India. *Engineering Geology, 293*, 106302.

Geomorphic Response to the Active Tectonics Along a Sub-Himalayan Basin in Kangra Valley, India

Swati Sharma, Suranjita Mohapatra, Chinmaya Mund, A. K. Mahajan, and Rohan Kumar

Abstract This study delves into the complex relationships between geomorphic indicators and relative tectonics that regulate geological parameters of a Sub-Himalayan drainage basin revealing the subtle but significant links that determine the topography. Using a comprehensive approach incorporating morpho-analytical tools, this study reveals the reasoning that emphasizes the interdependence of various geological events and analyses the tectonic and structural development of sub-watersheds in the Baner Basin in the Himalayan foothills of District Kangra, Himachal Pradesh. The valley floor width-to-height ratio (Vf), Stream length index (SL), Basin shape index, mountain front sinuosity etc. helped to access the impacts of active tectonics and erosional activities in the study area. The average of all statistical values of geomorphic indices, hypsometric and morphometric parameters were used to generate a relative tectonic map (RTM) of the sub-watersheds and their respective ranking for tectonic activity was done. The parameter results and insights can be used in regional planning initiatives, laying the groundwork for long-term development and hazard mitigation. In this study, it was analyzed how the tectonics have modified the overall geo-morphometry of the Baner basin which may affect the hazard potential at the sub-basin level.

Keyword Geomorphic indices · Sub-Himalayan basin · Active mountain fronts · Relative tectonic map

---

S. Sharma (✉)
Department of TransDisciplinary Research, Division of Research and Development, Lovely Professional University, Jalandhar, Punjab, India
e-mail: swati.9238@gmail.com; swati.33384@lpu.co.in

S. Mohapatra · C. Mund
Amity Institute of Geoinformatics and Remote Sensing, Amity University, Noida, Uttar Pradesh, India

A. K. Mahajan
Department of Geology, Central University of Himachal Pradesh, Shahpur, Himachal Pradesh, India

R. Kumar
Center for Space Research, Lovely Professional University, Jalandhar, Punjab, India

© The Author(s), under exclusive license to Springer Nature Switzerland AG 2025

S. Ghosh (ed.), *The Himalaya Dilemma*, Springer Natural Hazards,
https://doi.org/10.1007/978-3-031-95083-4_4

1 Introduction

The tectonically active Himalayan terrains cross the Indian subcontinent from east to northwest which are also drained by highly dense drainage networks in various Himalayan catchments along with the prominent thrust boundaries transversing these terrains (i.e. from Higher Himalaya to the Outer/Sub-Himalayan ranges) (Thakur, 2004). The north-western Himalayan region is particularly highly vulnerable to geological hazards and erosional activities, especially activated in the monsoon season which is evident by the records of various natural calamities triggered due to high precipitation or cloudbursts, etc. The river Beas, one of the several active sub-Himalayan River catchments, crosses the state of Himachal Pradesh's Rohtang Pass in the Higher Himalayan heading towards the lower and outer Himalayan ranges in the district Kangra. A substantial drainage network of river Beas passes through the cities of Kullu and Mandi and connects the eastern and northwest regions of the state of Himachal Pradesh (Sharma and Mahajan, 2018). Major drainage channels such as the Gaj-Khad (local name for drainage), Baner-Khad, Banoi-Khad, Dehar-Khad, and several other local drains merge into the River Beas in the District Kangra. These drainages have a dense network of tributaries ranging up to stream order 5 (Sharma and Mahajan, 2020) and they collectively make the Beas watershed in District Kangra which abodes the areas on the northern to southern margin of the lesser to outer/sub-Himalayan terrain, respectively. At the southernmost edge of the Beas catchment, the Pong Dam's reservoir receives the sediments plus the water from this profound network. A part of the Beas catchment which is the Baner Watershed was considered under this study, which is tectonically very active, as evidenced by the record of the frequency and magnitude of seismic events this overall area (District Kangra) has previously experienced. This is due to the presence of active thrust boundary namely main boundary thrust (MBT) between the lesser and outer Himalayan terrains and Himalayan frontal thrust (HFT) locally known as the Jwalamukhi Thrust which separates the sub-Himalayan ranges from the low-lying alluvium along with local thrust. Not only the longitudinal structural features but the transverse discontinuities also contribute to the tectonic vulnerability of the study area. The morphological patterns and terrain in this area have been instinctively built by tectonic activity over the years, and this has led to the current erosional activities and geological hazards (floods, debris flow, shallow landslides, etc.) in this region (Mahajan et al., 2022).

Morphotectonics is a crucial technique for analyzing basin tectonic deformation and linking landform responses to active deformation processes (Bull & McFadden, 1977; Keller, 1996, 2002; Chen, 2003; Goudie, 2004; Pérez-peña, 2010; Aier, 2011; Phartiyal, 2012 Kale, 2014; Kothyari, 2016; Dubey, 2017; Pancholi, 2017; Kothyari, 2018; Joshi, 2021, Kandregula, 2020; Misra, 2020; Pant, 2020). Quantifying landforms for tectonics activity involves calculating the geomorphic indicators (Melton, 1958; Strahler, 1964; Tandon, 1974; Jordan & Schott, 2005; Rudraiah, 2008; Alam et al., 2017, 2018; (Barman et al., 2019, 2021; Yadav, 2020). Geomorphic indices assess the link between tectonics and landforms, revealing anomalies in landform distribution, river courses, terrace profiles, local relief, and unique landforms like

Knick points, narrow or wide valleys etc. (Burbank & Anderson, 2011; Goudie, 2013; Kothyari, 2014, 2019; Flores-Prieto, 2015 Kothyari, 2016; Shankar & Singh, 2020; Thakkar et al., 2021; Taioor & Joshi, 2021; Taloor, 2017, 2019). Morphotectonic indices are sensitive to rock resistance, climatic changes, and tectonic processes, resulting in landscape evolution. They play an important role in understanding the underlying processes. The examination of morphometric parameters can detect geomorphic anomalies, which are crucial indicators of seismotectonic activity (Cox, 1994; Pérez-peña, 2010; Tsodoulos, 2008). River drainages are the most prominent natural features that are susceptible to structural changes and reflect any tectonic activity in an area. Drainage anomalies, such as unexpected deflections or stream catches, might indicate residual zones etc. (Abrahams, 1984).

Using topographical maps and aerial photographs (remote sensing data) to evaluate drainage networks can aid in identifying active tectonics and quickly assessing large and undulated relief areas (Strahler, 1952). The use of DEM and satellite photos can also help identify geomorphic anomalies and morphotectonic patterns. (Keller, 1986; Mahmood & Gloaguen, 2012; Shukla, 2014; Smith, 2003; Grohmann, 2004; Mesa, 2006; Magesh, 2011; Arabameri, 2020; Khakhlari, 2016). Quantitative morphotectonic analysis viz. hypsometric integral, drainage basin asymmetry, stream sinuosity index, the ratio of valley floor width to valley height, stream length gradient index, basin shape index, and mountain front sinuosity index for a river valley or a catchment can be evaluated on high-resolution digital elevation model which can be verified with major tectonic event records and the structural patterns on the optical imageries to prepare the tectonic variation map and the residual topography. Such geomorphological parameters that fall within the geologic stages of watershed development are hypsometric integral (HI), mountain front sinuosity, stream length index, asymmetry factor etc. These factors are significant when estimating the watershed's state of erosion and setting priorities for initiating soil and water conservation measures (Omvir Singh, 2008). The intricate structure of these hydrological and landform processes working on watershed systems makes it highly challenging to quantify and explain the topographical changes (Strahler, 1952) that have been studied previously (Bishop et al., 2002; Ritter, 2002).

The degree to which river systems have adapted to the evolution of landforms and the neotectonics processes of the area may be understood via the use of longitudinal profiles. Along with adapting to the geological structure, a stream profile also considers the varying climatic circumstances that the river experiences (Darton, 1950). A curve that represents the influence of gradient on channel behavior is called a longitudinal profile, and it is produced by analyzing the connection between height and distance downstream of a channel (Christofoletti, 1980; Hack, 1957; Martinez, 2011). In addition to providing information on hydrological, sedimentological, and geomorphological aspects that influence a drainage basin's design and evolutionary history, longitudinal profile parameters enable the morphometric characterization of drainage networks. The stream gradient index (Hack, 1973), makes it easier to pinpoint the elements that regulate the drainage network, including the presence of resistant rock, tectonic activity, variations in the local topography, tributary confluence, human interference, etc. The mountainous regions of the Himalayas, which are subject to tectonic uplift, weathering, and denudation have undergone evolution due to the deformation processes (Valdiya, 1993; Bookhagen et al., 2005; Pérez-peña, 2010; Pazzaglia, 2013; Joshi & Kotila, 2015). Deformation processes impact the shaping of landforms, drainage patterns, and geomorphology. Through drainage network geomorphological investigation, historical tectonic disturbances may be deduced (Silva, 2003).

In this study, geomorphic indicators have been used to assess and evaluate the relative tectonic activity in the Baner catchment in the district Kangra of Himachal Pradesh. The indices reflect the residual topography whereas the aerial and relief aspects of the catchment reflect the degree of landscape evolution which is a function of tectonic forces that are ongoing in the study area. This study has quantified geomorphic indices to indicate the active tectonic and has characterized the sub-watersheds in the Baner catchment based on relative tectonic activities.

2 Study Area

The Baner catchment extends from 32°7′56.83"N latitude to 76°17′23.47"E longitude in District Kangra, Himachal Pradesh, covering an entire catchment zone of 587 km^2 as shown in Fig. 1. The highest elevation of the watershed is at 4834 m above mean sea level. The Dhauladhar range lies to the north of the watershed and the middle section of the study area traverses through the Lesser Himalayas, sometimes also referred to as the Lower Himalayas, and toward the south, the Sub-Himalayan alluviums are exposed. The reservoir where water from the watershed area is stored is known as Maharana Pratap Sagar, also known as Pong Dam Lake or Pong Reservoir. Constructed in 1975, Pong Dam is the highest earth-fill dam in India and the Beas River is the main fluvial channel entering the reservoir which originates in the higher Himalayan range near Rohtang Pass and has several tributaries that flow towards the Kangra Valley in the west, passing through the district Mandi's patches in the middle sections. Because the Baner watershed area has a varied terrain with plains, valleys, and mountains, the climate can change depending on where you are in the area. In general, the region has a nice and warm summer with temperatures between 25 and 35 °C. Due to the sub-humid climate of the area, substantial rainfall occurs during the monsoon season on the southern slopes of the watershed, while heavy snowfall occurs during the winter in the higher elevation areas (https://hydro.imd.gov.in/hydrometweb/(S(uoo24555i03okymvdyjcmm45))/DistrictRaifall.aspx). Most of the rainfall in district Kangra falls during the monsoon season (July to September), which makes up a substantial portion of the region's precipitation. Due to the district's geographical characteristics, various parts may receive varying amounts of rainfall. Prolonged precipitation episodes trigger several landslip incidents, ultimately supplying the drainage segments with more sediment which ultimately drains into the Pong Dam Reservoir. The region has a very dense vegetation cover, which ranges in elevation from alpine meadows at higher elevations to dry scrub woods at lower altitudes. Different vegetational zones, including Mixed Deciduous Forests,

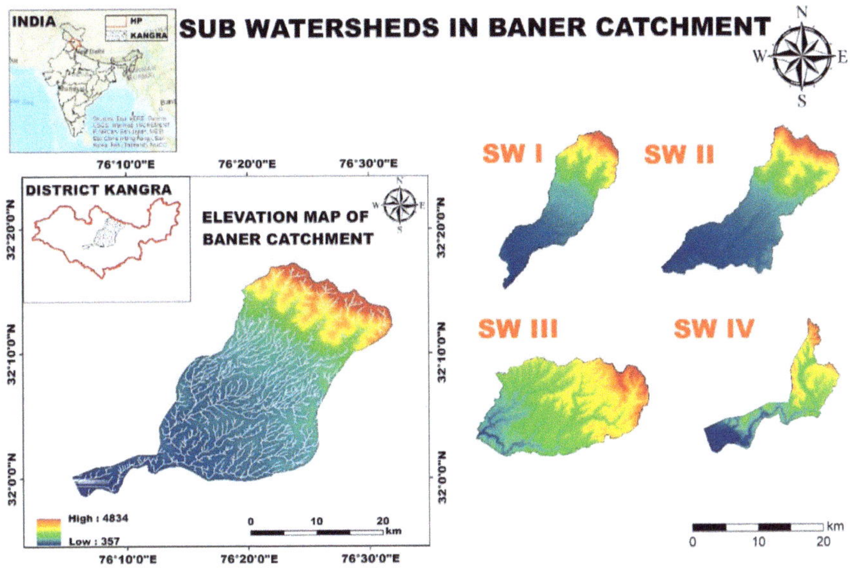

Fig. 1 Thematic maps shows the study area

Bamboo, Chirr Pine, Oak, Deodar, Kail, Fir, and Spruce, can be found between these two extremes (https://fsi.nic.in/isfr19/vol2/isfr-2019-vol-ii-himachal-pradesh.pdf). Within this study area, there are major longitudinal Himalayan Frontal Thrust (HFT), the Main Boundary Thrust (MBT), the local Drini Thrust along with numerous smaller transverse faults that drive the deformational activities and have led to the topographic restructuring of the basin.

2.1 Regional Setting

The Baner catchment is composed of a variety of lithological units, ranging from alluvium silt at the catchment's end to the middle Outer Himalayan rocks to the northern Lesser Himalayan rocks. The outer Himalayan rocks of the study area are weak and have led to many slope instability conditions in the past. Most of the road excavation and settlement lies in the outer Himalayas due to its position at the toe of high mountain ranges which make the area suitable for settlement and development. Due to this, the area experiences higher infrastructural and road damages when disturbed due to natural hazards owing to heavy rainfall or the deformation of the landscape. Geologically the southern part of the catchment lies in the outer Himalayas comprising the Siwalik (boulder Conglomerate) north to which the Muree Formation and the Dharmsala group (mudstone, claystone, sandstone) are detached by MBT from Lesser Himalayan rocks (limestone, Dhauladhar Granite, and Chails Formation) which are the low-grade metamorphic rocks.

In the study area, four major longitudinal Thrusts pass (Fig. 2) viz. Jwalamukhi Thrust (JT) at the southernmost edge between the lower Siwalik Formation and Dhokpathan Formation (Sandstone and Shale), Muree Thrust lying between Limestone Formation and Pinjore Boulder Conglomerate, Main Boundary Thrust (MBT) passing between Dhauladhar Granites and Limestone Formation and the northernmost Panjal Thrust (PT) between the Chails and Dhauladhar Granites. The presence of several major thrusts and lineaments in this area has led to the geological reworking of the rocks which has caused intensive weathering and shearing effectively contributing to the natural hazards in addition to sediment load and structural weaknesses. The study area i.e. Baner Catchment is important in terms of its geological dynamics evident through its geomorphic evolution with major signatures on the ground in the form of extreme topographical variations from its northern to southern edge as studied in this research work. In the mountains, earthquakes that cause landslides, rockslides, and snowstorms are detrimental and the study area is in the seismic zone V, which has quite a high damage risk, including the Kangra district (https://ddmakangra.org/earthquake). Across Himachal Pradesh, there have been significant earthquakes in past (Fig. 3); the massive one was the Kangra Earthquake in 1905.

2.2 Land Use and Land Cover in the Study Area

The LULC map was prepared from LISS-IV mx (5.8 m resolution) images with the application of a support vector machine (SVM) to classify signatures which was useful for large-scale data categorization. Multidomain applications in a large data environment can benefit most from the SVM classification (Suthaharan, 2016). The eight LULC classes that comprised the research region were built-up, dispersed vegetation, snow cover, cropland, flood plain, low relief forest cover, and high relief forest cover as shown in Fig. 4.

3 Materials and Methods

With the use of LISS IV-mx satellite imagery and ALOS PALSAR digital elevation model (DEM), thematic maps needed to analyze the relative tectonic activities and geomorphological evolution of the Baner catchment region were prepared (Fig. 5). Through the DEM data extraction, fill, flow accumulation, flow direction, stream order, slope, aspect, and contour map of the Baner catchment were prepared and further used to divide the study area into four sub-watersheds for in-depth individual analysis to determine the present tectonic setting and the erosional capacity from all the sections of the Baner catchment. Several variables were examined to discover the relationship between relative tectonics and geomorphic influences, which control the degree of impact from geological and geomorphological hazards in the study area which is ultimately shaping its present topography. Aerial, relief, and linear

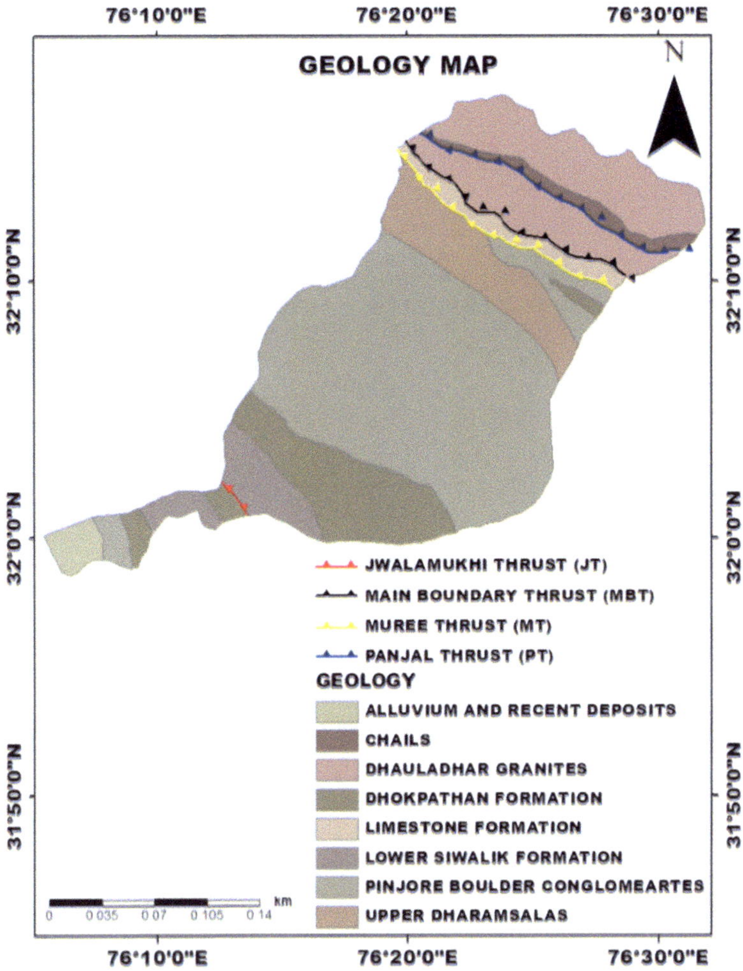

Fig. 2 Geological setting of the study area (modified after Survey of India Geology Map-52D)

aspects are the three categories of morphometric characteristics of a basin that were examined.

A total of thirty morphometric parameters were computed (Table 1) for the research region to gain insight into its drainage network characteristics and to understand its morphological, hydrological, and structural behaviours. Measures like hypsometric integrals (HI) were employed to display the region's topographic relief where for a heavily eroded terrain, the hypsometric integral value falls between 0 and 1.

Additionally, hypsometric curves were prepared to depict the topographic reliefs with respect to the denudational rates observed in drainage basins. Understanding the tilt direction and the degree of tectonic activity in the region is aided by the

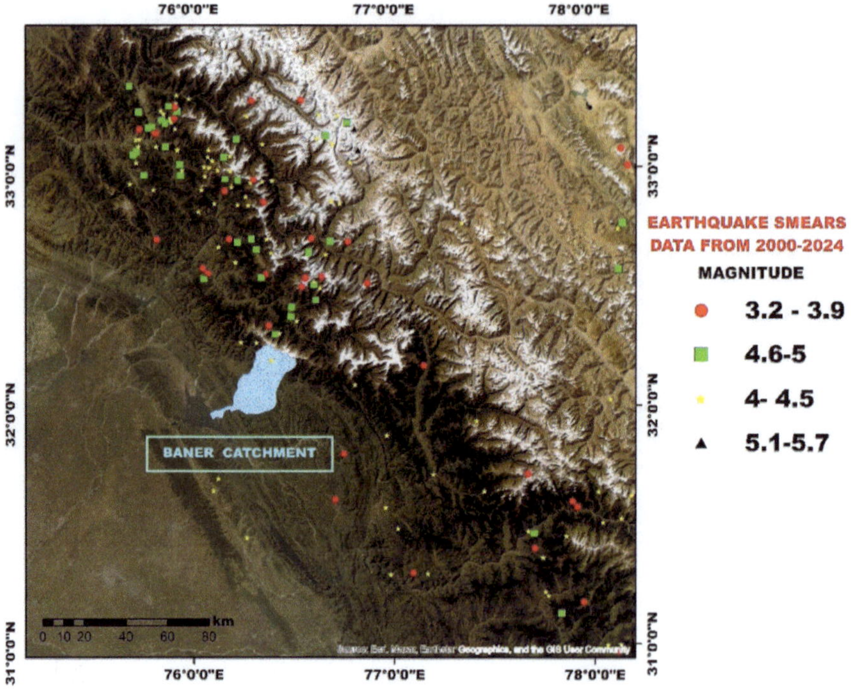

Fig. 3 Seismic epicenters data (2000–2024) overlain on the regional setup surrounding the study area inset (Baner catchment)

transverse topographic symmetry factor (Tt). Utilizing the midpoint tool in ArcMap, the midline of each sub-basin was generated to calculate the transverse topographic symmetric factor where Tt = 0 depicts a fully symmetric basin and if it approaches 1 the symmetry diminishes. The area right to the midline in a sub-watershed i.e. area of the right trunk and the overall area of the sub-watershed (Ar/At) were calculated to analyze the asymmetry factor (AF) of the drainage basin. Generally, if the AF values in the left and right halves of the basin are equal it indicates no or slight tilting which depicts less structural reworking i.e. AF is almost 50 (mid value), meaning that there is little to no tilting away from the basin's midline. AF factor of above or below 50 denotes the presence of active tectonics or lithological control. The longest flow path of drainages in each sub-watershed was interpolated using a 3D analyst tool for the stream length index (SL), and the path was classified into three categories depending on the geomorphological unit namely mountain, piedmont, and alluvial fan from top to the bottom of each sub-watershed respectively. This helped to extract the longitudinal profile of the longest drainage paths in each sub-watershed including a minimum of 8–9 points considered for calculating the SL value as per the formula given in Table 2. To calculate the ratio of valley floor width to valley height (Vf) 10 –15 Vf points were determined along the longest drainage paths in each sub-watershed. The Vf index characterized the steep and narrow valleys with a

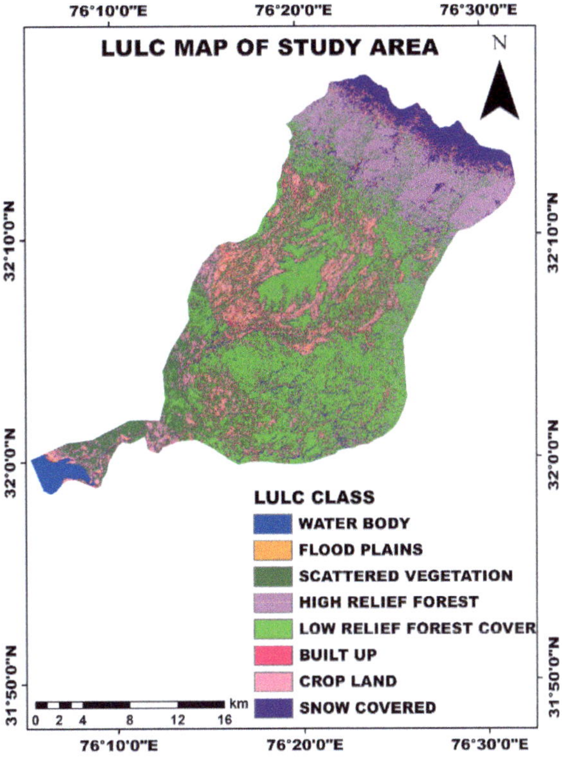

Fig. 4 Land use land cover map of study area

"V" form and broad valleys with a "U" shape that are somewhat taller than the valley walls. V-shaped valleys have relatively low values of Vf, whereas U-shaped valleys typically have large values. The basin's length-to-breadth ratio was derived using a shape index to determine the sub-watershed's shape which indicated the evolution stage of the basins (young or mature). Using the factor of mountain front sinuosity (Smf), tectonic fronts can be classified for the basin where the straight fronts with low Smf values indicate the uplift winning out over erosional processes in active mountain fronts and vice-versa (Keller, 2002; Silva, 2003). Erosion processes along less active fronts are indicated by the more sinuosity (Smf) between the front and the channel. Higher Smf values (>3) are typically associated with inactive fronts where the denudation has taken over the structural factors and, in such cases, the frontal ranges are sometimes more than 1 km away from the erosional front. In this study, Smf for the sub-watersheds were calculated using the longitudinal river profile. All the geomorphic indices were attributed to the respective sub-watersheds after calculating the quantitative values and were then spatially interpolated using the Krigging tool in ArcGIS software. Each parameter was classified into four classes based on their index values and the natural break classification method in ArcGIS to generate the geomorphic index variation maps for each sub-watershed such as Smf variation map, SL variation maps, Vf variation map etc.

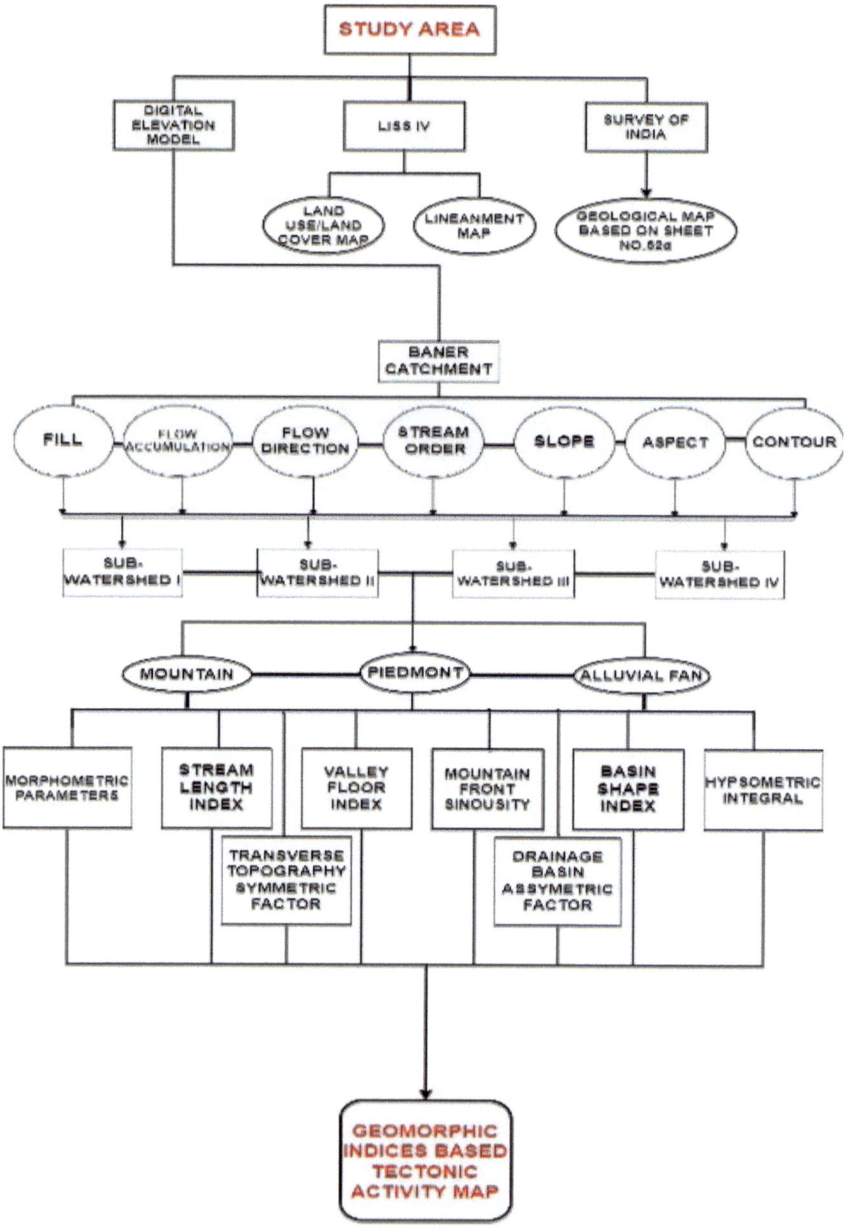

Fig. 5 Methodology flowchart

Table 1 Morphometric parameters calculated

Parameters	Formula	Description	References
Linear aspect			
Stream order (U)	Hierarchical rank	The process of determining a stream's place in the hierarchy of streams is known as the stream order	Strahler (1964)
Stream length (L_u)	$L_u = L_{u1} + L_{u2} + L_{u3} \cdots + L_{un}$	The mean length of a stream in each of the different orders within a watershed is defined as the stream length	Horton (1945)
Stream number (N_u)	$N_u = N_{u1} + N_{u2} + N_{u3} \cdots + N_{un}$	The number of streams in each catchment's sequence determines the total number of streams in that catchment	Horton (1945)
Stream length ratio (R_l)	$R_l = L_u/L_{u-1}$	The ratio of the mean stream length of the given order to the mean stream length of the next smaller order	Horton (1945)
Bifurcation ratio (R_b)	$R_b = S/N_{u+1}$	The ratio of streams of a certain order to streams of the next higher order; it explains the degree of a catchment's structural complexity and the impact of tectonic activity and geological structure	Schumm (1956)
Mean bifurcation ratio (R_{bm})	Average of bifurcation ratio (Rb) of all orders	The bifurcation ratio for each pattern is multiplied by the total number of streams to determine the weighted average ratio of bifurcation. The findings are then combined to provide a more representative figure	Strahler (1957)
Mean stream length (L_{sm})	$L_{sm} = (L_u/N_u)$	It is the ratio of stream length and number of streams	Horton (1945)
Mean stream length ratio (R_{lm})	Average of the steam length ratio of all orders	It is the proportion of stream length to total stream count	Schumm (1956)
Stream frequency (F_s)	$F_s = (\sum N_u)/A$	The quantity of stream segments for every order inside a fixed catchment area	Schumm (1956)
Drainage density (D_d)	$D_d = (\sum L_u)/A$	Defined as the ratio of the total length of all stream segments across all orders to the catchment area projected on the horizontal surface	Schumm (1956)
Drainage texture (D_t)	$D_t = (\sum N_u)/P$	It is the total number of streams within the drainage divides	Schumm (1956)
Length of overland flow (L_o)	$L_o = (1/(2D_d))$	It details each watershed's surface runoff	Schumm (1956)

(continued)

Table 1 (continued)

Parameters	Formula	Description	References
Drainage intensity (D_i)	$D_i = F_s/D$	As the stream frequency is divided by the drainage density, drainage intensity is determined	Faniran (1968)
Infiltration number (I_f)	$I_f = F_s \times D_d$	Combining drainage density with stream frequency gives infiltration number	Faniran (1968)
Constant of channel maintenance (C_{cm})	$Cc_m = 1/ Dd$	Quantity of catchment area required to sustain one route length unit	Schumm (1956)
Relief aspect			
Maximum elevation (H)	DEM based analysis	Highest Point of Catchment	
Minimum elevation (h)	DEM based analysis	Lowest Point of Catchment	
Relief (B_h)	$B_h = H–h$	Difference between maximum and minimum elevation	Strahler (1957)
Relief ratio (R_h)	$R_h = B_h/L_b$	The ratio of the minimum catchment length, which runs parallel to the primary catchment line, to the maximum catchment relief (Bh)	Schumm (1956)
Relative relief (R_{hp})	$R_{hp} = H \times 100/P$	Relative relief is determined using the watershed's elevation difference and its perimeter	Melton (1957)
Ruggedness number (R_n)	$R_n = B_h \times Dd$	It is the result of combining maximum catchment relief with drainage density	Strahler (1957)
Aerial aspects			
Area of the watershed (A)	Shape file-based analysis	The total amount of water can be directly impacted by the watershed's area	
Perimeter of the watershed (P)	Shape file-based analysis	The outside limit of the watershed enclosing its region	
Basin length (L_b)	$L_b = 1.312 \times A^{0.568}$	Determining each watershed's length from the highest to the lowest point	Nookaratnam et al. (2005)
Circulatory ratio (R_c)	$R_c = 4\pi A/P^2$; where $\pi = 3.14$	The catchment's area is divided by the area of a circle whose radius is equal to that of the catchment	Miller (1953)
Elongation ratio (R_e)	$R_e = (2\ (A/\pi)^{0.5})/(L_b)$	The ratio of the catchment's minimum length to the diameter of a circle with the same area covered by the catchment	Schumm (1956)
Form factor (F_f)	$F_f = (A/Lb^2)$	The form factor is the ratio of catchment area to catchment length squared	Horton (1945)

(continued)

Table 1 (continued)

Parameters	Formula	Description	References
Lemniscate ratio (K)	$K = (Lb^2/4A)$	It is employed to determine the slope of the watershed	Chorley et al. (1957)
Shape index (S_b)	$S_b = 1/F_f$	It is the reciprocal of the Form factor value	Horton (1932)
Compactness coefficient (C_c)	$C_c = (P/2(\pi A)^{0.5})$	The ratio of the catchment's perimeter to the circle of a corresponding circular region	Horton (1945)

3.1 Morphometric Indices

The arrangement of drainage geomorphology and its evolution through time on the surface of the earth is measured and quantitatively analyzed using the morphometric approach. (Agarwal, 1998; Vittala et al., 2004). To acquire and assess the morphometric characteristics, a surface runoff area map was used (Vijithand & Satheesh, 2006). Using this approach the parameters that were studied for the present study area included—Linear Aspect, Relief Aspect, and Aerial Aspect as shown in Table 1.

3.2 Geomorphic Indices

3.2.1 Stream Length Gradient Index (SL)

The SL index determines whether the drainage has attained equilibrium through its evolution history and studies the effects of environmental and geological factors on longitudinal stream profiles (Hack, 1973). The mathematical formula for SL is given in Eq. 1.

$$SL = (\Delta H/\Delta L)\,L \quad (1)$$

where ΔH represents the change in altitude between two points within a basin, ΔL denotes the length of a reach, and L is the horizontal length between the watershed split and the reach's midpoint. Relative tectonic activity in a basin can be measured using the SL index where the rivers and streams running across the active uplifts show high SL index value. When the drainage flows parallel to features like valleys created by strike-slip faulting, their SL index value may decrease (Keller, 2002).

Table 2 Geomorphic indices calculated

Geomorphic indices	Formula	Description	References
Stream length gradient index (SL Index)	$SL = (\Delta H / \Delta L) L$	**ΔH**-Change in Altitude **ΔL** Length of the reach **L** = Horizontal length from the watershed divide to the midpoint of the reach	Hack (1973)
Valley floor width to height ratio (Vf)	$Vf = 2V_{fw}/[(E_{ld} - E_{sc}) + (E_{rd} - E_{sc})]$	V_{fw} = Valley floor width E_{ld} = Elevation of left valley divide; E_{rd} = Elevation of right valley divide; E_{sc} = Elevation of the valley floor	Bull and McFadden (1977)
Basin shape (B_s)	$Bs = B_I/B_W$	B_I = Basin length from Head point to mouth Bw = Width of the Basin	Ramírez Herrera (1998)
Mountain front sinuosity index (S_{mf})	$S_{mf} = L_{mf}/L_s$	L_{mf} = Planimetric length of mountain front along the mountain Piedmont junction; L_s = length of the mountain front measured along a straight line	Bull (2007)
Asymmetric factor (A_f)	$Af = (A_r/A_t) * 100$	A_r = Area of the Basin to the right of the main channel facing downstream A_t = Total area of the basin	Hare (1985a, b)
Transverse topographic symmetry factor (T)	$T = Da/Dd$	Da is the length between the drainage basin's and the active meander belt's midline, Dd is the length between the basin's midline and the basin split	Cox (1994)
Hypsometric integral (H_i)	$H_i = (Ele_{avg} - Ele_{min})/(Ele_{max} - Ele_{min})$	Elev avg = Average of Elevation min = Minimum Elevation max = Maximum Elevation	Pike (1971), Mayer, 1990), Keller (2002)

3.2.2 Valley Floor Width to Valley Height (Vf)

Another metric that is susceptible to tectonic uplift is the ratio of valley floor width to valley height (Vf), (Bull & McFadden, 1977) which may be calculated as follows:

$$Vf = 2V_{fw}/[(Eld - Esc) + (Erd - Esc)] \quad (2)$$

The variables Eld, Erd, and Esc represent the elevation of the left divide and right divide of the valley, respectively, while V_{fw} the width of the valley floor, represents the ratio of the valley floor height to the valley floor breadth. By using the Vf index, it is possible to distinguish between narrow, steep valleys with a "V" form and valleys with a wide floor (U shaped). The V-shaped valleys often have relatively low values of Vf, and U-shaped valleys typically have large values of Vf. Because of the correlation between uplift and, low values of Vf are assumed to be indicative of higher rates of incision and uplift, serving as a benchmark for active tectonics. Despite at an equilibrium condition, incision and uplift are about equal, this index measures incision rather than uplift.

3.2.3 *Basin Shape Index (B_S)*

Relatively young drainage basins in active tectonic areas tend to be elongated in shape normal to the topographic slope of a mountain. With continued evolution or less active tectonic processes, the elongated shape tends to evolve to a more circular shape (Bull & McFadden, 1977). The elongation ratio, B_s (Cannon, 1976; Ramírez-Herrera, 1998) can be used to characterize the horizontal projection of a basin form, represented by the Eq. 3:

$$Bs = B_I/B_W \quad (3)$$

where B_w is the basin's width between its widest point and B_I is the basin's length, measured from the headwaters to the mouth. Extended basins are linked to elevated Bs values, which are often correlated with heightened tectonic activity. A basin with a more circular form and typically lower tectonic activity which is indicated by low Bs values. Elongated, steep basins are typically created by quickly raised mountain fronts; as tectonic activity decreases or stops, the basins expand (Ramírez Herrera, 1998). On the flip side, no evidence of pervasive rapid uplift allows the basins to enlarge (Mahmood & Gloaguen, 2012). Consequently, the basin's structure is considered as a sign of tectonic activity (Ramírez Herrera, 1998).

3.2.4 *Mountain Front Sinuosity Index (S_{mf})*

S_{mf} is a balance between active vertical tectonics, which tends to build straight mountain fronts frequently coinciding with active faults or folds, and erosive processes, which tend to erode a mountain front, making it more sinuous by streams that cut laterally and into the front (Bull & McFadden, 1977; Keller, 1986).

The mathematical equation for the mountain front sinuosity index (Smf) is

$$S_{mf} = L_{mf}/L_s \tag{4}$$

where Ls is the length of the mountain front measured along a straight line and L_{mf} is the planimetric length of a mountain front (Bull, 2007). The relative tectonic activity along mountain fronts is assessed using this index (Keller, 2002; Silva, 2003). A straight front with low S_{mf} values infers active mountains, whereas irregular or sinuous fronts with high S_{mf} values are produced by erosional processes along less active fronts. Tectonically active fronts are indicated by lower S_{mf} values (<1.4) (Keller, 1986), whereas inactive fronts are often linked with greater S_{mf} values (>3) (Bull, 2007).

3.2.5 Asymmetric Factor (Af)

The inclusion of tectonic tilting at the drainage basin scale is shown by the asymmetry factor (Af). The technique may be used across a somewhat wide region. The formula for parameter Af is (Hare, 1985a, b; Keller, 2002)

$$Af = (A_r/A_t) * 100 \tag{5}$$

where A_t is the overall area of the drainage basin and Ar is the area of the basin to the right of the mainstream facing downstream. If the AF values in the left and right halves of the basin are identical, there is either no tilting or very little tilting. If the value of the asymmetric factor (AF) is almost 50, that means there is little to no tilting away from the midline of the basin. On the other hand, an AF value of 50 or less suggests the impact of active tectonics or lithological control (El Hamdouni et al., 2008).

3.2.6 Transverse Topographic Symmetry Factor (T)

Together with the drainage basin asymmetry factor (Af), the transverse topographic symmetry factor (T) may also be utilized to interpret minor tectonic activity-related disturbances or reactions in basins at the basin scale. (Cox, 1994). It is calculated as

$$T = Da/Dd \tag{6}$$

is the formula, where Da is the length between the drainage basin's midline and the active meander belt's midline, and Dd is the length between the basin's midline and the basin split. The parameters Af and T serve as indices of deformation-induced ground tilting (Cox, 1994). The T factor may be thought of as a vector with direction and magnitude ranging from zero to one (T = 0–1), articulating, respectively, a perfectly symmetrical, asymmetrical, or tilted basin (Cox, 1994; Burbank, 2001; Keller, 2002).

3.2.7 *Hypsometric Integral (H_i)*

The relative distribution of height in a certain landscape feature, especially a drainage basin, is described by the hypsometric integral (HI) (Strahler, 1952). By expressing the relative area beneath the hypsometric curve, this index represents the volume of a basin that has not experienced erosion. An easy formula to roughly determine the hypsometric index is given by (Pike, 1971; Mayer, 1990 Keller, 2002).

$$H_i = H_i = \left(Ele_{avg} - Ele_{min}\right)/(Ele_{max} - Ele_{min}) \quad (7)$$

where mean, minimum and maximum elevations are represented, respectively, by the terms Elev mean, Elev min, and Elev max. High HI values typically indicate a younger terrain that may have been created by active tectonics, indicating that not many of the uplands have been eroded.

Significant details on the basin's state are provided by the hypsometric curves and hypsometric integral (Ritter, 2002; Sharma et al., 2013). The hypsometric curves (Fig. 6) are obtained by plotting the relative area denoted by a/A, where "a" refers to the relative area of a contour and "A" is the total area of the basin with the relative elevation h/H, where "h" is the maximum height of the same level curve, and "H" is the maximum height of the basin (Strahler, 1964; Wei, 1998; Sharma et al., 2013; Kusre, 2013).

The integrated HI, which shows the erosion cycle, is produced using results from the hypsometric analysis. There are three phases in an erosional cycle first one is the Monadnock phase, the second is the Mature stage (Equilibrium), and the last one is the young stage (Inequilibrium) (Martínez-Ramírez et al., 2017).

Variations in the hypsometric integral value and the curve's form (Fig. 6) correspond to the degree of instability in the tectonic and destructive force balance (Weissel et al., 1994; Dowling et al., 1998; Ritter, 2002). Langbein (1947) presented hypsometric analysis for the first time to represent the drainage basin's shapes and overall slope. The amount of soil mass in the basin and the degree of erosion that has taken place there in comparison to the residual mass are connected to the hypsometric curve (Hurtrez et al., 1999).

4 Results and Discussions

The Baner catchment with the highest elevation point at 4834 m traverses from the Lesser Himalayan rocks in the north towards the outer Himalayan Alluvium in the south and has well-developed geomorphological characteristics (Fig. 7) that have

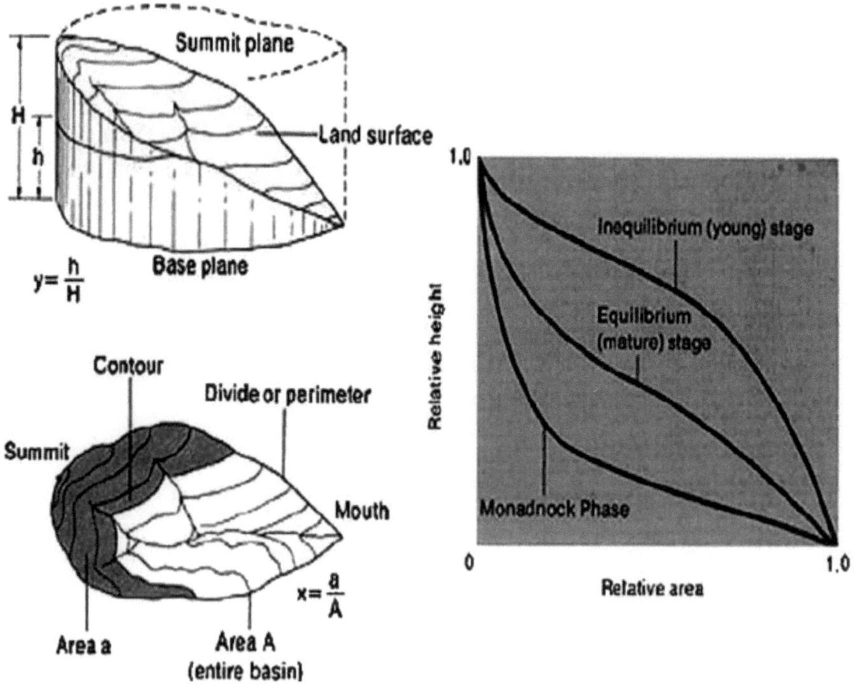

Fig. 6 Ritters model followed for hypsometric analysis (Ritter, 2002)

molded and evolved through decades of tectonic and erosional re-working. In this study, the most accepted mathematical parameters for analyzing the morphometric, residual topography and geomorphic indices were derived to establish the correlation between the progressive changes in the morphology of the study area and the prevailing tectonic impact of longitudinal and transverse faults, varying lithological formations, anthropogenic activities, and climatological processes.

In the very first steps of catchment analysis, the stream ordering was performed which indicated an order up to 5 in SW1, SW2 and SW3 whereas in the SW4 the stream order up to 4 resulted using the hydrology tool that uses the Strahler's classification in ARC-GIS. Further, the linear parameters helped to calculate the stream number (Nu) (Table 3), stream length (Table 4) and an average of stream lengths (Table 5) using which the mean bifurcation ratio ($\sum$Rb) of each sub-watershed (WS1, WS2, WS3 and WS4—Fig. 1) was calculated. Generally, the 'Rb' value indicates the structural control and, in this study, Rb ranged between 1.77 and 3.42 for all the four sub-watersheds (Table 6) which indicates overall a moderate to high structural control on their landforms. Also, Table 1 in Chap. 3 indicated a higher number of low-order streams (1st and 2nd order) in SW1 and SW 2 which is a representation of high mountainous area morphology and an increase in high-order streams i.e. 3rd order to 5th-order streams in SW3 and SW4 indicated a change in morphological units in the middle and low reaches of the study area mostly depicting a mature to

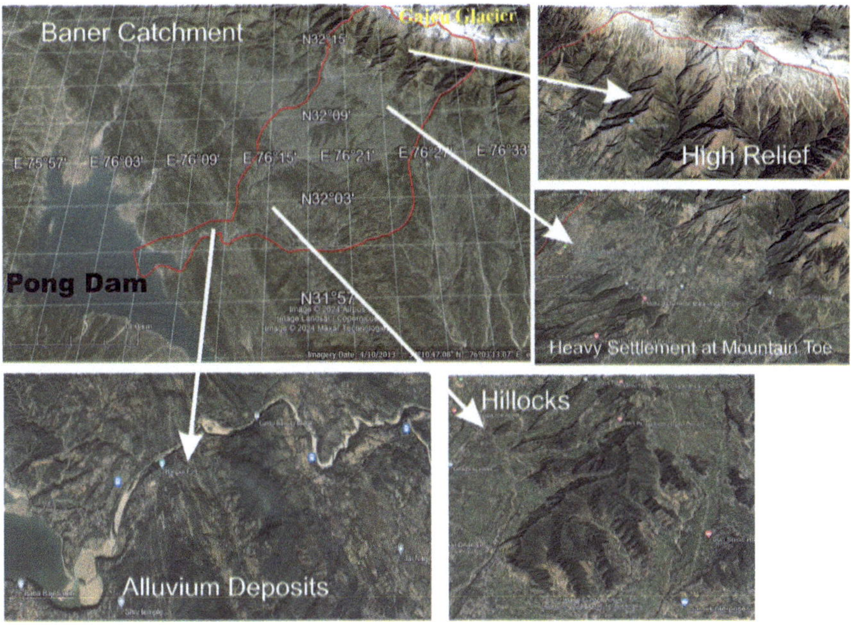

Fig. 7 Morphological features of the study area

Table 3 Order-wise stream no. in the sub-watershed

Sub-watershed	Stream number (Nu)					
	Order-wise stream no.					Sum
	1	2	3	4	5	
WS-1	742	327	146	147	65	1427
WS-2	1005	502	101	213	53	1874
WS-3	417	243	120	75	43	898
WS-4	817	69	145	105	0	1136
						5335

late mature evolution stage in the basins. The mean stream length remained low for the low-order streams whereas it increased for the high-order streams which is also indicative of a changing topography from high mountains towards low-lying areas.

Under the areal aspects of the study area the drainage density (Dd) (Table 7), length of overland flow (Table 8), stream frequency (Table 9), constant of channel maintenance (Table 10), circulatory ratio, elongation ratio, form factor (Tables 11 and 12) was calculated for all the sub-watersheds. The stream frequency and drainage density (Dd) in a landscape indicate dissection, runoff potential and permeability. Drainage density which is the stream length per unit area (Horton, 1945) infers the landform runoff potential as well as the dissections rate. The overall Dd value for the

Table 4 Linear aspects of watersheds showing stream length in the sub-watersheds of the study area

	Order-wise stream length in km						Length of stream in watershed
	Sub-watershed	1	2	3	4	5	
Stream length	WS-1	137.96	64.88	42.91	36.27	14.10	296.13
	WS-2	191.84	116.32	27.87	55.29	18.64	409.96
	WS-3	100.05	55.86	28.60	14.23	13.23	211.97
	WS-4	72.60	21.58	13.78	22.03	0.00	130.00
	Sum						**1048.06**

Table 5 Mean stream length in the sub-watersheds

Mean stream length					
Sub-watershed	Order-wise mean length in km				
	Stream order 1	Stream order 2	Stream order 3	Stream order 4	Stream order 5
WS-1	0.19	0.20	0.29	0.25	0.22
WS-2	0.19	0.23	0.28	0.26	0.35
WS-3	0.24	0.23	0.24	0.19	0.31
WS-4	0.09	0.31	0.10	0.21	

Table 6 Stream bifurcation ratio of the sub-watershed in the study area

Bifurcation ratio (Rb)						
Sub-watershed	Rb of 1st order	Rb of 2nd order	Rb of 3rd order	Rb of 4th order	The sum of Rb of each order	ΣRb
WS-1	2.27	2.24	0.99	2.26	7.76	1.94
WS-2	2.00	4.97	0.47	4.02	11.47	2.87
WS-3	1.72	2.03	1.60	1.74	7.09	1.77
WS-4	11.84	0.48	1.38		13.70	3.42

study area ranges from 1.71 to 1.95 which indicates steep to moderately steep slopes with high runoff potential and moderate infiltration capacity for the sub-watersheds. It was seen that the SW1 and SW2 had more Dd as they belonged to high relief areas where the slopes are steep and the lithology is hard (Dhauladhar Granites, Low-Grade Metamorphic) and non-porous which has also led to the increased length of overland flow (Table 7) in SW1 and SW2 compared to low elevation sub-watersheds SW3 and SW4. Form factor values up to 0.33 indicated a near circular to elongated basin which marked a moderate peak flow in limited time which can lead to chances of frequent flooding and bank failures leading to destructive episodes in peak monsoon season. The low-laying watershed SW4 has the lowest value of form factor which indicates an elongated basin inferring a sub-humid climate. Similarly, the elongation ratio which

Table 7 Drainage density in the sub-watershed

Drainage density								
Sub-watershed	Order wise length in km					Sum of the length of the watershed	Area (sq km)	DD in km/SQ km
	1	2	3	4	5			
WS-1	137.96	64.88	42.91	36.27	14.10	296.13	152.20	1.95
WS-2	191.84	116.32	27.87	55.29	18.64	409.96	224.73	1.82
WS-3	100.05	55.86	28.60	14.23	13.23	211.97	123.76	1.71
WS-4	72.60	21.58	13.78	22.03	0.00	130	71.41	1.82

also indicates the basin shape and the geomorphological control on the river channel has a value between 0.60 and 0.64 which means the overall sub-watersheds have near-circular to moderate elongated basins and the geomorphological control on river channels is higher (Strahler, 1964) which is also verified with high stream frequency values in Table 8 which pointed towards the development of a good drainage network. This also shows that all sub-watersheds have moderate to high flood characteristics with median discharge in peak season. The constant of channel maintenance (CCM) is the minimum area required for channel development and its low values may indicate the young morphology of the basin. In the study area, all sub-watersheds have shown a value of CCM ranging between 0.51 and 0.58 with a lower value for the high-elevation sub-watershed (SW1) which means low infiltration and higher overland runoff potential which is verified with steep slopes in the high points of the study area.

Table 8 Length of overland flow in the sub-watersheds

Sub-watershed	Total length of stream	Area (sq km)	DD in km/SQ km	1/(2*DD)
WS-1	296.13	152.20	1.95	3.89
WS-2	409.96	224.73	1.82	3.65
WS-3	211.97	123.76	1.71	3.43
WS-4	130	71.41	1.82	3.64

Table 9 Stream frequency values in the sub-watersheds

Stream frequency								
Sub-watershed	Order-wise stream no.					Sum	Area	Frequency
	1	2	3	4	5			
WS-1	742	327	146	147	65	1427	152.20	9.38
WS-2	1005	502	101	213	53	1874	224.73	8.34
WS-3	417	243	120	75	43	898	123.76	7.26
WS-4	817	69	145	105	0	1136	71.41	15.91

Table 10 Showing drainage intensity, infiltration number and constant of channel maintenance

Sub-watershed	Drainage intensity	Infiltration number	Constant of channel maintenance
WS-1	4.81	18.29	0.51
WS-2	4.58	15.18	0.55
WS-3	4.25	12.41	0.58
WS-4	8.74	28.96	0.55

Table 11 Other aerial aspects of watersheds

Watershed	Area	Perimeter	Basin length	Elongation ratio	Form factor	Lemniscate ratio	Shape index	Compactness co-efficient
WS-1	152.2	73	22.78	0.61	0.29	0.85	3.41	1.67
WS-2	224.73	91.93	28.42	0.60	0.28	0.90	3.59	1.73
WS-3	123.76	54.62	20.25	0.62	0.30	0.83	3.31	1.39
WS-4	71.41	71.41	14.82	0.64	0.33	0.77	3.08	2.38

Table 12 Relief aspects of watersheds

Sub-watershed	Max elevation	Min elevation	Relief	Relief ratio	Relative relief	Ruggedness
WS-1	4536	521	4.02	0.01	6213.70	7.83
WS-2	4834	521	4.31	0.01	5258.35	7.85
WS-3	1118	455	0.66	0.00	2046.87	1.13
WS-4	826	357	0.47	0.00	1330.76	0.85

Talking of the relief parameters (Table 12), an abrupt fall in the relief was seen from 4.02 to 0.47 from SW1 to SW4 respectively which is a typical indication of the mountain-plain front area which is prone to geomorphological hazards such as debris flows and landslides. Also, the very high values of relative relief ranging between 6213 and 1330 in SW1, SW2 and SW4 respectively indicate high erosion potentials in the overall study area. The ruggedness number 'R' which is derived from the Dd and relative relief product mainly represents the geology, vegetation cover and climate of a region. In this study, the high values of R in SW1 and SW2 show a late young to early mature stage of evolution and less denudational activities whereas the SW3 and SW4 have shown low R values indicating mature and highly eroded areas with more subsidence and weathering phenomenon.

Figure 8 represents the hypsometric integral values for all the Sub-Watersheds using the elevation-to-relief ratio which helped to calculate the cumulative surface area at various elevations in the sub-watersheds. Graphs were plotted between the surface area and the elevation points to get the hypsometric curves where the SW1 and SW2 have shown a concave upward curve in the middle parts of the sub-watershed indicating eroded and monadnock stage which is true to evident images (Fig. 7)

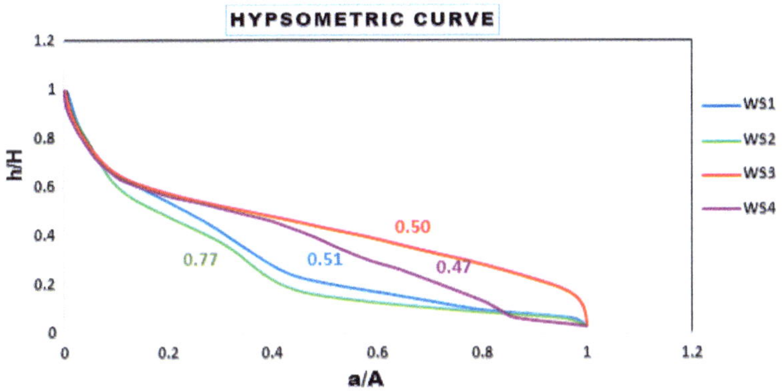

Fig. 8 Hypsometric curves and HI values for the sub-watersheds in Baner catchment

showing flat land in middle part of the study area, whereas the top and bottom parts of curves have a mature stage trend. The SW3 and SW4 have represented mature to young stages of basin evolution with S shape patterns. The hypsometric integral (HI) value of 0.77 for SW2 represents its late youthful stage of evolution subjecting it to future hazard-prone events and more morphological reworking. Overall, the Baner catchment from north to south shows a typical trend of morphological development in mountain-front to plains with HI values higher for high elevation and steeply sloping areas at the youthful stage whereas low HI for the low-lying and middle-flat portion of the study area which are near to mature or old stage.

Geomorphic Indicators

The stream length index (SL) mainly shows the impact of environmental processes on the drainages in the basin implying the movements driven by tectonic uplift or the dominance of erosional activities. If the SL values are higher, they mainly represent deflections of river channel profiles (kinking, rapids, turns, narrowing) that indicate the presence of tectonic impact and lithological factor dominance. In each sub-watershed SW1, SW2, SW3 and SW4 within each geomorphological unit i.e. mountain, piedmont and alluvial fans at least five to six SL values were calculated based on the length of main drainage channels (Tables 13, 14, 15, 16, 17 and 18) and their average were calculated for each sub-watershed and using the kriging method maps were prepared in Arc-GIS. In this study, the average of all the SL point values resulted ranging from 124 (SW4), 819 (SW1), 1023 (SW2) and 472 (SW3). The sub-watershed SW1 and SW2 belong to high-elevation areas and are traversed by the Main Boundary Thrust and the local Drini Thrust with comparatively hard lithology such as Dhauladhar Granites, Chails, Sandstones whereas if we consider the value 124 for the SW4 there is erosional dominance over the tectonics and geological variance. The SW3 with moderate values of SL indicate a balance in the erosional and tectonic processes where the presence of Jwalamukhi thrust and high-order streams presence cause erosion. The piedmont and the alluvial fan areas in the Baner catchment have shown moderate SL average values of 80–52 and 282–78 respectively reflecting less

Table 13 Stream length gradient index

Geomorphological unit	SW1	SW2	SW3	SW4	Formula
Mountain	499	356.06	111.76	5.96	SL = (ΔH/ΔL) L (Hack, 1973)
	363.41	574	97.46	21.39	
	712.28	950	1430.88	147.32	
	1252	1375.42	97.27	273.65	
	928.96	694.19	75.32	48.58	
	1162	2189	455.16	288.63	
			1051.65		
Avg of mountain	**819.60**	**1023.11**	**474.21**	**124.92**	
	77.21	68.95	0.11	15.4	
	126.21	204.15	19.65	66.44	
	103.68	520.58	6.49	29.14	
Piedmont	316.38	297.58	135.79	59.45	
	423.02	178.8	42.58	152.56	
	647.96	325.53	388.24	95.58	
	–	146.18	–	133.15	
	–	135	–	–	
Avg of piedmont	**282.41**	**234.59**	**98.81**	**78.81**	
	5.2	9.42	1.51	32.78	
	55.71	47.61	15.83	9.87	
	29.23	191.48	107.58	81.44	
Alluvial	54.55	32.8	49.52	98.53	
Fan	143.94	83.95	23.88	37.39	
	61.69	–	–	–	
	274.12	–	–	–	
Avg of alluvial fan	**89.20**	**73.05**	**39.66**	**52**	

structural control in the low-lying areas of the catchment with weak lithology such as alluvium deposits and low-strength conglomerates. Although the SW4 has shown anomalous values of high SL in the tail end of the watershed even if its position in the catchment area is at the low elevations which may be because of the presence of active Jwalamukhi Thrust passing through it and this region part is affected by the tectonic activities as well even if the lithology is not hard (alluvium deposits). The graphical representations of the SL values for all three sub-watersheds (Fig. 9) have shown the highest values for the mountain systems in each basin.

The Vf index generally marks the incision rates of the drainages passing through a valley where the high Vf values are associated with less tectonic uplift forming U-shaped valleys whereas the low Vf values are associated with high tectonic uplift and river incision forming a V-shaped valley. It is implied that the incision is directly

Table 14 Valley floor width to valley height ratio (Vf)

S. no.	SW1	SW2	SW3	SW4	Formula
1	3.28	1.75	4.92	3.65	Vf = 2Vfw/ [(Eld – Esc) + (Erd – Esc)]
2	3.42	1.98	5.06	3.32	
3	3.62	6.25	5.9	3.39	
4	4.66	2.65	4.71	2.97	
5	12.82	3.78	4.47	3.7	
6	27.28	4.72	0.53	3	
7	18.19	6.34	3.21	5.9	
8	25.7	6.95	4.05	4.54	
9	16.12	6.55	6.12	10.86	
10	18.1	10.35	3.75	–	
11	17.29	6.24	3.36	–	
12	22.83	4.83	3.93	–	
13	3.93	–	–	- –	
14	2.24	–	–	–	
Average Vf value	**12.82**	**5.19**	**4.16**	**4.59**	

Table 15 Basin shape index (BS)

SW1	3.02	Formula
SW2	2.84	Bs = B_I/B_W (Ramírez-Herrera, 1998)
SW3	1.75	
SW4	3.78	

Table 16 Mountain front sinuosity index (Smf)

Geomorphological unit	SW1	SW2	SW3	SW4	Formula
Mountain	0.94	0.98	0.97	0.98	Smf = Lmf/Ls (Bull, 2007)
	0.97	0.95	1	0.95	
	0.97	0.96	1	1.01	
	0.99	0.97	0.96	0.98	
Piedmont	0.96	0.94	0.86	0.97	
	0.95	0.99	0.98	1	
	0.99	0.97	0.96	1	
	0.97	0.94	0.99	0.98	
Alluvial fan	0.97	0.98	0.96	0.97	
	0.99	0.97	0.98	0.99	
	0.98	0.98	0.96	0.97	
	1	0.97	0.97	0.98	

Table 17 Asymmetric factor of sub-watersheds

Sub-watershed	SW1	SW2	SW3	SW4	Formula
AF values	34.77	32.74	75	56.95	$AF = (A_R/A_T) * 100$ (Hare, 1985a, b)

Table 18 Transverse topography symmetry factor (T) of sub-watersheds

	SW1	SW2	SW3	SW4	Formula
1	0.43	0.9	0.5	0.79	T = Da/Dd (Cox, 1994)
2	0.3	0.94	0.34	0.06	
3	0.7	0.56	0.54	0.55	
4	0.3	0.55	0.57	0.5	
5	0.3	0.46	0.14	0.73	
6	0.6	0.36	0.08	0.36	
7	0.58	0.14	0.72	0.83	
8	0.86	0.06	11.78	0.36	
9	0.64	0.25	–	0.69	
10	0.42	0.78	–	0.14	
11	0.54	0.78	–	0.68	
12	–	–	–	0.83	
Avg	**0.15**	**0.52**	**1.83**	**0.54**	

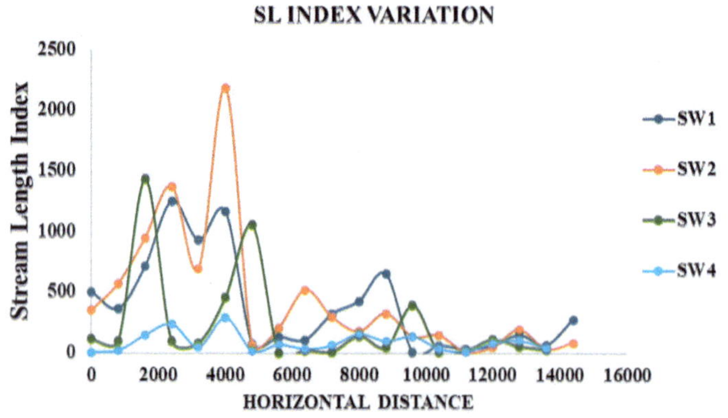

Fig. 9 Shows the comparative variations of SL-index in all the sub-watersheds

proportional to the type of lithology that can give a fair comparison of varying Vf values in an area therefore in this study Vf was analyzed at sub-watershed levels and through kriging the values were interpolated in the form of Vf variation maps for all the sub-watersheds. In this study after analyzing at least 9 Vf data points in SW4 and a maximum of 14 data points of SL in SW1 (Table 14) depending on the length of drainages of the sub-basins, the maximum Vf value resulted in the high elevation watershed SW1 and the lowest resulted for the SW3 whereas increased value of Vf in the low elevation SW4 again (Fig. 10) highlighted the presence of active thrust leading to active tectonics in the southern parts whereas decreased Vf in the northern parts of SW4 with much wider basin represents lateral erosion activities.

Table 15 represents the values of Basin shape index (Bs) with the highest and anomalous value for SW4 i.e. 3.78 whereas SW1 on second rank with a value of Bs 3.02 and both these sub-watersheds reflect high tectonic rates and elongated shape with more prominence of ongoing erosional activities in the active seasons. The average Bs values are graphically presented in Fig. 11.

The mountain front sinuosity index (Smf) generally refers to the tectonic activity along the mountain fronts; the straight and triangular fronts refer to more tectonic involvement represented with low Smf values and less denudation whereas the less active mountain fronts have higher sinuosity values and are more irregular in shape. In this study, the Smf values have been used as an indicator of the relative tectonic

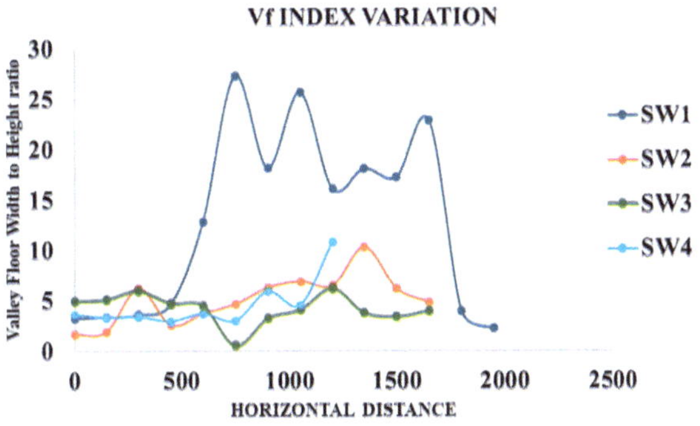

Fig. 10 Shows the comparative variations of Vf-index in all the sub-watersheds

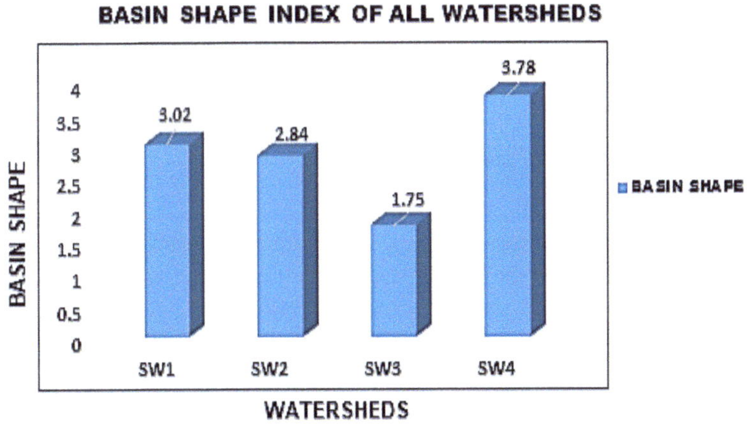

Fig.11 Basin shape index (Bs) values of sub-watersheds in Baner catchment

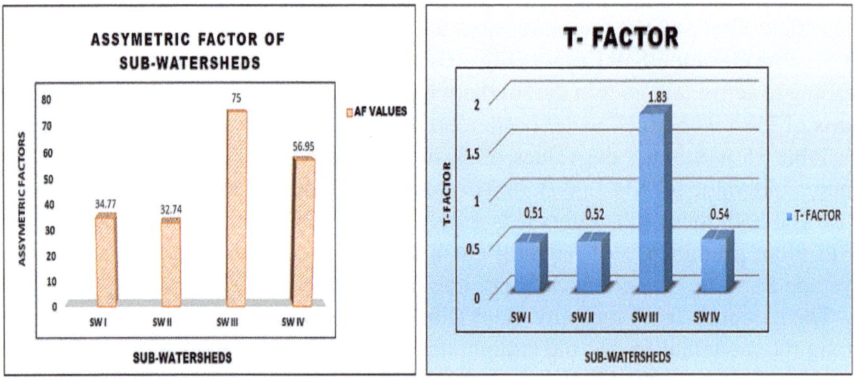

Fig. 12 Asymmetric factor (Af) and transverse topographic factor (T) factor variations in sub-watersheds of Baner catchment

activity shown as maps where all four sub-watersheds have shown a value ranging from a maximum of 1.01–0.94 (Table 16) with higher values for SW4 and SW3 inferring more sinuous fronts and more denudations along the river channels. Still, overall, the Smf values indicated active tectonic uplift in the Baner catchment area which is marked with the presence of active thrusts in this region.

To affirm the ongoing analysis the asymmetric factor (Af) and the transverse topographic factor (T) were used to further find the tilting in the sub-watersheds (Fig. 12) using the midline method and analyzing the area in right and left trunks. The sub-watershed SW3 particularly has shown more tilting with a maximum Af value of 75 (Table 17) showing a lot of deflection from a median value of 50, which is also validated with a maximum T value for SW3 i.e. 1.83 (Table 18). Nevertheless, the T factor and Af factor values for all other three sub-watersheds have also shown tilting indicating tectonic disturbances from north to south in the study area.

Finally, taking the average of the morphometric parameters along with hypsometric integral values and geomorphic indices they were attributed to the sub-watersheds and then classified for generating a relative tectonic map (RTM) (Fig. 13) of the study area. The relative tectonic index map was reclassified based on the derived average values of the total indices used in this study, and finally, the sub-watersheds were given the relative rankings (I, II, III and IV) as per their vulnerability to active tectonics. The SW1 and the SW2 have shown the highest index values in the relative tectonic map (RTM) following which SW4 ranked III and the SW3 secured the rank IV. This study has shown the parts of the sub-Himalayan catchment that are geomorphologically remodelled due to active tectonics and erosional activities.

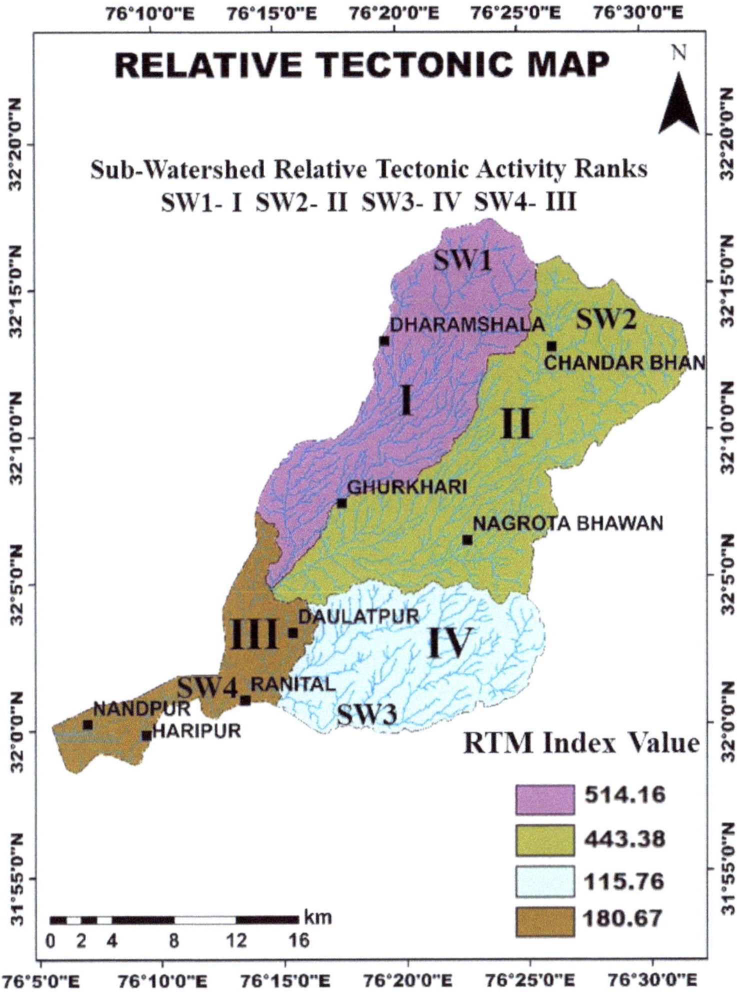

Fig. 13 Relative tectonic map of the study area

5 Conclusions

- Studying tectonic geomorphology through hypsometric, morphometric and geomorphic indicators is a common approach in which the relatively active areas with higher frontal activities and uplift can be classified and studied to see an absolute tectonic variation through the study area. The present study has given the spatial distribution of the geomorphic indices to classify the highly active tectonic fronts and active uplift areas where the findings were found in correspondence with the morphometric and hypsometric parameters.

- The study shows that the active tectonic uplift in the study area is mostly associated with the spatial zones having a high density of faults that especially fall in SW1, SW2 and SW4. These sub-watersheds are traversed by the major thrusts such as MBT, Panjal Thrust and the Muree Thrust in SW1 and SW2 whereas in SW4 Jwalamukhi Thrust is seen passing at the boundary of Dhokpathan Formation (Boulder Conglomerates and Alluvium) and Lower Siwalik Formation (Sandstone, claystone). These sub-basins have Vf and SL values indicating the drainages modified due to tectonic uplifts whereas in the SW3 and SW4, erosional activity-driven geomorphological evolution is more dominant inferred from the basin shape for SW3 and its least Vf value. But at the same time for SW3 the transverse topographic factor (T) and the asymmetry factor (Af) are maximum which shows that the drainage morphology was modified through the tectonics and the soft lithology i.e. sandstone and alluvium deposits and presently the erosional and lateral activities are dominant in the lower reaches of SW3 and SW4.
- This study also reflects the application of satellite data for tectonic and erosional process interpretations which if used as a hybrid with the statistical data can act as an absolute tectonic indicator for further geological activities in any study area.
- Overall if we look at the results, they indicate the tectonic reactivation in the SW4 i.e. the lower reaches of the Baner catchment which has an almost circular basin shape in its upper reaches but the Vf, SL and T factors have anomalous values that indicate towards dynamic tectonics. This part of the study area has witnessed many land subsidence activities and slow-moving landslides in the recent years (Mahajan et al., 2022).
- Looking at the overall results have pointed out the relative tectonic activity (RTA) in four sub-watersheds in this study where SW1, SW2 and SW4 are tectonically much more active than SW3. The Vf-SL factors for the sub-watersheds have shown an inverse relation as where the SL value has gone up there the Vf factor value has gone low or vice-versa in SW1, SW2 and SW4 whereas only in SW3 the Vf factor and SL factor show a smooth combination with not much relative variance indicating greater erosional control in the middle parts of the overall Baner catchment that has mostly flat areas.
- The Vf-Smf statistical values are true to their representation for active tectonics in SW1 and SW2 whereas, more erosional activity-driven morphological adjustments have resulted for SW3. The Vf-Smf variance in SW4 shows anomalous trends which highlights this area's complexity in terms of tectonic re-activation indicated by low Vf index and sinuous mountain fronts represented by higher Smf values.
- Overall, this study has outlined that how the tectonics have modified the overall geo-morphometry of the Baner catchment which may reflect the hazard potential at the sub-basin level.

Acknowledgements Amity Institute of Geoinformatics and Remote Sensing, Noida, India, provided the authors with the chance and technological facilities to undertake this study. We would also like to thank USGS and the European Space Agency for making the satellite data available for the present study. All the co-authors are acknowledged for their support.

Funding Statement The authors have received no funding for the research work and have submitted the article with mutual consent.

Data Availability Statement The data used in this work is freely available and anyone can use it by downloading it from sources. Further data may be shared upon reasonable request.

Disclosure Statement No potential conflict of interest was reported by the author(s).

References

Abrahams, A. (1984). Channel networks: A geomorphological perspective. *Water Resources., 20*, 161–188.

Agarwal, C. S. (1998). Study of drainage pattern through aerial data in). Study of drainage pattern through aerial data in Naugarh area of Varanasi district, U.P. *Journal of the Indian Society of Remote Sensing, 26*, 169–175.

Aier, I. L. (2011). Geomorphic evolution of Medziphemaintermontane basin and Quaternary deformation in the schuppen belt, Nagaland, NE India. *Geomorphology, 55*, 247–265.

Alam, A., Bhat, M. S., Kotlia, B. S., Ahmad, B., Ahmad, S., Taloor, A. K., & Ahmad, H. F. (2017). Coexistent pre-existing extensional and subsequent compressional tectonic deformation in the Kashmir basin, NW Himalaya. *Quaternary International, 444*, 201–208.

Alam, A., Bhat, M. S., Kotlia, B. S., Ahmad, B., Ahmad, S., Taloor, A. K., & Ahmad, H. F. (2018). Hybrid tectonic character of the Kashmir basin: response to comment on "Coexistent pre-existing extensional and subsequent compressional tectonic deformation in the Kashmir basin, NW Himalaya (Alam et al., 2017)" by Shah (2017). *Quaternary International, 468*, 284–289.

Arabameri, A. T. (2020). Morphometric analysis for soil erosion susceptibility mapping using novel GIS-based ensemble model remote sensing. *Remote Sensing, 12*, 874.

Barman, B. K., Gogoi, C., & Rao, K. S. (2019). Remote sensing and CIS application for morphometric analysis of upper Mat watershed, Serchhip District, Mizoram. *Journal of Applied Geochemistry, 21*(2), 185–195.

Barman, B. K., Rao, C. U. B., Rao, K. S., Patel, A., Kushwaha, K., & Singh, S. K. (2021). Geomorphic analysis, morphometric-based prioritisation and tectonic implications in Chite Lui River, Northeast India. *Journal of the Geological Society of India, 97*, 1–11.

Bishop, M. P., Shroder, J. F., Jr., Bonk, R., & Olsenholler, J. (2002). Geomorphic change in high mountains: A western Himalayan perspective. *Global and Planetary Change, 32*, 311–329.

Bookhagen, B., Thiede, R. C., & Strecker, M. R. (2005). Late Quaternary intensified monsoon phases control landscape evolution in the northwest Himalaya. *Geology, 33*(2), 149–152.

Bull, W. (2007). *Tectonic Geomorphology of Mountains: A New Approach to Paleoseismology* (p. 328). Wiley-Blackwell.

Bull, W. E., & McFadden, L. (1977). Tectonic geomorphology north and south of the Garlock fault, California. In: D. O. Doehring (Ed.), *Proceedings of the Eighth-Annual Geomorphology Symposium, 36, State University of New-York, Binghamton Creek Drainage Basin. Oklahoma Geology Notes* (pp. 2–16, 115–138).

Burbank, D. W. (2001). Tectonic geomorphology. In *Blackwell Science, Malden* (Vol. 274). ISBN 0-632-043865.

Burbank, D. W., & Anderson, R. S. (2011). *Tectonic Geomorphology*. Wiley.

Cannon, P. J. (1976). Generation of explicit parameters for a quantitative geomorphic study of Mill Creek drainage basin. *Oklahoma Geology Notes, 36*(1), 3–16.

Chen, Y. S. (2003). Along-strike variations of morphotec. *Geomorphology, 56*, 109–137.

Chorley, R. J., Malm, D. E. G., & Pogorzelski, H. A. (1957). A new standard for estimating drainage basin shape. *American Journal of Science, 225*, 138–141.

Christofoletti, A. (1980). *Geomorphology* (Vol. 188, 2nd ed.). Edgard Blucher.
Cox, R. (1994). Analysis of drainage-basin symmetry as a rapid technique to identify areas of possible quaternary tilt-block tectonics: An example from the Mississippi Embayment. *Geological Society of America Bulletin, 106*, 571–581.
Darton, N. (1950). Configuration of the bedrock surface of the District of Columbia and vicinity. *United States Geological Survey Professional, 217*, 42p.
Dowling, T., Walker, J., Richardson, D. P., O'Sullivan, A., & Summerell, G. K. (1998). *Application of the hypsometric integral and other terrain based metrics as indicators of the catchment health: A preliminary analysis.* CSIRO Land and Water.
Dubey, R. J. (2017). Evaluation of relative tectonic perturbation of the Kashmir Basin, Northwest Himalaya, India: An interated morphological approach. *Journal of Asian Earth Sciences, 148*, 153–172.
El Hamdouni, R., Irigaray, C., Fernandez, T., Chacón, J., & Keller, E. A. (2008). Assessment of relative active tectonics, southwest border of the Sierra Nevada (southern Spain). *Geomorphology, 96*, 150–173.
Faniran, A. (1968). The index of drainage intensity: a provisional new drainage factor. *Australian Journal of Science, 31*(9), 326–330.
Flores-Prieto, E. Q. (2015). Morphotectonic interpretation of the Makuyuni catchment in Northern Tanzania using DEm and SAR data. *Geomorphology, 248*, 427–439.
Goudie, A. (2004). *Encyclopedia of Geomorphology* (Vol. 2). *Phychology Press.*
Goudie, A. (2013). *Entries in A-Z in Encyclopedia of Geomorphology* (Vol. 35, p. 1157). Routledge.
Grohmann, C. (2004). Morphometric analysis in, geographic information systems: Applications of free software GRASS and R. *Computers & Geosciences, 30*, 1055–1067.
Hack, J. (1957). Studies of longitudinal stream profiles in Virginia and Maryland. United States. *Geological Survey Professional Paper, 294*, 45–97.
Hack, J. (1973). Stream profile analysis and stream gradient index. *Journal of Research of the United States Geological Survey, 1*(4), 421–429.
Hare, G. (1985a). Geomorphic indicators of vertical neotectonism along converging plate margins, Nicoya Peninsula, Costa Rica. In *Binghamton Symposia in Geomorphology: International Series* (Vol. 15, pp. 75–104).
Hare, G. (1985b). Geomorphic indicators of vertical neotectonism along converging plate margins, Nicoya Peninsula Costa Rica. In: M. Morisawa, & J. T. Hack (Eds.), *Tectonic. Geomorphology. Proceedings of the 15th Annual Binghamton Geomorphology Symposium* (pp. 123–134). Allen and Unwin, Boston, MA.
Horton, R. E. (1932). Drainage basin characteristics. *Transactions, American Geophysical Union, 13*, 350–361.
Horton, R. E. (1945). Erosional development of streams and their drainage basins; hydro physical approach to quantitative morphology. *Bulletin Geological Society of America, 56*, 275–370.
Hurtrez, J. E., Sol, C., & Lucazeau, F. (1999). Effect of drainage area on hypsometry from an analysis of small-scale drainage basins in the Siwalik hills (Central Nepal). *The Journal of the British Geomorphological Research Group, 24*, 799–808.
Jordan, G., & Schott, B. (2005). Application of wavelet analysis to the study of spatial pattern of morphotectonic lineaments in digital terrain models. A case study. *Remote Sensing of Environment, 94*, 31–38.
Joshi, N. K. (2021). Drainage conformation and transient response of river system in thrust segmentation of Northwest Himachal Himalaya, India. *Quaternary International, 575*, 37–50.
Joshi, L. M., & Kotlia, B. S. (2015). Neotectonically triggered instability around the palaeolake regime in Central Kumaun Himalaya, India. *Quaternary International, 37*, 219–231.
Kale, V. S. (2014). Tectonic controls upon Kaveri River drainage, cratoinc Penninsular India: Inference from Longitudinal profiles, morphotectonics indices, hanging Valleys and fluvial records. *Geomorphology, 227*, 153–165.
Kandregula, R. K. (2020). Evaluating the seismic hazard in the Kachchh region, western India using the river gradient length anomoly technique. *Journal of Earth System Science, 129*, 1–18.

Keller, E. (1986). *Investigation of Active Tectonics: Use of Surficial Earth Processes* (pp. 136–147). National Academy Press.
Keller, E. P. (1996). *Active Tectonics: Earthquakes, Uplift, and Landscape* (Vol. 338). Prentice Hall.
Keller, P. (2002). *Active Tectonics: Earthquakes and Landscape.*
Khakhlari, M. N. (2016). Morphometric analysis of Barapani River Basin in Karbi Anglong District, Assam. *International Journal of Scientific and Research Publications, 6*(10), 238–249.
Kothyari, G. (2014). Morphometric analysis of tectonically active Pindar and Saryu River basins: Central Kumaun Himalaya. *Geomorphology, 9*, 421–442.
Kothyari, G. L. (2016). Late quaternary tectonic landforms and fluvial aggradation in the Saryu River Valley: Central Kumaun himalaya. *Geomorphology, 268*, 159–176.
Kothyari, G. J. (2019). Landscape evolution and deduction of surface deformation in the soan Dun NW Himalaya India. *Quaternary International, 507*, 302–323.
Kothyari, G. S. (2018). *Evolution of drainage in response to brittle-ductile dynamics and surface processes in Kachchh Rift Basin, Western India, 10 Tectonics Problem Settings.*
Kusre, B. C. (2013). Hypsometric analysis and watershed management of Diyung watershed in north eastern India. *Journal of the Geological Society of India, 82*(3), 262–270.
Langbein, W. B. (1947). *Topographic characteristics of drainage basins (No. 968-C).* USGPO.
Magesh, N. C. (2011). Morphometric evaluation of Papanasam and Manimuthar watersheds, parts of Westemn Ghats Tirunelveli district, Tamil Nadu, India: A GIS approach. *Environment and Earth Science, 64*, 373–381.
Mahajan, A. K., Sharma, S., Patial, S., Sharma, H., Pandey, D. D., & Negi, S. (2022). A brief address of the causal factors, mechanisms, and the effects of a major landslide in Kangra valley, North-Western Himalaya, India. *Arabian Journal of Geosciences, 15*, 925.
Mahmood, S. A., & Gloaguen, R. (2012). Appraisal of active tectonics in Hindu Kush: Insights from DEM derived geomorphic indices and drainage analysis. *Geoscience Frontiers, 3*(4), 407–428.
Martinez, M. H. (2011). SL Index as indicator of anomalies in the longitudinal profile of Pirapo river. *Geosciences, 30*(1), 63–76.
Martínez-Ramírez, Á., Steinich, B., & Tuxpan, J. (2017). Morphometric and hypsometric analysis in the Tierra Nueva Basin, San Luis Potosí, México. *Environmental Earth Sciences, 76*, 444.
Mayer, L. (1990). *Introduction to Quantitative Geomorphology.* Prentice Hall.
Melton, M. A. (1957). *An analysis of the relations among elements of climate, surface properties and geomorphology* (Vol. 11). Project NR 389-042 [Technical report, Columbia University].
Melton, M. (1958). Correlation structure of morphometric properties of drainage systems and their controlling agents. *The Journal of Geology, 66*, 442–460.
Mesa, L. M. (2006). Morphometric analysis of a subtropical Andean basin (Tucuman Argentina). *Environmental Geology, 50*, 1235–1242.
Miller, V. C. (1953). *A quantitative geomorphologic study of drainage basin characteristics in the Clinch mountain area, Virginia and Tennessee* (Vol. 3, pp. 389–402). [Technical report 3, Columbia University, New York].
Misra, A. A. (2020). Quantitative geomorphic approach for identifying active deformation in the foreland region of central Indo-Nepal Himalaya. *Geotectonics, 54*, 543–562.
Nooka Ratnam, K., Srivastava, Y. K., Venkateswara Rao, V., Amminedu, E. K. S. R., & Murthy, K. S. R. (2005). Check dam positioning by prioritization of micro-watersheds using SYI model and morphometric analysis remote sensing and GIS perspective. *Journal of the Indian Society of Remote Sensing, 33*(1), 25–38.
Omvir Singh, A. S. (2008). Hypsometric integral estimation methods and its relevance on erosion status of North-Western lesser Himalayan watersheds. *Water Resources Management, 22*, 1545–1560.
Pancholi, K. (2017). Morphotectonic of Sabarmati-Cambay basin, Gujrat Western India. *Journal of the Indian Geophysical Union, 21*(5), 371–383.
Pant, N. D. (2020). Soil erosion and flood hazard zonation using morphometric and morphotectonic parameter in upper Alaknanda river basin. *Natural Hazards and Earth System Sciences, 1032*, 3263–3301.

Pazzaglia, F. J. (2013) Fluvial terraces. In *Treatise on Geomorphology* (pp. 379-412). Elsevier, Academic.

Pérez-peña, J. A. (2010). Active tectonic in the Sierra Neveda(Betic Cordillera, SE Spain): Insight form geomorphic indexes and drainage pattern analysis. *Geomorphology, 119*, 74–87.

Phartiyal, B. (2012). Impact of neotectonic on drainage network evolution reconstructed from morphometric indices case study from NW Indian Himalayan. *Geomorphology, 56*, 121.

Pike, R. W. (1971). Elevation–relief ratio, hypsometric integral and geomorphic area–altitude analysis. *Geological Society of America Bulletin, 82*, 1079–1084.

Ramírez-Herrera, M. T. (1998). Geomorphic assessment of active tectonics in the Acambay Graben, Mexican volcanic belt. *Earth Surface Processes and Landforms, 23*, 317–332.

Ritter, D. F., Kochel, R. C., & Miller, J. R. (2002). *Process Geomorphology*. Mc Graw-Hill.

Rudraiah, M. G. (2008). Morphometry using remote sensing and GIS techniques in the sub-basins of Kagna river basin, Gulburga district, Karnaraka, India. *Journal of the Indian Society of Remote Sensing, 36*, 351–360.

Schumm, S. A. (1956). Evaluation of drainage system and slopes in bad lands at Perth Amboy, New Jersey. *Geological Society of America Bulletin, 67*(5), 597–646.

Shankar, R., & Singh, A. S. (2020). Active rectonics influences in the Satlui river basin in and around Rampur, Himachal Pradesh Himalaya, India. *Arabian Journal of Geosciences, 13*, 1–21.

Sharma, S., & Mahajan, A. K. (2018). A comparative assessment of information value, frequency ratio and analytical hierarchy process models for landslide susceptibility mapping of a Himalayan watershed, India. *Bulletin of Engineering Geology and the Environment, 78*, 2431–2448.

Sharma, S., & Mahajan, A. K. (2020). GIS-based sub-watershed prioritization through morphometric analysis in the outer Himalayan region of India. *Applied Water Science, 10*, 163.

Sharma, S. K., Tignath, S., Gajbhiye, S., & Patil, R. (2013). Use of geographical information system in hypsometric analysis of Kanhiya nala watersed. *International Journal of Remote Sensing & Geoscience (IJRSG), 2*, 30–35.

Shukla, D. D. (2014). GIS-based morpho-tectonic studies of Alaknanda river basin: A precursor for hazard zonation. *Natural Hazards, 71*, 1433–1452.

Silva, P. G. (2003). Fault generated mountain fronts in Southeast Spain: Geomorphologic assessment of tectonic and earthquake activity. *Gemorphology, 250*, 203–226.

Smith, B. S. (2003). Accuracy and resolution of shuttle radar topography mission data. *Geophysical Research Letters, 30*(9).

Strahler, A. (1952). Hypsometric (area-altitude) analysis of erosional topography. *Geological Society of America Bulletin, 63*, 1117–1142.

Strahler, A. N. (1957). Quantitative analysis of watershed geomorphology. In: H. S. Schumm (Ed.), *Drainage Basin Morphometry, Benchmark Papers in Geology* (Vol. 41). *Transactions of American Geophysical Union, 38*(6), 913–920.

Strahler. AN. (1964). Part H. Quantitative geomorphology of dramage basins and channel networks. In *Handbook of Applied Hydrology* (pp. 4–39). McGraw-Hill, New York.

Suthaharan, S. (2016). Support vector machine. Machine learning models and algorithms for big data classification: Thinking with examples for effective learning 207–235.

Taloor, A. K. (2019). Tectono-climatic infuence landscape changes in te glaciated BurungDrung basin, Zanskar Himalaya. India. *Quaternary International, 507*, 262–273.

Taioor, A. K., & Joshi, L. K. (2021). Tectonic imprints of landscape evolution in the Bhilangana and Mandakini basin, Gariwal Himalaya, India: A geospatial approach. *Quaternary International,*575–576, 21–36.

Taloor, A. R. (2017). Active tectonic deformation along reactivated faults in Binta basin in Kumaun Himalaya of north India inference from tectonogeomorphic evaluation. *Zeitschrift fur Geomorphologie, 61*, 159–180.

Tandon, S. (1974). Litho-control of some seomorphic properties: An illustration from the Kumaun Hjmalaya, India. *Zeitschrift fur Geomorphologie, 18*, 460–471.

Thakkar, M. G., Kothyari, G. C., Jani, C., Chauhan, G., Lakhote, A., & Taloor, A. K. (2021). Time assessment of tectonic and ciimatic forcing on the formation of Khari bedrock gorge. Kachchh, western India: A mathematical approach. *Quaternary International, 575–576*, 328–337.
Thakur, V. C. (2004). Active tectonics of Himalayan frontal thrust and seismic hazard to Ganga Plain. *Current Science* 1554–1560
Tsodoulos, I. K. (2008). Tectonic gecmorphclogy of the easternmost extension of the Culf of Corinth (Beotia, Central Greecei). *Tectonophysics, 453*, 211–232.
Valdiya, K. S. (1993). Uplift and geomorphic rejuvenation of the Himalaya in the Quaternary period. *Current Science, 64*(11/12), 873–885.
Vijithand, H., & Satheesh, R. (2006). GIS based morphometric analysis of two mayor upland sub-watersheds of Meenachil river in Kerala. *Journal of the Indian Society of Remote Sensing, 34*(2), 181–185.
Vittala, S. S., Govindaiah, S., & Honne Gowda, H. (2004). Morphometric analysis of sub-watersheds in the Pavagada area of Tumkur district, South India using remote sensing and GIS techniques. *Journal of the Indian Society of Remote Sensing, 32*(4), 351–362.
Wei, L. (1998). Hypsometric analysis with a geographic information system. *Computers & Geosciences, 24*, 815–821.
Weissel, J. K., Pratson, L. F., & Malinverno, A. (1994). The length-scaling properties of topography. *Journal of Geophysical Research, 99*, 13997–14012.
Yadav, S. D. (2020). Spatiat regionalisation of morphometric characteristics of mini watershed of Northern Foreland of Peninsular India. *Arabian Journal of Geosciences, 13*, 1–16.

Geomorphic Assessment of Tectonic Influence on the Neora River Basin, Darjiling-Sikkim Himalaya

Bikramjit Roy, Sohini Neogy, Ujwal Deep Saha, and Rachita Chakraborty

Abstract The present study tried examining the role of tectonic perturbations on the fluvial processes through an array of morphotectonic indices coupled with channel longitudinal form analysis in the Neora River catchment. ALOS Palsar 12.5 m DEM is used to derive the channel longitudinal profile along with the Stream Length Gradient Index (SL), and thereafter, calculate the morphotectonic indices. Using the entropy weightage method (EWM), an aggregated tectonic index (ATI) has been prepared to map the degree of tectonic controls at sub-catchment scale. Furthermore, Principal Component Analysis (PCA) groupings and Kernel Density Estimate (KDE) plots have been performed on the sub-basins—indicating the localized effects of surface deformation—based on the morphotectonic parameters to evaluate the localised nature of tectonic activity. The tectonic deformities have been strongly reflected in the channel longitudinal profile with distinct differences in channel responses related to notable lithological variations. The knickpoint distribution and its correspondence with structural deformities and the nature of sub-catchments in-terms of morpho-tectonic signatures defines segment specific dominance of either tectonic deformation and lithological composition. Within the segment above Himalayan thrusts, lithology was seemed to be dominant force governing catchment forms while on the HFB, localised surface deformations and relative channel adjustments is the dominant phenomenon.

Keywords Catchments · Himalayan thrusts · Long profile · Morphotectonics · Tectonics

B. Roy (✉) · R. Chakraborty
Department of Geography, Vidyasagar College, Kolkata, India
e-mail: bikramjitroy405@gmail.com

S. Neogy
Department of Geography, Banwarilal Bhalotia College, Asansol, India

U. D. Saha
Department of Geoenvironmental Research, Institute of Geography and Spatial Organization, Polish Academy of Sciences, Krakow, Poland

© The Author(s), under exclusive license to Springer Nature Switzerland AG 2025

S. Ghosh (ed.), *The Himalaya Dilemma*, Springer Natural Hazards,
https://doi.org/10.1007/978-3-031-95083-4_5

1 Introduction

Rivers play a vital role in developing its immediate environment as the majority of the earth's surface exhibits traces of fluvial action on its landscapes (Saha et al., 2023). Climatic action coupled with underlying geological structure and lithology influences the development of fluvial systems (Das et al., 2022). Additionally, the natural evolution of drainage catchments and their fluvial processes are altered by landscape deformation, such as faults, thrusts, or any other lineament features (Thiede et al., 2005). Tectonism, an allogenic process, notably influences the development of major structural features. Large-scale uplift, subsidence, or deformation disrupts the channel network in tectonically active locations because fluvial systems are constantly searching for their equilibrium state, or dynamic equilibrium (Charlton, 2007). These processes not only affect the channel network system but also the overall form of the catchment (Simoes et al., 2021). Analyzing drainage patterns, catchment forms and their deviations from the idealized condition can reveal how tectonism shapes landscape evolution. Here, the quantification of morphotectonic parameters of the catchment aids in identifying the structural setup of the channel and tectonic response (Keller & Pinter, 1996). The intricate equilibrium of morpho-tectonic parameters in a river system indicates that the river is sensitive to any change within its given environmental setup where the degree of stress and the river's primary ability to absorb these changes determines the nature of the adjustments (Jiongxin, 1990). The study of morphotectonic indices, quantified using toposheets, satellite images, DEMs, and field investigation, helps to evaluate the zonation of tectonically active regions (Anand & Pradhan, 2019). Consequently, this helps in separating the localised effects of surface deformation (Saha et al., 2024). However, these morpho-tectonic indices have a few shortcomings and lithological variations or physiographic changes may lead to erroneous results. Thus, the inferences drawn from the morpho-tectonic approach need careful investigations in the post-training period (Holbrook & Schumm, 1999). These issues can significantly be solved if the channel longitudinal forms can be collaborated with the study (Lee & Tsai, 2009). The Stream Length Gradient Index (SL) analysis gives further insight into the hillslope development process and structural and lithological control over the regularity of the channel flow (Das et al., 2022; Saha et al., 2021).

The Himalayan Mountain Belt is one of the most tectonically active regions of the world. It is a product of multiple deformation of huge sedimentary mass over millions of years on the northern marginal platform of the Indian sub-continent (Valdiya, 1986). Three major thrust lines; Main Central Thrust (MCT) which divides the Greater and Lesser Himalayas, Main Boundary Thrust (MBT), which separates the outer Himalayas from the Lesser Himalayas and the Main Frontal Thrust (MFT) divides the Himalayas foothills and Indo-Gangetic plain; these primary thrusts limit the distribution of lithology and their associated translation (Jain et al., 2012) thereby dividing Himalayas into specific litho-tectonic domains (Saha et al., 2021). Surface deformation in the Eastern Himalayan region in the Quaternary period is the outcome of the continuous downward movement of the Indian plate at different rates over time

(Valdiya, 2002). This uneven and dynamic landscape impacts the fluvial activity as well as the whole region (Saha et al., 2021). The Neora River Catchment which is the main focus of this study flows across the thrusts and thus, encounters different lithological combinations and varying degrees of strains. The MBT and MCT are transverse to the catchment and packed close to each other due to crustal shortening in this part of the Himalayas (Mukul, 2010). Moreover, the Himalayan Frontal Catchment contains a deformed fan surface with the development of blind thrusts across it within the catchment boundary of the Neora River. These are often linked to the southward translation of the Himalayan thrusts (Kar et al., 2014). This landscape configuration packed with intense deformation activity provides a unique setup for the fluvial system to adjust and express itself. The river catchments on the eastern Himalayas have been studied earlier to understand their response to regional tectonic activities (Simoes et al., 2021). These works (Table 1) have mostly considered each catchment as individual entities (Saha et al., 2021; Ayaz & Dhali, 2019; Srivastava et al., 2017) or contain larger rivers (glacier-fed) that portray regional implications of tectonism on catchment development (Simoes et al., 2021; Kashyap et al., 2024). Analyzing relatively smaller rivers (mountain-fed) on complex structural setups draws fundamental interest and the results' resolution will be greater since sub-catchments are considered at the base of the spatial scale. This study was aimed towards answering, how diverse lithological composition and packed tectonic elements within a relatively smaller area affect river catchment evolution. It has applied the morphotectonic approach aided with channel longitudinal form analysis to ascertain the relative role of tectonic deformation on fluvial processes and catchment form in the Neora River catchment. The work has been done at the sub-catchment scale to separate the differential localized effects of tectonism within the catchment. The objectives of this study are as follows

i. To analyse the degree of tectonic influence at sub-catchment level
ii. To identify the relative role of lithology and tectonics on geomorphic properties of the sub-catchments.

2 Study Area

The Neora River, a mountain-fed channel originates from the Rechila Hill Reserve Forest located in the Neora Valley National Park and drains into River Chel after flowing southward. The river catchment lies between 26° 42' N to 27° 80' N and 88°42' E to 88°48' E and covers approximately 282 km^2 area. The length of the Neora River is 61 km (approx.) and flows within Darjiling and Kalimpong districts. This watershed is elongated in shape (Saha et al. 2022) and covers rugged mountains and the adjoining frontal plains (Biswas, 2015; Biswas & Chakraborty, 2016). The watershed area lies across the Main Central Thrust (MCT) and Main Boundary Thrust (MBT) (Fig. 1). The MBT runs along the mountain front of this watershed and on the frontal plains multiple blind thrusts are present; Samsing thrust, Matiali thrust, Chalsa thrust and Baradighi thrust. Since it covers multiple tectonic units, the

Table 1 Reviewing major literatures related to basin scale morphotectonic analysis

Author	Methods							
	Geomorphological analysis	Morphotectonic	ATI	Statistical tests	Tectonic and Stratigraphic analysis	Long profile analysis	Structural analysis using	MCDM
Manna and Ghosh (2023)	✓	✓				✓		
Sharma et al. (2022)	✓	✓						
Kothyari and Rastogi (2013)		✓			✓			
Ayaz and Dhali (2019)		✓				✓		
Kothyari et al. (2020)		✓						
Alipoor et al. (2011)		✓			✓		✓	
Srivastava et al. (2016)								
Bali et al. (2016)		✓						
Das and Gupta (2019)	✓	✓						
Altın (2011)		✓						
Ahmad and Bhat (2012)		✓			✓			
Arlegui and Soriano (1998)					✓			

(continued)

Table 1 (continued)

Author	Methods							
	Geomorphological analysis	Morphotectonic	ATI	Statistical tests	Tectonic and Stratigraphic analysis	Long profile analysis	Structural analysis using	MCDM
Azañón et al. (2012)		√			√			
Al-Attar et al. (2022)		√		√			√	√
Alizadeh et al. (2020)		√						
Saha et al. (2021)		√	√			√		
Alaei et al. (2017)		√						
Ali and Ikbal (2020)		√						
Saha et al. (2022)		√	√	√				

lithological composition is also diverse. Above the MCT, the watershed area contains Kanchenjunga Gneiss (Central Crystalline Gneissic Complex) of the Proterozoic era and its extent up to the mountain front contains the Daling group, composed of lower-grade sub-metamorphosed rocks of the Higher Himalayan crystalline. On the frontal plain, the entire watershed is composed of catchment fill Quaternary alluvium (Guha et al., 2007, Dhali & Mukhopadhyay, 2020). On the frontal plains, the watershed shares its extent on multiple tectonically deformed surfaces, the Matiali alluvial surface and the Gorubathan recess. The trunk stream bears significant development of alluvial terraces (mostly unpaired) on the plains (Fig. 2). The significant growth of the alluvial fan influences entrenched channel segments where a significant length on the frontal part flows parallel to the faults (Barman et al., 2022). This watershed area experiences tropical humid climatic conditions with an average annual rainfall of 400 mm and the giant share of rainfall is concentrated between June and September. Since this watershed is rainfed, the hydrograph forms a unimodal character with a significantly high rainfall-runoff relationship (Saha, 2023).

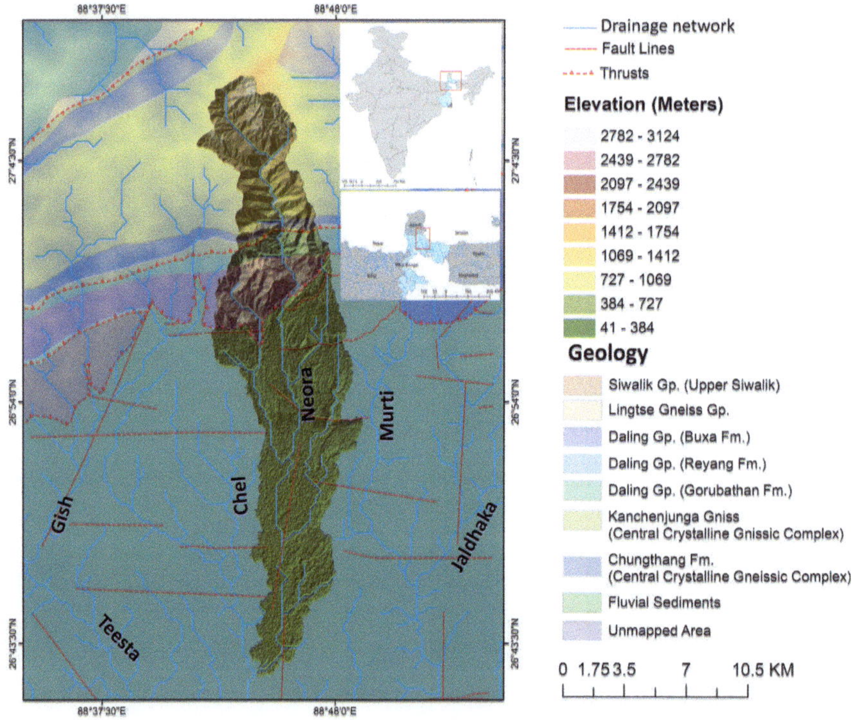

Fig. 1 (Left) Extent of the Neora River watershed. The river flows across five thrusts, from N to S—MCT, MBT (on the mountainous part), Samsing, Matiali and Chalsa thrust (on the HFB); the white dot represents the location of the photograph in Fig. 2. (Right) The entire HFB is composed of Quaternary deposits. This region does not contain any Siwalik formation

Fig. 2 Viewing the Neora River (blue arrow) near its entrance on the HFB (viewing North). On the HFB, the river showcases unpaired terraces with maximum Terrace Level 4 (T4) development. T1 and T2 are terrace level 1 and 2

3 Materials and Methods

3.1 Data Preparation

The watershed area was delineated using the ALOS Palsar 12.5 m DEM in HEC-HMS software. The DEM was downloaded from https://search.asf.alaska.edu/#/ and trained for radiometric and geometric corrections in ArcMap10.4 before employing for topographic computations. The drainage network has been delineated using the D8 algorithm where flow direction is traced down by the flow away from each pixel to its 8 neighborhood pixels. Strahler's stream order was computed for delineating the sub-watersheds. Watersheds of 3rd order channels were considered to demarcate the sub-watershed because the DEM-derived first-order channels are often found with error in terms of their real-world existence even on mountainous terrain (Schwanghart & Scherler, 2017). A total of 62 sub-catchments were prepared. The dataset bearing structural elements was downloaded from https://bhukosh.gsi.gov.in/Bhukosh/Public.

3.2 Channel Long Profile and Stream Length Gradient Index (SL)

The long profile provides a conceptual framework for understanding the influence of lithology, structure, tectonics, and watershed hydrology on river profiles. The SL index is the most useful tool to identify and understand the sensitive characteristics of channel slope and its role in identifying areas of potential tectonic activity, rock resistance, stream length, and stream power (Lee & Tsai, 2009). It also helps to demarcate the possible location of knickpoints. It can be calculated using Eq. 1 (Hack, 1973).

$$SL = \frac{\Delta H}{\Delta L} \times L \tag{1}$$

A low SL value indicates the lower resistance of the underlying rock and a high SL value suggests increased tectonic activity in the area (Taloor et al., 2023). Initially, we divided the course of the river into 100-m segments from source to confluence and then converted these points into a point vector format. Further, the spatial information of these values was used to calculate the distance between two points using Eq. (2).

$$d = \sqrt{\{(x_2 - x_1)^2 + (y_2 - y_1)^2\}} \tag{2}$$

The elevation of the points was further extracted from the DEM and the half-length was measured from the water divide.

3.3 Morphotectonics Indices

To evaluate the tectonic influence at the sub-catchment level, several morphotectonic indices were calculated. To determine the catchment shape, Form Factor (F_f), Elongation Ratio (R_e), and Circularity Ratio (C_r) were computed. To examine the catchment relief and anomalies according to the influence of tectonic activity, Relief ratio (R_f), Topographic Roughness Index (STD), Hypsometric integral (H_i) and Normalised Stream Length Gradient Index (NSLI) were calculated. To determine the catchment asymmetry, Topographic Symmetry Factor (T), Asymmetry Factor (A_f) and Valley width-to-height ratio (V_f) were evaluated (Saha et al, 2021; Taloor et al, 2023). The formulas used to derive all of these morphotectonic indices are given in Table 2.

Table 2 Mathematical derivations and explanations of morphotectonic indices applied in the study

Properties	Index	Mathematical derivation	Explanations	Source
Shape	Form factor (F_f)	F_f = A/L^2 A = Catchment Area, L = Catchment Length)	Higher values indicate bulging and oval-shaped catchment whereas lower values indicate elongated catchment	Horton (1932)
	Elongation ratio (R_e)	$R_e = 2*(\sqrt{A/\pi})/Lb$ (A = Catchment area, Lb = Maximum length of the catchment)	$R_e < 0.50$; tectonically active; R_e = 0.50–0.70; slightly active; $R_e >$ 0.70; tectonically inactive	Bull and McFadden (1977)
	Circularity Ratio (C_r)	$C_r = 4\pi A/P^2$ where A is catchment area, P = Area of a Circle with same catchment perimeter	High C_r value represents strong structural control on catchment whereas lower values represent no structural disturbance or circular form of catchment	Miller (1953), Strahler (1964)
Relief	Relief ratio (R_f)	R_f = (Emax-Emin)/ Lb Emax = Maximum elevation of the catchment, Emin = Minimum elevation of the catchment, Lb = Longest length of the catchment	High values of relief ratio indicate steep slope and high relief. In other side low values indicate small ridge with lower degree of slope	Panek (2004)
	Topographic roughness index (STD)	Standard deviation of the elevation value	Magnitude of topographic variability; higher the value greater is the roughness	Klinkenberg and Goodchild (1992)
	Hypsometric integral (H_i)	H_i = (Hmean − Hmin)/(Hmax − Hmin) where Hmean = Average of elevation, Hmax & Hmin = Maximum and Minimum elevation of area respectively	$H_i > 0.6$ characterise tectonic mobility $H_i < 0.6$ indicates mature stage of development	Strahler (1952)

(continued)

Table 2 (continued)

Properties	Index	Mathematical derivation	Explanations	Source
	Normalized stream length Gradient Index (SL)	$(\Delta H/\Delta L) * L / K$ ΔH = Elevation change, ΔL = Horizontal length for a given channel L = Horizontal distance from the midpoint of the reach of the drainage divide		Hack (1973)
Asymmetry	Topographic symmetry factor (T)	T = Da/Dd Da is the distance from the channel to the catchment midline. Dd is the distance from the lateral catchment margin to the catchment midline	T = 0; symmetric catchment with no ambiguities T > 0 asymmetric catchment	Cox (1994)
	Asymmetry factor (A_f)	$A_f = 100(Ar/At)$ Ar = Right hand side area of drainage looking downstream At = Total area of drainage catchment	$A_f = 50$; stable $50 > A_f > 50$ characterise tilt	Keller and Pinter (1996)
	Valley width to height ratio (V_f)	$V_f = 2Vfw/\{(Eld\text{-}Esc) + (Erd\text{-}Esc)\}$ Vfw = width of the valley floor, Eld & Erd = Elevation of left and right valley divides respectively Esc = Elevation of valley floor	$V_f < 1$ characterize active incision $V_f > 1$ active deposition into a broad valley	Bull and McFadden (1977)

3.4 Aggregated Tectonic Index (ATI)

Based on the morphotectonic indices computed at sub-catchment scale, an aggregation of sub-catchments using relative similarities is required to compartmentalise regions with different degrees of tectonic attractions. Here, since all the indices are important equally, the individual weightage of the morphotectonic indices was assigned using the Entropy Weighted Method (EWM) (Saha et al., 2021). In the pre-training phase, to reduce the covariance error, a correlation matrix was prepared (Pearson's) where indices having a correlation coefficient (r) more than 0.90 were excluded from computing ATI (Fig. 3). The EWM method generates a unique value that measures the magnitude of differences between variables. Where greater dispersion of measured value indicates a higher degree of differentiation for index resulting in more informative data and vice-versa. (Zhang et al., 2014). EMW is a 4-step computational method where Eq. 3 is used to standardise a model by eliminating the dimensionality and commensurability.

$$P_{ij} = X_{ij}/\Sigma X_{ij} \tag{3}$$

Here, Pij is the proportion indicator of the variables and i is the matrix location of the jth variable (here index). Next, the information entropy value (e) of each index is computed using the Eq. 2:

$$ei = \left[(1/\,ln\,(m))\Sigma P_{ij} ln\left(P_{ij}\right)\right] \tag{4}$$

Then, the differential coefficients (g) of the ith index, which exhibit a negative association with entropy properties, are calculated as:

$$gi = (1 - ei) \tag{5}$$

Finally, the weightage is determined using Eq. 6.

$$ai = (gi\,/\,\Sigma gi) \tag{6}$$

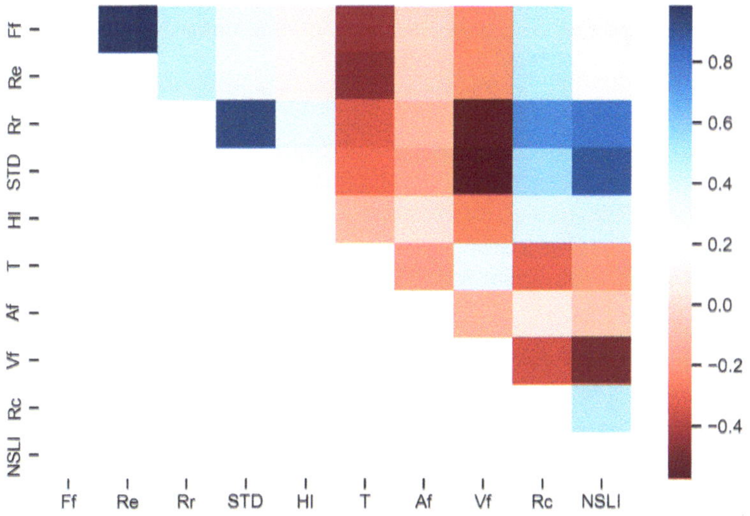

Fig. 3 Pearson's correlation matrix among the morpho-tectonic indices selected for preparing ATI. Since the Vf ratio is highly correlated with NSLI, STD and Rr while STD is highly correlated with Rr and NSLI

3.5 Principal Component Analysis

PCA is the most significant method to represent a set of data by removing the noise that occurred due to the presence of correlated variables in different observations (Yang et al., 2022). The main framework behind the concept of principal component analysis (PCA) is to measure several correlated variables, diminish the dimensionality of the data, and incorporate these while preserving a feasible variability of the data (Mishra et al., 2017). Here, the grouping of the sub-catchments based on the morphotectonic indices was achieved using PCA (Eq. 7–10) where a similarity plot of the sub-catchments was done based on varimax rotation (Eq. 11). In the pre-processing phase, no variable was excluded from the exercise.

PCA is generally computed using a four-step method. Firstly, the dataset is standardized using the following equation:

$$\chi_{new} = (\chi - \mu)/\delta \tag{7}$$

After standardizing the variables (where each variable has a mean of 0 and a standard deviation of 1); a covariance matrix table is generated using the following equation:

$$Cov_{(x,y)} = \{\Sigma(x_i - \overline{x}) * (y_i - \overline{y})\}/(N-1) \tag{8}$$

Then the eigenvalue is calculated using the following equation:

$$Av - \lambda v = 0;\ (A - \lambda I)v = 0 \tag{9}$$

After the calculation of the eigenvalue; it is multiplied by the matrix to get transformed data:

$$Feature\ matrix * top\ k\ eigenvectors = Transformed\ Data \tag{10}$$

4 Results

4.1 Morphometric Parameters

The average catchment area and catchment length above MCT is 3.71 sq km & 3.47 km, MCT to MBT is 6.35 sq. km & 4.44 km and on HFB it is 4.59 sq. km & 4.88 km. The average perimeter of the catchment above MCT is 12.11 km, between MCT and MBT is 15.71 km, and on HFB is 17.79 km (Fig. 4). Different zonal parameters are used to analyse the relief of the catchment; in this process, maximum elevation, minimum elevation, and mean elevation have crucial roles. The elevation of the study area ranges between 96 and 3164 m. The average elevation above MCT is 1891.12 m, MCT to mountain front is 625.59 m and 208.61 m on the HFB. To calculate the liner aspects of the catchment channel length is calculated. The average channel length above MCT is 3.04 km, between MCT and MBT is 4.54 km and 4.75 km on HFB. The elevation of the Right and Left valley divide is calculated to analyse the asymmetry of the catchment using the valley width-to-height ratio (V_f). The average elevation of the right valley divide above MCT is 1836.17 m, 605.82 m in between MCT and MBT, and 202.57 m on HFB. The average elevation of the Left valley divide above MCT is 1840.98 m, 617.32 m in between MCT to MBT, and 199.67 m on HFB. Also, the width of the valley floor is important to assess the Vf. The width of the valley floor is 43.52 m above MCT, 28.26 m in MCT to MBT, and 72.00 m on HFB. The average elevation of the valley floor is 1733.88 m above MCT, 500.95 m between MCT to MBT, and 193.05 m on HFB.

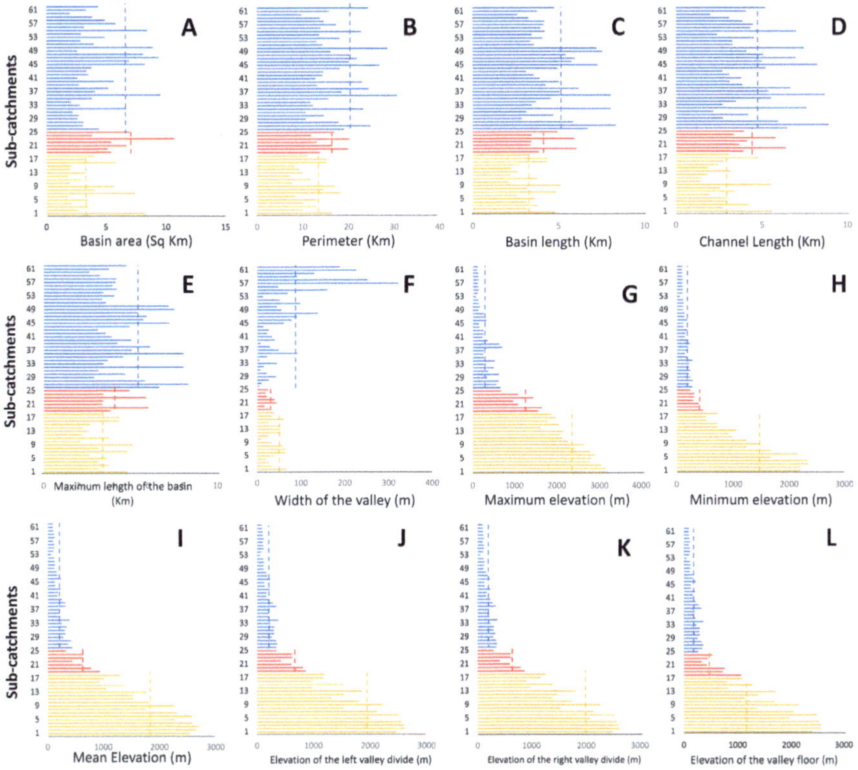

Fig. 4 Morphometric parameters: **a** catchment area, **b** perimeter, **c** catchment length, **d** channel length, **e** maximum length of the catchment, **f** width of the valley floor, **g** maximum elevation. **h** Minimum Elevation, **i** mean elevation, **j**, **k**, **l** elevation of left, right & lowest point of the valley. The yellow segment denotes the zone above MCT, the red segment denotes MCT to MBT and the blue segment denotes the zone on the HFB; Average values of the values of the respective segments are demarcated using dotted lines

4.2 Morphotectonic Indices

4.2.1 Form Factor (F_f)

Higher values of F_f indicate a bulging or oval shape of the catchment while lower values indicate an elongated catchment (Horton, 1932). The F_f is lowest at SB 22 (0.044) and highest at SB 21 (0.592) (Table 2). The values are divided into 5 classes: class 1 (0.04–0.12), class 2 (0.12–0.21), class 3 (0.21–0.29), class 4 (0.29–0.40) and class 5 (0.4 –0.59). 9.67% (6) fall under class 1 indicating the highest number of tectonic activities while 27.41% (17) in class 2, 29.03% (18) in class 3, 25.80% (16) in class 4, and 8.06% (5) in class 5 (Fig. 5). The interquartile range between the highest and lowest value of F_f is 0.146.

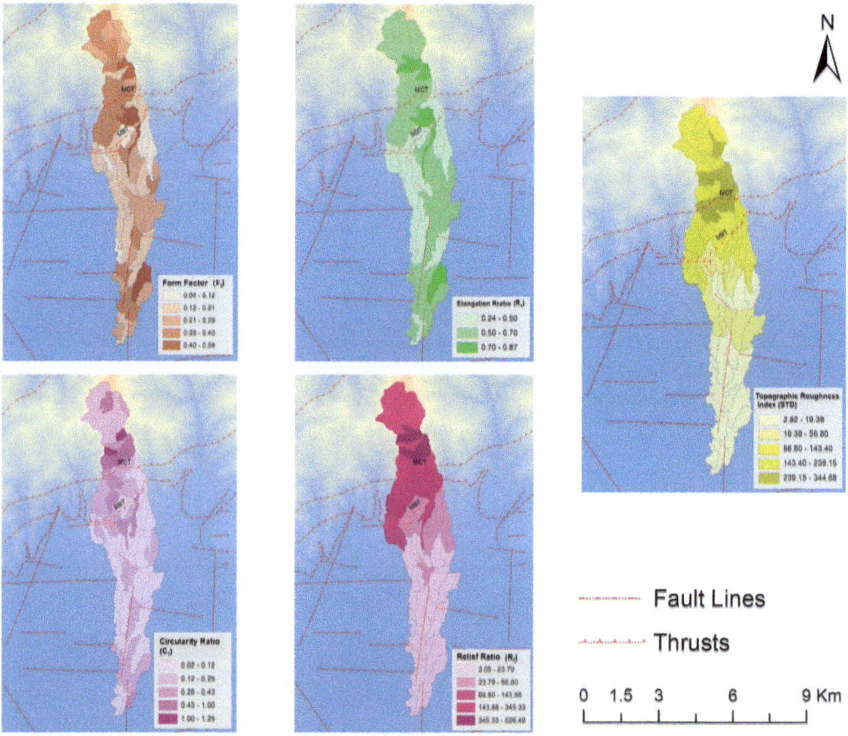

Fig. 5 Morpho-tectonic indices (from top left): From Factor (F_f), Elongation Ratio (R_e), Circularity Ratio (C_r), Relief Ratio (R_f), Topographic Roughness Index (STD), Hypsometric Integral (H_i), Normalized stream length gradient index (NSL), Topographic Asymmetry Factor (T), Valley width to Height ratio (V_f), Asymmetry Factor (A_f)

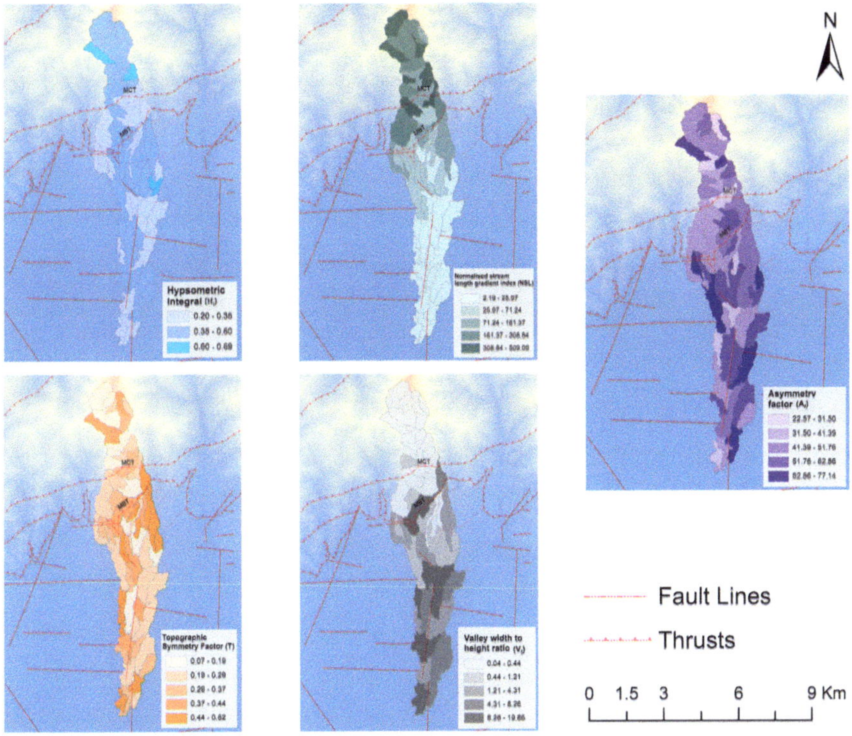

Fig. 5 (continued)

4.2.2 Elongation Ratio (R_e)

A higher value of R_e indicates a high elevation with a steep slope and the lowest value indicates an elongated shape of the catchment which is a proxy indicator of tectonic activity (Saha et al, 2021). The lowest R_e was recorded at SB 22 (0.24) and the highest value was recorded at SB 21 (0.87). R_e values of all sub-catchments have been divided into three classes: class 1 (0.24–0.50), class 2 (0.50–0.70), and class 3 (0.70–0.87). Among the 62 sub-catchments, 29.03% (18) falls in class 1 representing higher tectonic activity, 61.29% (38) in class 2 representing slightly active tectonic activity and 9.67% (6) in class 3 representing low tectonic activity (Fig. 5). It was also observed that most of the tectonically active region is located below the MBT indicating the neo-tectonics footprint on the alluvial reach.

4.2.3 Circularity Ratio (C_r)

The circularity ratio compares the area of a catchment to a circle that has the same area and perimeter. A lower circularity ratio suggests minimal structural disturbances and vice-versa. In this study, the circularity ratio values are divided into 5 classes: class 1(0.02–0.12), class 2(0.12–0.25), class 3 (0.25–0.43), class 4 (0.43–1.00) & class 5 (1.00–1.25) (Fig. 5). The highest value of C_r was recorded in SB 12 (1.25) while the lowest value was recorded in SB 36 (0.02). Among the 62 sub-catchments, 37.09% (23) are in class 1, 35.48% (22) are in class 2, 12.90% (8) are in class 3, 11.29% (7) in class 4, and 3.23% (2) in class 5.

4.2.4 Relief Ratio (R_f)

The Relief Ratio of a catchment is the ratio between catchment relief and longest length of the catchment. It measures the intensity of the erosion process and gauges the complete steepness of a drainage water-shade. In this study, one of the values of R_f ranges between lowest at sub-catchment number 49 (3.05) to highest at sub-catchment number 12 (526.49). The values are classified into 5 classes: Class 1 (3.05–23.79), Class 2 (23.79–66.80), Class 3 (66.80–143.88), Class 4 (143.88–345.33) and Class 5 (345.33–526.49). The whole of Class 1 (40.32%) and Class 2 (16.12%) and a major portion of Class 3 (8.06%) are located on HFB.

4.2.5 Topographic Roughness Index (STD)

The Topographic Roughness Index (STD) is used to measure the magnitude of the topographic variability of a surface (Saha et al., 2021). The higher value of STD represents a high upliftment rate and lower values indicate the opposite (Klinkenberg and Goodchild, 1992; Partabian et al.,2016; Saha et al., 2021). The values of STD range between 2.82 at SB 61 to 344.68 at SB 17. The calculated STD values are divided into 5 classes: Class 1 (2.82–19.39), Class 2 (19.39–56.80), Class 3 (56.80–143.40), Class 4 (143.40–239.15), and Class 5 (239.15–344.68). Based on these results 33.87% (21) fall under Class 1, 14.51% (9) fall under Class 2, 19.35% (12) under Class 3 & 4 each and 12.90% (8) under Class 5.

4.2.6 Hypsometric Integral (H_i)

The H_i is used to quantify the relative distribution of land surface at different elevations within a given drainage catchment. The values of H_i greater than 0.6 are represented as the erosional stage, the values between 0.6 and 0.35 as the mature stage, and less than 0.35 indicates the old stage (Strahler, 1952). The H_i values range between the lowest (0.2) at SB 57 and the highest (0.69) at SB 43. The calculated values of H_i of all 62 sub-catchments are classified into 3 classes: Class 1 (0.20–0.35), Class 2 (0.35–0.60), And Class 3 (0.60–0.69). Based on the result 33.87% (21) fall under class 1, representing a youthful stage of erosion; 61.29% (38) under class 2, representing a mature stage of erosion; and 4.84% (3) under class 3, representing an old stage of erosion.

4.2.7 Topographic Symmetry Factor (T)

The T is used to calculate the intensity of tectonism by measuring the lateral tilting of the catchment (Cox, 1994). The value of T close to 0 represents asymmetry and close to 1 represents topographic symmetry (Cox, 19,994). In this study area, the values of T range from lowest at SB 29 (0.07) to highest at SB 32 (0.82). The values are classified into 5 classes: Class 1 (0.07–0.19), Class 2 (0.19–0.29), Class 3 (0.29–0.37), Class 4 (0.37–0.44), and Class 5 (0.44–0.82). The result shows that 19.35% (12) fall under Class 1, 29.03% (18) under Class 2, 19.35% (12) under Class 3, 9.68% (6) under Class 4, and 22.58% (14) under Class 4.

4.2.8 Asymmetry Factor (A_f)

Changes in the A_f occur due to the variations in the inclination, perpendicular to the stream direction. The values that deviate significantly from 50, suggest the influence of active tectonic or lithological factors on differential erosion (Keller & Pinter, 2002). The values of A_f calculated for the sub-catchments are classified into 5 classes; class 1 (22.57–31.50), class 2 (31.50–41.40), class 3 (41.40–50.00), class 4 (50.00–62.86), class 5 (62.86–77.14). The highest value is recorded in SB 33 (77.14) and the lowest value is recorded in SB 44 (22.57). 8.05% (5) fall under Class 1, 24.19% (15) fall under Class 2, 17.74% (11), 32.26% (20), and 17.74% (11) in Classes 3, 4, and 5 respectively.

4.2.9 Valley Width-To-Height Ratio (V_f)

The ratio of valley width to height reflects the shift between V-shaped valleys (with V_f values close to 0), where rivers are carving downwards due to upliftment. This indicates significant tectonic activities and wider valley floors associated with slower uplift rates (Bull & McFadden, 1977). Higher values of V_f represent low tectonic activity. The V_f values range between lowest at SB 3 (0.03) to highest at SB 62 (19.86). The values (table 3) are divided into 5 classes: class 1 (0.04–0.44), class 2 (0.44–1.21), class 3 (1.21–4.31), class 4 (4.31–8.26), and class 5 (8.26–19.86). 4.83% (3) falls under Class 1, 46.78% (29) under Class 2, 17.74% (11) under Class 3, and 11.29% (7) &19.35% (12) under Class 4 and 5 respectively.

4.3 Channel Long Profile and Stream Length Gradient Index

Multiple breaks in the channel long profile are indicative of adjustments between surface upliftment and channel downcutting in the long profile of the Neora River (Fig. 6). These bumps in the long profile are observed when the channel encounters the MCT and MBT. Following this, knickpoints were also found corresponding to the

Table 3 Descriptive statistics of calculated values of Morpho-tectonic parameters

Index	Min	Max	Mean	Q1	Q3	IQR	SD
Form Factor (F_f)	0.04	0.59	0.26	0.17	0.32	0.15	0.11
Elongation Ratio (R_e)	0.24	0.87	0.56	0.47	0.64	0.17	0.13
Relief Ratio (R_f)	3.05	526.49	132.57	13.20	226.13	212.94	147.64
STD	2.82	344.68	102.76	10.83	194.22	183.39	100.88
Hypsometric integral (H_i)	0.20	0.69	0.41	0.31	0.50	0.19	0.12
Topographic symmetry factor (T)	0.07	0.82	0.33	0.23	0.41	0.18	0.16
Asymmetry factor (A_f)	22.57	77.14	48.32	37.59	58.05	20.46	12.86
Valley width to height ratio (V_f)	0.04	19.86	3.98	0.23	7.77	7.54	4.98
Circularity ratio (C_r)	0.02	1.25	0.24	0.08	0.29	0.22	0.25
SL index	2.19	509.09	123.06	13.33	227.66	214.33	136.75

Matiali, Chalsa and Baradighi thrusts. Moreover, the channel long profile displays an overall concave shape, accompanied by a gradual decrease in gradient with increasing distance downstream. The exponential curve with highest $R^2 = 0.96$ is found the best fit which indicates a possible balance between channel incision and structural perturbations.

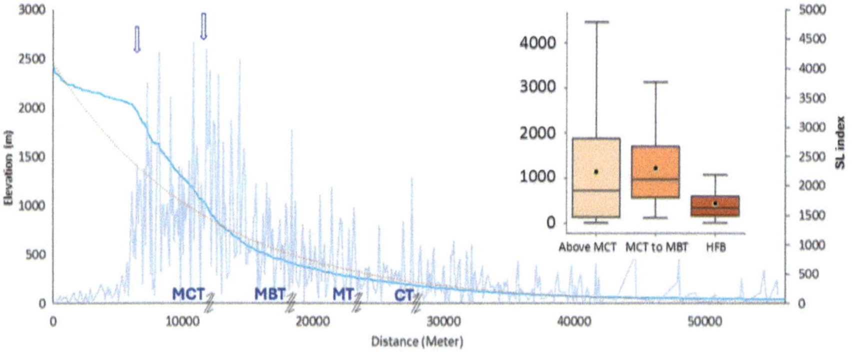

Fig. 6 The channel-long profile and the calculated SL index are represented in the figure, The channel long-profile is shown with a blue line and the SL index is indicated using a faded blue line. The best-fit line (exponential curve) for the channel long profile is shown with a dotted red line. Major knickpoints are pointed with black arrows. The places where major thrusts cut the river knickpoints are created on the channel long profile and the SL values fluctuate

For the Neora River only, the SL index ranges between 0.86 and 4443.13 m and most of the anomalies on the long profile were observed until the MBT. Between the MCT and MBT, the SL values mostly fluctuate registering a range of 4159.85 m to 109.01 m. The SL values at the Matiali, Chalsa, and Baradighi thrusts are 1977.97, 2151.60, and 842.76 m respectively which are 54.81, 53.06 and 16.59% higher than its immediate upstream values respectively. The categorical plot of SL values is calculated after dividing the whole channel flow into 3 major structural compartments. These are above MCT, MCT to MBT, and on the HFB to find out the outliers (Fig. 6). The channel segment above the MCT registered highest SL range of 4442.27 m. With respect to the SL distribution above MCT, three prominent knickpoints at 6202, 8179, and 10,839 m length of the river can be observed over the central crystalline gneiss complex. It is possibly a backward translation of Knickpoints through prolonged erosion of the basic metamorphosed rocks.

The Neora River has three major tributaries; the right bank tributary is the Mal River, and the two left bank tributaries are the Juranti and Kurti Rivers. The Mal River originates somewhere between the MCT and MBT at an elevation of 764.28 m and flows up to 72.53 m on the HFB (hill-fed river) while the SL index ranges between 1448.69 and 0.32 m (Fig. 7). The highest SL is recorded at a distance of 8860 m where it encounters MCT. The Juranti River originates at an elevation of 514 m and the entire channel is present on the HFB only. The SL values of this channel ranges between 1544.05 and 0.09 m. It crosses the Matiali thrust at 5980 m from its source where the SL value is 1187.38 m, recorded the highest. The Kurti River originates at an elevation of 412.43 m to 63 m. It crosses the Matiali and Chalsa Scarp at 3563 m and 9192 m length from its source respectively. The recorded highest SL value for Kurti River is 2736.92 m, found at the cross-over with a local fault (Figs. 7 and 8).

4.4 Aggregated Tectonic Index (ATI)

The ATI value ranges between 0.23 and 2.15 and it was subsequently categorized into five distinct classes (Fig. 9) spaced at equal intervals. Class 1 (0.23–0.61) represents very high tectonic sensitivity, class 2 (0.61–0.99) represents high tectonic sensitivity, class 3 (0.99–1.38) represents moderate tectonic sensitivity, class 4 (1.38–1.76) represents low tectonic sensitivity and class 5 (1.76–2.15) represents very low tectonic sensitivity. According to this classification, around 41.94% of the catchment area is in class 1 and 29.03% in class 2 which lies below the mountain-front. Other 17.74%, 4.84% and 6.45% are in classes 3, 4 and 5 respectively. The average ATI value for the sub-catchment above MCT (completely) is 1.38 with a range of 1.37. It signifies possibly a dominant effect of lithological control of the Higher Himalayan crystalline composition. Between the mountain front and the MCT, the average ATI is 0.96 and it ranges between 0.47 and 1.38. It is indicative of greater tectonic sensitivity due to the presence of Himalayan thrusts. Moreover, on the HFB the average ATI is 0.51 and it ranges between 0.23 and 0.95. Here, the presence of multiple thrusts which are southward translations of the HFB (Saha et al., 2024). This river drains a significant

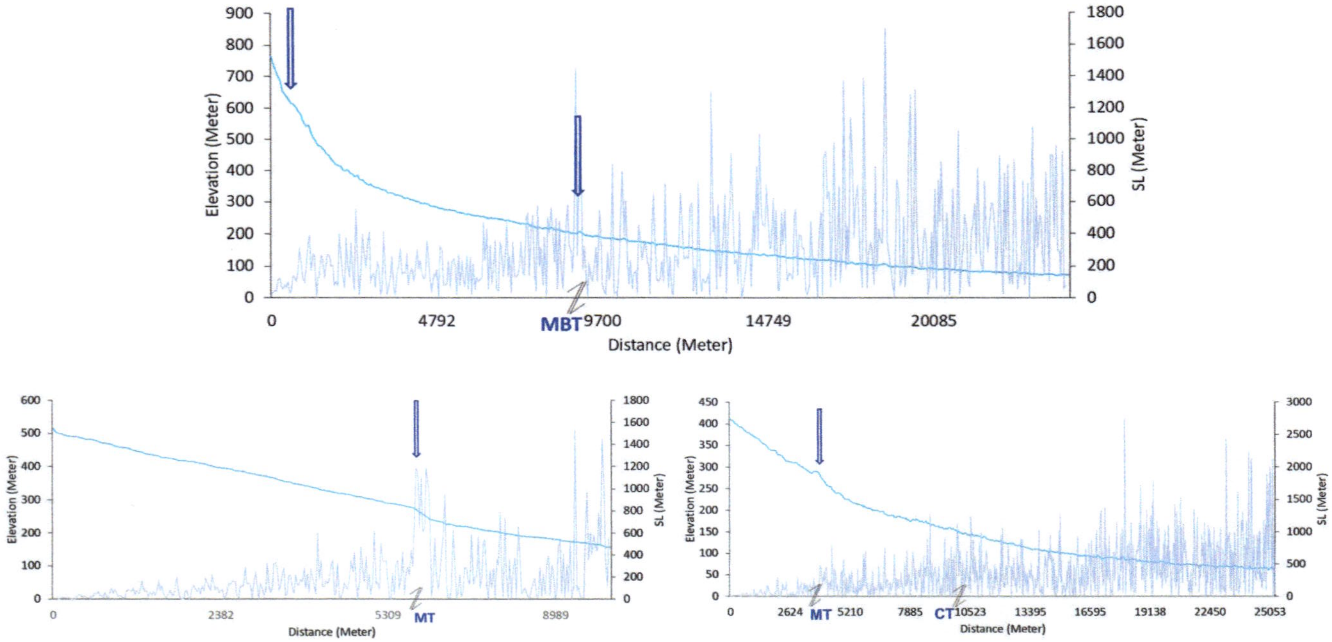

Fig. 7 Long profiles of (top) the Mal, Juranti (bottom-left) and Kurti River (bottom-right). All three tributaries are associated with major knickpoints while crossing the thrusts on the HFB

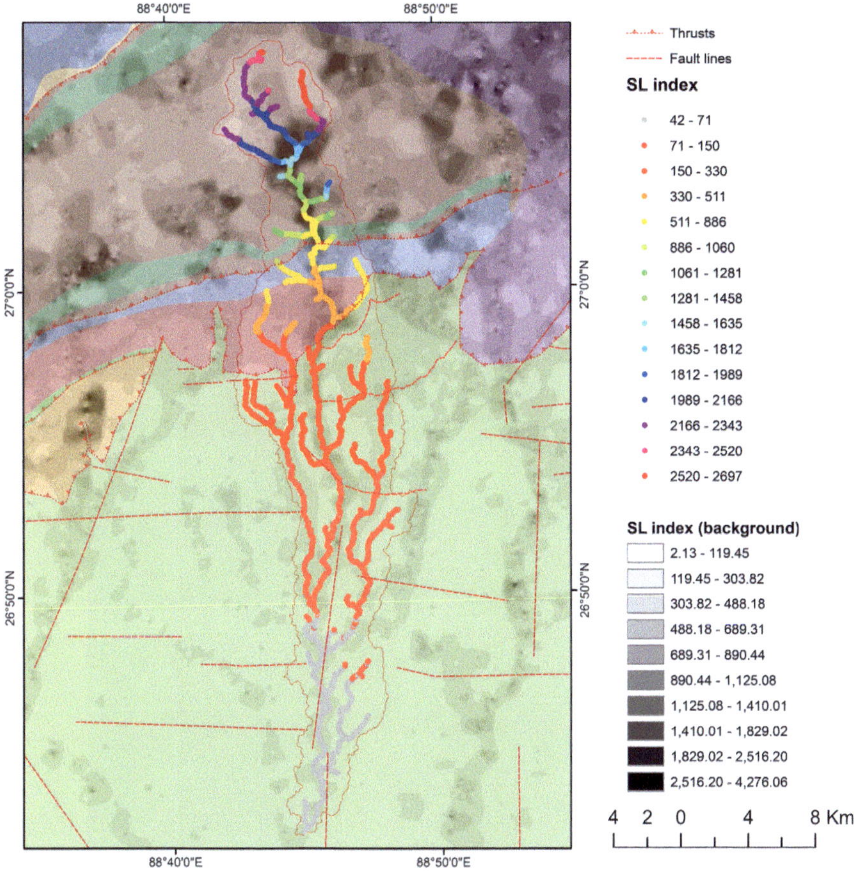

Fig. 8 Stream length gradient index map indicating major trunk streams of the Neora River catchment

part of the deformed Gorubathan surface and Matiali fan surface. On the HFB, the sub-watersheds that contain a connection with any of the thrusts record higher ATI compared to others within the same physiographic domain.

4.5 *Grouping of Sub-Catchments*

A four-group clustering was conducted based on the similarity of the sub-catchments as found in the PCA operation with the Varimax rotation component (Fig. 10 and Table 4). Here, the sub-watersheds under Quadrant 4 are characterized as elongated with high SL index values and higher topographic asymmetry. The sub-watersheds immediately on the south of the mountain front and thrusts located on the HFB;

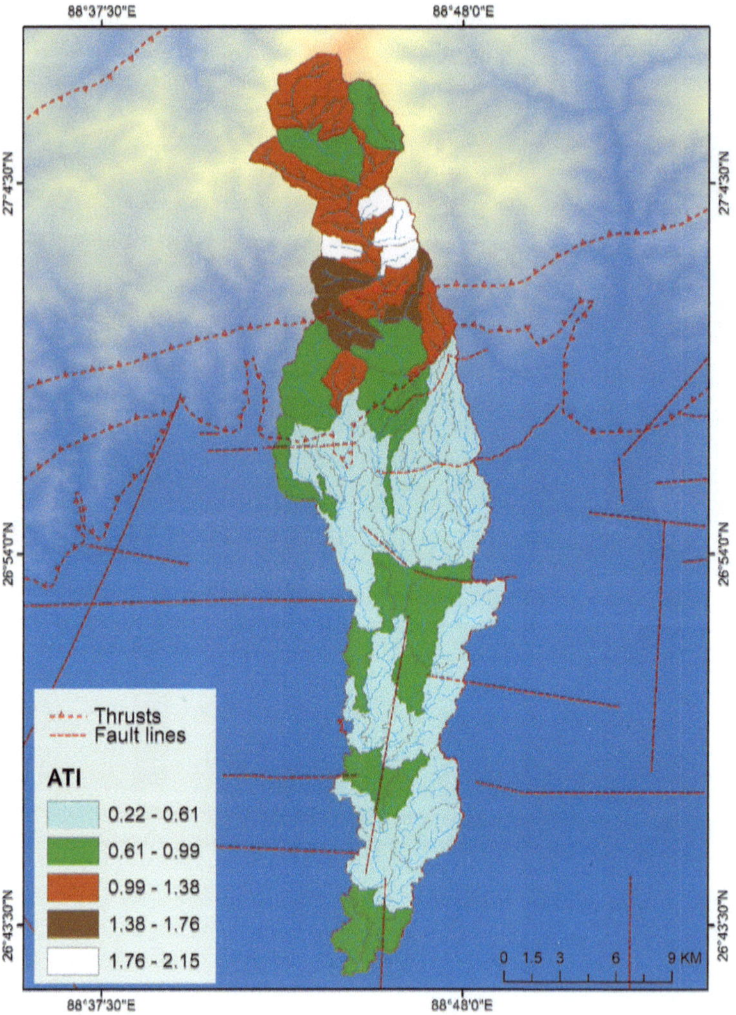

Fig. 9 Aggregated Tectonic Index (ATI), formulated through the Entropy Weightage Method (EWM); Higher values of ATI depict Low tectonic activity

Samsing scarp, Matiali Scarp, Chalsa Scarp, and Borodighi Thrust generate the Group 4. In quadrant 3, sub-watersheds are also quite sensitive to tectonic activities as the elongation ratio is high (low as compared to Q4), and the Circularity ratio is low (high as compared to Q4) with a higher concentration of SL. Sub-watersheds in Quadrants 1 and 2 are similar in terms of the catchment shape and asymmetric properties. These are mostly from the ATI class of 1–3 which are low in sensitivity to possible tectonic effects.

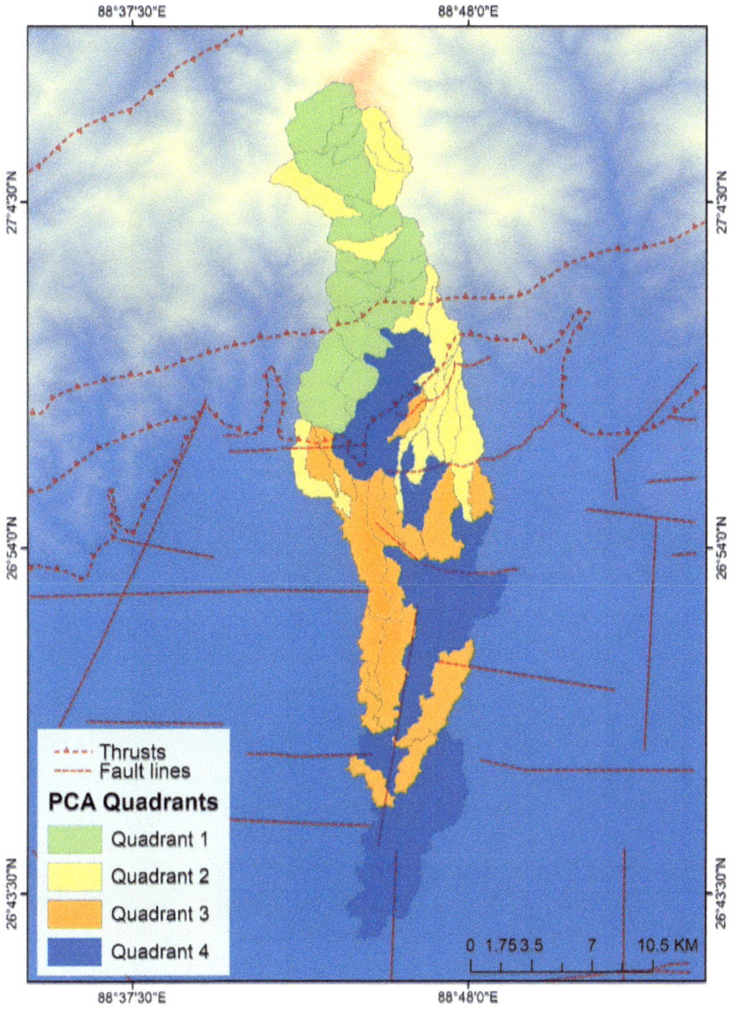

Fig. 10 Cluster mapping based on the PCA grouping shows each quadrant with similar morphotectonic characteristics

4.6 Distribution of Catchment Morphometric Values

Focusing on the sub-catchment scale, the values of different morphotectonic indices vary for three structural segments of the study area, which includes above MCT, MCT to MBT, and HFB. Kernel Density Estimate has been conducted to evaluate the local probability of the occurrence of tectonic activity and the degree of activity of the area with respect to the structural elements. The odds of different segments can be extracted from this estimation which would further give an insight into the

Table 4 PCA factor loadings and relative contribution of variables after Varimax rotation

Factor loadings after Varimax rotation:				Contribution of the variables (%) after Varimax rotation			
Variables	D1	D2	D3	Variables	D1	D2	D3
Form Factor (F_f)	0.160	0.948	−0.012	Form Factor (F_f)	0.744	35.599	0.012
Elongation Ratio (R_e)	0.153	0.953	−0.012	Elongation Ratio (R_e)	0.680	36.017	0.012
Relief Ratio (R_f)	0.903	0.345	−0.026	Relief Ratio (R_f)	23.584	4.706	0.053
STD	0.929	0.180	−0.076	STD	24.963	1.282	0.453
Hypsometric Integral (H_i)	0.498	−0.016	0.336	Hypsometric Integral (H_i)	7.155	0.011	8.823
Topographic Symmetry Factor (T)	−0.166	−0.586	−0.384	Topographic Symmetry Factor (T)	0.794	13.613	11.491
Asymmetry Factor (A_f)	−0.190	0.037	0.839	Asymmetry Factor (A_f)	1.039	0.055	55.002
Valley Width to Height Ratio (V_f)	−0.473	0.099	−0.532	Valley Width to Height Ratio (V_f)	6.454	0.392	22.104
Circularity Index (C_r)	0.601	0.458	0.161	Circularity Index (C_r)	10.445	8.319	2.016
SL index	0.914	0.014	0.020	SL index	24.142	0.007	0.033

structural control of the sub-catchments (Fig. 11). The relative concentration of the elongation ratio (R_e) values of the HFB is almost leptokurtic, followed by the above MCT region, whereas negative kurtosis can be seen with regard to the area between MCT and MBT. The lowest average elongation was observed on the HFB (Table 5). Congestion of parallel drainage systems and locally constrained tributaries on the HFB could be the reasons. The same pattern of concentration is observed for the form factor (F_f), thereby representing higher structural constraints on the alluvial plains. The concentration of the R_f is leptokurtic on the HFB with lowest mean, 26.5 compared to its platykurtic concentration on the mountainous part. This is due to the break of slope at the mountain front and low relief on the HFB with multiple thrust ridges across the river. Positively skewed distribution can be observed with regard to the H_i values concerning the area under HFB. The mean H_i between MCT to MBT, and on the HFB recorded mean values less than 0.40, lower compared to above MCT segment. For the Circularity Ratio (C_r), the concentration under HFB is low which represents positive skewness with a lowest mean 0.13 compared to other structural compartments, indicating more circular catchments with lack of valley confinity (Table 4). A very similar pattern was also witnessed in the case of the NSLI. The highest mean NSLI was observed above the MCT where the channel slope is highest with maximum relief concentration. A strikingly different pattern was revealed from the V_f values, wherein the concentration in the above MCT segment showcased a pointed rise depicting positive kurtosis. The mean values were quite high for HFB and

MCT to MBT segment, representing wider valleys with lack of valley confinement compared to the segment above MCT. The IQR of V_f is lowest in the zone above MCT due to similar valley configuration (deeply incised and confined) but the lowest CV of it on the HFB indicates occurrence of isolated catchments which are incised due to adjustment with respect to structural perturbation. In terms of T, the values in the segment above MCT and on the HFB are leptokurtic in nature where the distribution is bi-modal on the MCT. The maximum CV of it in the segment between MCT and MBT is the maximum due to presence of catchment with varied order of asymmetry. Similar results were also found in case of A_f, but the mean values of it in all the segments are less than 50, indicates dominant tilting towards right hand side of the rivers while looking downstream.

5 Discussion

In the Darjiling-Sikkim Himalayas, the MCT zone is marked by the localisation of high stress along with several protoliths, deformation mechanisms, and an assortment of imbricated lithologies (Mukhopadhyay et al., 2017). This differential lithological contact coupled with strains, faults, and thrusts has led to the development of typical controls on the fluvial environments in this particular region (Srivastava et al., 2017). The present study has applied a morphotectonic approach to examine the tectonic controls on the Neora River catchment whose area lies across the MCT and MBT. In the development of the fluvial environment here, past studies revealed the notable impact of litho-tectonic control owing to the presence of discontinuities and surface stresses (Saha et al., 2021). The presence of isolated pockets of active tectonism along with discontinuities has resulted in shear zones of varied intensity and nature (Mukhopadhyay et al., 2017). Furthermore, in this region, rock structures and rock resistances are highly related to the formation of the recent drainage networks. The Matiali scarp which represents the trace of MBT, adds to the dynamism of the landscape, with respect to its fan abandonment and incision which led to the possible orientation of the Neora river (Kar et al., 2014). Additionally, the Chalsa scarp can also be found in this region which represents the Himalayan Frontal Thrust (HFT) (Nakata, 1972). The interplay of the complex scenarios of variegated deformation mechanisms, stress factors, and active tectonism has been quite imperative in forming the watershed processes in the study region.

This region has undergone significant tectonic perturbations in addition to lithological variations. Subsequently, focussing in terms of the structural zones of the river catchment would provide a deeper insight into the litho-tectonic controls of the region. Low R_e and F_f mean values in the HFB point towards deformed pockets and associated structural confinement. However, the low deviation of C_r values in the HFB exhibits no structural disturbance which is contradictory. However, the sub-catchments located in the eastern part of the main channel of Neora River recorded higher values thereby representing structural disturbances in the R_f and T mean values. Some of the sub-catchments located on the alluvial reach are impacted by

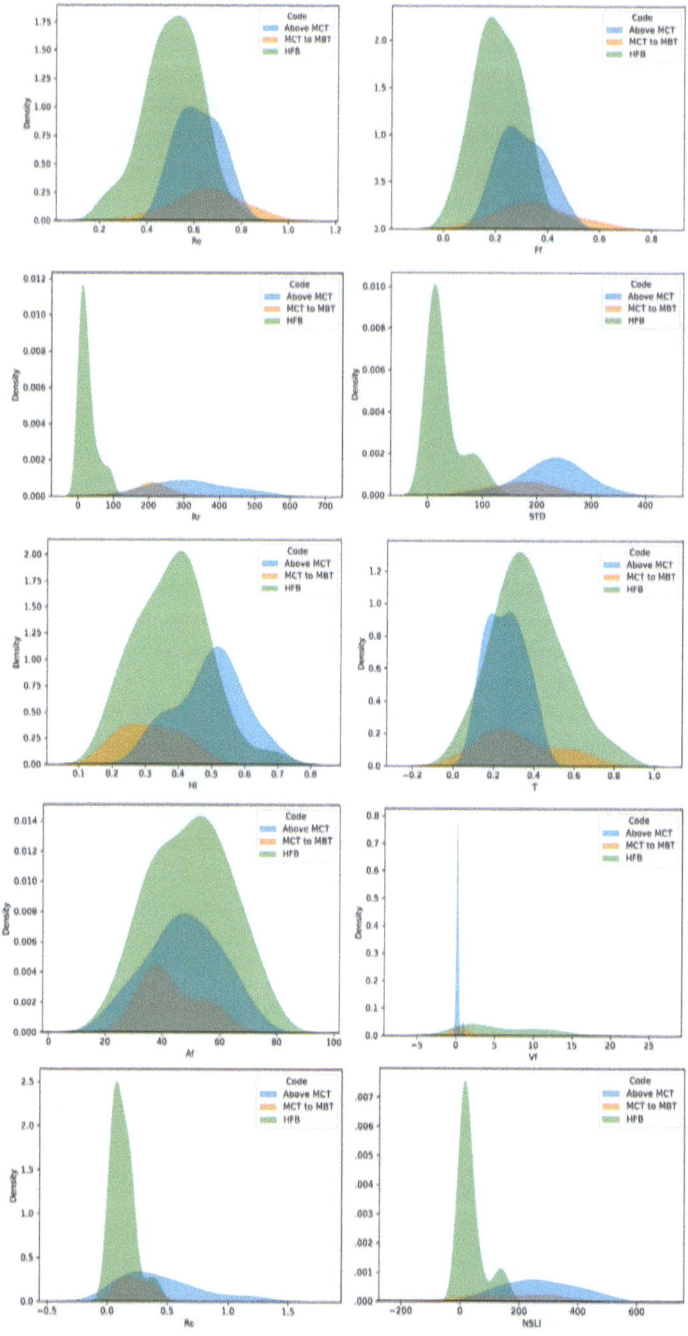

Fig. 11 Kernel Density Estimate (KDE) of the morpho-tectonic parameters for different structural elements. The volume shows the relative concentration of the values and it's both tails show the distributional estimates of positive and negative ends

Table 5 Descriptive statistics of the selected morphotectonic index concerning different structural segments

Index	Above MCT						MCT-MBT						HFB					
	Q1	Q3	IQR	Mean	SD	CV	Q1	Q3	IQR	Mean	SD	CV	Q1	Q3	IQR	Mean	SD	CV
F_f	0.23	0.37	0.13	0.30	0.08	0.27	0.28	0.46	0.17	0.35	0.13	0.37	0.14	0.27	0.12	0.21	0.09	0.42
R_e	0.54	0.68	0.14	0.61	0.08	0.13	0.60	0.76	0.16	0.65	0.12	0.19	0.43	0.59	0.15	0.50	0.11	0.22
R_f	224.85	402.2	177.38	323.75	106.05	0.32	192.66	227.30	34.63	201.2	53.22	0.26	8	34.5	26.59	26.5	24.71	0.93
STD	190.23	267.4	77.25	228.29	58.36	0.25	130.04	200.52	70.47	168.15	53.12	0.31	7.65	36.3	28.65	29.31	31.46	1.07
H_i	0.42	0.56	0.14	0.49	0.09	0.19	0.22	0.42	0.19	0.31	0.08	0.25	0.30	0.44	0.14	0.38	0.10	0.28
T	0.17	0.32	0.16	0.25	0.08	0.34	0.24	0.57	0.33	0.33	0.16	0.51	0.25	0.48	0.23	0.37	0.17	0.46
A_f	38.76	56.8	18.09	47.39	12.19	0.25	36.47	54.63	18.15	43.1	8.87	0.20	37.4	60.1	22.66	49.73	13.5	0.27
V_f	0.09	0.22	0.13	0.19	0.19	1.02	0.37	1.10	0.73	1.24	1.72	1.38	1.65	10.87	9.22	6.33	5.19	0.81

(continued)

Table 5 (continued)

Index	Above MCT						MCT-MBT						HFB					
	Q1	Q3	IQR	Mean	SD	CV	Q1	Q3	IQR	Mean	SD	CV	Q1	Q3	IQR	Mean	SD	CV
C_r	0.20	0.64	0.44	0.46	0.33	0.71	0.12	0.36	0.23	0.22	0.11	0.50	0.06	0.18	0.12	0.13	0.09	0.75
NSLI	189.8	370.6	180.7	277.6	118.9	0.42	48.3	308.8	260.4	181.9	120.9	0.66	6.62	43.8	37.2	36.7	44.04	1.20

thrust translations (Srivastava et al., 2017). Apart from that, multiple faults (parallel and sub-parallel) have affected the fluvial trajectory (Fig. 12a). H_i values less than 0.40 imply ongoing channel adjustments to surface warping. Looking at the sub-catchments located in the north of the MCT shows a growing stability in terms of drainage adjustments and catchment evolution since the Hi values are relatively higher. Similarly, isolated high values of V_f on the HFB indicate active channel entrenchment. Channels crossing the Matiali Thrust have adjusted their courses through downcutting (Fig. 12c) since the major flow arteries are antecedent to the structures (Chakraborty Goswami et al., 2012). Overall, the values have accurately reflected in the places where tectonic deformations have occurred. With regards to the topographic asymmetry, the values in the segment MCT to MBT vary majorly, chiefly in the alluvial reach zones. Distinct differences along the east and west valley divide can be noticed in a few of the sub-catchments and the presence of high peaked V_f values with their pertinent variations above the MCT indicate structural interferences. The difficult terrain that predominates in this area exhibits irregularities in terms of surface deformation and deviate from graded condition. This ultimately results in an elongated catchment shape, indicating a tectonically active environment.

The sub-catchments near the MCT and MBT are impacted by significant slope breaks, as the MCT and MBT are sensitive to higher strain properties (Seeber & Gorntiz, 1983). This, in turn, leads to fluctuations in thc morphotectonic values. The groupings of the sub-catchments put forth the fact that the catchments under the fourth quadrant display high SL values and low C_r values, thereby portraying high tectonic activity. The greater decline in the channel gradient on the valley side slopes of the watershed results in rougher terrain and more elongated sub-watersheds as parallel drainage networks predominate. This phenomenon suggests that the lithology along the valley side slopes is has active controls on the catchment properties. It controls the intensity of valley slope processes and determines the erodibility of the slopes that affects the lower order drainage development. Subsequently, active neo-tectonism can be observed in the sensitive zone of the third quadrant. Furthermore, knickpoint distribution and overall channel longitudinal form are regarded as signatures of active tectonism (Keller & Pinter, 1996). Many times, while using DEMs of high resolution on small watersheds, it happens that the effect of active tectonism gets subdued and the hillslope processes come to the forefront (Bishop, 2007). In this study, the usage of medium-resolution DEM pinpoints the presence of knickpoints in highlighting tectonic imprints. The SL index anomaly is linked to small deformations and lithological resistance, while no large depressions or structural deformations are visible on the surface (Fig. 3). There are SL outliers in the area between MCT and MBT as well as on the HFB, which is consistent with the location of the river at significant thrusts. Strong tectonic disturbances that are isolated in relation to the regions upstream and downstream are represented by a greater number of outliers. Low variation and a greater number of outliers are seen in the stretch on the HFB, which suggests topographic abnormalities brought about by neotectonic deformation mainly due to the presence of blind thrusts ridges (Fig. 13). Similar results were also seen in the work of (Saha et al., 2022) where rivers were found associated with significant breaks at their crossings (Neora and Murti River). Since major

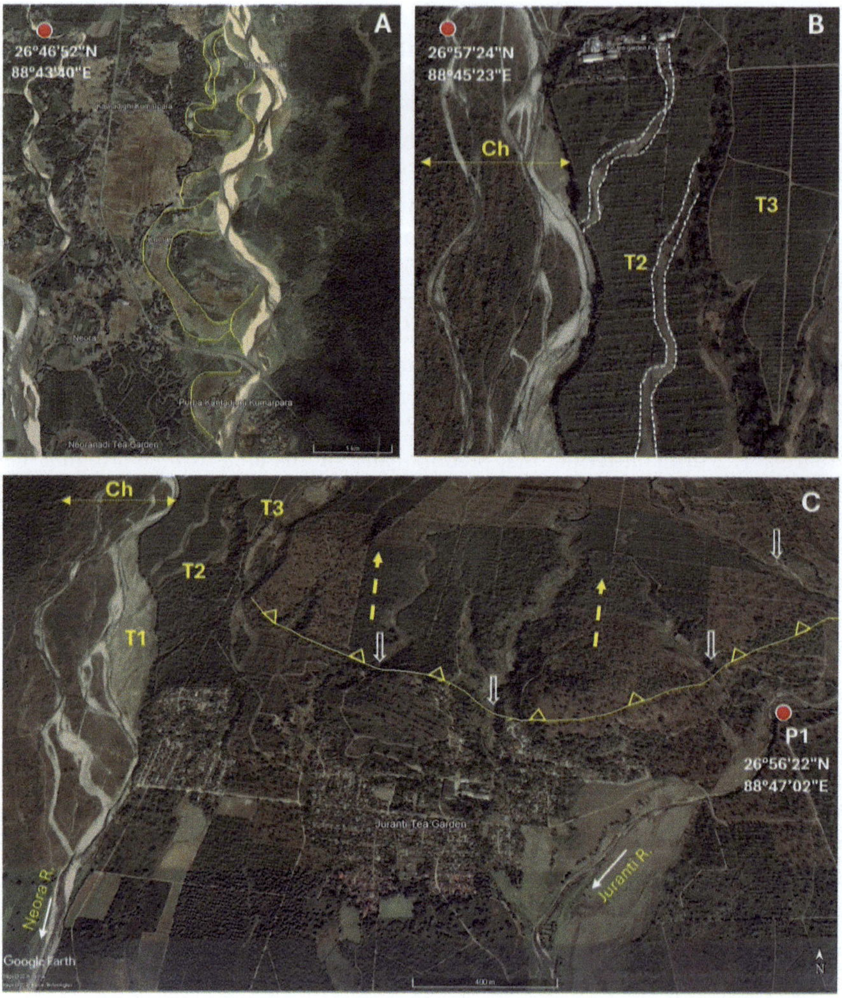

Fig. 12 **a** wandering channel properties of the Neora, following an N-S fault line in the downstream reach of its confluence with Kurti River. The westward restriction to channel shifting has created all the cut-offs along the right bank; **b** association of unpaired terraces along the river on the alluvial fan surface (roughly 4 km downstream of the mountain front); **c** tributary channels have developed incised valleys (white arrow) while they cross the Matiali thrust (yellow line with triangles). The immediately upper part of the scarp has with northward slope (yellow dashed arrow) because of back-tilting. The absence of rivers being developed in that direction was showcased as evidence for the channels to be antecedent to these thrusts. T1, T2 and T3 are terrace level 1, 2 and 3

rivers are mostly antecedent to the structures; on the Gorubathan recess (Srivastava et al., 2017) and Matiali fan surface (Kar et al., 2013) rivers have adjusted through downcutting and develop breaks at the frontal part of the scarps. The major break in slope is developed upstream of the MCT within the higher Himalayan crystalline surface which is possibly associated with headward migration of the knickpoints. A possible trace of knickpoint migration can be assumed for the Neora River on the Higher Himalayan Crystalline surface. Since, there is not much evidence of it, it is difficult to ascertain this phenomenon but it can be taken up for future research. The ATI and PCA groupings also indicate notable differences in terms of lithological and tectonic controls above the MBT. The sub-catchments above MBT are associated with higher ATI values compared to the sub-catchments on the HFB. The morpho-tectonic indices often portray a complex interaction of lithology and tectonics where it often becomes confusing to ascertain the distinctions of any of the processes only analysing the ATI (Diercks et al., 2023). Here, the PCA grouping plays a significant role. The sub-catchments above the mountain front (roughly guided by MBT) are grouped into Quadrant 1 and 2. Similarly, the sub-catchments on the eastern corner, just below the mountain front (on the Matiali fan surface) have registered in Quadrant 2. These catchments are part of the radial drainage systems (slope guided) above the Matiali Thrust (Chakrabarti Goswami et al., 2013). Thus, Quadrant 1 and Quadrant 2 possibly indicates dominant role of lithology at the catchment scale rather than forming isolated bumps on the river profiles. It is very much possible that the high-grade metamorphic composition with high sheer properties both along the MCT is responsible for lithographic controls to be more dominant while relatively more recent thick skinned thrust mechanism on the HFB has more dominant signatures of surface warping on the fluvial systems. Similar results were discovered by Pike et al. (2010) in Puerto Rico, Martins et al. (2017) in central-eastern Portugal, Guha and Jain (2020) in Western Ghats, India, where both lithological and tectonic features have a distinct impact on the fluvial landscape.

Apart from the lithology and tectonic perturbations, rainfall has a significant control on the Himalayan orogen as well as on its terrain evolution. The rainfall gradient is stronger on the eastern Himalayas but the spatiality of its variation is much larger than compared to a single river basin across the Himalayan thrusts. Thus, it would not be logical enough to venture for any correlation between climate and terrain evolution for a small areal extent as considered in this study.

6 Conclusions

The present study puts forward the peculiarity of the Neora river catchment (both at watershed and sub-watershed scale) and its complex response to active tectonics and rock structure. Here, the uniqueness lay in understanding the watershed dynamics of the Neora River, which portrayed a complex reaction of tectonic and rock structure,

Fig. 13 **a** different terrace surfaces at P1 (Fig. 12b), viewing North; **b** Matiali thrust as viewed from P2 (Fig. 12c); viewing North. T1, T2 and T3 are terrace level 1, 2 and 3. Date of Photograph—18.02.2023

is the main area of research attention. When viewed with respect to the entire watershed, the morphotectonic indices accurately depict that lithology-controlled hillslope processes and river erosion are dominant, making it less responsive to tectonic disturbances. Tectonic sensitiveness is focused on specific regions of differential stress caused by Himalayan thrust faulting. Moreover, the degree of tectonic impacts is confined to few pockets because of the presence of different structural elements, like the southern and central parts of the catchment, chiefly around the HFB. Additionally, the opted analyses also recommend future studies with better resolution using thorough geological descriptions of the main tectonic features and quantitative metrics for channel long profiles in accordance to structural units. The findings of this study could provide a solid foundation for comprehending the morphological processes occurring in the area. Although the conducted methodology has helped uncover the litho-tectonic responses to a great extent surrounding the Himalayan thrusts, the fieldwork related to the present study could not be performed exhaustively due to the rugged terrain of the river catchment leading to high inaccessibility. However, it can be asserted that intensive field study combined with advanced field logistics might provide better and more precise results.

Acknowledgements The authors have no one to acknowledge

Declarations The authors have no conflict of interest.

Funding No external funding has been utilised for conducting this work.

Data Availability The entire data is prepared by the authors and thus it will be made available on request. The data will not be made available for any commercial purpose where the authors are not a part of the venture.

References

Ahmad, S., & Bhat, M. I. (2012). Tectonic geomorphology of the Rambiara basin, SW Kashmir Valley reveals emergent out-of-sequence active fault system. *Himalayan Geology., 33*, 162–172.

Alaei, M., Dehbozorgi, M., Ghassemi, M. R., & Nozaem, R. (2017). Evaluation of relative tectonic activity of Buin Zahra-Avaj area, northern Iran. *Arabian Journal of Geosciences, 10*(10). https://doi.org/10.1007/s12517-017-3025-x

Al-Attar, Z. T., Othman, A. A., Al-Hakari, S. H. S., Obaid, A. K., Salar, S. G., & Liesenberg, V. (2022). A neotectonic statistical assessment through watershed geomorphic analysis: A case study in the Greater Zab River Basin, Turkey–Iraq. *Environmental Earth Sciences, 81*(13). https://doi.org/10.1007/s12665-022-10478-7

Ali, S. A., & Ikbal, J. (2020). Assessment of relative active tectonics in parts of Aravalli mountain range, India: implication of geomorphic indices, remote sensing, and GIS. *Arabian Journal of Geosciences, 13*(2). https://doi.org/10.1007/s12517-019-5028-2

Alipoor, R., Poorkermani, M., Zare, M., & Hamdouni, R. E. (2011). Active tectonic assessment around Rudbar Lorestan dam site, High Zagros Belt (SW of Iran). *Geomorphology, 128*(1–2), 1–14. https://doi.org/10.1016/j.geomorph.2010.10.014

Alizadeh, A., Moghadam, H. H., & Seraj, M. (2020). DEM-derived geomorphic indices for assessment of tectonic activity at the Dara anticlinal oil structure within the Zagros fold-thrust belt, southwestern Iran. *Arabian Journal of Geosciences, 13*(4). https://doi.org/10.1007/s12517-020-5156-8

Altın, T. B. (2011). Geomorphic signatures of active tectonic in drainage basins in the southern Bolkar Mountain, Turkey. *Journal of the Indian Society of Remote Sensing, 40*(2), 271–285. https://doi.org/10.1007/s12524-011-0145-8

Anand, A. K., & Pradhan, S. P. (2019). Assessment of active tectonics from geomorphic indices and morphometric parameters in part of Ganga basin. *Journal of Mountain Science, 16*(8), 1943–1961. https://doi.org/10.1007/s11629-018-5172-2

Arlegui, L. E., & Soriano, M. A. (1998). Characterizing lineaments from satellite images and field studies in the central Ebro basin (NE Spain). *International Journal of Remote Sensing, 19*(16), 3169–3185. https://doi.org/10.1080/014311698214244

Ayaz, S., & Dhali, M. K. (2019). Longitudinal profiles and geomorphic indices analysis on tectonic evidence of fluvial form, process and landform deformation of Eastern Himalayan Rivers,

India. *Geology Ecology and Landscapes, 4*(1), 11–22. https://doi.org/10.1080/24749508.2019.1568130

Azañón, J. M., Pérez-Peña, J. V., Giaconia, F., Booth-Rea, G., Martínez-Martínez, J. M., & Rodríguez-Peces, M. J. (2012). Active tectonics in the central and eastern Betic Cordillera through morphotectonic analysis: The case of Sierra Nevada and Sierra Alhamilla. *Journal of Iberian Geology, 38*(1). https://doi.org/10.5209/rev_jige.2012.v38.n1.39214

Bali, B. S., Wani, A. A., Khan, R. A., & Ahmad, S. (2016). Morphotectonic analysis of the Madhumati watershed, northeast Kashmir Valley. *Arabian Journal of Geosciences, 9*(5). https://doi.org/10.1007/s12517-016-2395-9

Barman, J., Biswas, B., & Das, J. (2022). *Active tectonics and associated channel shifting pattern of Neora River Basin, Darjeeling Himalaya* (pp. 19–31). Springer eBooks. https://doi.org/10.1007/978-3-031-15377-8_2

Bishop, P. (2007). Long-term landscape evolution: Linking tectonics and surface processes. *Earth Surface Processes and Landforms, 32*, 329–365. https://doi.org/10.1002/esp.1493

Biswas, M. (2015). Impact of Neotectonism in the discussion of geomorphological processes as a feedback system: North Bengal Foothills, West Bengal. *GSTF Journal of Geological Sciences, 2*(1). https://doi.org/10.5176/2335-6774_2.1.22

Biswas, R., & Chakraborty, S. (2016). Watershed prioritization based on geo-morphometry and land use parameters—An approach to watershed development using remote sensing and GIS, Neora Watershed, Darjeeling And Jalpaiguri Districts, West Bengal, India. *IOSR Journal of Applied Geology and Geophysics (IOSR-JAGG), 4*, 2321–2990. https://doi.org/10.9790/0990-0403013649

Bull, W. B., & McFadden, L. D. (1977). Tectonic geomorphology north and south of the Garlock fault, California. In *Proceedings 8th annual geomorphology symposium* (pp. 115–138). Binghamton, 23–24 September 1977. https://doi.org/10.4324/9780429299230-5

Chakrabarti Goswami, C., Mukhopadhyay, D., & Poddar, B. C. (2013). Geomorphology in relation to tectonics: A case study from the eastern Himalayan foothills of West Bengal, India. *Quaternary International, 298*, 80–92. https://doi.org/10.1016/j.quaint.2012.12.020

Charlton, R. (2007). *Fundamentals of fluvial geomorphology*. Routledge eBooks. https://doi.org/10.4324/9780203371084

Cox, R. T. (1994). Analysis of drainage basin symmetry as a rapid technique to identify areas of possible quaternary tilt block tectonics: An example from the Mississippi embayment. *Geological Society American Bulletin, 106*, 571–581. https://doi.org/10.1130/0016-7606(1994)106%3c0571:AODBSA%3e2.3.CO;2

Das, B. C., Barman, S. D., & Islam, A. (2022). Influence of neotectonics on channel evolution of Kameng River, North–East Himalaya. In *Society of earth scientists series* (pp. 397–417). https://doi.org/10.1007/978-3-030-95435-2_15

Das, S., & Gupta, K. (2019). Morphotectonic analysis of the Sali River basin, Bankura district, West Bengal. *Arabian Journal of Geosciences, 12*(7). https://doi.org/10.1007/s12517-019-4406-0

Dhali, M. K., & Mukhopadhyay, S. (2020). Formation, Migration, and morphodynamic alteration of 50 channel bars in Darjeeling Himalayan Piedmont Zone, India. *Air Soil and Water Research, 13*, 117862212094143. https://doi.org/10.1177/1178622120941436

Diercks, M. L., Mildon, Z. K., Boulton, S. J., Hussain, E., Alçiçek, C., Yıldırım, C., & Aykut, T. (2023). Constraining historical earthquake sequences with Coulomb stress models: An example from western Türkiye. *Journal of Geophysical Research: Solid Earth, 128*(11), Article e2023JB02662. https://doi.org/10.1029/2023JB026627

Guha, D., Bardhan, S., Basir, S. R., De, A. K., & Sarkar, A. (2007). Imprints of Himalayan thrust tectonics on the Quaternary piedmont sediments of the Neora-Jaldhaka Valley, Darjeeling-Sikkim Sub-Himalayas, India. *Journal of Asian Earth Sciences, 30*(3–4), 464–473. https://doi.org/10.1016/j.jseaes.2006.11.010

Guha, S., & Jain, V. (2020). Role of inherent geological and climatic characteristics on landscape variability in the tectonically passive Western Ghat, India. *Geomorphology, 350*, 106840.

Hack, J. T. (1973). Stream profile analysis and stream-gradient index. *Bulletin United States Geological Survey, 1*(4), 421–429.
Holbrook, J., & Schumm, S. A. (1999). Geomorphic and sedimentary response of rivers to tectonic deformation: A brief review and critique of a tool for recognizing subtle epeirogenic deformation in modern and ancient settings. *Tectonophysics, 305*(1–3), 287–306. https://doi.org/10.1016/S0040-1951(99)00011-6
Horton, R. (1932). Drainage basin characteristics. *Transactions, American Geophysical Union, 13*, 350–361. https://doi.org/10.1029/TR013i001p00350
Jain, A. K., Ahmad, T., Singh, S., Ghosh, S. K., Patel, R. C., Kumar, R. O. H. T. A. S. H., ... & Bhargava, O. N. (2012, September). Evolution of the Himalaya. In *Proceedings of the Indian national science academy* (Vol. 78, No. 3, pp. 259–275).
Jiongxin, X. (1990). An experimental study of complex response in river channel adjustment downstream from a reservoir. *Earth Surface Processes and Landforms, 15*(1), 43–54.
Kar, R., Chakraborty, T., Chakraborty, C., Ghosh, P., Tyagi, A. K., & Singhvi, A. K. (2014). Morpho-sedimentary characteristics of the Quaternary Matiali fan and associated river terraces, Jalpaiguri, India: Implications for climatic controls. *Geomorphology, 227*, 137–152.
Keller, E. A., & Pinter, N. (1996). *Active tectonics: Earthquakes, uplift and landscape.* Prentice Hall.
Klinkenberg, B., & Goodchild, M. F. (1992). The fractal properties of topography: A comparison of methods. *Earth Surface Processes and Landforms, 17*(3), 217–234. https://doi.org/10.1002/esp.3290170303
Kothyari, G. C., & Rastogi, B. K. (2013). Tectonic control on drainage network evolution in the Upper Narmada Valley: Implication to neotectonics. *Geography Journal, 2013*, 1–9. https://doi.org/10.1155/2013/325808
Kothyari, G. C., Kotlia, B. S., Talukdar, R., Pant, C. C., & Joshi, M. (2020). Evidences of neotectonic activity along Goriganga River, Higher Central Kumaun Himalaya, India. *Geological Journal, 55*(9), 6123–6146. https://doi.org/10.1002/gj.3791
Lee, C., & Tsai, L. L. (2009). A quantitative analysis for geomorphic indices of longitudinal river profile: A case study of the Choushui River, Central Taiwan. *Environmental Earth Sciences, 59*(7), 1549–1558. https://doi.org/10.1007/s12665-009-0140-3
Manna, D., & Ghosh, K. (2023). Comparison between two rivers and rivers' basin in the light of the evidence of tectonic activity: Chel and Neora. *Khoj an International Peer Reviewed Journal of Geography, 10*(1), 42–65. https://doi.org/10.5958/2455-6963.2023.00004.8
Martins, A., Cabral, J., Cunha, P. P., Stokes, M., Borges, J. F., Caldeira, B., & Martins, A. C. (2017). Tectonic and lithological controls on fluvial landscape development in central-eastern Portugal: Insights from long profile tributary stream analyses. *Geomorphology, 276*, 144–163. https://doi.org/10.1016/j.geomorph.2016.10.012
Mishra, S., Sarkar, U., Taraphder, S., Datta, S., Swain, D., Saikhom, R., Panda, S., & Laishram, M. (2017). Principal component analysis. *International Journal of Livestock Research, 7*, 60–78. https://doi.org/10.5455/ijlr.20170415115235
Mukhopadhyay, D. K., Chakraborty, S., Trepmann, C., Rubatto, D., Anczkiewicz, R., Gaidies, F., Dasgupta, S., & Chowdhury, P. (2017). The nature and evolution of the main central thrust: Structural and geochronological constraints from the Sikkim Himalaya, NE India. *Lithos, 282*, 447–463.
Mukul, M. (2010). First-order kinematics of wedge-scale active Himalayan deformation: Insights from Darjiling-Sikkim-Tibet (DaSIT) wedge. *Journal of Asian Earth Sciences, 39*, 645–657. https://doi.org/10.1016/j.jseaes.2010.04.029
Nakata, T. (1972). Geomorphic history and crustal movements of the foothills of the Himalayas. *Tohoku University Science Reports, 7th Series(Geography), 22*, 39–177.
Panek, T. (2004). The use of morphometric parameters in tectonics geomorphology (on the example of the Western Beskydy MTS). *Journal of Geographia, 1*, 111–126.

Partabian, A., Nourbakhsh, A., & Ameri, S. (2016). GIS-based evaluation of geomorphic response to tectonic activity in Makran Mountain Range, SE of Iran. *Geosciences Journal, 20*(6), 921–934. https://doi.org/10.1007/s12303-016-0106-x

Pike, A. S., Scatena, F. N., & Wohl, E. E. (2010). Lithological and fluvial controls on the geomorphology of tropical montane stream channels in Puerto Rico. *Earth Surface Processes and Landforms, 35*(12), 1402–1417.

Saha, U., Bhattacharya, H., & Islam, A. (2024). Neotectonic impact on drainage development in the Eastern Himalayan foreland basin. *Geomorphology, 462*, 109328. https://doi.org/10.1016/j.geomorph.2024.109328

Saha, U. D., Biswas, N., Mondal, S., & Bhattacharya, S. (2022). Morphotectonic expressions of the drainage basins and channel long profile forms on a selected part of Sikkim-Bhutan Himalayas. In *Society of Earth Scientists series* (pp. 175–206). https://doi.org/10.1007/978-3-030-95435-2_7

Saha, U. D., Bhattacharya, S., Islam, A., & Mondal, S. (2023). Allogenic and autogenic impact on morphodynamics of a tropical river on the Himalayan frontal plain, India. *International Journal of Environmental Science and Technology, 21*(3), 3289–3316. https://doi.org/10.1007/s13762-023-05187-y

Saha, U. D., Neogy, S., Kar, J., & Mukhopadhyay, U. (2021). Lithological and tectonic response on catchment characteristics of Rishi Khola, Sikkim, India. *Journal of Mountain Science, 18*(11), 3003–3024. https://doi.org/10.1007/s11629-021-6666-x

Schwanghart, W., & Scherler, D. (2017). Bumps in river profiles: Uncertainty assessment and smoothing using quantile regression techniques. *Earth Surface Dynamics, 5*(4), 821–839. https://doi.org/10.5194/esurf-5-821-2017

Seeber, L., & Gornitz, V. (1983). River profiles along the Himalayan arc as indicators of active tectonics. *Tectonophysics, 92*(4), 335–367.

Sharma, S., Baruah, S., & Sarma, J. N. (2022). Neotectonic evidences of some major rivers of North East India. *Current Science, 122*(8), 918. https://doi.org/10.18520/cs/v122/i8/918-933

Simoes, M., Sassolas-Serrayet, T., Cattin, R., Le Roux-Mallouf, R., Ferry, M., & Drukpa, D. (2021). Topographic disequilibrium, landscape dynamics and active tectonics: An example from the Bhutan Himalaya. *Earth Surface Dynamics, 9*(4), 895–921. https://doi.org/10.5194/esurf-9-895-2021

Srivastava, P., Ray, Y., Phartiyal, B., & Sundriyal, Y. P. (2016). Rivers in the Himalaya: Responses to neotectonics and past climate. *Proceedings of the Indian National Science Academy, 82*(3). https://doi.org/10.16943/ptinsa/2016/48483

Strahler, A. (1952). Dynamic basis of geomorphology. *Geological Society of America Bulletin, 63*, 923–938. https://doi.org/10.1130/0016-7606(1952)63[923:DBOG]2.0.CO;2

Taloor, A. K., Sharma, R., & Kothyari, G. C. (2023). Tectono-geomorphic and active deformation studies in the Ujh basin of Northwestern Himalaya. *Quaternary Science Advances, 12*, 100121. https://doi.org/10.1016/j.qsa.2023.100121

Thiede, R. C., Arrowsmith, J. R., Bookhagen, B., McWilliams, M. O., Sobel, E. R., & Strecker, M. R. (2005). From tectonically to erosionally controlled development of the Himalayan orogen. *Geology, 33*(8), 689–692. https://doi.org/10.1130/G21483AR.1

Valdiya, K. S. (1986). Neotectonic activities in the Himalayan belt. In *New tectonics in South Asia* (pp. 241–267).

Valdiya, K. S. (2002). Emergence and evolution of Himalaya: Reconstructing history in the light of recent studies. *Progress in Physical Geography Earth and Environment, 26*(3), 360–399. https://doi.org/10.1191/0309133302pp342ra

Yang, F., Ma, Z., & Xie, M. (2022). Image classification with parallel KPCA-PCA network. *Computational Intelligence, 38*(2), 397–415. https://doi.org/10.1111/coin.12503

Zhang, Q., Xu, W., & Zhang, J. (2014). Method for determining the weight of functional objectives on manufacturing system. *Scientific World Journal* 1–9

Himalayan Active Fault Zones: Basic Understanding, Review and Concerns

Arindam Sarkar and Shreya Basak

Abstract The location of the active fault is the most vulnerable to earthquakes. Integrated approaches to assessing fault activity in the Himalayan region are geologic, geomorphic, geodetic, and seismological data analysis. The most significant active continental megathrust fault (The Main Himalayan Thrust) is beneath the Himalayan range. Deformation of continental crust occurred in three different formats (Himalayan Frontal Thrust, Main boundary thrust, and Main Central thrust). Himalayan frontal thrust is a geologic fault that defines the boundary between the Indian and Eurasian plates. Three significant seismic gaps in the Himalayan region are the Assam gap, the Central gap, and the Kashmir gap. The Himalayan state of Uttarakhand lies in the Central Gap. It has the maximum potential to experience magnitude earthquakes very shortly, as per geological, geomorphological, and GPS data analysis. Significant stress accumulated in this area due to the subduction of the Indian plate into the Eurasian plate can be released anytime through an earthquake. Himalayan earthquakes can make about 1 million people from various large mountain cities. Previous major earthquakes with rupture area M_w 7.5 (1803, 1835, 1905, etc.) occurred along the shallowest portion of the megathrust boundary and the subduction earthquake zone. Due to the push of the Indian plate at the edge of the Eurasian plate, a mega-earthquake occurred in the Nepal basin of Central Himalaya. Vulnerability mapping has been done through high-end GPS data in this region to save human lives and property. The present study is focused on typically assessing the active fault zone in the Himalayan region and its vulnerability to contemporary and future concerns with the help of an intensive literature survey, cloud-based analysis of satellite data, and web-sourced GPS locations. The study's outcome revealed the region's possible location of high-magnitude future earthquakes. The result of the study helps to find high-risk areas for earthquakes by evaluating stress zones due to subduction plate movement. It has a significant chance to estimate the time and potential intensity of seismic events in the Himalayan landscape. This study makes

A. Sarkar (✉)
Department of Geography, Purash Kanpur Haridas Nandi College (University of Calcutta), Kolkata, West Bengal, India
e-mail: arindam.srkr1@gmail.com

S. Basak
Department of Geography, University of Calcutta, Kolkata, West Bengal, India

© The Author(s), under exclusive license to Springer Nature Switzerland AG 2025

S. Ghosh (ed.), *The Himalaya Dilemma*, Springer Natural Hazards,
https://doi.org/10.1007/978-3-031-95083-4_6

it easy to identify vulnerable inhabitant regions in the Himalayas with a high chance of earthquakes due to fault movement.

Keywords Active · Stress · Shake · Vulnerability · Subduction

1 Introduction

1.1 Himalayan Mountain System: Spatio-Temporal Aspects

The Himalayan Mountain System (HMS) is a young fold mountain in the world's mountain cluster, characterized by its lack of volcanic activity due to its lithospheric configuration. It is one of the most complex mountain systems in the world's tectonic history. HMS is across India, Nepal, Bhutan, China, and Pakistan. HMS also creates physical boundaries between those neighbouring countries. HMS is a significant driver of the climatic conditions and hydrological parameters in these nations. The HMS is classified into three parallel ranges: the Greater Himalayas (Himadri), the Lesser Himalayas (Himachal or Middle Himalayas), and the Outer Himalayas (Shivalik Range). The Greater Himalayas are rugged and are home to some of the world's highest peaks, including Mount Everest (8,848 m), Kangchenjunga (8,586 m), Lhotse (8,516 m), and Nanga Parbat (8,126 m). The Greater Himalayas are mainly composed of granite and several metamorphic rocks. The Lesser Himalayas, situated south of the Greater Himalayas, are elevated between 3,500 and 4,500 m. The Outer Himalayas are the youngest and least elevated (900–1500 m), as opposed to the Greater and Lesser Himalayas. The Himalayas are the source of three major transboundary sizeable tropical river systems: the Indus, the Ganga, and the Brahmaputra Rivers. The Indus Originates from the Tibetan Plateau near Lake Manasarovar and is associated with tributaries like the Jhelum, Chenab, Ravi, Beas, and Sutlej. The large Tropical River Ganga originates from the Gangotri Glacier. The Tibetan Plateau's geological structure resembles a mosaic of terranes that were gradually accreted northward from the Early to Late Palaeozoic and into the Mesozoic period (Zheng, 2025). Major tributaries of the Ganga system are Yamuna, Ghaghara, Gandak, Kosi, and Son. The Brahmaputra Originates from the Angsi Glacier in Tibet, known as the Yarlung Tsangpo Suture line. The sediment load of the Ganga and Brahmaputra River system produced the largest Ganga–Brahmaputra-Meghna (GBM) Delta in the world. A substantial share of the world's population, as well as the major ancient civilizations of the world, has been fed by the Himalayan River system.

1.2 Overview of the Himalayan Tectonic Setting

The Himalayan Mountain range is considered the most tectonically active and youngest in the world. It is a product of the collision between the Indian Plate and the Eurasian Plate. This region is characterized by active crustal deformation, seismic activity, and ongoing mountain-building processes. The India-Eurasia collision zone (Fig. 6.1) is the largest deforming region on the planet (Zheng et al., 2017). The Himalayas are the result of continental collision and orogeny. Major tectonic divisions of the Himalayan Mountain system are the Tibetan Plateau (North of the Himalayas associated with thick continental crust), Trans-Himalayan Belt (situated north of the Indus-Tsangpo Suture Zone), Tethyan Himalayas, Greater Himalayas (Himadri and oldest stable part of the Himalayas), Lesser Himalayas (Himachal), Sub-Himalayas (Shivalik) and Indus-Tsangpo Suture Zone (ITSZ). During the brief Early Cenozoic collision, these terranes and their sutures were reactivated variably rather than remaining as a single collisional entity throughout the entire Cenozoic (Zheng, 2025). The Indian Plate is moving northward at ~5 cm/year, colliding with the Eurasian Plate. This movement leads to continuous uplift (1–10 mm/year) and fault reactivation. Crustal stress buildup significantly contributes to medium to large-scale earthquakes in this area. Ongoing thrusting and folding lead to shortening, which accelerates fault slip. This continuous upward trend, driven by compressional forces, exemplifies the region's active tectonism. The belt of ophiolites following the Indus and Tsangpo valleys marks the boundary between the Eurasian Plate and the Indian Plate. To the North, the presence of acidic volcanic rocks indicates the subduction of oceanic lithosphere prior to the collision of the Indian and Eurasian landmasses (Summerfield, 1991).

1.3 Himalayan Geology

Himalayan geology is characterised by a series of thrust faults and several seismic gaps. Due to the subduction of the Indian Plate beneath the Eurasian Plate, severe stress accumulates in the majority of the Himalayan region. The Trans-Himalayan Belt is composed of igneous and metamorphic rocks of Eurasian origin. ITSZ contains remnants of the Tethys Ocean located south of the Trans-Himalayan Belt. The Tethyan Himalayas are composed of sedimentary rocks, including limestones, sandstones, and shales, which were deposited in the Tethys Sea. The Greater Himalayas (Himadri) are composed of high-grade metamorphic rocks, including gneisses, schists, and quartzites, as well as granites. Major fault system of the Himalayan Mountain chain is located at the Lesser Himalayas. Unconsolidated sediments transported from the Himalayas by the region's fluvial system are the main geological characteristic of the sub-Himalayas. The Himalayan Mountain system is considered an active seismo-tectonic region of the world where major thrust faults accommodate crustal shortening. The major thrust faults of the HMS are the Main

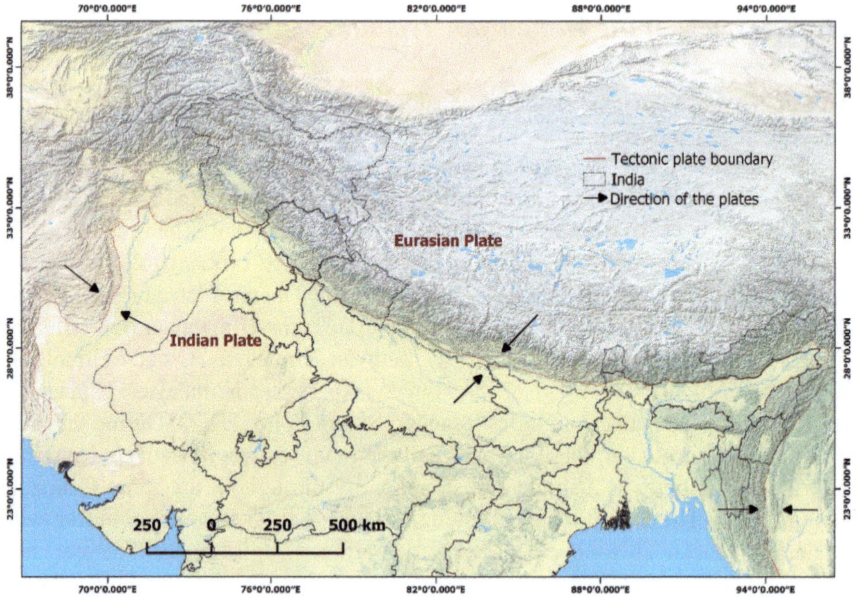

Fig. 6.1 Tectonic setting of the Himalaya, showing plate motions

Frontal Thrust (MFT), Main Boundary Thrust (MBT), Main Central Thrust (MCT), and South Tibetan Detachment System (STDS).

1.4 Evolution of the Himalayas

The Himalayan landscape evolved approximately 45–50 million Years Ago (Mya) due to the continent–continent collision between the Indian Plate and the Eurasian Plate. Approximately 50 million years ago, the collision between the Indian and South Eurasian plates resulted in the formation of the Himalayas (Hodges, 2000). Active tectonism and the ongoing mountain-building process are among the factors responsible for the world's highest elevation in this mountain system. The evolution of the Himalayan Mountain system has been classified into four significant phases: the Pre-Collision Phase (140–65 million Years Ago—Mya), the Initial Collision Phase (65–50 Mya), the Main Orogenic (Mountain-Building) Phase (50–23 Mya) and Post-Collision and Recent Activity (23 Mya–Present). More than 140 million years ago, the Indian plate (part of the Gondwana supercontinent) and the Eurasian plate were separated by the Tethys Sea. During the Mesozoic Era, the Indian Plate drifted northward at a rate of approximately 15 cm per year due to plate tectonic movement. In this phase, a substantial number of sediments is entrained and deposited in the Tethys Sea basin from the surface of the Indian and Eurasian lithospheric plates.

The initial collision between the Indian and Eurasian plates occurred in the Palaeocene–Eocene epoch (65–50 million Years Ago). This collision led to the subduction of the Indian Plate beneath the Eurasian Plate. For millions of years, the sediments deposited in the Tethys Sea were compressed, folded, and uplifted to form the initial Himalayan structures. This is the phase of birth of the incredible Himalayan Mountain landscape. The third and most crucial phase of the evolution and development of the Himalayan Mountain system is the Orogenic or mountain-building phase. This mountain-building phase began approximately 50–23 million Years Ago. The central Himalayan fault system (MCT, MBT, HFT) was formed during this phase. The last and contemporary phase of Himalayan Orogeny was initiated on 23 Mya. Post-collision and contemporary activities connected to the evolution and development of the Himalayan Mountain chain are continuous erosion and upliftment (2 cm/year), seismic activity, and geodynamic adjustment (Indian Plate continues to move northward at ~5 cm/year). The most significant impacts of the India-Asia collision are observed in the Himalayan and Gangdese orogens (Zheng, 2025). The evolution of the Himalayas is also influenced by the evolution and development of the Himalayan subsystems, including the Trans-Himalayas, Tibetan Plateau, Greater Himalayas, Lesser Himalayas, and Siwalik Hills. The dynamic evolution of the Himalayan Mountain System is also connected to the Ganga Basin, Brahmaputra Basin, Sindh Basin, and Bengal Basin.

1.5 Active Fault Zone: Key Characteristics of the Himalayan Fault Zone

An active fault zone refers to an area where a fault is currently moving and often produces earthquakes. This region exemplifies a tectonically active zone where earthquakes frequently occur due to the movement of lithospheric plates. The Himalayan active fault zone (HAFZ) is associated with medium- to large-scale seismic events. The ongoing movement of tectonic plates has caused measurable displacement along this fault. HAFZ is deeply related to the location of the epicenter of the quake. The potential energy accumulated in the HAFZ as a form of stress poses a possible threat to large-scale earthquakes in the Himalayan landscape. HAFZ refers to the response to the collision of two lithospheric plates, namely the Indian and Eurasian plates. This collision led to the ongoing uplift of the Himalayas and frequent seismic activity.

1.6 Methodological Review

A critical discussion of the methodologies used in previous fault zone assessments, particularly in tectonics, active fault zones, and seismic hazards, involves analyzing the strengths, limitations, and advancements of various approaches necessary for further discussion. Methodologies from the previous study on this topic can be classified into fault mapping, the application of different geospatial data through geoinformatics techniques, and the use of seismic data from historical and instrumental records. Several tool-based methods, including fault mapping and modelling, paleo-seismological parametric analysis, geodetic survey and interpretation, Numerical simulation, the application of vulnerability assessment tools (HAZUS), and risk matrix approaches, have been employed by different authors over the last few decades. All previous research is attributed to the findings on the spatial distribution of active faults, temporal aspects related to active faults, historical and contemporary seismic activity, its trends, and future possibilities of earthquake occurrence, as well as the identification and mapping of vulnerable earthquake regions and their mapping towards mitigation and sustainability.

1.7 The Foundation of the Research Gap and Scope

High-resolution data from remote sensing and GNSS is crucial for precisely identifying active faults, measuring plate movement rates, and estimating recurrence intervals. Our review of the literature uncovered a notable temporal gap in seismic records throughout the Himalayan region. Modern scientific research emphasizes the geological and geophysical aspects of the Himalayas, while socio-economic vulnerabilities and infrastructure risks remain underexplored. We have identified regional disparities in scientific research within the Himalayan zone, where most research efforts are concentrated in specific high-risk areas, such as Kashmir and the Nepal Himalayas. Meanwhile, regions such as Arunachal Pradesh and western Bhutan remain under-researched, resulting in significant gaps in spatial research. There is a lack of research focused on translating scientific findings into disaster risk reduction (DRR) policies, early warning systems, and community preparedness strategies for active fault zones in the Himalayas. The intricate relationship between climate change impacts and seismic vulnerabilities is a vital area that necessitates deeper exploration. Often ignored, these interactions play a significant role in the instability of these at-risk regions. Additionally, the inconsistent application of advanced geodetic technologies, such as interferometric synthetic aperture radar (INSAR) and light detection and ranging (LiDAR), poses considerable challenges. This lack of reliable technology hampers effective surveillance of the geologically diverse and rapidly evolving Himalayan arch, complicating efforts to monitor harmful activities. Consequently, our ability to understand and mitigate seismic risks in the region is severely restricted. It is crucial to utilize advanced remote sensing data in conjunction with high-end genetic tools to address the current research gap in the Himalayan active tectonic landscape. Multi-hazard assessments, the development of integrated frameworks for geophysical and socio-economic vulnerabilities, rapid policy implementation, and increased funding and research opportunities are also necessary to protect human society in this region.

1.8 Research Statement and Objectives

Active tectonics in the Himalayan region pose significant risks to the densely populated areas across South Asia. Half of the world's population is closely tied to the Himalayan Mountain system in terms of resource utilization. Himalayan active fault zones are responsible for seismically hazardous regions in this region. Understanding the dynamics of these fault zones is crucial for assessing seismic risks, predicting future earthquakes, and mitigating the socio-economic and environmental impacts. The objectives of this research are to assess the geological characteristics, spatial and temporal aspects related to the Himalayan active fault, seismic activity patterns, and the vulnerability of communities located near the Himalayan active fault zones. This study aims to provide a comprehensive evaluation of the risks and propose strategies for disaster preparedness, sustainable land-use planning, and fostering community resilience.

1.9 Location and Spatial Dimension of the Study

The location of an active fault zone has been determined by the movement of lithospheric plates and collision under different conditions, Intense compression and crustal shortening, geological weaknesses, such as older fault lines, suture zones, accumulation of higher tectonic stress, rugged topography with active upliftment and erosion process, geothermal gradients, sediment entrainment in adjacent basins. The location of an active fault zone has been categorized into seven significant concerns: the Main Central Thrust (MCT), the Main Boundary Thrust (MBT), the Himalayan Frontal Thrust (HFT), the Karakoram Fault, the Indus-Yarlung Suture Zone (IYSZ), the Kangra and Dehradun Faults, and the Arunachal Faults (Table 6.1).

Table 6.1 Localizing active fault zone in the Himalayan region

Active Fault Zone	Spatial dimension
Main Central Thrust (MCT)	Central and eastern Himalayas
Main Boundary Thrust (MBT)	Separating Siwalik Hills from the Lesser Himalayas
Himalayan Frontal Thrust (HFT)	Indo-Gangetic Plain
Karakoram fault	Northwestern Himalayas-Ladakh
Indus-Yarlung Suture Zone (IYSZ)	Southern Tibet
Kangra and Dehradun faults	Western Himalayas
Arunachal faults	Eastern Himalayas

2 Keywords for Basic Understanding of Active Tectonics in the Himalayan Region

Continental collisions: Two lithospheric plates and the continental crust (Indian vs. Eurasian) collide, an Important phase of mountain building (Orogeny), which involves the deformation of crustal material through faulting, folding, and thickening.

Active fault or Thrust system: Geological structures formed by tectonic stress, Fractures in rock strata, and compressive forces push a block of crust one over another. An active fault or Thrust system is the key factor in the occurrence of medium to large-scale earthquakes.

Main Himalayan Thrust (MHT): The MHT is a detachment fault formed by the subduction of the Indian Plate beneath the Eurasian Plate. It is linked to the rapid upliftment of the Himalayas and the occurrence of significant earthquakes. It is also a Strain accumulation zone, subjected to release as a surface feedback response, similar to an earthquake.

Main Central Thrust (MCT): MCT is a geological fault. This central fault system is characterized by the high-grade metamorphic rocks of the Higher Himalayas thrust over the Lesser Himalayas. This thrust plays a significant role in crustal shortening, which produces intense deformation during the Himalayan orogeny.

Main Boundary Thrust (MBT): The MBT demarcates the boundary between the Lesser Himalayas and the Siwalik Hills. MBT is an active thrust fault system. Here, Lesser Himalayan rocks have been pushed over the younger Siwalik sediments.

Accumulation of strain: Strain accumulation is subjected to tectonic stress. Strain accumulates as a form of energy, which is responsible for the formation of fault slips and the occurrence of earthquakes. In the Himalayan region, strain accumulates along major thrust systems.

Seismicity: Seismicity refers to the frequency, distribution, and intensity of earthquakes in the Himalayan region. The Himalayan landscape experiences high seismic activity due to the active movement of tectonic plates. Large earthquakes along the MHT, MCT, and MBT result from strain release that has accumulated over this area over Millions of years.

Crust and Tectonism: The Earth's crust is deformed by the operation of tectonic forces. Crustal material is compressed, folded, and thickened in different spatial settings in continental collision zones, such as the Himalayan Orogeny. Himalayan tectonism and the Crust system are associated with complex tectonic features such as thrust faults, shear zones, and crustal shortening.

Crustal Deformation and Shortening: The deformation of the earth's Crust due to tectonic forces is called Crustal Deformation. Crustal Deformation is associated with folding, faulting, and warping. In the Himalayan system, horizontal compression is responsible for crustal deformation and shortening.

Plate movement and Surface expression: Plate movement causes the upliftment of Young Fold Mountain, the formation of fault scarps, and land deformation. The northward movement of the Indian plate (4–5 cm/year) is responsible for active deformation along major thrusts.

Himalayan Orogeny: Orogeny is the process of mountain-building. Himalayan Orogeny is subjected to the collision of two continental crusts (tectonic plates). The Himalayan mountain-building process is associated with complex geological feedback mechanisms like folding, faulting, and crustal thickening.

Sub-process through plate tectonics: folding, faulting, and Crustal thickening (Horizontal compression leading to continuous upliftment of the Himalayan Mountain system).

3 Data and Methodology

The study was conducted through an extensive literature review, data analysis, and visual interpretation. Secondary data sources, including peer-reviewed journal articles, geological reports, and fault zone inventories, have been utilized to form the foundation of this review. Topographic and geological maps, along with satellite imagery, are analysed to identify fault traces, displacement patterns, and surface deformation features. Photographic evidence, sourced from existing publications and field documentation, is used to interpret surface ruptures, fault scarp morphology, and geomorphic indicators of tectonic activity. Diagrams and schematic models are examined to understand fault kinematics and their regional implications. The methodology integrates qualitative and quantitative analysis, enabling a comprehensive synthesis of seismotectonic characteristics. The visual representation of fault dynamics through maps and photographs enhances the spatial understanding of fault distribution, deformation patterns, and associated geological risks across the Himalayan region.

4 Result and Interpretation

4.1 Spatial Distribution of Active Fault in the Himalayan Region

The Himalayan active fault system (HAFS) is associated with four major southward-propagating Thrust systems. The Central/Main Himalayan Thrust (MHT) system is deeply associated with an active fault zone in this area. The most prominent aspects and mechanisms of the Himalayan fault system are active defamation and thrust faulting. Seismic response over this area is an active feedback mechanism that is prominent in HAFS. Generally, the thrust is demarcated by a tectonic boundary. Activation of the thrust system is a signature of tectonic activity in the area, characterized by faulting and the occurrence of earthquakes of varying magnitudes. Thrust is navigating along the weaker zones of the lithosphere. This area features extensive deep sedimentary formations, such as mudstone, sand deposits, and low-grade conglomerates. A reverse fault functions as a thrust fault, categorized as a dip-slip fault. Within the thrust fault system, the upper block lies above the fault plane and rises at the same rate as the regional uplift.

4.2 Temporal Aspect Related to the Himalayan Active Fault

The regional geology of the Himalayan landscape is characterized by a complex network of active fault systems. The Himalayan fault system was activated during the Paleogene-Neogene period of the Cenozoic era, when the Indo-Eurasian Plate collided with the Indian Plate as the latter moved beneath the Eurasian Plate. The evolution and development of the active fault system in the Himalayan region occurred during the Miocene-Pliocene Epoch (25–2.6 million Years Ago).

4.3 History of Seismicity in the Himalayan Region

Katmandu Valley in Nepal has experienced several devastating earthquakes (1255, 1833, 2023) in the past. Assam in northeast India has also experienced several earthquakes. Historical records of previous earthquakes have been devastating. In 1255, 30% of the population was killed by an earthquake in the Katmandu Valley, Nepal. In the Himalayan region, spatiotemporal variations in seismic activity indicate that periods of seismic quiescence often precede earthquakes with magnitudes ranging from 4.8 to 6.8 (Srivastava, 1992, 2004).

One of the largest earthquakes (M_w 8.2) in the Himalayas occurred in Western Nepal and Tibet in 1505 (Table 6.2). Severe damage occurred across northeast India due to the 1897 Assam earthquake (M_w 8.0). The devastating earthquake (M_w 8.00) took place in the year 1934 at the Bihar, India–Nepal Border, causing over 10,000 deaths and massive infrastructure damage. The Largest earthquake (M_w 8.6) in India occurred in 1950 at the Assam-Tibet Border (Arunachal) and caused extensive damage and landslides. Different places in Himachal Pradesh and Uttarakhand, India, have experienced several devastating earthquakes. An earthquake with Mw7.6 occurred at Muzaffarabad, Pakistan (Kashmir) in 2005, which was responsible for the death of 80 thousand people and massive infrastructural damage. Imphal, Manipur, India, and Myanmar, near the Indian border, also experienced medium-scale earthquakes. Evidence regarding past large earthquakes in the Himalayan landscape has been found in the fault scarps and deformed river terraces. Paleo-seismic evidence, data, and studies suggest that large-scale earthquakes in the Himalayan landscape have occurred approximately every 300–800 years in various regions. The most potent continental earthquake on record, measuring M_w 8.7, struck Assam near the India-China border on August 15, 1950. Between 1897 and 1952, the area experienced significant seismic activity, with five M_w 8 earthquakes, including the Shillong earthquake in 1897, the Kangra earthquake in 1905, the Bihar-Nepal earthquake in

1934, the Assam earthquake in 1950, and a 1952 aftershock of the Assam quake. Since 1952, no M_w 8 earthquake has been recorded in this region (Gupta, 2023). The Himalayan seismic belt ranks among the world's most seismically active zones, created by the collision of the Indo-Eurasian plates. Within just 53 years, four major earthquakes struck this region: Shillong in 1897, Kangra in 1905, Bihar–Nepal in 1934, and Assam–Tibet in 1950 (Srivastava et al., 2015).

Table 6.2 Chronology of significant earthquakes in the Himalayan region

Serial No.	Date/Year	Location	Magnitude (M_w)
Before 1900			
1	1255	Kathmandu Valley, Nepal	~7.8
2	1344	Kashmir, India	~7.5
3	1505	Western Nepal, Tibet	~8.2
4	1555	Kashmir, India	~7.6
5	1803	Garhwal, Uttarakhand, India	~7.5
6	1833	Kathmandu, Nepal	~7.6
7	1897	Assam, India	~8.0
1901–1950			
8	1905	Kangra, Himachal Pradesh, India	7.8
9	1916	Assam, India	7.5
10	1934	Bihar, India–Nepal Border	8.0
11	1947	Assam, India	7.3
1951–2000			
12	1950	Assam–Tibet Border (Arunachal)	8.6
13	1966	Nepal (Western Region)	6.3
14	1975	Kinnaur, Himachal Pradesh, India	6.8
15	1980	Dharamsala, Himachal Pradesh	6.4
16	1991	Uttarkashi, Uttarakhand, India	6.8
17	1997	Chamoli, Uttarakhand, India	6.8
18	1999	Chamoli, Uttarakhand, India	6.6
2001–Recent			
19	2005	Muzaffarabad, Pakistan (Kashmir)	7.6
20	2011	Sikkim, India	6.9
21	2015	Gorkha, Nepal	7.8
22	2015	Dolakha, Nepal	7.3
23	2016	Imphal, Manipur, India	6.7
24	2016	Myanmar (near the Indian border)	6.9
25	2021	Assam, India	6.4
26	2023	Nepal	6.4
27	2024	Afghanistan-Himalayan Belt	6.0

4.4 Region of Vulnerability in the Himalayas

The Himalaya is a highly tectonically active zone of global tectonics. The activation of the fault system in this region makes it highly vulnerable to severe hazards (Table 6.3).

Living near an active fault is vulnerable. This area is very prone to earthquakes and landslides. Living in a floodplain of the Himalayan River system and a high-sloped upper catchment area is associated with physical, biological, economic, and environmental vulnerability. The Himalayas and their foothills have been highly populated for a long time. Physical vulnerability in the Himalayan region is subjected to settlements, high-rise buildings (Hotels, administrative blocks, restaurants, etc.), roads, railways, and other ways of communication like power lines, telephone lines, wire bridges, suspension bridges, ropeways, etc.). High population density in the Himalayan region requires infrastructural development for communication and other public services as a principal concern to biological vulnerability. Economic vulnerability is associated with subsistence agriculture practices along the slope (Step cultivation) and in the floodplain of the Himalayan River. The development of tourism and its infrastructure in the Himalayan region is a crucial concern in terms of economic vulnerability. The construction of dams and hydropower projects is also a significant economic concern related to the economic vulnerability of the landscape. Environmental vulnerability in the Himalayas region is exacerbated by rapid deforestation, grazing, steep cultivation, and slash-and-burn practices. The Himalayan landscape is also a significant concern in terms of water diplomacy and international security.

Table 6.3 Mechanism, zonation, and feedback of Himalayan hazard connected to active fault and tectonism

Type of hazard	Mechanism and process	Response	Vulnerable zone
Seismic hazard	Tectonic fault Normal fault Thrust fault Cause: Due to the movement of lithospheric plates and deformation	Activation of Rupture Zone Seismic activity Frequent earthquake	Regional network of fault zone MCT MBT HFT
Geomorphic hazard	Escarpment, fault, mechanical weathering, mass movement	Landslides Rock falls Breakdown of glaciers Formation of glacier lake GLOF Flash Flood Melting of snow	Darjiling Himalaya Gangtok, Sikkim Joshi Math, Uttarakhand Har-Ki-Dun Gangatri glacier

5 Discussion on the Relevant Topics

5.1 *Himalayan Active Fault Zone (HAFZ)*

Activation of fault is subjected to tectonic movement over the area. Activation and movement of faults are common phenomena in the Active Fault Zone (AFZ) of the Himalayan landscape, resulting from rapid tectonic activity and the interaction of the Indian plate with the Eurasian plate. AFZ of the Himalayas experience large to shallow-scale earthquakes and related phenomena. The Himalayan region experiences frequent seismic activity along the active fault zone. Due to the active tectonic movement of the Indian and Eurasian lithospheric plates, faults are being activated, and they experience significant, measurable displacement over time. Active fault zones of the Himalayan system are epicenters of historical earthquakes. A significant level of stress accumulates along the AFZ in the Himalayas, posing a potential risk for major earthquakes in this region. The central fault system within the Himalayan Active Fault Zone includes the Himalayan Frontal Thrust (HFT)/Main Frontal Thrust (MFT), Main Boundary Thrust (MBT), Main Central Thrust (MCT), and South Tibetan Detachment System (STD).

5.1.1 Himalayan Frontal Thrust (HFT)

Himalayan Frontal Thrust (HFT) is also known as Main Frontal Thrust (MFT). HFT is the southernmost and youngest thrust fault as well as a low-angle reverse fault in the Himalayan orogenic belt. HFT demarcates the boundary between the Indian Plate and the Eurasian Plate, as well as between the sub-Himalaya (Siwalik) and the Indo-Gangetic alluvial plains. The present boundary separating India from Asia is determined by the thrust along the Main Himalayan Front (MFT). This line distinguishes rigid India from southern Tibet (Replumaz & Tapponnier, 2003). HFT is situated at the interface between the Himalayan Mountain System and the Indo-Gangetic Plain. Along the HFT, the Indian plate is actively subducting beneath the Eurasian plate, and deformation and rapid uplift are occurring. The India-Eurasia collision zone represents the most extensive deforming region on Earth (Molnar & Tapponnier, 1975, 1978; Zheng et al., 2017). HFT formed due to the compression and uplift produced by the subduction of the Indian Plate beneath the Eurasian Plate. HFT is a central considered part of the Himalayan complex fold system as well as the thrust belt. HFT is a southward-propagating thrust fault characterized by a steep fault plane. The dip of these steep fault planes is towards the north. In HFT, surface-breaking scarps, also known as fault splays, are located in numerous places. HFT extends across the entire Himalayas, from Pakistan to Arunachal Pradesh, India, through Nepal and Bhutan. Two important large tropical rivers, the Ganges, Brahmaputra, and Indus, cut across HFT, creating dip gorges. HFT is highly seismically active and has experienced significant earthquakes, including the 1934 Bihar-Nepal Earthquake (Mw > 7.0), the 1950 Assam-Tibet earthquake (Mw 8.0), and the 2015 Gorkha earthquake

(Mw 7.8), among others. Recent studies reveal that significant Crustal shortening (15–20 mm/year) is going on in HFT regions. These have the potential to trigger disastrous landslides, changes in river courses, ground ruptures, and other severe consequences. HFT is an important part of the Himalayan orogeny, which began 50 million years ago (Ma) when the Indian plate collided with the Eurasian plate. The Indian plate is subducting beneath the Eurasian plate at a rate of 5 cm per year. The subduction rate of the Indian plate beneath the Eurasian plate is responsible for the multiple thrust systems in the Himalayas. HFT represents the surface expression of the Main Himalayan Thrust. The Sub-Himalaya zone, situated between the HFT and the MBT, is characterized by Late Quaternary to Holocene rock deformation (Thakur, 2013).

5.1.2 Main Boundary Thrust (MBT)

MBT is the major tectonic fault and one of the important north-dipping thrust faults in the Himalayan orogeny. MBT is characterized by a high-angle fault plane, steep escarpment, and fault scarp. MBT is bearing the signature of active upliftment and deformation over the Himalayan landscape. MBT delineates the boundary between the Lesser Himalayas and the Sub-Himalayas, specifically the Siwalik. Along the MBT, the Lesser Himalayan metamorphic rock system is thrust over the Sub-Himalayan sedimentary rock system. The Main Boundary Fault (MBF) is an active fault that cuts through the Himalayan front, extending southeast into the alluvial plain (Nakata & Kumahara, 2002). Significant variation in dip and structure has been identified along the MBT fault system. MBT is a significant part of the Himalayan complex fold system and thrust belt, formed due to the southward propagation of Himalayan deformation as the Indian plate ceased to move northward along the Indian Proto-continents. Nevertheless, the faults connecting the MBT and HFT were reactivated in the Late Quaternary and likely during the Holocene period (Thakur, 2013). MBT is older than MFT and still active as well. It played a significant role in the structural characteristics of the Himalayas by accommodating a substantial portion of the Indo-Eurasian lithospheric collision. MBT is considered a crucial structural feature in Himalayan geology. Seismic activity in this area, the mountain-building process, denudational chronologies, and the evolution of the landscape system have been influenced by the MBT. MBT is also responsible for earthquakes and landslides over the Himalayan landscape. Several large-scale earthquakes (1905-Kaangra-Mw 7.8, 1991-Uttarkashi-Mw 6.8, 1999-Chamoli-Mw 6.6, etc.) have been experienced in the MBT region. MBT extends across the entire Himalayan Great Arc from Pakistan to Arunachal Pradesh (India), passing through Himachal Pradesh (India), Uttarakhand (India), Nepal, and Bhutan. MBT is situated parallel to MCT and MFT, experiencing Crustal shortening of 5–10 mm/year, which is responsible for the upliftment and folding of major rock systems in Lesser Himalaya. MBT is also an important factor in the occurrence of landslides in the Himalayas. Entrainment and channel incision in the fluvial system of the Himalayan foothills have been influenced by the Main Boundary Thrust (MBT). The MBT shapes the primary channels of the Ganga,

Yamuna, and Brahmaputra rivers as they traverse the thrust zones linked to it. Most of the active deformation occurs along the Himalayan front, which encompasses the HFT and the hanging wall frontal anticlines (Thakur, 2013).

5.1.3 Main Central Thrust (MCT)

The Himalayan fault system is deeply rooted in MCT. MCT is the significant tectonic signature in the Himalayan region. MCT demarcates the boundary between the Higher Himalayas and the Lesser Himalayas. According to litho-sedimentology, MCT separates Tethyan sedimentary and metamorphic rock systems into low-grade, relatively newer metamorphic rock. MCT is a significant fault zone formed due to the deformation mechanism of the Himalayan orogeny. Exhumation of the higher Himalayan Crystalline complex has been done by MCT. MCT represents the ductile shear zone. This north-dipping thrust fault extended towards Tibet. The structural features of MCT are associated with two major share zones: Lower MCT (MCT-I) and Upper MCT (MCT-II). The Lower MCT marks the transition from the Lesser Himalayas to the Higher Himalayas, and the Upper MCT denotes a deep shear zone within the metamorphic rock system. MCT is extensively deformed with Mylonite. The metamorphic gradient of MCT rises with temperature towards the higher Himalayas. Fault gouges and Breccias were identified in some areas along MCT, indicating Bittle faulting with an insignificant depth. The MCT stretches from Pakistan to Arunachal Pradesh (India), passing through Himachal Pradesh (India), Uttarakhand (India), Nepal, Bhutan, and Tibet. The Ganges, Brahmaputra, and Indus rivers traverse the MCT, exposing a significant fault section. In recent years, the MCT has undergone multiple seismic events.

5.1.4 South Tibetan Detachment System (STD)

The STD is a comprehensive low-angle normal fault system that demarcates the boundary between the Higher Himalayan Crystalline and Tethyan Himalayan Sedimentary systems. Geologically, an STD is considered an extensional detachment fault. In that case, the upper crust is allowed to move away. The STD accommodates a substantial part of the crustal shortening and exhumation resulting from the Indo-Eurasian plate collision and the subduction of the Indian plate beneath the Eurasian plate. STDs are extended from Pakistan to Arunachal Pradesh (India) and through Nepal, Bhutan, and Tibet. Metamorphic and structural signatures along the STD indicate vertical extension. Upper and Lower detachment are the two components of STD. The upper detachment is a brittle fault system exposed in unmetamorphosed sediment of the Tethys Sea. Lower detachment is a ductile share zone with high-grade metamorphic rock exposure. STDs are well-exposed in the Zanskar region, the Everest region, and along the Nepal-Tibet border, as well as in some areas of the Bhutan Himalayas. STD is less seismically active. STD plays a significant role in crustal thinning in the higher Himalayas.

5.1.5 Geotectonic Mystery of the Himalayan Fault System

Indus Yarlung Tsangpo Suture (TS) is a central geotectonic mystery related to the Himalayan collision. TS is the key tectonic boundary in the Himalayan formation. TS is also popular as Indus-Tsangpo Suture Zone (ITSZ). In the earth science discipline, suture refers to the area of convergent plate margins. Tsangpo Suture is a demarcating boundary between the Eurasian and Indian plates. The length of the Himalayan suture zone is approximately 2500 km (Gaisser, 1980). Himalayan suture is highly tectonic active. This is the largest suture in the tectonic history of the world (Tripathy, 2018). The suture zone is typically characterized by oceanic crustal materials (volcanic) and ultrabasic slabs (Gansser, 1980). The suture located in the Himalayas is highly sensitive to its destructive nature, and it is also significant for the study of active tectonics in this area. The Himalayan orogeny process made this region different from others (Tripathy, 2018). The Himalayan suture zone experiences large-scale earthquakes of varying magnitudes. For example, the Nepal earthquake occurred due to TS.

TS represents the collision zone of the Indo-Eurasian plate. The Indian Plate is subducted beneath the Eurasian Plate along the Indus-Yarlung Tsangpo Suture Zone (IYTS). TS extended from Pakistan to Namche Barwa across Tibet, Bhutan, Nepal, and India. TS is associated with the Indus-Tsangpo region, including Tibet and Ladakh; the western Himalayas, including Ladakh and Zanskar; and the eastern Himalayas, including Tibet, Bhutan, and Arunachal Pradesh. IYTS is named after the River Indus and the Yarlung Tsangpo. Yarlung Tsangpo River flows parallel to TS before entering Arunachal Pradesh as the most significant tropical river, Brahmaputra. TS is considered as remanent of the Tethys Ocean formed 55–50 Ma. TS comprises fragments of oceanic crust and upper mantle (Ophiolites), high-pressured metamorphic rock, and remnants of volcanic arc associated with the subduction zone.

5.2 Tectonic Process and Himalayan Active Fault

5.2.1 The Lithospheric Plate

The Himalayan mountain-building process is associated with two major lithospheric plates: the Indian and Eurasian plates. The Indian lithospheric plate is part of the Indo-Australian plate, which is moving in a northward direction at a rate of 4–5 cm per year. Indian plate started a northward movement during the Late Cretaceous (70–65 Ma). The Eurasian plate, a massive continental plate, moved south. The movement of the Eurasian plate is less than that of the Indian plate. Movement of the Indian Plate accelerated by the Kerguelen and Réunion hotspots near the pole and equator. The Indian crust is denser than the Eurasian crust. Lhasa is a microcontinental block that was part of the Eurasian Plate before colliding with the Indian Plate. The Yarlung Tsangpo Suture Zone demarcates the boundary between the Lhasa block of the Eurasian Plate in the north and the Indian Plate in the South. The Indian Plate is associated with both continental and oceanic crust. The oceanic crust of the Indian plate extended up to the Carlsberg ridge. The thickness of the continental crust of the Indian Plate ranges from 30 to 70 km. The thickness increases from the Peninsular part towards the Himalayas. The thickness of the oceanic crust of the Indian Plate ranges between approximately 5 and 15 km.

The peninsular part, which is a stable craton of the Indian Shield, is primarily composed of Precambrian rocks. The Indo-Gangetic Plain is composed of young Himalayan alluvial sedimentary deposits situated between the Himalayas and the Indian Peninsula. The Eurasian Plate comprises the continental crust, which encompasses Europe, Asia, and a portion of the Middle East, as well as the oceanic crust. The oceanic crust of the Eurasian plate extended into the Arctic Ocean along with the portion of the North Atlantic Ocean. The thickness of the Eurasian continental crust varies between 30 and 50 km, but it rapidly increases to 80 km under the Tibetan Plateau due to collision and subduction. In contrast, the thickness of the oceanic crust of the Eurasian plate varies between 5 and 15 km. The Eurasian crust is associated with Precambrian shields and Phanerozoic sedimentary basins, as well as younger orogenic belts. The Eurasian plate is associated with several tectonic boundaries, including the Himalayan Collision Zone (southern boundary), the Mid-Atlantic Ridge (Western Boundary), and the East African Subduction Zone (eastern boundary). The Siberian Craton is a stable continental block composed of Precambrian rock, situated at the center of the Eurasian Plate. The Indian Plate is associated with the *Dharwar* Craton (southern India), the *Singhbhum* Craton (eastern India), the Bundelkhand Craton (central India), the Aravalli Craton (Northwestern India), and the Bastar Craton (central India). The second aspect pertains to post-collisional reworking, stemming from the upwelling of the asthenospheric mantle due to the sinking of the lithospheric mantle. This mechanism results in crustal anatexis, the formation of leucogranites, and the emergence of metamorphic core complexes, all of which are characterized by domical uplift. This primarily occurred between 30 and 10 million years ago during the Late Cenozoic (Zheng, 2025).

5.2.2 Plate Driving Mechanism in Himalayan Geotectonic

The plate-driving mechanisms are responsible for the formation of the Himalayan orogeny. Mantle convection and slab-pull forces, Plate push and ridge-push forces, Slab break-off and rollback, Gravitational collapse and crustal flow, and Isostatic adjustments are plate-driving mechanisms related to the formation and development of the Himalayan Mountains. The circulation of materials within the mantle due to thermal variations is called mantle convection. Upwelling currents at mid-ocean ridges and Downwelling currents at subduction zones have been generated by Mantle convection. In the Late Cenozoic era, the Himalaya-Tibet tectonic collage functioned within an intracontinental setting, undergoing uplift driven by asthenospheric upwelling resulting from lithospheric foundering (Zheng, 2025). The generation of slab-pull forces is influenced by the sinking of denser oceanic lithosphere into the mantle, which drags the attached plate behind it. Slab-pull forces are the primary driver of the rapid northward movement of the Indian Plate at ~15–20 cm/year during the Palaeocene–Eocene. Slab-pull forces diminished as the Indian Plate encountered continental resistance in the post-collision phase. The northward movement of the Indian plate is due to ridge-push forces from the Indian Ocean Ridge. This force, combined with mantle drag, accelerated the plate velocity. Continuous folding and faulting lead to the formation of Anticlinal ridges and Synclinal valleys. Thrust faulting is most pronounced along the convergent plate boundaries (Tarbuck et al., 2010).

5.2.3 Fault and Thrust

The fault is a zone of fracture in the Earth's crust (lithospheric plate) that runs along the direction of plate movement. When the stress-building process occurs due to tectonic forces, the rocks on the other side of the fault slip, causing an earthquake. When hanging, the wall of earth's lithosphere moves upward relative to the footwall due to strong compressional forces. Thrust fault is a typical type of reverse fault. Many strike-slip faults occur in zones of collision, with the most prominent example being the continental collision between Asia and India (Molnar & Tapponnier, 1975, 1977, 1978, 1979). Dip-slip faults in which the hanging wall blocks move up relative to the footwall blocks are called Reverse faults (Tarbuck et al., 2016). A thrust fault is a type of reverse fault characterized by a dip of less than 45 degrees, allowing the overlaying block to move nearly horizontally over the underlying block (Tarbuck et al., 2016). The fault plane in the thrust fault remains low (<30°). The rock layer of the lithospheric plate pushes over another in the thrust faulting area. In the Himalayan region, thrust faults formed due to strong compression forces, which enhance the process of active mountain building. The mountain front of the central Himalayas experiences active thrusting and folding (Taylor & Yin, 2009). A thrust fault has formed due to the primarily horizontal operation of compressional forces, accompanied by some vertical movement. Reserve faults result from compressional tresses that produce horizontal shortening of the lithospheric crust (Tarbuck et al., 2016).

5.2.4 Compression and Himalayan Active Tectonics

Compressional tectonics is the principal dynamic factor of the formation of the Himalayan Mountains. Compressional stress produces fold-and-thrust faults (Tarbuck et al., 2016). In the Himalayan landscape, compression is the result of the convergence between the Indian plate and the Eurasian plate along the subduction zone. The Himalayan Mountain range is a product of continental collision and compressional tectonics resulting from the convergence of the Indian Plate and the Eurasian Plate. The compressional forces have led to Crustal shortening and thickening, the Development of significant thrust faults, the Formation of fold-and-thrust belts, Seismic activity due to stress accumulation, and Ongoing uplift and deformation of the Himalayas. The compressional forces have shaped the Himalayas through the formation of Thrust faults and shear zones, Fold-and-thrust belts, Fault-propagation folds, Crustal deformation and uplift. Significant thrust faults have displacements ranging from 10 to 100 km (Tarbuck et al., 2016). Firm compression is present in the active fault region of the Himalayas because of the pushing mechanism of the Indo-Eurasian plates against each other. Firm compression in the Himalayan region is responsible for crustal deformation, folding, and faulting throughout the entire landscape, with varying intensities. The operation of strong compressional forces is responsible for the occurrence of mega-earthquakes. Substantial and significant compressional forces in the Himalayan region are also responsible for folding. Compressional forces enhance plastic deformation of the rock under excessive pressure. That is why significant folding is occurring in the Himalayan Mountain belt. Compression is responsible for the shortening of the lithospheric crust. Compression squeezes the crust, making it shorter. Crustal shortening due to strong compressional forces initiates mountain-building processes throughout the entire Himalayan landscape, resulting in increased seismic activity. Compressional forces associated with colliding plates generally create fold as well as thrust faults that thicken and shorten the crust to produce mountain topography (Tarbuck et al., 2016). The active Main Frontal thrust positions the Miocene–Pliocene Siwalik Group atop Quaternary sediments (Lave et al., 2005).

5.2.5 Earthquake Rupture Zone

The occurrence of a slip along the fault line is a common geological phenomenon, and medium- to large-scale earthquakes are typically responsible for this. These incidents occurred within the Earthquake Rupture Zone (ERZ/RZ). ERZ is associated with the accumulation of tectonic stress. The sudden and massive release of tectonic stress through the fault zone is the primary factor in generating seismic waves and their subsequent propagation through the Earth's crust along an active fault. The distinctive character of RZ is positively related to the magnitude, intensity, and outbreak pattern of the earthquake. The occurrence of high-magnitude earthquakes depends on the length and width of the RZ. The horizontal and vertical dimensions of RZ along the fault line positively related with the magnitude of the earthquake. High-magnitude earthquakes (> Mw7) should be associated with a 350 km length of RZ. Rupture propagation is close to shear wave velocity. Rupture propagation direction should be uni or bi-directional with a speed of 2–5 km per second. Shaking of the ground during an earthquake is governed by the direction and speed of rupture propagation.

Heterogenous displacement of blocks along active fault is shared. An intense occurrence of slip is experienced in and around the hypocenter of the earthquake, and it decreases towards the edge. A large-scale earthquake is not associated with surface rupture. Rupture reaches the surface during shallow earthquakes by creating a fault scarp and offsets. A continuous aftershock of the earthquake is experienced in areas surrounded by the central rupture zone. ERZ is closely associated with fault type, rock strength, structure, the type and intensity of stress accumulation, and the depth of the earthquake. Horizontal displacement (Strike-Slip-Fault) is responsible for the longer and narrower RZ. The thrust fault is a perfect example of vertical displacement, which is responsible for a wider RZ with great shaking potential. A moderate rupture is caused by a normal fault. The horizontal extension of RZ is also related to the strength of the rock and the state of stress accumulation over the region. A larger RZ is characterized by high rock strength and a high potential level of stress. The gap between two RZ is called seismic gap, which is also the area of high potentiality of future earthquakes.

5.2.6 Seismic Gap in the Himalayan Region

The seismic gap (SG) refers to a segment of the Himalayan active fault that has not experienced a significant earthquake in a considerable amount of time. The seismic gap is also related to the tectonic boundary of the Indo-Eurasian plate, which has a high likelihood of experiencing a large-scale earthquake shortly. Seismic gaps in the Himalayas are also considered zones of stress accumulation on a large scale. Several seismic gaps (Kangra SG, Central SG, Western Nepal SG, Assam SG, etc.) have been identified in the Himalayan region. Kangra, SG, is located between the 1905 Kangra earthquake (Mw 7.8) and the 2005 Muzaffarabad earthquake (Mw 7.6). It is also known as Kashmir SG. Scientists anticipate a significant accumulation of seismic stress in this region, which has not previously experienced an earthquake of this magnitude. The Central Himalayan SG extends between Uttarakhand and western Nepal. The Central SG is positioned between the rupture zones of the 1905 Kangra earthquake and the 1934 Bihar-Nepal earthquake (Mw 8.0). Western Nepal SG lies between the 1934 Bihar-Nepal earthquake and the 2015 Gorkha earthquake (Mw 7.8). Eastern Himalayan SG is located between the 1950 Assam earthquake (Mw 8.6) and the Bhutan region. This region has not experienced a major earthquake since 1950, making it a potential seismic gap.

5.2.7 Himalayan Subduction Zone and the Zone of Earthquake

The zone of collision of the Indian plate with the Eurasian plate is called the subduction zone. A Himalayan subduction zone is closely related to the Himalayan mountain-building process. The length of the Himalayan subduction zone is approximately 2,500 km, extending from Nanga Parbat in Pakistan to Namche Barwa in Tibet. The 2000-km-long Himalayan orogenic belt stretches from the Nanga Parbat syntaxis in the west to the Namche Barwa syntaxis in the east (Yin, 2000, 2006). The Himalayan subduction zone is associated with a Continent–Continent collision, where the continental part of the Indian Plate is subducting beneath the relatively stable, large, and rigid continental mass of the Eurasian Plate at a rate of approximately 5 cm per year. The intensity of collision along the subduction zone is responsible for complex deformation in Himalayan regional geology. The Himalayan subduction is associated with Crustal Thickening and Shortening, Metamorphism, Magmatism, Surface Deformation, and Isostatic Rebound, among other phenomena. The occurrence of large-scale folding, the formation of nappes, the uplift of metamorphic rock, and the formation of fault scarps are associated with the Himalayan subduction zone. The continuous rise in elevation of the Himalayas is closely related to the Himalayan subduction zone.

5.3 Geological Structure and Himalayan Active Fault

The Siwalik range of the Sub-Himalayas was formed by the sediment load deposited in the Himalayan foreland basin during the Miocene to Pleistocene. The area is associated with the thrust fault system, characterized by a significant lithological composition, including sandstone, conglomerate, and other rock types. The Lesser Himalayas are separated from the Sub-Himalayas by the Main Boundary Thrust. The Lesser Himalayas are composed of sedimentary rocks, including phyllite, quartzite, and slate. The main central Thrust is the principal fault system of the greater Himalayas. The Greater Himalayas are composed of high-grade metamorphic rocks, including gneisses and schists, as well as intrusive granites. Trans-Himalaya (Tethyan Himalaya) is composed of sedimentary rocks of Palaeozoic to Mesozoic age. The South Tibetan Detachment System is located between the Greater Himalayas and the Trans-Himalayas. The Himalayan active fault is responsible for the uplift of the Himalayas, the formation of fault scarps and folds, the creation of deformation zones, frequent landslides and earthquakes, and changes in river courses.

5.4 Tectonic Geomorphology of the Himalayas

Himalayan tectonic geomorphology is characterized by the active tectonism in the Himalayas that shapes the landscape through folding, faulting, uplift, subsidence, deformation, erosion, and mass wasting. The Himalayan landscape is characterized by several fault-controlled landforms due to its active thrust system. Fault Scarps are created by vertical fault displacement in the Himalayan Mountain landscape. Uplifted river terraces along active faults have been located at the Kali Gandaki River terrace in Nepal. The Himalayan landscape features asymmetrical valley slopes along the rivers Kosi and Gandak, resulting from tectonic tilting. River incision, governed by tectonic uplift, is identified in the Indus, Sutlej, and Tsangpo river systems. In the Himalayan landscape, the course of the river is greatly influenced by tectonic upliftment and faulting. The capture of the Yarlung Tsangpo River by the Brahmaputra is a significant example of this. Several knickpoints are visible along the river's course. Teesta gives a clear signature of tectonic activity in this area. Himalayan rivers are characterized by steep gradients due to successive uplift.

Prominent glacial erosional features, such as hanging valleys, in the upper Himalayan landscape have been developed through tectonic uplift along faults. The Rupin Pass hanging valleys in Himachal Pradesh are a significant example of that. Moraines and glacial deposits in the Zanskar Valley of Ladakh have been influenced by the tectonic uplift of the region. Uplift-induced changes in precipitation and temperature gradients influence glacier mass balance. The retreat of the Gangotri Glacier is influenced by local tectonic activity. Landslides, rockfalls, and debris flows occur in the Himalayan region due to tectonic activity and the presence of active faults. Asymmetrical Drainage Patterns have been formed in the Himalayan region due to tectonic tilting and faulting. Gandak River drainage anomalies are the perfect example of tectonic deformation in the Himalayan region. Beheaded tributaries of the Teesta River formed due to tectonic-driven active fault displacement.

5.5 Geological Events (Time Scale) of the Himalayan Formation

The Himalayas have undergone complex geological formation and activity (Table 6.4).

5.6 Himalayan Orogeny

The Himalayan orogeny is associated with active tectonic processes occurring in this landscape, which are responsible for the uplift of the mountains. The Himalayan orogen consists of two main stages. The initial stage involves continental suturing, transitioning from soft to hard collision and deep subduction, which leads to crustal shortening through slice thrusting. This process primarily occurred during the Early Cenozoic, specifically between 55 and 45 million years ago (Ma) (Zheng, 2025). The Himalayan orogeny is characterized by Crustal shortening and thickening, Tectonic uplift and folding, Metamorphism and magmatism, Seismic activity and faulting. The Himalayan Mountain belt, running some 2500 km, is seismically one of the most active continental regions (Gupta, 2023). The Himalayan orogeny is characterized by High-Grade Metamorphism and the formation of leucogranite intrusions, as well as crustal melting. High-grade metamorphism is subjected to the formation of garnet, kyanite, and sillimanite. Radioactive heating and crustal thickening are responsible for the melting of the crust. Partial melting of the crust has produced leucogranite bodies, such as the Manaslu granite in Nepal. The Himalayan intercontinental collision orogen is situated in a southward-bending arc, spanning approximately 200–250 km in width, extending over 2500 km from the Indus River in the west to the Brahmaputra River in the east. It is separated from the Trans-Himalayan zone and the Tibetan Plateau to the north by the structurally controlled valleys of the Indus

Table 6.4 Briefing of geological incidents in context to the formation of the Himalayas

Eon	Era	Period	Epoch	Time range (Millions of Years Ago) MYA	Geotectonic activities of Himalayan formation
Phanerozoic	Cenozoic	Quaternary	Holocene	Present 0.01	Glacier advances (early Holocene: 11,700–9000 kaBP) Intensification of Indian summer monsoon (mid Holocene: 9–5 kaBP) Aridity in North Western Himalaya (late Holocene)
			Pleistocene	0.01 2.6	Rapid uplift Glacier activity Tectonic deformation Formation of distintives landscape High erosion and sedimentation (Brahmaputra Dessert)
		Tertiary (Neogene)	Pliocene	2.6 5.3	Less active upliftment Sedimentation changes in the Bengal Basin
			Miocene	5.3 23	Rapid upliftment Denudation and exhumation
		Tertiary (Paleogene)	Oligocene	23 33.9	The rise of the Himalayas Formation of the Tibetan Plateau Himalayan Orogen derived by India-Asia collision
			Eocene	33.9 56	Himalaya magnetism Subathu formation in Jammu
			Palaeocene	56 66	Initial stage of India-Asia collision Formation of Himalayan foreland basin Deposition of shallow marine sediment (Subathu formation)

(continued)

Table 6.4 (continued)

Eon	Era	Period	Epoch	Time range (Millions of Years Ago) MYA	Geotectonic activities of Himalayan formation
	Mesozoic	Cretaceous		66 145	Northward drift of the Indian plate Formation of deepwater sediment in areas like Spiti, Zanskar, and Garhwal Closing of Neo-Tethys Ocean
		Jurassic		145 201.3	The Himalayan region was a part of the southern margin of the Neo-Tethys Ocean Intensive marine sedimentation Formation of Spiti shale rock
		Triassic		201.3 251.9	Maximum depositional age in the Tethys Himalaya
	Palaeozoic	Permian		251.9 298.9	Supercontinent Pangea intact Tethys Ocean
		Pennsylvanian		298.9 323.2	Existed between the Indian and Eurasian plates
		Mississippian		323.2 358.9	Deposition of marine sediments in the Tethyan basin
		Devonian		358.9 419.2	Muth and Aishmuqam formations in Kashmir Basin
		Silurian		419.2 443.8	Deposition of sediment and formation of the Tethys Ocean
		Ordovician		443.8 485.4	Deposition of rock in a warm, shallow sea water and the formation of ancient sea flows
		Cambrian		485.4 541	Setting of passive continental margin Gradual drowning of carbonate platform

(continued)

Table 6.4 (continued)

Eon	Era	Period	Epoch	Time range (Millions of Years Ago) MYA	Geotectonic activities of Himalayan formation
Proterozoic	Precambrian			541 2000	Formation of the Indian Craton (part of the Gondwana supercontinent). Development of basement rocks (gneisses, schists, and granites)
Archean				2500	
Hadean				4000 4600	

and the Brahmaputra (in its upper reaches, called the Tsangpo), and to the south, it is bounded by the sediment-fitted peripheral foreland basin of the Indo-Gangetic Plain (Summerfield, 1991). Himalayan orogeny can be classified into several phases. The Indian plate moved northward at a rate of 15–20 mm per year along the Tethys Sea during the Pre-Collision phase (120–55 Ma). The subduction of the Tethys Oceanic Plate beneath the Eurasian Plate led to the formation of island arcs and accretionary formations. In the Collison phase (55–40 Ma), the Indian Plate collided with the Eurasian Plate by the closing of the Tethys Ocean. Crustal deformation and the initiation of thrust faulting began with the start of the continental collision. The tectonic development of the Himalayan–Tibetan orogen and surrounding regions throughout the Cenozoic feature's complex fault systems and folds (Taylor & Yin, 2009). The convergence of the Indian and Eurasian plates has continued at a rate of 4–5 cm per year in the Post-Collision Phase (40 Ma–Present). Major thrust faults of the Himalayas, such as the Main Frontal Thrust (MFT), Main Boundary Thrust (MBT), Main Central Thrust (MCT), and South Tibetan Detachment System (STDS), have developed during the Post-Collision Phase. An intercontinental collision orogen is subjected to no further development as a convergent boundary since subduction has halted (Summerfield, 1991). The Himalayan–Tibetan orogen comprises extensive and intricate fault systems, notably the Himalayan Main Frontal thrust zone, with some faults extending over 1,500–2,000 km (Tapponnier & Molnar 1977, 1979; Taylor & Yin, 2009).

6 Major Findings

6.1 Geological Findings

The concept of a Himalayan active fault becomes easily understandable after the discussion. Active fault is generally located along the frontal zone in the Himalayas. A major active fault is located at MCT and MFT. The location of active faults is influenced by the collision and subduction zones of the Indo-Eurasian plates. Several active fault like *Bangad* fault, *Kulekhani* fault (Chandra Giri Mountain range), *Chittang* fault (near Katmandu valley), *Malagiri* fault (Chandra Giri Mountain range), *Surkhet-Ghorahi* fault (central Bhutan Himalaya), Dras fault zone has also been located in the Himalayan region and several has to be identified near future (Cannon et al. 2018). Intensive scientific research, aided by modern technologies such as artificial intelligence and remote sensing, is ongoing throughout the entire Himalayan region. Active faults have been occurring along the MBF, MFT, and their associated faults (Nakata, 1989). The study reveals the current status of Himalayan deformation and the present scenario. The Himalayan main frontal thrust currently accommodates approximately 12–23 mm per year of convergence between the Indo-Eurasian litho-tectonic plates through uplift and deformation of the sub-Himalayas (Bouscary et al., 2023, 2024). Recent TL dating by Bouscary et al. (2024) indicates that the sub-Himalayan fold and thrust belt, specifically the Main Frontal Thrust, accommodates at least 62% of the Himalayan convergence since at least 200,000 years ago. Active tectonic deformation of the entire Himalayas is a result of the ongoing under-thrusting of the Indian Plate beneath the Eurasian Plate (Tiwari et al., 2021a, 2021b). Two significant ranges, the lesser Himalaya and the higher Himalaya, both of which exceed 800 m in height, are bounded by three central linear structure; from south to north these are the Main Boundary Thrust (MBT), the Main Central thrust (MCT), and the Indus-Tsangpo Suture Zone (ITSZ) (Summerfield, 1991).

6.2 Seismotectonic Findings

The Himalayas are one of the most tectonically active and seismically vulnerable regions in the world. The tectonic structure of the Himalayan region can be divided into several subdivisions, including the Trans-Himalaya and Tibetan Plateau, the Tethyan Himalaya, the Greater Himalaya, the Lesser Himalaya, the Sub-Himalaya (also known as the Siwalik Himalaya), and the Indo-Gangetic Plain (IGP). Strike-slip and oblique-slip faults have formed in the Trans-Himalaya region due to crustal extension and active deformation. The Greater Himalaya is associated with high-grade metamorphic rock, while the Lesser Himalaya is linked to Precambrian to Paleozoic metamorphic rock. The Main Boundary Thrust (MBT) is present in the Lesser Himalayan region. The Tethyan Himalaya is associated with the South Tibetan Detachment (STD). The Siwalik is the younger part of the Himalayan region, composed of Neogene sediment. The Indo-Gangetic Plain is a foreland basin comprising Quaternary sediments. The Himalayan region is seismically classified into Zones IV and V, according to the BIS Seismic Zonation Map of India, which indicates high to very high seismic hazards. Earthquakes in the Himalayas typically occur at shallow to moderate depths (10–30 km) and are associated with thrust faulting. Some deeper earthquakes (those with depths greater than 50 km) also occur. The Himalayan seismic cycle is associated with a prolonged interval of activity, characterized by the continuous accumulation of strain over centuries. Seismic gaps in Himalayan geotectonic provinces remain stacked and delinquent for a large earthquake. Contemporary research, equipped with high-end instruments and modern technology, predicts that high-magnitude earthquakes are possible to occur in western Nepal, Uttarakhand (India), and Arunachal Pradesh (India). The Himalayas experience a recurrence of large to moderate earthquakes over centuries, and certain areas of the Himalayas remain locked with a high accumulation of strain, posing a devastating seismic risk.

6.3 Geomorphological Findings

The Himalayan active fault plays a pivotal role in shaping the regional geomorphology and landscape evolution of the Himalayan region through continuous tectonic activity and deformation. The collision between the Indian and Eurasian tectonic plates led to crustal shortening, resulting in the uplift of Everest to its maximum elevation in the world. The typical mountain range, rugged topography, and complex terrain system of the Himalayas are a result of ongoing tectonic convergence. Fault-induced landforms are common in the Himalayan landscape. Active faulting in the Himalayan geotectonic region produces distinctive geomorphological features, such as fault Scarps, offset river terraces, and linear valleys. Several geomorphic signatures of active faulting and thrust faults have been identified in various locations throughout the Himalayan landscape. *Dras* is an oblique strike-slip fault in the northwestern Himalayas, bearing the signature of crustal deformation. Mass movement is a significant exogenetic landfarming process, as well as a denudation process, producing maximum sediment in the Himalayan River system. Earthquake-triggered landslides and mass movements are a significant mechanism in Himalayan land farming processes, sediment budgets, and denudation chronologies. Seismic activity along the active fault zone is a causing factor for landslides and rockfall. Active faults are responsible for the alternation of the landscape in the Himalayan region.

The *Tsergo Ri* landslide in Langtang Valley, Nepal, is the most significant mass movement of the earth due to the tectonic behavior of active faults. The local topography of Langtang Valley was transformed and reshaped by the mass movement. An active fault is a crucial controlling factor in river dynamics and the evolution of drainage systems. Tectonic upliftment and faulting have a more significant influence on the shaping of the river course, the process and mechanism of formation of the river terrace, and the alluvial fan. In the Kumaon sub-Himalayas, active fault has shaped the local landscape and regional geomorphology. The shaping of the landscape, river planforms, natural adjustment of the river, sediment load, and entrainment have been controlled by the active fault. The implication of seismic hazard in the Himalayan landscape has been controlled by active faults and tectonism. Active faults can be considered both driving sources and process mechanisms for identifying distinct geomorphic signatures in the Himalayan region. The active nature of these fault zones poses a significant seismic hazard, with the potential for large-magnitude earthquakes. The frontal zone of the Himalayas (MBF and Himalayan Frontal Front) is highly active in faulting in contemporary time. Active fault zones in the Himalayas are fundamental in driving the region's dynamic geomorphology, influencing mountain-building processes, creating distinct landforms, triggering mass movement, altering the river system, and contributing seismic hazards.

6.4 Himalayan Active Tectonics- Activation of Fault

Active faults are the mechanical drivers of dynamic Himalayan regional geomorphology and landscape evolution. Active faults and their behavior are positively related to the Himalayan mountain-building process, regional elevation and slope, diversity in the ruggedness of the Himalayan topography at different elevations, mass movement, sediment entrainment, and the natural alteration of the Himalayan drainage system. An active fault is a crucial factor in determining transverse topographic symmetry, mountain front sinuosity, and basin asymmetry. The formation, destruction, and reformation of significant landforms in the Himalayan landscape are closely related to the location and activity of faults within the landscape. Fault-related driving factors and process mechanisms include tectonic activity, crustal deformation, and lithological strain.

6.5 Vulnerability, Human Civilization, and Risk Assessment

Urban areas in the Himalayan region are highly seismically active due to the ongoing convergence of the Indian and Eurasian tectonic plates. Several seismic gaps have been identified in the Himalayan landscape, where significant earthquakes are likely to occur in the near future. The Indo-Nepal border, situated on the Central Himalayan seismic gap, is considered a high-risk zone for earthquakes. Similarly, the India-Pakistan border, located at the Kashmir seismic gap, is also considered a high-risk area for seismic activity. The large populations and infrastructures of prominent Himalayan cities, including Kathmandu, Srinagar, Muzaffarabad, and Lhasa, are associated with the highest seismic vulnerability. Cities like Dehradun, Shimla, Thimphu, and Islamabad are considered to be at moderate risk for earthquakes. In the western Himalaya Garhwal-Kumaon region of Uttarakhand, India, there has been no significant rupture in the last 500–600 years, making this area vulnerable to large-scale earthquakes. Western Nepal and the Uttarakhand segment of the Central Himalayas are associated with a large-scale accumulation of stress beneath this region. Certain areas of Arunachal Pradesh (India) and Bhutan, located on the Tawang, Mishmi, and Kopili faults, are overdue for a large to medium-sized earthquake.

Kathmandu, Dehradun, Srinagar, and Guwahati are densely populated urban agglomerations situated in valleys (Table 6.5) located at different tectonic segments, and seismic gaps in the Himalayas are considered high-risk zones for the occurrence of large-scale earthquakes. Those are the ancient civilizations associated with historic heritage constructions, poorly constructed old buildings, and new High-Rise buildings in a vulnerable state. Poor infrastructure and unplanned urbanization throughout the Himalayan landscape suffer from substantial threat to earthquakes. Ancient and small Himalayan cities are associated with significantly less earthquake-resistant structure. Connectivity network through the Himalayas is an international concern for security, geopolitics, and administration. Transportation networks, modes of transportation, and dams are in huge threat to earthquake and earthquake-induced landslides. Earthquakes can trigger glacial lake outburst floods in the Himalayan region, resulting in catastrophic flooding in the valleys of this region.

7 Conclusion

The findings of research associated with Himalayan tectonics and deformation are equally important for understanding the evolution mechanism of the Himalayan landscape and its vulnerability. All previous studies have highlighted the comprehensive aspects to understand the Himalayan tectonic behavior and consequences. The findings of previous research have the potential to significantly inform future research directions in the Himalayan landscape. It can also present nuanced arguments and offer actionable suggestions to policymakers, urban planners, and disaster management authorities. Implementation of Policy and Mitigation. The current research is significant for Disaster preparedness and mitigation strategies.

An effective early warning system should be implemented throughout the Himalayas with international cooperation. An assessment of the existing structure is required throughout the Himalayan region. A strong recommendation is required for the earthquake-resistant design of infrastructure and development. Evaluation of the existing preparation level by the regional government and various agencies related to central governments should be conducted at regular intervals, typically every 2–3 years. International collaboration among Himalayan countries (India, Nepal, Bhutan, China, and Pakistan) is necessary to develop effective seismic risk mitigation policies and ensure their regular application. Transparency in international collaboration should also be an important mitigation aspect. Countries associated with the Himalayan landscape must collaborate with various international agencies, such as the Red Cross, the World Bank, and UNICEF, to develop mitigation and rapid

Table 6.5 Seismic vulnerable Himalayan cities

Urban area	Seismic risk type	Geotectonic setting	Remarks
India			
Srinagar	Very high	Close to the Kashmir Seismic Gap, near the Main Frontal Thrust (MFT)	Area: 294 km^2 Population: 1,180,570 Population density: 703/km^2 2005 Kashmir earthquake (Mw 7.6)
Dehradun	High	Near the Himalayan Frontal Thrust (HFT)	Area: 196.5 km^2 Population: 569,578 Population density: 550/km^2 Potential risk from seismic gaps
Shimla	High	Near the Main Boundary Thrust (MBT)	Area: 25 km^2 Population: 814,010 Population density: 159/km^2 1905 Kangra earthquake (Mw 7.8)
Gangtok	High	Near the Darjiling-Sikkim seismic zone	Area: 25 km^2 Population: 283,583 Population density: 297/km^2 Frequent moderate earthquakes
Itanagar	Very high	Proximal to subduction zone near Arunachal	Area: 51.69 km^2 Population: 59,490 Population density: 1200/km^2 Highly vulnerable due to proximity
China			
Lhasa	High	Near the Indus-Yarlung Suture Zone	Area: 29518 km^2 Population: 867,900 Population density: 890/km^2 1950 Assam-Tibet earthquake (Mw 8.6)

(continued)

Table 6.5 (continued)

Urban area	Seismic risk type	Geotectonic setting	Remarks
Shigatse	High	Close to the Yadong-Gulu rift	Area: 182,000 km^2 Population: 798,153 Population density: 4.4/km^2 Vulnerable to seismic events
Nyingchi	Very high	Near the India-Tibet collision zone	Area: 116,175 km^2 Population: 238,936 Population density: 1.7/km^2 High seismic hazard
Xigaze	High	Close to the Brahmaputra suture zone	Area: 182,000 km^2 Population: 798,153 Population density: 4.4/km^2 Vulnerable to future events
Nepal			
Kathmandu	Very High	Near the Main Himalayan Thrust (MHT)	Area: 50.7 km^2 Population: 2,041,587 Population density: 20,288/km^2 2015 Gorkha earthquake (Mw 7.8)
Pokhara	Very high	Near MHT and MBT	Area: 464.24 km^2 Population: 513,504 Population density: 1106/km^2 Significant ground shaking risk
Biratnagar	High	Close to the Nepal-India border	Area: 58.48 km^2 Population: 243,927 Population density: 3200/km^2 Vulnerable to future events
Lalitpur	Very high	Part of the Kathmandu Valley seismic zone	Area: 385 km^2 Population: 551,667 Population density: 1433/km^2 Highly vulnerable
Bhutan			

(continued)

Table 6.5 (continued)

Urban area	Seismic risk type	Geotectonic setting	Remarks
Thimphu	High	Close to main frontal thrust	Area: 26.1 km^2 Population: 138,736 Population density: 67.1/km^2 2009 Bhutan earthquake (Mw 6.1)
Paro	High	Near MBT and MFT	Area: 1293 km^2 Population: 46,316 Population density: 36/km^2 High vulnerability
Punakha	High	Close to active fault zones	Area: 1110 km^2 Population: 28,740 Population density: 26/km^2 Seismic history indicates potential risk
Pakistan			
Islamabad	High	Near Hazara thrust and MFT	Area: 906.5 km^2 Population: 1,108,872 Population density: 2609/km^2 2005 Kashmir earthquake (Mw 7.6)
Muzaffarabad	Very high	Close to the Kashmir seismic gap	Area: 1642 km^2 Population: 171,959 Population density: 396/km^2 Epicenter of 2005 Kashmir earthquake
Gilgit	Very high	Near the Karakoram Fault	Area: 72971 km^2 Population: 2,300,000 Population density: 21/km^2 Frequent moderate earthquakes
Skardu	Very high	Proximal to Kohistan-Ladakh Suture Zone	Area: 77 km^2 Population: 214,848 Population density: 14/km^2 Seismically vulnerable

action plans. Community-based disaster risk reduction, along with ecosystem-based adaptation approaches, should be implemented in the Himalayan region to mitigate earthquake hazards.

Synthetic Aperture Radar Interferometry (InSAR) can measure ground displacement. The extensive spatial coverage of InSAR enables this remote sensing technology to achieve very accurate and precise detection of surface deformation. The detection of surface deformation is crucial for mapping inter-seismic strain accumulation, identifying seismic gaps, verifying potential rupture zones, detecting areas with unstable slopes and zones of mass movement, and estimating slip rates along significant fault lines. Prominent offset hinterland strain accumulation was collected with cross-orogen strike-slip and extensional fault systems. Monitoring deformation near large dams, roads, and urban agglomerations is important for assessing vulnerability to earthquakes and landslides. The monitoring of the retreat of Himalayan glaciers and their impact on deformation should be studied in greater detail.

References

Bouscary, C., Avouac, J. P., Grujic, D., Burgess, P. M., & Goral, T. (2024). Recent thermoluminescence dating reveals active deformation along the Himalayan Main Frontal Thrust. *Tectonophysics, 878*, 229–247.

Bouscary, C., King, G. E., Grujic, D., Lavé, J., Almeida, R., Hetényi, G., & Herman, F. (2023). Sustained deformation across the Sub-Himalayas since 200 ka. *Geology, 52*(1), 72–76. https://doi.org/10.1130/G51656.1

Cannon, J. M., Murphy, M. A., & Taylor, M. (2018). Segmented strain accumulation in the High Himalaya expressed in river channel steepness. *Geosphere, 14*(3), 1131–1149. https://doi.org/10.1130/GES01508.1

Gupta, H. K. (2023). Himalayan seismic belt, seismic gaps and related issues. *Journal of Geological Society of India, 99*, 1187–1190. https://doi.org/10.1007/s12594-023-2450-6

Gaisser, M. (1980). Length of the Himalayan suture zone: Approximately 2500 km. *Journal of Geotectonics and Geodynamics, 12*(3), 245–260.

Gansser, A. (1980). *The geology of the Himalayas.* Wiley-Interscience.

Hodges, K. V. (2000). Tectonics of the Himalayas and southern Tibet from two perspectives. *Geological Society of America Bulletin, 112*, 324–350.

Lave, J., Yule, D., Sapkota, S., Basant, K., Madden, C., Attal, M., & Pandey, R. (2005). Evidence for a great Medieval earthquake (~1100 AD) in the Central Himalayas, Nepal. *Science, 307*(5713), 1302–1305. https://doi.org/10.1126/science.1104804

Molnar, P., & Tapponnier, P. (1975). Cenozoic tectonics of Asia: Effects of a continental collision. *Science, 189*(4201), 419–426.

Molnar, P., & Tapponnier, P. (1977). Relation of the tectonics of eastern China to the India-Eurasia collision: Application of slip-line field theory to large-scale continental tectonics. *Geology, 5*(3), 212–216.

Molnar, P., & Tapponnier, P. (1978). Active tectonics of Tibet. *Journal of Geophysical Research, 83*(B11), 5361–5375. https://doi.org/10.1029/JB083iB11p05361

Molnar, P., & Tapponnier, P. (1979). The collision between India and Eurasia and its consequences. *Annual Review of Earth and Planetary Sciences, 7*(1), 219–256.

Nakata, T. (1989). Active faults of the Himalaya of India and Nepal. *Geological Society of America Special Papers, 232*, 243–264.

Nakata, T., & Kumahara, Y. (2002). Active faulting across the Himalayas and its significance in the collision tectonics. *Act Fault Research, 22*, 7–16.

Replumaz, A., & Tapponnier, P. (2003). Reconstruction of the deformed collision zone between India and Asia through the backward motion of lithospheric blocks. *Journal of Geophysical Research, 108*(B6), 2285. https://doi.org/10.1029/2001JB000661

Srivastava, H. N. (1992). Earthquake prediction studies in Himalaya: Critical evaluation, Himalayan seismicity (Editor G.D. Gupta). *Memoir Geological Society of India, 23*, 151–170.

Srivastava, H. N. (2004). *Earthquakes: Forecasting and mitigation* (p. 344). National Book Trust.

Srivastava, H. N., Vermaz, M., Bansal, B. K., & Sutar, A. K. (2015). Discriminatory characteristics of seismic gaps in Himalaya. *Geomatics, Natural Hazards and Risk, 6*(3), 224–242. https://doi.org/10.1080/19475705.2013.839483

Summerfield, M. A. (1991). *Global geomorphology: An introduction to the study of landforms* (pp. 57–81). Pearson Prentice Hall.

Thakur, V. C. (2013). Active tectonics of Himalayan frontal fault system. *International Journal of Earth Science, 102*, 1791–1810. https://doi.org/10.1007/s00531-013-0891-7

Tapponnier, P., & Molnar, P. (1977). Active faulting and tectonics in China. *Journal of Geophysical Research, 82*, 2905–2930.

Tapponnier, P., & Molnar, P. (1979). Active faulting and Cenozoic tectonics of the Tian Shan, Mongolie, and Baikal regions. *Journal of Geophysical Research, 84*, 3425–3459.

Tarbuck, E. J., Lutgens, F. K., & Tasa, D. G. (2010). *Earth: An introduction to physical geology* (10th ed.). Pearson Education.

Tarbuck, E. J., Lutgens, F. K., & Tasa, D. G. (2016). *Earth: An introduction to physical geology* (pp. 366–367), Pearson Publication.

Taylor, M., & Yin, A. (2009). Active structures of the Himalayan-Tibetan orogen and their relationships to earthquake distribution, contemporary strain field, and Cenozoic volcanism. *Geosphere, 5*(3), 199–214. https://doi.org/10.1130/GES00217.1

Tripathy, R. (2018). *Tectonic history and geodynamics of the Himalayan region.* Springer.

Tiwari, J., Goswami, P. K., Bhakuni, S. S. (2021). Pattern of active crustal deformation in a part of a Lesser Himalayan tectonic window, Himachal Pradesh, India. *Quaternary International, 592*, 111–120. https://doi.org/10.1016/j.quaint.2021.03.043.

Tiwari, V. M., Rai, S. S., Gahalaut, V. K., & Mishra, P. (2021). Active deformation and seismicity in the Himalayan region: A geophysical perspective. *Geophysical Research Letters, 48*(9), e2020GL091578.

Yin, A. (2000). Mode of Cenozoic east-west extension in Tibet suggesting a common origin of rifts in Asia during the Indo-Asian collision. *Journal of Geophysical Research, 105*(B9), 21745–21759. https://doi.org/10.1029/2000JB900168

Yin, A., (2006). Cenozoic tectonic evolution of the Himalayan orogen as constrained by along-strike variation of structural geometry, exhumation history, and foreland sedimentation: *Earth-Science Reviews, 76*, 1–131. https://doi.org/10.1016/j.earscirev.2005.05.004.

Zheng, Y. F. (2025). A revisit to continental collision between India and Asia. *Earth-Science Reviews, 264*, 105087. https://doi.org/10.1016/j.earscirev.2025.105087, ISSN 0012-8252

Zheng, G., Wang, H., Wright, T. J., Lou, Y., Zhang, R., Zhang, W., Shi, C., Huang, J., & Wei, N. (2017). Crustal Deformation in the India-Eurasia Collision Zone From 25 Years of GPS Measurements. *Journal of Geophysical Research: Solid Earth, 122*(11), 9290–9312. https://doi.org/10.1002/2017jb014465

Landslide Susceptibility and Hazard Assessment

Advances in Understanding Mass Movement and Landslides in the Jammu and Kashmir Region, Western Himalaya: An Integrated Approach from Geomorphological, Geological, and Geotechnical Perspectives

Subhajit Sinha and Surajit Kundu

Abstract India ranks as the second most landslide-prone country globally. The tectonically active and geomorphologically diverse Western Himalaya, particularly the Jammu and Kashmir region, is highly susceptible to landslides due to natural and anthropogenic factors. This study focuses on its ten districts, with six of them being considered as India's most landslide-vulnerable areas. An integrated geomorphological, geological, and geotechnical approach is employed to assess the landslide dynamics in these districts. Twenty important causal factors derived from field-based geospatial and secondary data were taken into consideration, which was later classified into geological, geomorphological, hydrological, and environmental categories. A comprehensive landslide inventory is compiled using the data from authoritative sources. Spatial analysis was performed under the ArcGIS and MATLAB platforms, and the frequency ratio model was applied for unbiased landslide susceptibility assessment. The results produced a susceptibility map that identified three zones: high, moderate, and low risk. Critical hotspots include northern Anantnag, northeastern Riasi and Ramban, and southeastern Udampur. Probabilistic likelihood ratios distinguished positive and negative sub-factor correlations, while spatial multi-criteria evaluation, normalized and integrated into an adaptive neuro-fuzzy inference system, refined vulnerability classification. The findings categorized the region into three distinct risk levels: all ten districts showed moderate susceptibility, eight districts out of ten exhibited moderate to high vulnerability, and four were classified as having moderate to low risk. This study provides critical insights for landslide

S. Sinha (✉)
Department of Geology, University of Calcutta, 35 Ballygunge Circular Road, Kolkata 700019, India
e-mail: ssgeol@caluniv.ac.in

S. Kundu
Department of Geography, Trivenidevi Bhalotia College, Paschim Bardhaman, Raniganj 713347, West Bengal, India
e-mail: surajitkundu@tdbcollege.ac.in

© The Author(s), under exclusive license to Springer Nature Switzerland AG 2025
S. Ghosh (ed.), *The Himalaya Dilemma*, Springer Natural Hazards,
https://doi.org/10.1007/978-3-031-95083-4_7

hazard zonation in the Western Himalaya, emphasizing data-driven zonation for further decision-making.

1 Introduction

The downslope movement of surface material is a prevalent and significant phenomenon in the dynamic processes which is one of the essential agents for shaping the Earth's surface. The downslope movement of individual particles is relatively unusual in phenomenological understanding. Therefore, the phrase 'mass movement' is appropriate for describing the downslope movement of earth's surface materials. The mass movement although confined to very specific regions, defined as the gravitationally-driven downslope displacement of material which has a frequent occurrence. Various geomorphologists have provided insights into the nature of mass movements (Selby, 1993a, 1993b, 1993c; Selby & Hodder, 1993; Turner, 1996). In the Encyclopaedia of Sediments and Sedimentary Rocks, Michael J. Bovis characterised mass movement as the process by which soil and rock materials move downward under the influence of their own weight or potential energy (Middleton et al., 2003). He identified six primary types of movements; namely, creep, sliding, flow, toppling, free-fall, and rolling specifically. The materials resulting from mass movements are typically poorly sorted irrespective of fluvial, glacial, and fluvio-glacial environments. This heterogeneity of material character often complicates the precise identification of the specific type of mass movement and its associated features. Consequently, it is often more practical to refer to these events as "landslides," focusing on the overall mechanism rather than the specific details of the dominant process and outcome. Richard Dikau, in the Encyclopaedia of Geomorphology (Goudie, 2004), further elaborated on mass movement where he describes it as the downward and outward movement of materials that shape slopes under the influence of gravity. Dikau also used the term "landslide" synonymously with mass movement, underscoring its general applicability to a range of downslope processes.

> …in a pure sense the term landslide is used as a generic term describing those downward movements of slope-forming material as a result of shear failure occurring along a well-defined shear plane.
>
> [Encyclopedia of Geomorphology (page 644) by Andrew S. Goudie, 2004.]

J. N. Hutchinson stated

> *Mass movements exhibit great variety, being affected by geology, climate and topography, and their rigorous classification is hardly possible.*
>
> [The Encyclopedia of geomorphology (page 688) by Fairbridge, and Rhodes Whitmore, 1968.]

Mass movement is often perceived as a complex phenomenon due to its variable characteristics. However, the rapid and slow processes involved can be distinguished from one another. Identifying mass movement across terrestrial environments is challenging, as the mechanisms driving these processes often resemble one another with the exception of subsidence. This similarity in underlying mechanisms complicates the clear differentiation of mass movement types which making it difficult to precisely characterise them in the broader context of Earth's surface dynamics.

Karl Terzaghi (1950) described

> The term landslide refers to a rapid displacement of a mass of rock, residual soil, or sediments adjoining a slope, in which the center of gravity of the moving mass advances in a downward and outward direction. (page 83)

Terzaghi (1950) also said that

> If the difference between landslides and continuous creep resided only in the velocity of the movement, it would hardly be justifiable to consider landslides and creep as different types. (page 85)
>
> [Application of Geology to Engineering Practice by Sidney Paige, Geological Society of America, 1950]

The term "landslide" has gained widespread recognition and acceptance within both the scientific community focused on surface processes and disaster management. It refers to a specific type of mass movement which is characterised by the downslope displacement of Earth materials driven by gravitational forces and often augmented by fluid action. The materials involved in landslides are described using various terms depending on the specific Earth science discipline and related research with the common descriptors including debris, soil creep, and rock fragments etc. Notably, landslide materials differ significantly from water as their density is typically at least ten percent greater than that of water. Consequently, the role of wet soil and gravity-driven flows is crucial in the occurrence of landslides worldwide, contributing to the complexity and variability of these events (Fabio Vittorio de Blasio, 2011).

U. S. Geological Society stated that…

> A landslide is a downslope movement of rock or soil, or both, occurring on the surface of rupture—either curved (rotational slide) or planar (translational slide) rupture—in which much of the material often moves as a coherent or semicoherent mass with little internal deformation. It should be noted that, in some cases, landslides may also involve other types of movement, either at the inception of the failure or later, if properties change as the displaced material moves downslope.
>
> [The Landslide Handbook—A Guide to Understanding Landslides; Page 5 (Highland & Bobrowsky, 2008)]
>
> *Landslide and avalanche occurred due to the fast or slow movement of earthy materials (rocks, boulder, soil*etc.*) or snow/ice blocks in mountains or hilly regions under the influence of gravity.*
>
> [National Disaster Management Authority, 2023]

'Land' is a geographical space on the earth surface comprising 3-dimensional spatial property (latitude, longitude and elevation) and 'slide' is the downward movement of the solids or fluid mixed solids resting on that geographical space whose angle of repose is greater than the critical angle of that particular material.

Landslides are the downward movement of solids or fluid mixed solids that moves under the influence of gravity.

Landslides have been categorised by various researchers and organizations across different countries. The classification system proposed by David J. Varnes being one of the most widely regarded. Varnes developed an initial general classification in 1958 which he subsequently refined in 1978 and again in 1996. His material- and movement-based classification system is valuable for theoretical understanding though it is less suited for practical applications. Region-specific classifications can prove to be more appropriate for more effective standardisation of classification schemes. In the context of the Kashmir Himalaya, we focused on the primary types of material movement and the predominant complex movement patterns (Table 1). We modified the existing classification system by adjusting the complex movement types to better align with the region's characteristics based on spatial data we gathered and insights from our field visits to the landslide-prone areas.

Landslides are a widespread natural hazard that can occur on any terrain irrespective of the type of landscape. They are particularly prevalent and frequent in mountainous regions across the globe and new fold mountains are especially prone to such events due to their inherent instability. The Himalayas are no exception to this pattern. Landslides are frequent occurrence in various regions of the range. Data from NASA's Landslide Catalogue spanning from 2007 to 2018 reveals that India ranks second worldwide after the United States in the total number of landslide events. Also, India holds the second position globally in terms of fatalities resulting from these events.

Table 1 Types of mass movement (modified after Varnes, 1978)

Type of movement		Type of material		
		Bedrock	Engineering soils	
			Predominantly course	Predominantly Fine
Falls		Rock fall	Debris fall	Earth fall
Topples		Rock topple	Debris topple	Earth topple
Slides	Rotational	Rock slide	Debris slide	Earth slide
	Translational			
Lateral Spreads		Rock spread	Debris spread	Earth spread
Flows		Rock flow (deep creeps)	Debris flow	Earth flow
			(soil creep)	
Complex		Combination of two or more principal types of movement (Rock fall, Rock fall Debris topple, Rock fall Debris fall, Soil creep, Soil creep Debris fall, Soil creep Debris spread, Debris fall, Debris topple Soil creep Rockfall, Debris slide Soil creep Earth spread, Debris flow Earthflow, Earthflow, Unrecognizable)		
Water Content		Water content $\leq$ 10% is Debris flow		
		Water content more than $\geq$ 10% and < 50% is Mudflow		
		Water content > 50% is Stream flow		

According to the Centre for Research on the Epidemiology of Disasters (CRED) and its EM-DAT database, India is classified as one of the most vulnerable countries to landslides due to its geological conditions, rapidly growing population, and inadequate disaster management infrastructure in certain regions. Landslide study involves understanding causes (natural and human-induced), assessing risks, implementing early warning systems, adopting mitigation measures (engineering, land use planning, and vegetation), preparing for emergencies, addressing economic and environmental impacts, establishing policies, raising community awareness, advancing research, and adapting to climate change impacts. Landslide research faces challenges due to the complexity of mechanisms, variability in triggers, difficulties in data collection and monitoring, impacts of climate change, human activities, limitations in modeling and technology, socioeconomic and political barriers, regional variability, and the need for sustainable long-term monitoring and maintenance. This combination of factors makes India highly susceptible to the destructive impacts of landslides which urge the need for focused research aimed at improved early-warning systems and enhanced mitigation strategies to reduce the risks associated with these frequent natural disasters.

Human settlements are increasingly expanding toward high-altitude landscapes of the Himalayan region that are prone to landslides. The significant driving forces behind these are the growing global interest in tourism, cultural diffusion, and the allure of these majestic mountains mainly resulting an intensified human presence throughout the region. Climate change has emerged as a critical factor exacerbating environmental stresses is standing alongside. Concurrently, modern societal transformations entail an influx of amenities and infrastructure which are contributing to substantial pressure on these mountainous ecosystems rapidly. This intensified human encroachment in this mountainous region is forcing the construction of essential infrastructures including roads, tunnels, buildings etc. Such rapid progress is driven by an expanding demand for access and services that often leads to unintended consequences leading to clash between the hastily implemented developmental projects and the fragile natural environment. The unique geological conditions of the region characterised by a new fold mountain belt and associated surface processes are further complicated by climate change added to the human interventions. These combined forces significantly undermine the stability of the Jammu and Kashmir region, heightening vulnerability to a range of natural hazards and disasters. Landslides have emerged as a particularly prominent and concerning issue following the ongoing environmental instability in the region. This study aims to understand landslide causes and mechanisms in the region, develop risk assessment, enhance overall predictability, address inclusive climate and human-induced challenges by providing vulnerability zone mapping.

2 Study Area

2.1 Location

The accessibility of the Himalayan region towards Srinagar and the further northerly regions is through Jammu and Kashmir roadways. Hence, neglecting the region's hazard vulnerability will affect the ease of accessibility to natural beauty, declining substantial foreign investment, and strategic position towards the approachability of the higher mountainous terrains. Specifically, the former Jammu & Kashmir state or particularly the Kashmir region is synonymous with the epithet 'Heaven on Earth'. This so-called 'Heaven' is nestled within the Kashmir Basin and framed by the majestic Greater Himalayan Range to the north and the Pir Panjal Range to the south.

2.2 Climate

The India Meteorological Department (IMD) reports significant temperature variations in Jammu and Kashmir region due to its diverse topography and global trend of climate change. Summer season (May–August) experiences temperature ranging from 20 to 30 °C in the valleys of Jammu, while winters (December–February) can drop to sub-zero levels, particularly in higher regions like Srinagar, Pahalgam etc. The Kashmir Valley often experiences snowfall with temperatures ranging from 0 to 7 °C in winters and these variations are influenced both by the Indian Summer Monsoon and the Westerlies (IMD Reports, 2001–2023). In this connection, different scientific studies have identified significant warming trend in the region with the higher minimum temperatures and the higher precipitation patterns throughout the year with significant annual variation observed especially in the first phase of twenty-first century. The average annual temperature of Jammu and Kashmir has increased approximately 1.2–1.4 °C over the last century which is higher than the global average (ICIMOD, 2022). In tandem the average summer temperatures has also increased by 0.5–1.0 °C over the past few decades (ICIMOD, 2022). Impacts of climate change on Western Himalayan ecosystems in India is due to changes in temperature variation, increased precipitation and biodiversity deterioration particularly (Rasul & Molden, 2019). This has also implications for glacial melt and uncontrolled water availability which is creating significant problems for this region. The spatial heterogeneity in climate trends with higher-altitude regions experiencing more pronounced changes. These trends have significant impact for the physical landscape like region of glaciers, rivers, and ecosystems which are potentially affecting the hydrological cycle and water management along with the anthropogenic environment including socioeconomic conditions and local livelihood (Pai et al., 2014).

2.3 Geology

The region is anchored in a Pre-Cambrian hard rock foundation with the Plio-Pleistocene (Karewa) and Late Pleistocene (Alluvium) deposits overlaying the ancient bedrock (Wadia, 1953). The Kashmir Basin (Table 2) lies within the embrace of the young fold mountain system which is geologically and tectonically unstable (Ahmad et al., 2009; Mukhopadhyay & Dasgupta, 2015). Despite the inherent instability of the region's geology, human habitation has been proliferating at an accelerated pace. The seismically active zone is susceptible to tectonic activity coupled with its natural allure which has led to rapid urbanisation. The juxtaposition of environmental beauty and geological volatility presents a complex challenge to sustainable development in this area (Bhat et al., 2024; Burbank et al., 1983; Valdiya, 2016; Wadia, 1953).

2.4 Geomorphology

The Western Himalayas, particularly in the Jammu and Kashmir region, exhibit a diverse and complex terrain characterised by dramatic variations in elevation, geology, and geomorphology. The region is categorised into three principal geomorphological zones, namely, the Outer Himalayas or Siwaliks or Jammu Region, Middle Himalayas or Kashmir Valley and Trans-Himalayas or Ladakh Region (Sarkar & Gupta, 2012). The Jammu Region occupying the southernmost extent of Jammu & Kashmir is characterised by low-relief topography, encompassing hills showing cuesta structure, steep-walled valleys, and foothills of the Siwalik range. It is composed of young, unconsolidated sedimentary deposits including gravel, sand, and clay which rendering this region highly susceptible to erosional processes, landslides, and flash floods due to its geologically fragile structure. Middle Himalayas is a fertile intermontane basin which is bounded by the Pir Panjal and Greater Himalayan ranges and formed through a combination of tectonic and glacial processes (Ganjoo, 2014). This valley's geomorphology is defined by elevated plateaus originating from ancient fluvial and lacustrine deposits. The Trans-Himalayas are the high-altitude cold desert regions characterised by rugged, mountainous terrain, deep gorges, and glacial valleys with prominent geomorphic features include moraines, glacial lakes, and alluvial fans. This tripartite division underscores the region's geomorphic diversity, driven by tectonic, glacial, and fluvial processes, which are collectively define its dynamic landscape.

Table 2 The stratigraphic sequence of Kashmir Basin (after Iqbal et al., 2018)

Rock formations	Age	Thickness in meters	Description
Quaternary alluvium	Recent	–	Unconsolidated deposits of gravel, silt and clays
Unconformity			
Mirpur formation	Pleistocene	20	Conglomerates having cobbles & pebbles of igneous, metamorphic and sedimentary rocks
Unconformity			Clays, claystone and grey sandstone. Clays are brown and yellowish grey
Soan formation	Pliocene	600	Dominantly consists of sandstone, siltstone and clays. Sandstone is grey, fine to medium grained and medium to thick bedded
Dhok Pathan formation	Late Miocene	1720	Dominantly it consists of greenish grey sandstone, siltstone and mudstone. Sandstone is massive and medium to coarse grained. Sandstone and clay ratio is 60%
Nagri formation	Late Miocene	1160	Red to purple, greenish grey, ash grey sandstone and siltstone. The mudstone is purple and reddish brown. The Formation has 70% clays and 30% sandstone
Kamlial formation	Middle to Late Miocene	933	Mainly sandstone, clays and intraformational conglomerates
Murree formation	Early to Middle Miocene	261	Mostly clays, shales and sandstone. Sandstone is red to purple red and fine to medium grained
Kuldana formation	Early Miocene	682	Variegated shales with subordinate sandstone. Shales are arenaceous
Chorgali formation	Middle to Late Eocene	150	Mostly calcareous shale, limestone and dolomitic limestone
Margalla Hill limestone	Early Eocene	50	Main nodular fossiliferous limestone with shales
Patala formation	Early Eocene	22	Mainly shales interbedded with marl and limestone
Lockhart formation	Late Paleocene	70	Grey to dark grey limestone with subordinate shales
Hangu formation	Early Paleocene	10	Mainly laterite, bauxite and fireclay
Unconformity			
Muzaffarabad formation	Cambrian	254	Mainly dolomitic limestone with cherty dolomite and chert bands

(continued)

Table 2 (continued)

Rock formations	Age	Thickness in meters	Description
Unconformity			
Dogra formation	Precambrian	1000	Slates

2.5 *Land Use and Land Cover*

Anthropogenic activities have emerged as the primary driver of Landuse and Land-cover (LULC) changes globally and significantly altering ecosystems every day. The modifications are governed by a string of forces that are mutually interconnected with each other. LULC changes also impacted profoundly on ecological balance which increasing vulnerability to landslides and other hazards in any region. The western Kashmir Himalayan regions are no exception and is experiencing widespread LULC shifts that have led to numerous environmental challenges. Since 2005, the National Remote Sensing Centre (NRSC), under ISRO's Natural Resources Census (NRC) Programme, has conducted annual LULC assessments for India using multi-temporal and multi-sensor satellite data, primarily AWiFS which shows that over the past three decades, the western Kashmir Himalayas have witnessed significant LULC transformations including a notable decline in water bodies (~18.21%), forests (~13.56%), snow/glacial cover (~29.32%), and agricultural land (~22.37%). Conversely, urban areas (~22.33%), barren land (~37.32%), plantations (~14.53%), and marshes (~13.21%) have expanded (Hamid et al., 2024). These shifts highlight the growing influence of human activities on the region's landscape, underscoring the need for sustainable land management to mitigate environmental risks like landslides which are most frequent hazardous event. The global climate change, uncontrolled population growth, diverse economic compulsion, and unplanned development among others are the major causes or drivers of rapid LULC changes in the study region.

3 Background

The Jammu & Kashmir region situated in the northern expanse of the Indian subcontinent which approximately spans an area of 15,297.80 square kilometres. Our study specifically encompasses ten selected districts of this region, namely Srinagar, Anantnag, Pulwama, Badgam, Shupiyan, Kulgam, Riasi, Ramban, Udhampur, and Jammu. All these districts are of paramount significance and lies within the Kashmir basin. These districts are playing key roles in tourism, cultural activities, economic ventures, and serving as vital transportation link in the region with the other parts of Indian subcontinent. These districts are home to over eight million residents at present and 7,003,727 population were there according to the 2011 Indian Census.

The majority of the population of the former Jammu & Kashmir state resides within these ten districts. The National Highway 44 (NH-44) is a crucial route that stretches over 240 km from Jammu in the south to Srinagar in the north which facilitates the primary connectivity with rest of India as well as this route is currently facing frequent landslide events. This highway is lifeline of this region and it provides essential access to different historical landmarks, economically important zones, and facilitates travel for both domestic and international visitors. The livelihood and economic prosperity of the local population not only depends on the rich cultural heritage but also on the diverse local industries like cash-crop agriculture, mineral extraction, and transportation etc. This basin is highly susceptible to natural hazards and landslides is the most frequent natural disaster. The Landslide Atlas 2023 compiled by the National Remote Sensing Centre (NRSC) and the Indian Space Research Organisation (ISRO) has identified six of the ten districts selected for this study as landslide-prone areas out of the 147 districts throughout India. Specifically, the districts of Jammu (ranked 14th), Udhampur (17th), Pulwama (27th), Anantnag (52nd), Srinagar (98th), and Badgam (119th) are most vulnerable due to extensive landslide occurrences. The remaining four districts also experiences frequent landslide events emphasizing the widespread susceptibility of the region to such hazards. The prevalent landslide activity in these districts poses substantial challenges to both the safety of the resident population and the stability of the economic infrastructure.

3.1 Rationale of Selecting the Study Area

The examination of this particular region is crucial for advancing sustainability and achieving sustainable development objectives. Several key factors were considered in selecting this study area (Fig. 1). *First*, the diverse landscape includes both mountainous and valley terrains which makes the study area an exemplary case for assessing landslide vulnerability. This region is characterised by intricate geological formations consisting of slate, sandstone, quartzite etc. with dip varying between 45° and 55° N-NE sometimes reaching very steep angle that makes the region considerably prone to devastating landslides.

Second, the area has a well-established history of landslides with frequent occurrences particularly prevalent during monsoon seasons and following different seismic events throughout the year. Analysing these historical events can provide invaluable insights into the patterns and triggers of landslides to enhance our understanding of the underlying causes. *Third*, the effects of climate change manifested through intensified rainfall and accelerated glacier melt further compound the landslide risk in this sensitive region. This underscores the importance of investigating how climate change-induced shifts affect the geological stability of the area which emphasising the need for vulnerability assessments in response to the current environmental changes. *Fourth*, since the region holds a significant socio-economic value, the growing anthropogenic encroachment further added a factor to the intensification of landslides. *Fifth*, while considerable research has been conducted on landslides in general, there is a

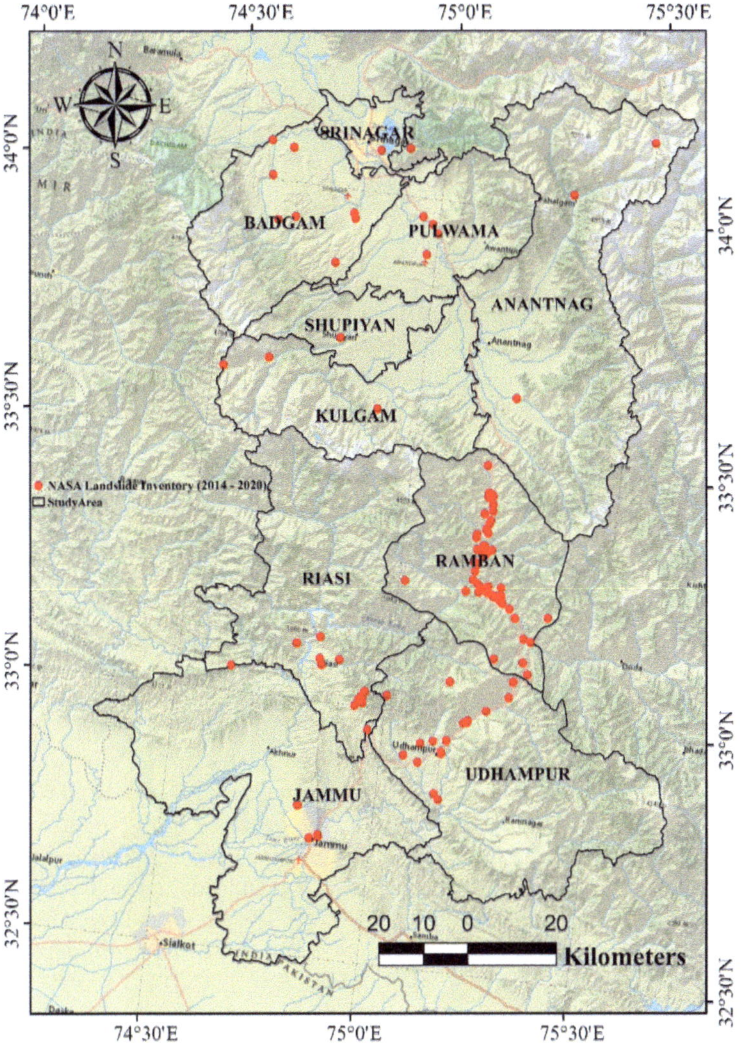

Fig. 1 Map of the study area showing the historical occurrences of landslides between 2014 and 2020

distinct paucity of studies that leverage advanced geospatial technique within this particular region. The use of geospatial technique adds visuality to the inaccessible areas, facilitates larger spatial coverage and high-resolution temporal data deliverables. *Sixth*, the region's intrinsic vulnerability to natural disasters necessitates the development of effective risk assessment frameworks and disaster management strategies.

Addressing the above-mentioned issues and finding the scientific solutions for developing targeted risk management strategies will not only contribute to the academic understanding of landslide dynamics but also foster more sustainable and resilient development practices in the region.

4 Materials and Methods

4.1 Database

Our approach involved consulting a range of published studies and relevant literature on landslides. We recognised the limitations of relying solely on satellite-based inventory data, hence, supplemented this information by gathering data from national and local archives as well as databases containing historical records of landslide occurrences within the study area. Subsequently, many field visits were conducted to various targeted locations where active landslides are meticulously identified and spatial data were collected using hand-held Global Positioning System (GPS) devices. The fieldworks covered the ten selected districts excluding some challenging areas with minimal anthropogenic influences. The landslide data were primarily collected from regions with significant human-nature interactions where at least one instance of human activity was evident. Approximately ten percent of the data was also gathered from areas with no anthropogenic influence as a control group for validation and cross-verification of the collected data to ensure data accuracy and reliability.

A total of nine different types of specific data were collected from reputed government and non-government sources for further processing. *First*, Digital elevation void-filled data with 1° spatial resolution image collected from NASA Earth Data website. *Second*, Lithology data collected as raster map from USGS archive following U. S. Geological Survey Open File Report 97-470C and Geological Survey of India. *Third*, Gridded Rainfall data with 0.25×0.25 spatial resolution collected from Indian Meteorological Department database. *Fourth*, FAO coded Soil world data accessed from Food and Agricultural Organisation archive. *Fifth*, LANDSAT 9 satellite images collected from USGS earth explorer during 1st January 2023 to 31st December 2023. *Sixth*, District shapefile of the study area downloaded from Survey of India (Order ID: OD24022253165). *Seventh*, peak ground acceleration data with 2.5-min global resolution data collected from NASA socio-economic data and application centre (SEDAC). *Eighth*, Lineament data collected from BHUVAN: Indian Geo-Platform of ISRO. Last, the *ninth* i.e., landslide occurrence point collected from field survey using Garmin Global Positioning System eTrex30.

The collected nine types of data were segregated into four major categories considered as prime factors for landslide occurrence and susceptibility. Those are the geological factor, geomorphological factor, hydrological factor and environmental factor. Each major factor is a combination of more than one type of data. Geological factor constitutes lithology (L), lineament (LM) and distance from fault line (FL). Geomorphological factor comprises of elevation (E), slope (S), aspect (A), plan curvature (PC), profile curvature (PRC), topographic position index (TPI), surface area ratio (SAR), and soil cover (SC). Hydrological factor includes annual rainfall (R), drainage density (DD), stream power index (SPI), topographic wetness index (TWI), and distance to drainage (DDR). The environmental factor is constituted of normalised difference vegetation index (NDVI), land use & land cover (LULC), peak ground acceleration (PGA), and distance from roads (DRR). Six particular variables among the twenty variables are identified as independent variables for triggering landslide occurrences. These independent triggering mechanisms are annual rainfall, distance to drainage, normalised difference vegetation index, land use & land cover, peak ground acceleration, and distance from roads. The other factors work in association to initiate a slide.

4.2 *Inventory Creation*

Substantial research has been conducted in various countries, including India, to identify landslide-prone zones and to develop effective strategies for managing the disastrous consequences both during and after landslide events. The creation of spatial databases is fundamental to any landslide study as it forms the foundation for identifying vulnerable areas. The preparation of a landslide inventory is crucial considering both historical and current landslide occurrences. This inventory serves as a prerequisite for pinpointing regions at risk regardless of their spatial scale, which can range from several square kilometers to thousands of square kilometers. Long-term data collection and validation against various sources—such as global databases, national-level records, regional archives, and local information—are essential for the creation of a precise inventory. This multiple source database ensures the accuracy and reliability of the landslide inventory. A sufficient number of point data for smaller regions is typically acceptable for landslide studies, while polygon data offers greater accuracy for larger areas. However, there are no standardised guidelines regarding the minimum or maximum number of data points required, nor are there universally applicable statistical methods for determining the suitability of data. But, any form of sampling method is strictly prohibited in data collection for detailed and accurate study purposes. As a result, the notion of long-term data collection and ensuring an adequate sample size remains somewhat ambiguous. Although collecting data for every individual landslide event is virtually impossible due to the varying filtration processes in different databases, efforts were made to analyse both point and polygon data for accuracy assessment. The collected data was further classified based on its nature and characteristics, providing a clearer understanding of the types of landslides present in the region. The process of gathering comprehensive landslide data is challenging. Though a detailed field survey remains highly recommended for the best outcome. Such surveys would be invaluable in obtaining both point and polygon data for this methodology. But this kind of data set is beneficial for historical trend analysis and reliability of locational information along with the future intensive research purposes.

4.3 Choice of Causal Factors

Understanding the causal factors behind landslides is crucial for assessing their occurrence, intensity, and implementing appropriate mitigation measures. The causal factors of landslides vary across different regions and even for the same location in consecutive years. These factors can be classified into two categories. The first category comprises spatial factors which are directly related to regional geology, local geomorphology, hydrology, vegetation cover, and similar elements. These factors are generally static or change at an extremely slow pace often requires centuries to detect significant alterations. The second category includes triggering factors which have the potential to intensify landslide events by influencing whether they occur as 'high magnitude and low frequency' or 'low magnitude and high frequency' events.

The factor of safety (FS) is a stability measure, defined as the ratio between the actual strength of a material and the hypothetical mobilised strength needed to achieve a state of limit equilibrium. This equilibrium condition assumes the equal mobilisation of both cohesion and friction forces within the material. The FS serves as a critical indicator of the material's capacity to resist failure under specific conditions, with values greater than one indicating stability and values less than one suggesting potential instability (Klar et al., 2011); that is,

$$FS = \frac{\tau_f}{\tau_m} = \frac{c + \sigma\, tan\phi}{c_m + \sigma\, tan\phi_m} = \frac{tan\phi}{tan\phi_m}\frac{c}{c_m}$$

where ϕ and c are the true strength properties of the slope,

ϕ_m and c_m are the mobilised strength properties required for the limit (equilibrium) state,

τ_f is the strength and

τ_m is the mobilised strength with a factor of safety FS.

A factor of safety greater than 1 (FS > 1) indicates stability, meaning the system possesses more strength than required for limit equilibrium. Conversely, a factor of safety less than 1 (FS < 1) is physically unrealistic as the system cannot mobilise more strength than its inherent capacity. However, applying this concept to landslides is not straightforward; rather, the process is inherently complex. The intricacies of various contributing factors necessitate simplification to effectively describe and analyse each individual component influencing stability, making the overall system's behaviour challenging to assess accurately (Klar et al., 2011).

True strength of any landscape can be influenced by a multitude of factors such as geomorphological characteristics, landform conditions, surface cover, water availability, climatic conditions etc. Likewise, the mobilised strength of a landscape is contingent on various parameters encompassing both endogenic (internal) and exogenic (external) processes. Many of these factors are not readily available in quantitative terms, which highlights the necessity of quantification to comprehensively understand the landslide phenomenon. We have identified numerous parameters that contribute either directly or indirectly to the instability of the landscape through our study. Some of these factors are inherently qualitative in nature and can be observable on the Earth's surface, which requires precise numerical assessment for effective analysis.

Lithology encompasses the physical, geochemical, and mineralogical characteristics of geological materials. Global datasets, such as those provided by the US Geological Survey, alongside regional data sources like the Geological Survey of India (GSI) for the Indian subcontinent, are invaluable for lithological analysis. It is crucial to examine the surrounding region of different lithological units with particular attention, as these regions tend to exhibit greater instability compared to other sections of the terrain. The stability of the landscape is also influenced by its susceptibility to weathering, erosion and the seismo-structural makeup along with other destabilising forces. In the study area, twelve different types of lithological units are identified constituting undivided Paleozoic rocks, Triassic metamorphic, Neogene sedimentary rocks and the Quaternary sediments (Wandrey and Law, 1998).

The term "*lineament*" was first introduced in the field of geology by William Herbert Hobbs (Hobbs, 1904). Simply, a lineament refers to linear or curvilinear features present on the Earth's surface that are typically identifiable and mappable. These features can be either simple or composite, comprising one or more distinct linear elements. Lineaments are products of purely natural geological processes often considered to reflect subsurface phenomena in areas unaffected by human activity, thus being. Lineaments can also serve as distinct landforms within a specific region and also provide valuable insight into the underlying geological structure. The identification and verification of lineaments from satellite imagery may sometimes be compromised by image contrast issues leading to the potential misinterpretations. Nevertheless, lineament analysis remains a crucial tool in the study of geological features as it aids in identifying fault systems and fractures within the Earth's crust. Lineament data has been widely utilised since 1972 for a variety of applications, including structural investigations, resource exploration, and geohazard assessment. This method continues to play a significant role in advancing our understanding of Earth's geological processes. This helps to inform strategies for natural resource management and risk mitigation related to geohazards (Gabrielsen & Olesen, 2024). Diversified lineament composition exists throughout the whole region except the low elevation plain lands of Jammu. Pulwama, Ramban, Kulgam and Anantnag comprises most lineament.

Distance from fault line is another important factor for stable and risk-free zone identification. Fault lines are significantly unstable areas in the new fold mountains. Any kind of regional planning without considering fault lines can be a disastrous situation in near future. Distance from fault line should be measured because it can influence the frequency and intensity of the landslide (Zhuang et al., 2010). The study area contains many regional thrusts mainly the Balapur Thrust that runs through Badgam, Shupian, Kulgam; the Main Boundary Thrust and Main Central Thrust passes through Ramban and Riasi; the Panjal Thrust runs through Ramban and the Jwalamukhi thrust through Udhampur along with many other minor faults and thrusts (Ansari et al., 2022). The landslides sites are located at varying distances from the faults and thrusts present in the area.

Elevation data refers to the altitude of a particular place above mean sea level. Digital elevation data is the most suitable for elevation calculation and mapping any large region where manual altitude determination is either hectic task or possible to face errors. Shuttle radar topographic mission of NASA provides void filled 1-arc second 30-m resolution digital elevation all over the world. GPS use during field verification is recommended for verification of altitude. Elevation class determination depends on the researcher's choice and understanding of the problem (Dai & Lee, 2002; Meunier et al., 2008). Elevation value ranges from 233 to 5,347 m and Ramban, Riasi, Kulgam, Udhampur, Anantnag covers the most complex elevation landscape which are experiencing maximum number of landslide events.

Slope is the steepness or the degree of inclination of a surface. Slope computation from either a raster or Triangulated irregular network surface is essential.

The slope for a slide location is the maximum rate of change of elevation between that location and its surroundings (Fig. 2). The data collection from large place of interest is near to impossible which can be overcome by using advanced GIS platform. Landslide depends on the degree of slope and detachment of the material from a place depends on the shear stress. Mohr–Coulomb equation stated that the successful slide depends on the downslope component of the bulk weight of the material that equals shear strength along a plane of detachment. It can be described as:

$$s = c + (\sigma + \mu) tan\emptyset$$

where s is shear strength, c is cohesion, σ is total normal stress, μ is pore water pressure, and $\emptyset$ the angle of shearing resistance (Barton & Choubey, 1977). Maximum slope angle is 74° and most of the region comprises rugged topography with slope variety from 35° to 57°.

Aspect is the direction or orientation of slope faces generally measured clockwise in degrees or cardinal direction from 0 to 360. The 0 is north-face, 90 is east-face, 180 is south-face, and 270 is west-face. This direction is determined as the direction of any person would be looking downhill toward downhill at the line of steepest descent. Michael de Smith states Aspect as the directional component of the gradient vector and is the direction of maximum gradient of the surface at any given point. Aspect calculation with slopes from the partial derivatives are (Smith et al., 2018):

$$\frac{\partial z}{\partial x}, \frac{\partial z}{\partial y}$$

where,

z is the elevation or gradient

x and y is the position vector or direction

Aspect is computed in degrees from north which is an azimuth in degrees not radians. The expression is (Skidmore, 1989):

$$A = 270 - \frac{360}{2\pi} tan^{-1}\left(\frac{\partial z}{\partial x}, \frac{\partial z}{\partial y}\right)$$

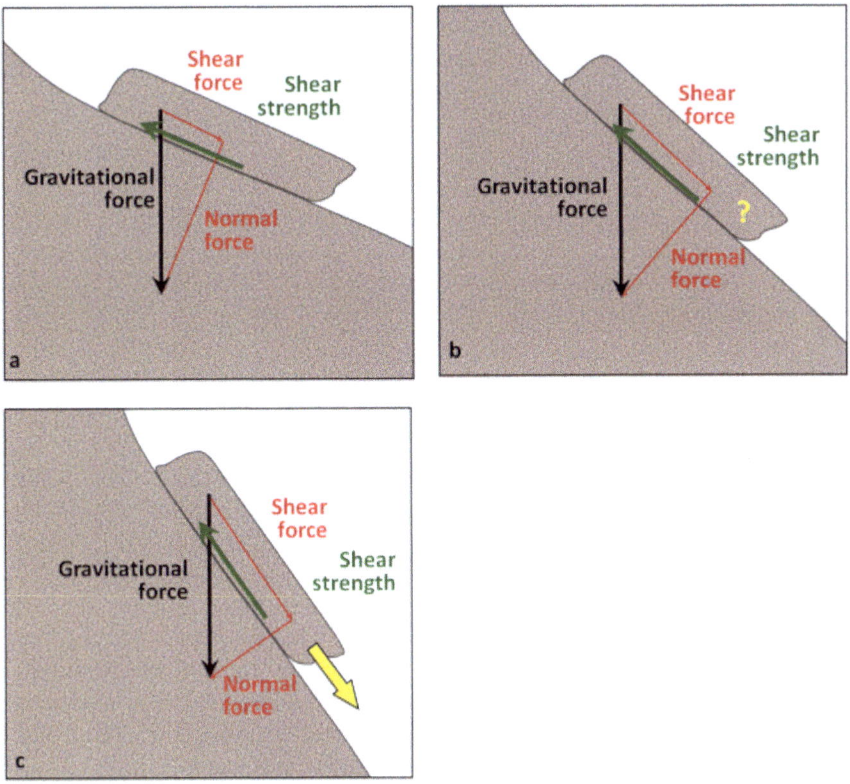

Fig. 2 The gravitational force is the same in all three cases. In **a** the shear force is less than the shear strength, so the block should be stable. In **b** the shear force and shear strength are about equal, so the block may or may not move. In **c** the shear force is greater than the shear strength, so the block is very likely to move (After Earle, 2015)

The Curvature means to displays the shape or curvature of the slope. A part of a surface can be concave or convex. The curvature calculation computes the second derivative of the surface. The curvature may be profile curvature or plan curvature. Simply, profile curvature is parallel to the slope and indicates the direction of maximum slope which affects the acceleration and deceleration of flow across the surface. Planform curvature is perpendicular to the direction of the maximum slope which affects the convergence and divergence of flow across a surface.

Profile curvature is the rate of surface slope change, S, with the movement in the direction of the unit vector $(-\nabla f/S)$, (i.e. following downstream). It can be expressed as (Peckham, 2011):

$$\kappa_p = -S^{-1}(\nabla S \cdot \nabla f)$$

Plan Curvature or contour curvature is the rate of flow direction change (given by $\phi + \pi$), with the movement in the direction of ($\nabla^{\perp} f / S$), (i.e. following a contour line). It can be expressed as (Peckham, 2011):

$$\kappa_c = S^{-1}(\nabla\phi \cdot \nabla^{\perp} f)$$

The $\kappa_c\,(x, y)$ measures the curvature of the contour line (a curve in the xy plane) that goes through the point (x, y). Thus, Plan curvature is negative for valleys and positive for ridges. Therefore, Plan curvature is closely related to the tangential curvature, given by (Peckham, 2011):

$$\kappa_t = \frac{f_y^2 f_{xx} - 2 f_x f_y f_{xy} + f_x^2 f_{yy}}{-S^2\sqrt{1 + S^2}}$$

The key difference is that κ_t is a normal curvature. The extra function of S in the denominator is due to the angle between the surface normal and the xy plane. That is, $\kappa_t\,(x, y)$ measures the curvature of the 3D curve formed by the intersection of the surface $f\,(x, y)$ and the plane that contains both the surface normal vector at (x, y) and $\nabla^{\perp} f$. Expressions for plan curvature should not have extra functions of S in the denominator.

Longitudinal profiles in valleys are always concave up. The $\kappa_c < 0$ (valley) implies that $\kappa_p < 0$. Narrower valleys have higher profile curvatures. For a fixed S, $|\kappa_p|$ increases linearly with $|\kappa_c|$. Steeper valleys have higher profile curvatures. For a fixed $\kappa_c < 0$ (valley), $|\kappa_p|$ is a rapidly increasing function of slope, S. In fact, if $\kappa_c = 0$ everywhere, as it is for a sheet-like surface with parallel contour lines (Fig. 3).

Curvature value calculated in radians per meter and maximum curvature showing in Ramban, Riasi, Udhampur and Anantnag. Most of the Jammu, Srinagar, eastern portion of Shupian, Badgam, Kulgam and western stretch of Pulwama experiencing less curvature.

Topographic position index or TPI generally used to distinguish landscape features or terrain classification. Andrew Weiss presented a very interesting and useful poster at the 2001 ESRI International User Conference about the concept of Topographic Position Index (TPI) and how it could be calculated. Proper inclusion of TPI at different scales with slope can classify the landscape into both slope position (i.e. ridge top, valley bottom, mid-slope, etc.) and landform category (i.e. steep narrow canyons, gentle valleys, plains, open slopes, mesas, etc.). The algorithms are simple. The TPI is the basis of the classification system and is simply the difference between a cell elevation value and the average elevation of the neighbourhood around that target cell. Positive value signifies the cell is higher than its surroundings while negative value means it is lower. Slope position calculation combines the degree to which it is higher or lower, along with the slope of the cell. If it is significantly higher than the surrounding neighbourhood, then it is likely to be at or near the top of a hill or ridge. Significantly low values suggest the cell is at or near the bottom of a valley. TPI values near zero could mean either a flat area or a mid-slope area, so the cell

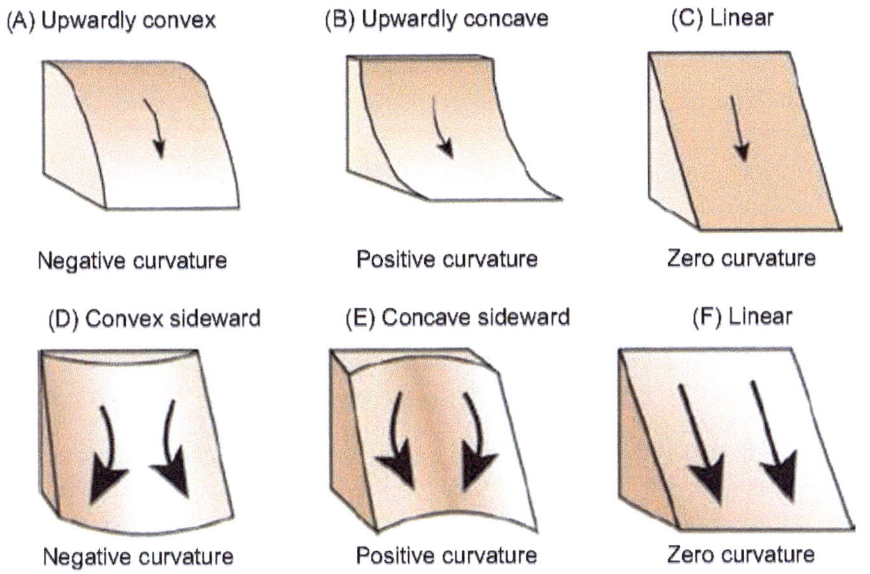

Fig. 3 Schematic diagram showing the different types of profile curvatures where (A) indicates upwardly convex surface indicates negative value where flow will be decelerated, (B) a positive value indicates upwardly concave surface and flow will be accelerated and (C) having zero value indicating a linear surface. The different types of plan curvatures with (D) a positive value indicates laterally convex surface, (E) a negative value indicates laterally concave surface and (F) zero value indicating a linear surface (After Kimerling et al., 2016)

slope can be used to distinguish the two (Sørensen et al., 2006). Highest TPI value Ramban, Riasi, Kulgam, Anantnag and Udhampur. Ramban comprises highest TPI value throughout the whole district.

Surface area ratio is a basis for a useful measure of landscape topographic roughness. The surface area ratio of any particular region on the landscape can be calculated by dividing the surface area of that region by the planimetric area. The early computation method found in Hobson (1972). Ruggedness calculation by Beasom et al. (1983), fractal dimension by Mandelbrot (1982), contour line density by Jenness (2010) created the simplification idea of this. Hodgson (1995) and Berry (2002) method can be thought of as simply using the slope to calculate an adjustment factor for the cell planimetric area. If the slope is flat, then the adjustment factor is equal to 1 and therefore the surface area is equal to the planimetric area. As the slope increases, the adjustment factor is equal to: Slope is expressed here unit per unit. Our case, it is meter per meter.

$$Adjustment\ factor = \frac{1}{cos(slope)}$$

Here, slope is in Radians calculated as

$$Slope(radians) = \frac{[slope(degree)]\pi}{180}$$

Assuming cell size is defined as the length of one edge of the target cell and slope is measured in degrees, then

$$Surface\ area = \frac{c^2}{cos\left(S\left(\frac{\pi}{180}\right)\right)}$$

Here, c is the cell size and S is the slope in degree.

Soil cover is essential for any region on earth and is the end product of rocks and minerals which comprises genetic traits of different earth forming elements that is capable of sustaining life. The scientific study of soil is an assessment of the genetic relationship between soils and landforms (Gerrad, 1992). The water holding capacity, nutrient richness, grain sizes, structure and texture, porosity and permeability etc. defines the actual condition of soil in any region. Modern soil science has emerged in different countries and regional organisations. USDA, FAO like international organisations initiated the world-wide study. ICAR, NBSS&LUP working nicely in India. Therefore, different national and international organisation are responsible for creating and classifying soil cover maps for whole world as well as country and region specific. Genetic characteristics-based naming and code name is the most suitable and scientifically justified (USDA, 2014). FAO code used with the NBSS&LUP classification and output result is showing eight different categories of soils are covering the whole region. More than 95% area is covered with clay, loamy soil and less than 5% area comprises sandy soils.

Annual rainfall is the total quantity of water from the clouds in a particular year. Condensation of atmospheric water vapor that returns to the Earth's surface as rain, sleet, drizzle, hail etc. All these known as precipitation which returns from higher atmosphere by the influence of gravitational force. We do not consider the all-precipitation forms. Only rainfall considered here for some specific triggering characteristics. Rainfall enhances the surface erosion by directly striking the water droplet from higher altitude. It also increases the water quantity in landforms which helps to slide or move any surface over a sloping region. Sufficient amount of rainfall also helps local vegetation to grow properly and stood healthy. A healthy vegetation cover reduces the soil erosion. But, huge amount of rainfall creates the flood prone situation as well as the flood water erodes the soil (Gariano & Guzzetti, 2016). India meteorological department gridded data showing a huge variation in annual precipitation range from 456 mm in north-eastern districts to 2,596 mm in southern districts of Jammu, Udhampur, Ramban, Riasi. Maximum precipitation recorded in the southern part of the study area and only rainfall, snowfall considered as major input of precipitation.

Drainage density is a simple mathematical calculation where the total length of all existing streams (L_t) within a drainage network divided by the total area of the watershed (A_d).

$$\textit{Drainage Density} = \frac{L_t}{A_d}$$

Drainage density is an important geometric parameter for channel networks as it determines the spacing of channels, length of hill slopes, maximum length scale of slope failures and reflects the processes governing landscape dissection (Schumm, 1997). The high value of drainage density refers to the more dissected the basin created by the existing stream network. Drainage density value can vary from single digit to three-digit number. Drainage density differences among the neighbouring drainage basins reflects variations in the effectiveness of some factors that controls the development of channel flow. These factors include climate, past climate conditions, biomass, parent materials, lithology, relief, time and land use primarily. Interaction among these factors in different environmental conditions makes it difficult to determine any independent effects. Generally, drainage density increases with the increasing of precipitation, decreasing vegetation cover, increasing altitude, increasing slope angle, and decreasing surface resistance to erosion. However, drainage density can decrease with increase of slope angle when hillslope erosion is dominated by mass wasting activity (Talling & Sowter, 1999). Interestingly, overland flow does not necessarily originate at the top of hillslope (Horton, 1945; Schumm, 1956). Therefore, water addition in the hillslope results to move the surface material due to the availability of drainage network. Increase of drainage density is an important factor for the slope failure and slide in the hillslope when the hillslope fails to accommodate the channel flow.

Stream power index is a GIS derived calculation of stream power which is devised by Bagnold (1966). Stream power is a measure of the energy of a stream channel which is related to ability to transport the sediment. Equation expressed (Bagnold, 1966) as

$$\Omega = \rho g Q S$$

Here,

Ω is the power, ρ is the density of water (1000 kg/m^3), g is the acceleration of gravity (9.81 m/s^2), Q is the whole discharge of the stream, S is the stream slope.

Drainage density value calculated from GIS platform and ranges from 3.65 km per square kilometers to 31.79 km per square kilometers. Ramban, Riasi and Udhampur comprises highest drainage density regions among the ten districts.

Topographic wetness index or TWI idea is described by Moore et al. (1991). This tool is developed and extensively used to calculate the water accumulation using digital elevation model. TWI signifies the water accumulation potentiality in a given landscape. It depends on the slope and existing depression along with porous-pervious layers on the surface and near surface layer. Material property and slope exhibits a complex form which express the possibility to retain water and make wet. This can be generated from GIS platform using the equation:

$$TWI = ln\left(\frac{A}{tan\beta}\right)$$

Here,

A is the contributing area calculated using the flow accumulation algorithm using a DEM which specifically reflects the total area drains into a given point.

tanβ is the tangent of the slope in degree value which quantifies the steepness of the terrain *A*.

A steeper slope has the potential to reduce the water accumulation in any given region. Therefore, TWI determines the hydrological behaviour of any surface in terms of the water movement and retention. Soil moisture condition, wetness pattern and flooding condition prediction can be done using this GIS tool. TWI value showing the Ramban, Anantnag, northern part of Pulwama, Udhampur and eastern part of Riasi are most wet places.

Distance to drainage is an essential factor for slope stability study, geotechnical understanding and landslide hazard condition. This factor governs the average slope condition, water input probability and surface moisture which are the complex form of hydrology, geology and geomorphology of any area. Water plays an important role behind the mechanism of mass movement and drainage is an essential input for landslide occurrence influenced by water.

Normalised difference vegetation index or *NDVI* is introduced in the 1990s as the modified version of transformed vegetation index (Jackson, 1983; Rouse et al., 1973). Initially, NDVI was useful to differentiate vegetation region from non-vegetation region. Later, NDVI calculation incorporated in spatial science discipline for forest monitoring and plant health identification. Spectral reflectance of near infrared band and red band is useful to calculate the green wave effect. Multispectral remotely sensed (MSS) images are required for the calculation. The equation is

$$NDVI = \frac{NIR - RED}{NIR + RED}$$

where, NIR is the spectral value near infrared band of MSS.

RED is the spectral value red band of MSS.

This equation generates values between -1 and $+1$, where higher values indicate healthier and denser vegetation, and lower values suggests sparse vegetation or non-vegetated surfaces like water or bare soil. NDVI minimises the effects of varying lighting conditions and sensor differences which provides a consistent measure of vegetation cover.

This mathematical calculation is very useful for the large area coverage, high frequency data analysis, cost benefited, less operational complexity and low weather dependency by normalizing the difference between the NIR and RED bands.

NDVI class determination and colour coded image output depends on the researcher solely. Though, in general, red to blue colour variation is most popular and red signifies less healthy, whereas blue indicates healthy vegetation. Some researcher uses green instead of blue colour for better colour variation and catchy outlook.

Most healthy vegetation covers found in the Jammu, southern stretch of Udhampur and western portion of Riasi; whereas, barren land and poor vegetation health identified in the northern part of Anantnag, western fringe of Shupiyan, Badgam and Kulgam.

Land use and land cover (LULC) maps are essential tools for depicting regional conditions. Land cover refers to the physical attributes observed on the Earth's surface, including natural vegetation and other naturally occurring features. Whereas, land use is defined by the ways in which humans interact with the naturally occurring land, encompassing the activities, arrangements, and inputs employed to alter, maintain, or utilise a particular land cover type. These maps provide valuable insight into human activities, the extent, frequency, and intensity of alterations made to the natural environment over time. LULC map offers a comprehensive view of how natural landscapes are modified for various purposes, such as agriculture, urbanisation, or infrastructure development. Organisations like the US Geological Survey (USGS) and other spatial mapping agencies regularly produce and update these maps to track and illustrate the degree of human impact on the environment. These maps serve as vital resources for understanding environmental changes and informing land management strategies by capturing the dynamic relationship between land use and land cover. This study can be classified into five broad classes and near about 45% of whole region identified as barren land, 25% area has vegetation cover, 15% is built-up area, about 10% is agricultural land and less than 5% area covers water bodies.

Peak ground acceleration or PGA is an essential fact for ground shaking in any region. PGA is not similar to peak ground velocity (PGV). PGA identification and measurement stations are now available in different earthquake susceptible zones throughout the world. PGA is important than PGV. PGV is directly associated to endogenic processes like earthquake due to plate movement and PGA is the result of the exogenic localised effect of PGV (Wu et al., 2003). The PGA value ranges from 8 to 1, where 8 is the maximum intensity and 1 is the minimum intensity. North-eastern part of Ramban, Southern part of Anantnag, south-eastern part of Kulgam and eastern most part of Udhampur falls in the high PGA. Lowest PGA found in the southernmost part of Jammu.

Distance from roads is another important variable for landslide occurrence in terms of anthropogenic factor. Road density is very in response to the whole area. NH-44 is the main highway and several small distributary roads are distributed in the Jammu and Kashmir urban connected regions only. Historical landslide database showing the high frequency of landslide occurrence near the NH-44 within the Ramban and Udhampur mostly.

4.4 Machine Processing

All nine individual data set performed in ArcGIS desktop version 10.8.1 (Fig. 4). Here are two types of inputs. First one is collected from different sources and second

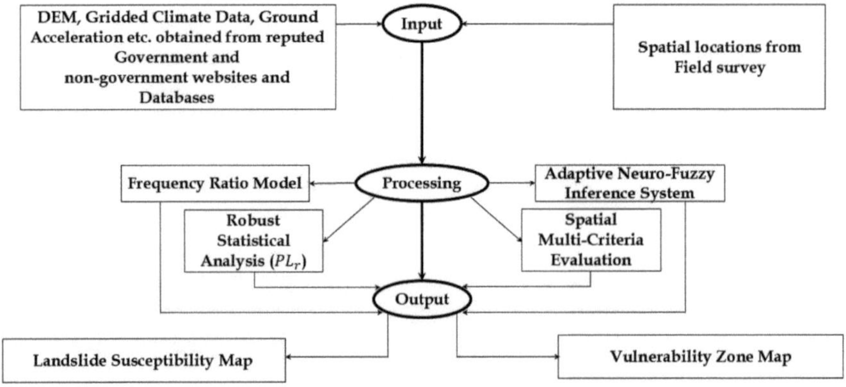

Fig. 4 Flow-chart of the data input and processing performed in ArcGIS 10.8.1, MatlabR2019a platform

one is collected from field survey. Initially, all data set modified and converted to the feasible geographic information system (GIS) following the standard GIS processing rules. After that, we made our choice to process the whole data set using four different kinds of calculation to follow the common analysis process, modified analysis process to generate the most accurate result.

First one is frequency ratio method. Frequency ratio method is the most common method used in the statistical analysis of landslide susceptibility (Aditian et al., 2018; Lee & Sambath, 2006; Yalcin et al., 2011; Yilmaz, 2009). This method is based on the relationship between the spatial distribution of landslides and each conditioning parameter. We estimate this relationship with the calculation of landslide susceptibility index (LSI). This method calculates the LSI for each category of all conditioning factors (e.g., L, LM, FL, E, S, A, PC, PRC, TPI, SAR, SC, R, DD, SPI, TWI, DDR, NDVI, LULC, PGA, and DRR), which are selected for this study. Thus, supposing that j is a category within the factor i, then the LSI for this category ($LSI_{i,j}$) is defined as follows (He & Beighley, 2008):

Frequency ratio F_r is:

$$F_r = \frac{f}{f_n}$$

Here,

f is the operating or driving frequency (e.g., the input or external frequency);

f_n is the natural frequency of the system (e.g., the system's inherent frequency);

Landslide Susceptibility Index (LSI) is:

$$LSI_{i,j} = ln\left(\frac{N_{i,j}}{A_{i,j}} \Big/ \frac{N_T}{A_T}\right)$$

Here,

$N_{i,j}$ is the number of landslides in category j of the factor i;
$A_{i,j}$ the area of this category;
N_T is the total number of landslides; and
A_T is the total area under investigation.

Thus, frequency ratio method helped us to identify the region with landslide susceptibility in the study area following three different classes as highly susceptible region, moderately susceptible region, and low susceptible region (Fig. 5).

Selected study area classified into three different landslide susceptibility zones i.e. low LSI zone, moderate LSI zone, and high LSI zone. Low zone depicted here in green colour, moderate zone depicted in yellow colour and high susceptible zone in red colour. All nine landslide factors combined and LSI calculated using the LSI properties of the conditioning factors following minimum value, maximum value, range of values and standard deviation. Final output calculated using the weighted overlay method. Weight or assigned values for individual factors depends on the researcher's ability and understanding of landslide science. Scientific understanding, logical explanation and justification is more important than biasness. In some cases, only one bias input could change the whole result for the study using less than ten causal factors. Similarly, assigning same weightage for every input is another kind of biasness which should be strictly prohibited.

The LSI depends on the number of data availability in category and sub-category of calculation. Hundred percent data cannot be gathered by any means of satellite technology or human support due to extensive region and difficult terrains. Therefore, probability calculation using surveyed data would be a fruitful process to capture all susceptible regions exist in the study area. Probability of landslide prone region could be identified by probabilistic likelihood ratio method.

Second, Probabilistic likelihood ratio created using empirical formula of probability calculation. This method applied to refine the data set of frequency ratio result. 'Yes' algorithm applied on the minimum one landslide occurrence regions and 'no' algorithm used to exclude the regions with zero landslide events.

Probabilistic Likelihood Ratio, PL_r is:

$$PL_r = \frac{P(L \cap C)}{P(L \cap \neg C)}$$

Here,

$\boldsymbol{L}$ is the landslide event;
$\boldsymbol{C}$ is the conditioning factor;
$\boldsymbol{\neg C}$ is the negation of the conditioning factor;
$\boldsymbol{P(L \cap C)}$ is the probability of the landslide event L occurring in the given condition C (elevation, lithology type, annual rainfall etc.);
$\boldsymbol{P(L \cap \neg C)}$ is the probability of the landslide event L occurring in the given condition C is **not** present.
$\boldsymbol{PL_r} > 1$ is the condition C increases the likelihood of the event E;
$\boldsymbol{PL_r} = 1$ is the condition C has no effect on the likelihood of the event E;
$\boldsymbol{PL_r} < 1$ is the condition C decreases the likelihood of the event E;

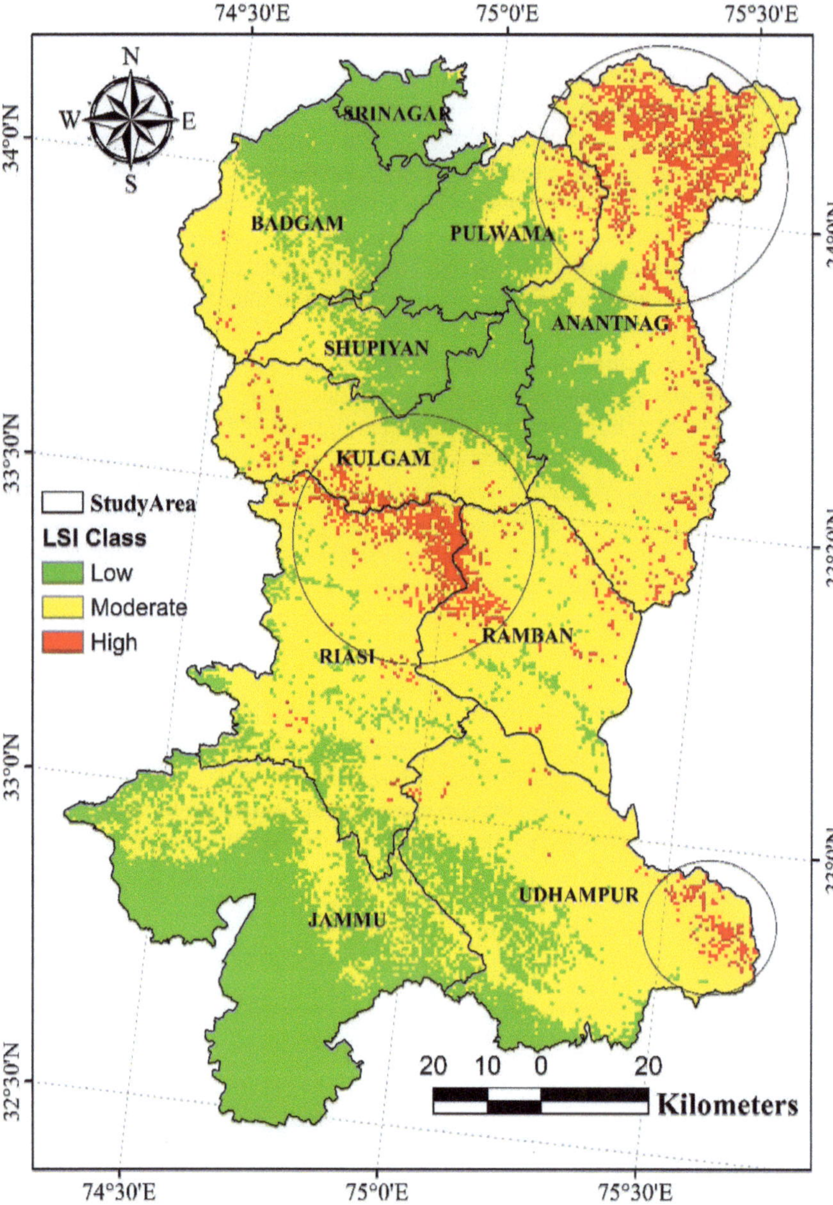

Fig. 5 Landslide susceptibility index map showing the various landslide prone zones where high-LSI indicated in red, moderate in yellow and low in green

All twenty conditioning factors are classified as per requirement of analysis depending on data suitability and availability. Each classified factors are calculated using the frequency ratio category classes of previous method. Similarity of category classes among the applied method is most acceptable without further refinement to get the justified spatial location-based results. Results collected through either yes algorithm or no algorithm which is varies from less than 1 to more than 1. Combined factor calculation among the conditioning factors needed to get the probabilistic likelihood ratio of individual factors of landslide events.

Alone probability can only provide the answer in two means either yes or no. But, twenty individual conditioning factors are independent in nature. On the other hand, combined effect of conditioning factors could be resulted to more accurate situation in terms landslide vulnerability. Only mathematical calculation is not sufficient to study these factors. Human input through rational understanding and justified process will help to get criteria-based analysis depending on the spatial data. Criteria dependent spatial analysis is popular term today. When the region is diversified and different type of landform exist, then, multi-criteria evaluation would be justified instead of any single criteria analysis.

The Spatial Multi-Criteria Evaluation (SMCE) methodology integrates multiple criteria along with their respective weights to generate a comprehensive Suitability Score for each spatial unit (Nsengiyumva et al., 2018; Pourghasemi et al., 2014). All PL_r result further adjusted with weightage and suitability score calculated accordingly. Suitability calculation method as

$$S = \sum_{i=1}^{n} w_i.C_i$$

Here,

S is the Suitability score for the spatial unit;
n is the Total number of criteria (elevation, lithology type etc.);
w_i is the Weight assigned to the ith criterion, reflecting its significance;
C_i is the Score of the ith criterion, represented as a normalized value.
Weighting Criteria (Analytic Hierarchy Process considered here) as

$$\sum_{i=1}^{n} w_i = 1$$

The weights should sum to 1.

Analytic Hierarchy Process (AHP) is one of the most popular terms in computer dependent work environment in spatial science as well as data science (Pourghasemi et al., 2012). This is a simple method to analyse complex decision using mathematics and psychology. AHP helps to identify suitability among complex situation rather than pin pointed decision. Comprehensive knowledge and rational understanding of causal factors enhances the decision finding mechanism properly. The event like landslide requires multiple criteria to understand the processes. Selecting one criterion for landslide among the alternatives is very difficult due to the limited scientific exposures till date. All twenty criteria considered for AHP following the prioritization, management and conflict resolution. Consideration of twenty criteria minimises the erroneous prioritization among the factors of landslide occurrence.

Score normalisation is applied to ensure the comparability of scores across distinct criteria is essential to normalising the data. Data normalisation refers to the process of transforming the values within a dataset to a consistent scale by enabling fair comparison among different variables. Various normalisation methods exist in landslide related literature, such as the minimum–maximum method, maximum absolute value method, mean normalisation method, Z-score method, and log transformation etc. Here, the minimum–maximum normalisation method is commonly employed for simplicity and ease of interpretation. This method provides a direct approach to standardising the data and making it useful for practical applications. The minimum–maximum normalisation method involves two primary components. The denominator represents the range of values in the dataset, which is determined by subtracting the minimum value from the maximum value of the dataset. The numerator calculates the difference between the specific value of a given factor and the minimum value within that set. This transformation ensures that all data points are scaled to fall within a fixed range, typically between 0 and 1. As a result, each value is proportionally adjusted relative to the overall range of the dataset. However, if the normalisation process results in a negative value for any conditioning factor, then the equation must be modified for this discrepancy. This adjustment ensures that all values remain within the desired range, typically preserving the integrity of the analysis while accommodating potential variations in the dataset. The minimum–maximum normalisation method is particularly advantageous when the goal is to maintain a clear, interpretable scale that facilitates comparison between multiple factors, avoiding the complexities associated with more advanced methods. Therefore, normalisation plays a crucial role in standardising scores and ensuring that they are comparable across different criteria. While there are several methods available for this process, the minimum–maximum method is widely adopted due to its simplicity and ease of calculation. This technique enables effective comparison and interpretation of data, provided that any issues with negative values are properly addressed. By applying normalisation, analysts can ensure that their datasets are on a consistent scale that enhances the accuracy and reliability of the conclusion (Juszczak et al., 2002).

$$C_i' = \frac{C_i - C_{min}}{C_{max} - C_{min}}$$

Here,

C_i' is the Normalised score of the ith criterion;
C_{min} is the Minimum score for that criterion;
C_{max} is the Maximum score for that criterion.

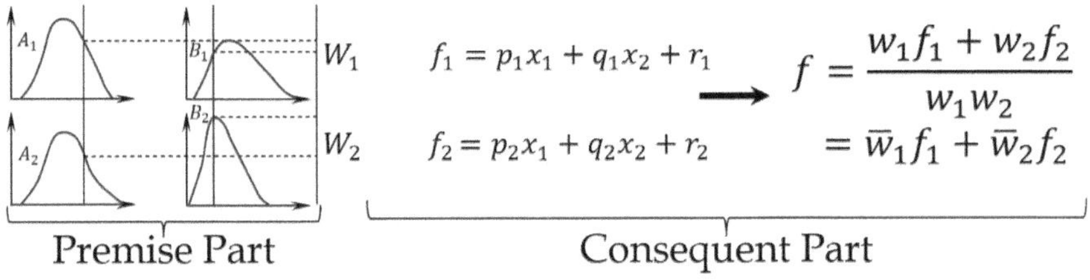

Fig. 6 Adaptive neuro-fuzzy inference system (ANFIS) showing the dependencies of the members as the input and output functions

Adaptive neuro-fuzzy inference system (ANFIS) is one of the most accepted method for landslide analysis (Kanungo et al., 2006; Shahabi & Hashim, 2015). All normalised score for each sub-categories are transferred to MATLAB for further processing with the objective to generate the vulnerability map of the study area (Fig. 6). ANFIS is a strong method for multi-layer data analysis. Specifically, ANFIS is a multi-layer feed forward structure, which is formed by a number of nodes connected through directional links. Two basic rule applied to continue the ANFIS (Sugeno, 1985; Takagi & Sugeno, 1983, 1985):

Rule 1: if x_1 is A_1 and x_2 is B_1, then $f_1 = p_1x_1 + q_1x_2 + r_1$

Rule 2: if x_1 is A_2 and x_2 is B_2, then $f_1 = p_2x_1 + q_2x_2 + r_2$

where,

A_1, A_2, B_1, B_2 are the membership functions (MFs) for input x_1 and x_2; p_1,q_1,r_1,p_2,q_2,r_2 are the parameters of the output function.

A fuzzy inference system consists of five layers and each layer is formed by several nodes and node function. There are two types of nodes: Adaptive nodes and Fixed nodes. Adaptive nodes are marked by Squares that represent the parameter sets, which may be adjustable. Fixed nodes are marked by Circles, and their parameter sets are fixed in the system (Fig. 7).

Layer 1: All nodes in this layer are adaptive nodes.

Layer 2: Every node of this layer is fixed node. The output of each node is the product of all the incoming signals.

Layer 3: Every node of this layer is fixed node. The output is calculated by the ratio of the *i*-th rule's Fs over the sum of all rule's Fs.

Layer 4: Every node is an adaptive node.

Layer 5: This is a single fixed node. Output value is the sum of all incoming signals.

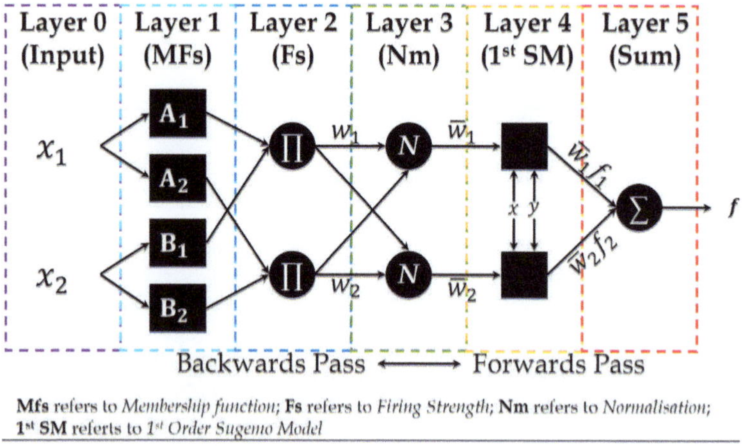

Fig. 7 Flow-chart showing the ANFIS membership functions among the variables

5 Results

The ANFIS process is a typical feeder-food relationship. Result of the ANFIS study showing the three different vulnerability zones within the study area (Fig. 8) namely, high vulnerability zone, moderate vulerability zone and low vulnerability zone specifically. Though, this vulnerable map is comparable to susceptibilty map, but, also showing the differences depending on the different feed-forward structure of ANFIS. Less than 10 percent area of the whole study region is directly falls within the most vulnerability zone. On the other hand, moderate and low vulnerability zone are sharing equal quanity of area. Interstingly, no such direct relationship identified as the most important factor for the landslide occurrences, Though, the trigerring factors are the most concerning issue for the vulnerability condition. Red marked susceptible areas are vulnerable for landslide events along with some other regions, specifically, along the NH-44 within the Ramban and Udhampur district. These two pockets are showing most vulnerable following the output colour density and coarseness.

Southern portion of Kulgam district, northeastern portion of Riasi, eastern most pockets of Udhampur, northern pockets of Anantnag and a few points of Pulwama has been identified as most vulnerable region. The southern flat land of Jammu district and plain lands of Srinagar including more than the half of Badgam, Shupiyan, Pulawama clearly showing the low landslide vulnerability zone. Moderately vulnerable zone is the second most concerning issue in the whole study region. These zones are most unpredictable due to their volatile nature for all causal factors. Juxtaposed association with trigerring factors and non-trigerring factors are liable to recongnise these areas as moderately vulnerable as well as highly probable zones for future landslide vulnerablity.

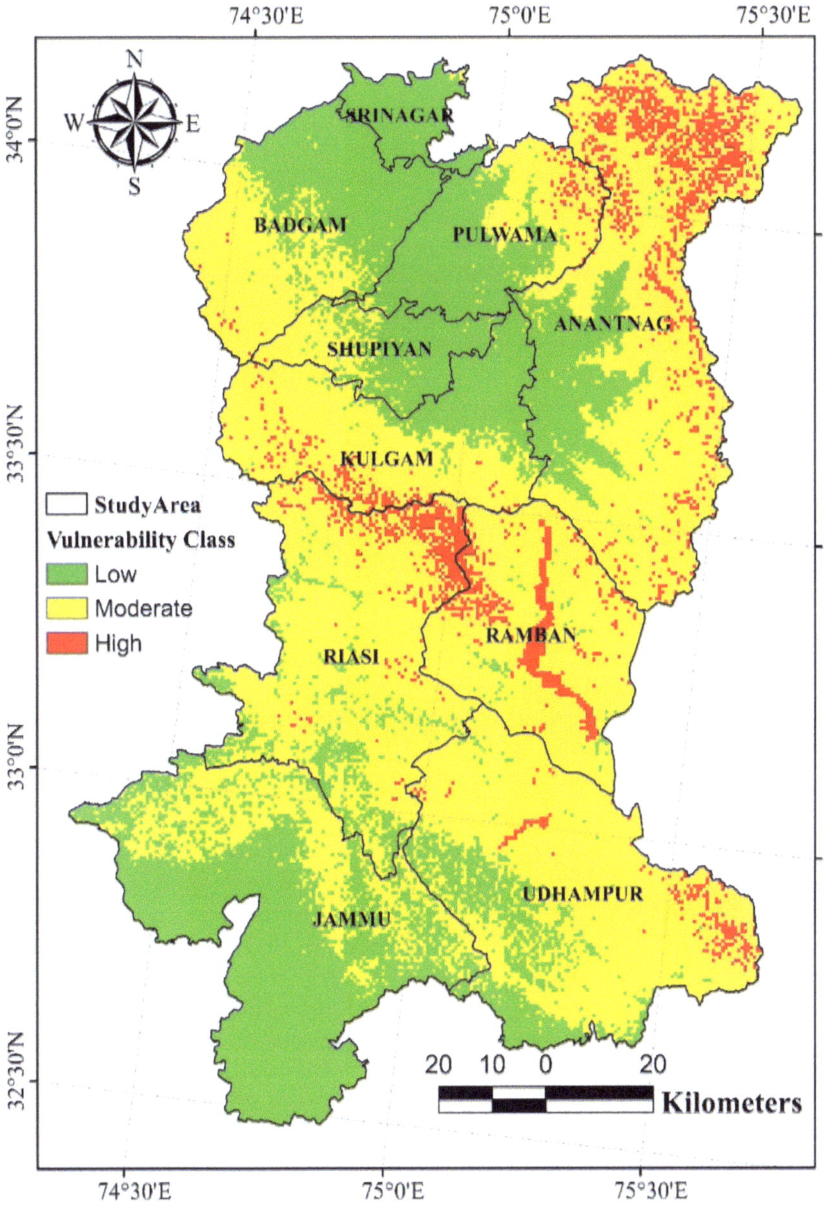

Fig. 8 Map showing the result of the ANFIS study where the output shows three different prominent landslide vulnerability zones

6 Conclusion

This research intended to address the landslide issue in terms of identification of the susceptible zones and assess the vulnerable regions in the study area that comprises of ten districts within the Kashmir basin. There are several methods available worldwide to address both the landslide susceptibility zones and vulnerability assessment. But, data dependent study is the main constraints to assess the issue with authenticity. Therefore, we used the frequency ratio method which is the most popular and simplest of all the statistical methods applied for landslide study. The probabilistic likelihood method along with the spatial multi-criteria analysis is further applied to support the results. Finally, we got the highest accuracy through adaptive neuro-fuzzy method to obtain the vulnerability zone. The results indicate that the landslide occurrences in these districts have been escalating every day and requires urgent concerns about the long-term sustainability of the region. The analytical results indicating three zones within the study area: high, moderate, and low vulnerability. Less than 10% of the area falls into the high vulnerability zone, while moderate and low zones cover near about equal portions. Key landslide-triggering factors are highlighted, but no single factor is identified as the most critical. High-risk areas include parts of Ramban, Udhampur, Kulgam, Riasi, Anantnag, and Pulwama districts, particularly along NH-44. These areas undergoing rapid and ongoing land use and land cover changes are identified as the most vulnerable zones. These regions are mostly confined within the Ramban, Udhampur, Raisi and Anantnag district and features steep slopes, loose soil cover, and highly porous subsurface layers along with uncontrolled development panorama which includes road expansion, built-up concrete structures and most concerning issue is the deforestation for settlement, industrialisation and related purposes. In contrast, the low-elevation flatlands of Jammu and the high-elevation flatlands of Srinagar are less vulnerable due to their broad, flat surfaces and the presence of less permeable rocks. Moderate vulnerability zones are unpredictable due to their volatile nature, influenced by both triggering and non-triggering factors, making them potential future landslide hotspots. However, the highland areas of Anantnag, Badgam, and Shupiyan, characterised by steep slopes, snow-covered mountainous terrain, and limited topsoil development, exhibit moderate to high vulnerability. The cause-and-effect relationship between various indicators of landslides are needed using Driver-Pressure-State-Impact-Response (DPSIR) assessment framework for spatial location-based management strategy framing.

Acknowledgements Authors are thankful to the University of Calcutta, Kolkata and Trivenidevi Bhalotia College, Raniganj, West Bengal for providing the necessary facilities to carry out this piece of research work. Authors also gratefully acknowledge Dr. Sandipan Ghosh, Chandrapur College, Purba Bardhaman, for providing the opportunity to include our research work in this book volume. Authors would also like to thank all those persons who helped during field and laboratory analysis. Authors also thankful to the anonymous reviewers for fruitful suggestions to improve and finalise this chapter.

Conflict of Interest This research carried out as a part author's personal interest on modern-day hazards in Indian Himalaya. The whole research is carried by author's personal fund and no other government and non-government funds utilised for this. All data are either collected from open archives or directly collected from author's field visit. No human and other biological organisms are harmed during the visits to the study area.

References

Aditian, A., Kubota, T., & Shinohara, Y. (2018). Comparison of GIS-based landslide susceptibility models using frequency ratio, logistic regression, and artificial neural network in a tertiary region of Ambon, Indonesia. *Geomorphology, 318*, 101–111. https://doi.org/10.1016/j.geomorph.2018.06.006

Ahmad, B., Bhat, M. I., & Bali, B. (2009). Historical record of earthquakes in the Kashmir Valley. *Himalayan Geology, 30*, 75–84.

Ansari, A., Rao, S., Jain, A., & Ansari, A. (2022). Deep learning model for predicting tunnel damages and track serviceability under seismic environment. *Modeling Earth Systems and Environment, 9*, 1–20. https://doi.org/10.1007/s40808-022-01556-7

Bagnold, R. A. (1966). *An approach to the sediment transport problem from general physics (Report 422I; Professional Paper)* (43p.). USGS Publications Warehouse. https://doi.org/10.3133/pp422I

Barton, N., & Choubey, V. (1977). The shear strength of rock joints in theory and practice. *Rock Mechanics, 10*, 1–54.

Beasom, S. L., Wiggers, E. P., & Giardino, J. R. (1983). A technique for assessing land surface ruggedness. *The Journal of Wildlife Management, 47*(4), 1163–1166. https://doi.org/10.2307/3808184

Berry, J. K. (2002). Use surface area for realistic calculations. *Geoworld, 15*(9), 20–21.

Bhat, N. A., Bali, B. S., Mir, S. A., & Kumar, P. (2024). Seismotectonics and seismogenesis of Kashmir Valley, NW Himalaya, India from a local seismic network. *Journal of Earth System Science, 133*(2), 61. https://doi.org/10.1007/s12040-024-02270-x

Burbank, D. W., & Johnson, G. D. (1983). The late cenozoic chronologic and stratigraphic development of the Kashmir intermontane basin, Northwestern Himalaya. *Palaeogeography, Palaeoclimatology, Palaeoecology, 43*(3), 205–235. https://doi.org/10.1016/0031-0182(83)90012-3

Dai, F. C., & Lee, C. F. (2002). Landslide characteristics and slope instability modeling using GIS, Lantau Island, Hong Kong. *Geomorphology, 42*(3), 213–228. https://doi.org/10.1016/S0169-555X(01)00087-3

De Blasio, F. V. (2011). *Introduction to the Physics of Landslides—Lecture Notes on the Dynamics of Mass Wasting* (418p.). Springer Dordrecht. https://doi.org/10.1007/978-94-007-1122-8

Earle, S. (2015). *Physical Geology*. BCcampus Open Publishing. https://opentextbc.ca/geology

Fairbridge, R. W. (1968). *The Encylopedia of Geomorphology* (1320p.). Reinhold Book Corporation.

Gabrielsen, R. H., & Olesen, O. (2024). The concept of lineaments in geological structural analysis; principles and methods: A review based on examples from Norway. *Geomatics, 4*(2), 189–212. https://doi.org/10.3390/geomatics4020011

Ganjoo, R. K. (2014). The vale of Kashmir: Landform evolution and processes. In V. S. Kale (Ed.), *Landscapes and Landforms of India* (pp. 125–133). Springer Netherlands. https://doi.org/10.1007/978-94-017-8029-2_11

Gariano, S. L., & Guzzetti, F. (2016). Landslides in a changing climate. *Earth-Science Reviews, 162*, 227–252. https://doi.org/10.1016/j.earscirev.2016.08.011

Gerrad, A. J. (1992). *Soil Geomorphology: An Integration of Pedology and Geomorphology* (292p.). Chapman & Hall.
Goudie, A. S. (2004). *Encyclopedia of Geomorphology* (1200p.). Routledge
Hamid, I., Dar, L. A., & Akintug, B. (2024). Assessing land use land cover change using remote sensing and GIS techniques: A case study of Kashmir Valley. *Journal of Earth System Science, 133*(3), 168. https://doi.org/10.1007/s12040-024-02369-1
He, Y., & Beighley, R. E. (2008). GIS-based regional landslide susceptibility mapping: A case study in southern California. *Earth Surface Processes and Landforms, 33*(3), 380–393. https://doi.org/10.1002/esp.1562
Highland, L. M., & Bobrowsky, P. (2008). *The Landslide Handbook—A Guide to Understanding Landslides* (Report 1325; Version 1.0, Circular, p. 147). USGS Publications Warehouse. https://doi.org/10.3133/cir1325.
Hobbs, W. H. (1904). Lineaments of the Atlantic Border region. *GSA Bulletin, 15*(1), 483–506. https://doi.org/10.1130/GSAB-15-483
Hobson, R. D. (1972). Surface roughness in topography: a quantitative approach. In R. J. Chorley (Ed.), *Spatial Analysis in Geomorphology* (pp. 221–246). Methuen Publisher.
Hodgson, M. (1995). What cell size does the computed slope/aspect angle represent? *Photogrammetric Engineering & Remote Sensing, 61*, 513–517.
Horton, R. E. (1945). Erosional Development of Streams and their Drainage Basins; Hydrophysical Approach to Quantitative Morphology. *GSA Bulletin, 56*(3), 275–370. https://doi.org/10.1130/0016-7606(1945)56[275:EDOSAT]2.0.CO;2
ICIMOD Annual Report 2022. (2023). https://doi.org/10.53055/icimod.1038
Iqbal, O., Baig, M., Pervez, S., & Siddiqi, M. (2018). Structure and stratigraphy of Rumbli and Panjar Areas of Kashmir and Pakistan with the aid of GIS. *Journal of the Geological Society of India, 91*, 57–66. https://doi.org/10.1007/s12594-018-0820-2
India Meteorological Department (IMD). (2020). *Climate variability and trends in Jammu and Kashmir*. IMD.
Jackson, R. D. (1983). Spectral indices in N-Space. *Remote Sensing of Environment, 13*(5), 409–421. https://doi.org/10.1016/0034-4257(83)90010-X
Jenness, J. (2010). *Surface Area and Ratio for ArcGIS (surface_area.exe) v. 1.0.138*. Jenness Enterprises. https://www.jennessent.com/arcgis/surface_area.htm
Juszczak, P., Tax, D., & Duin, R. (2002). Feature scaling in support vector data description. In *Proceedings of 8th Annual Conference on Advanced School for Computing and Imaging* (pp. 25–30).
Kanungo, D. P., Arora, M. K., Sarkar, S., & Gupta, R. P. (2006). A comparative study of conventional, ANN black box, fuzzy and combined neural and fuzzy weighting procedures for landslide susceptibility zonation in Darjeeling Himalayas. *Engineering Geology, 85*(3), 347–366. https://doi.org/10.1016/j.enggeo.2006.03.004
Kimerling, A. J. (2016). Map Use: Reading, Analysis, and Interpretation (620p.). ESRI Press.
Kimerling, A. J., Buckley, A. B., Muehrcke, P. C., & Muehrcke, J. O. (2016). *Map Use: Reading, Analysis, Interpretation* (908p.). Esri Press. ISBN: 1589484428; 9781589484429.
Klar, A., Aharonov, E., Kalderon-Asael, B., Katz, O. (2011). Analytical and observational relations between landslide volume and surface area. *Journal of Geophysical Research (Earth Surface)*. https://doi.org/10.1029/2009JF001604
Lee, S., & Sambath, T. (2006). Landslide susceptibility mapping in the Damrei Romel area, Cambodia using frequency ratio and logistic regression models. *Environmental Geology, 50*(6), 847–855. https://doi.org/10.1007/s00254-006-0256-7
Mandelbrot, B. B. (1982). *The Fractal Geometry of Nature* (p. 504p). W.H. Freeman.
Meunier, P., Hovius, N., & Haines, J. A. (2008). Topographic site effects and the location of earthquake induced landslides. *Earth and Planetary Science Letters, 275*(3), 221–232. https://doi.org/10.1016/j.epsl.2008.07.020
Middleton, G. V., Church, M. J., Coniglio, M., Hardie, L. A., & Longstaffe, F. J. (2003). *Encyclopedia of Sediments and Sedimentary Rocks* (p. 928p). Springer.

Moore, I. D., Grayson, R. B., & Ladson, A. R. (1991). Digital terrain modelling: A review of hydrological, geomorphological, and biological applications. *Hydrological Processes, 5*(1), 3–30. https://doi.org/10.1002/hyp.3360050103

Mukhopadhyay, B., & Dasgupta, S. (2015). Seismic hazard assessment of Kashmir and Kangra valley region, Western Himalaya, India. *Geomatics, Natural Hazards and Risk, 6*, 832405. https://doi.org/10.1080/19475705.2013.832405

Nsengiyumva, J. B., Luo, G., Nahayo, L., Huang, X., & Cai, P. (2018). Landslide susceptibility assessment using spatial multi-criteria evaluation model in Rwanda. *International Journal of Environmental Research and Public Health, 15*(2), 243.

Pai, D., Rajeevan, M., Sreejith, O., Mukhopadhyay, B., & Satbha, N. (2014). Development of a new high spatial resolution (0.25° × 0.25°) long period (1901–2010) daily gridded rainfall data set over India and its comparison with existing data sets over the region. *MAUSAM, 65*(1), 1–18. https://doi.org/10.54302/mausam.v65i1.851

Peckham, S. (2011). Profile, plan and streamline curvature: A simple derivation and applications. In *Conference: Geomorphometry 2011, Proceedings of the International Society for Geomorphometry* (pp. 27–11).

Pourghasemi, H. R., Moradi, H. R., Fatemi Aghda, S. M., Gokceoglu, C., & Pradhan, B. (2014). GIS-based landslide susceptibility mapping with probabilistic likelihood ratio and spatial multi-criteria evaluation models (North of Tehran, Iran). *Arabian Journal of Geosciences, 7*(5), 1857–1878. https://doi.org/10.1007/s12517-012-0825-x

Pourghasemi, H. R., Pradhan, B., & Gokceoglu, C. (2012). Application of fuzzy logic and analytical hierarchy process (AHP) to landslide susceptibility mapping at Haraz watershed, Iran. *Natural Hazards, 63*(2), 965–996. https://doi.org/10.1007/s11069-012-0217-2

Rasul, G., & Molden, D. (2019). The global social and economic consequences of mountain cryospheric change. *Frontiers in Environmental Science, 7*, 91. https://doi.org/10.3389/fenvs.2019.00091

Rouse, J., Haas, R., Schell, J., & Deering, D. (1973). Monitoring vegetation systems in the great plains with ERTS. In *Proceedings of the Earth Resources Technology Satellite Symposium* (Vol. 1, pp. 309–317). NASA.

Sarkar, S. C., & Gupta, A. (2012). The Himalaya. In *Crustal Evolution and Metallogeny in India* (pp. 679–716). Cambridge University Press.

Schumm, S. A. (1956). Evolution of drainage systems and slopes in badlands at Perth Amboy, New Jersey. *GSA Bulletin, 67*(5), 597–646. https://doi.org/10.1130/0016-7606(1956)67[597:EODSAS]2.0.CO;2

Schumm, S. A. (1997). Drainage density. In *Process and Form in Geomorphology* (Vol. 15).

Selby, M. J. (1993a). Mass wasting of soils In: *Hillsope: Materials & Process* (Vol. 2, 480p.). Oxford University Press.

Selby, M. J. (1993b). Mass wasting of soils. *Hillslope Materials and Processes* (2nd ed., pp. 249–355). Oxford: Oxford University Press.

Selby, M. J. (1993c). *Hillslope Materials* (451p.). Oxford University Press.

Selby, M. J., & Hodder, A. P. W. (1993). Rock-slope processes. In *Hillslope Material and Processes* (pp. 320–355).

Shahabi, H., & Hashim, M. (2015). Landslide susceptibility mapping using GIS-based statistical models and Remote sensing data in tropical environment. *Scientific Reports, 5*(1), 9899. https://doi.org/10.1038/srep09899

Skidmore, A. K. (1989). A comparison of techniques for calculating gradient and aspect from a gridded digital elevation model. *International Journal of Geographical Information Systems, 3*(4), 323–334.https://doi.org/10.1080/02693798908941519

Smith, M. J. D., Goodchild, M. F., Longley, P. A., & Associates. (2018). *Geospatial Analysis—A Comprehensive Guide to Principles Techniques and Software Tools* (6th ed., 618p.). Winchelsea Press.

Sørensen, R., Zinko, U., & Seibert, J. (2006). On the calculation of the topographic wetness index: Evaluation of different methods based on field observations. *Hydrology and Earth System Sciences, 10*(1), 101–112. https://doi.org/10.5194/hess-10-101-2006

Sugeno, M. (1985). An introductory survey of fuzzy control. *Information Sciences, 36*(1), 59–83. https://doi.org/10.1016/0020-0255(85)90026-X

Takagi, T., & Sugeno, M. (1983). Derivation of fuzzy control rules from human operator's control actions. *IFAC Proceedings Volumes, 16*(13), 55–60. https://doi.org/10.1016/S1474-6670(17)62005-6

Takagi, T., & Sugeno, M. (1985). Fuzzy identification of systems and its applications to modeling and control. *IEEE Transactions on Systems, Man, and Cybernetics, 15*(1), 116–132. https://doi.org/10.1109/TSMC.1985.6313399

Talling, P. J., & Sowter, M. J. (1999). Drainage density on progressively tilted surfaces with different gradients, Wheeler Ridge, California. *Earth Surface Processes and Landforms: THe Journal of the British Geomorphological Research Group, 24*(9), 809–824.

Terzaghi, K. (1950). Mechanism of landslides. In S. Paige (Ed.), *Application of Geology to Engineering Practice* (Berkey Volume, pp. 83–123). Geological Society of America.

Turner, A. K. (1996). *Landslides: Investigation and Mitigation* (Chap. 20-Colluvium and Talus) (No. 247).

USDA. (2014). *Keys to Soil Taxonomy by Soil Survey Staff* (12th ed., 372p.). United States Department of Agriculture, Natural Resources Conservation Service.

Valdiya, K. S. (2016). The Making of India—Geodynamic Evolution. Springer International Publishing Switzerland. 960p.

Varnes, D. J. (1978). Slope movement types and processes. In: R. L. Schuster, & R. J. Krizek, (Eds.), *Landslides, Analysis and Control* (pp. 11–33). Transportation Research Board, Special Report No. 176, National Academy of Sciences.

Wadia, D. (1953). *Geology of India* (p. 592p). Macmillan and Co. Limited.

Wandrey, C. J., & Law, B. E. (1998). *Maps showing geology, oil and gas fields and geologic provinces of South Asia (Report 97-470C; Version 2.0, Open-File Report).* USGS Publications Warehouse. https://doi.org/10.3133/ofr97470C

Wu, Y., Teng, T., Shin, T. C., & Hsiao, N. C. (2003). Relationship between peak ground acceleration, peak ground velocity, and intensity in Taiwan. *Bulletin of the Seismological Society of America, 93*, 386–396. https://doi.org/10.1785/0120020097

Yalcin, A., Reis, S., Aydinoglu, A. C., & Yomralioglu, T. (2011). A GIS-based comparative study of frequency ratio, analytical hierarchy process, bivariate statistics and logistics regression methods for landslide susceptibility mapping in Trabzon, NE Turkey. *CATENA, 85*(3), 274–287. https://doi.org/10.1016/j.catena.2011.01.014

Yilmaz, I. (2009). Landslide susceptibility mapping using frequency ratio, logistic regression, artificial neural networks and their comparison: A case study from Kat landslides (Tokat—Turkey). *Computers & Geosciences, 35*(6), 1125–1138.

Zhuang, J., Cui, P., Hu, K., Chen, X., & Ge, Y. (2010). Characteristics of earthquake-triggered landslides and post-earthquake debris flows in Beichuan County. *Journal of Mountain Science., 7*, 246–254. https://doi.org/10.1007/s11629-010-2016-0

Rockfall Hazard Assessment and Mitigation Strategies Along NH-7, Garhwal Himalaya: An Integrated RHRS and Numerical Simulation Approach

Tariq Siddique and Sarada Prasad Pradhan

Abstract Rockfall hazards pose significant threats to infrastructure and public safety in mountainous terrains, necessitating comprehensive assessment and mitigation strategies. This study presents a comprehensive rockfall hazard analysis along the National Highway-7 (NH-7). Rockfall Hazard Rating System (RHRS) and numerical simulation approach have been applied to evaluate rockfall potential and mitigation strategies. RHRS was applied to categorize the engineered slopes based on their vulnerability, considering parameters such as slope geometry, geo-environmental factors, and traffic exposure. Numerical modeling has been performed using RocFall software to simulate possible rockfall trajectories and quantify several parameters, including bounce height, maximum run-out distance and kinetic energy of falling blocks. Rockfall simulations provided detailed insights into the behavior of falling rocks under varying slope conditions. To optimize, mitigation strategies and site-specific protective barriers have been proposed and evaluated based on simulation outputs. Barrier placement was optimized to maximize energy absorption and prevent rocks from reaching the road. Results indicate that proposed barrier installations can arrest all plausible falling blocks.

Keywords Rockfall · Rockfall modeling · Rockfall hazard rating system · Himalaya · National Highway-7

1 Introduction

Rockfall hazards are a major geotechnical concern in mountainous regions, posing significant risks to transportation networks, infrastructure and public safety. Rockfall involves detachment of rock blocks from slopes and travel downslope by falling,

T. Siddique (✉)
Department of Geology, Aligarh Muslim University, Aligarh, India
e-mail: *tariqsiddiqueiitr@gmail.com

S. P. Pradhan
Department of Earth Sciences, Indian Institute of Technology Roorkee, Roorkee, India

© The Author(s), under exclusive license to Springer Nature Switzerland AG 2025

S. Ghosh (ed.), *The Himalaya Dilemma*, Springer Natural Hazards,
https://doi.org/10.1007/978-3-031-95083-4_8

rolling, sliding or any combination of these, primarily under gravity (Lakherwal et al., 2024; Roul et al., 2023). The initial detachment of blocks depends upon several intrinsic weaknesses in the rockmass and external triggering factors (Ahmad et al., 2013; Pandey et al., 2025; Vishal et al., 2017). In orogenic belts such as the Himalaya, slope failures and rockfall events are especially frequent due to rugged terrain with steeply inclined slopes, persistent neotectonism, freeze and thaw cycles, extreme meteorological activities, etc. (Vishal et al., 2017; Verma et al., 2019; Siddique et al., 2022; Singh et al., 2023; Dahiya et al., 2024; Khan et al., 2024; Varshney et al., 2024). Generally, the meta-sedimentary rocks of the Lesser Himalaya bear three or more sets of discontinuities, often forming kinematically unstable wedge geometry. Among these three sets, the relict bedding discontinuity has maximum persistence and is often involved in most structurally controlled landslides and rockfalls. Altered natural topography for road alignment in the hilly region exacerbates the potential for rockfall by manifolds (Aher et al., 2024; Jamir et al., 2024; Kumar et al., 2024; Singh et al., 2016). Due to such excavations, the existing geomechanical and geomorphological equilibrium of slopes is greatly hampered, leading to rockfalls along transportation routes (Maheshwari & Bhowmik, 2024; Singh et al., 2021a). Such hazards threaten vital roadways such as the NH-7, a critical transportation corridor that experiences recurring rockfall incidents, often causing fatal road accidents, traffic delays and infrastructural damage. Rockfall incidents along NH-7, particularly between Devprayag and Srinagar are one of the major geotechnical challenges. The situation has worsened due to extensive road-widening activities undertaken over the last decade. Understanding and mitigating rockfall risks in such regions is vital to ensuring safety and resilience.

RHRS serves as a preliminary screening tool for assessing and quantifying the potential risk of rockfalls. Moreover, various commercial and freeware applications for rockfall assessment are often integrated with conventional rating systems to enhance hazard evaluation and risk quantification. RocFall software, developed by Rocscience is widely used to analyze rockfall trajectories, bounce heights, velocities, and impact energies of potential falling rock blocks. This study focuses on two rockfall hazard-prone slope sections along NH-7 with the following objectives: (i) To conduct a detailed RHRS assessment, (ii) Perform simulations using RocFall software to model rockfall trajectories, evaluate bounce heights and quantify kinetic energy impacts, (iii) Propose protective barriers based on simulation.

This integrated approach combining empirical assessments and numerical modeling provided a scientifically robust framework for rockfall hazard mitigation. By focusing on a case study in the Himalaya, this research contributes to the broader understanding of slope stability and offers feasible solutions for reducing risks in similar geotechnical settings.

2 Study Area

The study area encompasses two road cut sections along NH-7 between Devprayag to Srinagar of Tehri Garhwal district of Uttarakhand (Fig. 1) and lies in 53 J/12 toposheet of Survey of India. The confluence of the Bhagirathi and Alaknanda Rivers form the Ganga River at Devprayag. The studied section has great importance due to several socio-economic and cultural activities. Many devotees travel through this route to Char Dham pilgrimage sites every year. The scenic and spiritual appeal of the area promotes general tourism, which has been steadily increasing over the years. Furthermore, this is one of the important strategic routes that lead towards the India-China border. Consequently, NH-7 is a critical lifeline for regional connectivity, spiritual activities, bolstering the economy.

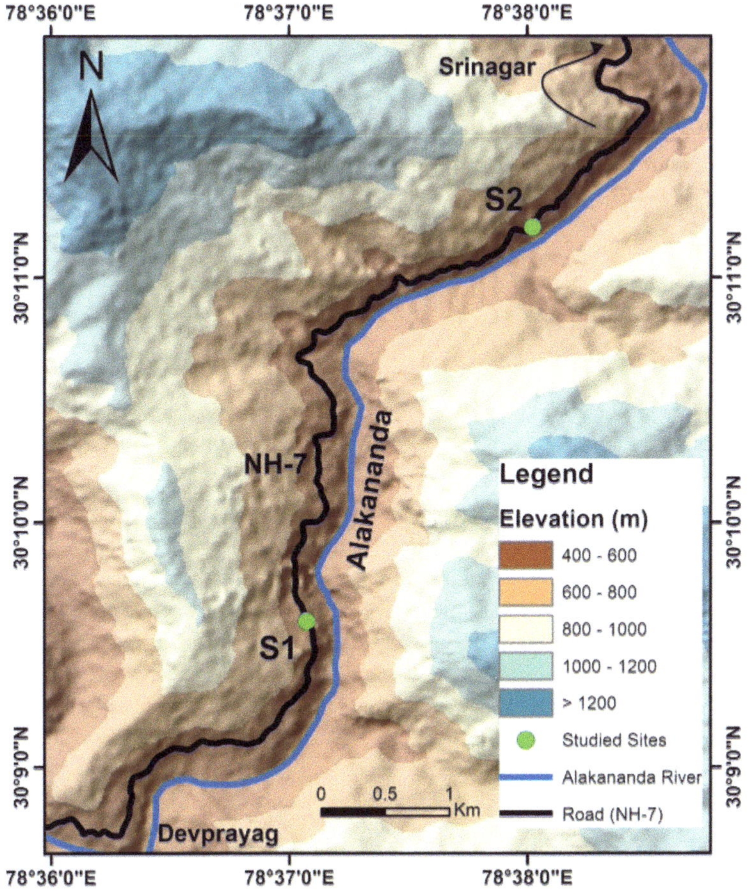

Fig. 1 Digital elevation model depicting topography and studied rockfall sites along NH-7

The selected section along NH-7 has a history of rockfall incidence. In September 2019, a fatal rockfall incident occurred near Teen Dhara of Devprayag city that claimed six lives and four persons were severely injured (https://www.ndtv.com/india-news/six-killed-4-injured-after-boulders-fall-on-their-car-in-uttarakhand-2108835). According to Mey et al. (2024), around 250 km long section of NH-7 from Rishikesh to Joshimath has landslide density of 1.25 landslide per km. In this 250 km long section, Devprayag city lies in a heavily clustered landslide zone having maximum landslide counts. Ansari et al. (2020a) conducted rockfall appraisal and rock mass characterization in phyllite near Devprayag and concluded that improper blasting for road construction had escalated the rockfall vulnerability along NH-7 near Devprayag. Many other geotechnical assessments, including numerical simulation studies also highlighted the concerns of frequent rockfalls and landslides along NH-7 near Devprayag (Ansari et al., 2020b; Dahiya et al., 2024; Kumar et al., 2020; Kundu et al., 2022; Singh et al., 2021b).

The studied section lies between Devprayag and Srinagar and comprises of meta-sedimentary sequence having steep slopes and rugged terrain. The section lies in temperate climate zone and high-altitude regions receive moderate snowfall from December to February. Geologically, it belongs to Chandpur Formation, Jaunsar Group of Neoproterozoic age (Ansari et al., 2020b). The studied section lies between the Main Boundary Thrust (MBT) and North Almora Thrust (NAT). The studied engineered slopes comprise of meta-sedimentary rocks, including quartzite (Slope-S1, Location: 30° 9′35.76" N and 78°37′4.43" E) and phyllite (Slope-S2, Location: 30°11′12.33" N and 78°38′1.29" E). The studied slope sections have three sets of discontinuities forming a blocky mass with wedge geometry. During the field survey, some fallen rock blocks have been reported in the limited ditch portion at the toe parts of the slope.

3 Methodology

The present study evaluates rockfall along a road-cut section of NH-7 by integrating the Rockfall Hazard Rating System (RHRS) and numerical simulations. A comprehensive field investigation was conducted to identify rockfall-prone sites and various controlling parameters like lithology, structural discontinuities, slope geometry, geotechnical parameters and historical rockfall occurrences were collected during the field survey. RHRS is applied to systematically classify and prioritize hazardous slopes based on several quantitative and qualitative parameters, including slope height, block size, geotechnical characteristics, etc. Numerical simulations were also undertaken using RocFall application to determine possible rockfall trajectories. Simulations also determined key rockfall-related metrics, such as bounce height, kinetic energy and maximum run-out of fallen blocks. The adopted methodology has been depicted using a flowchart (Fig. 2).

An extensive geotechnical field survey was conducted along NH-7 from Devprayag to Srinagar and two potential rockfall sites were selected for detailed

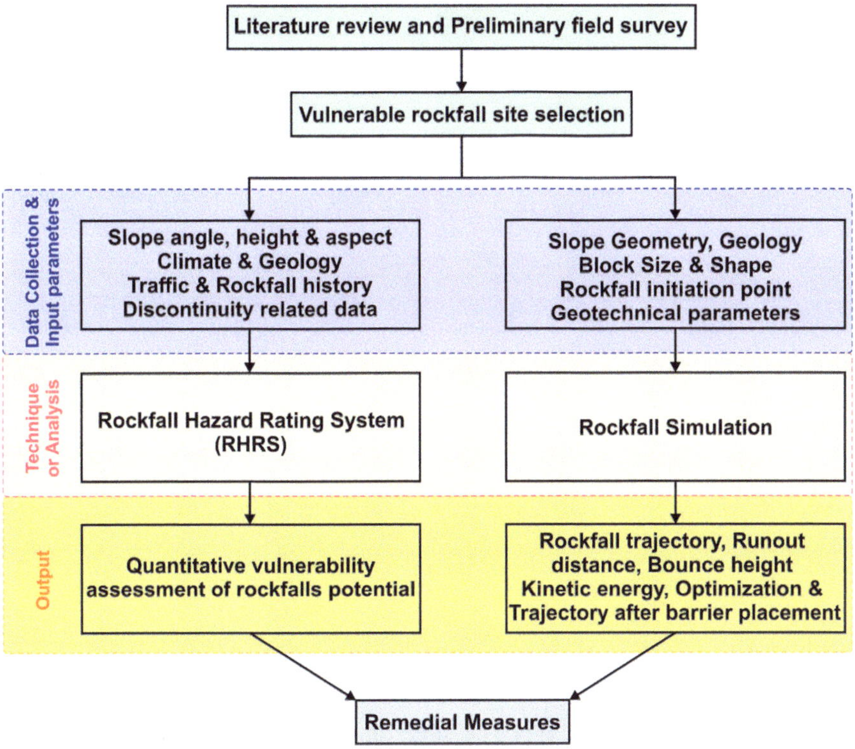

Fig. 2 Flowchart depicting the adopted methodology

rockfall assessment. The selected locations (S1 and S2) have very narrow ditches or roadside pavements, where blocks of different sizes have fallen from the engineered slopes. Although there is no settlement downstream to these slope faces, the live traffic is always vulnerable to rockfalls from these sections. Required geological, geotechnical, geometrical, structural data were collected to apply RHRS for Indian rockmass and to perform numerical simulations.

The cut slope height at location S1 is around 38 feet and comprises weathered quartzite forming blocky rockmass (Fig. 3a). Several rock blocks of different sizes were found on the bottom of the slope, ranging from about five to twenty kg in weight, indicating previous rockfall occurrences. Vegetation in the form of certain isolated plants is found, but it is limited to the upper section of the slope. The rocks are exposed in the cut face, allowing blocks to fall freely from the cut slope face. The ditch is almost flat and narrow, which cannot arrest rockfalls initiated from the cut slope face. The engineered slope at location S2 has a height of 21 feet and comprises phyllite (Fig. 3b). It has large persistent rock foliations in the rockmass, which intersect with other discontinuities to form blocks for rockfalls. Like S1, this section has no vegetation and bears a flat-narrow ditch, where rockfall cannot be

Fig. 3 Field photographs depicting the geotechnical condition of rock mass and potential to rockfalls: **a** blocky and weathered quartzite with fallen blocks near toe portion at location S1; **b** highly persistent rock foliation within phyllite and fallen blocks at S2 along NH-7

prevented. Almost every rockfall event will hit the road and ultimately reach the downstream valley.

4 Results

The present rockfall hazard appraisal study utilized the Rockfall Hazard Rating System (RHRS) and numerical simulation. Rockfall hazard rating is widely applied for preliminary assessment of rockfall risk potential. The initial systematic and well-recognized attempt to characterize rockfalls was proposed by Brawner and Wyllie (1976). Subsequently, several researchers (Andrew, 1994; Pierson & Van Vickle, 1993; Pierson et al., 1990; Santi et al., 2009) have developed different rockfall hazard rating-based systems. Ansari et al. (2013) modified existing frameworks to accommodate local geo-climatic conditions and developed a Rockfall Hazard Rating System for Indian Rockmass (RHRSI). RHRSI is based on 24 controlling parameters related to slope geometrical characteristics, ability to catch or arrest rockfalls, hydrology, hydrogeology, durability, geological structures, geotechnical factors, traffic vulnerability, etc. Based on existing site conditions, scores were assigned to all parameters, and rockfall hazard rating values were determined for both sites (Table 1). Slope S1 is relatively more hazardous, with a cumulative score of 779, while Slope S2 has 615 score.

4.1 Rockfall Simulation Using RocFall Simulator

Numerical modeling was conducted using RocFall 4.0 simulator tool, which provided a detailed and dynamic behavior of plausible rockfalls from both sites along NH-7.

Table 1 Rockfall hazard rating system of rock mass vulnerable to rockfalls along NH-7 from Devprayag to Srinagar

Parameter and category			Slope 1		Slope 2	
Category	Parameter		Description	Score	Description	Score
Slope	Slope height (m)		11.62	5	6.4	3
	Average slope angle		D	81	B	9
	Vegetation		Isolated plants	27	Isolated plants	27
	Launching features		Many (0.6–1.8 m surface variation)	27	Many (0.6–1.8 m surface variation)	27
	Ditch catchment		Moderate	9	Limited	27
Climate and geology	Annual precipitation (mm)		508	9	508	9
	Seepage or Water		Damp-Wet	9	Dry	3
	Slope aspect		NE	27	SE	27
	Sedimentary rock	Degree of undercutting (m)	0–0.3	3	0.6–1.2	27
		Slake durability index (SDI)	99%	3	98.90%	3
		Degree of interbedding	1–2 weak interbeds with thickness > 15 cm	9	1–2 weak interbeds with thickness < 15 cm	3
	Discontinuities	Block size (m)	0.6	9	0.6	9
		Block shape	Blocky to angular	27	Blocky to angular	27
		Number of sets	3	81	3	81
		Persistence/orientation	>3 m and daylight out of the slope	81	>3 m and daylight out of the slope	81
		Aperture (mm)	>5	81	0.1–1	9
		Weathering Condition	Highly	27	Moderately	9
		Friction	Rough	3	Planar	27
		Infilling material	None	0	None	0
Traffic	Percent Decision sight distance (DSD) in %		40	81	40	81
	Average vehicle risk (AVR) in %		100	81	100	81
	Roadway width (m)		5	81	7	27
	No. of accidents		3–5	9	3–5	9
Rockfall history/Frequency			Occasional	9	Occasional	9
Total score			779		615	

Depending upon the potential rock blocks and plausible site of initiations, simulations were performed on different weights and imitation sites. Various inputs for modeling, including slope geometry, rock density, block size variations, release point, etc., were determined during the field survey. Coefficient of restitution has been consulted from the existing literature. The performed simulations predicted bounce heights, velocity and maximum runout distance, suggesting that such areas are prone to high-impact rockfall events. From site S1, three possible rockfall sources were identified during the field and simulated. The predicted rockfall trajectories show that large numbers of rock blocks are hitting the road and bouncing off to travel further in the valley (Fig. 4). Kinetic energies and bounce height (Figs. 5 and 6) of falling blocks indicate that the falling rocks hit the road again for a secondary bounce, thereby increasing the chances of damaging any structure, vehicle or person over the road.

Two plausible rockfall sources were identified and simulated for slope S2 (Fig. 7). The predicted trajectories of the rocks show that rocks that start from the top of the slope land initially on the ditch area and then hit the road for the secondary bounce.

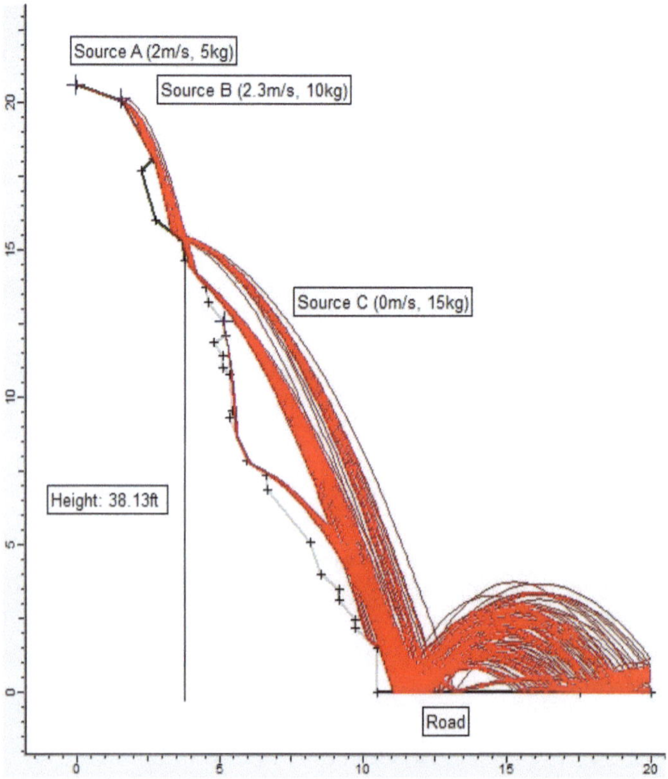

Fig. 4 Simulation of rockfall from slope S1

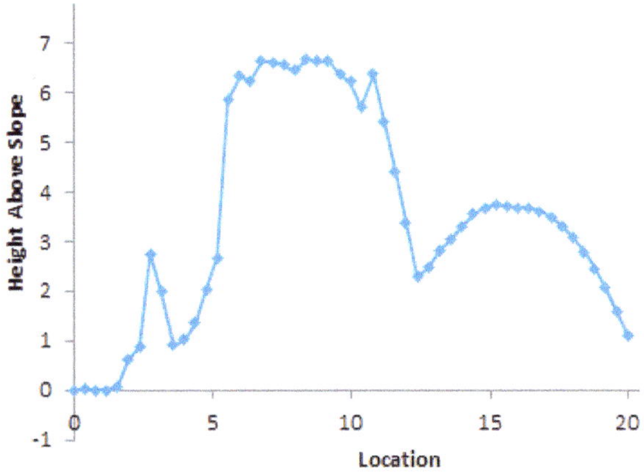

Fig. 5 Graph showing bounce heights for slope S1

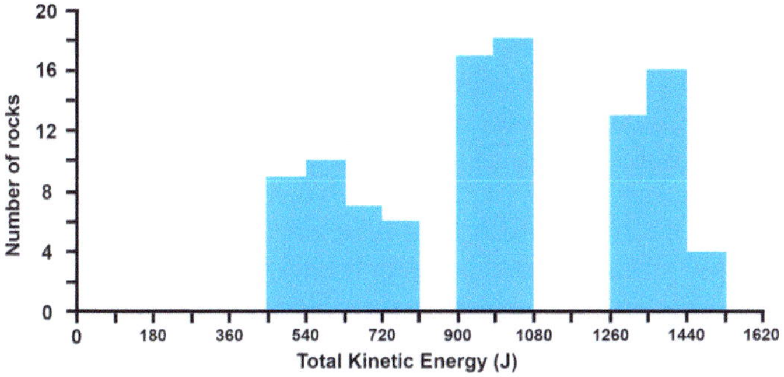

Fig. 6 Graph showing the distribution of total kinetic energy for slope S1

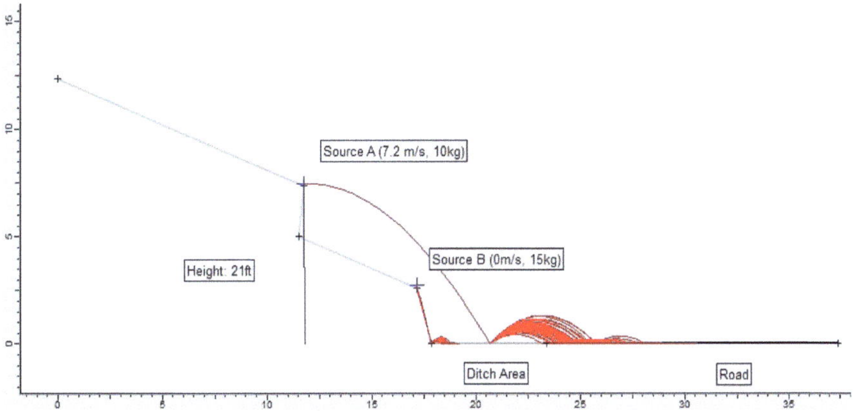

Fig. 7 Simulation of rockfall for slope S2

5 Discussion

In Slope S2, the bounce height and kinetic energies are relatively lower than S1 (Figs. 8 and 9). This is because of the relatively lower slope gradient and the initial landings of rockfalls within the ditch area, which is comprised of soil. Soil mass in the ditch area absorbs the kinetic energy much more efficiently than relatively harder asphalt. The mechanical behavior of soil differs significantly from that of asphalt or exposed hard rock surfaces, leading to enhanced energy absorption and attenuation. Soil is relatively softer and has a greater tendency for plastic deformation and energy dissipation upon impact. This allows it to absorb and attenuate the kinetic energy of falling rock blocks more effectively than rigid asphalt. The absorption capacity of the soil is influenced by factors such as compaction, moisture content and grain size distribution. In contrast, asphalt or exposed bedrock surfaces exhibit higher coefficients of restitution, resulting in greater rebound heights and increased energy retention in rockfall trajectories. It is also evident that a group of blocks can be arrested within the ditch area, thereby emphasizing the importance and effectiveness of a ditch. Depending on the bounce heights and run-out distances, protective barriers can be placed in positions that could prevent the rocks from falling over the road. The mechanical strength and quality of these barriers can be estimated from the energies and velocities of the rocks, which have been calculated through the simulations (Hearn et al., 1995). Due to relatively higher bounce heights, slope S1 need a barrier of 5.5 m, while a barrier of 1.5 m is sufficient to block the rockfalls from S2 (Figs. 10 and 11). The findings emphasize the crucial role of slope geometry and interacting material or terrain type in controlling rockfall dynamics. Strategic designing of the studied rockfall sites by modifying ditches and placing barriers can significantly reduce the hazard posed by rockfalls.

Fig. 8 Graph showing bounce heights for slope S2

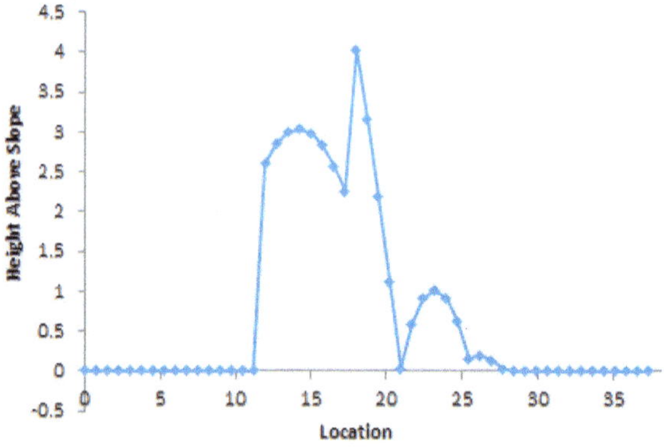

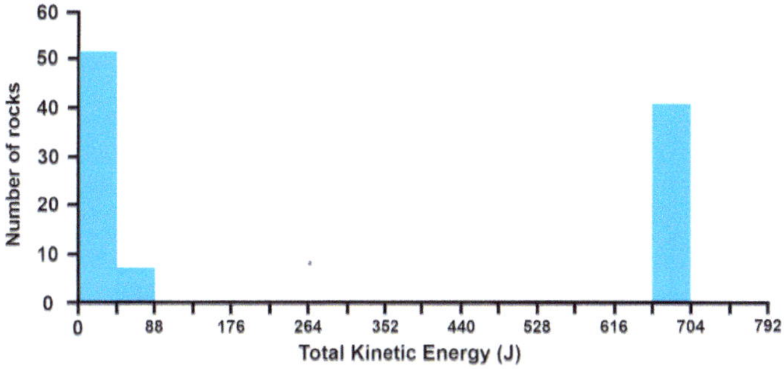

Fig. 9 Graph showing the distribution of total kinetic energy for slope S2

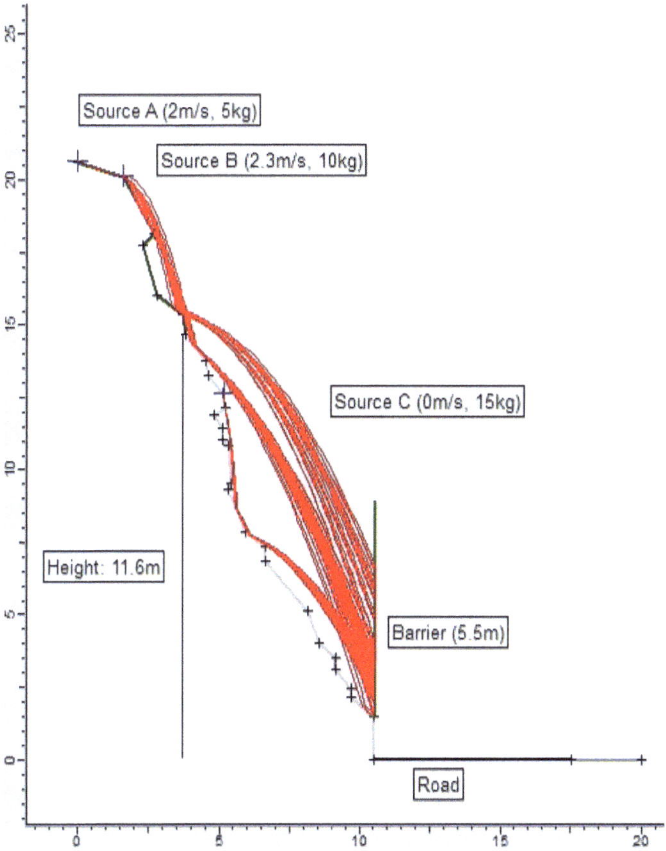

Fig. 10 Simulated trajectories after placing a barrier of 5.5 m for slope S1

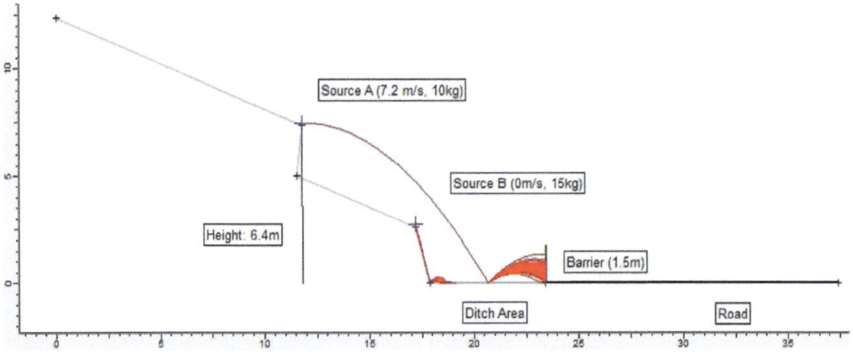

Fig. 11 Simulated trajectories after placing a barrier of 1.5 m for slope S2

6 Conclusions

An extensive rockfall analysis of two engineered slopes along NH-7 using empirical rockfall hazard rating and computer-based simulations was conducted. From both techniques, S1 is more hazardous than S2. It was observed that the significant source of rockfall in the studied sections is due to the adverse orientation of discontinuities. Numerical simulation confirmed that most of the fallen blocks impact the road before continuing their trajectory into the Alaknanda Valley, demanding immediate attention to install protective barriers. As per the rockfall optimization, 5.5 and 1.5 m barriers can prevent the road from potential rockfalls. Apart from protective barriers, re-sloping and trimming the loose rock blocks should also be undertaken to reduce the rockfall initiation from the cut slopes. Increasing the ditch catchment area and the drapery system are other measures that can rule out rockfall occurrences at the studied sites. This study highlights the necessity of geotechnical appraisal and timely hazard assessment to mitigate rockfall risks along strategic transportation corridors. Implementing proposed mitigation measures can significantly reduce the likelihood of hazardous rockfall incidents, ensuring safer and more resilient road infrastructure in tectonically active regions like Himalaya.

References

Aher, S. P., Mohanty, D. P., Bhapkar, P. V., Pradhan, S. P., & Vishal, V. (2024). Understanding the rock slope stability along Malshej Ghat, India, using rock mass characterisation and block shape classification techniques. *Journal of Earth System Science, 133*(2), 1–16.

Ahmad, M., Umrao, R. K., Ansari, M. K., Singh, R., & Singh, T. N. (2013). *Assessment of rockfall hazard along the road cut slopes of state highway-72.* Maharashtra.

Andrew, R. D. (1994). The colorado rockfall hazard rating system [Doctoral dissertation, University of Colorado at Colorado Springs].

Ansari, M. K., Ahmad, M., Singh, R., & Singh, T. N. (2013). Rockfall hazard rating system for Indian rock mass. *International Journal of Earth Science and Engineering, 6*(01), 18–27.
Ansari, T. A., Singh, H. O., Singh, T. N., & Singh, K. H. (2020a). Rock slope vulnerability assessment Near Devprayag, Uttarakhand, India. In *ARMA US Rock Mechanics/Geomechanics Symposium* (pp. ARMA-2020). ARMA.
Ansari, T. A., Singh, K. H., Singh, T. N., & Tripathi, J. N. (2020b). Stability analysis of road cut slope near Devprayag in Lesser Himalaya, India. *Journal of Earth System Science, 129*, 1–13.
Brawner, C. O., & Wyllie, D. (1976). Rock slope stability on railway projects. *Area Bulletin, 77* (Bulletin 656).
Dahiya, N., Pandit, K., Sarkar, S., & Pain, A. (2024). Various aspects of rockfall hazards along the mountain roads in India: a systematic review. *Indian Geotechnical Journal* 1–23.
Hearn, G., Barrett, R. K., & Henson, H. H. (1995). Testing and modeling of two rockfall barriers. *Transportation Research Record, 1504*, 1–11.
Jamir, I., Kumar, V., Ojha, A. K., Gupta, V., Martha, T. R., & Griffiths, D. V. (2024). Evaluating failure regime of an active landslide using instability and rockfall simulation, NW Himalaya. *Environmental Earth Sciences, 83*(8), 256.
Khan, I., Bahuguna, H., & Kainthola, A. (2024). Regional landslide susceptibility zonation utilizing bivariate statistical techniques in the northwestern Himalayas, Jammu and Kashmir, India. *Journal of Earth System Science, 133*(3), 157.
Kumar, A., Pradhan, S. P., Singh, S., & Tripathi, K. (2024). Stability analysis for rock slope using finite element modeling near Darekasa Railway Station: A case study. *Journal of Intelligent Construction, 2*, 9180005.
Kumar, K., Joshi, V., & Biswakarma, P. (2020). Mapping of vulnerable landslide zones by large scale mapping in and around Devprayag Area along National Highway 58, Uttarakhand, India. *Environment & We an International Journal of Science & Technology, 15*, 145–154.
Kundu, S., Mehta, H. A., Sarkar, N. S., & Bhowmik, R. (2022). Rockfall hazard in himalayan region: assessment and review of trajectory analysis parameters for mitigation strategies. In *Indian Geotechnical Conference* (pp. 101–115). Springer Nature Singapore.
Lakherwal, M., Dhiman, R. K., Thakur, M., & Kumar, M. (2024). Kinematic assessment of a rockfall disaster: a case study from Batseri Village, Sangla Valley, Himachal Pradesh, India. *Landslides* 1–14.
Maheshwari, S., & Bhowmik, R. (2024). Performance assessment of a protection embankment designed for a rockfall-prone slope along lesser Himalayas of Jammu and Kashmir, India. *Natural Hazards* 1–39.
Mey, J., Guntu, R. K., Plakias, A., Silva de Almeida, I., & Schwanghart, W. (2024). More than one landslide per road kilometer–surveying and modeling mass movements along the Rishikesh-Joshimath (NH-7) highway, Uttarakhand, India. *Natural Hazards and Earth System Sciences, 24*(9), 3207–3223.
Pandey, V. H. R., Kainthola, A., Yadav, V., Kundu, J., Mazzanti, P., Singh, R. P., & Singh, T. N. (2025). 3D Stochastic Simulation of Rockfall Mechanism and Mitigation in the Batseri Zone. *Geotechnical and Geological Engineering, 43*(1), 1–24.
Pierson, L. A., & Van Vickle, R. (1993). *Rockfall hazard rating system: participant's manual* (No. FHWA-SA-93-057). Federal Highway Administration.
Pierson, L. A., Davis, S. A., & Van Vickle, R. (1990). *Rockfall Hazard rating system: implementation manual* (No. FHWA-OR-EG-90-01).
Roul, A. R., Pradhan, S. P., Tulsawadekar, K., & Vishal, V. (2023). A Python-based optimization method to define the trajectory of rockfalls: A case study of Malshej Ghat, India. *Bulletin of Engineering Geology and the Environment, 82*(11), 421.
Santi, P. M., Russell, C. P., Higgins, J. D., & Spriet, J. I. (2009). Modification and statistical analysis of the Colorado rockfall hazard rating system. *Engineering Geology, 104*(1–2), 55–65.
Siddique, T., Haris, P. M., & Pradhan, S. P. (2022). Unraveling the geological and meteorological interplay during the 2021 Chamoli disaster, India. *Natural Hazards Research, 2*(2), 75–83.

Singh, A. K., Kundu, J., Sarkar, K., Verma, H. K., & Singh, P. K. (2021a). Impact of rock block characteristics on rockfall hazard and its implications for rockfall protection strategies along Himalayan highways: A case study. *Bulletin of Engineering Geology and the Environment, 80*(7), 5347–5368.

Singh, J., Pradhan, S. P., Vishal, V., & Singh, M. (2023). Characterization of a fractured rock mass using geological strength index: A discrete fracture network approach. *Transportation Geotechnics, 40*, 100984.

Singh, P. K., Kainthola, A., Panthee, S., & Singh, T. N. (2016). Rockfall analysis along transportation corridors in high hill slopes. *Environmental Earth Sciences, 75*, 1–11.

Singh, R. A., Negi, R., Singh, P. K., & Singh, T. N. (2021b). Landslide Studies Between Devprayag and Pali along National Highway-7, Tehri District, Garhwal Lesser Himalaya. *Journal of Scientific Research, 65*(1), 32–38.

Varshney, H., Siddique, T., Ahamad, A., & Akhtar, W. (2024). Correlating circular failure charts, limit equilibrium and finite element method based safety factor for circular failure type landslides. In *Landslide: Susceptibility, Risk Assessment and Sustainability: Application of Geostatistical and Geospatial Modeling* (pp. 695–712). Springer Nature Switzerland.

Verma, A. K., Sardana, S., Sharma, P., Dinpuia, L., & Singh, T. N. (2019). Investigation of rockfall-prone road cut slope near Lengpui Airport, Mizoram, India. *Journal of Rock Mechanics and Geotechnical Engineering, 11*(1), 146–158.

Vishal, V., Siddique, T., Purohit, R., Phophliya, M. K., & Pradhan, S. P. (2017). Hazard assessment in rockfall-prone Himalayan slopes along National Highway-58, India: Rating and simulation. *Natural Hazards, 85*, 487–503.

Estimating Terrain Feasibility Using BIS Code and AHP Methods for Landslide Susceptibility Zonation: A Case Study of Mao-Maram Region Manipur

Kaikho Khusulio, Rohan Kumar, and Swati Sharma

Abstract A feasibility study was conducted to ascertain the scope of the landslide hazard evaluation factor using the Bureau of Indian Standards (LHEF—also a BIS code) and the Analytical Hierarchy Process (AHP) approach for landslide susceptibility zonation (LSZ). LHEF guidelines recorded in the BIS code have been slightly modified considering the scale of the study. Multi-source geospatial data has been used to extract landslide causative factors in addition to the slope facet. A total of 153 slope facets have been identified for the study area Moa-Maran region of Manipur. For each slope facet, an estimate of landslide propensity has been made using ratings derived from the LHEF scheme and the AHP method. The prediction rate curve and Receiver operating characteristic (ROC) have been evolved for LHEF and AHP methods and further used for establishing feasibility. The prediction rate curve estimated 70.9 and 72.63% accuracy for the LHEF/BIS method and the AHP method respectively, whereas ROC curve estimate was found to be 77.2% for LHEF/BIS and 79.4% for the AHP method. This study has found both methods reasonably feasible and can be used in similar terrain conditions found in other regions.

Keywords Landslide susceptibility zonation · BIS code · AHP · GIS · Slope facet and remote sensing

K. Khusulio (✉) · R. Kumar
Lovely Professional University, Phagwara, Punjab, India
e-mail: Khusiilio16@gmail.com

R. Kumar
e-mail: rohan.25322@lpu.co.in

S. Sharma
Department of Transdisciplinary Research, Division of Research and Development, Lovely Professional University, Jalandhar, India
e-mail: swati.33384@lpu.co.in

© The Author(s), under exclusive license to Springer Nature Switzerland AG 2025

S. Ghosh (ed.), *The Himalaya Dilemma*, Springer Natural Hazards,
https://doi.org/10.1007/978-3-031-95083-4_9

1 Introduction

Landslides have caused property damage and the loss of countless lives all across the world. According to disaster research on a worldwide scale, emerging nations for example Nepal, China and India have experienced numerous deaths and destruction because of landslide vulnerability in the current decade (Kumar & Anbalagan, 2015). According to estimates, India accounts for 18% of worldwide landslide deaths, even though industrialized nations account for 80% of landslide fatalities (Froude & Petley, 2018). In the hilly terrain of India, landslides are one of the frequently occurring disasters which cost lives and damage properties. It is said that landslide activities cannot be completely averted but the power and severity of their power may be mitigated by picking up well-organized measures and suitable planning in the high landslide zone (Sarkar et al., 1995). Prediction and occurrence of landslides are thought to be tough and complicated therefore evaluation of landslide susceptibility is regarded to be one of the essential problems and with the development of remote sensing and GIS, there is an increase in landslide susceptibility research (Chen & Li, 2020). Landslide zonation, on the other hand, may be defined as "the broad classification of a land surface into zones and the ranking of those zones rendering to the degree of actual and possible threat from other mass movements or landslides on the slope" (Varnes, 1984). Anthropogenic activities, for example, the building of infrastructures in steep regions such as roads, canals, dams, tunnel townships and so on have also enhanced the effect of disasters as indicated by several incidents occurring in recent few years (Kumar & Anbalagan, 2019). Landslides are caused by several variables, including geological, geomorphologic, and meteorological influences. Geomorphology (grade of the slope, aspect, and relative relief), bedrock geology (lithology, grade of weathering, structure), soil (deepness, structures, porousness, and permeability), LULC, and geohydrologic conditions are often the most critical components. We can analyze and anticipate landslide-prone places using scientific research, and thereby limit landslide-induced devastation by applying appropriate mitigation measures (Kundu et al., 2013; Galve et al., 2016; Jemec Auflič et al., 2018; Mahajan et al., 2022). The multidimensionality of geo-environmental data collection, which is diverse from nature and comes from a variety of sources, such as air photos, satellite images, historical records, and field surveys, has made it a challenge for research scientists to develop accurate and effective models from geo-environmental data (Nhu et al., 2020; Pham et al., 2019).

In Manipur increased landslide activity is being caused by the unstable topography, severe precipitation, quick infrastructure development, and construction of the large structure in a mountainous region. Although the landslide situation in Manipur is dreadful, no efficient slope stability estimation is used throughout the state of Manipur, India, although 90% of Manipur is highly susceptible to landslide activity every year. It has also not yet been explored in many ways, including using available advanced technology, availability of details ancillary data, factors causing landslides, cause of increasing landslides activities in the region, and minimizing/preventive mitigation for landslides. From the literature review in general for

Manipur state, there are only a few landslide susceptibility studies carried out so far even though it is considered a highly landslide-prone area and fatalities incidents are reported every year. Few studies which are carried out in Manipur are found to only focus along the roadside leaving behind other important regions and also lack terrain feasibility assessment. In this study, two approaches that are the BIS code and Analytical Hierarchy Process methods were used for the study region. The BIS code (BIS 1998—14,496 part 2) was used to apply the usual slope facet-based technique. The grading of geo-environmental elements impacting every facet of the slope was used for determining using the landslide hazard evaluation factor (LHEF) approach described in the code (BIS code). In general terms, the LHEF rating is considered to fit the best for the heuristic approach even though it is an empirical technique method from an engineering geological viewpoint (Anbalagan, 1992). AHP is a measuring theory that has been applied to numerous domains, including decision theory and dispute resolution, for dealing with quantitative and intangible factors. "AHP is a cross multi-criteria decision-making method that allows the user to choose a scale of preference from a collection of options" (Yalcin, 2008). Six landslide-affecting elements offered by the BIS code, such as lithology, structural morphometry, hydrological condition, slope, relative relief, and LULC, were determined and graded based on the supplied LHEF rating for the geo-environmental condition. For comparison of the research region, LSZ maps were constructed using two methods: Analytic Hierarchy Process (AHP) and BIS code with the same six conditioning causal factors mentioned in the BIS code. Landslide hazard assessments include evaluating alternative scenarios based on the kind and severity of the triggering mechanism, as well as the direct and indirect implications of the triggering mechanism. As a result, extensive mapping and analysis of regions prone to landslides are critical, particularly for the use of spatial planning, decision-making, and disaster management procedures. The scenarios must be taken into account in regional and local making plans, whether through prevention measures like prohibiting construction in highly exposed sites, mitigation measures like building drainage channels, taking precaution measures like installing an alert system and organizing evacuation procedures (Kumar et al., 2017; Vojtekova & Vojtek, 2020).

2 Study Area

The research study area (Fig. 1) is a part of the Mao-Maram subdivision region located at latitude 25.42° E and longitude 93.98° N in Senapati, Manipur India. In general, the study area is a part of the Northeast region that is mostly affected by landslide incidents of a confusing variety. Landslides in this region cause constant problems, causing loss of lives, properties and economic losses worth billions of rupees. During strong and prolonged rains, natural hazards like landslide disasters commonly afflict Manipur's hilly roads, particularly the national highways. National highways NH-37 and NH-2 connect the state to the rest of the nation and serve as the state's backbone which also runs along the present study area region. Every year, especially during the rainy season, the mountainous territory of Manipur, notably along the NH-2, the state lifeline, is prone to landslides (Singh et al., 2011). The current research region Mao-Maram comprises fragile lithology due to the Disang and Barail groups of rock of the tertiary period which is a steep landscape with degraded and fractured geology. Manipur's soil types may be divided into two major categories: alluvium in the lowlands and red ferruginous soil in the mountainous areas. The topsoil on the steep slopes is quite thin compared to the lowlands, which are much heavier, particularly in flood plains. Sand, sandy clay, and tiny rock pieces of various types are often found in soils. The soil on a steep slope is thought to be prone to a lot of erosion. In the state of Manipur, extremely widespread weathering, tectonic collision, and human activity due to slope cutting and road expansion for development create frequent landslides. The study area is prone to landslides and every year, notably during monsoon, it encounters landslides and obstructs the roads for days. In general, Manipur has a tectonic landscape encompassing the Arakan-Yoma mountains that are characterized by exceptionally steep, rough, and weak slopes (Balamurugan et al., 2016). The present study area also belongs to the tectonic setting of the subduction zone taking place between the Indian plate and the Eurasian plate where the Indian plate is subducting beneath the Eurasian. The study area which is under Manipur state falls in the highest seismic active zone V of India proposed by "the Bureau of Indian Standard (BIS)" of the Trans-Asiatic earthquake belt. It also comes within Purvanchal Himalaya which is mostly formed of geological shale and sandstone formations.

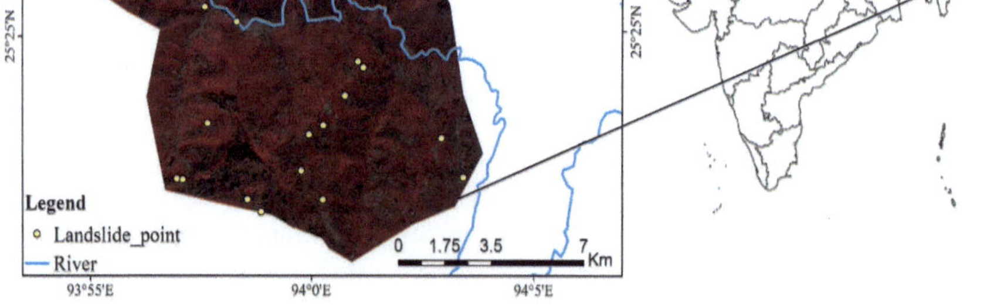

Fig. 1 Base map of the study area obtained from Sentinel 2B satellite

3 Landslide Inventory and Data Preparation

Recognizing and mapping past landslide sites (Fig. 2) is essential to investigating the link between landslide distribution and causative parameters. In this study, 36 landslides location was considered for validation purposes. Landslides observed in the study area are mostly 50–120 m^2 in size which was noted during field study and using remote sensing techniques. The study area Mao-Maram belongs to Purvanchal Himalaya and is primarily made of mostly sandstone and shale with few siltstones of geological origin.

Geospatial databases and remote sensing have become important factors for the mapping of landslide susceptibility zones. In this paper, six causative factors

Fig. 2 Field photographs **a** rotational type of landslide **b** rotational type of landslide **c** earthflow type of landslide **d** earth flow type of landslide **e** complex landslide **f** rotational landslide

Table 1 Geospatial data of the causative factor

Type of data	Detector sensor	Resolution or scale	Derived	Source
Multispectral data	Sentinel 2 B	10 m	LULC	Copernicus open-access hub
	Landsat series	15 m–15 cm	Landslide inventory	Google earth pro
DEM	ALOS PALSAR	12.5 m	Slope morphometry	Alaska satellite facility
			Relative relief	
			Hydrogeological condition	
			Structure lineament	
Ancillary data	Lithology map of the study area	1:50,000	Lithology	Bhukosh (GSI)

proposed by BIS 1998–14,496 part-2, were used for the study and creation of landslide susceptibility zonation of the study area. Data used for geo-environmental parameter extraction are discussed in Table 1.

4 Proposed BIS Code for a Causative Factor

In this paper, six causative factors proposed by the Bureau of Indian Standard (BIS) namely slope angle, lithology, structural favorability, relative relief and hydrogeological condition were used for LSZ mapping. These causative factors were provided with a rating based on the given BIS code called "landslide hazard evaluation factor rating" (LHEF). The slope facet for each factor was selected as a mapping unit. In this paper, the slope facet was derived from the DEM of ALOS PALSAR which has a resolution of 12.5 m (Fig. 3). The shaded relief map which is derived from ALOS PALSAR DEM was digitized according to the slope facet for analysis.

4.1 Slope Morphometry

A region with a steep slope gradient (Fig. 4) is considered to be highly prone to landslides whereas a region with a lower grade angle is less hazardous to landslides (Kumar & Sandhu, 2020). According to the BIS code, "slope angle with 0–15° is very low, 16–25° low, 26–35° moderate, 36–45° high and >45° very high" are indicated. The slope angle of the study area was generated using DEM along with the support of

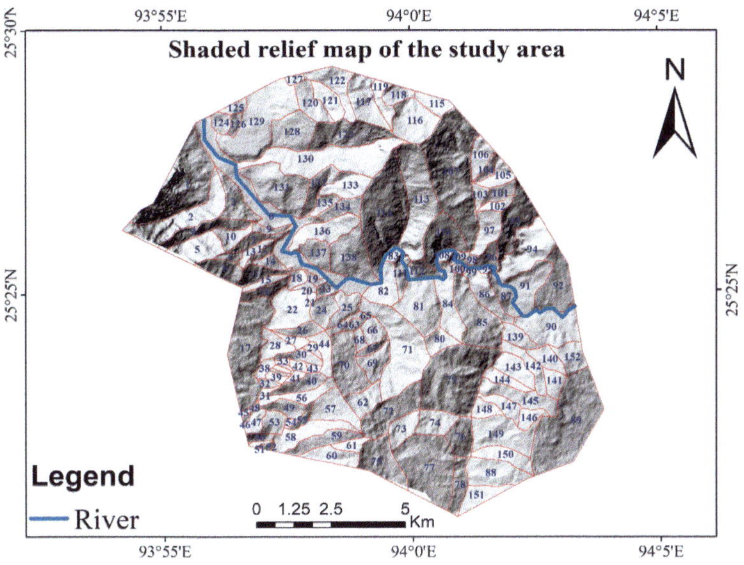

Fig. 3 Shaded relief map of the study area

the GIS software and classified into five groups based on the slope angle mentioned in the BIS code.

4.2 Relative Relief

It is the difference between the highest point and lowest point of the region. Relative relief of the study area was detected to be ranging from 44 to 235 m. Relative relief for all the facets was classified into three classes and rated according to the provided BIS code for rating.

4.3 Lithology

In the research region, two significant rock groups were discovered that is Disang and Barial groups of rock (Fig. 4). In these two groups of rock 3 rock formation was observed i.e. Jenam and Laisong formation of Barial group which consist of flaggy sandstone, subordinate shale, carbonaceous sandy shale, sandstone and coal seams. Upper Disang formation of Disang group with shale, siltstone, greywacke rhythmite and sandstone. During the ground study and investigation, it is also detected that the rock is somewhat eroded and weathered. All three forms of rock were rated according

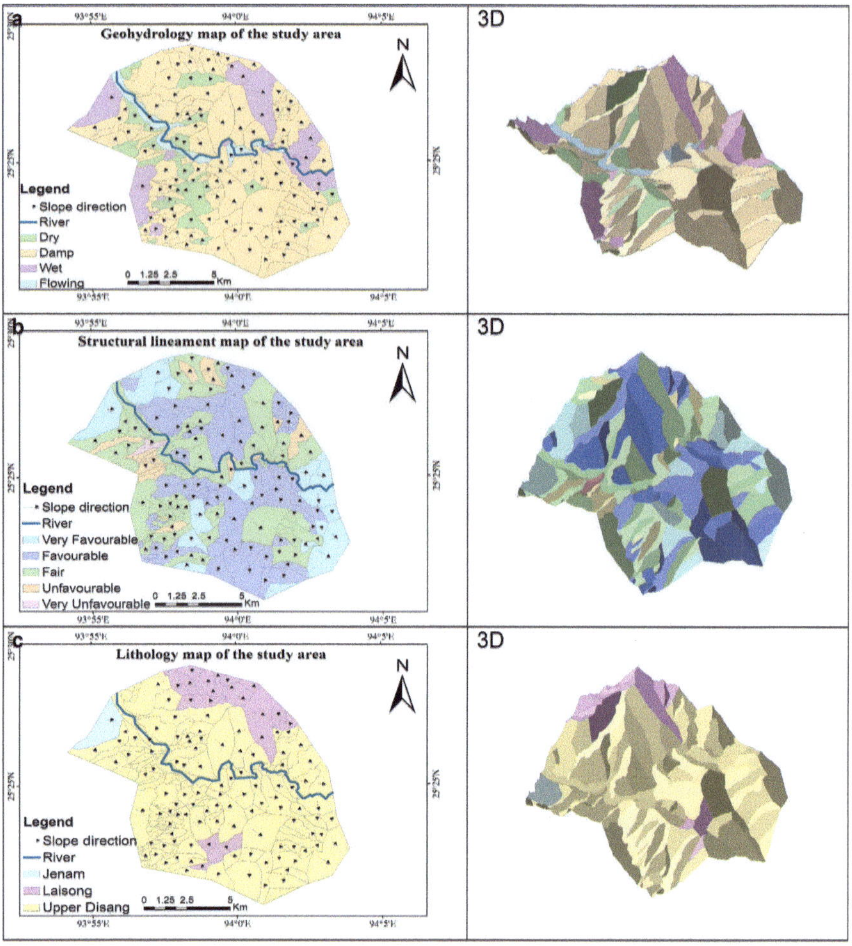

Fig. 4 Causative factors of landslide: **a** geohydrology **b** structural lineament **c** lithology **d** lulc **e** relative relief and **f** slope morphology

to the BIS code. A lithology map of the research study region is a kind of ancillary data obtained from the Bhukosh (GSI) portal.

4.4 Structural Lineament

Lineament is considered as a linear landscape feature result of a primary geological structure, which includes fault, fold, thrust, ridge, shoreline or a combination of all characteristics. In this paper, the structural favorability plot of the research area was done utilizing the structural photo-lineament approach which is less time taking

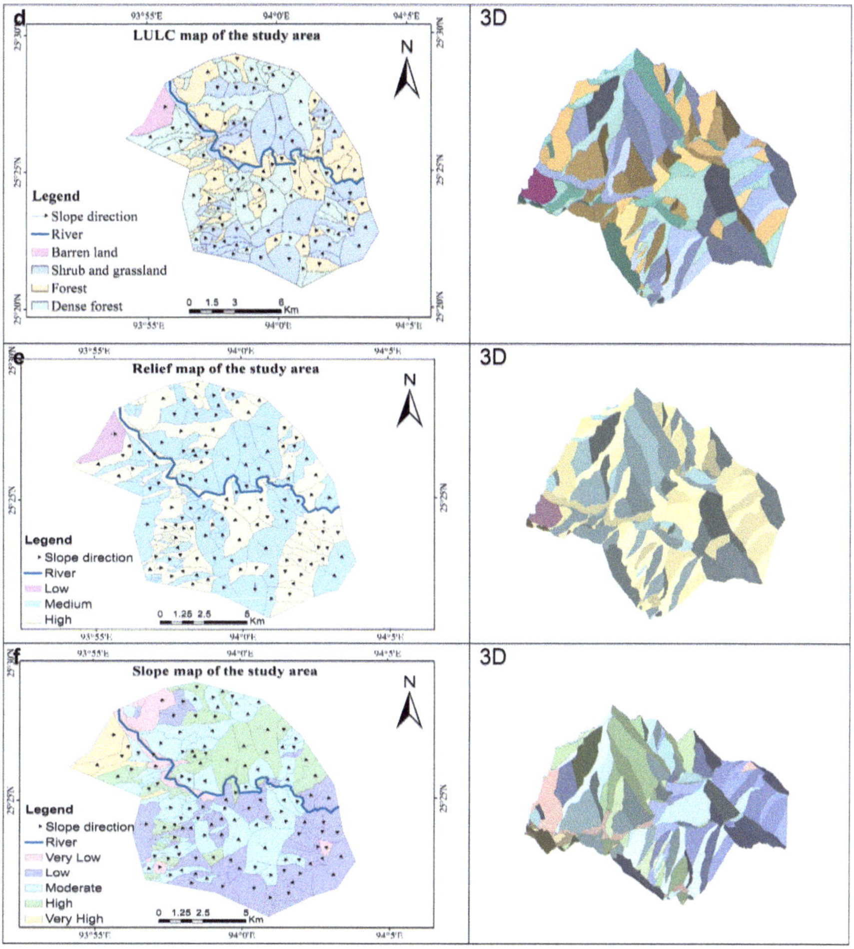

Fig. 4 (continued)

and convenient with consideration of the given BIS code. Lineament features were generated using remote sensing techniques and GIS software. With the help of the GIS software, photo-lineament density was obtained for all the facets and classified into five classes. Rating for all five classes was given with consideration to the given BIS code and a structural favorability map was generated for the study area.

4.5 *Land Used/Land Cover (LULC)*

Land use land cover of the study zone was obtained using the sentinel 2B image of 10 m resolution. Visible light (red) and Near-infrared (NIR) was used for generating the LULC map of the research region. Apparently to the given BIS code differentiation of each facet was analysed depending on the LULC status and classified into four classes and rated accordingly.

4.6 Hydrogeological Condition

The existence of streams, groundwater, and the moisture state of slope-forming materials establish an area's hydrogeological condition, which is an influential concern in landslide initiation. For the hydrogeological study, a stream order of drainage density was created for the study area. Each facet of the study area was calculated concerning the drainage density and classified into four classes according to the BIS code. The rating of each class was given according to the threshold values and names given in BIS code 1998.

5 Methodology for Generating Landslide Susceptibility Zonation

5.1 BIS Code Method

The landslide susceptibility zones were created using the "landslide hazard evaluation factor" (LHEF) rating method as per the BIS code. In this paper to calculate landslide hazard zones, a modified BIS technique was used. The modification was done for the causative factor during extracting data and analysis concerning the proposed BIS code. With advances in technologic, software and remote sensing, some slight modification was done which was convenient, advanced, and less time-consuming. The slope facet for six causative variables is determined and graded according to the LHEF rating, after which the total of the six parameters, also known as a total estimated hazard (TEHD), is computed to create the final landslide susceptibility zone map for the research area. After factor classes have been assigned ratings, the total estimated hazard (TEHD) may be calculated by aggregating the ratings assigned to each factor class in a facet (Kumar & Sandhu, 2020).

Rating of all the six causative factors (Table 2) was rated according to the proposed BIS code. A slight modification was done during the extraction of the causative factor with advancing technology and resources but for the LHEF rating (Table 3), it was rated regarding the "Preparation of landslide hazard zonation in mountainous terrains- guideline (BIS code IS 14496 part 2 1998)".

Table 2 Max LHEF to be rated according to BIS code for landslide parameters

Variables	Max rating
Structural lineament	2.0
Relative relief	1.0
Lithology	2.0
LULC	2.0
Hydrogeological	1.0
Slope morphometry	2.0
Total	10

5.2 *Analytical Hierarchy Process (AHP) Method*

AHP is a "multi-criteria decision-making method that allows the user to choose a scale of preference from a collection of options and is commonly used in site selection, suitability analysis, and LHZ mapping" (Yalcin, 2008). AHP is a logical instrument for generating the weight of geo-environmental factors according to their relevance in producing landslides considering LHZ mapping. The AHP model uses the professional expertise of specialists or groups of experts regarding the landslide influencing factor as an input, making the AHP method even more subjective. In this study, AHP was used for the determination of the Six landslide geo-environmental factors for rating and weighting of the conditioning landslide parameters and their sub-classes by using a heuristic approach and quantitative matric calculation based on eigenvector.

"Saaty (1980) developed AHP, a multi-criterion decision-making approach that allows objective and subjective factors to be considered in the decision-making process". AHP separates the complexity judgement issue into a hierarchy of variables and alternately, each of these is given weight on a 9-point ordinal scale based on pair-wise comparisons. The pair-wise comparative study was carried out by generating a matrix with assessed scores along one side of diagonal and diagonal cells with a value of 1 (Kumar & Anbalagan, 2016; Devara et al., 2021. The professional's judgment on the variables is impacted by his or her experience. Subjective judgment was used to give relative ratings to components in the matrix in this study. Scores range from 1 to 9 if the elements in the column are more significant than those in the row, and from 1/2 to 1/9 if the ones in the row are more essential (Kumar & Anbalagan, 2016; Sur et al., 2020). For Matrix calculation, a matrix's eigenvector is used to determine factor or class weights. The eigenvectors, maximal eigenvalue, consistency index (CI), and consistency ratio must all be calculated (CR). The conventional matrix computation approach was used to determine the eigenvalue and eigenvector. The factors were rated using Eigenvector. The transitivity rule may be violated by the subjective decision rule, resulting in contradiction in judgment. To determine the consistency of a judgement matrix, CI and CR calculations were performed (Feizizadeh & Blaschke, 2012; Kumar & Sandhu, 2020). The formula for consistency by Saaty (1980)

Table 3 Based on the LHEF system, ratings were given to the factor classes detected in the region

Causative factors	Categories	Description	Rating
Lithology	Laisong	Flaggy sandstone with subordinate shale rare coal	1.3
	Jenam	Carbonaceous sandy shale, sandstone & coal seams	1.3
	Upper Disang	Shale, siltstone, greywacke rhythmite & sandstone	1.8
Structural lineament	Scale ≤ 0.86	Very Favorable	0.7
	0.86 < Scale ≤ 1.72	Favorable	1.05
	1.72 < Scale ≤ 2.58	Fair	1.4
	2.58 < Scale ≤ 3.44	Unfavorable	1.7
	Scale > 3.44	Very Unfavorable	2
Slope morphometry	Very Low	<15°	0.5
	Low	16°–25°	0.8
	Moderate	26°–35°	1.2
	High	36°–45°	1.7
	Very High	>45°	2
LULC	1	Barren land	2.0
	2	Shrub and grassland	1.5
	3	Forest	1.2
	4	Dense forest	0.8
	5	Populated flat land	0.6
Relative relief	Low	>100 m	0.3
	Medium	100–200 m	0.45
	High	200–300 m	0.6
	Very high	>300 m	1.0
Hydrogeological condition	1	Dry	0.0
	2	Damp	0.2
	3	Wet	0.5
	4	Dripping	0.8
	5	Flowing	1.0

$$\mathrm{CI} = \frac{(\lambda \max - N)}{N - 1}$$

Here, λ_{max} denotes the largest eigenvalue, and N is the number of entries in the matrix. The ratio CI/RI may be used to derive CR, and RI is a randomized index (Saaty, 1980). The "random index was created by Satty (1980)" using a variety of randomly chosen data. If the CR is less than 0.1, the matrices are regarded to be consistent. This also shows that the weight distribution is appropriate (Tables 4 and

Table 4 Ordinal scale judgement by Saaty (1977)

Ordinal scale	Preference degree	Remarks description
1	"Equally"	"Factors inherit equal contribution"
3	"Moderately"	"One factor moderately favoured over other"
5	"Strongly"	"Judgement strongly favours one over other"
7	"Very strongly"	"One factor very strongly favoured over other"
9	"Extremely"	"One factor favoured over others in the highest degree"
2,4,6,8	"Intermediate"	"Compensation between weights 1,3,5,7 and 9"
Reciprocals	"Opposite"	"Refers inverse comparison"

Table 5 Random consistency index (RI) Saaty (1980)

Nos	1	2	3	4	5	6	7	8	9	10	11	12	13	14	15
RI	0	0	0.58	0.9	1.12	1.24	1.32	1.41	1.45	1.49	1.51	1.53	1.56	1.57	1.59

5). If the CR value is more than 0.1, the judging matrix is deemed to be inconsistent for judgment and must be changed to attain appropriate consistency.

Factors were categorized using the ratings collected from the AHP matrix to create the LSZ map. In the GIS domain, classified feature classes were submitted to arithmetic sum using the given below formula:

$$\mathrm{LHI} = \sum_{\mathrm{J}=1}^{J=n} \text{Weight of factor } (\mathrm{W_J}) \times \text{Weight of factor classes } (\mathrm{W_{iJ}})$$

The weight of the *i*th class of component J is denoted by WIj, while the number of classes is denoted by n. It produced a facet-by-facet LSI map with susceptibility indexes for each facet.

6 Results and Discussion

6.1 BIS Code Approach

The slope facet was chosen as the mapping unit for landslide susceptibility zonation, as per the BIS code. Overall, 153 facets were extracted using a shaded relief map obtained from a DEM. All the extracted facets were then applied to all the six-causative factors for evaluation and rating using LHEF (BIS code). The total estimated hazard (TEHD) proposed by BIS code 1998 for the generation of landslide hazard zonation is mentioned below in Table 6.

Table 6 TEHD based on landslide hazard zonation/ landslide susceptibility zonation (BIS code 1998 part 2)

Zone	TEHD value	Description of zone
I.	<3.5	Very low hazard
II.	3.5–5.0	Low hazard
III.	5.1–6.0	Moderate hazard
IV.	6.1–7.5	High hazard
V.	>7.5	Very high hazard

Equation (1) was used to determine TEHD values for each facet, and then the LSZ map was shaped and set on the cut (threshold) values in Table 3. It indicated that TEHD value for the study zone v ranged from 4.4 to 7.85. According to the TEHD value obtained from the Six causative factors by the LHEF rating of the study region, LSZ was generated and classified into four classes low hazard, moderate hazard, high hazard and very high hazard (Fig. 5). Very low hazard for the study area was not observed as the TEHD value was ranging from 4.4 to 7.85 based on the given BIS code for Landslide hazard zonation (Table 7).

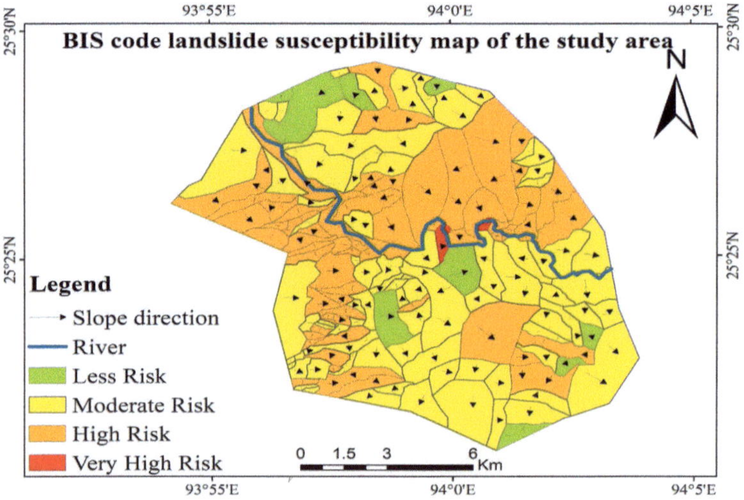

Fig. 5 BIS code landslide susceptibility zonation of the study region

Table 7 The number of facets present in each BIS code LSZ class

Classes	Scaling	Number of facets
Low risk	4.4–5	9
Moderate risk	5.05–6	80
High risk	6.05–7.35	62
Very high risk	>7.5	2

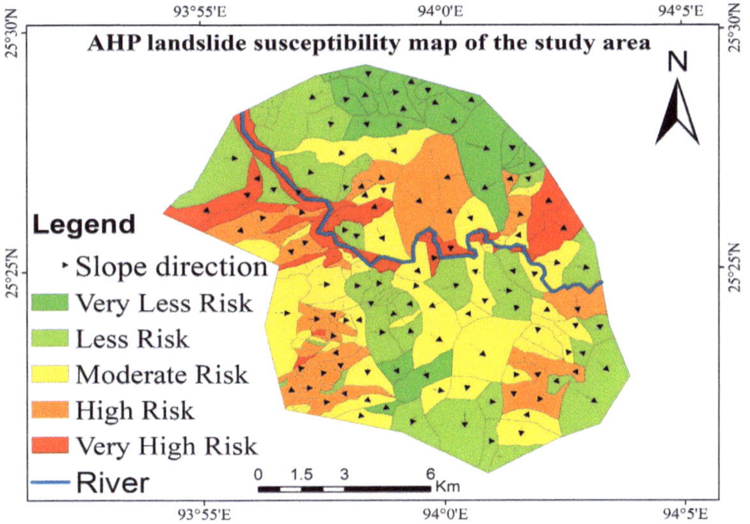

Fig. 6 AHP landslide susceptibility zonation of the study region

6.2 Analytical Hierarchy Process Approach

Preparation of LSZ was done using the rating obtained from the AHP method. For the AHP approach slope facets as used in BIS code were used as a mapping unit. 153 slope facets were retrieved and applied for six causative factors mentioned in the BIS code for geo-environmental factors. On a slope-by-slope facets basis, weight values were added to vector mappings of each component. Weighted vector map integration was achieved using Equation (2). The outcome was a Landslide susceptibility index (LSI) map containing numerical susceptibility data, with greater LSI values showing higher susceptibility and lower values suggesting low susceptibility. LSI values range from 1.714 to 2.910. The Natural Breaks classifier was utilized to categorize the continuous LSI map into five categories (Fig. 6): very low vulnerability (01.714–2.032), low vulnerability (2.032–2.238), medium vulnerability (2.238–2.368), high vulnerability (2.368–2.520), and very high vulnerability (2.520–2.910) (Tables 8, 9 and 10).

6.3 Validation

The cumulative percentage curve was used to verify LSZ. The obtained landslide susceptibility index (LSI) for the cumulative percentage curve was plotted on the X and Y axis against the cumulative percentage of landslides to construct success rate curves. To estimate the AUC curve, 36 testing landslide inventory locations were

Table 8 AHP scores of six causative factors, CR, maximum eigenvalue and eigenvector

Factor		Structural lineament	Relative relief	Lithology	LULC	Hydrological condition	Slope morphometry	Factor pairwise weight
Nos		1	2	3	4	5	6	
1	Structural lineament	1	2	3	1/2	1/3	1/3	0.112
2	Relative relief	0.50	1	1/2	1/3	1/5	1/5	0.051
3	Lithology	0.33	2	1	1/3	1/4	1/4	0.066
4	LULC	2	3	3	1	1/3	1/3	0.150
5	Geo-Hydrological condition	3	5	4	3	1	1	0.311
6	Slope morphometry	3	5	4	3	1	1	0.311

CR: 0.032 Max Eigen value = 6.200

Table 9 AHP scores of all the classes of the factors, CR, maximum eigenvalue and eigenvector

Lithology							
Nos	Classes	Laisong	Jenam	Upper Disang	Classes weight		
1	Laisong	1	1/2	1/4	0.137		
2	Jenam	2	1	1/3	0.238		
3	Upper Disang	4	3	1	0.625		
CR: 0.019 Max eigen value = 3.018							
Structural lineament							
Nos	Classes	Very Favorable	Favourable	Fair	Unfavourable	Very Unfavorable	Classes weight
1	Very Favourable	1	1/2	1/3	1/4	1/5	0.062
2	Favourable	2	1	1/2	1/3	1/4	0.097
3	Fair	3	2	1	1/2	1/3	0.160
4	Unfavourable	4	3	2	1	1/2	0.263
5	Very Unfavourable	5	4	3	2	1	0.419
CR:0.015 Max eigen value = 5.068							
Slope morphometry							
Nos	Classes	Very Low	Low	Moderate	High	Very High	Classes weight
1	Very Low	1	1/2	1/3	1/4	1/5	0.062
2	Low	2	1	1/2	1/3	1/4	0.097
3	Moderate	3	2	1	1/2	1/3	0.160
4	High	4	3	2	1	1/2	0.263

(continued)

Table 9 (continued)

Lithology

Nos	Classes	Laisong	Jenam	Upper Disang	Classes weight		
5	Very High	5	4	3	2	1	0.419

CR:0.015 Max eigen value = 5.068

LULC

Nos	Classes	Barren land	Rock and settlement	Shrub and grassland	Vegetation	High vegetation	Classes weight
1	Barren land	1	2	3	4	5	0.419
2	Settlement	0.50	1	2	3	4	0.263
3	Shrub and grassland	0.33	0.50	1	2	3	0.160
4	Forest	0.25	0.33	0.50	1	2	0.097
5	Dense forest	0.20	0.25	0.33	0.50	1	0.062

CR:0.015 Max eigen value = 5.068

Relative relief

Nos	Classes	Low	Medium	High	Very High	Classes weight
1	Low	1	1/2	1/3	1/4	0.095
2	Medium	2	1	1/2	1/3	0.160
3	High	3	2	1	1/2	0.277
4	Very High	4	3	2	1	0.467

CR:0.011 Max eigen value = 4.031

Geo-Hydrogeological Condition

(continued)

Table 9 (continued)

Lithology							
Nos	Classes		Laisong	Jenam	Upper Disang	Classes weight	
Nos	Classes	Flowing	Dripping	Wet	Damp	Dry	Classes weight
1	Flowing	1	2	3	4	5	0.419
2	Dripping	0.50	1	2	3	4	0.263
3	Wet	0.33	0.50	1	2	3	0.160
4	Damp	0.25	0.33	0.50	1	2	0.097
5	Dry	0.20	0.25	0.33	0.50	1	0.062
CR:0.015 Max eigen value = 5.068							

Table 10 The number of facets present in each AHP LSZ class

Classes	Scaling	Number of facets
Very low hazard	1.714–2.032	17
Low hazard	2.033–2.238	36
Moderate hazard	2.239–2.368	47
High hazard	2.369–2.520	35
Very high hazard	2.521–2.910	18

used to generate a cumulative percentage curve based on a trapezium technique. Validation accuracy for two models under the AUC curve for the cumulative percentage curve was found to be 70.09 and 72.63% for the BIS code model and AHP model respectively.

ROC curve (Figs. 7 and 8) was also used for authentication of the study region, 36 landslide location and 36 non-landslide location was taken into account. A total of 72 location was taken for the ROC curve for validation. The false positive rate (FP) on the x-axis and the true positive rate (TP) on the y-axis for a collection of ROC scores specified by single confusion matrixes are shown on the ROC curve. TP stands for sensitivity, which is defined as the ratio of TP to TP and FP. The ratio of FP to FP and TP is known as FP. FP stands for the difference between the true negative rate (specificity), where specificity is described as the ratio of true negative to false positive and true negative. The accuracy yield for ROC under the AUC curve for the BIS code was 77.2% and for AHP was 79.4%.

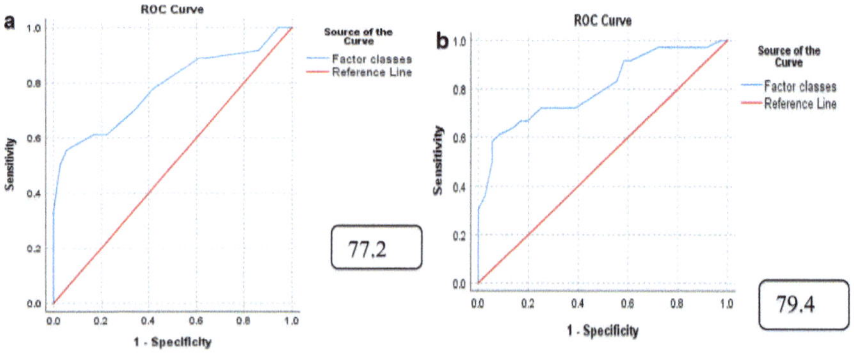

Fig. 7 ROC curve for **a** BIS code approach and **b** AHP approach

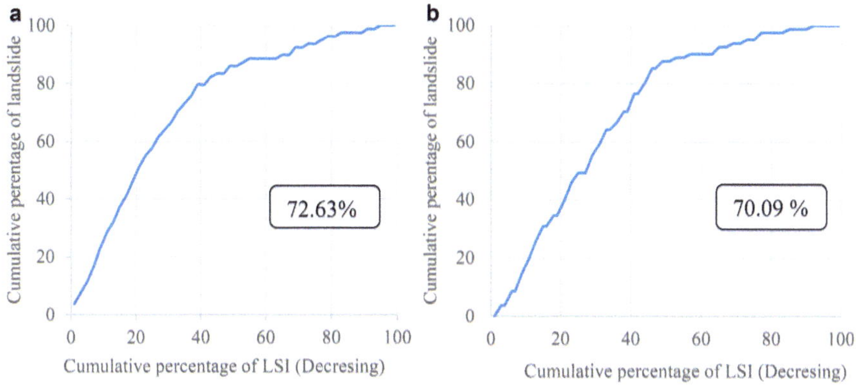

Fig. 8 Cumulative percentage curve for **a** AHP model **b** BIS code model

7 Conclusion

Landslides have grown widespread in mountainous terrain like the Himalayas, especially during the rainy season when there is a lot of rain. The LSZ might be extremely beneficial to long-term planning and growth. In this research, BIS/LHEF and AHP grading methods have been utilized to develop a landslide vulnerable map of the research study region. The slope facet was chosen as the mapping unit for both the model (BIS code, AHP) and it was outlined using a high-resolution DEM. The geo-environmental parameters described in the LHEF grading method were derived using remote sensing technology and auxiliary data. The LHEF technique was used to give ratings to factors for the BIS code approach. For AHP rating for geo-environmental factor confusion matric used in the AHP calculator was used to generate the factor weight for grading. LSI maps for both models were classed into relative hazard zone for landslide susceptibility zonation. This paper also shows the importance of slope facets used as mapping units for identifying landslide susceptibility zonation. Slope facets and geo-environmental factors were extracted from the high-resolution satellite image, DEM (ALOS PALSAR 12.5 m), and ancillary data. Generated landslide susceptibility zone map of the Mao-Maram region (study area) has been further validated using two validation processes, cumulative percentage curve and receiver operating characteristic and obtained an accuracy of 77.2% (BIS code) and 79.4% (AHP) using ROC curve. The cumulative percentage curve also obtained an accuracy of 70.09% for the BIS code and 72.63% for the AHP approach. In this study, it is observed that the AHP model yield better performance as compared to the BIS code model by using both validation approach. This study also tries to compare the two heuristics (BIS code) and semi-heuristic (AHP) approach by using the same fixed geo-environmental factor proposed by BIS code 1998 part 2, with the slope facets as the mapping unit. During the study, it also came across that advancing in technology and remote sensing tools with GIS has given the advantage and feasibility of doing research. Finally, this research also yields a good result for both models using slope facets as a mapping unit. The landslide susceptibility map may be used as a guide for planning and developing appropriate stabilizing measures, as well as to choose the safest area for construction operations.

Acknowledgements The authors greatly thank the Lovely Professional University for providing software and ancillary data for this research. Authors are also grateful to the Mao Council for its support in data collection during their field visit to the study area and for their hospitality throughout the research.

Statement and Declarations

Competing Interest The authors declare no competing interest.

Funding No funds or grants were received for this research.

Informed Consent All co-authors of this research paper are informed of every step of the publication process, and everything has been done with their common approval.

References

Anbalagan, R. (1992). Landslide hazard evaluation and zonation mapping in mountainous terrain. *Engineering Geology, 32*(4), 269–277. https://doi.org/10.1016/0013-7952(92)90053-2

Balamurugan, G., Ramesh, V., & Touthang, M. (2016). Landslide susceptibility zonation mapping using frequency ratio and fuzzy gamma operator models in part of NH-39, Manipur, India. *Natural Hazards, 84*(1), 465–488. https://doi.org/10.1007/s11069-016-2434-6

Chen, W., & Li, Y. (2020). GIS-based evaluation of landslide susceptibility using hybrid intelligence models. *CATENA, 195*, 104777. https://doi.org/10.1016/j.catena.2020.104777

Devara, M., Tiwari, A., & Dwivedi, R. (2021). Landslide susceptibility mapping using MT-InSAR and AHP enabled GIS-based multi-criteria decision analysis. *Geomatics, Natural Hazards and Risk, 12*(1), 675–693. https://doi.org/10.1080/19475705.2021.1887939

Froude, M., & Petley, D. (2018). Global fatal landslide occurrence from 2004 to 2016. *Natural Hazards and Earth System Sciences, 18*(8), 2161–2181. https://doi.org/10.5194/nhess-18-2161-2018

Feizizadeh, B., & Blaschke, T. (2012). GIS-multicriteria decision analysis for landslide susceptibility mapping: Comparing three methods for the Urmia lake basin, Iran. *Natural Hazards, 65*(3), 2105–2128. https://doi.org/10.1007/s11069-012-0463-3

Galve, J., Cevasco, A., Brandolini, P., Piacentini, D., Azañón, J., Notti, D., & Soldati, M. (2016). Cost-based analysis of mitigation measures for shallow-landslide risk reduction strategies. *Engineering Geology, 21*, 142–157. https://doi.org/10.1016/j.enggeo.2016.09.002

IS 14496 (Part 2) Preparation of landslide hazard zonation maps in mountainous terrains—Guidelines, part 2: Macro-Zonation: Bureau of Indian Standards (BIS): Free Download, Borrow, and Streaming: Internet Archive (2022). Retrieved March 18, 2022, from https://archive.org/details/gov.law.is.14496.2.1998

Jemec Auflič, M., Kumelj, Š, Peternel, T., & Jež, J. (2018). Understanding of landslide risk through learning by doing: Case study of Koroška Bela community, Slovenia. *Landslides, 16*(9), 1681–1690. https://doi.org/10.1007/s10346-018-1110-1

Kumar, R., & Anbalagan, R. (2015). Landslide susceptibility zonation in part of Tehri reservoir region using frequency ratio, fuzzy logic and GIS. *Journal of Earth System Science, 124*(2), 431–448. https://doi.org/10.1007/s12040-015-0536-2

Kumar, R., & Anbalagan, R. (2016). Landslide susceptibility mapping using analytical hierarchy process (AHP) in Tehri reservoir rim region, Uttarakhand. *Journal of the Geological Society of India, 87*(3), 271–286. https://doi.org/10.1007/s12594-016-0395-8

Kumar, R., & Anbalagan, R. (2019). Landslide susceptibility mapping of the Tehri reservoir rim area using the weights of evidence method. *Journal of Earth System Science, 128*(6). https://doi.org/10.1007/s12040-019-1159-9

Kumar, R., & Sandhu, H. (2020). Landslide susceptibility zonation mapping of Keteshwar reservoir area using modified BIS code and AHP methods. *European Journal of Molecular & Clinical Medicine, 7*, 3552–3573.

Kumar, N. T., Hindayar, J., Mohan, M., Dasarwar, P., Ibrahim, M., & Som, S. (2017). Population based binary reclassification of Indian standard landslide hazard model. *Journal of the Geological Society of India, 89*(2), 175–182. https://doi.org/10.1007/s12594-017-0581-3

Kundu, S., Saha, A., Sharma, D., & Pant, C. (2013). Remote sensing and GIS based landslide susceptibility assessment using binary logistic regression model: A case study in the Ganeshganga Watershed, Himalayas. *Journal of the Indian Society of Remote Sensing, 41*(3), 697–709. https://doi.org/10.1007/s12524-012-0255-y

Mahajan, A. K., Sharma, S., Patial, S., Sharma, H., Pandey, D. D., & Negi, S. (2022). A brief address of the causal factors, mechanisms, and the effects of a major landslide in Kangra Valley, North-Western Himalaya, India. *Arabian Journal of Geosciences, 15*(9), 925.

Nhu, V., Hoang, N., Nguyen, H., Ngo, P., Thanh Bui, T., & Hoa, P. (2020). Effectiveness assessment of Keras based deep learning with different robust optimization algorithms for shallow landslide susceptibility mapping at tropical area. *CATENA, 188*, 104458. https://doi.org/10.1016/j.catena.2020.104458

Pham, B., Prakash, I., Singh, S., Shirzadi, A., Shahabi, H., Tran, T., & Bui, D. (2019). Landslide susceptibility modeling using Reduced Error Pruning Trees and different ensemble techniques: Hybrid machine learning approaches. *CATENA, 175*, 203–218. https://doi.org/10.1016/j.catena.2018.12.018

Sarkar, S., Kanungo, D., & Mehrotra, G. (1995). Landslide hazard zonation: A case study in Garhwal Himalaya, India. *Mountain Research and Development, 15*(4), 301. https://doi.org/10.2307/3673806

Singh, C., Behera, K., & Rocky, W. (2011). Landslide susceptibility along NH-39 between Karong and Mao, Senapati district, Manipur. *Journal of the Geological Society of India, 78*(6), 559–570. https://doi.org/10.1007/s12594-011-0120-6

Saaty, T. L. (1977). A scaling method for priorities in hierarchical structures. *Journal of Mathematical Psychology, 15*(3), 234–281.

Saaty, T. L. (1980). *The analytic hierarchy process planning, priority setting, resources allocation*. Mcgraw-Hill, New York—References—Scientific Research Publishing. (2022). Retrieved March, 18, 2022, from https://www.scirp.org/(S(lz5mqp453edsnp55rrgjct55))/reference/ReferencesPapers.aspx?ReferenceID=1895817

Sur, U., Singh, P., & Meena, S. (2020). Landslide susceptibility assessment in a lesser Himalayan road corridor (India) applying fuzzy AHP technique and earth-observation data. *Geomatics, Natural Hazards and Risk, 11*(1), 2176–2209. https://doi.org/10.1080/19475705.2020.1836038

Varnes, D. (1984). *Landslide hazard zonation: A review of principles and practice*. UNESCO.

Vojteková, J., & Vojtek, M. (2020). Assessment of landslide susceptibility at a local spatial scale applying the multi-criteria analysis and GIS: A case study from Slovakia. *Geomatics, Natural Hazards and Risk, 11*(1), 131–148. https://doi.org/10.1080/19475705.2020.1713233

Yalcin, A. (2008). GIS-based landslide susceptibility mapping using analytical hierarchy process and bivariate statistics in Ardesen (Turkey): Comparisons of results and confirmations. *CATENA, 72*(1), 1–12. https://doi.org/10.1016/j.catena.2007.01.003

GIS-Based Landslide Susceptibility Mapping of the Darjiling Himalaya, India Using Frequency Ratio (FR) and Fuzzy Logic (FL) Models

Sonam Lama, Abhishek Ghosh, Sunil Kumar De, and Balai Chandra Das

Abstract Landslides are the most common and devastating natural calamities in the Darjiling Himalaya. It cripples the daily life of inhabitants for three to four monsoonal months of the year. Hence, the present study aimed to compare the predictive performance of two bivariate models, namely, the frequency Ratio (FL) and fuzzy Logic (FL), for assessing the susceptibility of landslides in the region. A total of 1556 landslide locations were identified in the study area from the inventory reports and then verified with Google Earth images and field surveys. Of these 1556 identified landslide points, (70%) 1089 were randomly used for data training, and the remaining (30%) 467 points were used for data validation. Fifteen landslide-causing parameters, namely, elevation, slope, rainfall, distance to river, distance to lineament, distance to road, aspect, geomorphology, curvature, plan curvature, profile curvature, seismic zone, LULC, TWI (topographic wetness index) and NDVI (Normalized Difference Water Index), were used as independent variables in the modeling process and to construct the spatial database. Subsequently, geographical information systems (GIS) and two predictive models were ensembled to assess landslide susceptibility. The resulting landslide susceptibility maps (LSMs) were subsequently classified into four classes, namely, low, medium, high, and very high susceptibility to landslide occurrence, using the natural breaks classification method in the GIS environment. Finally, validation results obtained from the receiver operating characteristics (ROC) curve showed that the landslide susceptibility map constructed with the FR model (76.1%) was slightly more accurate than that produced with the FL model (73.6%). Landslide susceptibility assessments using comparative predictive models are always better than those using any single model. The findings of this study could lead to potential resources for future planning and development in the study area.

S. Lama (✉)
Department of Geography, Darjiling Government College, Darjiling, West Bengal, India
e-mail: sonamgeo@yahoo.co.in

A. Ghosh · S. K. De
Department of Geography, North-Eastern Hill University (NEHU), Shillong, Meghalaya, India

B. C. Das
Department of Geography, Barasat Government College, Kolkata, West Bengal, India

© The Author(s), under exclusive license to Springer Nature Switzerland AG 2025
S. Ghosh (ed.), *The Himalaya Dilemma*, Springer Natural Hazards,
https://doi.org/10.1007/978-3-031-95083-4_10

Keywords Landslide susceptibility · LSM · Frequency ratio · Fuzzy logic · Darjiling Himalaya

1 Introduction

The term landslide denotes "the movement of a mass of rock, debris or earth down a slope" (Cruden, 1991). Landslide is a complex natural phenomenon influenced by many geo-environmental factors such as geology, topography, morphology, vegetation, and land use pattern (Chen et al., 2017; Varnes, 1984). Extreme rainfall, earthquake, excavation for roads, and drainage are some of the triggering factors responsible for landslides (Das et al., 2010). Landslides are the most common and devastating disaster in the Indian Himalaya. About 80% of the landslide events in India belong to this region (Onagh et al., 2012). West Bengal has 2980 km^2 area prone to Landslides and major portion of this total landslide prone area falls under Darjiling Himalaya for being the only hilly regions in the state (www.gsi.gov.in). According to GSI, during 2015–2020, West Bengal registered 374 landslide events which is next to Kerala and second highest number among the landslide registered states of India. The region has long history of disasters. Landslides and flash floods (1968), Ambootia landlside (October 1968), Ghayabari Landslide (8th July, 2003), 02.07.2015 landslide (Soureni G.P.-20 people killed, Kalimpong-7 killed and Rangbhang bustee, Jorebunglow-1 killed, other deaths took place in Lava, Sukhia and Gorubathan areas) are some of the major landslides happened in the region (Table 1). Most of the landslides are shallow translational failure in nature (NASA global landslide catalogue). The recent Teesta flash flood of 04.10.2023 have once again exposed the very high level of vulnerability in the region. Landslide research has gained worldwide attention, primarily because of growing awareness of the socioeconomic effects of these disasters as well as mounting urbanisation pressure on the alpine environment (Aleotti & Chowdhury, 1999; Verma et al., 2016). Incidents of landslides and consequent scale of damages in Himalayan region are much likely to increase in consistent with climate change, population growth, urbanization and infrastructure development. Therefore, identification of potential areas of landslide occurrence becomes imperative for safer strategic planning of future developmental activities in the region (Dikshit et al., 2020; Kanungo et al., 2006). This can be achieved through detailed Landslide Susceptibility mapping (LSM) of an area wherein the area may be divided into near-homogeneous domains and ranked according to degrees of potential hazard due to mass movements (Varnes, 1984).

LSM is the first stage (pre-disaster stage) of landslide disaster management and therefore it should give important information aiding authorities to take appropriate decisions for reducing considerable potential landslide damage. Landslide susceptibility mapping shall be done prior to landslide assessment planning, management and disaster mitigation (Kanwal et al., 2016). LSM aids in identifying the landslide-prone zone in advance, helping planners in making decisions for the future development and disaster mitigation measures (Chawla et al., 2018). Preparation of LSM is based

Table 1 Landslide prone areas in the study area

Sub-divisions with landslide event years	Blocks/Municipality/ Municipal Area	Number of Gram Panchayats/ Wards	No. of Landslide prone GPs/ Wards	Probable causes of landslide
Darjiling-Sadar 1968, 1980, 1998, 1999, 2000, 2003, 2006, 2007, 2008, 2009, 2012, 2013, 2015, 2017, 2018	Darjiling Municipality	32	24	Flash floods, Cyclone Aila, heavy rainfall, steep slopes, earthquakes, highly dissected topography, piping action and toe cutting, ubiquity of fragile Darjiling Gneiss and Daling series rocks, slope cutting, unplanned and poorly maintained drainage, irresponsible dumping of slope cutting materials
	Darjiling-Pulbazar	23	16	
	Jorebunglow-Sukhia Pokhari	16	10	
	Rangli-Rangliot	11	5	
Kurseong 1934, 1950, 1968, 1985, 1999, 2008, 2009, 2010, 2010, 2013, 2015, 2017, 2018	Kurseong municipality	20	12	Earthquake, Incessant Rainfall, Cyclone Aila, slope cutting, steep slopes, toe erosion by excess water in nala, highly dissected topography, coming from Giddapahar, Coal Quarrying in Tindaharia area, proximity to MCT, ubiquity of weak Darjiling Gneiss, Daling and Damuda series rocks
	Kurseong	14	10	
Mirik 1950, 2003, 2008, 2013, 2015, 2018	Mirik municipality	9	6	Cyclone Aila, heavy rainfall, steep slopes, earthquakes, moderately dissected topography, Piping action and toe cutting, slope cutting, unplanned and poorly maintained drainage, irresponsible dumping of slope cutting materials, pressence of fragile Darjiling Gneiss rocks
	Mirik	6	3	
Siliguri	Siliguri Municipal Corporation	47 (14 falls in Jalpaiguri District)	–	–

(continued)

Table 1 (continued)

Sub-divisions with landslide event years	Blocks/Municipality/ Municipal Area	Number of Gram Panchayats/ Wards	No. of Landslide prone GPs/ Wards	Probable causes of landslide
–				
	Matigara	5	–	–
	Naxalbari	6	–	–
	Kharibari	4	–	–
	Phansidewa	7	–	–
Kalimpong 1968, 1993, 1996, 2006, 2009, 2015, 2019, 2021	Kalimpong Municipality	23	17	Incessant rainfall, flash floods, cyclone Aila, proximity to teesta river, slope cutting, earthquake, piping action and toe erosion, highly dissected topography, unplanned and poorly maintained drainage, irresponsible dumping of slope cutting materials, ubiquity of fragile Darjiling Gneiss, Daling and Damuda series rocks
	Kalimpong-I	18	9	
	Kalimpong-II	13	8	
	Gorubathan	11	4	

Source Compiled from District Disaster Management Plan report of West Bengal (2018, 2019, 2020–2021), NASA Global Landslide Catalogue and old newspapers

on assumption that the future landslide is expected on those locations which has same set of geo-environmental conditions as that of past and present landslide locations (Kanungo et al., 2009; Ram et al., 2020; Varnes, 1984).

Historically, the LSM of the Himalayan region has been achieved largely manually through field work, superimposition of different thematic layers of causative factors and interpretation (Anbalagan, 1992; Gupta et al., 1993; Mehrotra et al., 1996; Pachauri & Pant, 1992; Panikkar & Subramanyan, 1996; Sarkar et al., 1995). The method was laborious, time taking, uneconomical, less accurate, biased and largely irreplicable. However, with the advent of fast computing personal computers, availability of wide range of Remote sensing data, GIS models, machine learning and easy availability of other ancillary data in digital format, has made preparation of LSM much easier and accurate (Gupta, 2003; Gupta & Joshi, 1990; Kanungo et al., 2006; Nagarajan et al., 1998; van Westen, 1994). In fact, all these developments have propelled the landslide hazard research to a new high. Today there are scores of methodologies in practice for the landslide susceptibility mapping. These can be broadly categorized into qualitative, semi-quantitative and quantitative approaches. In the qualitative approach weights are assigned to thematic layers according to their relative importance in a GIS environment for generation of landslide susceptibility map (Gupta et al., 1999; Saha et al., 2002, 2005; Sarkar & Kanungo, 2004). Heuristic, Boolean, fuzzy logic, multiclass overlay, and spatial multicriteria evaluation are the primary techniques used in qualitative research. However, qualitative approach suffers from subjective biasness of the expert towards the assumptions made and weights given (Kanungo et al., 2006). In semi quantitative methods weights and rankings are assigned on the basis of logical tools such as AHP, fuzzy logic and weighted linear combination (Anbalagan et al., 2015). Quantitative methods are landslide inventory driven statistical methods wherein relationship of landslide factors with landslide inventory is established. quantitative methods are statistical, deterministic, distribution free approaches or probabilistic in nature. Semi-quantitative and quantitative methods largely reduce the subjectivity and biasness in the weight assignment process (Kanungo et al., 2006).

Few notable literatures relating to causes of landslide and landslide susceptibility mapping using different models in the Darjiling Himalaya are Sinha et al., 1975, Sarkar, 1999, Starkel & Basu, 2000, Basu & Dey, 2003, Kanungo et al., 2006, Ghosh et al., 2009, Chakraborty et al., 2011, Ghosh et al., 2011, Mondal & Maiti, 2012, Sarkar et al., 2013, Chawla et al., 2018, Chawla et al., 2019. None of these studies have used FR and FL bivariate models for mapping the landslide susceptibility of the study area.

Various scholars have used Frequency Ratio (FR) and Fuzzy Logic (FL) bivariate statistical models for mapping landslide susceptibility (Anbalagan et al., 2015; Khan et al., 2018, Sahana & Sajjad, 2017, Pradhan, 2010, Rasyid et al., 2016) in other parts of the world. The frequency ratio model takes into account how each landslide conditioning parameter affects the likelihood of a landslide, but it does not take into account how parameters relate to one another (Lee & Pradhan, 2007). Whereas in fuzzy logic approach weights can be assigned to landslide conditioning factors through expert knowledge (Lee, 2007). In this study landslide susceptibility was evaluated by GIS technique using the Frequency Ratio (FR) and Fuzzy Logic (FL) modelling with respect to the bivariate statistical approach.

Thus, the main objective of the present study is to identify landslide prone areas and prepare landslide susceptibility map of the Darjiling Himalaya. The objective is achieved through application of Frequency Ratio (FR) and Fuzzy Logic (FL) bivariate models coupled with remote sensing and other ancillary data in GIS. The results can be helpful for researchers, local authorities and planners for devising effective landslide mitigation plans.

2 Study Area

The hilly parts of Darjiling and Kalimpong districts of West Bengal, located above 300 m contour line is generally considered as Darjiling Himalayas. But in the present study, for the ease of boundary delimitation and availability of secondary data, we have considered the administrative boundary of Darjiling and Kalimpong districts as the Darjiling Himalaya. Therefore, in this chapter, along with the hill sub-divisions of Darjiling Sadar, Kalimpong Sadar and Kurseong, mostly flat sub-division of Siliguri together is considered as Darjiling Himalaya (Fig. 1). Henceforth, Darjiling Himalaya will be synonym to combined administrative extent of Darjiling and Kalimpong districts. Located in the northernmost edge of the West Bengal, Darjiling Himalya have a geographical area of 3146 km^2 which is 3.5% of state's total area. The geographical extension of the study area is 26° 27" to 27° 13"N and 87° 59" to 88° 53"E (Fig. 1). It is surrounded by Sikkim in the north, Bhutan in the east, Jalpaiguri district (Southeast), North Dinajpur district (south), Bangladesh (Southeast) and Nepal in the west. The hill region of the study area experiences warm humid mild summer (maximum temperature rarely crosses 25 °C) whereas the plains experience hot humid condition (maximum temperature reaches as high as 42 °C). The average annual rainfall ranges from 2400 to 3900 mm, 90% of which happens during four-five months of the high-sun season (May–September) whereas very little precipitation happens during the low-sun season. During winter, average temperature ranges between 5 and 9 °C which occasionally plummets below freezing point inducing rare snowfall. During monsoon and winter, the hilly region of the study area often remains shrouded in mist and fog (District Disaster Management Plan report of Darjiling and Kalimpong, West Bengal, 2020–21). Almost 75% of the study area have undulating hilly terrain. The elevation ranges from 61 to 3616 m. Geologically, the region is composed of Precambrian (Darjiling gneiss, Daling series), Gondwana (Damuda series), Tertiary (Siwaliks), and recent Pleistocene (Alluvium) lithologies (Gansser, 1964) (Fig. 2). Geomorphologically the area is composed of alluvial plains, flood plains, moderately dissected hills and valleys, highly dissected hills and valleys, and piedmont fan plains. The study area is characterized by various soil texture classes; namely, gravelly-loamy, fine-loamy to coarse-loamy, gravelly-loamy to loamy skeletal, and gravelly-loamy to coarse-loamy (Pramanik, 2016). The study area has very high drainage density owing to high rainfall and undulating topography. Tista, Mahananda, Balason, Relli, Rohini, Jaldhaka along with their numerous tributaries drains the districts. As per 2011 census, it has 18,46,823 population with density of 586 inhabitants per square kilometer. The study area features many landslide conditioning factors like steep slopes, highly dissected hills and valleys, heavy rainfall, high earthquake intensity zone, very high drainage density, proximity to MCT and abundance of fragile Darjiling gneiss, Daling and Damuda series rocks. Landslide is a reoccurring annual phenomenon in the study area. Most of the landslides are rainfall induced, happens during monsoon season (June–September) and cripple general life in the district. Pagla Jhora landslide along the old Siliguri-Kurseong Highway and Sweti Jhora landslide along the NH-10 are the most infamous ones for their almost perennial nature. District Disaster Management Plan report of West Bengal (2018, 2019, 2020–2021) shows that majority of gram panchayats in the hilly community development blocks of Darjeeing-Kalimpong districts are prone to landslides (Table 1). During last few decades the districts have seen unprecedented increase in anthropogenic activities like growth in settlement areas, new roads and widening of existing ones, deforestation, land use change, unplanned construction, dams over river Tista and most recently Sevoke-Rangpo railway project has been added. The undergoing Sevoke-Rangpo rail project is 44.96 km long, of which 41.55 km passes through Darjiling Himalaya and only 3.41 km falls in Sikkim Himalaya. Further, out of the total length of 44.96 km, 38.623 km (86%) is in the form of tunnels, 2.24 km (5%) in Bridges and 4.79 km (9%) length in open cutting or filling of station yards. The proposed line consists of 14 tunnels, 13 major bridges, 10 minor bridges in the project and five stations viz. Sivok, Riyang, Melli, Rangpo and one underground halt station viz. Teesta Bazar (https://nfr.indianrailways.gov.in/view_detail.jsp?lang=0&dcd=2256&id=0,4,268). This railway project is already gathering debates and disputes, as people including experts have expressed concerns that it will further aggravate the slope instability along NH-10. NH-10 and NH-110 are the two major highways in the study area. NH-10 is of great economic and strategic significance as it connects Sikkim with West Bengal and rest of India via Teesta Bazar, Rambi, Ranpo, Singtam whereas NH-110 connects Darjiling with Siliguri Via Ringtong, Tindharia, Mahanadi, Kurseong and Sonada. Rockfalls along these two major highways are perennial. Panglajora landslide and subsidence zone between Tindharia and Mahanadi is a major sliding zone along NH-110. Due to its perennial nature, the Darjiling bound vehicles usually avoid the Siliguri-Kurseong section of NH-110 and follow either Rohini road or Pankhabari road. Landslide points at Sweti jhora, 29th mile and Birrik Dara are the major landslide susceptible points along NH-10 highway. The study area has very popular hill stations namely—Darjiling, Kalimpong, Mirrik, Kurseong along with ever growing rural tourism. The districts attract tourists seeking tranquility of nature with view of snowclad mountains and be amidst lush green tea gardens and sub-tropical to sub alpine forests. It is also a paradise for adventure lovers as plenty of diverse adventure sports options are available, such as mountaineering, trekking, trail running, Paragliding, mountain biking etc. Thus, apart from natural annual population growth, Darjiling Himalaya also have to accommodate the increasing inflow of both domestic and foreign tourists. This is leading to ever growing increase in settlement areas, transport routes and infrastructures. Under this background, preparation of detailed landslide susceptibility mapping becomes imperative for timely preparedness to avoid potential large socio-economic, strategic and environmental damage in the region.

3 Materials and Methods

All the necessary data were procured from different sources and based on it; spatial database was created for preparing landslide susceptibility maps (Table 2). Susceptibility maps were prepared through the integration of pixels of past landslide inventory location and thematic layers of all selected parameters using the weights derived through frequency ratio and fuzzy logic models. Details of the steps followed are given in methodology flow chart (Fig. 3).

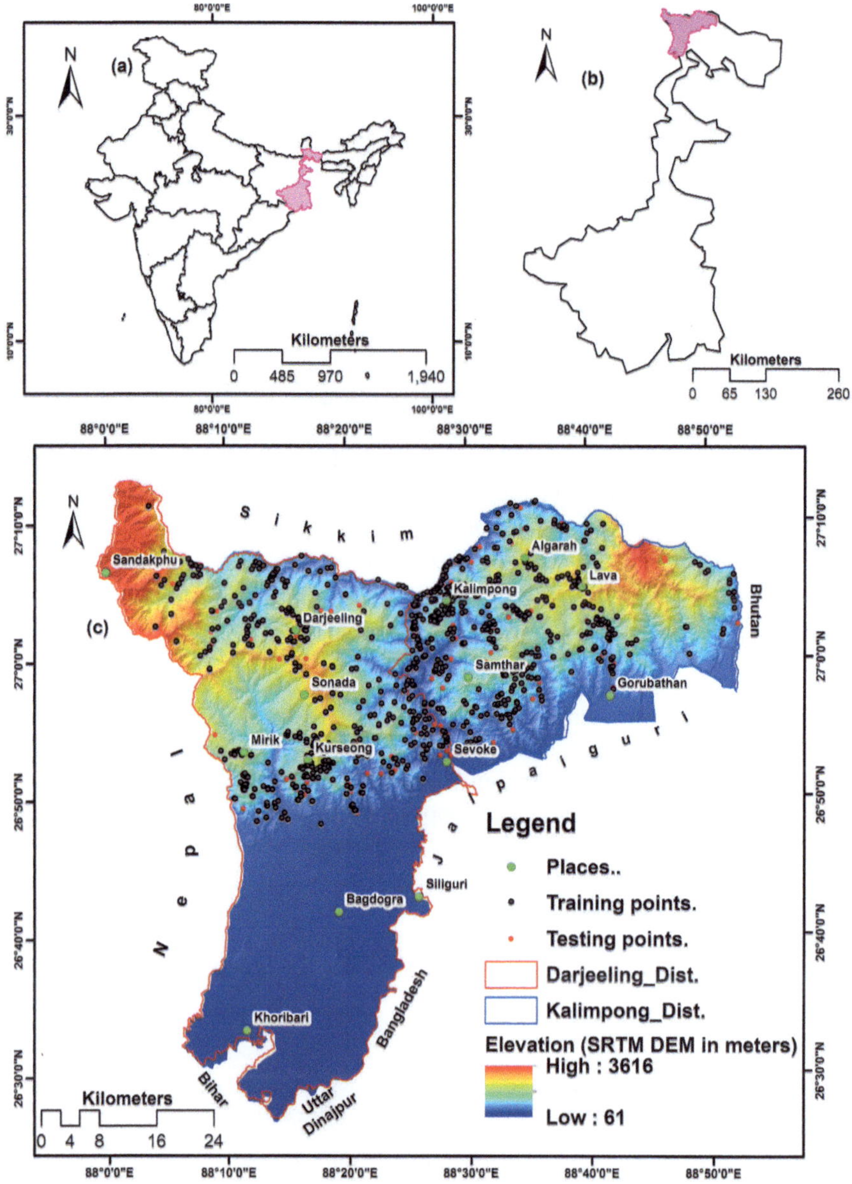

Fig. 1 Location of study area and landslide points

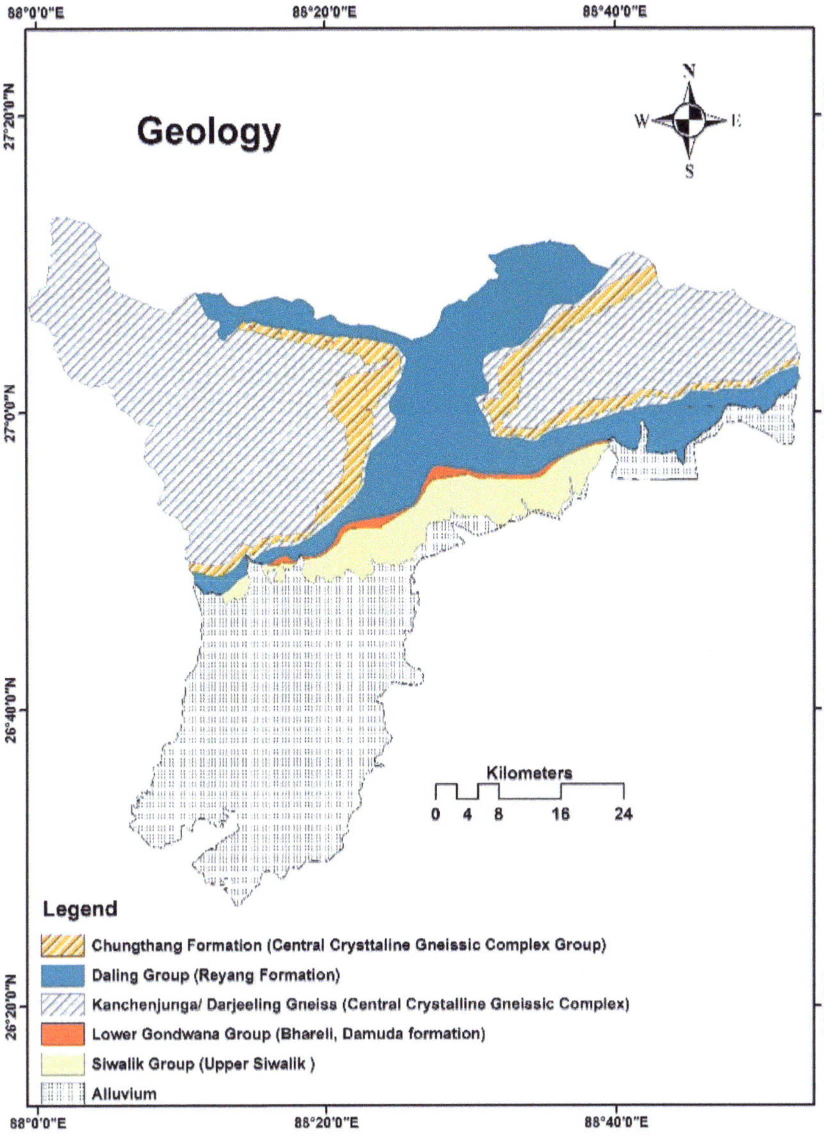

Fig. 2 Geological structure of the study area. Based on Gansser (1964) and Geological and Mineral map of West Bengal, Govt. of India, 1990

Table 2 Details of input parameters used in landslide susceptibility prediction

Sl. No.	Parameters	Data format	Resampled resolution (m)	Source	Technique	Functional relation with landslide occurrence
1	Elevation	Raster (SRTM DEM 30 m × 30)	30	SRTM DEM (September 23, 2014) extracted from USGS earth explorer	30 m × 30 m digital elevation model	Higher the slope higher is the frequency and intensity of landslide
2	Slope	Raster (SRTM DEM 30 m × 30)	30	SRTM DEM (September 23, 2014) extracted from USGS earth explorer	$\tan\theta = N \times i/636.6$ N = No. of contour crossings; i = contour Interval	Rainfall has a positive relation with landslide
3	Rainfall	Annual average rainfall data of different stations in the last 5 years-Gridded Rainfall (0.25 × 0.25) NetCDF File	30	Indian Metrological Department (IMD)	Kriging interpolation method	Higher the rainfall higher is the intensity of landslide
4	Distance to lineament	Vector layers extraction and conversion into raster	30	https://bhuvan-app1.nrsc.gov.in/thematic/thematic/index.php	Euclidian distance buffering	Lower the distance to lineament higher is the intensity of landslide

(continued)

Table 2 (continued)

Sl. No.	Parameters	Data format	Resampled resolution (m)	Source	Technique	Functional relation with landslide occurrence
5	Distance to river	Vector layers extraction and conversion into raster	30	Raster (Landsat 8 OLI/TIRS 30 m × 30, March, 2022. Path and row 139, 41 and 139, 42), SRTM DEM (September 23, 2014) extracted from USGS earth explorer	Euclidian distance buffering	Lower the distance to river higher is the intensity of landslide
6	Distance to road	Vector layers extraction and conversion into raster	30	Geological Survey of India, Government of India, Kolkata, India on 20.02.2023, https://extract.bbbike.org/	Euclidian distance buffering	Lower the distance to road higher is the intensity of landslide
7	Aspect	Raster (SRTM DEM 30 m × 30)	30	SRTM DEM (September 23, 2014) extracted from USGS earth explorer	Aspect = 57.29 × αtan2([dz/dy] − [dz/dx]) Where, dz/dx = ((c + 2f + i) − (a + 2d + g))/8 dz/dy = ((g + 2 h + i) − (a + 2b + c))/8 Here, a to i indicates the cell value of 3*3 window	South, south east and south west facing slopes are more prone to landslides
8	Geomorphology	Vector layers extraction and conversion into raster	30	Geological Survey of India, Government of India, Kolkata, India on 20.02.2023	Digitization process	Dissected Hills and valleys are more prone to landslides
9	Curvature	Raster (SRTM DEM 30 m × 30)	30	SRTM DEM (September 23, 2014) extracted from USGS earth explorer	30 m × 30 m digital elevation model	Convex slopes are more prone to landslides

(continued)

Table 2 (continued)

Sl. No.	Parameters	Data format	Resampled resolution (m)	Source	Technique	Functional relation with landslide occurrence
10	Plan curvature	Raster (SRTM DEM 30 m × 30)	30	SRTM DEM (September 23, 2014) extracted from USGS earth explorer	31 m × 30 m digital elevation model	Plan curvature is generally inversely related to landslide occurrence
11	Profile curvature	Raster (SRTM DEM 30 m × 30)	30	SRTM DEM (September 23, 2014) extracted from USGS earth explorer	32 m × 30 m digital elevation model	Plan curvature is generally positivelyrelated to landslide occurrence
12	Seismic zone	Last 200 years point data of earthquake 30 m × 30	30	USGS Earthquake catalogue https://earthquake.usgs.gov/earthquakes/search/	Spline Interpolation	Seismicity has a positive relation with landslides
13	LULC	Raster (SRTM DEM 30 m × 30)	30	2021 Esri landcover map https://livingatlas.arcgis.com/landcover/	Digitization process	Fallow land and builtup areas are more prone to landslides

(continued)

Table 2 (continued)

Sl. No.	Parameters	Data format	Resampled resolution (m)	Source	Technique	Functional relation with landslide occurrence
14	TWI	Raster (SRTM DEM 30 m × 30)	30	SRTM DEM (September 23, 2014) extracted from USGS earth explorer	TWI = In(As/tanθ) Where α is the cumulative upslope area draining through a point (per unit contour length), and β is the slope gradient (in degree)	Higher the TWI higher the intensity of landslide
15	NDVI	Raster (Landsat 8 OLI/TIRS 30 m × 30, March, 2022. Path and row139, 41 and 139, 42)	30	Raster (Landsat 8 OLI/TIRS 30 m × 30, March, 2022. Path and row139, 41 and 139, 42) extracted from USGS earth explorer	NDVI = NIR − IR NIR + IR Where NIR is the near infrared band or band 4 and IR is the infrared band or band 3	NDVI has negative relation with landslides

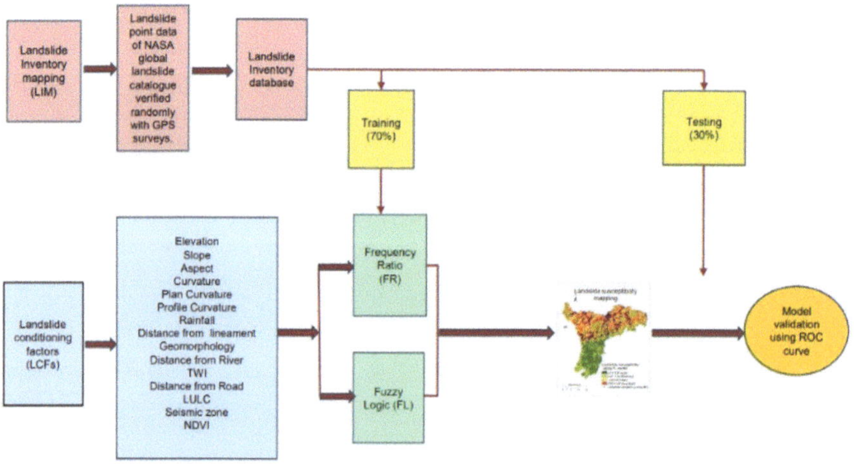

Fig. 3 Methodological flow chart of landslide susceptibility mapping

3.1 Preparation of Landslide Inventory Maps

Landslide inventory means the spatial distribution of landslide locations in an area. Preparation of landslide inventory map is vital for determining the spatial links of historic earth slips with factors of geomorphological, hydrological, climatic factors, and anthropogenic activities (Hong et al., 2017; Bui et al., 2017). It aids us to have an idea about the relationship between the factors affecting the landslide incidences and determine the importance of factors for landslide occurrence as well (Yalcin, 2011). Landslide inventory map of the study area was prepared using published landslide point data of the NASA Global Landslide Catalogue (landslide.nasa.gov), accessed on 22.02.2023. Sonker et al., 2021 has similarly used the landslide point data from Bhukosh site (Geological Survey of India) to prepare landslide inventory map of Sikkim Himalaya. Most of the landslides in the area are rainfall-induced in the study area. In the present study, a total of 1556 landslide locations were identified which were randomly verified through extensive field investigations using a global positing system (GPS) and Google Earth imagery. Out of 1556 landslides, 1089 (70%) landslides were chosen randomly for landslide modeling purposes, and remaining 467 (30%) landslides were used to validate the prepared landslide susceptibility maps. The landslide inventory map (LIM) was prepared in a GIS environment and is shown in Fig. 1c. Field photographs of some landslides in the study area are shown in Fig. 4.

Fig. 4 Photographs of a few landslides in the study area. **a** Landslides in Shivkhola gram panchayat under Kurseong Block seen as white and grey patches (Photographed-15.03.2020). **b** Sherpa gaon landslide, Lava, Kalimpong of 19.10.2021. Courtesy-India.com News Desk. **c** Sittong Village Landslide (Photographed 22nd November, 2015). Incessant rain triggered this landside in July, 1999 and since then it is gradually enlarging in lack of controlling measures (Courtesy-savethehills.blogspot.com). **d** A major landslide that occurred at 14th Mile, NH-55 near Paglajhora, Kurseong on 16.06.2010 (Courtesy-savethehills.blogspot.com)

3.2 Determination of Landslide Causative Factors

In susceptibility mapping, it is assumed largely that future landslides will occur in the same condition that caused the past landslides (Rasyid et al., 2016). Therefore, selection of the conditioning factors is essential and a mandatory step in landslide susceptibility assessment (Hong et al., 2015). These factors can be divided into four groups: Physical geographic factors, hydrological factors, climate factors, and anthropogenic activity related factors (Chang et al., 2019).

In this study the conditioning factors were mainly derived based on earlier established works on the region and field survey based on certain attributes such as easy obtainability, representativeness and practicality (Ghosh et al., 2022). A total of 15 explanatory factors were resampled at 30-m resolution for analyzing both the models and to generate their respective outputs. Some of the causative factors selected are continuous data whereas others are categorical in nature. Thematic maps of 15 conditioning factors, viz. elevation, slope, rainfall, distance to river, distance to lineament, distance to road, aspect, geomorphology, curvature, plan curvature, profile curvature, seismic zone, LULC, TWI (Topographic Wetness Index) and NDVI (Normalized Difference Water Index) were produced in the GIS environment (Fig. 5). The details of each created dataset and their relation with landslide are listed in Table 2.

3.2.1 Elevation

Surface elevation is one of the most significant determinants of landslides. Incident of landslide and its magnitude are strongly influenced by this explanatory factor. Higher the land surface from river bed level, the higher will be the landslide susceptibility and vice versa. Elevation also affects landslide by controlling the drainage direction, surface runoff, velocity, infiltration rate and soil characters. The study area has been separated into four elevation zones (61–546, 546–1233, 1233–2049 and 2049–3616 m) using SRTM DEM (2014) (Fig. 5a). The highest elevation is in the north-west and north-east region of the study area.

3.2.2 Slope

On a slope of uniform isotropic material, increased slope correlates with increased likelihood of failure (Rasyid et al., 2016). SRTM DEM (2014) has been utilized to obtain and extract statistics on slope. Landslide happens more on steeper slopes, as the angle of slope is one of the primary factors in inducing the slope instability. The slope map (Fig. 5a) was generated from the DEM of the study area in the GIS environment. Based on slope, the study area was divided into 0–8.6, 8.6–18.8, 18.8–27, 27–35.9- and 35.9–84-degree angles of slope (Fig. 5b).

3.2.3 Aspect

Similarly, the aspect map is important for evaluating slope stability (Chauhan et al., 2010). In this study, aspect is divided into nine classes namely, flat (−1), north (0–22.5; 337.5–360), northeast (22.5–67.5), east (67.5–112.5), southeast (112.5–157.5), south (157.5–202.5), southwest (202.5–247.5), west (247.5–292.5), northwest (292.25–337.5) based on SRTM DEM, 2014 (Fig. 5c). To show the variation among classes, aspect maps display the distribution of each direction in the topography by using different colors to each cell in the study area (Quan & Lee, 2012).

3.2.4 Curvature

Curvature is a geomorphological factor which controls the flow of water on the surface and influences the runoff and storage of water (Pham et al., 2017, Ayalew

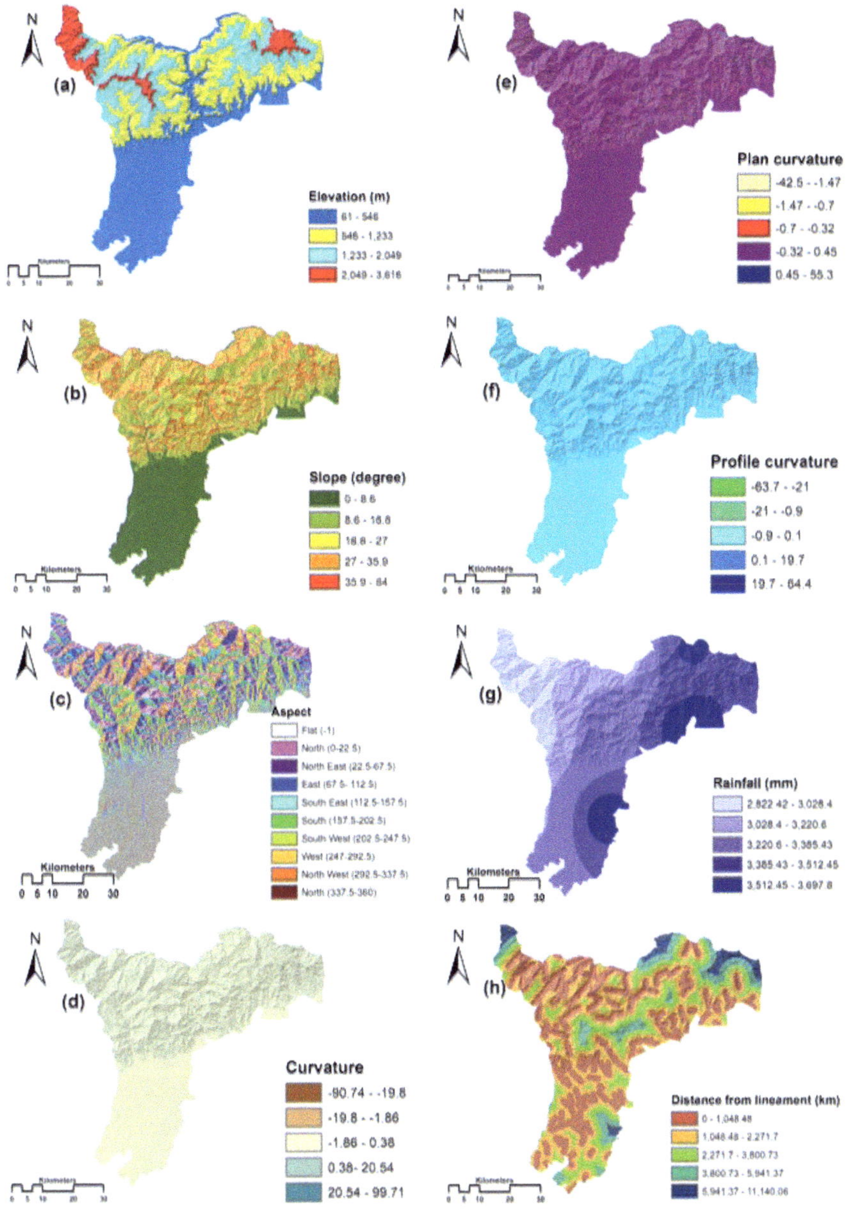

Fig. 5 Landslide conditioning factors—**a** elevation, **b** slope, **c** aspect, **d** rainfall, **e** geology, **f** soil texture, **g** distance from river, **h** distance from lineament, **i** land use/land cover (LULC), **j** normalized differential vegetation index (NDVI), **k** distance from road, **l** topographic wetness index (TWI), **m** stream power index (SPI), **n** sediment transportation index (STI), **o** geomorphology, **p** seismic map

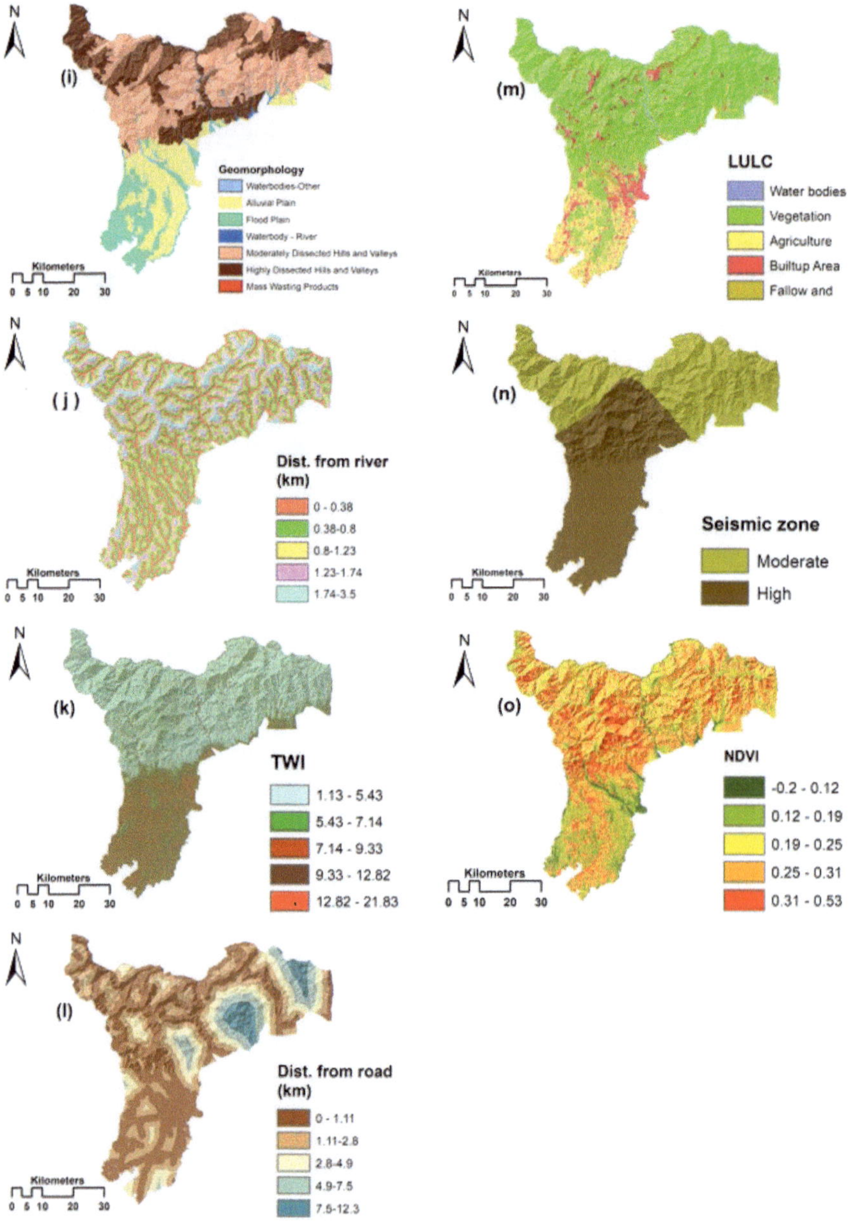

Fig. 5 (continued)

et al., 2004). A curvature map was generated from SRTM DEM (2014) with 30 m resolution, and classified into three classes: Concave (<−0.05), flat (−0.05–0.05) and convex (>0.05) (Fig. 5d).

3.2.5 Plan Curvature

Plan curvature is considered an important factor affecting landslide as it controls the convergence or divergence of landslide material and water in the direction of landslide motion (Carson & Kirkby, 1972). It actually is a curvature of a contour line formed by intersecting a horizontal plane with the surface (Wilson & Gallant, 2000). Again SRTM DEM (2014) with 30 m resolution was used to prepare plan curvature map and was then grouped into five classes. The plan curvature values range from −42.5 to 55.28 (Fig. 5e).

3.2.6 Profile Curvature

Profile curvature affects the driving and resisting stresses within a landslide in the direction of motion. Profile curvature has implications for changes in flow velocity and sediment transport processes which in turn influences the landslide processes (Wilson & Gallant, 2000). SRTM DEM (September 23, 2014) was used to prepare the profile curvature map and was grouped into five classes. The profile curvature values range from −63.7 to 64.40 (Fig. 5f).

3.2.7 Rainfall

Rainfall plays a critical role in the unforeseen events like landslides. It is a temporal external trigger, and too much of it might cause the slope to become heavy from an increase in pore water pressure, which would lead to slope slips. The rainfall map was prepared based on annual average rainfall data of different stations in the last 5 years-Gridded Rainfall (0.25 × 0.25) in NetCDF File format downloaded from Indian Metrological Department (IMD) website. Kriging Interpolation method was followed in ArcGIS to get the desired map. The map divides the study area in five rainfall zones of 2822.42–3028.4, 3028.4–3220.64, 3220.64–3385.43, 3385.43–3512.45 and 3512.45–3697.82 mm (Fig. 5g).

3.2.8 Distance from Lineament

Lineaments are the structural features, which describe the plane of weakness, fractures, and faults along which landslide susceptibility is higher (Kanungo et al., 2006). Lineament density and distance from lineaments is considered pivotal when studying landslide. It is well known that locations near lineaments have a higher chance of experiencing landslides, which not only impact the surface material structures but also make contribution to terrain permeability causing slope instability. The proximity distance from lineaments were identified by buffering from lineament or fault map in Arc GIS following Rasyid et al., 2016. For this purpose, lineaments were extracted as vector layers from Bhuvan (https://bhuvan-app1.nrsc.gov.in/thematic/thematic/index.php) and vectorized in ArcGIS. Lastly, multiple ring buffers were demarcated around each lineaments using Euclidian Distance buffering technique (0–1048.48, 1,048.48–2271.7, 2271.7–3800.73, 3800.73–5941.3 and 5941.37–11,140.06 m) and converted into raster format (Fig. 5h).

3.2.9 Geomorphology

Geomorphologically, the research area consists of water bodies-other than rivers, alluvial plain, flood plain, water bodies-rivers, moderately dissected hills and valleys, highly dissected hills and valleys and mass wasting products (Fig. 5i). Geomorphology map of the study area was produced in the GIS environment based on the geomorphology map published by the Geological Survey of India, Kolkata (accessed on 22.02.2023). Highly dissected hills and valley regions are prone to landslide events.

3.2.10 Distance from River

Distance from river is an important factor affecting landslides as this conditioning factor is directly related to landslide. Landsat 8 (OLI/TIRS, 30 m, March, 2022. Path and row-139, 41 and 139, 42) and SRTM DEM (September 23, 2014) were used to demarcate major rivers of the study area. The vectorized stream lines were prepared in ArcGIS, followed by demarcation of concentric buffers (0–0.38, 0.38–0.8, 0.8–1.23, 1.23–1.74 and 1.74–3.5 m). The drainage proximity map thus produced was converted into raster format (Fig. 5j).

3.2.11 Distance from Road

Landslide occurrence and distance from roads have indirect correlation. Lesser the distance from the road, higher is the probability of landslides and vice-versa. Major transport routes or roads were demarcated by consulting data from GSI and https://extract.bbbike.org/. Cutting of natural slopes during road construction leaves the slopes unstable and coupling of any other triggering phenomenon like high rainfall, earthquake, constructions along the roads etc. leads to landslides. Just like in distance from river parameter, we used euclidian distance buffering technique along all the roads in ArcGIS and five classes namely 0–1.11, 1.11–2.8, 2.8–4.9, 4.9–7.5 and 7.5–12.3 m were created (Fig. 5l).

3.2.12 TWI

TWI (Topographic Wetness Index) is an important topographic factor and theoretically refers to the accumulation of flow at any point within a watershed (Ghosh et al., 2022). TWI was calculated using the following formula (Beven & Kirkby, 1979):

$$\text{TWI} = \text{In}\ (a/\tan\beta) \tag{1}$$

Here a represents cumulative upslope area draining through a point and β represents slope angle at that point. The TWI = In (a/tan β) index reflects water's tendency to accumulate at any point in the catchment (in terms of a) and gravity's tendency to transfer that water down slope (expressed in terms of tan β as an approximate hydraulic gradient) (Poudyal et al., 2010). TWI map was prepared in ArcGIS using SRTEM DEM to detect spatial distribution of wetness which were divided into five classes and the values ranges from 1.12 to 21.83 (Fig. 5k).

3.2.13 LULC

Land use is an indirect measure for the strength of the slope, as it commands the rate of weathering and erosion (Chawla et al., 2018). Although cultivated land impacts slope stability due to covered soil saturation, forest land controls continuous water flow and regular infiltration of water (Devkota et al., 2013). The 2021 Esri landcover map was accessed from https://livingatlas.arcgis.com/landcover/ and through digitization process, the LULC of study area was prepared in GIS environment. The map thus produced consists of five different land use types; namely, Water bodies, Vegetation, Agriculture, Built-up Area and Fallow land (Fig. 5m).

3.2.14 Seismic Zone

Seismicity has a positive correlation with the landslide occurrences. The ground shaking owing to earthquakes not only produce fresh landslides rather they also reactivate the older and dormant ones. The seismic zone map for the study area has been prepared by using the point data of last 200 years from the official website of National Centre for Seismology, New Delhi, India. The map was generated using Spline interpolation technique in GIS environment. Based on it, the study area has been sub-divided into two zones of moderate and high for past earthquakes having seismic intensity of 3–4 and 4–5 in the Richter scale respectively (Fig. 5n).

3.2.15 NDVI

One of the crucial elements in landslide assessment is the natural differential vegetation index, or NDVI, i.e. Normalized Difference Vegetation Index (Roy & Saha,

2019). Usually, high normalized differential vegetation index areas are less prone to landslide occurrence and vice-versa. Landsat 8 OLI/TIRS 30 m, March, 2022, path and row-139, 41 and 139, 42 were used in ArcGIS10.5 software to calculate the NDVI from bands 4 (red band) and 5 (infrared band). The NDVI map was calculated using Eq. (2):

$$NDVI = IR - R/IR + R, \tag{2}$$

where IR is the infrared part of the electromagnetic spectrum and R is the red part of the electromagnetic spectrum.

The NDVI map thus produced has been classified into five classes such as very low (−0.2–0.12), low (0.12–0.19), medium (0.19–0.25), high (0.25–0.31) and very high (0.31–53) (Fig. 5o).

3.3 Landslide Susceptibility Mapping

3.3.1 LSM Using Frequency Ratio (FR) Model

The frequency ratio (FR) shows how likely it is for a particular attribute to occur. It is the ratio of the area where landslides occurred to the total study area, and also, is the ratio of the probabilities of a landslide occurrence to a non-occurrence for a given factor's attribute (Lee & Pradhan, 2006). In this study we have used FR model to show the relationship between the spatial distribution of landslides and each selected causative factor, i.e., elevation, slope, rainfall, distance to river, distance to lineament, distance to road, aspect, geomorphology, curvature, plan curvature, profile curvature, seismic zone, LULC, TWI and NDVI. This model is an effective bivariate method for evaluating the impact of each unit of conditioning factor (here 15 causative factors are used) on the landslide occurrences. Landslide factors were compared with landslide inventory and a correlation between them were quantitatively analyzed by landslide frequency ratio method.

FR model can be expressed as:

$$\text{FR} = f(i) / \sum f / p(i) / \sum p \tag{3}$$

where *f(i)* is the number of landslide points for each class of each factor, $\sum f$ is the total number of landslide points in the study area, *p(i)* is total number of pixels in each class of the factor, and $\sum$p is the total number of pixels in the entire study area.

Generally, any factor classes with a frequency ratio value of >1 will have higher probability landslide occurrence and vice versa.

Landslide susceptibility mapping (LSM) using FR model was accomplished using the following steps:

(i) Firstly, each sub-class of selected fifteen causative variables were reclassified using reclass tool.
(ii) Secondly, separate column of each FR value against all sub-classes of fifteen separate raster layers were added in ArcGIS.
(iii) Thereafter, another separate raster layers were created using those FR values of each sub-class of all selected causative variables with the help of 'Lookup' tool of ArcGIS.
(iv) Finally, raster calculator tool was used to combine (summation) newly generated fifteen separate raster layers. Here, the sum of landslide frequency value of the entire set of factors of a particular pixel denotes the landslide probability of that pixel.

The composite landslide susceptibility map was articulated as:

$$\mathrm{LSM} = \sum \mathrm{FR} \tag{4}$$

where LSM denotes landslide susceptibility map and $\sum$ FR denotes total of FRs of all classes in a factor.

FR model is easy to compute, easy to understand and a very popular bivariate method for evaluating the impact of each unit of conditioning factor on the occurrences of a phenomenon like landslide, flood, forest fire etc. It is generally considered as a base model to make comparison with the results of other advanced models (Ghosh et al., 2022). Therefore, this model was used here in this chapter.

3.3.2 LSM Using Fuzzy Logic (FL) Model

Fuzzy logic set theory introduced by Zadeh (1965) is one of the oldest tools used to handle complex problems. Therefore, we find its wide application in different disciplines. Fuzzy logic considers the spatial distribution of features on a map as members of a set (Pradhan, 2010). In the classical set theory, the membership of an object is in binary form of either yes or no. It means an object will have a membership value of 1, if it is a member of a set, or a membership value of 0 if is not a member. Whereas in case of fuzzy set theory, membership can take on any value between 0 and 1 reflecting the degree of certainty of membership. Thus, fuzzy set theory employs the idea of a membership function that expresses the degree of membership with respect to some attribute of interest (Chung & Fabbiri, 2001; Pradhan, 2010).

A sample of fuzzy set is shown in the following equation (Mcbratney & Odeh, 1997).

$$\mathrm{A} = \{\mathrm{X}, \mu\mathrm{A}(\mathrm{X})\} \text{ for each } \mathrm{x} \ \varepsilon \ \mathrm{X} \tag{5}$$

where, μA is the MF (membership of x in fuzzy set A) so that:

If x does not belong to A then $\mu A = 0$.

If x belongs completely to A then $\mu A = 1$.

If x x belongs in a certain degree to A then
$0 < \mu_A(x) < 1$.

Application of fuzzy logic in landslide susceptibility takes into account the pixels on any layer of conditioning parameters susceptible to landslides (Pradhan, 2010; Sahana & Sajjad, 2017). The pixel values of these parameters can range anywhere between 0 and 1 where absolute 0 and 1 indicates non-susceptibility and susceptibility respectively. The fuzzy membership values can be extracted from a frequency ratio model (Lee, 2007; Pradhan & Lee, 2009) or through AHP process (Saaty, 1977). Many scholars have also derived fuzzy membership values by using qualitative knowledge-based approach (Champatiray et al., 2007; Chung & Fabbiri, 2001; Kayastha et al., 2013). Here we have computed fuzzy membership values based on frequency ratio values (Table 3).

The computed frequency ratio values were first normalized between 0 and 1 to describe the fuzzy membership functions. The level of support or confidence among paired members is determined by a fuzzy set. The members of the set are figured after the slash (/) and the membership (confidence) is represented by the value before the slash (/). Fuzzy sets represent the values more consistently and properly by setting them in language terms (Juang et al., 1992; Kanungo et al., 2006; Kayastha, 2012). Values of linguistic variables like phrases, words or sentences constitute fuzzy sets. Following Pourghasemi et al., (2012) and Sahana and Sajjad (2017), the fuzzy sets are expressed in this study in linguistic forms where very low is illustrated as (0/5, 0.25/4, 0.5/3, 0.75/2, 1/1), low as (0/5, 1/4, 0.75/3, 0.25/2, 0/1) moderate as (0/5, 0.5/4, 1/3, 0.5/2, 0/1), high as (0/1, 1/2, 0.75/3, 0.25/4, 0/5) and very high as (1/5, 0.75/4, 0.5/3, 0.25/2, 0/1).

Landslide susceptibility was grouped based on output of each parameter. The outputs obtained from the normalized values in frequency ratio model represent the degrees of fuzzy memberships. The fuzzy set of the landslide parameters for landslide susceptibility were obtained as follows:

(i) Elevation = (0.15/1, 0.73/2, 1.00/3, 0.54/4, 0.00/5)
(ii) Slope = (0.00/1, 0.42/2, 0.64/3, 0.87/4, 1.00/5)
(iii) Slope Aspect = (0.00/1, 0.34/2, 0.71/3, 0.80/4, 1.00/5, 0.82/6, 0.65/7, 0.92/8, 0.57/9, 0.75/10)
(iv) Curvature = (0.00/1, 1.00 /2, 0.50/3, 0.71/4, 0.00/5)
(v) Plan Curvature = (1.00/1, 0.41/2, 0.34/3, 0.00/4, 0.30/5)
(vi) Profile Curvature = (0.00/1, 1.00/2, 0.50/3, 0.73/4, 0.00/5)
(vii) Rainfall = (0.51/1, 1.00/1, 0.72/3, 0.43/4, 0.30/5)
(viii) Distance from Lineament = (0.03/1, 0.49/2, 1.00/3, 0.09/4, 0.00/5)
(ix) Geomorphology = (0.02/1, 0.00/2, 0.02/3, 0.41/4, 1.00/5, 0.56/6, 0.00/7)
(x) Distance from River = (0.11/1, 0.00/2, 0.11/3, 0.72/4, 1.00/5)
(xi) TWI = (1.00/1, 0.64/2, 0.26/3, 0.15/4, 0.00/5)
(xii) Distance from Road = (1.00/1, 0.07/2, 0.13/3, 0.00/4, 0.55/5)
(xiii) LULC = (0.91/1, 0.58/2, 0.00/3, 0.87/4, 1.00/5)
(xiv) Seismic Map = (1.00/1, 0.00/2)
(xv) (0.96/1, 1.00/2, 0.52/3, 0.00/4, 0.02/5).

Table 3 Frequency ratio and fuzzy membership values of each unit of each causative parameter

Factors/ Parameters	Sub-units	Pixel	Ratio	Percentage	Landslide points	Ratio	Percentage	FR (Frequency Ratio)	FR-a	b-a	Fuzzy Membership (FR-a/b-a)
1. Elevation (m)	**61–450**	1,310,058	0.37497	37.50	155	0.142332	14.23	**0.38**	0.2721	1.7975	**0.15**
	450–996	776,780	0.22233	22.23	345	0.316804	31.68	**1.42**	1.3175	1.7975	**0.73**
	996–1577	698,936	0.20005	20.01	415	0.381084	38.11	**1.90 (b)**	1.7975	1.7975	**1.00**
	1577–2293	499,015	0.14283	14.28	167	0.153352	15.34	**1.07**	0.9662	1.7975	**0.54**
	2293–3616	209,014	0.05982	5.98	7	0.006428	0.64	**0.11 (a)**	0.0000	1.7975	**0.00**
2. Slope (degree)	**0–8.57**	1,140,856	0.32654	32.65	20	0.018365	1.84	**0.06**	0.0000	2.0047	**0.00**
	8.57–18.78	570,461	0.16328	16.33	161	0.147842	14.78	**0.91**	0.8492	2.0047	**0.42**
	18.78–27.02	815,897	0.23353	23.35	343	0.314968	31.50	**1.35**	1.2925	2.0047	**0.64**
	27.02–35.91	678,596	0.19423	19.42	380	0.348944	34.89	**1.80**	1.7403	2.0047	**0.87**
	35.91–84.01	287,993	0.08243	8.24	185	0.169881	16.99	**2.06**	2.0047	2.0047	**1.00**
3. Aspect	**Flat (−1)**	11,653	0.00334	0.33	0	0.000000	0.00	**0.00**	0.0000	1.3161	**0.00**
	North (0–22.5)	209,761	0.06004	6.00	29	0.026630	2.66	**0.44**	0.4436	1.3161	**0.34**
	North-East 22.5–67.5	403,459	0.11548	11.55	118	0.108356	10.84	**0.94**	0.9383	1.3161	**0.71**
	East 67.5–112.5	493,905	0.14137	14.14	162	0.148760	14.88	**1.05**	1.0523	1.3161	**0.80**
	South East (112.5–157.5)	516,781	0.14791	14.79	212	0.194674	19.47	**1.32**	1.3161	1.3161	**1.00**
	South (157.5–202.5)	478,515	0.13696	13.70	160	0.146924	14.69	**1.07**	1.0727	1.3161	**0.82**
	South West (202.5–247.5)	435,734	0.12472	12.47	117	0.107438	10.74	**0.86**	0.8615	1.3161	**0.65**

(continued)

Table 3 (continued)

Factors/ Parameters	Sub-units	Pixel	Ratio	Percentage	Landslide points	Ratio	Percentage	FR (Frequency Ratio)	FR-a	b-a	Fuzzy Membership (FR-a/b-a)
	West (247.5–292.5)	396,665	0.11353	11.35	149	0.136823	13.68	**1.21**	1.2051	1.3161	**0.92**
	North West (292.5–337.5)	368,878	0.10558	10.56	87	0.079890	7.99	**0.76**	0.7567	1.3161	**0.57**
	North (337.5–360)	178,452	0.05108	5.11	55	0.050505	5.05	**0.99**	0.9888	1.3161	**0.75**
4. Curvature	**−90.73–−19.79**	48	0.00001	0.00	0	0.000000	0.00	**0.00**	0.0000	1.7195	**0.00**
	−19.79 to−1.86	151,131	0.04326	4.33	81	0.074380	7.44	**1.72**	1.7195	1.7195	**1.00**
	−1.86–0.38	2,421,287	0.69302	69.30	655	0.601469	60.15	**0.87**	0.8679	1.7195	**0.50**
	0.38–20.54	921,296	0.26369	26.37	353	0.324151	32.42	**1.23**	1.2293	1.7195	**0.71**
	20.54–99.71	41	0.00001	0.00	0	0.000000	0.00	**0.00**	0.0000	1.7195	**0.00**
5. Plan curvature	**−42.5 to−1.47**	60,510	0.01732	1.73	46	0.042241	4.22	**2.44**	1.6551	1.6551	**1.00**
	−1.47 to−0.7	227,182	0.06502	6.50	103	0.094582	9.46	**1.45**	0.6707	1.6551	**0.41**
	−0.7 to−0.32	404,001	0.11563	11.56	170	0.156107	15.61	**1.35**	0.5661	1.6551	**0.34**
	−0.32 to 0.45	2,259,281	0.64665	64.67	552	0.506887	50.69	**0.78**	0.0000	1.6551	**0.00**
	0.45 to 55.28	542,829	0.15537	15.54	218	0.200184	20.02	**1.29**	0.5046	1.6551	**0.30**
6. Profile curvature	**−63.7 to−21**	32	0.00001	0.00	0	0.000000	0.00	**0.00**	0.0000	1.6056	**0.00**
	−0.9	223,795	0.06405	6.41	112	0.102847	10.28	**1.61**	1.6056	1.6056	**1.00**
	−0.9 to 0.10	1,898,698	0.54345	54.34	476	0.437098	43.71	**0.80**	0.8043	1.6056	**0.50**
	0.10 to 19.69	1,371,244	0.39248	39.25	501	0.460055	46.01	**1.17**	1.1722	1.6056	**0.73**
	19.69 to 64.40	34	0.00001	0.00	0	0.000000	0.00	**0.00**	0.0000	1.6056	**0.00**
7. Rainfall (mm)	**2822.42–3028.4**	459,334	0.13147	13.15	123	0.112948	11.29	**0.86**	0.8591	1.6857	**0.51**

(continued)

Table 3 (continued)

Factors/ Parameters	Sub-units	Pixel	Ratio	Percentage	Landslide points	Ratio	Percentage	FR (Frequency Ratio)	FR-a	b-a	Fuzzy Membership (FR-a/b-a)
	3028.4–3220.64	481,520	0.13782	13.78	253	0.232323	23.23	**1.69**	1.6857	1.6857	**1.00**
	3220.64–3385.43	1,141,523	0.32673	32.67	430	0.394858	39.49	**1.21**	1.2085	1.6857	**0.72**
	3385.43–3512.45	975,612	0.27924	27.92	214	0.196511	19.65	**0.70**	0.7037	1.6857	**0.42**
	3512.45–3697.82	435,814	0.12474	12.47	69	0.063361	6.34	**0.51**	0.5079	1.6857	**0.30**
8. Distance from lineament (km)	**0–1.49**	1,294,001	0.37037	37.04	325	0.298439	29.84	**0.81**	0.0168	0.6006	**0.03**
	1.49–2.27	1,130,264	0.32351	32.35	382	0.350781	35.08	**1.08**	0.2953	0.6006	**0.49**
	2.27–3.8	607,205	0.17379	17.38	263	0.241506	24.15	**1.39**	0.6006	0.6006	**1.00**
	3.8–5.94	320,008	0.09159	9.16	84	0.077135	7.71	**0.84**	0.0532	0.6006	**0.09**
	5.94–11.14	142,325	0.04074	4.07	35	0.032140	3.21	**0.79**	0.0000	0.6006	**0.00**
9. Geomorphology	**Water Bodies-other**	53,984	0.01545	1.55	6	0.005510	0.55	**0.36**	0.3566	1.7852	**0.20**
	Alluvial Plain	623,008	0.17832	17.83	1	0.000918	0.09	**0.01**	0.0051	1.7852	**0.00**
	Flood Plain	382,992	0.10962	10.96	5	0.004591	0.46	**0.04**	0.0419	1.7852	**0.02**
	Waterbody—River	60,994	0.01746	1.75	14	0.012856	1.29	**0.74**	0.7364	1.7852	**0.41**
	Moderately Dissected Hills and Valleys	1,329,925	0.38065	38.07	740	0.679522	67.95	**1.79**	1.7852	1.7852	**1.00**
	Highly Dissected Hills and Valleys	1,038,984	0.29738	29.74	323	0.296602	29.66	**1.00**	0.9974	1.7852	**0.56**
	Mass Wasting Products	3916	0.00112	0.11	0	0.000000	0.00	**0.00**	0.0000	1.7852	**0.00**

(continued)

Table 3 (continued)

Factors/ Parameters	Sub-units	Pixel	Ratio	Percentage	Landslide points	Ratio	Percentage	FR (Frequency Ratio)	FR-a	b-a	Fuzzy Membership (FR-a/b-a)
10. Distance from river (Km)	**0–0.38**	980,611	0.28067	28.07	267	0.245179	24.52	**0.87**	0.1092	0.9860	**0.11**
	0.38–0.8	965,406	0.27632	27.63	230	0.211203	21.12	**0.76**	0.0000	0.9860	**0.00**
	0.8–1.23	754,690	0.21601	21.60	205	0.188246	18.82	**0.87**	0.1071	0.9860	**0.11**
	1.23–1.74	536,492	0.15356	15.36	247	0.226814	22.68	**1.48**	0.7127	0.9860	**0.72**
	1.74–3.5	256,604	0.07345	7.34	140	0.128558	12.86	**1.75**	0.9860	0.9860	**1.00**
11. TWI	**1.12–5.43**	1,207,131	0.34551	34.55	556	0.510560	51.06	**1.48**	1.2846	1.2846	**1.00**
	5.43–7.14	1,183,127	0.33864	33.86	373	0.342516	34.25	**1.01**	0.8184	1.2846	**0.64**
	7.14–9.33	691,740	0.19799	19.80	115	0.105601	10.56	**0.53**	0.3403	1.2846	**0.26**
	9.33–12.82	328,732	0.09409	9.41	40	0.036731	3.67	**0.39**	0.1973	1.2846	**0.15**
	12.82–21.83	83,073	0.02378	2.38	5	0.004591	0.46	**0.19**	0.0000	1.2846	**0.00**
12. Distance from road (km)	**0–1.11**	1,637,206	0.46860	46.86	723	0.663912	66.39	**1.42**	0.8870	0.8870	**1.00**
	1.11–2.8	966,009	0.27649	27.65	178	0.163453	16.35	**0.59**	0.0614	0.8870	**0.07**
	2.8–4.9	447,383	0.12805	12.81	90	0.082645	8.26	**0.65**	0.1157	0.8870	**0.13**
	4.9–7.5	278,589	0.07974	7.97	46	0.042241	4.22	**0.53**	0.0000	0.8870	**0.00**
	7.5–12.3	164,616	0.04712	4.71	52	0.047750	4.78	**1.01**	0.4837	0.8870	**0.55**
13. LULC	**Water bodies**	24,931	0.00714	0.71	12	0.011019	1.10	**1.54**	1.5107	1.6666	**0.91**
	Vegetation	2,487,605	0.71200	71.20	779	0.715335	71.53	**1.00**	0.9711	1.6666	**0.58**
	Agriculture	382,420	0.10946	10.95	4	0.003673	0.37	**0.03**	0.0000	1.6666	**0.00**
	Builtup Area	338,430	0.09687	9.69	156	0.143251	14.33	**1.48**	1.4453	1.6666	**0.87**
	Fallow land	260,417	0.07454	7.45	138	0.126722	12.67	**1.70**	1.6666	1.6666	**1.00**

(continued)

Table 3 (continued)

Factors/ Parameters	Sub-units	Pixel	Ratio	Percentage	Landslide points	Ratio	Percentage	FR (Frequency Ratio)	FR-a	b-a	Fuzzy Membership (FR-a/b-a)
14. Seismic map	**Moderate**	2,062,683	0.59038	59.04	720	0.661157	66.12	**1.12**	0.2927	0.2927	**1.00**
	High	1,431,120	0.40962	40.96	369	0.338843	33.88	**0.83**	0.0000	0.0000	**0.00**
15. NDVI	**−0.2–0.12**	151,929	0.04349	4.35	62	0.056933	5.69	**1.31**	0.4733	0.4939	**0.96**
	0.12–0.19	501,806	0.14363	14.36	208	0.191001	19.10	**1.33**	0.4939	0.4939	**1.00**
	0.19–0.25	964,104	0.27595	27.59	328	0.301194	30.12	**1.09**	0.2555	0.4939	**0.52**
	0.25–0.31	1,182,053	0.33833	33.83	308	0.282828	28.28	**0.84**	0.0000	0.4939	**0.00**
	0.31–0.53	693,911	0.19861	19.86	183	0.168044	16.80	**0.85**	0.0101	0.4939	**0.02**

Fuzzy operation is the next stage in the fuzzy logic technique. There exists Several fuzzy operators for combining membership functions such as Fuzzy OR, fuzzy AND, fuzzy algebraic sum, fuzzy algebraic product and fuzzy gamma (Chung & Fabbiri, 2001). In case of fuzzy OR and fuzzy AND, only one of the contributing fuzzy sets has an effect on the resultant value. The resulting set shall be larger than, or equal to the maximum value and smaller than, or equal to the minimum value of all the fuzzy sets, by means of fuzzy algebraic sum and fuzzy algebra product operator (Chi et al., 2002). Values that fall between fuzzy algebraic product and fuzzy algebraic sum are computed via the Fuzzy gamma (γ) operator. The range of a gamma (γ) value is 0 (no compensation) to 1 (complete compensation). Determination of optimum γ value is dependent on the degree of compensation between two extreme confidence levels (Anbalagan et al., 2015). Choice of suitable fuzzy operator for the data integration is required to achieve optimum result in landslide prediction studies. The kind of spatial data that needs to be integrated determines which fuzzy operator should be used (Choi et al., 2000). In the present study, Fuzzy gamma operator was chosen to combine factors using the formula given below:

$$\mu_\gamma = (\mu_{sum})k^\gamma \cdot \left(\mu_{product}\right)^{.1-\gamma} \tag{6}$$

where, exponent γ is a number from < 0, 1 > interval and it allows optimization of the membership combination. Either the fuzzy algebraic product (γ = 0) or the fuzzy algebraic sum (γ = 1) is obtained when it is set to the extremes of the interval. To perform fuzzy gamma operation, several gamma operator (k) values, i.e. 0.50, 0.70, 0.80, 0.90, 0.95, and 0.975 are there. In the present study, fuzzy gamma operator (k) value of 0.975 has been applied for producing the landslide susceptibility map following Roy and Saha (2019).

3.4 Validation of the Landslide Susceptibility Maps

3.4.1 Validation of Susceptibility Maps Receiver Operating Characteristics (ROC)

FR and FL models were used to prepare the landslide susceptibility maps of the study area. The two LSMs produced were then validated using the ROC (receiver operating characteristics) curve and through frequency analysis of landslide occurrences on different susceptible zones to evaluate the accuracy level of the models. The X and Y axes were used to prepare the ROC curve. The X-axis indicates the false positive rate (1-specificity) and the Y-axis indicates the true positive rate (sensitivity) (Chung & Fabbri, 2003). ROC curves have been extensively used for the assessment of susceptibility maps (Kanungo et al., 2006; Pham et al., 2019; Pradhan, 2010; Ram et al., 2020; Roy & Saha, 2019). One common technique for validating landslip models is the ROC curve (Beguería, 2006). In this study, out of total 1556 landslide locations, 467 landslides (30%) were used to validate the landslide susceptibility maps. The prediction capacity of AUC has been categorized as: poor (0.5–0.6), average (0.6–0.7), good (0.7–0.8), very good (0.8–0.9) and excellent (>0.9). The AUC value of 1 indicates model is perfect whereas value of 0.5 indicates incorrect models (Bui et al., 2016).

3.4.2 Validation of Susceptibility Maps by Frequency Ratio Model/ method

The next validation process of showing the level of accuracy of the LSM is by calculating the ratio of landslide data for validation that falls into each susceptibility class. It is assumed that most of the landslides used for validation should fall in the high to greater susceptibility class (Rasyid et al., 2016). Pham et al., 2019 have also used frequency analysis of landslide occurrences on susceptible zones for validation of LSMs.

4 Results and Discussions

4.1 Landslide Susceptibility Mapping Using FR Model

The relationship between Landslide susceptibility and the classes of each conditioning factor was established through FR analysis. A high value of FR (FR > 1) indicates a strong relationship between landslide susceptibility and selected explanatory factors and vice-versa. The calculated FR value for every parameter has varied from 0 to 2.44 in the study area (Table 3). The elevation of the study area ranges from 61 to 3616 m. FR analysis suggests that areas with moderate elevation (996–1577 m) and very steep slope (35.91–84.01 degrees) have very close relation (FR-1.9 and FR-2.06 respectively) with landslide susceptibility. The south-east facing slopes (FR-1.32) are highly susceptible to landslides. Generally, the south facing slopes in northern hemisphere receives most sunlight and rainfall, are prone to landslides. Likewise slopes with Curvature (−19.79 to −1.86), Plan curvature (−42.5 to −1.47) and Profile curvature (−21 to −0.9) are most susceptible to landslides with FR values of 1.72, 2.44 and 1.61 respectively. The analysis doesn't show direct relationship of higher rainfall with higher landslide susceptibility which is a popular opinion. Rather moderate rainfall zone (3028.4–3220.64 mm) shows highest susceptibility to landslide (FR-1.69). Similarly, areas moderately far from the lineaments (2.27–3.8 kms) are highly susceptible to landslide (FR-1.39). Geomorphologically, moderately dissected hills and valleys of the districts show very high susceptibility to the landslides (FR-1.79) whereas highly dissected hills and valleys are moderately susceptible (FR-1) and the alluvial plains are the least susceptible (FR-0.01). Contrary to the popular opinion that distance from rivers and landslide susceptibility are highly correlated, FR analysis in fact shows that the areas farthest from the rivers

(1.74–3.5 kms) are the most landslide susceptible with highest FR value of 1.75. Least TWI zones (1.12–5.43) have scored the highest FR value of 1.48. Areas closest to roads (<1.1 kms) are very highly susceptible to landslides as suggested by FR-1.42. The model suggests that Fallow land is highly susceptible to landslides (FR-1.7). Whereas water bodies and built-up areas are also significantly susceptible to landslides with FR values of 1.54 and 1.48 respectively. Seismologically the study area was grouped into two classes Moderate (3–4) and high (4–5) intensity. FL analysis depicts that moderate intensity areas are much prone to landslides (FL-1.1) whereas high intensity areas are also significantly susceptible with FL value of 0.83. The NDVI map of the study area has been classified into five classes such as very low (−0.2–0.12), low (−0.12–0.19), medium (0.19–0.25), high (0.25– 0.31) and very high (0.31–0.53). According to FL model, the low sub-class (−0.12–0.19) of NDVI is very highly susceptible to landslides with FL-1.33.

From the FR model, we can conclude that moderately dissected hill and valleys with moderate elevation, steep south facing slopes, negative curvature, moderate rainfall, moderate seismic intensity and areas moderately far from lineaments are highly landslide susceptible. There is no direct correlation between amount of rainfall and TWI with landslide susceptibility. Moreover, the areas closest to roads, fallow, settlement with low NDVI are also prone to landslides. Hence, the landslide events in the districts are consequence of coupled impact of both natural hydro-geomorphic setup and anthropogenic alterations at play (Fig. 6a).

Finally, FR of each sub-class was combined together using raster calculator tool of GIS to prepare the landslide susceptibility map. The FR in the landslide susceptibility model varies from 7.65 to 23.48. The greater FR value indicates a higher probability

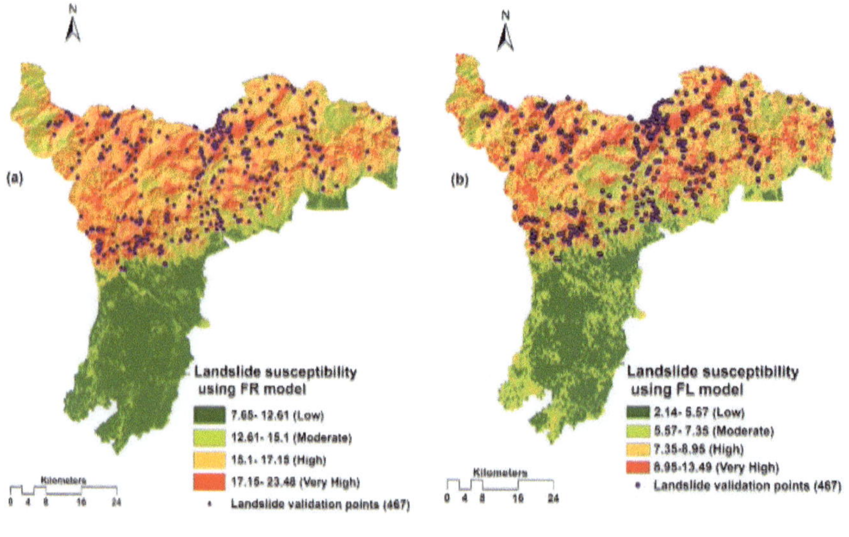

Fig. 6 Landslide susceptibility map of Darjiling Himalaya based on **a** frequency ratio (FR), and **b** fuzzy logic (FL) models

Table 4 Areal distribution of FR and FL model landslide susceptibility maps (LSMs)

Landslide susceptibility classes	Frequency Ratio (FR)		Fuzzy Logic (FL)	
	Area in Sq.km	Area (in %)	Area in Sq.km	Area (in %)
Low	818	26.2	608	19.5
Moderate	483	15.5	735	23.6
High	1086	34.8	1123	36
Very high	731	23.5	652	20.9

of landslide. The developed database was classified into four different susceptible zones: low (7.65–12.61), moderate (12.61– 15.1), high (15.1–17.15) and very high (17.15–23.48) in the GIS environment using the natural break method (Fig. 6a). In the Frequency Ratio (FR) landslide susceptibility map, the four landslide susceptibility classes of low, moderate, high, and very high covered 818 km^2 (26.2%), 483 km^2 (15.5%), 1086 km^2 (34.8%), and 731 km^2 (23.5%) area of the districts, respectively (Table 4 and Fig. 8).

4.2 *Landslide Susceptibility Mapping Using FL Model*

Fuzzy logic uses fuzzy membership functions, where all raster layers of landslide conditioning factors were normalized between '0' and '1', where '0' indicates no flood susceptibility and '1' indicates significant flood susceptibility. Thus, the fuzzy membership value becomes the relative weights of each sub-unit of selected causative variables. Finally, gamma overlay operator allows to prepare landslide susceptible zone of the study area assembling all membership values. The landslide susceptibility map prepared by FL has been divided into four susceptibility classes: low (low lying alluvial plains of Siliguri sub-Division and parts of foothills), moderate (much of flood plains of Siliguri Sub-division and parts of foothills), high (areas above 500 m elevation in both moderately and highly dissected hills and valleys topography) and very high (areas above 500 m elevation in partly in moderately and heavily on highly dissected hills and valleys topography). Landslide susceptibility map prepared by fuzzy logic shows that maximum parts of moderately and highly dissected hills and valleys topography of the districts are highly susceptible to landslides whereas areas below 500 m elevation belongs to low landslide susceptibility zone (Fig. 6b).

The FL index values in the landslide susceptibility model varies from 2.14 to 13.49. The greater FL value indicates a higher probability of landslide. The developed database was classified into four different susceptible zones: low (2.14–5.57), moderate (5.57– 7.35), high (7.35–8.95) and very high (8.95–13.49) in the GIS environment using the natural break method (Fig. 6b).Meanwhile, in the Fuzzy Logic (FL) landslide susceptibility map, the classes of low, moderate, high, and very high landslide susceptibility covered 608 km^2 (19.5%), 735 km^2 (23.6%), 1123 km^2 (36%) and 652 km^2 (20.9%) of the area, respectively (Table 4 and Fig. 8).

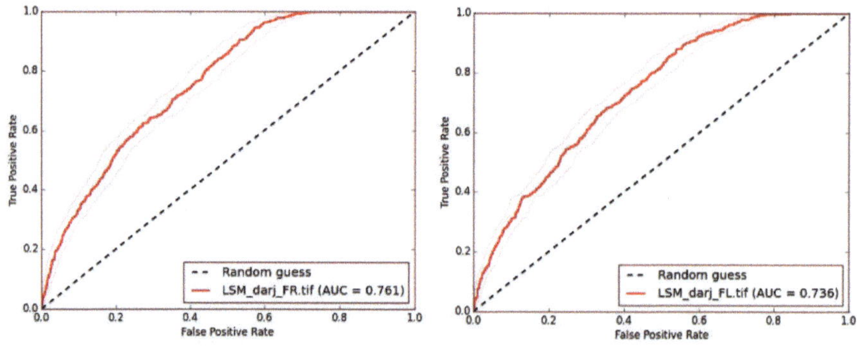

Fig. 7 Validation of LSMs using the ROC curve showing the area under curve (AUC)—**a** Frequency Ratio (FR) and **b** Fuzzy Logic (FL)

4.3 Validation of Landslide Susceptibility Maps

The results of AUC were used to validate the two Landslide Susceptibility Maps (LSMs). In the area under curve (AUC), values of the Frequency Ratio (FR) and Fuzzy Logic (FL) models are 76.1% and 74%, respectively, indicating that they are good models for the identification of landslide hazard zones (Fig. 7a and b). AUC results shows that the FR model gives slightly better performance than the FL model in Landslide susceptibility prediction.

Frequency analysis of landslide occurrences on susceptible zones suggests that majority of landslides were observed on very high and high zones. More specifically, in FR model, the highest frequency is in very high landslide susceptible zone (59.31%) followed by high landslide susceptible zone (35.55%) and moderate landslide susceptible zone (5.14%). Likewise in FL model too, the frequency is highest in very high landslide susceptible zone (46.25%) followed by high landslide susceptible zone (44.11%) and moderate landslide susceptible zone (9.42) (Table 5 and Fig. 9). When we compare the two models, the FR model has slightly better performance in terms of AUC = 76.1% and the frequency ratio of landslides in very high and high zones on landslide susceptibility map produced by the FR model is also the highest. In general, the results show that both the landslide susceptibility maps have good prediction accuracy; however, the map prepared from the FR model has better performance.

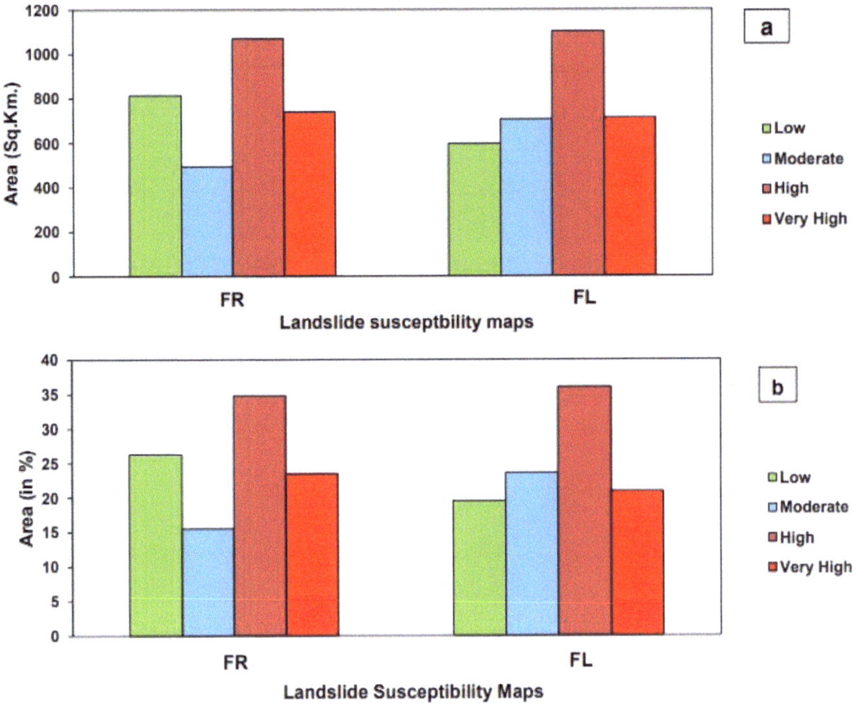

Fig. 8 Areal distributions of LSMs by—**a** area distribution of LSMs, **b** percentage of area distribution of LSMs

Table 5 The characteristics of susceptibility classes on LSMs

Class number	Reclassified index value	Landslide susceptibility class	Number of pixels	Area (in %)	Number of landslide validation pixels	Area of landslide validation covered (in %)
Frequency ratio model						
1	7.65–12.61	Low	905,011	26.1	0	0
2	12.61–15.10	Moderate	550,030	15.9	24	5.14
3	15.10–17.15	High	1,187,859	34.3	166	35.55
4	17.15–23.48	Very high	821,458	23.7	277	59.31
Fuzzy logic model						
1	2.14–5.57	Low	665,063	19.1	1	0.21
2	5.57–7.35	Moderate	829,974	22.6	44	9.42
3	7.35–8.95	High	1,229,934	35.4	206	44.11
4	8.95–13.49	Very high	739,387	22.9	216	46.25

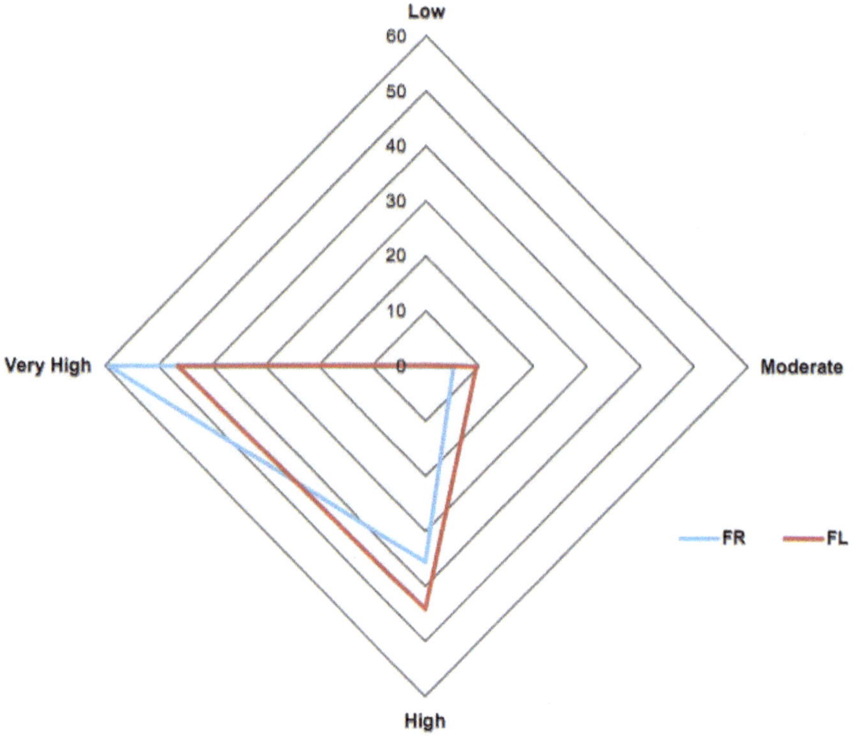

Fig. 9 Frequency analysis of landslides on the susceptibility maps

5 Conclusion

The Darjiling Himalaya is witnessing increasing pressure of both natural phenomenon and anthropogenic activities in the wake of climatic fluctuations and increasing population. Hence the frequency and severity of landsides are much likely to increase. Therefore, preparation of updated landslide susceptibility maps is needed. In this paper, landslide susceptibility maps (LSMs) of Darjiling Himalaya, West Bengal, India was achieved through Frequency Ratio (FR) and Fuzzy Logic (FL) models. A total of 1556 landslide locations were identified of which 70% were randomly used for data training and remaining 30% were used for data validation. Fifteen landslide causing parameters namely elevation, slope, rainfall, distance to river, distance to lineament, distance to road, aspect, geomorphology, curvature, plan curvature, profile curvature, seismic zone, LULC, TWI (Topographic Wetness Index) and NDVI (Normalized Difference Water Index) were used as independent variables in the modeling process and to prepare the spatial database. Both the LSMs suggests that almost entire study area except the Siliguri Sub-division of Darjiling District falls under moderate to very high landslide susceptible zone owing to prevalence of steep slopes, fragile rocks, high rainfall and highly dissected topography.

The very high landslide susceptibility zone falls in the middle and western portion of the hilly section (approximately above 500 m) of the districts. The susceptibility maps also indicate that the land sliding in the study area is controlled mainly by elevation, plan curvature, slope, slope aspect, geomorphology and land use–land-cover. More specifically, the areas closer to roads are very highly susceptible. In the entire study area, the areas closer to NH-10 which connects Gangtok and Kalimpong with Siliguri is most critical and falls under very highly vulnerable as it bears the brunt of being close to both NH-10 (very high traffic) and the river Teesta (mightiest river of the region). The ongoing Sevoke-Rangpo railway project has made this zone further vulnerable. Likewise, area closer to other roads such as NH-110, Darjiling-Sukhia pokhari road, Tista Bazar-Kalimpong Road, Rishi Road, Darjiling-Jorethang Road are also susceptible. Area-wise all the prominent hilly towns of the districts namely Darjiling, Kalimpong, Kurseong and Mirik falls under very high susceptibility zones. Anthropogenic activities such as high road density, slope cutting for house construction, changing LULC and local geomorphology coupled with other conditioning factors is resulting in very high susceptibility scores. Very high landslide susceptibility zones of the FR and FL models cover areas of 731 km^2 (23.4%) and 652 km^2 (20.9%), respectively. The validation results obtained from ROC (receiver operating characteristics) curve shows that landslide susceptibility map prepared by FR model (76.1%) is slightly more accurate in its prediction with the data at hand as compared to FL model (73.6%). The concerned agencies can use these produced landslide susceptibility map to create and implement landslip mitigation strategies in the research area.

Acknowledgements We would like to thank various agencies like USGS, GSI, IMD, Bhuvan, Esri and Save the Hills for various data used in this study. We convey our gratitude to Rippandi Lepcha, Assistant Professor, Department of English, Darjiling Govt. College, West Bengal for enhancing the readability and overall quality of the language. Authors also acknowledge Geography departments of Darjiling Government College, Barasat Government College, West Bengal, India and North-Eastern Hill University (NEHU), Shillong, Meghalaya, India for their support in making this book chapter.

Authors declaration We wish to confirm that there are no known conflicts of interest associated with this publication and there has been no financial support for this work that could have influenced its outcome.

Funding No funding was received for this work.

Availability of Data and Materials The datasets generated during and/or analyzed during the current study are available from the corresponding author on reasonable request.

References

Aleotti, P., & Chowdhury, R. (1999). Landslide hazard assessment: Summary, review and new perspectives. *Bulletin of Engineering Geology and the Environment, 58*, 21–44. https://doi.org/10.1007/s100640050066

Anbalagan, R. (1992). Landslide hazard evaluation and zonation mapping in mountainous terrain. *Engineering Geology, 32*, 269–277.

Anbalagan, R., Kumar, R., Lakshmanan, K., Parida, S., & Neethu, S. (2015). Landslide hazard zonation mapping using frequency ratio and fuzzy logic approach, a case study of Lachung Valley, Sikkim. *Geoenvironmental Disasters, 2*, 6. https://doi.org/10.1186/s40677-014-0009-y

Ayalew, L., Yamagishi, H., & Ugawa, N. (2004). Landslide susceptibility mapping using GIS-based weighted linear combination, the case in Tsugawa area of Agano River, Niigata Prefecture, Japan. *Landslides, 1*, 73–81.

Basu, S. R., & Dey, S. K. (2003). Causes and consequences of landslides in the Darjeeling-Sikkim Himalaya, India. *Geographia Polonica, 76*(2), 37–52.

Beguerıa, S. (2006). Validation and evaluation of predictive models in hazard assessment and risk management. *Natural Hazards, 37*(3), 315–329.

Beven, K. J., & Kirkby, M. J. (1979). A physically based, variable contributing area model of basin hydrology. *Hydrological Sciences, 24*, 43–69. https://doi.org/10.1080/02626667909491834

Bui, D. T., Tien-Chung, H., Pradhan, B., Pham, B. T., Nhu, V. A., & Revhaug, I. (2016). GIS-based modeling of rainfall-induced landslides using data mining-based functional trees classifier with AdaBoost, bagging, and MultiBoost ensemble frameworks. *Environmental Earth Sciences, 75*(14), 1–22. https://doi.org/10.1007/s12665-016-5919-4

Bui, D. T., Tuan, T. A., Hoang, N. T., Thanh, N. Q., Nguyen, D., Liem, N. V., & Pradhan, B. (2016). Spatial prediction of rainfall-induced landslides for the Lao Cai area (Vietnam) using a hybrid intelligent approach of least squares support vector machines inference model and artificial bee colony optimization. *Landslides*. https://doi.org/10.1007/s10346-016-0711-9.

Carson, M. A., & Kirkby, M. J. (1972). *Hillslope Form and Process* (p. 475). Cambridge University Press.

Chakraborty, I., Ghosh, S., Bhattacharya, D., & Bora, A. (2011). Earthquake induced landslides in the Sikkim-Darjeeling Himalayas—An aftermath of the 18th September 2011 Sikkim earthquake. *Geological Society of India.*

Champatiray, P. K., Dimri, S., Lakhera, R. C., & Sati, S. K. (2007). Fuzzy-based method for landslide hazard assessment in active seismic zone of Himalaya. *Landslides, 4*, 101–111.

Chang, K. T., Merghadi, A., Yunus, A. P., Pham, T., & Dou, J. (2019). Evaluating scale effects of topographic variables in landslide susceptibility models using GIS-based machine learning techniques. *Science and Reports, 9*, 1–21.

Chauhan, S., Sharma, M., & Arora, M. K. (2010). Landslide susceptibility zonation of the Chamoli region, Garhwal Himalayas, using logistic regression model. *Landslides, 7*, 411–423. https://doi.org/10.1007/s10346-010-0202-3

Chawla, A., Chawla, S., Pasupuleti, S., Rao, A. C. S., Sarkar, K., & Dwivedi, R. (2018). Landslide susceptibility mapping in Darjeeling Himalayas, India. *Advances in Civil Engineering, 1*, 6416492. https://doi.org/10.1155/2018/6416492

Chawla, A., Chawla, S., Pasupuleti, S., Rao, A. C. S., Sarkar, K., & Dwivedi, R. (2019). Landslide susceptibility zonation mapping: A case study from Darjeeling District, Eastern Himalayas, India. *Journal of the Indian Society of Remote Sensing*. https://doi.org/10.1007/s12524-018-0916-6.

Chen, W., Pourghasemi, H. R., & Naghibi, S. A. (2017). Prioritization of landslide conditioning factors and its spatial modeling in Shangnan County, China using GIS-based data mining algorithms. *Bulletin of Engineering Geology and the Environment., 77*, 611–629. https://doi.org/10.1007/s10064-017-1004-9

Chi, K. H. , Park, N. W., & Lee, K. (2002). Identification of landslide area using remote sensing data and quantitative assessment of landslide hazard. In *Proceedings of IEEE International Geosciences and Remote Sensing Symposium*, 19 July 2002, Toronto, Canada.

Choi, S. W., Moon, W. M., & Choi, S. G. (2000). Fuzzy logic fusion of W-Mo exploration data from Seobyeog-ri, Korea. *Geosciences Journal, 4*(2), 43–52.

Chung, C. F., & Fabbiri, A. G. (2001). Prediction models for landslide hazard zonation using a fuzzy set approach. In M. Marchetti & V. Rivas (Eds.), *Geomorphology & Environmental Impact Assessment* (pp. 31–47). Balkema Publishers.

Chung, C. J. F., & Fabbri, A. G. (2003). Validation of spatial prediction models for landslide hazard mapping. *Natural Hazards, 30*(3), 451–472. https://doi.org/10.1023/b:nhaz.0000007172.62651.2b

Cruden, D. M. (1991). A simple definition of a landslide. *Bulletin of the International Association of Engineering Geology, 43*, 27–29.

Das, I., Sahoo, S., van Westen, C., Stein, A., & Hack, R. (2010). Landslide susceptibility assessment using logistic regression and its comparison with a rock mass classification system, along a road section in the northern Himalayas (India). *Geomorphology, 114*, 627–637.

Devkota, K. C., Regmi, A. D., Pourghasemi, H. R., Yoshida, K., Pradhan, B., Ryu, I. C., Dhital, M. R., & Althuwaynee, O. (2013). Landslide susceptibility mapping using certainty factor, index of entropy and logistic regression models in GIS and their comparison at Mugling-Narayanghat road section in Nepal Himalaya. *Natural Hazards, 65*, 135–165.

Dikshit, A., Sarkar, R., Pradhan, B., Jena, R., Drukpa, D., & Alamri, A. M. (2020). Temporal probability assessment and its use in landslide susceptibility mapping for Eastern Bhutan. *Water, 12*, 267. https://doi.org/10.3390/w12010267

Froude, M. J., & Petley, D. N. (2018). Global fatal landslide occurrence from 2004 to 2016. *Natural Hazards and Earth Systems Sciences, 18*, 2161–2181. https://doi.org/10.5194/nhess-18-2161-2018

Gansser, A. (1964). *Geology of the Himalayas*. Wiley Interscience.

Ghosh, S., Carranza, E. J. M., Westen, C. J. V., Jetten, V. G., & Bhattacharya, D. N. (2011). Selecting and weighting spatial predictors for empirical modeling of landslide susceptibility in the Darjeeling Himalayas (India). *Geomorphology, 131*(2011), 35–56. https://doi.org/10.1016/j.geomorph.2011.04.019

Ghosh, A., Dey, P., & Ghosh, T. (2022). Integration of RS-GIS with frequency ratio, fuzzy logic, logistic regression and decision tree models for flood susceptibility prediction in lower Gangetic Plain: A study on Malda District of West Bengal, India. https://doi.org/10.1007/s12524-022-01560-5.

Ghosh, S., Van Westen, C. J., Carranza, E. J. M., Ghoshal, T. B., Sarkar, N. K., & Surendranath, M. (2009). A quantitative approach for improving the bis (indian) method of medium-scale landslide susceptibility. *Journal Geological Society of India, 74*, 625–638.

Gupta, R. P. (2003). *Remote Sensing Geology* (2nd ed.). Springer.

Gupta, R. P., & Joshi, B. C. (1990). Landslide hazard zonation using the GIS approach—A case study from the Ramganga Catchment, Himalayas. *Engineering Geology, 28*, 119–131.

Gupta, R. P., Saha, A. K., Arora, M. K., & Kumar, A. (1999). Landslide Hazard Zonation in a part of Bhagirathy Valley, Garhwal Himalayas, using integrated Remote Sensing—GIS. *Journal of Himalayan Geology, 20*(2), 71–85.

Gupta, V., Sah, M. P., Virdi, N. S., & Bartarya, S. K. (1993). Landslide hazard zonation in the Upper Satlej Valley, District Kinnaur, Himachal Pradesh. *Journal of Himalayan Geology, 4*, 81–93.

Hong, H., Liu, J., Zhu, A. X., Shahabi, H., Pham, T., Chen, W., Pradhan, B., & Bui, D. T. (2017). A novel hybrid integration model using support vector machines and random subspace for weather-triggered landslide susceptibility assessment in the Wuning area (China). *Environment and Earth Science, 76*, 652.

Hong, H., Pradhan, B., Xu, C., & Bui, D. T. (2015). Spatial prediction of landslide hazard at the Yihuang area (China) using two-class kernel logistic regression, alternating decision tree and support vector machines. *CATENA, 133*, 266–281.

Juang, C. H., Lee, D. H., & Sheu, C. (1992). Mapping slope failure potential using fuzzy sets. *Journal of Geotechnical Engineering., 118*(3), 475–494. https://doi.org/10.1061/(ASCE)0733-9410(1992)118:3(475)

Kanungo, D. P., Arora, M. K., Sarkar, S., & Gupta, R. P. (2006). A comparative study of conventional, ANN black box, fuzzy and combined neural and fuzzy weighting procedures for landslide susceptibility zonation in Darjeeling Himalayas. *Engineering Geology, 85*, 347–366. https://doi.org/10.1016/j.enggeo.2006.03.004

Kanungo, D. P., Arora, M. K., Sarkar, S., & Gupta, R. P. (2009). Landslide Susceptibility Zonation (LSZ) mapping—A review. *J South Asia Dis Stu, 2*, 81–105.

Kanwal, S., Atif, S., & Shafiq, M. (2016). GIS based landslide susceptibility mapping of northern areas of Pakistan, a case study of shigar and shyok basins. *Geomatics, Natural Hazards and Risk, 8*, 348–366. https://doi.org/10.1080/19475705.2016.1220023

Kayastha, P., Bijukchhen, S. M., Dhital, M. R., & Smedt, F. D. (2013). GIS based landslide susceptibility mapping using a fuzzy logic approach: A case study from Ghurmi-Dhad Khola area, Eastern Nepal. *Journal of Geological Society of India, 82*, 249–261. https://doi.org/10.1007/s12594-013-0147-y

Khan, H., Shafique, M., Khan, M. A., Bacha, M. A., Shah, S. U., & Calligaris, C. (2018). Landslide susceptibility assessment using Frequency Ratio, a case study of northern Pakistan. *The Egyptian Journal of Remote Sensing and Space Sciences, 22*, 11–24. https://doi.org/10.1016/j.ejrs.2018.03.004

Lee, S. (2007). Application and verification of fuzzy algebraic operators to landslide susceptibility mapping. *Environmental Geology, 50*, 847–855. https://doi.org/10.1007/s00254-006-0491-y

Lee, S., & Pradhan, B. (2006). Probabilistic landslide hazards and risk mapping on Penang Island, Malaysia. *Journal of Earth System Science, 115*(6), 661–672.

Lee, S., & Pradhan, B. (2007). Landslide hazard mapping at Selangor, Malaysia using frequency ratio and logistic regression models. *Landslides, 4*, 33–41. https://doi.org/10.1007/s10346-006-0047y

Mcbratney, A., & Odeh, I. O. (1997). Application of fuzzy sets in soil science: Fuzzy logic, fuzzy measurements and fuzzy decisions. *Geoderma, 77*(2–4), 85–113. https://doi.org/10.1016/s0016-7061(97)00017-7

Mehrotra, G. S., Sarkar, S., Kanungo, D. P., & Mahadevaiah, K. (1996). Terrain analysis and spatial assessment of landslide hazard in parts of Sikkim Himalaya. *Journal of the Geological Society of India, 47*, 491–498.

Mondal, S., & Maiti, R. (2012). Landslide susceptibility analysis of Shiv-Khola Watershed, Darjiling: A remote sensing & GIS based Analytical Hierarchy Process (AHP). *Journal of the Indian Society of Remote Sensing, 40*(3), 483–496. https://doi.org/10.1007/s12524-011-0160-9

Nagarajan, R., Mukherjee, A., Roy, A., & Khire, M. V. (1998). Temporal remote sensing data and GIS application in landslide hazard zonation of part of Western Ghat, India. *International Journal of Remote Sensing, 19*, 573–585.

Onagh, M., Kumra, V., & Rai, P.K. (2012). Landslide susceptibility mapping in a part of Uttarkashi district (India) by multiple linear regression method. *International Journal of Geology, Earth and Environmental Sciences*. ISSN: 2277–2081.

Pachauri, A. K., & Pant, M. (1992). Landslide hazard mapping based on geological attributes. *Engineering Geology, 32*, 81–100.

Panikkar, S. V., & Subramanyan, V. (1996). A geomorphic evaluation of landslides around Dehradun and Mussoorie, Uttar Pradesh, India. *Geomorphology, 15*, 169–181. https://doi.org/10.1016/0169-555X(95)00121-K

Petley, D. (2012). Global patterns of loss of life from landslides. *Geology, 40*, 927–930.

Pham, B. T., Bui, D. T., Prakash, I., & Dholakia, M. B. (2017). Hybrid integration of multilayer perceptron neural networks and machine learning ensembles for landslide susceptibility assessment at Himalayan area (India) using GIS. *CATENA, 149*, 52–63.

Pham, B. T., Bui, D. T., & Prakash, I. (2019). Landslide susceptibility modelling using different advanced decision trees methods. *Civil Engineering and Environmental Systems*. https://doi.org/10.1080/10286608.2019.1568418

Poudyal, C. P., Chang, C., Oh, H. J., & Lee, S. (2010). Landslide susceptibility maps comparing frequency ratio and artificial neural networks: A case study from the Nepal Himalaya. *Environment and Earth Science, 61*, 1049–1064.

Pourghasemi, H. R., Pradhan, B., & Gokceoglu, C. (2012). Application of fuzzy logic and analytical hierarchy process (AHP) to landslide susceptibility mapping at Haraz watershed, Iran. *Natural Hazards, 63*, 965–996. https://doi.org/10.1007/s11069-012-0217-2

Pradhan, B. (2010). Landslide Susceptibility mapping of a catchment area using frequency ratio, fuzzy logic and multivariate logistic regression approaches. *Journal of the Indian Society of Remote Sensing, 38*, 301–320.

Pramanik, M. K. (2016). Site suitability analysis for agricultural land use of Darjeeling district using AHP and GIS techniques. *Modeling Earth Systems and Environment, 2*, 1–22.

Ram, P., Gupta, V., Devi, M., & Vishwakarma, N. (2020). Landslide susceptibility mapping using bivariate statistical method for the hilly township of Mussoorie and its surrounding areas, Uttarakhand Himalaya. *Journal of Earth System Science, 129*, 167. https://doi.org/10.1007/s12040-020-01428-7

Rasyid, A. R., Bhandary, N. P., & Yatabe, R. (2016). Performance of frequency ratio and logistic regression model in creating GIS based landslides susceptibility map at Lompobattang Mountain, Indonesia. *Geoenvironmental Disasters, 3*, 19. https://doi.org/10.1186/s40677-016-0053-x

Roy, J., & Saha, S. (2019). Landslide susceptibility mapping using knowledge driven statistical models in Darjeeling District, West Bengal, India. *Geoenvironmental Disaster, 6*, 11. https://doi.org/10.1186/s40677-019-0126-8

Saha, A. K., Gupta, R. P., & Arora, M. K. (2002). GIS-based landslide hazard zonation in the Bhagirathi (Ganga) Valley. Himalayas. *International Journal of Remote Sensing, 23*, 357–369. https://doi.org/10.1080/01431160010014260

Saha, A. K., Gupta, R. P., Sarkar, I., Arora, M., & Csaplovics, E. (2005). An approach for GIS-based statistical landslide susceptibility zonation—With a case study in the Himalayas. *Landslides, 2*, 61–69. https://doi.org/10.1007/s10346-004-0039-8

Sahana, M., & Sajjad, H. (2017). Evaluating effectiveness of frequency ratio, fuzzy logic and logistic regression models in assessing landslide susceptibility: A case from Rudraprayag district, India. *Journal of Mountain Science, 14*(11), 2150–2167. https://doi.org/10.1007/s11629-017-4404-1

Sarkar, S. (1999). Landslides in Darjiling Himalayas. *Transactions, Japanese Geomorphological Union, 20*(3), 299–315.

Sarkar, S., & Kanungo, D. P. (2004). An integrated approach for landslide susceptibility mapping using remote sensing and GIS. *Photogrammetric Engineering and Remote Sensing, 70*(5), 617–625.

Sarkar, S., Kanungo, D. P., & Mehrotra, G. S. (1995). Landslide hazard zonation: A case study in Garhwal Himalaya, India. *Mountain Research and Development, 15*(4), 301–309.

Sarkar, S., Roy, A. K., & Tapas, R. M. (2013). Landslide susceptibility assessment using information value method in parts of the Darjeeling Himalayas. *Journal Geological Society of India, 82*, 351–362.

Sinha, B. N., Varma, R. S., & Paul, D. K. (1975). Landslides in Darjeeling district (West Bengal) and adjacent areas. *Bulletins of the Geological Survey of India, Series A, 36*, 1–45.

Sonker, I., Tripathi, J. N., & Singh, A. K. (2021). Landslide susceptibility zonation using geospatial technique and analytical hierarchy process in Sikkim Himalaya. *Quaternary Science Advances, 4*, 100039. https://doi.org/10.1016/j.qsa.2021.100039

Starkel, L., & Basu, S. R. (2000). *Rains, Landslides and Floods in the Darjiling Himalayas*. Indian Science Academy.

Van Westen, C. J. (1994). GIS in landslide hazard zonation: A review, with examples from the Andes of Colombia. In M. Price & I. Heywood (Eds.), *Mountain Environments and Geographic Information System* (pp. 135–165). Taylor & Francis.

Varnes, D. J. (1984). *Landslide Hazard Zonation: A Review of Principles and Practice* (pp. 1–63). UNESCO.

Verma, A. K., Singh, T. N., Chauhan, N. K., & Sarkar, K. (2016). A hybrid FEM–ANN approach for slope instability prediction. *Journal of the Institution of Engineers India Series A, 97*(3), 171–180.

Wilson, J. P., & Gallant, J. C. (2000). Digital terrain analysis. In J. P. Wilson & J. C. Gallant (Eds.), *Terrain Analysis: Principles and Applications* (pp. 1–27). John Wiley & Sons.
Yalcin, A. (2011). A geotechnical study on the landslides in the Trabzon Province, NE, Turkey. *Applied Clay Science, 52*(1–2), 11–19. https://doi.org/10.1016/j.clay.2011.01.015
Zadeh, L. A. (1965). Fuzzy sets. *Information and Control, 8*, 253–338.

Mapping Landslide Potential and Assessing Susceptibility in the Darjiling Himalaya Using GIS and Bivariate Statistics

Arnab Ghosh and Pankaj Chakraborty

Abstract The northern West Bengali districts of Darjiling and Kalimpong are located in the Darjiling Himalayas, which have been a landslide hotspot for centuries. The landslide risk is rising here due to unstable geological conditions, high rainfall, and the unskilled urbanization of the hill slopes near the fault. This article uses remote sensing, GIS, and bivariate statistical analysis (FR technique) to investigate the potential for future landslide disasters in the Darjiling Himalayas by identifying locations based on landslide susceptibility blocks. We used Arc GIS software to construct twelve factor and thematic layers to analyse remotely sensed and ground data. The all-thematic layer weights were estimated using the FR model to assess the region's landslide vulnerability. A landslide susceptibility map was created by combining the weights of all factor and theme layers using ArcGIS software. The study area is divided into five groups based on landslide susceptibility. The most vulnerable are Kalimpong II, Kurseong, Garubathan, Kalimpong I, and Darjiling-Pulbazar blocks in the Darjiling and Kalimpong districts. We evaluated the research field's ROC curve technique with a 95% level of correctness. Locating areas likely to experience landslides requires a more comprehensive approach, which this paper explores.

Keywords Darjiling Himalaya · Landslide susceptibility mapping · Remote sensing · Frequency ratio · AUROC curve

1 Introduction

Landslides are regarded as one of the most visible and destructive natural disasters in hilly locations. Although there are different geological natural factors, growing human development on unstable hillsides is now one of the primary causes of landslides (Basu & Pal, 2019; Kumar et al., 2018; Wu et al., 2016). Due to gravity, rocks

A. Ghosh (✉) · P. Chakraborty
21st Century GIS Academy, Kolkata, West Bengal, India
e-mail: drarnab.swre@gmail.com

© The Author(s), under exclusive license to Springer Nature Switzerland AG 2025

S. Ghosh (ed.), *The Himalaya Dilemma*, Springer Natural Hazards,
https://doi.org/10.1007/978-3-031-95083-4_11

and soil fall down the mountain slopes, causing financial, social and socio-economic damage to the local people, including loss of life (Aleotti & Chowdhury, 1999; Malamud et al., 2004). Not only are lives lost, but communication systems in mountainous areas are entirely damaged by landslides and cut off from the rest of the world, which imposes a financial burden on the government (Negi et al., 2013; Parkash, 2015). Unscientific urban expansion, uncontrolled felling of trees and excessive rainfall due to extreme weather conditions are increasing the number of landslides on the unstable slopes of the Darjiling Himalayas in the Eastern Himalayas. Darjiling district experienced significant landslides in 1993, 2009 and 2011 and Mirik, Kalimpong and Kurseong in 2015 (Panikkar & Subramanyan, 1996; Roy & Saha, 2019). Currently, in 2024, a total of 83 landslides occurred in the Jorebungalow-Sukhiapokhri block, and 57 houses collapsed, causing loss of life and property.

Researchers can prevent the harmful occurrence of landslides through various engineering methods, but first of all, it is necessary to identify landslide-prone areas. Landslide susceptibility mapping is needed to determine the actual landslide-prone areas. This sensitivity mapping has evolved into GIS using sophisticated mathematical methods that extract data from field surveys (Chung & Fabbri, 2008; Cross, 1998; Erener & Düzgün, 2010; Fernández et al., 2003; Gokceoglu & Sezer, 2009; Pradhan et al., 2010a, 2010b). However, by identifying landslide-sensitive zones in different countries of the world through the use of GIS and susceptibility mapping, various measures are being taken to control landslides S (Chen et al., 2019; Duc, 2013; Thanh & De Smedt, 2014; Westen & Terlien, 1996; Zhao & Chen, 2020; Mandal & Sarkar, 2021; Naito & Cairns, 2011). When assessing a region's landslide risk through GIS, current methodologies include physically based approaches, statistical methods, and advanced machine learning methods. A physical-based method or AHP is mainly feasible to apply successfully in a small area. In that case, it is possible to investigate the susceptibility of landslides through slope failure through a perfect and accurate geological analysis of that region (Gökceoglu & Aksoy, 1996; Westen & Terlien, 1996). It is possible to measure the sensitivity of landslides significantly through statistical methods and various bivariate and multivariate statistical methods. In this case, through FR (Aditian et al., 2018; Ding et al., 2017; Sonker et al., 2022), IV, CF, LR (Pourghasemi & Kerle, 2016) and EBF (Althuwaynee et al., 2014) methods, it is possible to establish the relationship between the landslides that occurred in the past and the existing factors for the current landslides. However, more accurate and sophisticated machine learning methods than these two methods are currently used to measure the sensitivity of landslides. ANN (Wang et al., 2016), NB (Lei et al., 2020), SVM (Ada & Sun, 2018), RF (Xu et al., 2024) and fuzzy logic (Wang et al., 2016) models eliminate the shortcomings of bivariate statistical models and calculate the sensitivity. Machine learning models are widely applied in susceptibility modelling due to their ability to handle complex, nonlinear relationships and large datasets, enabling accurate predictions and identification of high-risk areas. These models are particularly advantageous in geospatial and environmental applications, where they can analyse diverse factors such as terrain, climate, and environment. However, their reliance on high-quality, extensive training data can be a limitation, as data scarcity or bias may compromise results. Additionally, the interpretability of machine learning models in susceptibility studies is often limited, making it challenging to fully understand the driving factors behind predictions specially for qualitative data. Overfitting issues can further reduce reliability, especially when applied to new or unseen regions. Despite these challenges, machine learning models provide valuable tools for susceptibility analysis when properly validated and combined with domain knowledge.

The Darjiling Himalaya is prone to fatal landslide events. According to the available records (Griesbach, 1899–1900), a catastrophic landslide that only killed 72 people in Darjiling town and caused extensive property damage happened on September 24, 1899, after an unusually heavy downpour (Bandopadhyay, 1980; Basu & Sarkar, 1985, 1987; Ghosh, 1950; Nautiyal, 1951; Sarkar, 1995). For five days in 1950 (June 11 and 12), the Darjiling town area, Kalimpong, Mahanadi Paglajhora, Tindharia, Takdah, and hill regions are isolated from the rest of Bengal. Thousands of people were left homeless and 127 people lost their lives (Table 1). The areas of Rimbik, Lodhama, Bijanbari, Darjiling, Sukhiapokri, Manebhanjan, and Sonada Tindharia were once more severely impacted by landslides on September 3 and 4, 1980. There were more than 215 fatalities and 100 million rupees worth of property damage. There were several landslides on July 13, 1993. A total of fifteen people lost their lives in the Mongpoo area, Peshok, Pangkhabari, Mahanadi, and Gayabari. In 2009, a sequence of landslides struck the Kalimpong and Darjiling-Kurseong Road areas between May 26 and August 16. There were 500 houses destroyed, 37 people killed, and a great deal of property and animal damage. The hills were cut off for a long time. The USGS reports that the focal depth of the September 2011 earthquake was 19 kms. This destructive earthquake triggered 421 landslides of various sizes in Sikkim Himalayan region. These slides damaged roads and bridges and disrupted relief operations to towns and villages that were completely cut off, some for over three weeks. A series of geo-environmental factors were employed to evaluate landslide susceptibility, which was subsequently transformed into a hazard map by integrating it with temporal and magnitude data from landslide occurrences (Sarkar et al., 2010). The findings indicate that with a temporal probability of a 10-year return period, 2306 individuals residing in 691 structures across a 2.07 km^2 area are located within a high-risk zone. In contrast, 6798 individuals in 37,266 buildings (spanning 6.02 km^2) are situated in a moderate-risk zone. Moreover, the analysis reveals that a 5.44 km stretch of national highway and a 1.83 km stretch of state highway traverse areas highly prone to landslides. The risk assessment suggests that over the course of a year, approximately INR 4210 lakhs (equivalent to 65,88,650 USD) could be incurred due to road damage resulting from future landslide events.

While field surveys are essential for measuring landslide susceptibility, Himalayan hilly areas are inaccessible and risky to rely on GIS for data collection (Gupta et al., 2022; Zhao & Lu, 2018). However, with the advancement of technology and better satellite imagery, it has become possible to determine landslide susceptibility even in remote areas (Kannaujiya et al., 2019). Various natural and anthropogenic factors can highlight regional variation in landslide susceptibility in any remote area. By reviewing all the above issues seriously, the main objective of our present article is to map the landslide-prone areas of Darjiling Himalaya by the FR method by looking

Table 1 Current exceptional research works

Author	Title	Objective	Methodology	Finding
Dey et al. (2025)	Landslide susceptibility mapping using GIS	To identify landslide-prone zones using GIS and machine learning techniques	Machine learning models (random forest, SVM) integrated with GIS using parameters like slope, lithology, land use, and rainfall	Machine learning models provided high accuracy for identifying landslide-prone areas
Upadhay (2024)	Rainfall-induced landslides in Darjiling	Examine the impact of rainfall intensity on landslide triggering	Used rainfall intensity-duration (ID) thresholds and GIS spatial analysis	Intense rainfall events with specific thresholds strongly correlate with landslides in high-slope regions
Saha et al. (2023)	Climate change impacts on landslide occurrence	Assess the role of climate change on increasing landslide risks	Combined long-term climate data, GIS mapping, and landslide inventories	Climate change shifts rainfall patterns, increasing landslide frequency in vulnerable areas
Mondal et al. (2024)	Multi-hazard risk mapping	Develop a multi-hazard zonation map integrating landslide risk	Multi-criteria decision analysis (MCDA) combined with GIS	Produced comprehensive hazard maps, useful for disaster management and planning
Halder et al. (2024)	Early warning systems for landslides	Design GIS-based early warning systems integrating IoT sensors	Integrated GIS with IoT sensor data (rainfall and soil moisture)	Demonstrated effective real-time monitoring and reduced response times to landslide events
Goswami and Kundu (2024)	Remote sensing-based landslide detection	Utilize satellite data for landslide inventory mapping	High-resolution satellite images analyzed using GIS	Improved landslide detection accuracy using remote sensing and GIS

at the weight score of all the factors and to review the possibility of future landslides through block-based review. Since FR is a statistical method and it is possible to realize the weightage of each factor's sensitivity with the FR model's help, we have used the FR model here.

2 Study Area

The Darjiling Himalayan region, situated in the northernmost section of West Bengal, comprises the districts of Darjiling and Kalimpong (Fig. 1). The area of the entire region, including Darjiling and Kalimpong districts, is 3150 km^2. Among them, nine blocks of the Darjiling district and three blocks of the Kalimpong district are considered landslide-sensitive areas (Table 2). Although the northern and upper parts of Darjiling and Kalimpong districts are mountainous, the southern and lower parts are known as Landslide plains due to Teesta, Mahananda, Mechi and Balasan rivers. The region's altitude ranges from 15 to 3602 m above sea level, with slopes ranging from 0° to 80°. The entire region is divided into the Singalila range and the Darjiling-Kurseong range, and Sandakphu in the Singalila range is known as the highest peak in West Bengal. Geologically, the region is surrounded by the Triassic mantle, and the soil varies from region to region. The region's average annual rainfall is 3051 mm, and the maximum rainfall occurs during monsoon. Due to the lack of rainwater retention, most water flows down the slopes, creating landslides. Due to the tourism industry's popularity, most people in this region are attached to various hotels and homestays. In most cases, they are built unscientific on the slopes of the hills, which increases the possibility of human casualties due to landslides.

The primary investigations identified debris flow, rock and debris slides, topples, rockfalls, and rock slides as the primary types of rain-induced landslides. Darjiling experiences an average annual temperature of about 14.9 °C, with winter temperatures dropping as low as 1 °C due to the steep terrain. The lowest recorded temperature is 5 °C. The region's cold climate contributes to substantial annual rainfall, particularly from mid-April to late August, with an average annual precipitation of 2074.08 mm. Major towns within the area include Kurseong, Darjiling, Ghum, and Sonada. Heavy monsoon rains between June and August saturate the soil and rocks, often triggering slope failures and resulting in numerous landslides. These landslides significantly impact the livelihoods of residents, causing property damage, infrastructure destruction, casualties, and financial losses due to disrupted agriculture and trade, such as rotting produce caused by transport route closures. These challenges are compounded by factors like the area's geologic instability, rugged terrain, intense rainfall, deep weathering, high seismic activity, and human activities on unstable slopes, which together likely contribute to the frequent occurrence of landslides.

3 Materials and Methods

3.1 Database

Spatial mapping of landslide susceptibility requires understanding the causative factors for landslides. Sensitivity mapping is possible depending on the interplay of these factors. However, these factors differ in all regions and vary from region to

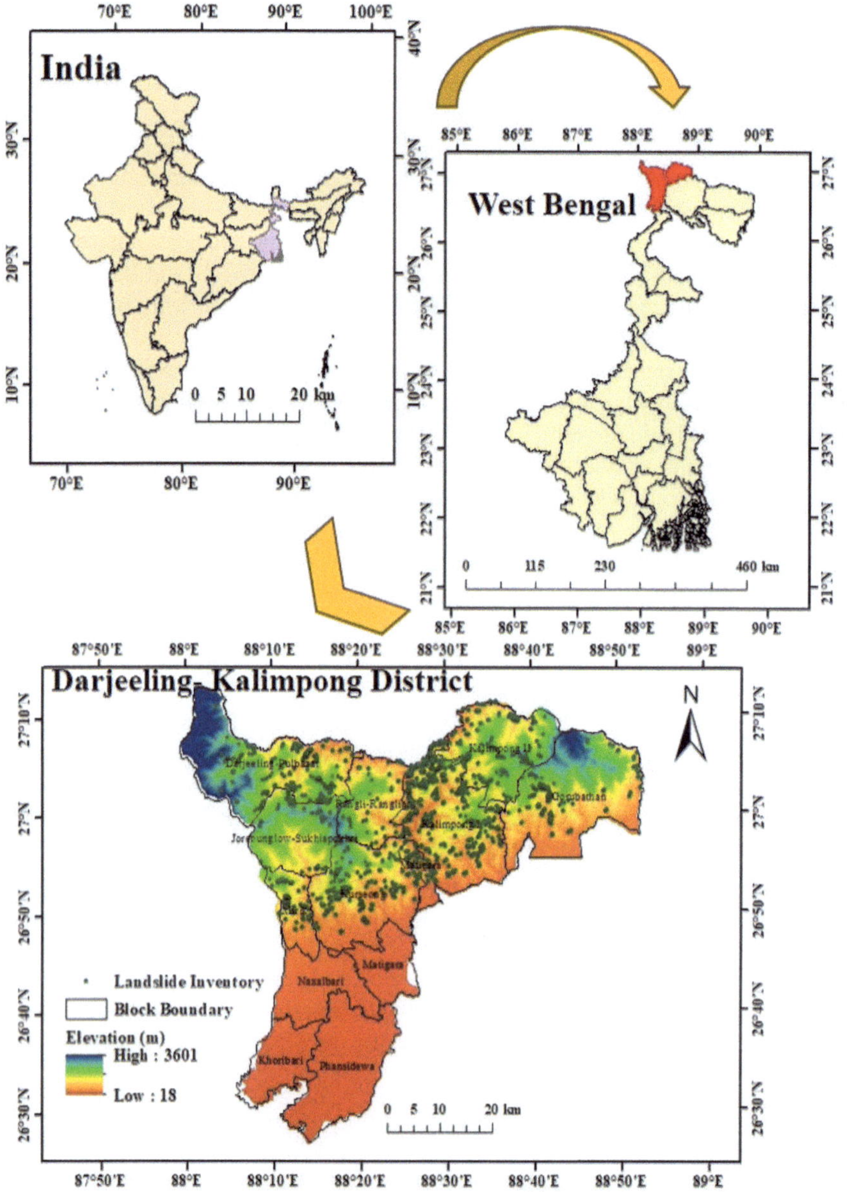

Fig. 1 Study area map with block boundary and landslide inventory points

Table 2 Block list of study area

Block_Name	District	Area (km^2)
Kalimpong II	Kalimpong	273.91
Gorubathan	Kalimpong	445.49
Kalimpong I	Kalimpong	372.33
Naxalbari	Darjiling	190.64
Khoribari	Darjiling	148.78
Darjiling-Pulbazar	Darjiling	415.06
Kurseong	Darjiling	334.18
Mirik	Darjiling	100.41
Matigara	Darjiling	168.69
Phansidewa	Darjiling	316.13
Jorebunglow-Sukhiapokhri	Darjiling	212.09
Rangli-Rangliot	Darjiling	172.92

region. For spatial mapping of landslide susceptibility of Darjiling Himalayas, we used rainfall as triggering factor, slope, elevation, aspect, distance from the road, TRI and geomorphology as topo-geomorphological factors, geology, soil, distance from the fault line, distance from lineament as lithological factors as a hydrological factor, and LULC as a human-induced factor. The FR approach evaluated all these elements using the Arc GIS 10.4.1 interface for landslide susceptibility mapping and was validated against landslide inventory data.

The Frequency Ratio (FR) model has been widely used in landslide susceptibility studies, including in the Darjiling Himalaya region, due to its simplicity, efficiency, and suitability for analyzing the relationship between landslide occurrences and contributing factors. This model quantifies the correlation between landslide-prone areas and specific causative factors such as slope, geology, rainfall, land use, and proximity to faults. The FR method calculates the likelihood of landslide occurrences by evaluating the spatial distribution of landslides against these factors, making it particularly effective in regions like Darjiling, where complex terrain and varied environmental conditions prevail.

The Darjiling Himalaya region is highly vulnerable to landslides due to its steep slopes, fragile geology, heavy monsoonal rainfall, and anthropogenic activities. The FR model is justified in this context as it provides a straightforward statistical approach to map landslide-prone zones without requiring extensive computational resources or advanced expertise. Moreover, it can integrate diverse geospatial datasets and deliver results that are interpretable and actionable for policymakers and local authorities. The model's ability to incorporate both continuous and categorical data ensures that the intricate interplay of natural and human-induced factors is adequately captured. Given the critical need for effective landslide mitigation in the Darjiling Himalaya, the FR model serves as a reliable and accessible tool for generating accurate susceptibility maps, thereby aiding in disaster preparedness and risk reduction strategies.

3.2 Landslide Inventory Map

Landslide inventory maps are used to measure the susceptibility and severity of landslides, where previous landslides are identified through some points. We have selected 800 landslide locations from the Geological Survey of India's Bhukosh site, where landslides have occurred before. Out of this, 560 (70%) data were used for training purposes, i.e., FR model construction and its validation, and the remaining 240 data (30%) were used by us for the model's authentication purposes (Fig. 1). The landslide and non-landslide points data has been placed in the Arc GIS interface with a raster format of 30 × 30 pixels spatial resolution. Landslide pixels were designated a value of 1, while non-landslide pixels were assigned a value of 0. We extracted the values of landslide-determining factors from these pixels and analyzed them using R Statistical Software (3.2.3). Landslide inventory maps revealed a greater density of landslide points in a specific section of the study area, clearly defining the area of interest.

3.3 Landslide Conditioning Parameters

Various geo-environmental and anthropogenic factors contribute to landslides in a given area (Tehrany et al., 2015). Effective mapping of landslide susceptibility in a region necessitates the consideration of relevant factors. Selection is essential, as pertinent factors in one area may not be appropriate for another region (Kia et al., 2012). Twelve factors have been identified in this article (Table 3).

3.3.1 Elevation

Elevation is critical in evaluating landslide susceptibility as it relates to the difference between a region's maximum and minimum altitudes. Elevation affects a region's geological features and vegetation characteristics. The elevation data was obtained from the ASTER-DEM 3 arc second void-filled DEM (https://search.earthdata.nasa.gov/search) and processed using ArcGIS raster tools for mosaicking. In contrast to floodplain regions, the elevation in this mountainous area is significantly higher, ranging from 18 to 3601 m (Fig. 3a).

3.3.2 Slope

The slope is a significant factor in the assessment of landslide hazards. The slope of a region influences surface runoff, infiltration processes, and various hydrological activities. The elevation contour illustrates the gradient of a specific area. This condition accelerates flood wave velocity and raises the probability of landslides.

Table 3 FR analysis with each factor

Factor	Range	No. of LSD	% of LSD	Pixels	% of pixels	FR	FRn
Elevation (m)	18–141	0	0.00	711,373	20.12	0.00	0.00
	141.01–509	271	17.34	707,165	20.00	0.87	0.49
	509.01–1,003	471	30.13	706,468	19.98	1.51	0.85
	1,003.01–1,585	556	35.57	706,393	19.98	1.78	1.00
	1,585.01–3,601	265	16.95	704,883	19.93	0.85	0.48
Slope (^{0})	0–9.24	57	3.65	1,115,790	31.55	0.12	0.05
	9.25–18.5	258	16.51	650,330	18.39	0.90	0.39
	18.6–26.9	471	30.13	801,184	22.66	1.33	0.57
	18.6–26.9	469	30.01	668,816	18.91	1.59	0.68
	36.3–69.3	308	19.71	300,161	8.49	2.32	1.00
Aspect	Flat (−1)	8	0.51	1845	0.05	9.81	1.00
	North (0–22.5)	51	3.26	212,158	6.00	0.54	0.06
	Northeast (22.5–67.5)	156	9.98	441,103	12.47	0.80	0.08
	East (67.5–112.5)	253	16.19	489,866	13.85	1.17	0.12
	Southeast (112.5–157.5)	293	18.75	497,948	14.08	1.33	0.14
	South (157.5–202.5)	220	14.08	468,450	13.25	1.06	0.11
	Southwest (202.5–247.5)	183	11.71	435,598	12.32	0.95	0.10
	West (247.5–292.5)	176	11.26	400,059	11.31	1.00	0.10
	Northwest (292.5–337.5)	167	10.68	398,984	11.28	0.95	0.10
	Northwest (292.5–337.5)	56	3.58	190,271	5.38	0.67	0.07
TRI	0.11–0.43	293	18.75	698,415	19.68	0.95	0.86
	0.44–0.48	323	20.67	727,646	20.51	1.01	0.91
	0.49–0.52	340	21.75	741,639	20.90	1.04	0.94
	0.53–0.57	331	21.18	681,678	19.21	1.10	1.00
	0.58–0.89	276	17.66	698,950	19.70	0.90	0.81
LULC	Water body	18	1.15	214,312	0.68	1.69	0.47
	Vegetation	1106	70.76	25,764,397	81.91	0.86	0.24
	Agriculture	237	15.16	3,328,060	10.58	1.43	0.40
	Built-up area	53	3.39	295,170	0.94	3.61	1.00
	Bare land	149	9.53	1,851,142	5.89	1.62	0.45
Geomorphology	Active flood plain	4	0.26	55,047	1.58	0.16	0.01

(continued)

Table 3 (continued)

Factor	Range	No. of LSD	% of LSD	Pixels	% of pixels	FR	FRn
	Highly dissected structural hills and valleys	1	0.06	618,630	17.76	0.00	0.00
	Mass wasting products	3	0.19	87,134	2.50	0.08	0.01
	Moderately dissected structural hills and valleys	0	0.00	295,960	8.50	0.00	0.00
	Older alluvial plain	2	0.13	996	0.03	4.48	0.38
	Older flood plain	31	1.98	62,539	1.80	1.10	0.09
	Piedmont alluvial plain	1023	65.45	1,328,417	38.13	1.72	0.15
	River	473	30.26	1,029,890	29.56	1.02	0.09
	Younger alluvial plain	26	1.66	4924	0.14	11.77	1.00
Geology	CHUNGTHANG Fm	3	0.19	58,000	1.66	0.12	0.05
	DALING Gp. (BUXA Fm.)	113	7.23	204,583	5.85	1.23	0.52
	DALING Gp. (REYANG Fm.)	24	1.54	1,042,140	29.82	0.05	0.02
	DALING Gp. (GORUBATHAN Fm.)	22	1.41	48,010	1.37	1.02	0.43
	DALING Gp. (REYANG Fm.)	82	5.25	196,693	5.63	0.93	0.39
	KANCHENJUNGA GNEISS (CENTRAL CRYSTALLINE Fm.)	1	0.06	7039	0.20	0.32	0.13
	LINGTSE GNEISS Gp	440	28.15	415,107	11.88	2.37	1.00
	LOWER GONDWANA Gp. (BHARELI, DAMUDA Fm.)	16	1.02	27,770	0.79	1.29	0.54
	SIWALIK Gp. (UPPER SIWALIK)	180	11.52	246,328	7.05	1.63	0.69
	UNDIFF.FLUVIAL/ AEOLIAN/COASTA & GLACIAL SEDIMENTS	682	43.63	1,248,747	35.74	1.22	0.52
Soil	Ah12-2bc	687	43.95	1,280,039	36.63	1.20	0.44

(continued)

Table 3 (continued)

Factor	Range	No. of LSD	% of LSD	Pixels	% of pixels	FR	FRn
	Bd32-2bc	437	27.96	894,164	25.59	1.09	0.40
	Bh10-2a	22	1.41	33,604	0.96	1.46	0.53
	I-Bh-U-c	0	0.00	43,071	1.23	0.00	0.00
	Je77-1/2a	0	0.00	852,896	24.41	0.00	0.00
	Rd28-1a	6	0.38	56,230	1.61	0.24	0.09
	Rd29-1a	411	26.30	334,713	9.58	2.75	1.00
Distance from road (m)	0–500	530	33.91	4866	16.65	2.04	1.00
	500–1000	400	25.59	7413	25.37	1.01	0.50
	1000–2000	206	13.18	5734	19.62	0.67	0.33
	2000–5000	218	13.95	5698	19.50	0.72	0.35
	>5000	209	13.37	5514	18.87	0.71	0.35
Distance from fault (m)	<1000	243	15.55	5064	17.33	0.90	0.65
	1000–2000	344	22.01	6494	22.22	0.99	0.72
	2000–3000	407	26.04	6159	21.07	1.24	0.90
	3000–5000	424	27.13	5783	19.79	1.37	1.00
	>5000	145	9.28	5725	19.59	0.47	0.35
Distance from lineament (m)	<500	500	31.99	5731	19.61	1.63	1.00
	500–1000	395	25.27	6088	20.83	1.21	0.74
	1000–2500	255	16.31	6188	21.17	0.77	0.47
	2500–5000	295	18.87	5660	19.37	0.97	0.60
	>5000	118	7.55	5558	19.02	0.40	0.24
Rainfall (mm)	2,041.76–2,290.27	123	7.87	169,816	4.80	1.64	0.37
	2,290.28–2,384.53	87	5.57	727,274	20.57	0.27	0.06
	2,384.54–2,475.94	354	22.65	1,641,299	46.41	0.49	0.11
	2,475.95–2,550.21	692	44.27	842,013	23.81	1.86	0.42
	2,550.22–2,770.15	307	19.64	155,878	4.41	4.46	1.00

Numerous researchers have noted a significant correlation between slope and landslide occurrences, as assessed through the Surface Analysis Tool in ArcGIS utilizing DEM. The slope in this area is defined by the angle, ranging from 0° to 69.3° (Fig. 3b).

3.3.3 Aspect

The slope aspect denotes the landslide direction, pertinent to topographic wetness influenced by solar exposure and precipitation. A slope aspect map was generated from the DEM and classified into ten categories. The directions included flat, north, northeast, east, southeast, south, southwest, west, northwest, and north (Fig. 3c).

3.3.4 TRI

TRI serves as a geomorphic indicator that characterizes the topographical features of the study area, indicating whether the terrain is flat or undulated and demonstrating the variability of the land surface. This method evaluates the elevation differences between neighbouring spatial pixels of the DEM data (Nakileza & Nedala, 2020). This study utilized DEM data to create the TRI map through ArcGIS software (Fig. 3d).

3.3.5 LULC

LULC is a region that affects that region's runoff, evaporation, and evapotranspiration (Mahmoud & Gan, 2018). The type and nature of land use of a particular area determine the flow dynamics of water and the effective application over landslide. Changes in the LULC overtime period play a key role in determining landslide events. The effect of landslide is inversely proportional to the areas covered by urban development, but in vegetative lands, the effect of landslide is much lower due to the presence of root cover (Bai et al., 2015). The LULC map for this research area was generated utilizing Sentinel 2 imagery with a spatial resolution of 10 m. Land use and land cover (LULC) in this region has been categorized into five distinct classes: water body, vegetation, agriculture, built-up area, and barren land (Fig. 3e). The region is predominantly mountainous with significant vegetation cover; however, urban development on sloped terrain heightens the risk of landslides.

3.3.6 Geomorphology

The geomorphologic map depicts surface landform features corresponding to subsurface geological formations, including lithology, which play a role in landslide occurrences. The region's notable geomorphic landscape comprises elevated, moderate, and low scattered hills and valleys. This research developed a geomorphology map utilizing the Landsat 8 OLI band and ArcGIS software (Fig. 3f).

3.3.7 Geology

Different geological formations have different lithological and structural features. The evaluation of landslide vulnerability relies heavily on this function. Therefore, the likelihood of landslides depends on factors such as the permeability of the rock, the type of rock, and the strength variations (Basu & Pal, 2019). Using the Bhukosh site, overseen by the Geological Survey of India, a geological map was created at a scale of 1:50,000. Various lithological formations, including intrusive rocks, schist, and gneisses, were included in a lithological map that was created using Arc GIS software using this map (Figs. 2 and 3g).

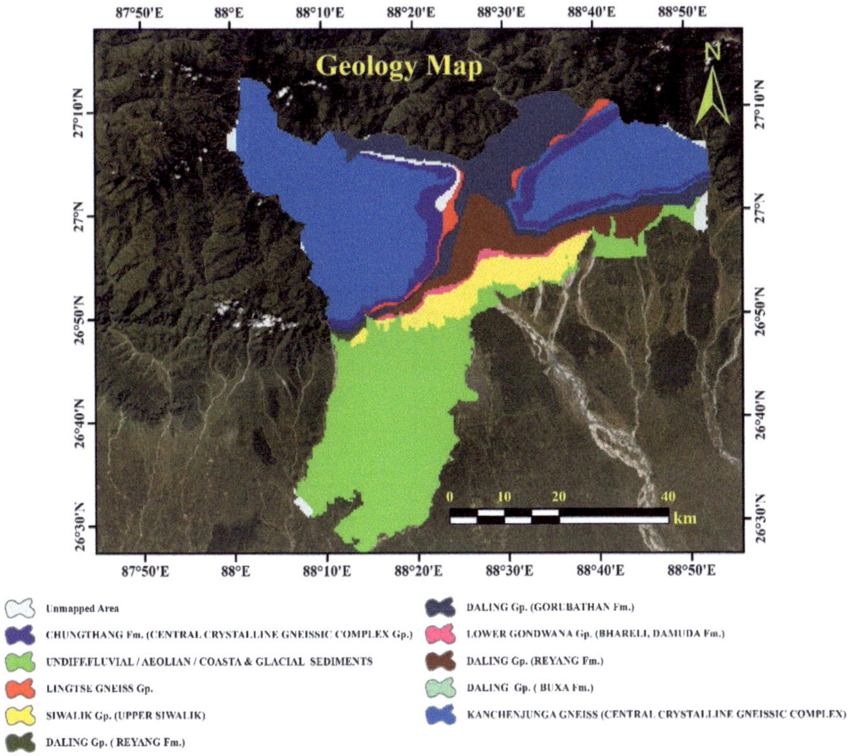

Fig. 2 Geological map of study area

3.3.8 Soil

Soils exhibit variability in their nature, composition, characteristics, and infiltration capacity, resulting in distinct differences in water-holding capacity. The volume of surface runoff is determined by the amount of water that surpasses the soil's water-holding capacity and subsequently moves through the soil layers. A reduction in infiltration rate increases surface runoff, consequently raising the probability of landslides (Fig. 3h). Excessive rainfall that surpasses the percolation rate in a specific area leads to the downslope movement of surplus water, which may cause a landslide. The soil in the study area has been sourced from the FAO soil map (https://www.fao.org/soils).

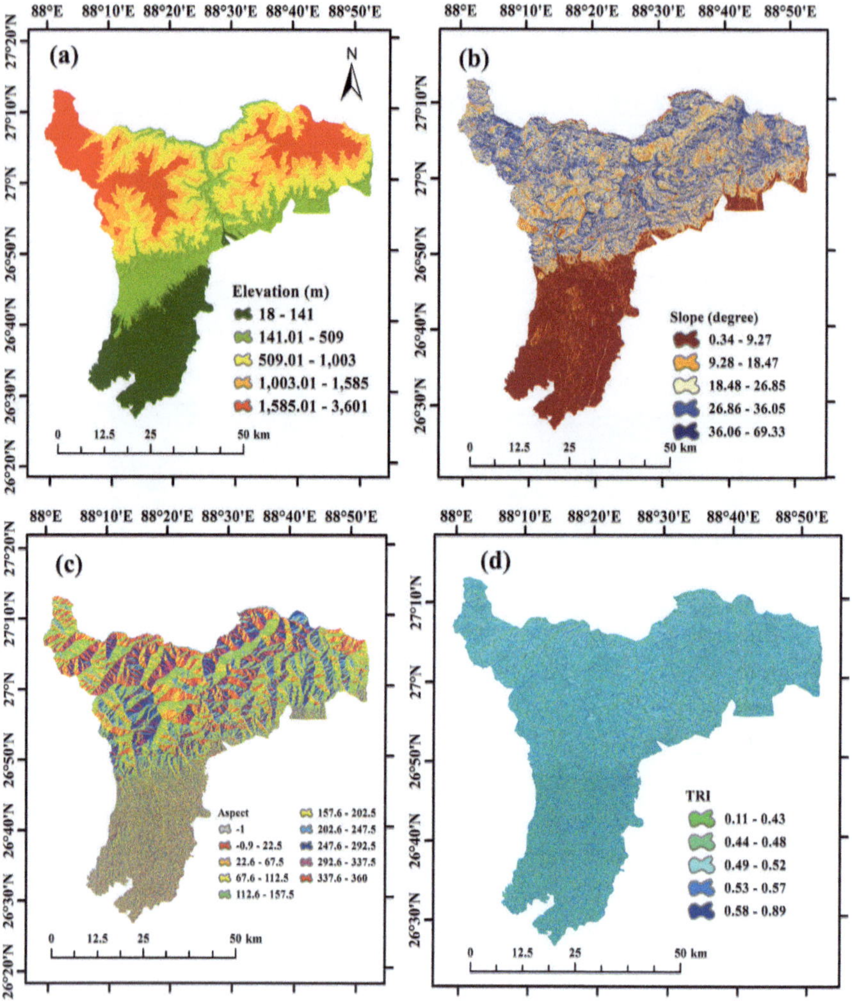

Fig. 3 Spatial variable factors used for FR model

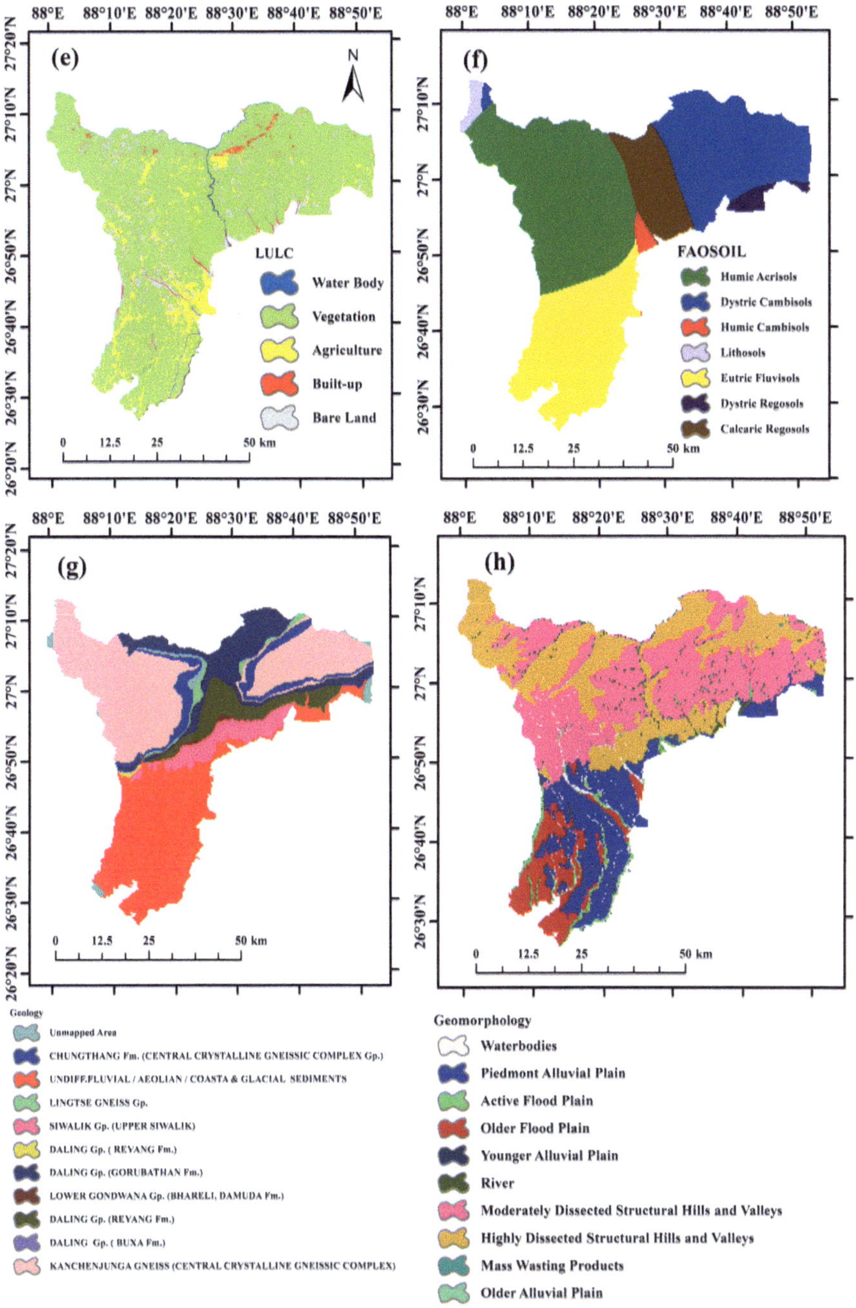

Fig. 3 (continued)

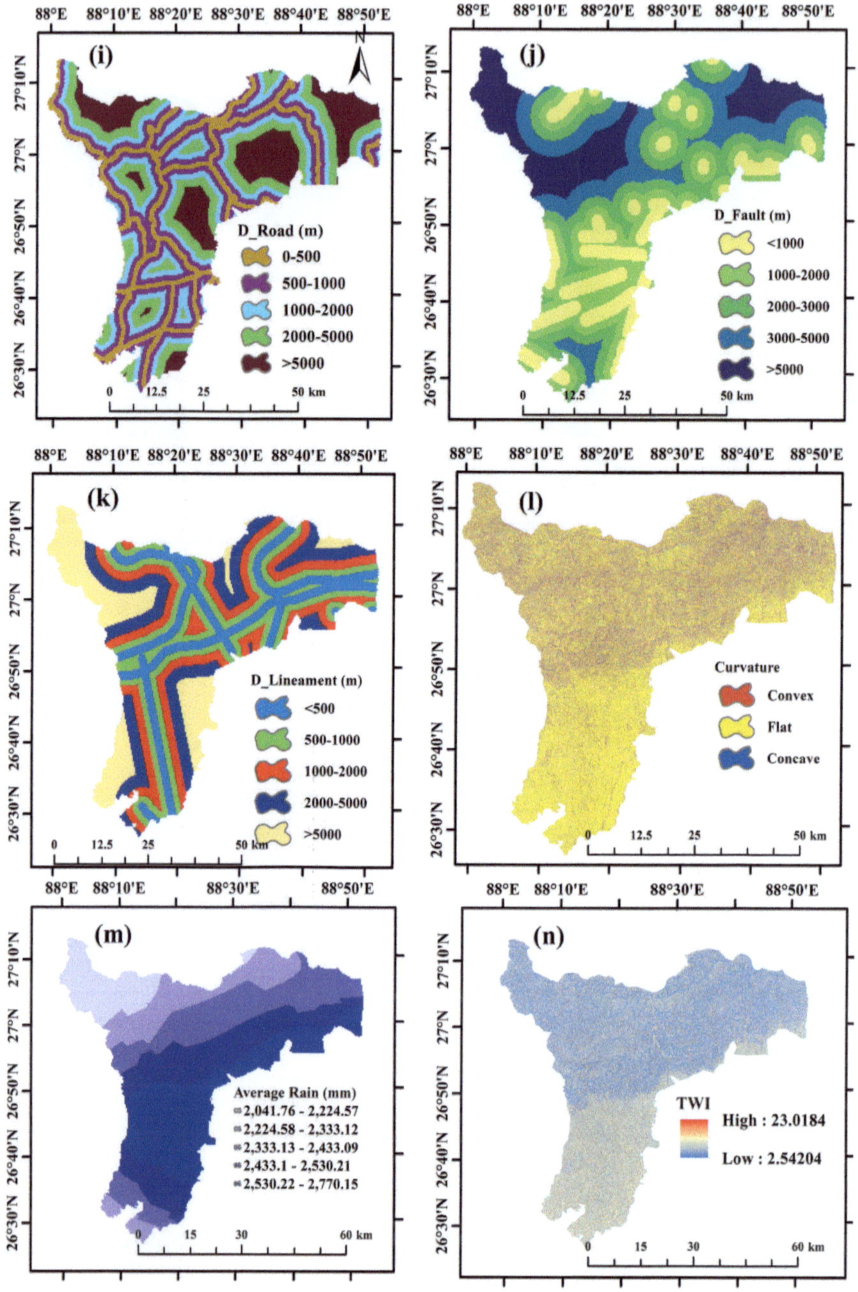

Fig. 3 (continued)

3.3.9 Distance from Road

The road network of the study area has been obtained from https://www.openstreetmap.org/#map=9/25.8110/89.3513. The presence of major roads is critical during landslides, particularly in post-landslide rescue operations (Fig. 3i). If the connectivity of the metal road is established, the disaster area can be accessed quickly, and the metal road exhibits durability against damage. The probability of disconnection in a communication system is low.

3.3.10 Distance from Faults

Most landslides transpire along faults, which trigger numerous landslides due to tectonic fractures that compromise the structural integrity of rock (El Jazouli et al., 2019). This element is taken into account in the landslide mapping of this study. Tectonic faults were delineated utilizing a geological map of the Darjiling Himalaya with Arc GIS. This map was classified into many buffers utilizing ArcGIS software (Fig. 3j).

3.3.11 Distance from Lineament

A Landsat 8 OLI image was used in conjunction with ENVI 4.7 and GEOMATICA software to derive the lineament of the research region. We used the Euclidean distance buffering tool in ArcGIS 10.4.1 to get the distance from the lineament map (Fig. 3k).

3.3.12 Rainfall

One of the reasons for landslides in hilly regions is excessive rainfall. After the runoff from a region with above-average rainfall has subsided, the soil's ability to absorb water declines. When creating a map of potential landslide areas, rainfall is essential. The consequent sloppy area or fault structure speeds up the landslide process. This study measures a 35-year rainfall average from 1985 to 2020 (https://pypi.org/project/imdlib/). About 2586 mm of rain falls here every year on average (Fig. 3l).

3.4 Methodology

3.4.1 Landslide Susceptibility Analysis Through Multicollinearity

Initially, a multicollinearity test was conducted using the SPSS statistical package to analyze the relationships among the conditioning variables. Multicollinearity arises

Table 4 Factor wise multicollinearity

Factor	VIF	TOL
Elevation (m)	2.43	0.456
Slope (degree)	1.23	0.675
Aspect	3.33	0.298
TRI	2.09	0.605
LULC	1.56	0.398
Geomorphology	3.98	0.288
Geology	4.22	0.785
Soil	4.37	0.492
Distance from road (m)	2.04	0.288
Distance from fault (m)	1.53	0.473
Distance from lineament (m)	2.47	0.277
Rainfall (mm)	4.22	0.311

when independent variables in a regression model exhibit correlation. Fitting the model and interpreting the results becomes problematic when there is a high correlation between the variables. All components undergo testing for multicollinearity by applying tolerance (TOL) and variance inflation factor (VIF). The VIF and TOL coefficients should be below 10.0 and 0.01, respectively.

$$\mathrm{TOL} = \frac{1}{\mathrm{VIF}} \tag{1}$$

$$\mathrm{VIF} = \frac{1}{1 - R_j^2} \tag{2}$$

where, R_j^2 represents the coefficient of determination in the regression analysis, encompassing all variables. No multi-collinearity issues were identified among the 12 variables; consequently, all 12 conditioning factors were utilized as inputs for the LSM analysis. The landslide susceptibility map produced using the FR method categorizes the probability into five distinct classes, ranging from very low (0) to very high (1) (Table 4).

3.4.2 Bivariate Statistical Analysis with FR

FR is a well-established and validated quantitative method to map landslide hazard inventories. Several research studies have utilized the FR technique to map landslide susceptibility (Sonker et al., 2022; Xu et al., 2024). Estimating historical landslide records is critical for forecasting future flooding events in a designated area. FR is a bivariate statistical method quantifying the relationship between landslide occurrences and the influencing factors.

$$\mathrm{FR} = (\mathrm{E}/\mathrm{F})\big/(\mathrm{M}/\mathrm{L}) \tag{3}$$

E is the landslide event of each factor, F depicts the total no of the landslide event, M refers to the number of pixels of each class of any aspect, and L is the actual number of pixels of the study area. The value of FR indicates the relationship between the incidence of a landslide and the various conditioning factors that affect it. The FR method analyzes the contributions of subdivisions within each landslide-forming element, assessing the ratio of landslide organization to the specific area of a given region. The normalized FR is analyzed and calculated using the following methods.

$$\mathrm{FR_n} = \frac{\mathrm{FR}}{\sum \mathrm{Class\ FR}} \tag{4}$$

Finally, Landslide Susceptibility Index (LSI) is calculated through the summation of the normalization of FR for every factor class,

$$\mathrm{LSI} = \sum \mathrm{FR_n} \tag{5}$$

3.4.3 Model Performance Validation

The Area Under Curve (AUC) is a standard metric for assessing overall performance. The plot is represented in two dimensions: specificity along the X-axis and sensitivity along the Y-axis. The value ranges from 0.5 to 1, indicating a low to high performance spectrum, respectively (Yesilnacar & Topal, 2005; Roy et al., 2020).

$$\mathrm{S_{AUC}} = (\mathrm{X_{k+1}} + \mathrm{X_k})\left(\mathrm{S_k} + 1 - \mathrm{S_{k+1}} - \frac{\mathrm{S_k}}{2}\right) \tag{6}$$

S_{AUC} is Area Under Curve, X_k refers to the 1-Specificity and S_k is 1-Sensitivity of ROC. Assessing validity and accuracy is essential for analyzing the predictive outcomes of landslide susceptibility mapping using the FR method.

4 Results and Discussions

4.1 Parameter Wise Clarification

In the present article, we have used 12 parameters to quantify the landslide hazard in the Darjiling Himalayas. These parameters make it easy to determine the landslide susceptibility of the Darjiling Himalayas. The FR method makes it possible to determine the probability of landslides in percentage form according to each parameter. Regarding elevation, 1003–1585 m elevation landslides accounted for 35.57% and affected about 20% of the area. According to the land slope, 18.6–26.9 slope accounts for 30.13% of landslides and 23% of the affected area. Landslides in the Southeast region are about 18.74%, affecting about 14.08% of the area of Kalimpong I, Kalimpong II, and Gurubathan blocks. About 21.75% of the previous landslides occurred, while the land roughness was between 0.49 and 0.52, affecting about 21% of the area. The number of landslides was about 70.76%, even under standard forest cover, and before that, it affected about 82% of the area. However, before this, about 65% of the landslides occurred in the Piedmont Alluvial Plain and about 38% of the area in the foothills experienced landslides. Geologically, the Lingtse Gneiss formation is responsible for 28% of landslides and has previously affected about 12% of the area. Humic acrisols in the soil increased the number of landslides by about 44%, affecting 37% of the total landslides in the entire region. 26% of landslides occurred within a distance of 500–1000 m from the main road and accounted for about 25% of the total landslide occurrences. Landslides are more abundant within a distance of 3000–5000 m from the faultline, affecting 22% of the total landslides. The number of landslides decreased by 500 m distance from the lineament by about 32%, affecting 20% of the total landslides. The range of rainfall between 2476 and 2550 mm increases the possibility of landslides by about 44%, which covers 24% of total landslide occurrences in this area.

4.2 Susceptibility Zonation Through FR Method

In our article, landslide susceptibility zonation is based on previously occurring landslides and blocks. The Darjiling Himalayan part of West Bengal is divided into two districts: Darjiling and Kalimpong. Among them are nine blocks of Darjiling district and three blocks of Kalimpong district. However, there has never been any landslide in the Darjiling district's Naxalbari, Kharibari, and Phansidewa blocks. Floodplains with disastrous floods are more common than landslides here. Landslides occurred earlier in the Jorebunglow-Sukhiapokhri and Rangli-Rangliot blocks to the north or upper part of the Matigara block. However, these landslides are significantly less and limited only to the steep slopes. Mirik, Garubathan, and Darjiling-Pulbazar blocks experienced moderate landslides earlier, with high damage. However, Kalimpong I, Kalimpong II, and Kurseong blocks have seen the highest number of landslides. Several hotels, settlements and roads built on the slopes of the mountains have been badly damaged by the landslides in the region.

We determined the landslide susceptibility of Darjiling and Kalimpong districts with the help of the FR method according to the Arc GIS interface and divided it into five classes. These classes are very high (32%), high (18%), moderate (16%), low (22%) and very low (12%) (Fig. 4). To determine the LSZ using the FR method, we used previous landslides and 12 parameters. Out of the total 3150 km^2 of the entire region, the 1008 km^2 area is very highly landslide-prone, and among these,

Kalimpong II and Kurseong block are notable. The high landslide potential zone covers an area of 567 km^2 and includes Garubathan, Kalimpong I and Darjiling-Pulbazar blocks. The moderate landslide potential zone consists of an area of 504 km^2 and includes Jorebunglow-Sukhiapokhri, Mirik and Rangli-Rangliot blocks. The low landslide potential zone consists of an area of 693 km^2, and the Matigara block is significant. The very low landslide potential zone includes an area of 378 km^2 and the Naxalbari, Kharibari, and Phansidewa blocks (Fig. 5). However, according to the inventory map, we took 800 examples of past landslides based on past landslides. No such landslides have occurred in the past or present in the very low landslide potential zone. In the low landslide potential zone, 2.43% of landslides have occurred in the past and present. In the past and present, 10.83% of the total landslides were in the moderate potential zone. In very high and high landslide potential zones, 56.27% and 30.47% of total landslides have occurred in the past and present, respectively. Excessive rainfall, overpopulation, instability in the Himalayan region, and unscientific urban encroachment are primarily responsible for these landslides (Fig. 6).

4.3 Validation of the Model

Landslide susceptibility of the Darjiling Himalayas is reviewed in the present article through the FR model and its validation by the AUROC curve. We previously took 800 points based on previous landslides, of which 560 points (70%) were used for training and 240 points (30%) for testing. Through the FR method, the future landslides' location and the current landslides' review have been highlighted, and its validation has been done (Fig. 7). Overall, the functionality of the entire model is reviewed here. The model's success is reviewed through the training dataset, and the future potential, i.e. validation, is done through the testing dataset. We all know that the value of AUROC lies between 0.5 and 1, whereas as the value approaches 1, the model's success rate is higher. A thorough review of the FR model has determined success and validation rates to be 85.4%, respectively.

In this study, a detailed statistical analysis among FR model has been done, showing that FR model is fitted with high goodness fit. FR has the good sensitivity (0.934), specificity (0.934), accuracy (0.920) and Kappa value (0.841). The predictive capability of the FR is assessed by the testing data shown in Table 5. FR will have the good sensitivity (0.951), specificity (0.941), accuracy (0.912) and Kappa value (0.839). Therefore, the local government should adopt the map obtained from these classifiers to reduce possible damage due to flood events.

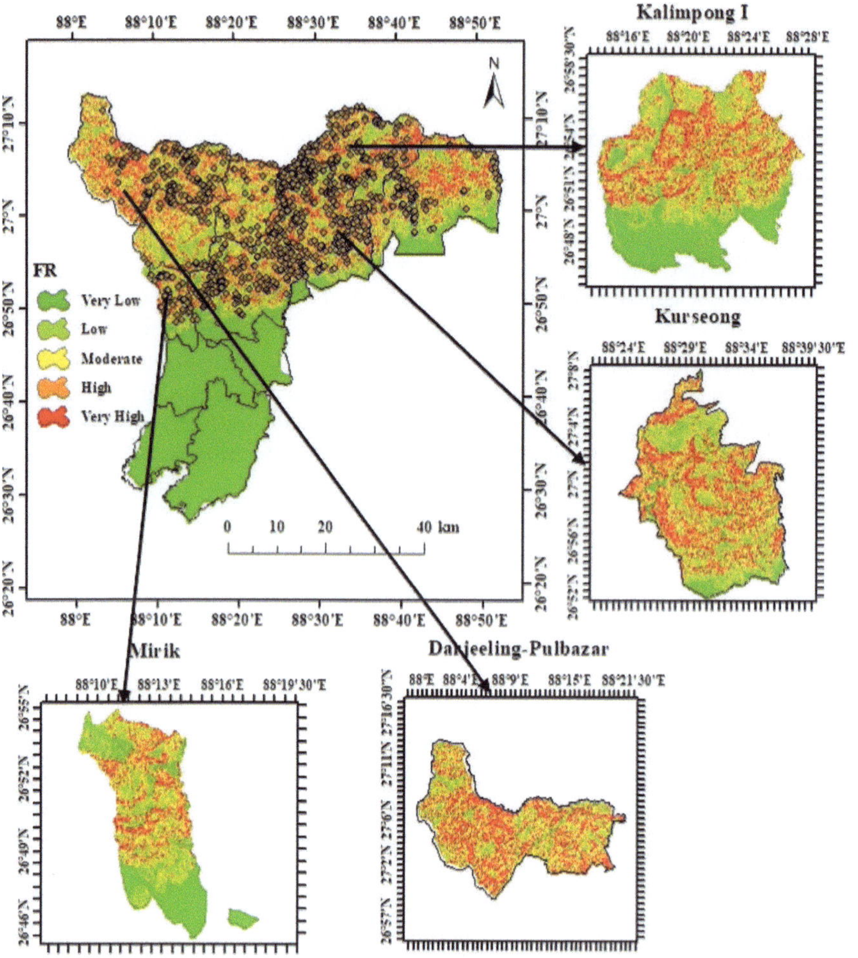

Fig. 4 Landslide susceptibility map developed by FR model

5 Conclusion and Recommendations

In our article, we want to understand the current and future situation of landslides in the Darjiling Himalayas, especially in the Darjiling and Kalimpong districts, through the FR model. Based on 12 factors or parameters, we have done landslide susceptibility zones and tried to highlight their micro-scale distribution on a block basis. Among the 12 parameters rainfall, fault location, elevation, slope, TRI, geology and soil—these factors play a significant role in determining the magnitude of landslides in the region. According to the LSZ map prepared by the FR method, the entire area is divided into five zones: very high (32%), high (18%), moderate (16%), low (22%) and very low (12%). Among the combined 12 blocks of the two districts, Kalimpong

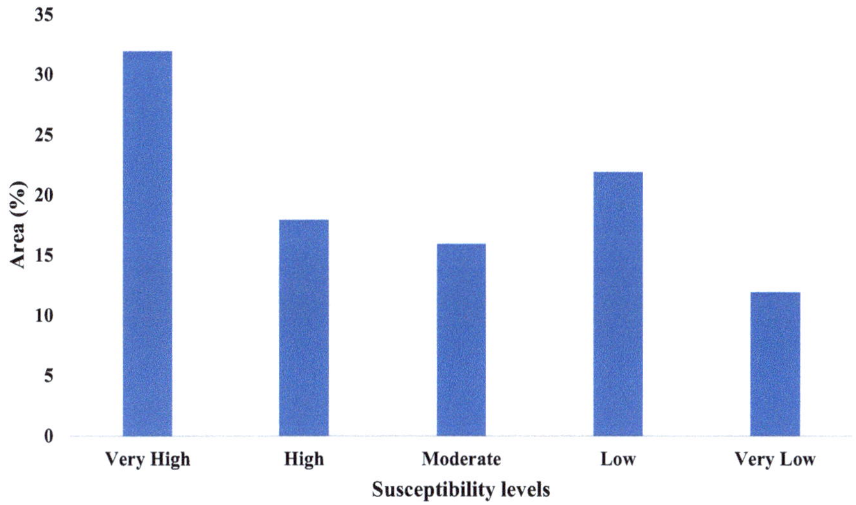

Fig. 5 Landslide susceptibility area in %

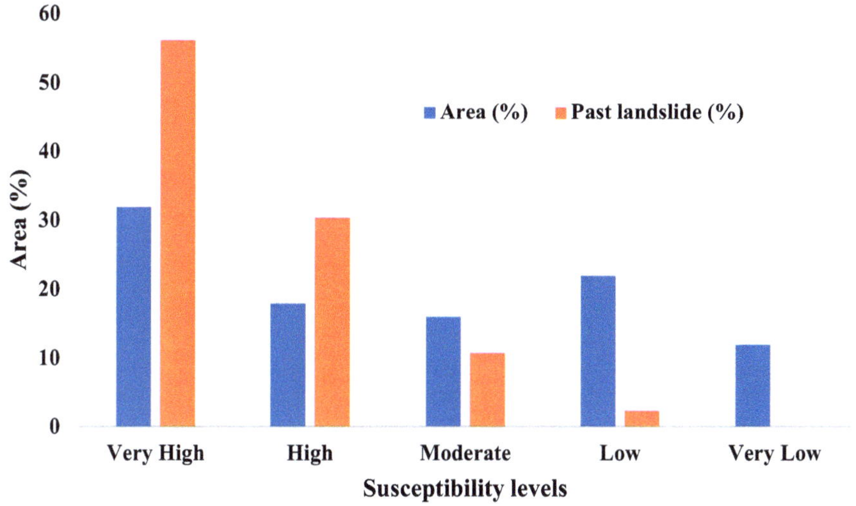

Fig. 6 Susceptible area in comparison to past landslide

II, Kurseong, Garubathan, Kalimpong I and Darjiling-Pulbazar blocks are known for the most landslides. According to the future probability of the FR method, the likelihood of future landslides in these regions is about 85%. According to the AUROC curve, the LSZ based on the FR method agrees with the current situation in Darjiling Himalaya in 95% of cases.

However, to eliminate the tendency of this landslide, the government has to formulate several laws and policies facing the difficult situation. Although we have not

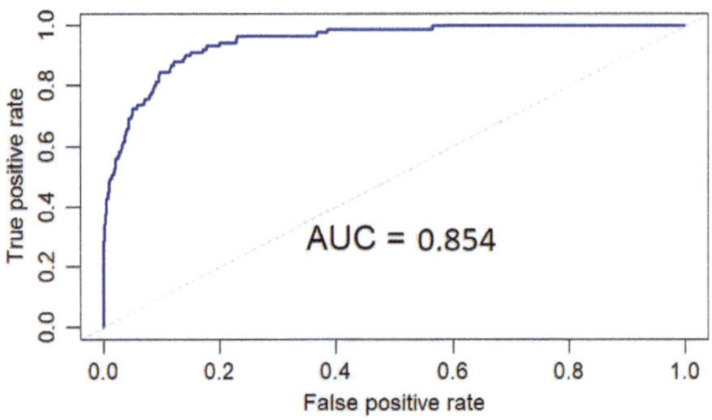

Fig. 7 ROC curve for validation of the model

Table 5 Evaluation of model performance based on selected parameters

Model	Stage	Parameters						
		Sensitivity	Specificity	PPV	NPV	Accuracy	F-score	Kappa
FR	Training	0.934	0.934	0.900	0.710	0.920	0.927	0.841
	Testing	0.951	0.941	0.963	0.789	0.912	0.968	0.839

considered NDVI a parameter, tree plantation programs are necessary to prevent landslides. The construction of homestays or hotels on mountain slopes should be declared illegal and prevented. After all. With the help of various scientific and engineering technologies, it can be assumed that it will be possible to prevent landslides through dyke, spur or rip-rap formation on the mountain slopes.

References

Ada, M., & San, B. T. (2018). Comparison of machine-learning techniques for landslide susceptibility mapping using two-level random sampling (2LRS) in Alakir catchment area, Antalya, Turkey. *Natural Hazards, 90*, 237–263. https://doi.org/10.1007/s11069-017-3043-8

Aditian, A., Kubota, T., & Shinohara, Y. (2018). Comparison of GIS-based landslide susceptibility models using frequency ratio, logistic regression, and artificial neural network in a tertiary region of Ambon, Indonesia. *Geomorphology, 318*, 101–111. https://doi.org/10.1016/j.geomorph.2018.06.006

Aleotti, P., & Chowdhury, R. (1999). Landslide hazard assessment: Summary review and new perspectives. *Bulletin of Engineering Geology and the Environment, 58*, 21–44. https://doi.org/10.1007/s100640050066

Althuwaynee, O. F., Pradhan, B., Park, H.-J., & Lee, J. H. (2014). A novel ensemble bivariate statistical evidential belief function with knowledge-based analytical hierarchy process and multivariate statistical logistic regression for landslide susceptibility mapping. *CATENA, 114*, 21–36. https://doi.org/10.1016/j.catena.2013.10.011

Bai, Y., Feng, M., Jiang, H., Wang, J., & Liu, Y. (2015). Validation of land cover maps in China using a sampling-based labelling approach. *Remote Sensing, 7*, 10589–10591. https://doi.org/10.3390/rs70810589

Bandopadhyay, M. (1980) Progress report on slope stability of Toonsoong area, Darjeeling Town (8 p.). Unpublished G.S. I report.

Basu, S. R., & Sarkar, S. (1985). Some considerations on recent landslides at Tindharia and their control. *Indian Journal of Power and River Valley Development* 188–194.

Basu, S. R., & Sarkar, S. (1987). Ecosystem vis-a vis landslide: A case study in Darjeeling Himalaya. Proce. Impact of Development on Environment. *Geographical Society of India, 2*, 45–53.

Basu, T., & Pal, S. (2019). RS-GIS based morphometrical and geological multi-criteria approach to the landslide susceptibility mapping in Gish River Basin, West Bengal, India. *Advances in Space Research, 63*(3), 1253–1269. https://doi.org/10.1016/j.asr.2018.10.033

Chen, W., Hong, H., Panahi, M., Shahabi, H., Wang, Y., Shirzadi, A., Pirasteh, S., Alesheikh, A. A., Khosravi, K., & Panahi, S. (2019). Spatial prediction of landslide susceptibility using GIS-based data mining techniques of ANFIS with Whale Optimization Algorithm (WOA) and Grey Wolf Optimizer (GWO). *Applied Sciences, 9*, 3755. https://doi.org/10.3390/app9183755

Chung, C.-J., & Fabbri, A. G. (2008). Predicting landslides for risk analysis—Spatial models tested by a cross-validation technique. *Geomorphology, 94*, 438–452. https://doi.org/10.1016/j.geomorph.2006.12.036

Cross, M. (1998). Landslide susceptibility mapping using the matrix assessment approach: A Derbyshire case study. *Geological Society, London, Engineering Geology Special Publications, 15*, 247–261. https://doi.org/10.1144/GSL.ENG.1998.015.01.26

Dey, S., Das, S., & Saha, A. (2025). Exploring uncertainty analysis in GIS-based landslide susceptibility mapping models using machine learning in the Darjeeling Himalayas. *Earth Science Informatics, 18*, 42. https://doi.org/10.1007/s12145-024-01561-7

Ding, Q., Chen, W., & Hong, H. (2017). Application of frequency ratio, weights of evidence and evidential belief function models in landslide susceptibility mapping. *Geocarto International, 32*, 619–639. https://doi.org/10.1080/10106049.2016.1165294

Duc, D. M. (2013). Rainfall-triggered large landslides on 15 December 2005 in Van Canh district, Binh Dinh province, Vietnam. *Landslides, 10*, 219–230. https://doi.org/10.1007/s10346-012-0362-4

El Jazouli, A., Barakat, A., & Khellouk, R. (2019). GIS-multicriteria evaluation using AHP for landslide susceptibility mapping in Oum Er Rbia high basin (Morocco). *Geoenvironmental Disasters, 6*(1), 3. https://doi.org/10.1186/s40677-019-0119-7

Erener, A., & Düzgün, H. S. B. (2010). Improvement of statistical landslide susceptibility mapping by using spatial and global regression methods in the case of More and Romsdal (Norway). *Landslides, 7*, 55–68. https://doi.org/10.1007/s10346-009-0188-x

Fernández, T., Irigaray, C., El Hamdouni, R., & Chacón, J. (2003). Methodology for landslide susceptibility mapping by means of a GIS. Application to the Contraviesa area (Granada, Spain). *Natural Hazards, 30*, 297–308. https://doi.org/10.1023/B:NHAZ.0000007092.51910.3f

Ghosh, A. M. N. (1950) *Observation of landslips of the 11 and 12 June 1950, in the Darjeeling Himalaya.* Unpublished G.S.I report.

Gökceoglu, C., & Aksoy, H. (1996). Landslide susceptibility mapping of the slopes in the residual soils of the Mengen region (Turkey) by deterministic stability analyses and image processing techniques. *Engineering Geology, 44*, 147–161. https://doi.org/10.1016/S0013-7952(97)81260-4

Gokceoglu, C., & Sezer, E. (2009). A statistical assessment on international landslide literature (1945–2008). *Landslides, 6*, 345. https://doi.org/10.1007/s10346-009-0166-3

Goswami, A., & Kundu, G. (2024). Geological and geomorphological interaction of the landslides in the Himalayan Region. In: U. Chatterjee, K. Lalmalsawmzauva, B. Biswas, & S. C. Pal (Eds.), *Landslides in the Himalayan Region. Disaster Risk Reduction.* Springer, Singapore. https://doi.org/10.1007/978-981-97-4680-4_1.

Griesbach, G. L. (1899–1900) General report. G. S. I, India.

Gupta, N., Pal, S. K., & Das, J. (2022). GIS-based evolution and comparisons of landslide susceptibility mapping of the East Sikkim Himalaya. *Spatial Science, 3*, 359–384. https://doi.org/10.1080/19475683.2022.2040587

Halder, K., Srivastava, A. K., Ghosh, A., Das, S., Banerjee, S., Pal, S. C., ... & Gaiser, T. (2024). Improving landslide susceptibility prediction through ensemble Recursive Feature Elimination and meta-learning framework.

Kannaujiya, S., Chattoraj, S. L., Jayalath, D., Champati Ray, P. K., Bajaj, K., Podali, S., & Bisht, M. P. S. (2019). Integration of satellite remote sensing and geophysical techniques (electrical resistivity tomography and ground penetrating radar) for landslide characterization at Kunjethi (Kalimath), Garhwal Himalaya, India. *Natural Hazards, 97*(3), 1191–1208. https://doi.org/10.1007/s11069-019-03695-0

Kia, M. B., Pirasteh, S., Pradhan, B., Mahmud, A. R., Sulaiman, W. N. A., & Moradi, A. (2012). An artificial neural network model for flood simulation using GIS: Johor River Basin Malaysia. *Environment and Earth Science, 67*(1), 251–264. https://doi.org/10.1007/s12665-011-1504-z

Kumar, V., Gupta, V., & Jamir, I. (2018). Hazard evaluation of progressive Pawari landslide zone, Satluj valley, Himachal Pradesh, India. *Natural Hazards, 93*(2), 1029–1047. https://doi.org/10.1007/s11069-018-3339-3

Lei, X., Chen, W., & Pham, B. T. (2020). Performance evaluation of GIS-based artificial intelligence approaches for landslide susceptibility modelling and spatial patterns analysis. *ISPRS International Journal of Geo-Information, 9*, 443. https://doi.org/10.3390/ijgi9070443

Mahmoud, S. H., & Gan, T. Y. (2018). Urbanization and climate change implications in food risk management: Developing an efficient decision support system for food susceptibility mapping. *Science of the Total Environment, 636*, 152–167. https://doi.org/10.1016/j.scitotenv.2018.04.282

Malamud, B. D., Turcotte, D. L., Guzzetti, F., & Reichenbach, P. (2004). Landslide inventories and their statistical properties. *Earth Surface Processes and Landforms, 29*, 687–711. https://doi.org/10.1002/esp.1064

Mandal, P., & Sarkar, S. (2021). Estimation of rainfall threshold for the early warning of shallow landslides along National Highway-10 in Darjeeling Himalayas. *Natural Hazards, 105*, 2455–2480. https://doi.org/10.1007/s11069-020-04407-9

Mondal, A., Chowdhury, I., Mukherjee, S., & Goswami, A. (2024). Assessment of climate change impact on landslides in Darjeeling District of West Bengal: A geospatial, geostatistical and ecosystem service based approach. In: G. K. Panda, R. Shaw, S. C. Pal, U. Chatterjee, A. Saha (Eds.), *Landslide: Susceptibility, Risk Assessment and Sustainability. Advances in Natural and Technological Hazards Research* (Vol. 52, 15(1), pp. 98–110). Springer, Cham. https://doi.org/10.1007/978-3-031-56591-5_13.

Naito, A. T., & Cairns, D. M. (2011). Relationships between Arctic shrub dynamics and topographically derived hydrologic characteristics. *Environmental Research Letters, 6*, 045506.

Nakileza, B. R., & Nedala, S. (2020). Topographic influence on landslides characteristics and implication for risk management in upper Manafwa catchment, Mt Elgon Uganda. *Geoenvironmental Disasters, 7*, 27. https://doi.org/10.1186/s40677-020-00160-0

Nautiyal, S. P. (1951) *A geological report on the hill slopes stability in and around Darjeeling, W. B.* Unpublished G.S.I report.

Negi, I. S., Kumar, K., Kathait, A., & Prasad, P. S. (2013). Cost assessment of losses due to recent reactivation of Kaliasaur landslide on National Highway 58 in Garhwal Himalaya. *Natural Hazards, 68*(2), 901–914. https://doi.org/10.1007/s11069-013-0663-5

Panikkar, S. V., & Subramanyan, V. (1996). A geomorphic evaluation of the landslides around Dehradun and Mussoorie, Uttar Pradesh, India. *Geomorphology, 15*, 169–181. https://doi.org/10.1016/0169-555X(95)00121-K

Parkash, S. (2015). Some socio-economically significant landslides in Uttarakhand Himalaya: Events. Consequences and Lessons Learnt. In H. Nibanupudi, & R. Shaw, (Eds.), *Mountain Hazards and Disaster Risk Reduction. Disaster Risk Reduction* (pp. 211–232). Springer, Tokyo. https://doi.org/10.1007/978-4-431-55242-0_12.

Pourghasemi, H. R., & Kerle, N. (2016). Random forests and evidential belief function-based landslide susceptibility assessment in Western Mazandaran Province, Iran. *Environmental Earth Sciences, 75*, 185. https://doi.org/10.1007/s12665-015-4950-1

Pradhan, B., Lee, S., & Buchroithner, M. F. (2010a). A GIS-based back-propagation neural network model and its cross-application and validation for landslide susceptibility analyses. *Computers, Environment and Urban Systems, 34*, 216–235. https://doi.org/10.1016/j.compenvurbsys.2009.12.004

Pradhan, B., Sezer, E. A., Gokceoglu, C., & Buchroithner, M. F. (2010b). Landslide susceptibility mapping by neuro-fuzzy approach in a landslide-prone area (Cameron Highlands, Malaysia). *IEEE Transactions on Geoscience and Remote Sensing, 48*, 4164–4177. https://doi.org/10.1109/TGRS.2010.2050328

Roy, J., & Saha, S. (2019). Landslide susceptibility mapping using knowledge driven statistical models in Darjeeling District, West Bengal, India. *Geoenvironmental Disasters, 6*, 11. https://doi.org/10.1186/s40677-019-0126-8

Roy, P., Chakrabortty, R., Chowdhuri, I., Malik, S., Das, B., & Pal, S. C. (2020). Development of different machine learning ensemble classifier for gully erosion susceptibility in Gandheswari Watershed of West Bengal, India. In *Machine Learning for Intelligent Decision Science* (pp. 1–26). https://doi.org/10.1007/978-981-15-3689-2_1

Saha, A., Villuri, V. G. K., & Bhardwaj, A. (2023). Development and assessment of a novel hybrid machine learning-based landslide susceptibility mapping model in the Darjeeling Himalayas. *Stochastic Environmental Research and Risk Assessment, 11*, 1–24. https://doi.org/10.1007/s00477-023-02528-8

Sarkar, S. (1995) Landslips in the Mahananda Basin and its control. In *Perspective in the Environmental Management in Developing Countries*, Concept, v-10.

Sarkar, S., et al. (2010). *Geo-Hazards in Sub Himalayan North Bengal*. University of North Bengal.

Sonker, I., Tripathi, J. N. & Swarnim. (2022). Remote sensing and GIS-based landslide susceptibility mapping using frequency ratio method in Sikkim Himalaya. *Quaternary Science Advances, 8*, 100067. https://doi.org/10.1016/j.qsa.2022.100067

Tehrany, M. S., Pradhan, B., & Jebur, M. N. (2015). Flood susceptibility analysis and its verification using a novel ensemble support vector machine and frequency ratio method. *Stochastic Environmental Research and Risk Assessment, 29*(4), 1149–1165. https://doi.org/10.1007/s00477-015-1021-9

Thanh, L. N., & De Smedt, F. (2014). Slope stability analysis using a physically based model: A case study from A Luoi district in Thua Thien-Hue Province. Vietnam. *Landslides, 11*, 897–907. https://doi.org/10.1007/s10346-013-0437-x

Upadhyay, V. (2024). Landslide hazard risk and vulnerability monitoring—GIS based approach. In: G. K. Panda, R. Shaw, S. C. Pal, U. Chatterjee, & A. Saha, (Eds.), Landslide: Susceptibility, Risk Assessment and Sustainability. Advances in Natural and Technological Hazards Research (Vol. 52, 8(2), pp. 112–125). Springer, Cham. https://doi.org/10.1007/978-3-031-56591-5_3

Wang, L.-J., Guo, M., Sawada, K., Lin, J., & Zhang, J. (2016). A comparative study of landslide susceptibility maps using logistic regression, frequency ratio, decision tree, weights of evidence and artificial neural network. *Geosciences Journal, 20*, 117–136. https://doi.org/10.1007/s12303-015-0026-1

Westen, C. V., & Terlien, M. (1996). An approach towards deterministic landslide hazard analysis in GIS. A case study from Manizales (Colombia). *Earth Surface Processes and Landforms, 21*, 853–868. https://doi.org/10.1002/(SICI)1096-9837(199609)21:9%3c853::AID-ESP676%3e3.0.CO;2-C

Wu, Y., Li, W., Wang, Q., Liu, Q., Yang, D., Xing, M., Pei, Y., & Yan, S. (2016). Landslide susceptibility assessment using frequency ratio, statistical index and certainty factor models for the Gangu County, China. *Arabian Journal of Geosciences, 9*(2), 1–16. https://doi.org/10.1007/s12517-015-2112-0

Xu, K., Zhao, Z., Chen, W., Ma, J., Liu, F., Zhang, Y., & Ren, Z. (2024). Comparative study on landslide susceptibility mapping based on different ratios of training samples and testing samples by using RF and FR-RF models. *Natural Hazards Research, 4*, 62–74. https://doi.org/10.1016/j.nhres.2023.07.004

Yesilnacar, E., & Topal, T. A. M. E. R. (2005). Landslide susceptibility mapping: A comparison of logistic regression and neural networks methods in a medium scale study, Hendek region (Turkey). *Engineering Geology, 79*(3–4), 251–266. https://doi.org/10.1016/j.enggeo.2005.02.002

Zhao, C., & Lu, Z. (2018). Remote sensing of landslides-A review. *Remote Sensing, 10*(2), 8–13. https://doi.org/10.3390/rs10020279

Zhao, X., & Chen, W. (2020). Optimization of computational intelligence models for landslide susceptibility evaluation. *Remote Sensing, 12*, 2180. https://doi.org/10.3390/rs12142180

Assessment of Landslide Susceptibility in the Reshi River Basin, West Bengal and Sikkim and Highlighting the Impact of Changing Vegetation Cover

Arpita Saha, Prabir Kumar Kundu, and Arup Kr. Saha

Abstract Landslides are significant natural hazards that alter the environment and negatively impact human lives and economies globally. Studying these phenomena is crucial in hilly regions, including the high hillsides of the Reshi River Basin, which are particularly susceptible to landslides due to climatic, geological, and geomorphological conditions. Initially, landslides in the Reshi River Basin were identified using Google Earth imagery. Thus, it is essential to create landslide susceptibility maps for the area. This study details the process of modelling landslide susceptibility using remote sensing data and GIS tools. Fourteen factors contributing to landslides—such as elevation, slope, curvature, lineament density, rainfall, drainage density, soil, lithology, relative relief, road density, settlement density, land use/land cover, and the topographic wetness index (TWI)—were analyzed. These factors were extracted from a spatial database created with remotely sensed data and topographic maps. The analysis involved weighting raster thematic maps, multiplying them by their respective weights, and combining them to generate a landslide susceptibility index (LSI) value for each cell. The final susceptibility map, after reclassification, includes four categories: low, moderate, high, and very high. According to the Linear Compositing Model (LCM), 90% of the identified landslide zones fall into the high and very high susceptibility categories. Vegetation plays a crucial role in preventing landslides by stabilizing the soil and reinforcing slopes. The roots of plants intertwine soil particles, which enhances the soil's cohesion and tensile strength. This diminishes the likelihood of the soil breaking away and sliding down, particularly in the event of heavy rain or earthquakes. For this, the study also observed changes in vegetation cover between 2001 and 2016.

A. Saha (✉)
Department of Geography, Chanchal College, Malda, West Bengal, India
e-mail: arpita.saha.mldt@gmail.com

P. K. Kundu
Department of Geography, Siliguri Mahila Mahavidyalaya, Siliguri, Darjiling, West Bengal, India

A. Kr. Saha
Chandrapur College, Chandrapur, West Bengal, India

© The Author(s), under exclusive license to Springer Nature Switzerland AG 2025
S. Ghosh (ed.), *The Himalaya Dilemma*, Springer Natural Hazards,
https://doi.org/10.1007/978-3-031-95083-4_12

Keywords Landslides · Natural hazards · Hillsides · Susceptibility · Topographic wetness index · Landslide susceptibility index · LCM models

1 Introduction

Landslides are among the most destructive natural disasters in hilly environments (Ayalew et al., 2005). Effective planning and management can reduce social and economic losses due to landslides (Rajakumar et al., 2007). Most landslide hazard analyses incorporate a current landslide inventory, which serves as a key tool in identifying the factors that cause hill-slope instability and trigger landslides (Lee & Sambath, 2006).

Mapping the areas that are susceptible to landslides is essential for proper land use planning and disaster management for a particular locality or region. Throughout the years, different techniques and methods have been developed and applied in the literature for landslide susceptibility mapping. Landslide susceptibility maps can be produced using both quantitative and qualitative approaches (Park et al., 2013). According to Schuster (1996), landslides are expected to increase in the coming decades due to ongoing deforestation and changing climatic patterns in landslide-prone regions. Three main approaches are used to identify areas susceptible to landslides: qualitative, deterministic, and quantitative methods. In particular, the concept of a geomorphic threshold is crucial for quantitatively analysing the stability of a given spatial area.

The landslide inventory map, providing information for the assessment of the influence of different causative factors on landslide occurrence, is the backbone of landslide susceptibility studies (Van Westen et al., 2006; Youssef et al., 2014a, b). The reliability and accuracy of the collected data related to landslides affect the success of landslide susceptibility analysis. Landslide inventory maps can be prepared either by collecting information related to landslides or by analysing satellite imagery and aerial photographs coupled with field surveys using GPS (Pradhan & Kim, 2014).

2 Literature Review

According to Schuster and Fleming (1986), in many countries, the economic losses and causalities due to landslides are more significant than commonly recognized and generate a yearly loss of property more significant than that from any other natural disaster, including earthquakes, floods and windstorms. Guzzetti et al. (1999) summarized many landslide hazard evaluation studies based on geomorphological relationships between landslide types, patterns, and the morphological, lithological and structural settings. Recently, studies on landslide susceptibility assessment have made use of GIS, and many probabilistic models have been applied, such as frequency ratio and logistic regression methods. Landslide researchers have also used quantitative methods in bivariate, multivariate, and nonlinear approaches. Most published literature regarding LSM using bivariate methods use weights of evidence and frequency ratio (FR) (Constantin et al., 2011; Yalcin et al., 2011; Conforti et al., 2012; Tehrany et al., 2013; Althuwaynee et al., 2014a). Pradhan and Lee (2009) assessed landslide susceptibility analysis in Penang, Malaysia, using frequency ratio, logistic regression, and the ANN model. They concluded that the logistic regression model showed slightly better prediction accuracy than the latter. Lee and Pradhan (2007) argued that the frequency ratio model provides a correlation between the historical slide locations and various influencing factors under consideration. Intarawichian and Dasananda (2011) applied the frequency ratio model to analyze slope instability and ascribed the model as a popular quantitative method. The frequency ratio (FR) model has become very popular as a realistic quantitative approach in landslide susceptibility mapping. This approach is related to the historical landslide events and their overall coverage. Akgun (2012) produced LSMs using logistic regression (LR), frequency ratio (FR), and analytical hierarchy process (AHP) methods. He showed that the prediction rates of these methods were equal to 81.0%, 76.0%, and 71.0%, respectively, and that LR and FR methods (both are quantitative methods) are significantly more accurate than the AHP. However, the prediction rates of all three methods are categorized as fair quality (Abul Hasanat et al., 2010).

3 Research Gap

The Darjiling Himalayas are part of the Eastern Himalayan range and are drained by several streams and rivers, including Tista, Rangeet, Mahannda, Jaldhaka, Balason, Mechi, Lish, Gish, and Murti. Major towns in this region include Darjiling, Kalimpong, Kurseong, and Mirik. The area is made up of soft phyllite, schists, and gneiss, making it highly susceptible to landslides, which have been a significant issue throughout its history. Many parts of the region, including the Siliguri-Kalimpong railway line and various roads, have experienced considerable property damage due to landslides. Each of the five main roads connecting Darjiling to Siliguri and Kalimpong has faced 15–20 landslides in different stretches (Table 1). The study area was selected due to its predisposition to landslides, influenced by its climatic conditions, geological features, and geomorphological characteristics. This area has not been studied before, but it is expected to be highly susceptible to landslides in the future.

3.1 Objectives

Here, two major objectives are set forth to carry out the work:

Table 1 Major landslides in the Darjiling Hill region during 1998–2004

Year	Block/Municipality	Area	Damage
1998	Kurseong	Tindharia	250 houses damaged, NH-55 damaged, 10 people died
		Sitting T. E	13 people died
1999	Kurseong	Sittong III & I GP	9 people died
		Tindharia	3 people died
2000	Sukhia pokhari	Pusumbing	2 people died
	Bijanbari	Relling Basty	1 people died
		Hathale Basty	2 people died
	Darjiling sadar	Botanical Garden	1 people died
		Jawar Basty	1 people died
	Kurseong	Sepoydhura	2 people died
	Darjiling Sadar	Dali	11 people died
		Harishatta	1 people died
2001	Kurseong	Rohini	2 people died
2002	Kurseong	Giddapahar	2 people died
2003	Mirik	Gayabari	24 people died
2004	Sukhia Pokhari	Mim T. E.	1 people died

Source D. M. Office, Relief Department, Darjiling, 2004

1. To evaluate the performance of the Linear Compositing Model (LCM) in identifying landslide susceptibility in the Reshi River basin; and
2. To assess the impact of changes in vegetation coverage on landslide occurrences in the study area.

4 Study Area

Reshi, also known as Rishi or Rishi Khola, is situated in the Kalimpong sub-division on the border of West Bengal and Sikkim. The name Rishi Khola comes from the Reshi River, with "Khola" meaning a small stream in Nepali. The study area encompasses the drainage basin of Rishi Khola, which flows into the Rongli River, a left-bank tributary of the Rongpo River. The Reshi River basin is situated between 27° 11′ 15″N and 27° 05′ 15″ latitude and 88° 35′ 15″ E to 88° 43′ 30″E longitude, covering approximately 119.302 km^2 along the West Bengal and Sikkim border (Fig. 1). The primary river in this basin is the Reshi River, which flows north and takes a north-westward bend near Reshi. This basin is located on the northern slope of the Lava-Pankhasari ridge. The Rishi River serves as the boundary between West Bengal and Sikkim in this area. Altitude in the study area varies between 433 and 2375 m. The landform of the area is hilly track which results in large elevation difference.

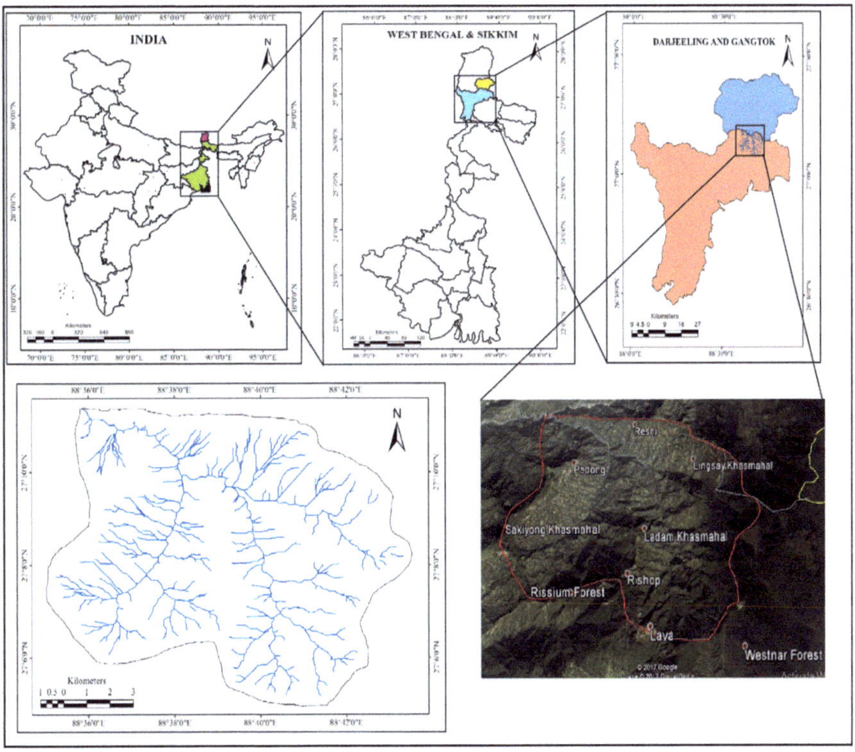

Fig. 1 Study area (Reshi river basin)

5 Methodology

5.1 Data Source

We had collected remote sensing data from satellite imagery, aerial photography, and other sources to analyze landslide characteristics, such as land use, land cover, and topography (Tables 2 and 3). Alongside the geological and geotechnical data were collected from existing reports, maps, and databases to analyze landslide causes and triggers. Review existing research and literature on landslides in the study area were very important part to collect data on landslide characteristics, causes, and impacts.

Table 2 Spatial databases for landslide susceptibility modelling

Category	Layer	Source	Data type
Topographic map	Elevation (derived)	DEM	GRID
	Slope aspect (derived)	DEM	GRID
	Slope angle (derived)	DEM	GRID
	Curvature (derived)	DEM	GRID
	Relative relief (derived)	DEM	GRID
Drainage map	Drainage density	DEM	GRID
Settlement map	Settlement density	Google Earth image	Polygon
Road map	Road density	Google Earth image	Line and polygon
Soil map	Soil texture	Soil survey atlas-(Dept. of Agriculture)	Polygon
Lithological map	Lithology	Geological Survey of India	Polygon
Lineament map	Lineament density	ArcGIS 10.2	Polygon
Precipitation map	Rainfall intensity	TRMM (Tropical Rainfall Measuring Mission)	GRID
Land use/land cover map	Land use/land cover	Remote sensing data	GRID
NDVI map	NDVI	Remote sensing data	GRID

Source Computed by researchers

Table 3 Details of satellite remote sensing data

Satellite	Path/Row	Date of acquisition	Spatial resolution	Bands	Source
Landsat 4–5 thematic mapper™	Path 139 Row 041	09/04/2001	30 × 30	1,2.3,4,5,7	USGS (United States Geological Survey)
		06/03/2006	30 × 30	1,2.3,4,5,7	
		20/03/2011	30 × 30	1,2.3,4,5,7	
Landsat 8	Path 139 Row 41	19/03/2016	30 × 30	2.3,4,5,6,7	

Source Computed by researchers

5.2 *Linear Compositing Model (LCM)*

To assess the impact of various factors on landslide hazard, the current distribution of landslides has been compared with different thematic data layers. A multi-criteria decision approach in a spatial environment has been a successful addition to GIS. Instead of solely focusing on the quantitative value of different parameters representing countable spatial units, this method considers spatial information and induces quantitative information of the smallest spatial unit. Before the introduction of this method, we used to gather information about different parameters from various articles and studies. With this method, we can analyze different parameters pixel by pixel in an automated manner. As a result, the output to achieve the objective is much more reliable and accurate. Moreover, a good number of spatial parameters can be taken into consideration to generate effective results. This study aims to identify the zones with higher susceptibility to landslides over the area of study. To perform this work, it is important to consider several relevant parameters. The specific numbers for these parameters may vary depending on the level of accuracy we are aiming for. In this study, I have taken into account thirteen parameters: **elevation, slope (in degrees), curvature, lineament density, rainfall, drainage density, soil type, lithology, relative relief, aspect, road density, settlement density, and land use/ land cover as well as TWI.**

The parameters do not all influence the objective in the same direction, so it's important to standardize them on a single scale. We've chosen a 10-point scale to rate the data. Each parameter is divided into 10 subclasses, and the scale points are distributed accordingly. The assumption is that the highest influencing parameter will be assigned a scale point of 10 to indicate the suitability for installing a new school on the site. After scaling the parameters, the composite of all these parameters in a GIS spatial analysis tool is created. The resulting weighted thematic maps are overlaid and numerically added to generate a Landslide Susceptibility Index (LSI) map.

$$\mathrm{LSI} = \mathrm{Sl} + \mathrm{As} + \mathrm{Rl} + \mathrm{El} + \mathrm{Li} + \mathrm{Soil} + \mathrm{Rd} + \mathrm{Sd} + \cdots$$

where, Sl, As, Rl, El, Li, Soil, Rd, and Sd are the product of derived weights and weight factors for slope, aspect, relative relief, elevation, lithology, lineament density, drainage density, land use and land cover, soil, road density, settlement density respectively.

6 Results and Discussion

6.1 Analysis of LCM Parameters

To create landslide susceptibility models, we developed a spatial database that takes into account 13 factors, including slope angle, slope aspect, slope curvature, altitude, relative relief, lineament density, drainage density, road density, TWI, lithology, rainfall, road density, settlement density, land use and land cover. All of this data was formatted as raster with a pixel size of 30 × 30 square meters to match the spatial resolution. A brief overview of the preparation process for each data layer is provided in the following section.

6.1.1 Lithology

The main types of rocks in the area are slate, schists, and quartzite. In the eastern part of Sikkim, the rocks are more gneissic at higher elevations in the northeastern area. Quartzite and schists are present in the central-western part of the Reshi River basin. Phyllites are exposed, while quartzites alternating with phyllites are prevalent along the northern boundary of the study area. Minerals such as coal, copper, and iron can be found throughout the study area. An analysis of landslide distribution based on the types of rocks revealed that the majority of landslides (60%) occur within shaly formations.

6.1.2 Soil

The topsoil covering a slope has been found to affect the occurrence of landslides, as observed in a previous study. The soils in the study area can be broadly classified into four major types: coarse-loamy, fine-skeletal cumulic hapludolls, gravelly loamy, and gravelly loamy-coarse loamy. The region is predominantly covered by gravelly, loamy-coarse soil, with smaller patches of skeletal soil and fine-skeletal cumulic hapludolls. Gravelly loamy-coarse loamy soil is found in hill areas and along major river courses.

6.1.3 Lineament Density

The LISS III images were used to interpret fractures, discontinuities, and shear zones. Figure 2c shows the lineament density map of the study area. The map indicates that the centre of the study area has a high lineament density, ranging from 0 to 4.6 sq. km. The lineament density for each 200-m by 200-m cell was calculated using the line density analyst extension (Saraf, 1999) of ArcGIS 10.2. The densities were then classified as very low (0–0.77 sq. km), low (0.77–1.53 sq. km), moderate (1.53–2.99 sq. km), moderately high (2.99–3.07 sq. km), high (3.07–3.83 sq. km), and very high (3.82–4.59 sq. km).

6.1.4 Drainage Density

The drainage density of the study area was calculated using the line density analysis tool in ArcGIS software. The study area was divided into six classes, each representing a range of drainage density. These classes were defined as very poor (0–1.72 km/sq.km), poor (1.72–3.45 km/sq.km), moderate (3.45–5.18 km/sq.km), moderately high (5.18–6.90 km/sq.km), high (6.90–8.63 km/sq.km), and very high (8.63–10.36 km/sq.km). The northeastern parts of the study area exhibited high drainage density, ranging from 8.63 to 10.36 km/km^2. (Fig. 2d).

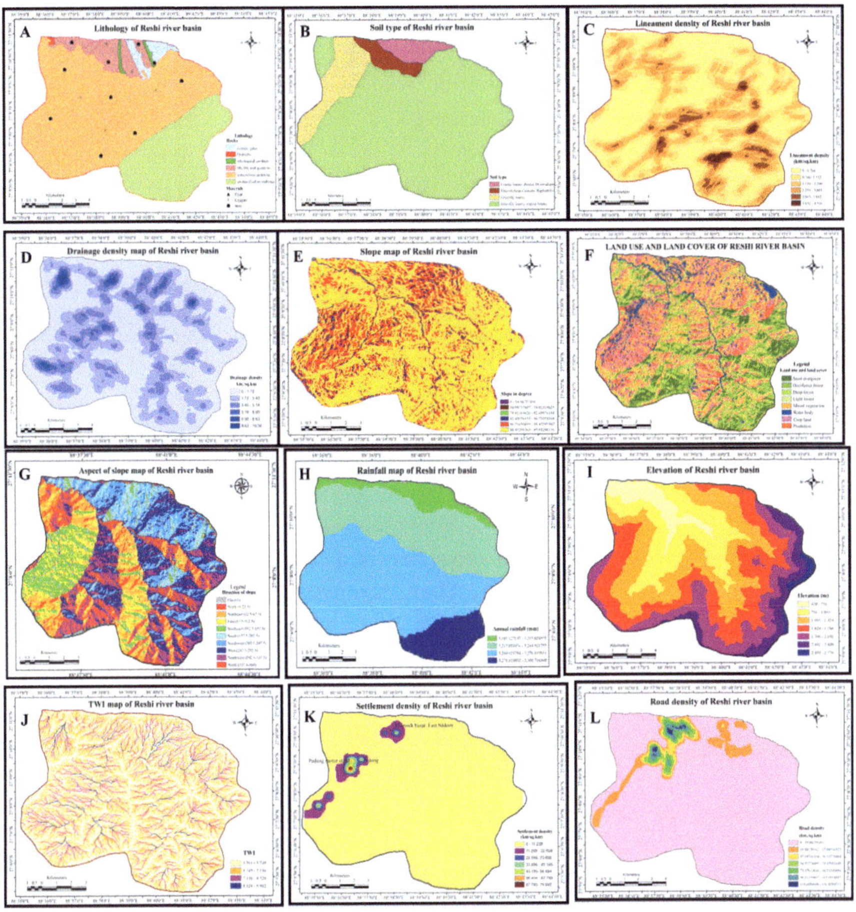

Fig. 2 **a** lithology map, **b** soil map, **c** lineament density map, **d** drainage density map, **e** slope map, **f** land use and land cover map, **g** aspect map, **h** rainfall map, **I** elevation map, **j** TWI map, **k** settlement density map, **l** road density map

6.1.5 Slope

The study area can be divided into six slope classes based on the degree of slope. Slope values range from 0° to 89°. Slopes with values between 82° and 86° are considered moderately steep, while slopes between 86° and 88° are classified as steep. Areas with escarpments of 88° to 89° can cause damage due to landslides. Generally, landslides are not expected to occur on gentle slopes due to lower shear stress. Figure 2e illustrates the slope map of the study area.

Table 4 Calculate the individual land use and land cover area in percentage

Land use and land cover	Pixel	Area in percentage
Semi evergreen	11,342	8.54
Deciduous forest	12,660	9.53
Deep forest	13,502	10.16
Light forest	14,637	11.02
Mixed vegetation	40,484	30.48
Water body	8502	6.40
Crop land	20,946	15.77
Plantation	10,760	8.10
Total	132,833	100

Source Calculated by researchers

6.1.6 Land Use/Land Cover

The land-use map was created using the supervised classification technique with the maximum likelihood parametric rule. After creating the map, it was cross-checked using Google Earth. Several errors were identified, especially in the areas marked as tea plantations and agricultural land (Table 4).

6.1.7 Slope Aspect

Aspect regions are categorized into nine classes based on aspect class: flat (−1°), north (0°–22.5°; 337.5°–360°), northeast (22.5°–67.5°), east (67.5°–112.5°), southeast (112.5°–57.5°), south (157.5°–202.5°), southwest (202.5°–247.5°), west (247.5°–292.5°), and northwest (292.5°–337.5°) (Fig. 2g).

6.1.8 Precipitation

The rainfall map was divided into four classes with different mean monthly rainfall rates: (a) 3181–3213 mm, (b) 3213–3244 mm, (c) 3244–3276 mm, and (d) 3276–3380 mm (Fig. 2h). Generally, higher rainfall rates indicate more favorable conditions for landslides.

6.1.9 Elevation

The relationship between elevation and landslides is often shown indirectly or through other factors. The elevation thematic map was classified into seven different ranges: (a) 420–756 m, (b) 756–1093 m, (c) 1093–1429 m, (d) 1429–1766 m, (e) 1766–2102 m, (f) 2102–2439 m, and (g) 2439–2776 m (Fig. 2I).

6.1.10 Relative Relief

The term "relative relief" refers to the actual variation in height, which is the difference between the maximum and minimum heights per grid. Relative relief is a technique that effectively presents the relief characteristics without considering sea level (Singh, 1992). The thematic map of relative relief was divided into six classes with different ranges: (a) 150–241 m, (b) 241–332 m, (c) 332–423 m, (d) 423–515 m, (e) 515–606 m, (f) 606–697 m. Major landslides occurred in areas with relative reliefs between 400 and 697 m.

6.1.11 TWI

The topographic wetness index (TWI) characterizes the impact of the terrain on the location and extent of saturated source areas for generating runoff, representing another topographic factor in the runoff model (Pourghasemi et al. 2013b; Pradhan & Kim, 2014). TWI is calculated as:

$$\text{TWI} = \text{Ln}\,(A_S/\tan\beta)$$

6.1.12 Settlement Density

The study area's settlement density is generally low, but there are clusters of settlements in the eastern part. The highest density is near the Pedong region and relatively high in Rhenock bazaar in East Sikkim. Villages such as Pedong Bazaar, Lingsay Village, and Renak are present throughout the area, primarily located along the roads.

6.1.13 Road Density

In the northeastern part of the Reshi river basin, road density ranges from 20 to 131 km/sq.km. Roads like Pedong-Reshi and Reshi-Rhenak are found in the eastern and western parts, respectively. The overall road density is low due to the hilly terrain.

6.2 Linear Compositing Model

In GIS, the multi-criteria decision approach considers spatial information and the quantitative information of the smallest spatial units. Its authors proposed this model. A 10-point scale is used to scale the data layer, with each parameter being subdivided into 10 subclasses (Table 5). The 10 points on the scale are distributed accordingly. It is assumed that point 10 on the scale will be assigned to the parameter with the maximum influence to make the landslide in the hilly region. The weights of these conditioning factors are selected based on their influence on landslides by authors' perception.

To visually interpret LSI maps, the data must be divided into distinct susceptibility classes (Jaafari et al., 2014). In this research, the landslide susceptibility map created with the Linear Compositing Model (LCM) was classified into four categories—low, moderate, high, and very high—using the natural breaks method in ArcGIS 10.2. (Fig. 3).

(A) **Very High Susceptibility Zone**

The area with very high susceptibility to landslides is mainly located in the southern and southeastern parts of the study area. This zone covers 23.33% of the total study area (Table 6). Geologically, this zone is precarious and is constantly threatened by landslides, especially during and after heavy rainfall. This is because the area consists of steep slopes with loose and unconsolidated materials and includes areas where active or past landslides have been observed. Additionally, it encompasses areas located near faults and tectonically weak zones.

(B) **High Susceptibility Zone**

The area described is generally stable as long as its current status is maintained. It consists of regions with moderately dense vegetation, moderate slope angles, and relatively compact, hard rocks. Although it may include areas with steep slopes exceeding 50 degrees, the orientation of the rock bed and the absence of loose debris and human activity make them less hazardous. This zone covers approximately 36.37% of the total study area, as indicated in Table 6. Due to the high landslide susceptibility of the High Hazard Zone, it is recommended that no human activity be carried out in this zone. The High Susceptibility Zone is well distributed throughout the study area.

(C) **Moderate Susceptibility Zone**

Due to various controlling parameters, the area in this zone is not expected to significantly impact slope stability. The altitudes between 750 and 1400 m are moderately vulnerable to landslides. The vegetation is relatively dense, and the slope angles are generally low. There is no evidence of instability observed within this zone in terms of risk factors, and mass movement is not expected unless major changes to the site occur. This zone covers approximately 31.59% of the total study area.

(D) **Low Susceptibility Zone**

This zone primarily consists of valley-fill areas, water bodies, low-lying land, and agricultural fields, which are notable features of the region. Due to its low slope angles and rock aspects, it is generally considered to be free from both current and future landslide hazards. Areas with gentle slopes are categorized within this zone. It is mostly confined to regions with minimal or no human activity, particularly in the northern part of the main town, and accounts for 8.71% of the total study area.

Table 5 Ranking of all parameters according to their importance

Parameters	Subclass of parameters	Subclass rank
Elevation (m)	420–655.6	1
	655.60–891.20	2
	891.20–1,126.80	3
	1,126.80–1,362.40	4
	1,362.40–1,598	5
	1,598–1,833.60	6
	1,833.60–2,069.20	7
	2,069.20–2,304.80	8
	2,304.80–2,540.40	9
	2,540.40- 2,776	10
Slope (degree)	0–41.95	1
	41.95–63.79	2
	63.79- 74.02	3
	74.02–78.95	4
	78.95–82.48	5
	82.48–85.30	6
	85.30–87.41	7
	87.41–88.82	8
	88.82–89.88	10
Lineament density (km/sq.km)	0–0.45	1
	0.459- 0.91	2
	0.91- 1.37	3
	1.37–1.83	4
	1.83–2.29	5
	2.29–2.75	6
	2.75–3.21	7
	3.21–3.67	8
	3.67–4.13	9
	4.13–4.59	10
Annual rainfall (mm)	3180–3190	1
	3190–3200	3
	3200–3294	4
	3294–3221	5
	3221–3249	6
	3240- 3250	7
	3250- 3270	9
	3270–3280	8

(continued)

Table 5 (continued)

Parameters	Subclass of parameters	Subclass rank
	3280–3290	10
	3290- 3300	2
Drainage density (km/sq.km)	0–1.036454	1
	1.036454–2.072909	2
	2.072909–3.109363	3
	3.109363–4.145817	4
	4.145817–5.182271	5
	5.182271–6.218726	6
	6.218726–7.25518	7
	7.25518–8.291634	8
	8.291634–9.328089	9
	9.328089–10.364543	10
Soil	Coarse-loamy Humic Dystrudepts	2
	Fine-skeletal Cumulic Hapludolls	4
	Gravelly loamy	7
	Gravelly loamy-coarse loamy	10
Lithology	Lithological contract	1
	Granite gniss	3
	Phyllite and quartzite	5
	Quarizite	6
	Slate, schists, quarteite	8
	Unclassified, crystallines	10
Relative relief (m)	150.343674–205.074065	1
	205.074065–259.804456	2
	259.804456–314.534846	3
	314.534846–369.265237	4
	369.265237–423.995628	5
	423.995628–478.726019	6
	478.726019–533.45641	7
	533.45641–588.186801	8
	588.186801–642.917192	9
	642.917192–697.647583	10
Aspect	Flat (−1)	1
	North (0–22.5)	2
	Northeast (22.5–67.5)	3
	East (67.5–112.5)	4
	Southeast (112.5–157.5)	5

(continued)

Table 5 (continued)

Parameters	Subclass of parameters	Subclass rank
	South (157.5–202.5)	6
	Southwest (202.5–247.5)	7
	West (247.5–292.5)	8
	Northwest (292.5–337.5)	9
	North (337.5–360)	10
Road density (km/sq.km)	0–13.189807	1
	13.189807–26.379614	2
	26.379614–39.569421	3
	39.569421–52.759229	4
	52.759229–65.949036	5
	65.949036–79.138843	6
	79.138843–92.32865	7
	92.32865–105.518457	8
	105.518457–118.708264	9
	118.708264–131.898071	10
Settlement density (km/sq.km)	0–8.787971	1
	8.787971–17.575941	2
	17.575941–26.363912	3
	26.363912–35.151883	4
	35.151883–43.939853	5
	43.939853–52.727824	6
	52.727824–61.515795	7
	61.515795–70.303765	9
	70.303765–79.091736	10
Landuse/Land cover	Semi ever green	8
	Decidious forest	10
	Deep forest	9
	Forest	7
	Mixed vegetation	6
	Water body	4
	Cropland	3
	Plantation	1
TWI	4.361429–4.91753	1
	4.91753–5.473632	2
	5.473632–6.029733	3
	6.029733–6.585835	4
	6.585835–7.141937	5

(continued)

Table 5 (continued)

Parameters	Subclass of parameters	Subclass rank
	7.141937–7.698038	6
	7.698038–8.25414	7
	8.25414–8.810241	8
	8.810241–9.366343	9
	9.366343–9.922444	10

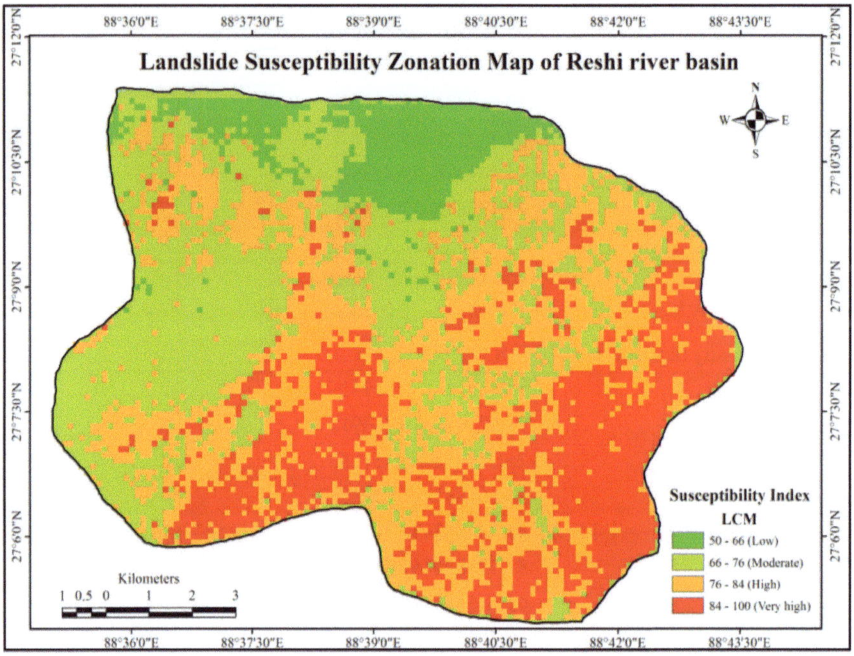

Fig. 3 Landslide susceptibility map derived from LCM model

Table 6 Susceptibility hazard zone and landslide area using LCM model

Landslide susceptibility category	LCM Index	Pixels	% of area
Low	50–66	691	8.71
Moderate	66–76	2504	31.59
High	76–84	2883	36.37
Very high	84–100	1849	23.33

Source Calculated by researchers

6.3 Changing Vegetation Coverage on the Occurrences of Landslides

Land use change plays a key role in the occurrence and movement of landslides triggered by rainfall (Kingsbury 1994; Glade, 2003 and Karsli et al., 2009), as landslide frequency is directly linked to land use variations (Glade, 2003). However, there is no consensus on how vegetation impacts slope stability. Some studies suggest that vegetation may not contribute to slope stability (Greenway 1987 and Wu & Sidle, 1995), while others argue that roots strengthen slopes (Gökceoglu and Aksoy 1996 and Jakob, 2000). The normalized difference vegetation index (NDVI) is a useful graphical tool for analyzing remote sensing data, often from satellites, to assess the presence of live green vegetation. By measuring the reflection of visible and near-infrared light, researchers can determine the density of green vegetation in an area. Chlorophyll in plant leaves absorbs visible light (0.4–0.7 μm) for photosynthesis, while the leaf structure reflects near-infrared light (0.7–1.1 μm). The more leaves present, the more these wavelengths are influenced.

Vegetation cover significantly affects landslides for several reasons. Firstly, the roots of vegetation help to hold soil particles together, which improves the soil's cohesion and stability. Secondly, plants regulate water movement by capturing rainfall and facilitating transpiration, which lowers the amount of water that penetrates the soil and reduces pore-water pressure. Additionally, vegetation protects the soil surface from direct rainfall impact, which minimizes erosion and helps prevent the formation of gullies that could result in slope failure. Finally, the organic material contributed by plants improves soil structure and drainage, which reduces the risk of landslides. In contrast, areas without vegetation are more susceptible to landslides because of reduced soil cohesion, heightened erosion, and more significant water infiltration.

The Normalized Difference Vegetation Index (NDVI) is a standardized index that allows the generation of an image displaying greenness (relative biomass). The NDVI equation is as follows (ArcGIS 10.2).

$$\text{NDVI} = (\text{NIR} - \text{Red})/(\text{NIR} + \text{Red})$$

where NIR = pixel values from the infrared band and Red = pixel values from the red band. This index outputs values between -1.0 and 1.0, primarily representing greenness, where any negative values are mainly generated from clouds, water, and snow, and values near zero are mainly generated from rock and bare soil. Very low values of NDVI (0.1 and below) correspond to barren areas of rock or sand. Moderate values (0.2–0.3) represent shrub and grassland, while high values (0.6–0.8) indicate temperate and tropical rainforests (Ahmed et al., 2013). In this research, OLI images and Landsat 4–5 TM images from the same season (Pre-monsoon) were acquired from the official website of the USGS. Finally, the NDVI map of CMA was prepared by analyzing band 3 and band 4 for Landsat 4–5 and band 4 and band 5 for the OLI image.

6.4 Analysis

The NDVI maps for four different periods (2001, 2006, 2011, and 2016) were created using satellite images and the NDVI equation. The images were classified into six categories: moist land, water bodies, plantation/mixed vegetation, light forest, deep forest, and deciduous forest. Significant changes in land use between 2001 and 2016 were made. The average NDVI values for the entire river basin area indicate an overall increase from 2001 to 2016 (see Figs. 4, 5, 6 and 7). In 2001, the deciduous forest area was 2.23%, but it has now decreased to 1.26%. Both light forests and deep forests have decreased. The percentage of light forest was 42.83% in 2001, but in 2016 it dropped to 31.48%. Almost 10% of the deep forest has been converted into light and mixed forest. Additionally, a significant proportion (27.94%) of mixed vegetation areas has been turned into plantation areas. These results indicate a continuous loss of forests and other vegetated areas from 2001 to 2016 (Table 7 and Fig. 8).

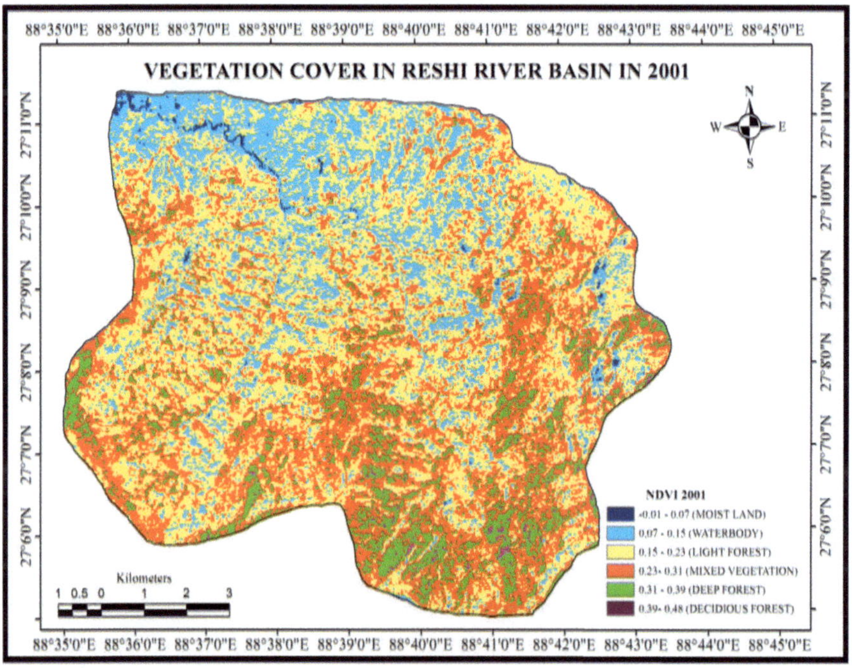

Fig. 4 Vegetation covers in 2001

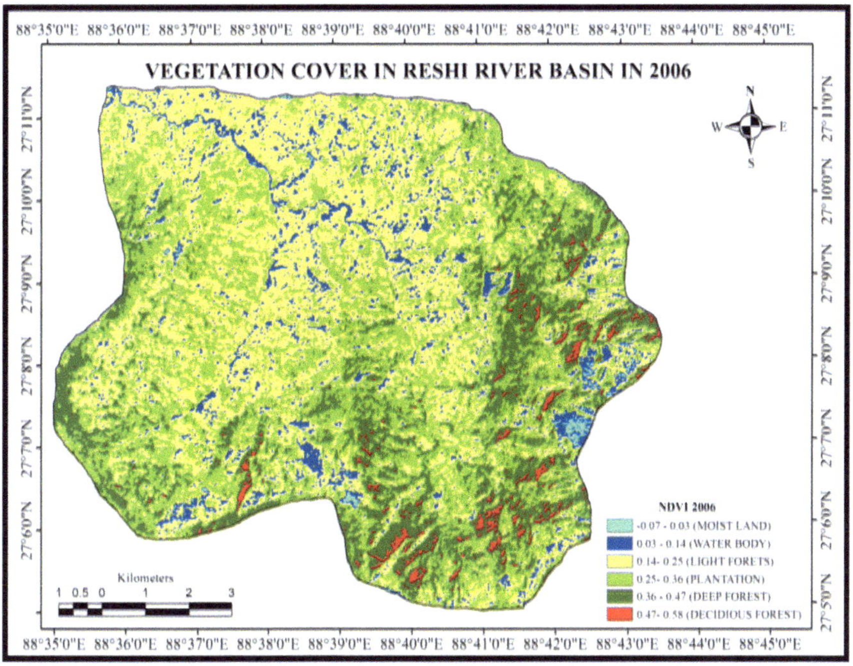

Fig. 5 Vegetation covers in 2006

6.5 *Major Findings*

(a) NDVI effectively links vegetation density with landslide susceptibility. Denser vegetation mitigates landslide risk.

(b) The study confirmed that landslides are more frequent on steep slopes near rivers, trails, and active faults. Vegetation cover helps stabilize the region. Landslides are most common in the natural forests, which cover the largest area in the Reshi River basin.

(c) In 2001, about 28% of mixed vegetation areas had become plantations by 2016. Due to the hilly terrain, this reduction of forested land is a critical issue in Darjiling, Gangtok, and surrounding areas. It leads to the threat of landslides impacting the environment and economy.

(d) Landslides related to natural forests are similar to all landslides in the basin. Natural forests cover 60% of the area and experience the highest number of landslides. Most natural forests are located on steep slopes prone to landslides during rainstorms

(e) In the last 16 years, step cultivation has rapidly increased, particularly in the eastern parts of the study area. Data analysis shows a 27% decline in forest cover from 2001 to 2016, accompanied by a 19% increase in step cultivation.

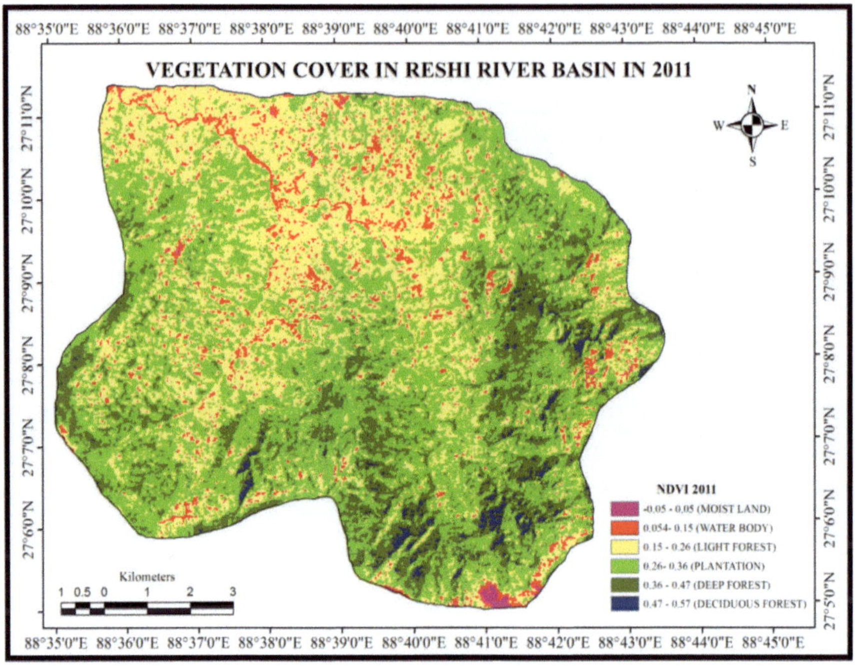

Fig. 6 Vegetation covers in 2011

This suggests an overall decline of almost 30% in forest cover during the study period.

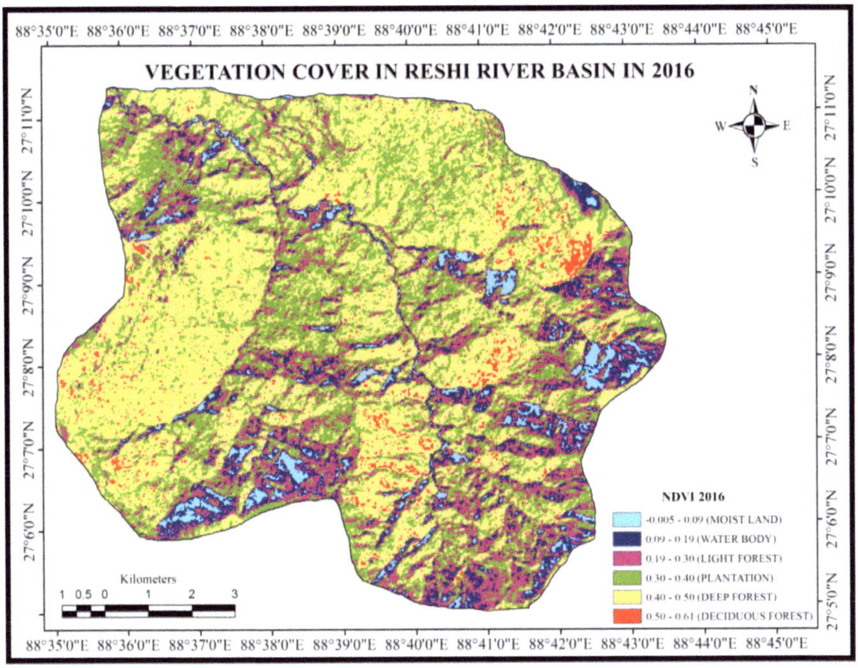

Fig. 7 Vegetation covers in 2016

Table 7 Calculate the individual land use land cover area in percentage in different periods

	2001	2006	2011	2016
Class	Area (%)	Area (%)	Area (%)	Area (%)
Moist land	0.48	0.48	0.30	3.07
Waterbody	13.00	6.71	5.02	3.21
Plantation/ Mixed vegetation	27.94	35.89	41.35	44.88
Light forest	42.83	37.90	34.72	31.48
Deep forest	13.53	17.41	17.23	16.09
Deciduous forest	2.23	1.61	1.38	1.26
Total	100	100	100	100

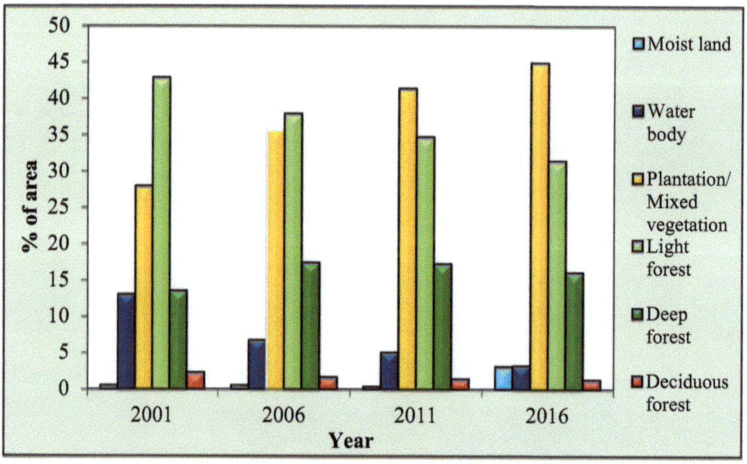

Fig. 8 Percentage of land use/land cover area in different periods (2001–2016)

7 Conclusion

Landslides are complex natural phenomena influenced by various known and unknown factors. Scientific analysis plays a crucial role in predicting and mitigating the damage caused by these events. This research utilized the GIS-based Landslide Susceptibility Mapping (LCM) approach to produce reliable susceptibility maps for the Reshi River basin. The study incorporated 13 thematic layers, including a drainage map, digital elevation model (DEM), slope angle map, land use/land cover map, relative relief map, lineament density map, and lithological map, to delineate landslide hazard zones using DEM-derived causative factors and remote sensing data. The analysis revealed that 70% of observed landslide points were situated in high and very high susceptibility zones, indicating the reliability of the generated map. These maps serve as a vital tool for identifying areas prone to landslides, providing a strong foundation for assessing landslide risks in the research area. Furthermore, the GIS-based methodology demonstrated significant potential in processing and analyzing spatial data, reaffirming its importance in natural disaster assessment.

The susceptibility maps generated in this study hold immense value for decision-makers in regional planning and disaster management. By outlining high-risk zones, these maps can guide the implementation of effective strategies to mitigate landslide hazards and support sustainable development in the Reshi River basin.

References

Ayalew, L., Yamagishi, H., Marui, H., & Kanno, T. (2005). Landslides in Sado Island of Japan: Part II. GIS-based susceptibility mapping with comparisons of results from two methods and verifications. *Engineering Geology, 81*(4), 432–445.

Glade, T. (2003). Landslide occurrence as a response to land use change: A review of evidence from New Zealand. *CATENA, 51*(3), 297–314.

Gökceoglu, C., & Aksoy, H. (1996). Landslide susceptibility mapping of the slopes in the residual soils of the Mengen region (Turkey) by deterministic stability analyses and image processing techniques. *Engineering Geology, 44*(1–4), 147–161.

Jakob, M. (2000). The impacts of logging on landslide activity at Clayoquot Sound, British Columbia. *CATENA, 38*(4), 279–300.

Karsli, F., Atasoy, M., Yalcin, A., Reis, S., Demir, O., & Gokceoglu, C. (2009). Effects of land-use changes on landslides in a landslide-prone area (Ardesen, Rize, NE Turkey). *Environmental Monitoring and Assessment, 156*(1), 241–255.

Lee, S., & Sambath, T. (2006). Landslide susceptibility mapping in the Damrei Romel area, Cambodia using frequency ratio and logistic regression models. *Environmental Geology, 50*, 847–855.

Park, S., Choi, C., Kim, B., & Kim, J. (2013). Landslide susceptibility mapping using frequency ratio, analytic hierarchy process, logistic regression, and artificial neural network methods at the Inje area, Korea. *Environmental Earth Sciences, 68*(5), 1443–1464.

Pradhan, A. M. S., & Kim, Y. T. (2014). Relative effect method of landslide susceptibility zonation in weathered granite soil: A case study in Deokjeok-ri Creek, South Korea. *Natural Hazards, 72*(2), 1189–1217.

Rajakumar, S., Nagarajan, R., & Rajamanickam, G. V. (2007). GIS-based landslide hazard zonation in part of Western Ghats, India. *Journal of the Indian Society of Remote Sensing, 35*(2), 163–175. https://doi.org/10.1007/BF02990784

Van Westen, C. J., Van Asch, T. W., & Soeters, R. (2006). Landslide hazard and risk zonation—why is it still so difficult? *Bulletin of Engineering Geology and the Environment, 65*(2), 167–184.

Wu, W., & Sidle, R. C. (1995). A distributed slope stability model for steep forested basins. *Water Resources Research, 31*(8), 2097–2110.

Landslide Inventory in Parts of Sikkim, Eastern Indian Himalaya: Navigating Spatial and Temporal Patterns

Kuldeep Dutta, Nishchal Wanjari, and Anil Kumar Misra

Abstract Landslides are one of the most prevalent and destructive disasters, causing substantial damage to human lives, infrastructure, and the environment. Their occurrence is governed by complex interactions of geological, topographical, climatic, and anthropogenic factors. With the increasing impacts of climate change and expanding human settlements into vulnerable regions, landslide occurrences are expected to intensify in both frequency and magnitude. This study focuses on developing a comprehensive landslide inventory for Gangtok and Pakyong districts in East Sikkim, integrating historical records and remote sensing data. Analysis of historical records revealed 478 landslides, predominantly shallow debris flows, with some spanning up to 30,000 sq. m. Landslides are concentrated along critical transportation corridors like NH-10 and NH-310A. Geological controls, such as the presence of Phyllite and Quartzite lithologies and zones of structural weakness like the MCT, significantly influence landslide dynamics. Between 2019 and 2023, an additional 113 landslides were recorded, raising the total to 591. Recurrence patterns indicated a shift from predominantly new landslides to a higher recurrence rate in recent years. The spatial distribution of high-risk areas also shifted during this period due to land-use and land-cover changes, with rising built-up areas correlating with increased landslide densities. The inventory developed combining historical records and semi-automatic mapping method provides detailed insights into landslide occurrence, recurrence, and spatial distribution. It highlights the significant role of natural and anthropogenic factors. This comprehensive inventory would serve as a vital tool for informed land-use planning, disaster preparedness, and the development of targeted mitigation strategies in East Sikkim.

K. Dutta (✉)
Centre for Earth Sciences and Himalayan Studies, Itanagar, Arunachal Pradesh, India
e-mail: kuldeepdutta1994@gmail.com

K. Dutta · N. Wanjari · A. K. Misra
Department of Geology, Physical Science Block, Sikkim University, Gangtok, Sikkim, India
e-mail: nwanjari@cus.ac.in

A. K. Misra
e-mail: akmisra@cus.ac.in

© The Author(s), under exclusive license to Springer Nature Switzerland AG 2025
S. Ghosh (ed.), *The Himalaya Dilemma*, Springer Natural Hazards,
https://doi.org/10.1007/978-3-031-95083-4_13

Keywords Google Earth engine · Sentinel 2 · Spatiotemporal inventory · Recurring landslides · Eastern Himalayas · Semi-automatic mapping

1 Introduction

Landslides are among the most prevalent natural hazards globally, posing significant risks to human lives, infrastructure, and the environment (Chen & Chen, 2021; Dilley, 2006; Froude & Petley, 2018; Mansour et al., 2011; Petley et al., 2005). Their occurrence is influenced by a multitude of factors including geological setting, topography, climate, and human activities (Beven & Kirkby, 1979; Hoek & Bray, 1991). With the expansion of populations into increasingly vulnerable areas and the changing patterns of precipitation due to climate change, the frequency and magnitude of landslides are expected to escalate (Fidan et al., 2024). Therefore, understanding the spatial and temporal distribution of landslides is imperative for developing effective hazard mitigation and risk management strategies (Gariano & Guzzetti, 2016; Jones et al., 2021; Muñoz-Torrero Manchado et al., 2021).

Landslides pose a significant global threat, responsible for approximately 72,000 fatalities and over tens of billions in damages worldwide from 1900 to 2022, as documented by the Emergency Events Database (EM-DAT 2023). These figures are projected to rise due to human-induced climate change (Gariano & Guzzetti, 2016). However, our ability to mitigate and predict landslide hazards is severely hindered by the lack of comprehensive and precise data regarding their occurrence in both time and space. This scarcity of information greatly hampers geoscientists' understanding of landslide dynamics and their correlation with triggering factors (Morgan & Wieczorek, 1996). Despite their critical importance for assessing landslide hazard and risk, maps depicting landslide-prone areas are surprisingly rare, with less than 1% of global slopes being adequately mapped (Guzzetti et al., 2012).

Recent advancements in remote sensing technology have revolutionized the ability to detect and monitor landslides. Remote sensing techniques, particularly those leveraging satellite data, have emerged as invaluable tools for mapping and analysing landslide occurrences across vast spatial extents. These techniques offer enhanced accuracy, efficiency, and cost-effectiveness compared to traditional ground-based surveys (Fiorucci et al., 2019; Jones et al., 2021; Kincey et al., 2021). By harnessing the power of remote sensing, researchers can identify and monitor landslides with unprecedented detail, thereby facilitating better-informed decision-making and proactive measures for landslide risk reduction (Behling et al., 2014, 2016; Chen & Chen, 2021; Chen et al., 2022; Yunus et al., 2020).

A landslide inventory serves as a comprehensive repository of landslide data, containing crucial information such as location, type, volume, run-out distance, and date of occurrence etc. (Fell et al., 2008; Fiorucci et al., 2019; Guzzetti et al., 2012; Lee & Jones, 2023; Marcelino et al., 2009; Moosavi et al., 2014). The compilation of such an inventory is essential for assessing landslide susceptibility, hazard, and risk within a given study area. The techniques employed in preparing landslide inventory maps depend on several factors, including the quality and accessibility of desired information, the scope and extent of the study area, the scales of base maps, and the available resources to conduct the work.

To create a robust landslide inventory, researchers must utilize a combination of historical records and remote sensing data. Historical records provide valuable insights into past landslide events, offering a foundation for understanding the characteristics and distribution of landslides within the study area. These records may include archival documents, field surveys, aerial photographs, and satellite imagery from earlier periods. However, the compilation of historical records alone does not capture the full extent of landslide occurrences, particularly those that are less conspicuous or occur in remote or inaccessible areas. This is where remote sensing techniques play a pivotal role. By harnessing satellite imagery and other remote sensing datasets, researchers can systematically detect and map landslides across large spatial extents. Remote sensing offers several advantages, including the ability to acquire data in a timely manner, cover vast areas efficiently, and detect subtle changes in the landscape indicative of landslide activity.

Various remote sensing techniques can be employed for landslide inventory mapping, depending on the specific objectives of the study and the characteristics of the study area. These techniques may include optical remote sensing (Fiorucci et al., 2018; Jones et al., 2021; Kincey et al., 2021), which utilizes visible and near-infrared wavelengths to identify surface changes associated with landslides, as well as radar remote sensing (Confuorto et al., 2023; Miele et al., 2022; Ventisette et al., 2014), which can penetrate cloud cover and vegetation to detect terrain deformations indicative of landslide activity. Furthermore, advanced image processing and analysis algorithms can be applied to remote sensing data to automate the detection and mapping of landslides, thereby improving efficiency and consistency. Machine learning techniques, such as supervised and unsupervised classification, can be trained to recognize landslide features based on their spectral and spatial characteristics, enabling rapid identification and mapping of landslide events (Behling et. 2014; Behling et al., 2016; Yunus et al., 2020; Chen et al. 2022).

The development of a comprehensive landslide inventory is fundamental for assessing landslide susceptibility, hazard, and risk. This inventory serves as a repository of landslide data, cataloguing the spatial distribution, characteristics, and temporal trends of landslide events within a specific study area. By compiling historical landslide records and utilizing remote sensing data, a detailed inventory can be created to facilitate informed decision-making and proactive measures for landslide risk reduction.

Darjiling Sikkim Himalaya has a history of several major landslides which caused loss of human life and destruction of infrastructure on a large scale. The first major occurrence of a landslide in Darjiling hills was reported in the year 1899 that led to the death of 72 people. Another report of major landslides from 1950 (127 fatalities) and 1968 (667 fatalities) (Mandal & Mondal, 2019) are the records of severe landslide disasters. Significant landslides have also occurred in the area in the 20th and 21st centuries (1991, 1993, 2004, 2005, 2009, and 2015) (Dikshit et al., 2020). In Sikkim major landslides in Gyalshing occurred during the third and last week of August 2000 claiming eleven lives. Other major slides such as the landslides along J N Marg Road, NH-10 & 310, and in Bardang, Martam, Sangkhola, Pacheykhani, Byang, Tintek, Pakoyng area in the East district are reported as frequently recurring. Again, Dzongu in North Sikkim, and many landslides in different parts of West Sikkim and South Sikkim are reported to affect the region severely. Despite the severity of this hazard, the efficacy of current assessment and management efforts remains compromised by significant scarcity in methodologies and data availability. One critical deficiency lies in the absence of an up-to-date and comprehensive landslide inventory, hindering accurate hazard assessment and effective management formulation.

In this chapter the creation of a comprehensive historical as well as a year-wise rainy season (April–August) landslide inventory for the period 2019–2023 in East Sikkim (Fig. 1) (now known as Gangtok and Pakyong district) is presented. This April to August period of the year is chosen for the study because rainfall and most landslide initiation/recurrence/remobilization incidences occur during this spell in the study area. The methodology involves the integration of historical landslide records with satellite-derived imagery to map and document landslide occurrences on an annual basis. By employing remote sensing techniques, the study aims to capture both the visible and hidden landslides, thereby enhancing the completeness and accuracy of the inventory.

The primary objectives of this chapter are to:

- Compile and analyse historical landslide records to establish an understanding of landslide distribution and characteristics in the study area.
- Utilize optical remote sensing satellite data (Sentinel 2) & Satellite Radar Topographic Mission (SRTM) Digital Elevation Model (DEM) to detect and map landslides occurring during the rainy seasons from 2019 to 2023 to develop a year-wise landslide inventory.

The creation of a detailed landslide inventory serves as a crucial step towards enhancing understanding of landslide dynamics, occurrence/recurrence, and nature. Again, there are many studies that utilize historical landslide records such as Geological Survey of India (GSI) or National Aeronautics and Space Administration (NASA) landslide inventory to analyse and prepare landslide susceptibility/risk/hazard maps. An updated landslide inventory that can be prepared through a rapid mapping technique can greatly assist in accurate landslide susceptibility/risk/hazard evaluation. By putting forward a replicable methodology and documenting the spatial and temporal trends of landslides, this inventory will provide valuable insights for land use planning and disaster preparedness efforts in East Sikkim.

2 Landslide Inventory, Background, and Context

A landslide inventory refers to a systematic record or database documenting the occurrence, characteristics, and impacts of landslides within a particular area or region. A new landslide inventory that is generated after a major triggering event,

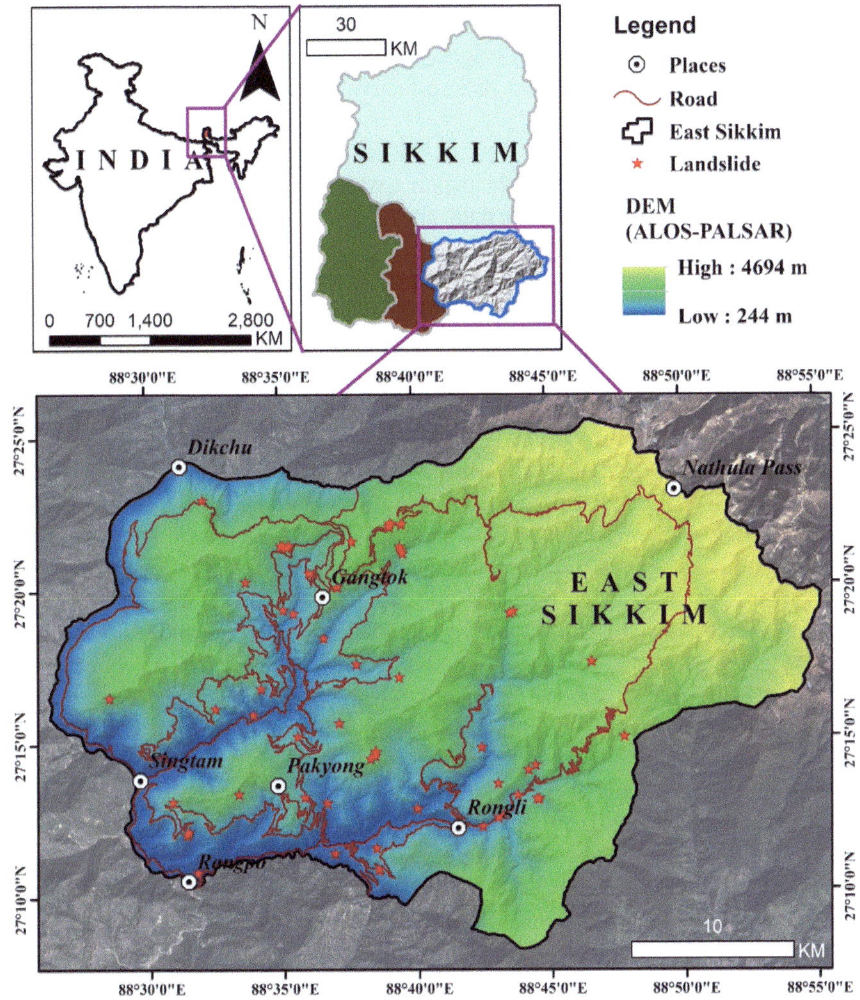

Fig. 1 Location of the study area i.e., East Sikkim, in the eastern Indian Himalayas. The region features an elevation range from 244 to 4694 m and shares its eastern border with Tibet, connected via the strategic Nathula Pass

e.g., extreme rainfall, an earthquake, storm, snowmelt, Glacial Lake Outburst Flood (GLOF) or volcanic eruptions, is referred to as an event-based landslide inventory (Galli et al., 2008; Segoni et al., 2018; Rosi et al., 2018; Fiorucci et al., 2019). Conducting field studies immediately after a hazardous triggering event has been instrumental in assessing and mapping landslides. Preparation of catalogues from news reports, other media, and personal communication (Kirschbaum et al., 2010, 2015; Petley, 2012) has also been practiced by researchers. However, in the

Himalayas, various factors such as adverse weather conditions, obstructed accessibility, and other logistical challenges significantly hinder the feasibility and efficiency of field visits as the quickest and most effective method for assessing landslides following such events.

Before the twenty-first century, a comprehensive review of the full potential of Earth observation (EO) for landslide mapping was lacking due to the limited spatial and temporal resolution of available data. Up to that point, most landslide inventories were created through the interpretation of topographic maps and aerial photographs (Parise, 2001). Advancements in earth observation technologies, like multi-temporal high-resolution optical satellite remote sensing, have facilitated the acquisition of remote sensing imagery for both pre- and post-landslide events. Consequently, this has led to an increase in the creation of landslide inventory maps (Casagli et al., 2017; Hao et al., 2020; Santangelo et al., 2015; Solari et al., 2019; Travelletti et al., 2012).

As the scenario of data availability improved (Belward and Skøien 2015), researchers have used optical remote sensing and demonstrated its effectiveness in the multitemporal analysis of landslides. Several studies have employed expert knowledge to visually interpret landslides using time-series satellite data, allowing for the detection of evolving landslide patterns (Fiorucci et al., 2019; Jones et al., 2021; Kincey et al., 2021). Automated methods of mapping landslides, such as thresholding vegetation indices from satellite time series to identify likely landslide footprints, have been developed to create multi-temporal landslide inventories and evaluate surface recovery (Behling et al., 2014, 2016; Yunus et al., 2020; Chen et al. 2022). By examining changes in vegetation index, studies can estimate the timing of landslide initiation and recurrence (Muñoz-Torrero Manchado et al., 2021). In recent years, with the rapid development of machine learning (ML) as a data science branch, and its spread over various fields, many researchers have started looking into disciplinary or thematic applications of ML methods (Huang & Zhao, 2018; Saha et al., 2020; Fang et al., 2021; Naemitabar & Zanganesh Asadi, 2021; Bhuyan et al., 2024).

The availability of free multi-sourced and multi-temporal High-Resolution Satellite Images (HRSIs) through platforms like Google Earth, which also offer 3D viewing capabilities (Crosby et al., 2012; Fisher et al., 2012), has significantly enhanced landslide inventory mapping (Mohammadi et al., 2018). Numerous studies have utilized the Google Earth platform for generating landslide inventories (Sato & Harp, 2009; Fiorucci et al., 2011; Borrelli et al., 2015; Rabby & Li, 2019; Lindsay et al., 2022; Notti et al., 2023). Additionally, the Google Earth (GE) history viewer can be employed to map event-based landslides by comparing pre- and post-event images.

In recent years, the Sentinel satellite constellation of the Copernicus programme has provided medium to high-resolution images, approximately 10 m in resolution, in both multi-spectral (Sentinel-2) and Synthetic Aperture Radar (SAR) (Sentinel-1) formats, freely accessible and with a high-frequency revisit rate. Google Earth Engine (GEE) cloud processing (Gorelick et al., 2017) has also been leveraged to generate time series of several satellite data types (optical/SAR), facilitating the detection of changes necessary for identification and mapping of landslides (Yu et al., 2018; Scheip & Wegmann, 2021; Handwerger et al., 2022; Lindsay et al., 2022; Ganerød et al., 2023; Notti et al., 2023).

In the Himalayas researchers have long practiced different methods to prepare landslide inventories. Van Westen et al. (2012, 2013) have addressed many aspects, such as the increasing concern of landslides in India due to population expansion in hilly areas, the role of the GSI in landslide studies, the methods used for landslide susceptibility assessment, the need for a detailed landslide inventory, challenges in landslide hazard and risk assessment, and the suggestion for a National Landslide Database in India. Which mission was later launched by GSI as the National Landslide Susceptibility Mapping (NLSM) (https://www.gsi.gov.in/webcenter/portal/OCBIS/pageQuickLinks/pageNLSM) in 2014–2015. Martha et al., (2016), Timilsina et al., (2017), Bhambri et al., (2017), Zhang et al., (2017), Jharendra et al., (2018), Roberts et al. (2019), Fayne et al., (2019), Ghosh et al., (2020), Martha et al., (2021), Bera et al., (2021), Roy et al., (2022) etc. have utilised EO satellite data coupled with other topographic data and field survey to study, prepare and analyse landslide inventory in many areas across Himalaya covering Western, Central and Eastern parts.

In Darjiling Sikkim Himalayas, the literature on comprehensive landslide inventory preparation is relatively sparse. Ghosh et al., (2012a) have analysed landslide inventory in Kurseong town in Darjiling, West Bengal. Ghosh et al., (2012c) studied event-based landslide inventory preparation in Darjiling Himalaya for determination of temporal and magnitude probability. This article applied the methodology that involved generating event-based landslide inventory maps using various source data spanning from 1968 to 2007. A series of studies (Ghosh et al., 2012b; Martha et al., 2015 and Joshi, 2022) have focused on the aftermath of the 2011 Sikkim earthquake to assess the damage and map the landslides triggered by the earthquake.

Amidst this there is a rising need of utilizing EO data to prepare rainfall induced landslide inventory in the study area. The preparation of landslide susceptibility mapping/maps (LSM) has attained a lot of interest to the researchers in this region. These studies to analyse susceptibility and risk in this region of Himalaya, mostly utilise GSI landslide inventory or NASA global landslide inventory data. For precise analysis of landslide susceptibility and hazard an updated inventory is necessary.

3 Description of the Study Area

Administratively, Sikkim is currently divided into six districts: Gangtok, Pakyong, Gyalshing, Namchi, Soreng, and Mangan. There were only four districts prior to 2022 and formerly Gangtok & Pakyong constituted one district, East Sikkim, covering an area of 954 km^2 (Fig. 1).

East Sikkim presents a complex interplay of geological, climatic, and geomorphological factors, significantly influencing its land use and land cover (LULC). Geologically, the region is situated within the Himalayan orogen, characterized by distinct tectono-stratigraphic units (Figs. 2 and 3, II). The study area primarily encompasses the Central Crystalline Gneissic Complex (CCGC) of the Higher Himalayan Sequence (HHS) in the east and the Daling group of rocks within the Lesser Himalayan Sequence (LHS) in the west, separated by the Main Central Thrust (MCT). The CCGC comprises lithologies such as banded migmatite, garnet biotite gneiss, and mica schist, while the LHS consists of chlorite sericite schist and quartzite. The presence of Lingtse Gneiss and numerous minor lineaments further contribute to the geological complexity and associated hazard potential, including mass movements along the MCT.

Climatically, East Sikkim exhibits significant variations over short distances due to elevation and geography. At lower elevations, such as Rangpo (315 m), the region experiences hot summers and supports deciduous and evergreen vegetation. However, within a mere 30 km, higher elevations like Zuluk and Nathang valley transform into sub-alpine dry lands with seasonal snowfall. The average rainfall (Fig. 3, III) is approximately 20 mm/5 days, with peaks reaching 190 mm. The mean daily temperature is around 13 °C, with observed increasing trends for mean, maximum, and minimum daily temperatures from 1994 to 2024. These climatic variations significantly impact the geomorphology, shaping the landscape through processes like erosion, weathering, and mass movements.

Geomorphologically, the terrain is predominantly mountainous and rugged, characterized by steep slopes, deep valleys, and numerous snow-fed lakes in the eastern higher altitudes. The region's geomorphology is heavily influenced by the active tectonics of the Himalayas, including the MCT, and the diverse climatic conditions.

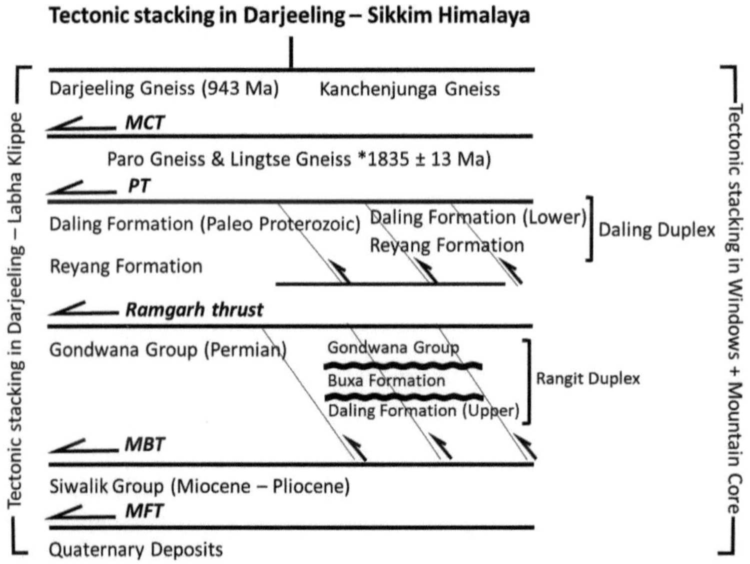

Fig. 2 Regional simplified stratigraphy of the Sikkim Darjiling Himalaya (modified after Bhattacharyya et al., 2015)

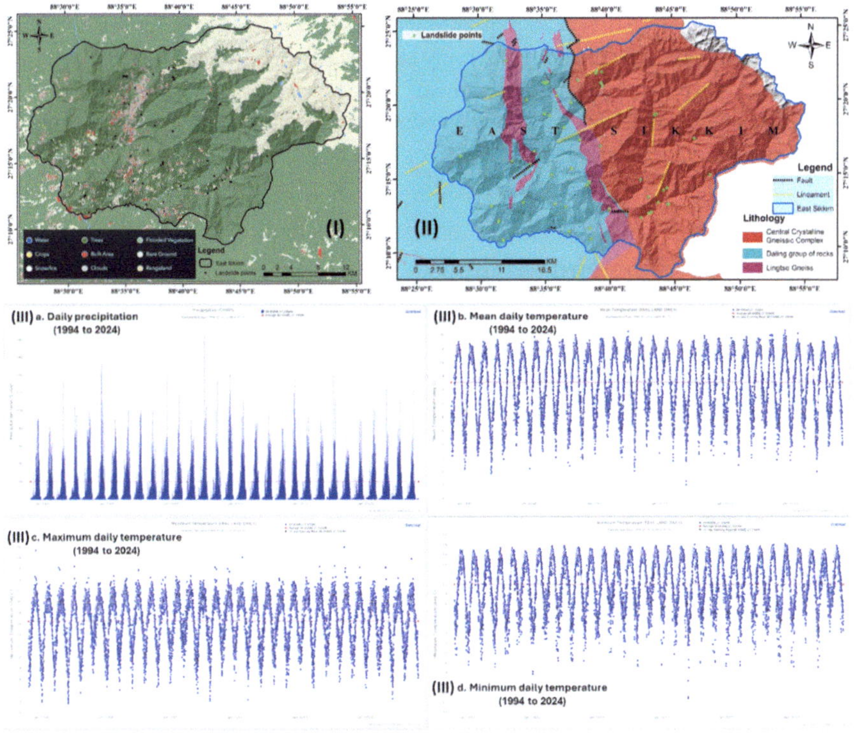

Fig. 3 **I** Land Use and Land Cover map of East Sikkim derived from Sentinel-2; **II** Shows distribution of different lithologic groups, fault, and lineament (after GSI, Bhukosh 2023); **III** Precipitation (CHIRPS) and temperature (ERA5) climatology of the study area for the last 30 years i.e., 1994–2024 for the study area, **a** daily precipitation, **b** mean daily precipitation, **c** maximum daily precipitation, **d** minimum daily precipitation

This leads to a high susceptibility to natural hazards such as landslides, particularly along the MCT and in areas with concentrated mass movements.

The LULC of East Sikkim (Fig. 3, I) is characterized by a low population density, with settlements concentrated around Gangtok, Pakyong, Singtam, and Rongli. A significant portion of the region is covered by forests, interspersed with small villages. Agricultural land, primarily used for cultivating maize, buckwheat, barley, and potatoes, also occupies a considerable area. Gangtok, a major tourist destination, has witnessed infrastructure development to support the tourism industry, along with the establishment of pharmaceutical manufacturing facilities. Recent development initiatives, such as the Sevoke-Rangpo railway project and the road corridor to Nathula Pass, have altered the LULC, particularly in the forest-covered eastern region.

The economy of East Sikkim is primarily driven by agriculture, forestry, and tourism. The region's natural resources support a thriving tourism industry, which is a vital source of income. However, increasing population growth, urbanization, and infrastructure development are exerting pressure on the natural landscape, leading to environmental degradation. The region relies heavily on roadways for connectivity, which are often disrupted during the rainy season. Efforts are underway to improve connectivity through the ongoing railway project, expected to be completed in mid-2025. The delicate balance between economic development and environmental preservation remains a critical challenge for East Sikkim.

4 Data Collection and Methods

For this study an ensemble of methods has been applied as shown in Fig. 4. The study commences with compilation of historical records of landslides in the study area. Ground information of some landslides is collected to cross verify the compiled historical inventory. Further, EO satellite imageries from GE, Sentinel 2 (S2) and Landsat 5/7/8/9 (LT 5/7/8/9) were analysed for creation of a yearly rainy season landslide inventory for East Sikkim for the period of 2019–2023.

4.1 Historical Records and Geospatial Database Generation

Historical records of landslide events in the study area have been compiled from different sources such as GSI inventory, National Disaster Management Authority (NDMA) and Sikkim State Disaster Management Authority (SSDMA) reports on landslides, NASA global landslide inventory data set, print and digital news articles, and Boarder Roads Organization (BRO) reports etc. All possible details such as location, time/date of initiation, cause, type, material, and damage caused etc. are noted for individual landslides and an inventory is prepared from these secondary sources.

4.2 Field Observations

A limited number of landslides were assessed along NH-10 and NH-310 between Rangpo, Singtam, Tashi viewpoint and Tsomgo lake. Moreover, some other landslides in Rumtek, Burtuk, Pakyong, Rangpo-Zuluk road, and Tintek were visited to collect information about their failure type, lithology, structural disposition, geology, soil cover material, hydrological condition, and joint condition etc. (Fig. 5).

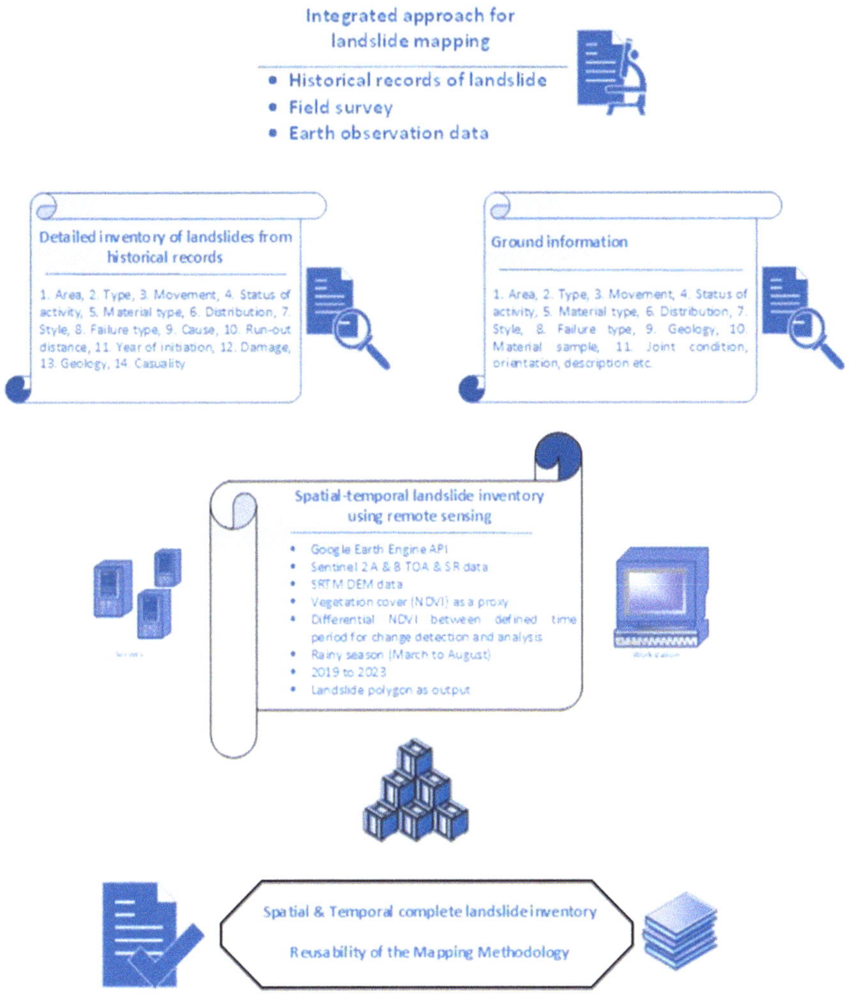

Fig. 4 Summary flowchart of integrated approach taken in the study to map and analyse landslide inventory for East Sikkim by evaluating historical records, field observations and Earth Observation data

4.3 Use of RS and GIS for Preparation of Landslide Inventory

A semi-automatic mapping technique utilizing Sentinel-2 surface reflectance data and the SRTM DEM serves as the foundation for generating yearly rainy season landslide inventories in East Sikkim.

Semi-automatic methods of mapping involve combination of automated processes and human input to analyse and interpret remote sensing data for mapping purposes.

Fig. 5 The manual mapping of landslides from GE imagery is performed for the known landslide polygon to use as training dataset for semi-automatic mapping

These methods leverage computer algorithms to perform initial data processing and analysis, followed by human intervention to refine results, validate findings, and interpret patterns.

The workflow, delineated in Figs. 6a, b, and 7, are executed within the GEE Application Programming Interface (API), in conjunction with ArcMap and QGIS software platforms, to integrate and analyse the data.

The process commences by overlaying the pre-prepared historical landslide inventory onto imagery sourced from GE and S2 to meticulously delineate landslide polygons (Fig. 5). Subsequently, a year-wise landslide polygon dataset (in KML/SHP format) is generated for the study period spanning 2019–2023. This dataset is later utilized to determine the threshold of spectral raster in semi-automatic detection.

The year-wise analysis of Sentinel-2 data is conducted within the GEE API, leveraging Java language for command execution. Within GEE, the workflow begins with the selection of datasets, including Sentinel-2 surface reflectance (S2-SR), SRTM DEM, and global surface water body data. The study area is delineated using a shapefile. Pre-processing functions, such as calculating slope from DEM and masking water and clouds, are defined.

Following this, pre- and post-event imagery are filtered from the collection based on location, date, and cloud cover criteria. Masks are then applied to the imagery, and bands 8 and 4 (NIR and Red) are selected to calculate the Normalized Difference Vegetation Index (NDVI) for both pre- and post-event images. The relative difference NDVI (RdNDVI) between the pre- and post-event imagery is computed using the

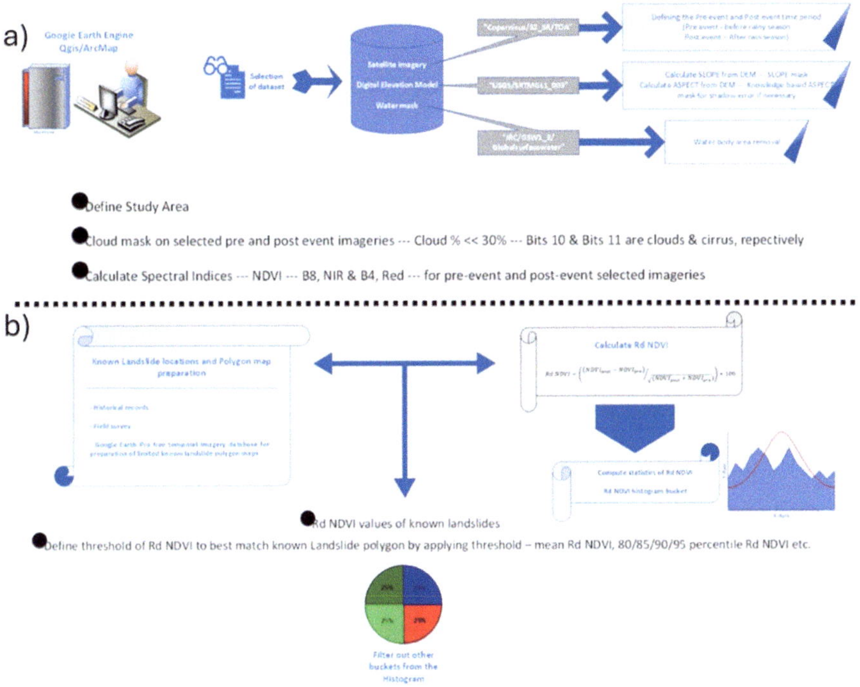

Fig. 6 **a** Initial stage of the GEE processing for semi-automatic mapping, flowchart showing the steps of data selection and pre-processing done in the study. **b** After selecting data and pre-processing, this flowchart shows the raster calculation and defining the threshold based on training dataset for mapping active landslides in GEE

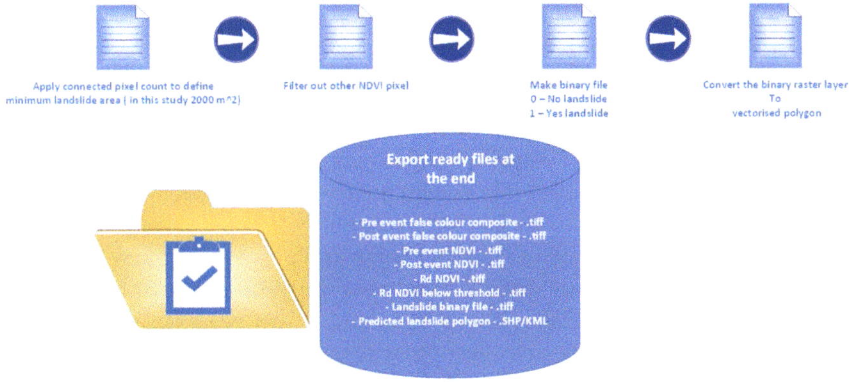

Fig. 7 This flowchart represents the workflow for post-processing steps to create the binary landslide raster and to successfully import the results in desirable format

equation given below,

$$\text{Rd NDVI} = \left(\left(\text{NDVI}_{\text{post}} - \text{NDVI}_{\text{pre}}\right)/\text{sqrt}\left(\text{NDVI}_{\text{post}} + \text{NDVI}_{\text{pre}}\right)\right) * 100\ldots \quad (1)$$

where:

$\text{NDVI}_{\text{post}}$ is the Normalized Difference Vegetation Index calculated from post-event satellite imagery.

NDVI_{pre} is the NDVI calculated from pre-event satellite imagery.

sqrt () represents the square root function.

The result is scaled by 100 to make it more interpretable.

The RdNDVI raster is exported as a TIFF file and utilized in ArcMap/QGIS to determine the threshold from known landslides obtained from the year-wise polygon dataset prepared earlier from GE/S2 imagery. Subsequently, within GEE, the threshold is applied to the RdNDVI raster, generating a binary raster where 'Yes' denotes landslide presence (1) and 'No' denotes absence (0). To mitigate any miscalculations or scattered pixels, the 'connected pixel minimum count function' is employed for filtering. Detailed information regarding the imagery, masks, and thresholds is presented in Table 2.

This workflow effectively maps landslides that have occurred within a defined timeframe and area. Additionally, the GEE code enables the export of various outputs, including:

1. Slope (TIFF)
2. Pre-event false colour composite (FCC) (TIFF)
3. Post-event FCC (TIFF)
4. Pre-event Normalized Difference Vegetation Index (NDVI) (TIFF)
5. Post-event NDVI (TIFF)
6. Relative difference NDVI (RdNDVI) (TIFF)
7. RdNDVI after threshold (binary TIFF)
8. RdNDVI after threshold (filtered) (TIFF)
9. Mapped/detected landslide polygons (KML/SHP).

This comprehensive set of outputs provides valuable information for further analysis and interpretation of landslide occurrences within the specified region and period.

After exporting the year-wise landslide polygon file from GEE, they are subsequently analysed in ArcMap/QGIS to determine their occurrence and recurrence status. The overlapping landslide polygons from different years are identified and marked out, providing insights into areas prone to repeated landslide events. Furthermore, the year-wise inventory of landslides is overlaid onto S2 derived Land Use and Land Cover (LULC) maps within East Sikkim. This overlay analysis allows for the evaluation and discussion of landslide density and hazard susceptibility across different parts of the region. By integrating the landslide inventory data with LULC information, it becomes possible to assess the relationship between land cover types and landslide occurrence, aiding in the identification of areas at higher risk of landslides.

5 Results and Discussion

5.1 *Historical Landslide Geodatabase*

The landslide inventory conducted for East Sikkim, through the compilation of historical data, identifies 478 landslides. This compilation unveils notable insights into the features, prevalence, causes, and geological composition of landslides in East Sikkim.

Most landslides in East Sikkim are identified as active (82%), planar failure or slide type (83%), shallow translation landslides (94%), and involve debris (64%). Most landslides are relatively small in area, with the majority below 5000 sq. m, although some are as large as 30,000 sq m. The runout distance of landslides is mostly less than 100 m, but some extend up to 200 m. Landslides are predominantly located in the lowly dissected hills geomorphological category (36%), followed by moderate and highly dissected categories. Landslides are less common on very steep slopes. A significant portion of landslides are concentrated along National Highways NH-10 and NH-310A. Additionally, approximately 48% of the landslides are situated in proximity to state highways. Debris are the most common material involved in landslides (64%), followed by rock (28%) and a combination of rock and debris (8%). Most landslides occur suddenly (rapid) (93%) and exhibit shallow failures (94%). The majority of landslides are slides (83%), with smaller percentages attributed to falls, topples, and subsidence. The hydrological condition of landslide sites varies, with 67% exhibiting wet or dripping conditions, followed by damp (8%), dry (15%), and flowing water (10%). Many landslides are enlarging and widening (61%) or retrogressive and widening (30%), while remaining are advancing and widening or confined. Rainfall is the predominant triggering factor for landslides (70%). More than half of the landslides (56%) occur on Phyllite & Quartzite lithology.

The analysis of landslide inventory data in East Sikkim until 2023 reveals several noteworthy patterns and attributes (Fig. 8). It is evident that landslides in the region are predominantly active, planar failures, primarily shallow landslides involving debris. This indicates a propensity for smaller-scale, relatively rapid movements in the landscape, with debris being a common component of these events. Despite many landslides being of smaller area, there are notable exceptions, with some extending up to 30,000 sq. m, underscoring the potential for significant geomorphic changes under aggravated erosion and weathering with the changing climate of Himalaya (Gabet et al., 2010). Many small landslides along national and state highways account for most of the landslides in the study area. These small landslides are very unpredictable and hazardous for moving traffic and habitation downslope.

Geomorphologically, landslides are concentrated in lowly dissected hills, suggesting a susceptibility to slope instability in these terrains. This correlation aligns with the findings that landslides are less common on very steep slopes, or on transportation slopes, where natural predisposing force or human intervention may have minimized instability to some extent. The geographical distribution of landslides, particularly along National Highways NH-10 and NH-310A, indicates a vulnerability of transportation infrastructure to landslide hazards. Furthermore, the

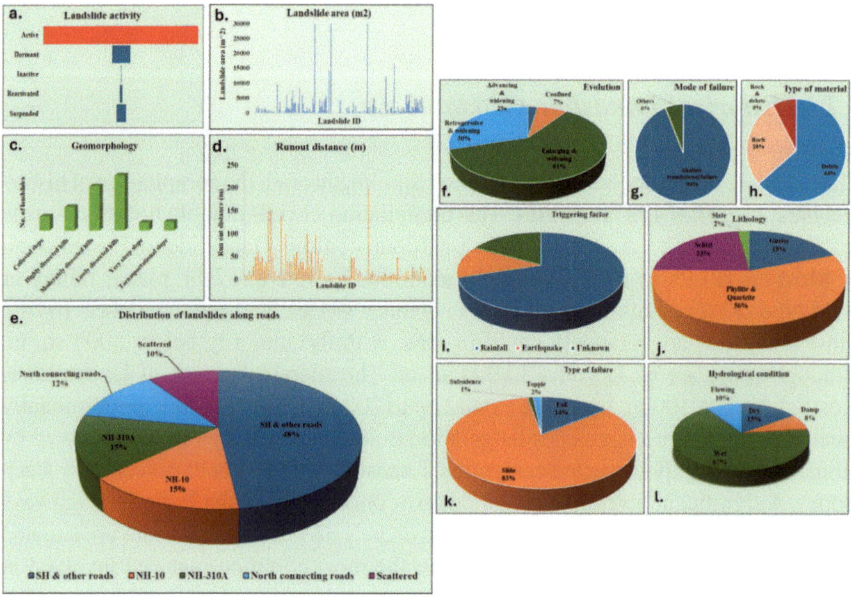

Fig. 8 The analysis of historical landslide database. **a** shows a major portion of landslides are active, **b** area wise most landslides are small but there are some large landslides, **c** geomorphology classes and landslide distribution, **d** 37 landslides have runout distance more than 100 m, **e** most of the landslides are distributed along state highways and NH-10 & NH-310, **f** most of the landslides are widening and enlarging, **g** shallow translation failure is dominant, **h** more than 60% landslides are debris material involved, **i** rainfall mostly initiate or trigger landslides in the region, **j** more than 50% landslides are distributed over Phyllite and Quartzite lithology of LHS, **k** Planar slide is the prominent type of failure, **l** wet hydrological condition contribute to slope failure in the study area in more than 65% cases

high percentage of landslides occurring in proximity to state highways underscores the potential implications for regional connectivity and transportation networks. The state administration is escalating its construction efforts to build a sustainable road network in the study area. As the area shares an international border with Tibet (China), it is a very strategic geographical area of the Indian nation. In recent times the construction and development activities have increased multi-fold across Indian Himalayas and the study area as well falls under the same. New road construction and excavation for road extension initiates many landslides. Sometimes it is because of poor planning an execution, and/or inherent instability prone of the slope. Thus, the presence of landslides along highways increases risk and demand proper evaluation and planning beforehand going for development activities.

A notable aspect is the predominant involvement of debris in landslides (Figs. 8 and 9), which suggests a significant influence of weathering and erosion processes in shaping slope instability. This is verified by the high proportion of landslides triggered by rainfall, highlighting the role of precipitation as a key triggering factor in landslide events in East Sikkim. The western part of the study area is covered with low grade metamorphic rocks of LHS, these rocks are more prone to weathering and erosion and consequent degradation. Micaceous minerals, clay mineral formed because of alteration of the feldspathic and mica group of minerals, high density of joints that are widened or degraded because of rainfall (Fig. 10) and spring water etc. contribute to the generation of debris (Fig. 9).

Moreover, the area is reported to have many paleo debris slopes that are more than thousands of years old. In many instances (Fig. 9) the presence of angular consolidated debris in the field indicates rapid movement conditions. Hence, a slope that is treated as liveable in the present time may have inherited debris flow initiation probability because of the presence of paleo debris and sediments. Thus, the evaluation of surface condition is necessary to fully understand bearing capacity of an area. The areas thought to have hard rock beneath ground probably do not have hard rock to a considerable depth instead filled with paleo debris. Hydrological conditions during landslides also appear to play a crucial role, with a majority exhibiting wet or dripping conditions. This suggests that saturation of slopes due to rainfall may contribute to increased landslide susceptibility, particularly in areas with poor drainage or high permeability lithologies.

The East Sikkim region receives heavy monsoonal and orogenic precipitation. The rainy season in the area extend from starting of April to the August and September. Prolonged exposure to rainfall is one of the factors that degrades the slopes (Fig. 10).

Fig. 9 Field photographs of unstable slopes along National Highway-10. **a** dormant debris flows become active during monsoon, **b**, **d**, and **e** semi-consolidated angular debris from paleo-deposit (sign of rapid and less distance deposition), **c** rock slabs and debris broken along foliation deposited on slope, **f** mixed lithology of foliated rock and highly weathered rock and debris, a susceptible slope along NH-10

Fig. 10 The Gaidhara, S-turning sinking zone stable during dry wintertime (left) and active during rainy season

Moreover, this region of Himalaya is known for sub surface network of springs. Natural springs in this part of the world serves for daily use of fresh water but also cause degradation of slopes. Continuous flow of water dampens the ground, increases pore pressure, enhance alteration, and contribute to landslide. Figure 11a shows an example of the mentioned processes acting in the crown part of Bardang landslide along NH-10 near 32 no. between Singtam and Majhitar. The highly deformed quartzite rock in the area is fractured in a systematic fashion under tectonic stress (part of Lesser Himalayan Duplex (LHD), intrinsic lithological properties and thrust effect), climatic (rainfall, seasonal temperature difference) and hydrogeologic processes.

The exhumed rocks of the LHD system displays history of intensive deformation and sulphide mineralization in the south-western part of East Sikkim in Rangpo area (Fig. 11). These mineralized zones of East (as well as South) Sikkim shows intense aggravated weathering processes.

Jha and Sharma (2022) have reported characteristic oxide and sulphide minerals in the study area with presence of magnetite, chalcopyrite, sphalerite, galena, pyrrhotite, and cobaltite along with galena, pyrrhotite chalcopyrite, arsenopyrite, sphalerite, and pyrite and the alteration minerals (Fig. 11c and d) malachite and azurite.

Ground water natural flow and seepage through existing fractures are the most common cause for slope failure in mining vicinity. The failure of slope in the mining vicinity can be influenced by various factors; viz., geometry, groundwater conditions, lithology, geological structures, cohesion, angle of internal friction, blasting effects, mining methods, and equipment selection. It is crucial to consider these factors during the planning and design phase of mining operations, as they not only impact design but also affect engineering assessments of slope stability. (Kolapo et al., 2022).

Regarding lithology, the prevalence of landslides on phyllite & quartzite lithology indicates a potential correlation between geological substrate and slope stability. Geologically phyllites are highly foliated rock with presence of flaky mica minerals. When slate to phyllite grade metamorphic rocks comprising quartzofeldpathic and micaceous minerals alter to clay minerals through degradation process in the presence

Fig. 11 Geological control over landslides in the study area is notable, **a** rock block naturally broken along the joints from the crown part of Bardang landslide, **b** folded strata and quartz veins in Martam landslide, **c** and **d** sulphide mineralisation in Rangpo, **e** weathered quartzite in Bardang, **f**. weathered Schist and Quartzite on slope dipping towards the road, **g** highly deformed folded layers, active landslide beside NH-10 near Rangpo

of water, they also decrease internal strength of the rock. Quartzite occurs as very highly jointed lithology in the study area (Figs. 9 and 11). Highly jointed rocks are more vulnerable to slope failure than other massive rocks (Bewick et al., 2019). Hence more than 50% of the landslide are observed over phyllite & quartzite lithology in the study area. Nearly half of East Sikkim is covered by phyllite, schist and quartzite of LHS. Moreover, the distribution of population and infrastructure is dense in the western part of East Sikkim in LHS region. The gneissic terrain of HHS containing CCGC group of rocks in the study area covers a major part in the high altitude, eastern part of East Sikkim. Landslides in this part are relatively low debris rock slope failures. The cold climate in this region forces the rock to undergo frost heave and thaw weakening contributes to degradation and opening of fractures, joints and fissures in the gneissic rocks. Moreover, the MCT that runs through the study area significantly affects landslide initiation in its vicinity such as in Tathengchen debris flow, Chandmari sinking zone within Gangtok, and 13th mile landslide along JN road.

The analysis of landslide point dataset compiled for 478 landslides in East Sikkim are used to generate a landslide density map (Fig. 12), revealing high landslide density areas.

The stretch of NH-310A passing through J N Marg Road, which links Gangtok to Nathula Pass and Baba Harbhajan Singh Mandir, exhibits a landslide density ranging from 2.2 to 2.8 landslides per square kilometres. This road is currently undergoing development into a circuit link network connecting Rangpo to Zuluk, Nathang Valley to Tsomgo, and Gangtok, with Pakyong at its centre. The aim is to ensure year-round accessibility for locals, tourists, and the military, highlighting the critical importance of this segment to the overall circuit. Geologically, this section comprises CCGC rocks with the presence of minor lineament, as mapped by GSI. Additionally, the South to North MCT buffer zone of the Himalayan orogeny intersects this West to East trending road section. Geomorphologically, the area is classified as low to moderately dissected hills. These intrinsic characteristics render this road segment susceptible to various types of mass movements. The 13th-mile area of this section is particularly active and has been studied extensively (Hindayar et al., 2016). The combination of inherent slope conditions and external influences makes this part of the road highly hazardous, posing risks to traffic. Military and tourist vehicles frequently experience delays and difficulties, necessitating urgent attention, planning, and engineering to mitigate mass movement hazards. It is crucial to conduct thorough investigations before any slope cutting for road widening, moreover identification of active slopes,

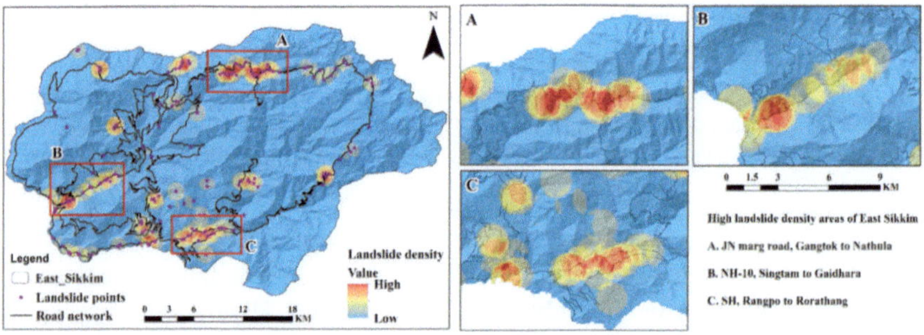

Fig. 12 Landslide density map of East Sikkim analysed for the compiled record of 478 historical landslides, high density of landslides along roads **a** J N Marg, **b** Singtam to Gaidhara, **c** Rangpo-Rorathang-Rongli

continuous monitoring of these slopes, and establishment of early warning systems is pertinent for uninterrupted traffic and safety of locals and tourists in this road section.

The segment of State Highway connecting Rangpo to Rorathang and Rongli experiences a landslide density ranging from 1.7 to 2.5 landslides per square kilometer. This highway runs parallel to the Rangpo-chu river, which originates at an altitude of approximately 4500 m from the eastern border of Sikkim/India and Tibet/China, forming part of the Nathu La Nathang Valley natural reserve. Descending from its source, the Rangpo-chu river drops to an altitude of 700 m in Rongli over 40 kms. As the river progresses downstream from Rongli, it is fed by surface runoff from various sub-watersheds such as Biring & Parkha and Zuluk & Pademchen, as well as from the Rongli and Rhenok forest blocks. Consequently, the river widens, and its water volume increases from Rorathang to Rangpo. The proximity of the road to the river leads to toe cutting and erosion, exposing the slopes and adding to their susceptibility to failure. Furthermore, the low-lying areas adjacent to the river are characterized by well-vegetated forests that experiences hot, humid conditions and significant precipitation during the monsoon season. These factors, coupled with the high curvature of the slopes and poorly planned construction activities, contribute to the elevated landslide density in this part of the study area.

Along NH-10 between Rangpo to Singtam and Gaidhara landslide density ranged from 1.1 to 2.4 landslides per square kilometres (Fig. 13). Though comparatively this is a low landslide density road section in East Sikkim but risk to the infrastructure and traffic is highest along this road. This only national highway (as of now) is the heart of the road network that connects the state with the outside world. This part of the road spread over the LHS of rocks consisting of phyllite and quartzite. Moreover, in many places along the roads, channel/river terrace deposits are present along with angular debris signifying paleo debris locations. These locations are characteristics of loosely consolidated debris materials, as vegetation cover over these kinds of areas is removed, rainfall easily affects them by percolating and degrading inherent strength properties, thus, making slopes vulnerable (Clague 2022) to sinking or mudflow/debris flow or landslides (Figs. 8 and 9).

The information of landslide location (from historical inventory), stream network [created from Advanced Land Observation Satellite (ALOS) Phased Array L-type Synthetic Radar (PALSAR) DEM], runout distance (from historical inventory), landslide distance to stream (calculated), landslide length (from historical inventory), along with other information from the compiled inventory is used to analyse type of landslide viz., base failure, face failure and toe failure. 124 landslides are situated in a distance less than 100 m from the streamline network. 37 landslides have runout distance more than 100 m, and 115 landslides are more than 100 m in length. Using this information, the landslides inventory of 478 landslides in East Sikkim were evaluated and found that 7% landslide in the area are base failure. 77% landslides are facing failure and 16% landslides are toe failure (Table 1).

Fig. 13 Unstable slopes along NH-10. **a** debris accumulated at the toe of Bardang landslide, **b** and **e** toe erosion and active landslide near Rangpo mining area, water seepage from mine shaft leading to slope degradation, **c** toe erosion of Teesta River and failed retaining wall, NH-10, **d** toe erosion and sinking zone remediation at Gaidhara, **f** toe erosion and crown weathering/erosion coupled process at Bardang location

Table 1 Type of slope failure base, face, and toe for the study area, based upon landslide location, runout distance, landslide distance to stream and landslide length

Type of slope failure	Base failure	Face failure	Toe failure
% of landslide	7%	77%	16%

Table 2 Year wise landslide inventory details from 2019 to 2023, pre- and post-event imagery collection dates, cloud mask, RdNDVI threshold, slope threshold, connected pixel count and number of mapped landslides

Year	Pre-event filter date range	Post-event filter date range	Cloud mask	RdNDVI threshold	Slope Threshold	Connected pixel minimum count/area (m^2)	No. of mapped landslides
2019	2017-01-01–2019-03-01	2019-08-31–2020-02-10	Bits 10 & 11	−20	15	20/2000	54
2020	2017-01-01–2020-03-01	2020-08-31–2021-02-10	Bits 10 & 11	−25	15	20/2000	98
2021	2017-01-01–2021-03-01	2021-08-31–2022-02-10	Bits 10 & 11	−60	15	20/2000	49
2022	2021-09-01–2022-03-01	2022-08-31–2023-02-10	Bits 10 & 11	−60	15	20/2000	84
2023	2021-09-01–2023-03-01	2023-08-31–2024-02-10	Bits 10 & 11	−47	15	20/2000	111

5.2 Spatial and Temporal Landslide Inventory

The year wise (2019–2023) landslide inventory for the rainy season (April to August) is prepared for the study area. This specific period is chosen because most of the landslide in the study occur or reoccur during this period. Moreover, major part (>80%) of the year wide precipitation of the study area is concentrated in these months.

The details of the utilized S2 imagery dates, masking, RdNDVI and slope threshold are given year wise in the Table 2.

The number of landslides occurring during the rainy season is highest in the year 2023 (111) and least (49) in the rainy season of 2021.

The Figs. 14 and 15 presents the outcomes of analysis for landslide polygon mapping. Figure 14a Mulukey landslide (old landslide) in Pakyong locality (Fig. 16 and Fig. 14—1 was earlier a landslide with two distinct flanks. During the rainy season of 2020 the landslide flanks joined and turned into a lens shaped landslide with active crown and body. The pre-event FCC (Fig. 14 A 1) and post-event FCC (Fig. 14 B 1) shows a notable change in the RdNDVI continuous raster (Fig. 14 E 1) and binary tiff (Fig. 14 F 1). The landslide area is successfully mapped and extracted as polygon (Fig. 13 G 1) for further analysis.

The Tareythang landslide (Fig. 14—2) is a new landslide. The pre-event FCC shows no sign of landslide in the area. But during the rainy season of 2020 the landslide initiated, caused by toe erosion and failure of debris slope developed into an elongated and widening landslide.

The 13th mile area along the J N Marg is notably active for more than a decade now. During the rainy season of 2019 the landslide increase in size with small, scattered portions failed in the vicinity. The landslide again aggravated in size during the rainy season of 2020 adding a new flank to the east of its older active flank. The post FCC (Fig. 14—3) and RdNDVI successfully mapped out active landslide area.

In the Chochenpery area road construction to connect Nathula through a new road commenced in the year 2019. After the period the area is exposed to high vegetation cover loss and high landslide density. The Fig. 14—4 shows the pre-event FCC and post-event FCC for this area. The change is not visible in the FCC but is visible clearly in the post-NDVI and RdNDVI.

Bojoghari hill north of Gangtok saw a rise in landslides during the rainy season of 2023. This area also saw new road construction and an increase in landslide activity during this period. There were small landslide scars present in the Lokdara area before the year 2022. During the rainy season of 2022 Lokdara landslide triggered and flowed through the road and settlement. In Changey-senti a new landslide initiated in the year 2019. There was no sign of landslide scars before the year 2022 in this area (Fig. 15). Distinct landslide lenses can be seen to have initiated during this period (Fig. 15—7). Tumlabong was a channel debris flow with a small active crown part before the rainy season of 2019. During the rainy season of 2019 it grew into a large landslide with a big crown and active body of landslide. Byang landslide was dormant during the rainy season of 2020 but triggered during the rainy season of

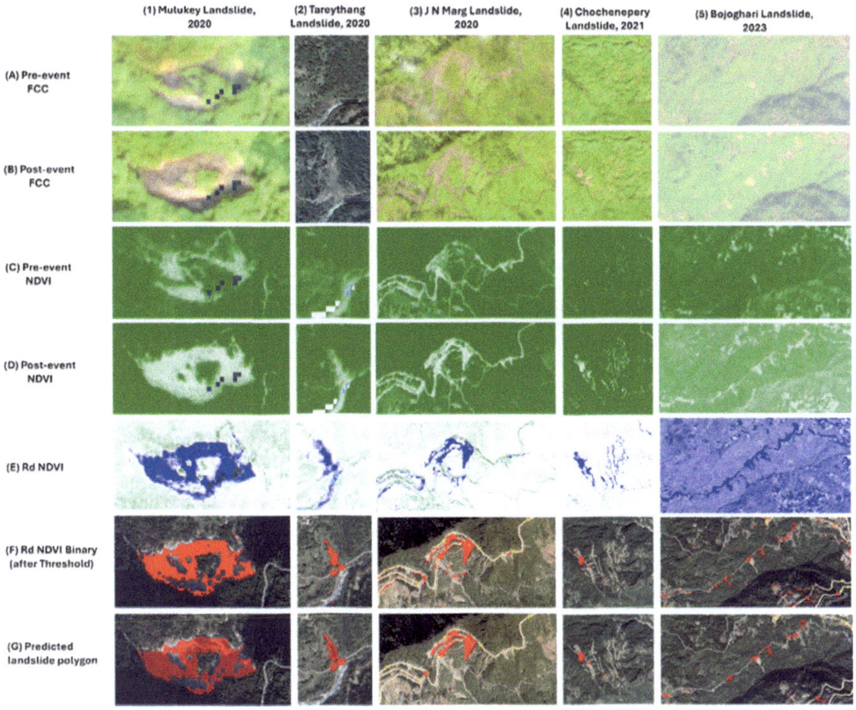

Fig. 14 The outputs of GEE analysis for year wise inventory **a** pre-event FCC, **b** post-event FCC, **c** pre-event NDVI, **d** post-event NDVI, **e** rdNDVI, **f** RdNDVI binary, and **g** active landslide polygon for the 1. Mulukey landslide, 2020; 2. Tareythang landslide, 2020; 3. Landslide along J N Marg, 2020; 4. Chochenpery landslide, 2021; 5. Bojoghari landslide, 2023

2021 and expanded. Likewise, the information on landslides in the area are extracted and used in landslide density and recurrence/occurrence status analysis.

The predicted landslide polygons for each year are compared to the known or mapped landslide polygons to determine the number of new landslides and recurring landslides between 2019 and 2023 (as shown in Table 3). The analysis reveals that approximately 70% of landslides occur during the rainy season in the study area. The highest persistence or recurrence of landslides from the previous year is observed in 2023, accounting for 46% of all landslides. For the years 2020, 2021, and 2022, the persistence or recurrence of landslides from the previous year is approximately 20%. The occurrence of new landslides is more than 50% for the years 2020–2022, while in 2023, there are fewer new landslides and a higher recurrence of landslides from the previous year.

During the period of 2020–2022, numerous construction activities in the southeast aspect of Sichey, Burtuk hill north of Gangtok, Southeast slope aspect of Chochenperi hill, areas of Rongli-Rorathang and Pakyong have been undertaken. In these areas the new landslides are concentrated. The overlapping polygons from different year (Fig. 16 and Table 4) are analysed and it is found that 15% landslide occur once in

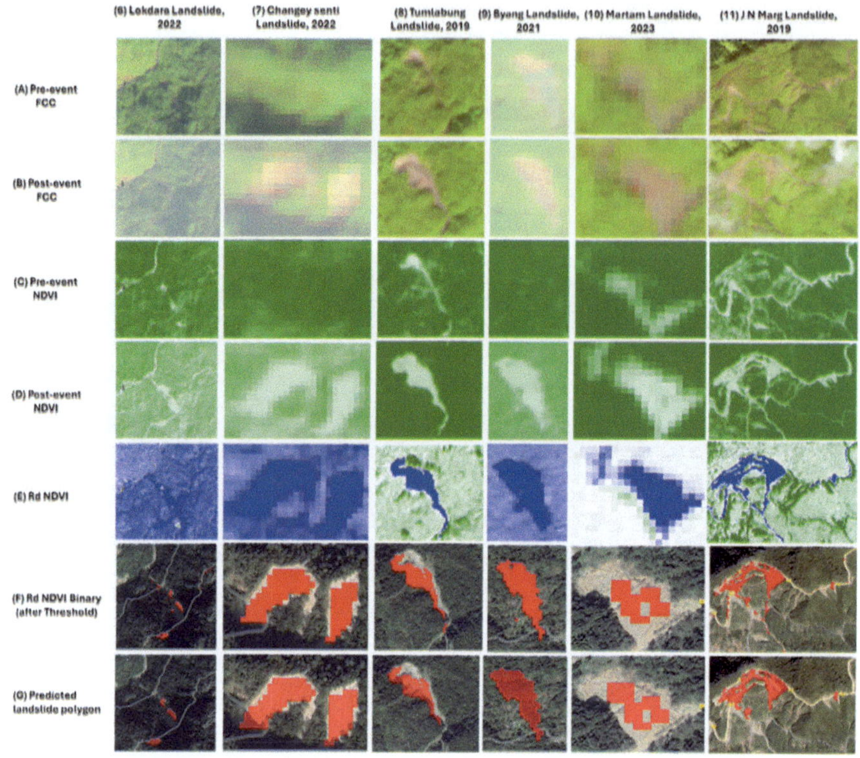

Fig. 15 The outputs of GEE analysis for year wise inventory **a** pre-event FCC, **b** post-event FCC, **c** pre-event NDVI, **d** post-event NDVI, **e** RdNDVI, **f** Rd NDVI binary, and **g** active landslide polygon for the 6. Lokdara landslide, 2022; 7. Changey-senti landslide, 2022; 8. Tumlabung landslide, 2019; 9. Byang landslide, 2021; 10. Martam landslide, 2023; 11. Landslide along J N Marg, 2019

5 year, 13% landslides once in 4 years, 11% landslides once in 3 years, 11% landslide once in 2 years, 15% landslides occur once in a year. Whereas 9% landslides occur twice a year and 4% landslides occur more than twice a year. The updated landslide inventory summed up 591 landslides in East Sikkim as of 2023. The historical landslide record included 478 landslides. During the period of 2019–2023, 113 newly occurred landslides are mapped and added to the inventory.

The intersection analysis of the newly mapped inventory and historical inventory shows that 78% landslides are persistent from the historical record and remained active to some extent throughout the study period of 2019–2023 (Figs. 17 and 18). 4% landslides are found to be occurring above twice a year during the study period and are scattered across East Sikkim (Fig. 18). Notable recurring and persistent landslides are Mulukey landslide, Byang landslide, 13th mile area along J N Marg, Martam landslide along NH-10, 2 landslides in Padamchey area, and landslides in Chochenperi area. The year wise landslide density map prepared for the study area is shown in the Fig. 18(right) b–f. In the year 2019 the active landslide density

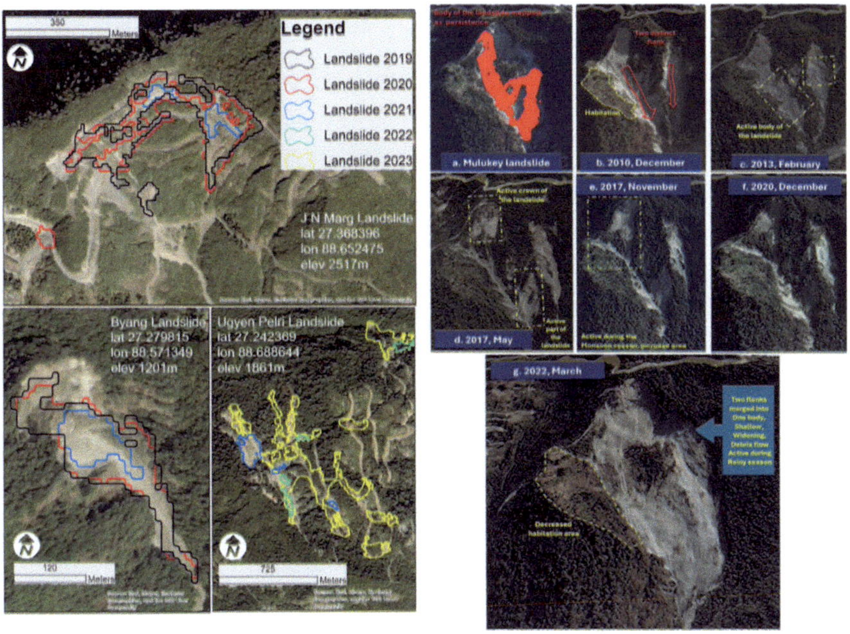

Fig. 16 Overlapping active landslide polygon for different years for J N Marg landslide, Byang landslide and Ugyenperi landslide; evolution of Mulukey landslide **a** recurring part of the landslide (red highlighted area), **b** 2010, two distinct flanks, **c** 2011, active body, **d** 2017, active crown and new area, changes between **d** May 2017 and **e** November 2017 shows activity during the rainy season, **e** and **f** widening of landslide area from 2017 to 2020, **g** transformed into a single landslide in 2022

Table 3 Year wise number of active landslides in the study area for the period of 2019–2023, total active landslides, new landslides, and recurring/persistence of landslides

Year	2019	2020	2021	2022	2023
Predicted landslide (total)	54	98	49	84	111
Dormant landslide from previous year (no.)	–	29	12	13	25
Dormant landslide from previous year (%)	–	29.5%	25%	15%	22.5%
Recurrence/Persistence of old/previous year landslide (no.)	–	17	12	17	51
Recurrence/Persistence of old/previous year landslide (%)	–	17.5%	25%	21%	46%
New landslide (no.)	–	52	25	54	35
New landslide (%)	–	53%	50%	64%	31.5%

Table 4 Status of landslide recurrence over different period in East Sikkim

<table>
<tr><td>Once in 5 years</td><td>Once in 4 years</td><td>Once in 3 years</td><td rowspan="2">Above twice a year</td><td rowspan="2">Not occurred during the period</td></tr>
<tr><td>15%</td><td>13%</td><td>11%</td></tr>
<tr><td>Once in 2 years</td><td>Once in a year</td><td>Twice a year</td><td rowspan="2">4%</td><td rowspan="2">22%</td></tr>
<tr><td>11%</td><td>15%</td><td>9%</td></tr>
</table>

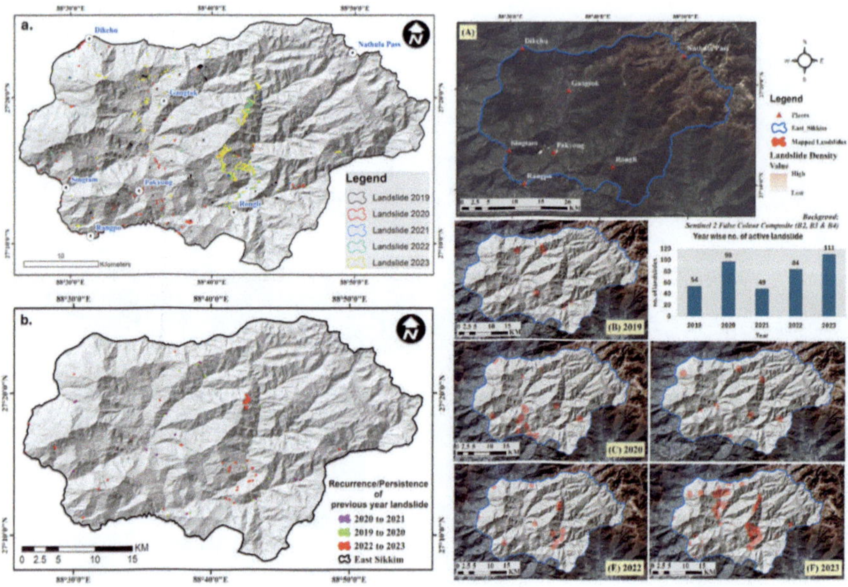

Fig. 17 (Left) **a** Year wise active landslides from 2019 to 2023, **b** landslides active for successive years comparison for East Sikkim. (right) **A** East Sikkim and towns; active landslide, and density maps for the years **B** 2019; **C** 2020; **D** 2021; **E** 2022; **F** 2023

was high along the J N Marg, Upper Burtuk, Pakyong and Rhenock area of East Sikkim. The distribution shifted to the southwest part of the study area during the rainy season of 2020 with more landslide density in the Rorathang and Pakyong area and in the Singtam-Dikchu connecting road. Moreover, new landslide occurrence in the Rakdong- Tintek area was also observed. In the year 2021 high active landslide density is observed in the J N Marg, Rorathang-Rongli, Ranipool vicinity and parts of Rhenock area. In the year 2022, landslides were concentrated in the Upper Assam Lingzey area within the Khedi forest and Chochenperi-Ugyenperi area. The active landslide numbers increased, and high-density area also increased in the year 2023. During this time many landslides in the eastern slope of Rumtek hill and southeast slope of Burtuk hill saw numerous active landslides. Moreover, Assam Lingzey, Rongli, Chochenperi and parts of Singtam-Dikchu road saw high density of active landslides.

The study area has undergone significant change in LULC in between the period of 2019–2023. Roads and infrastructure build-up in the recent time in these mentioned areas has significantly influenced landslide occurrence and reoccurrence. Once snow covered area of the eastern most part of the study area already turned into a dry, cold, rugged alpine terrain before 2019. The area has undergone further change after 2019 with more areas falling in the class of bare ground and range land with less or no vegetation.

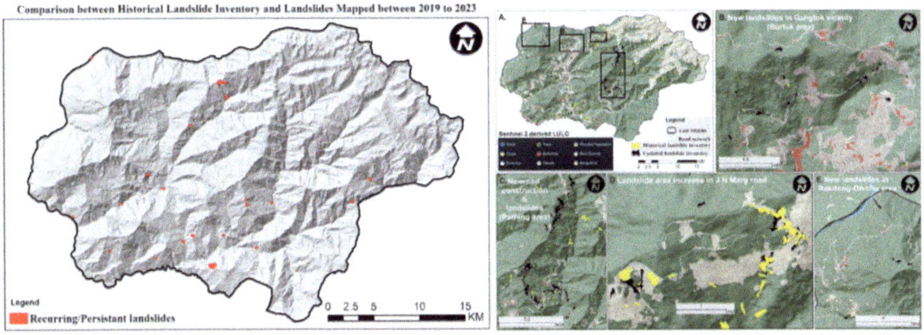

Fig. 18 The appraisal of historical landslide inventory and active landslides mapped between 2019 and 2023 show that 4% landslides are recurring/persistent (left), **a** LULC map of East Sikkim showing historical landslides and updated landslide inventory, **b** new landslides in Burtuk hill, **c** new landslide in Pathing, **d** landslide area increase along J N Marg, **e** new landslide in Rakdong-Tintek area

Seasonal snow, day/night and overall seasonal high temperature differences and tectonic forcing in this area has made the deformed rock more prone to weathering contributing to high landslide susceptibility.

The populated areas such as Gangtok, Pakyong, Rangpo and Rongli has also seen rise in built up and bare ground/rangeland. The rising built up and rising active landslide density in these regions of the study area show positive correlation. The mentioned areas exhibit a significant prevalence of landslides, primarily triggered by anthropogenic activities. Most of the landslides in the East Sikkim are relatively small in scale but pose a considerable threat due to their proximity to built-up areas and human habitations. These activities, ranging from deforestation to construction, create conditions conducive to slope instability.

The high susceptibility to landslide hazards in East Sikkim can be significantly attributed to anthropogenic activities, manifested across a spectrum of operational scales. This contribution comes from multi-level disarray, characterized by a lack of scientific consideration in land management practices. At the local level, the population's engagement in unscientific slope cutting for building construction makes slopes instable. This practice, often lack geotechnical considerations, disrupts natural equilibrium and predisposes slopes to failure. Infrastructure development agencies, both public and private, play a important role in escalating landslide risk. The implementation of new road construction and the broadening of existing road networks, alongside large-scale construction projects, frequently involve extensive slope modifications. The cumulative impact of these activities, ranging from localized interventions to large-scale infrastructure projects shows the complex nature of anthropogenic influence on landslide hazard.

Addressing this escalating risk necessitates a multi-level cooperative framework. A bottom-to-top approach is crucial for effective landslide disaster risk reduction and sustainable development. This framework should integrate local knowledge with scientific expertise, helping planning and implementation. At the local level, community education and capacity building are essential to promote responsible land use practices. Concurrently, infrastructure development agencies must adopt rigorous environmental impact assessments, incorporating geotechnical investigations and sustainable engineering solutions.

However, amidst these anthropogenic factors contributing to the predisposed conditions and landslide triggers so on by rainfall etc., there are notable exceptions represented by larger recurring landslides. The Sirwani landslide, for instance, is indicative of a probable paleo debris zone, suggesting a geological predisposition towards instability. Similarly, the Byang landslide appears to be primarily influenced by factors such as stream channel dynamics, erosion, and debris flow, underlining the disparities between geomorphological processes and human activities as landslide influencing factors. Again, the 13th mile landslide stands out due to its distinct geological and climatic context. Situated in the vicinity of the MCT, its favourable slope for failure, coupled with factors such as lack of vegetation, seasonal snow, and freeze–thaw action, shows the significant role played by natural processes in driving landslide occurrences. Similarly, the Mulukey landslide is characterized by geological factors such as slope steepness and high curvature, compounded by the presence of low-grade metamorphic rock formations. These geological attributes, rather than anthropogenic influences, emerge as primary drivers of instability in these cases.

6 Conclusion

The historical landslide inventory for East Sikkim has documented 478 occurrences, primarily shallow and debris-involved, with some extending up to 30,000 sq. m. Small landslides along national and state highways pose significant hazards, particularly for traffic and downslope habitations. A high geographical distribution of landslides is noted along NH-10, NH-310A, and state highways, with a density ranging from 2.2 to 2.8 landslides per square kilometre. Landslides are concentrated in lowly dissected hills, with fewer occurrences on very steep slopes. Detailed analysis indicates that 7% of failures are base failures, 77% are face failures, and 16% are toe failures. These events predominantly occur on phyllite and quartzite lithology within the Lesser Himalayan Sequence, with debris flows being more frequent during the rainy season. Mineralized areas, particularly in southwestern East Sikkim, exhibit increased erosion, weathering, and landslide activity.

The landslide inventory for East Sikkim from 2019 to 2023 reveals a significant increase in occurrences, with the total number of landslides rising to 591 by 2023, compared to the historical inventory of 478. A noticeable shift in the distribution of new versus recurring landslides is evident; while 50% of landslides between 2019 and 2022 were new, this proportion dropped to 37% in 2023, indicating a rise in recurrence to 46%. Persistent recurrence of 4% of landslides throughout the study period highlights areas of chronic instability. High-risk zones have shifted over time, with locations like J. N. Marg, Upper Burtuk, Pakyong, and Rhenock being prominent in 2019, and Rumtek Hill and the southeast slope of Burtuk Hill emerging as critical areas by 2023, influenced by land use and land cover changes. The rising density of landslides in populated regions such as Gangtok, Pakyong, Rangpo, and Rongli correlates with increasing construction activities in built-up areas.

Analysis shows that 22% of historical landslides did not recur during the study period. Annual recurrence of landslides is observed in 15% of cases, while others occur at varying intervals, including 15% once in 5 years, 13% once in 4 years, 11% once in 3 years, 11% once in 2 years, 9% twice a year, and 4% more than twice a year. Geological controls significantly influence landslide occurrences, as seen in areas like the 13th Mile and Mulukey associated with the Main Central Thrust zone, where minimal anthropogenic interventions exist. Landslides in the Lesser Himalayan sequences, such as Bardang and Martam, demonstrate the impact of natural forces, including exhumation, shear, rainfall, and groundwater dynamics. Larger recurring landslides, such as Sirwani and Byang, are influenced by paleo-debris zones and stream dynamics, respectively, and differ significantly from anthropogenic-induced landslides.

Maintaining a detailed and updated spatial and temporal landslide inventory is crucial for a landslide affected region like East Sikkim. The updated inventory of East Sikkim would help in identifying areas of concern, analysing susceptibility or hazard, and in systematic planning to minimize risk.

7 Data Availability

The corresponding author will provide details of all the figures and data representation upon request.

References

Behling, R., Roessner, S., Golovko, D., & Kleinschmit, B. (2016). Derivation of long-term spatiotemporal landslide activity—A multi-sensor time series approach. *Remote Sensing of Environment, 186*, 88–104.

Behling, R., Roessner, S., Kaufmann, H., & Kleinschmit, B. (2014). Automated spatiotemporal landslide mapping over large areas using rapid eye time series data. *Remote Sensing, 6*(9), 8026–8055.

Belward, A. S., & Skøien, J. O. (2015). Who launched what, when and why; trends in global land-cover observation capacity from civilian earth observation satellites. *ISPRS Journal of Photogrammetry and Remote Sensing, 103*, 115–128.

Bera, S., Upadhyay, V. K., Guru, B., & Oommen, T. (2021). Landslide inventory and susceptibility models considering the landslide typology using deep learning: Himalayas, India. *Natural Hazards, 108*(1), 1257–1289.

Beven, K. J., & Kirkby, M. J. (1979). A physically based, variable contributing area model of basin hydrology/Un modèle à base physique de zone d'appel variable de l'hydrologie du bassin versant. *Hydrological Sciences Journal, 24*(1), 43–69.

Bewick, R. P., Kaiser, P. K., & Amann, F. (2019). Strength of massive to moderately jointed hard rock masses. *Journal of Rock Mechanics and Geotechnical Engineering, 11*(3), 562–575.

Bhambri, R. A. K. E. S. H., Mehta, M., Singh, S., Jayangondaperumal, R., Gupta, A. K., & Srivastava, P. (2017). Landslide inventory and damage assessment in the Bhagirathi Valley, Uttarakhand, during June 2013 flood. *Himalayan Geology, 38*(2), 193–205.

Bhattacharyya, K., Mitra, G., & Kwon, S. (2015). Geometry and kinematics of the Darjeeling-Sikkim Himalaya, India: Implications for the evolution of the Himalayan fold-thrust belt. *Journal of Asian Earth Sciences, 113*, 778–796.

Bhuyan, K., Rana, K., Ferrer, J. V., Cotton, F., Ozturk, U., Catani, F., & Malik, N. (2024). Landslide topology uncovers failure movements. *Nature Communications, 15*(1), 2633.

Borrelli, L., Cofone, G., Coscarelli, R., & Gullà, G. (2015). Shallow landslides triggered by consecutive rainfall events at Catanzaro strait (Calabria–Southern Italy). *Journal of Maps, 11*(5), 730–744.

Casagli, N., Frodella, W., Morelli, S., Tofani, V., Ciampalini, A., Intrieri, E., ... & Lu, P. (2017). Spaceborne, UAV and ground-based remote sensing techniques for landslide mapping, monitoring and early warning. *Geoenvironmental Disasters, 4*, 1–23.

Chen, X., & Chen, W. (2021). GIS-based landslide susceptibility assessment using optimized hybrid machine learning methods. *CATENA, 196*, 104833.

Chen, Y., Ming, D., Yu, J., Xu, L., Ma, Y., Li, Y., ... & Zhu, Y. (2022). Susceptibility-guided landslide detection using fully convolutional neural network. *IEEE Journal of Selected Topics in Applied Earth Observations and Remote Sensing, 16*, 998–1018.

Clague, J. J. (2022). Paleo-landslides. In *Landslide hazards, risks, and disasters* (pp. 335–363). Elsevier.

Confuorto, P., Casagli, N., Casu, F., De Luca, C., Del Soldato, M., Festa, D., ... & Raspini, F. (2023). Sentinel-1 P-SBAS data for the update of the state of activity of national landslide inventory maps. *Landslides, 20*(5), 1083–1097.

Crosby, C. J., Whitmeyer, S. J., De Paor, D. G., Bailey, J., & Ornduff, T. (2012). Lidar and Google Earth: Simplifying access to high-resolution topography data. *Special Paper Geological Society of America, 492*, 37–47.

Del Ventisette, C., Righini, G., Moretti, S., & Casagli, N. (2014). Multitemporal landslides inventory map updating using spaceborne SAR analysis. *International Journal of Applied Earth Observation and Geoinformation, 30*, 238–246.

Dikshit, A., Satyam, N., Pradhan, B., & Kushal, S. (2020). Estimating rainfall threshold and temporal probability for landslide occurrences in Darjeeling Himalayas. *Geosciences Journal, 24*, 225–233.

Dilley, M. (2006). Setting priorities: Global patterns of disaster risk. *Philosophical Transactions of the Royal Society a: Mathematical, Physical and Engineering Sciences, 364*(1845), 2217–2229.

Fang, Z., Wang, Y., Peng, L., & Hong, H. (2021). A comparative study of heterogeneous ensemble-learning techniques for landslide susceptibility mapping. *International Journal of Geographical Information Science, 35*(2), 321–347.

Fayne, J. V., Ahamed, A., Roberts-Pierel, J., Rumsey, A. C., & Kirschbaum, D. (2019). Automated satellite-based landslide identification product for Nepal. *Earth Interactions, 23*(3), 1–21.

Fell, R., Corominas, J., Bonnard, C., Cascini, L., Leroi, E., & Savage, W. Z. (2008). Guidelines for landslide susceptibility, hazard and risk zoning for land use planning.

Fidan, S., Tanyaş, H., Akbaş, A., Lombardo, L., Petley, D. N., & Görüm, T. (2024). Understanding fatal landslides at global scales: A summary of topographic, climatic, and anthropogenic perspectives. *Natural Hazards* 1–19.

Fiorucci, F., Ardizzone, F., Mondini, A. C., Viero, A., & Guzzetti, F. (2019). Visual interpretation of stereoscopic NDVI satellite images to map rainfall-induced landslides. *Landslides, 16*, 165–174.

Fiorucci, F., Cardinali, M., Carlà, R., Rossi, M., Mondini, A. C., Santurri, L., ... & Guzzetti, F. (2011). Seasonal landslide mapping and estimation of landslide mobilization rates using aerial and satellite images. *Geomorphology, 129*(1–2), 59–70.

Fiorucci, F., Giordan, D., Santangelo, M., Dutto, F., Rossi, M., & Guzzetti, F. (2018). Criteria for the optimal selection of remote sensing optical images to map event landslides. *Natural Hazards and Earth System Sciences, 18*(1), 405–417.

Fisher, G. B., Amos, C. B., Bookhagen, B., Burbank, D. W., Godard, V., & Whitmeyer, S. J. (2012). Channel widths, landslides, faults, and beyond: The new world order of high-spatial resolution Google Earth imagery in the study of earth surface processes. *Geological Society of America Special Papers, 492*(01), 1–22.

Froude, M. J., & Petley, D. N. (2018). Global fatal landslide occurrence from 2004 to 2016. *Natural Hazards and Earth System Sciences, 18*(8), 2161–2181.

Gabet, E. J., Wolff-Boenisch, D., Langner, H., Burbank, D. W., & Putkonen, J. (2010). Geomorphic and climatic controls on chemical weathering in the High Himalayas of Nepal. *Geomorphology, 122*(1–2), 205–210.

Galli, M., Ardizzone, F., Cardinali, M., Guzzetti, F., & Reichenbach, P. (2008). Comparing landslide inventory maps. *Geomorphology, 94*(3–4), 268–289.

Ganerød, A. J., Lindsay, E., Fredin, O., Myrvoll, T. A., Nordal, S., & Rød, J. K. (2023). Globally vs. locally trained machine learning models for landslide detection: A case study of a glacial landscape. *Remote Sensing, 15*(4), 895.

Gariano, S. L., & Guzzetti, F. (2016). Landslides in a changing climate. *Earth-Science Reviews, 162*, 227–252.

Ghosh, G. K., Ghosh, K., De, S. K., Rawat, M. S., Joshi, V., & Ayala, I. A. (2012a). An inventory of recent (18th September 2011) earthquake triggered landslides in the Sikkim Himalaya. *J Geo-Environmental Observations, 1*, 1–94.

Ghosh, S., van Westen, C. J., Carranza, E. J. M., & Jetten, V. G. (2012b). Integrating spatial, temporal, and magnitude probabilities for medium-scale landslide risk analysis in Darjeeling Himalayas, India. *Landslides, 9*, 371–384.

Ghosh, S., van Westen, C. J., Carranza, E. J. M., Jetten, V. G., Cardinali, M., Rossi, M., & Guzzetti, F. (2012c). Generating event-based landslide maps in a data-scarce Himalayan environment for estimating temporal and magnitude probabilities. *Engineering Geology, 128*, 49–62.

Ghosh, T., Bhowmik, S., Jaiswal, P., Ghosh, S., & Kumar, D. (2020). Generating substantially complete landslide inventory using multiple data sources: A case study in Northwest Himalayas, India. *Journal of the Geological Society of India, 95*, 45–58.

Gorelick, N., Hancher, M., Dixon, M., Ilyushchenko, S., Thau, D., & Moore, R. (2017). Google Earth Engine: Planetary-scale geospatial analysis for everyone. *Remote Sensing of Environment, 202*, 18–27.

Guzzetti, F., Mondini, A. C., Cardinali, M., Fiorucci, F., Santangelo, M., & Chang, K. T. (2012). Landslide inventory maps: New tools for an old problem. *Earth-Science Reviews, 112*(1–2), 42–66.

Handwerger, A. L., Huang, M. H., Jones, S. Y., Amatya, P., Kerner, H. R., & Kirschbaum, D. B. (2022). Generating landslide density heatmaps for rapid detection using open-access satellite radar data in Google Earth Engine. *Natural Hazards and Earth System Sciences, 22*(3), 753–773.

Hao, L., Rajaneesh, A., Van Westen, C., Sajinkumar, K. S., Martha, T. R., Jaiswal, P., & McAdoo, B. G. (2020). Constructing a complete landslide inventory dataset for the 2018 monsoon disaster in Kerala, India, for land use change analysis. *Earth System Science Data, 12*(4), 2899–2918.

Hindayar, J. N., Dasarwar, P., Srivastava, S. P., Thrideep Kumar, N., Mohan, M., & Som, S. K. (2016). Dynamicity of the Himalayan landslide—a tectono-geotechnical appraisal of the 13th mile landslide, Sikkim. *Journal of the Geological Society of India, 88*(2), 197–205.

Hoek, E., & Bray, J. W. (1991). *Rock slope engineering* (3rd (reprinted), p. 358). The Institute of Mining and Metallurgy.

Huang, Y., & Zhao, L. (2018). Review on landslide susceptibility mapping using support vector machines. *CATENA, 165*, 520–529.

Jha, V., & Sharma, R. (2022). Formation of polymetallic ores in the metasedimentary rocks of Rangpo area, Sikkim Lesser Himalaya, India: Mineralogical and geochemical attributes. *Geological Journal, 57*(12), 5061–5082.

Jharendra, K. C., Gautam, D., Neupane, P., & Paudyal, K. R. (2018). Landslide inventory mapping and assessment along the Ramche-Jharlang area in Dhading, Rasuwa and Nuwakot districts, Lesser Himalaya Central Nepal. *Journal of Nepal Geological Society, 55*(1), 103–108.

Jones, J. N., Boulton, S. J., Stokes, M., Bennett, G. L., & Whitworth, M. R. (2021). 30-year record of Himalaya mass-wasting reveals landscape perturbations by extreme events. *Nature Communications, 12*(1), 6701.

Joshi, M. (2022). Co-seismic landslides in the Sikkim Himalaya during the 2011 Sikkim Earthquake: Lesson learned from the past and inference for the future. *Geological Journal, 57*(12), 5039–5060.

Kincey, M. E., Rosser, N. J., Robinson, T. R., Densmore, A. L., Shrestha, R., Pujara, D. S., ... & Swirad, Z. M. (2021). Evolution of coseismic and post-seismic landsliding after the 2015 Mw 7.8 Gorkha earthquake, Nepal. *Journal of Geophysical Research: Earth Surface, 126*(3), e2020JF005803.

Kirschbaum, D. B., Adler, R., Hong, Y., Hill, S., & Lerner-Lam, A. (2010). A global landslide catalog for hazard applications: Method, results, and limitations. *Natural Hazards, 52*, 561–575.

Kirschbaum, D., Stanley, T., & Zhou, Y. (2015). Spatial and temporal analysis of a global landslide catalog. *Geomorphology, 249*, 4–15.

Kolapo, P., Oniyide, G. O., Said, K. O., Lawal, A. I., Onifade, M., & Munemo, P. (2022). An overview of slope failure in mining operations. *Mining, 2*(2), 350–384.

Lee, E. M., & Jones, D. K. (2023). *Landslide risk assessment.* ICE publishing.

Lindsay, E., Frauenfelder, R., Rüther, D., Nava, L., Rubensdotter, L., Strout, J., & Nordal, S. (2022). Multi-temporal satellite image composites in google earth engine for improved landslide visibility: A case study of a glacial landscape. *Remote Sensing, 14*(10), 2301.

Mandal, S., & Mondal, S. (2019). *Statistical approaches for landslide susceptibility assessment and prediction* (Vol. 27). Springer International Publishing.

Mansour, M. F., Morgenstern, N. R., & Martin, C. D. (2011). Expected damage from displacement of slow-moving slides. *Landslides, 8*, 117–131.

Marcelino, E. V., Formaggio, A. R., & Maeda, E. E. (2009). Landslide inventory using image fusion techniques in Brazil. *International Journal of Applied Earth Observation and Geoinformation, 11*(3), 181–191.

Martha, T. R., Govindharaj, K. B., & Kumar, K. V. (2015). Damage and geological assessment of the 18 September 2011 Mw 6.9 earthquake in Sikkim, India using very high resolution satellite data. *Geoscience Frontiers, 6*(6), 793–805.

Martha, T. R., Kamala, P., Jose, J., Vinod Kumar, K., & Jai Sankar, G. (2016). Identification of new landslides from high resolution satellite data covering a large area using object-based change detection methods. *Journal of the Indian Society of Remote Sensing, 44*, 515–524.

Martha, T. R., Roy, P., Jain, N., Khanna, K., Mrinalni, K., Kumar, K. V., & Rao, P. V. N. (2021). Geospatial landslide inventory of India—an insight into occurrence and exposure on a national scale. *Landslides, 18*(6), 2125–2141.

Miele, P., Di Napoli, M., Novellino, A., Calcaterra, D., Mallorqui, J. J., & Di Martire, D. (2022). SAR data and field surveys combination to update rainfall-induced shallow landslide inventory. *Remote Sensing Applications: Society and Environment, 26*, 100755.

Mohammadi, A., Shahabi, H., & Bin Ahmad, B. (2018). Integration of InSAR technique, Google Earth images and extensive field survey for landslide inventory in a part of Cameron Highlands, Pahang, Malaysia. *Applied Ecology & Environmental Research, 16*(6).

Moosavi, V., Talebi, A., & Shirmohammadi, B. (2014). Producing a landslide inventory map using pixel-based and object-oriented approaches optimized by Taguchi method. *Geomorphology, 204*, 646–656.

Morgan, B. A., & Wieczorek, G. F. (1996). Debris flows and landslides resulting from the June 27, 1995, storm on the North Fork of the Moormons River, Shenandoah National Park, Virginia (No. 96–503). US Geological Survey.

Muñoz-Torrero Manchado, A., Allen, S., Ballesteros-Canovas, J. A., Dhakal, A., Dhital, M. R., & Stoffel, M. (2021). Three decades of landslide activity in western Nepal: New insights into trends and climate drivers. *Landslides, 18*, 2001–2015.

Naemitabar, M., & Zanganeh Asadi, M. (2021). Landslide zonation and assessment of Farizi watershed in northeastern Iran using data mining techniques. *Natural Hazards, 108*(3), 2423–2453.

Notti, D., Cignetti, M., Godone, D., & Giordan, D. (2023). Semi-automatic mapping of shallow landslides using free Sentinel-2 images and Google Earth Engine. *Natural Hazards and Earth System Sciences, 23*(7), 2625–2648.

Parise, M. (2001). Landslide mapping techniques and their use in the assessment of the landslide hazard. *Physics and Chemistry of the Earth, Part c: Solar, Terrestrial & Planetary Science, 26*(9), 697–703.

Petley, D. (2012). 14 Remote sensing techniques and landslides. In *Landslides: Types, mechanisms and modeling, 159*.

Petley, D. N., Higuchi, T., Petley, D. J., Bulmer, M. H., & Carey, J. (2005). Development of progressive landslide failure in cohesive materials. *Geology, 33*(3), 201–204.

Rabby, Y. W., & Li, Y. (2019). Landslide inventory (2001–2017) of Chittagong hilly areas, Bangladesh. *Data, 5*(1), 4.

Roberts, S., Jones, J. N., & Boulton, S. J. (2021). Characteristics of landslide path dependency revealed through multiple resolution landslide inventories in the Nepal Himalaya. *Geomorphology, 390*, 107868.

Rosi, A., Tofani, V., Tanteri, L., Tacconi Stefanelli, C., Agostini, A., Catani, F., & Casagli, N. (2018). The new landslide inventory of Tuscany (Italy) updated with PS-InSAR: Geomorphological features and landslide distribution. *Landslides, 15*, 5–19.

Roy, D., Das, S., & Mitra, R. (2022). An application of geospatial-based multi-criteria decision-making technique to identify landslide susceptibility zones in the Ragnu Khola River Basin of Darjeeling Himalayan region, India. *Applied Geomatics, 14*(4), 731–749.

Saha, A., Mandal, S., & Saha, S. (2020). Geo-spatial approach-based landslide susceptibility mapping using analytical hierarchical process, frequency ratio, logistic regression and their ensemble methods. *SN Applied Sciences, 2*(10), 1647.

Santangelo, M., Gioia, D., Cardinali, M., Guzzetti, F., & Schiattarella, M. (2015). Landslide inventory map of the upper Sinni River valley, Southern Italy. *Journal of Maps, 11*(3), 444–453.

Sato, H. P., & Harp, E. L. (2009). Interpretation of earthquake-induced landslides triggered by the 12 May 2008, M7. 9 Wenchuan earthquake in the Beichuan area, Sichuan Province, China using satellite imagery and Google Earth. *Landslides, 6*, 153–159.

Scheip, C. M., & Wegmann, K. W. (2021). HazMapper: A global open-source natural hazard mapping application in Google Earth Engine. *Natural Hazards and Earth System Sciences, 21*(5), 1495–1511.

Segoni, S., Rosi, A., Fanti, R., Gallucci, A., Monni, A., & Casagli, N. (2018). A regional-scale landslide warning system based on 20 years of operational experience. *Water, 10*(10), 1297.

Solari, L., Del Soldato, M., Montalti, R., Bianchini, S., Raspini, F., Thuegaz, P., ... & Casagli, N. (2019). A Sentinel-1 based hot-spot analysis: landslide mapping in north-western Italy. *International Journal of Remote Sensing, 40*(20), 7898–7921.

Timilsina, M., Bhandary, N. P., Dahal, R. K., & Yatabe, R. (2017). Large-scale landslide inventory mapping in lesser Himalaya of Nepal using geographic information system. In *GIS landslide* (pp. 97–112).

Travelletti, J., Delacourt, C., Allemand, P., Malet, J. P., Schmittbuhl, J., Toussaint, R., & Bastard, M. (2012). Correlation of multi-temporal ground-based optical images for landslide monitoring: Application, potential and limitations. *ISPRS Journal of Photogrammetry and Remote Sensing, 70*, 39–55.

van Westen, C. J., Ghosh, S., Jaiswal, P., Martha, T. R., & Kuriakose, S. L. (2013). From landslide inventories to landslide risk assessment; an attempt to support methodological development in

India. In *Landslide science and practice: Volume 1: Landslide inventory and susceptibility and hazard zoning* (pp. 3–20)
van Westen, C. J., Jaiswal, P., Ghosh, S., Martha, T. R., & Kuriakose, S. L. (2012). Landslide inventory, hazard and risk assessment in India. Terrigenous mass movements: Detection, modelling, early warning and mitigation using geoinformation technology (pp. 239–282)
Yu, B., Chen, F., & Muhammad, S. (2018). Analysis of satellite-derived landslide at Central Nepal from 2011 to 2016. *Environmental Earth Sciences, 77*, 1–12.
Yunus, A. P., Fan, X., Tang, X., Jie, D., Xu, Q., & Huang, R. (2020). Decadal vegetation succession from MODIS reveals the spatio-temporal evolution of post-seismic landsliding after the 2008 Wenchuan earthquake. *Remote Sensing of Environment, 236*, 111476.
Zhang, J., Regmi, A. D., Liu, R., Khanal, N. R., Schenato, L., Gurung, D. R., & Wahid, S. (2017). Landslides inventory and trans-boundary risk management in Koshi River Basin, Himalaya. In *Land cover change and its eco-environmental responses in Nepal* (pp. 409–426).

Landslide Susceptibility Mapping Along the Critical NH-10 Corridor of the Darjiling–Sikkim Himalaya: A Comparative Analysis of Analytical Hierarchy Process, Frequency Ratio, and Random Forest Models

Kabirul Islam

Abstract Landslides pose a significant threat to infrastructure and human life along the National Highway 10 (NH10) corridor in the Indian Himalayas. This study employed cutting-edge spatial modeling techniques—Analytical Hierarchy Process, Frequency Ratio, and Random Forest—to map landslide susceptibility and inform risk mitigation efforts. A comprehensive geo-database was created incorporating 11 conditioning factors, including slope, elevation, rainfall, distance to rivers, distance to roads, distance to lineament, topographic wetness index, normalized difference vegetation index, land use, geology, and aspect. A total of 304 historical landslide locations were identified and randomly divided into 70% training and 30% validation datasets. The resulting landslide susceptibility map categorizes the road into zones of varying landslide susceptibility levels, ranging from very low to very high. Approximately 33%, 44.64% and 47.51% of the road falls within the high and very high susceptibility categories according to the Analytical Hierarchy Process, Frequency Ratio, and Random Forest methods respectively. The susceptibility maps generated by the three methods were validated using the area under the receiver operating characteristic (ROC) curve and the success rate curve. The results showed that the RF model outperformed the AHP and FR models, with an area under the ROC curve of 0.859 for the validation dataset. The finer granularity and more conservative high-risk delineation of the RF model suggest it may be the most suitable approach for targeted risk mitigation planning and infrastructure development along the critical NH10 corridor in the Indian Himalayas. These findings can empower data-driven decision-making to safeguard lives, livelihoods, and transportation links in this treacherous region.

Keywords Landslide · Susceptibility mapping · Analytical hierarchy process · Frequency ratio · Random forest · National highway 10 · Indian Himalayas

K. Islam (✉)
Karajgram High School, Purba Bardhaman, West Bengal, India
e-mail: kabir.islam1987@gmail.com

© The Author(s), under exclusive license to Springer Nature Switzerland AG 2025

S. Ghosh (ed.), *The Himalaya Dilemma*, Springer Natural Hazards,
https://doi.org/10.1007/978-3-031-95083-4_14

1 Introduction

Landslide is "a downward and outward movement of slope forming materials under the influence of gravity" (Varnes, 1978). Landslides are among the most destructive natural hazards, particularly in mountainous regions, causing loss of life, property damage, and disruption of economic activities (Guzzetti et al., 2007; Youssef & Pourghasemi, 2020). Mass movements of earth material result from a combination of natural and human-induced factors. Natural triggers like heavy rainfall, earthquakes, and volcanic activity can destabilize slopes, while human activities such as deforestation and construction can worsen these conditions. Landslide risk is highest in areas with steep topography, particularly canyon bases, regions affected by wildfires, human-altered landscapes, river channels, and zones of concentrated water runoff or ground saturation. These locations are inherently vulnerable to earth movements due to compromised ground stability (Das et al., 2022; Achour & Pourghasemi, 2020; Dai & Lee, 2002). Landslides not only result in the destruction of infrastructure—such as roads, bridges, and buildings—but also disrupt communication, transportation, and economic lifelines, particularly in developing countries. The impact of landslides on communities is multifaceted: they can lead to the displacement of populations, interruption of trade, loss of agricultural productivity, and long-term environmental degradation. Between 1998 and 2017, landslides affected an estimated 4.8 million people and cause more than 18,000 deaths with billions of dollars in damage each year (Sreelakshmi et al., 2022). In the context of developing countries like India, where mountainous regions are frequently inhabited and infrastructure development is ongoing, landslides pose a serious challenge to sustainable development. Approximately 12.6% of India's total land area, equivalent to 0.42 million km^2, is susceptible to landslides. Of this, 43% (0.18 million km^2) is concentrated in the Darjiling–Sikkim Himalayas, 33% (0.14 million km^2) in the Northwest Himalayas, 22% (0.09 million km^2) in the Western Ghats, and only 2% (0.01 million km^2) in the Eastern Ghats (Nirmala et al., 2023). The Indian Himalayas represents one of the world's primary landslide hotspots, accounting for approximately 16% of global rainfall-triggered landslide events (Froude & Petley, 2018a, 2018b). Between of 2010 and 2020, over 1.7 million people have been affected by landslide events alone (EM-DAT 2020) and an estimated that India incurs annual losses ranging between INR 150–200 crores due to landslide disasters (NDMA, 2019). These statistics clearly demonstrate how landslides constitute a major hazard in the Himalayan region, posing significant threats to human lives, economic stability, and the region's biodiversity, including both flora and fauna (Das et al., 2022).

Mountain highways, which serve as vital lifelines connecting remote areas to the outside world, are often constructed along steep slopes, disrupt the ecological and geological stability of these areas (Sun et al., 2020). During construction, human engineering activities leads degree of slope surface fragmentation and fissures become more pronounced, the likelihood of landslides significantly rises. Landslides along mountain highways critically affect the safety of the ecological environment and hinder the social and economic development of the impacted regions. The Himalayan region, characterized by its dynamic geotectonic setting, complex lithological formations, and intense meteorological conditions, is particularly susceptible to landslide occurrences (Kumar & Anbalagan, 2015). Within this context, National Highway 10 (NH-10), extending from Siliguri to Gangtok, traverses through tectonically active Lesser and Higher Himalayan sequences, making it exceptionally vulnerable to slope failures and mass movements. National Highway 10 (NH-10), a major transportation corridor linking the states of West Bengal and Sikkim, is a critical lifeline for the region. The highway is not only vital for regional connectivity but also plays a crucial role in tourism and trade between India and neighboring countries like Bhutan and China. However, this highway passes through geologically fragile areas of the Eastern Himalayas, which are highly susceptible to landslides due to steep slopes, heavy rainfall, deforestation, and infrastructure development (Ghose et al., 2016). Landslides along NH-10 have repeatedly led to road closures, accidents, and economic losses, severely affecting the livelihood of local communities and disrupting supply chains. During the annual monsoon season, rainfall-triggered landslides consistently cause widespread devastation throughout corridors and in the recent past years i.e. from 2015 to 2024 cut off Sikkim from the rest of the country (Hindustan Times, 2024). Landslides have become more severe and frequent in recent years, prompting increased research into mapping areas susceptible to landslides to identify at-risk locations and improve disaster preparedness.

Landslide Susceptibility Mapping (LSM) is a specialized analytical technique that evaluates and maps the spatial probability of landslide events occurring under given conditions. The rapid advancement of computing power and Geographic Information System (GIS) technology has led to the development of numerous quantitative and statistical methods for assessing landslide susceptibility (Wang et al., 2013). Landslide hazard assessment studies generally categories into three: quantitative, qualitative, and artificial intelligence-based methods (Arabameri et al., 2021). Qualitative methods is a Knowledge-based techniques such as Analytic Hierarchy Process (Kumar & Anbalagan, 2016; Zhang et al., 2016; El Jazouli et al., 2019; Leonardi et al., 2022; Liu et al., 2024) based on experts' judgment and experience to determine factors weights according to their significance. Quantitative Methods for landslide susceptibility mapping are objective approaches such as Frequency Ratio (Silalahi et al., 2019; Abedini et al., 2019; Pham et al., 2015; Kumar & Annadurai, 2015), statistical index (Pourghasemi et al., 2013; Razavizadeh et al., 2017), Weight of Evidence (Batar & Watanabe, 2021; Ozdemir & Altural, 2013); information value (Chen et al., 2013; Kumar and Singh, 2019; Wubalem & Meten, 2020), Shannon Entropy (Rahaman et al., 2021; Shadman et al., 2016) based on mathematical relationships between past landslides and their contributing factors. These methods rely heavily on quality historical landslide data and environmental factor datasets, with the fundamental assumption that future landslide events will occur under similar conditions that triggered past landslides (Batar & Watanabe, 2021). Among the commonly used algorithms, machine learning algorithms have gained prominence for their superior accuracy compared to traditional statistical methods (Tran et al., 2024). Machine Learning approaches in landslide susceptibility mapping represent advanced computational techniques that offer superior accuracy and the ability to handle complex, non-linear relationships between landslide occurrence and various environmental factors (Reichenbach et al., 2018). These methods, including Random Forest (Taalab et al., 2018; Arabameri et al., 2019; Sun et al., 2020; Li et al., 2024), Support Vector Machines (SVM) (Ballabio & Sterlacchini, 2012; Lee et al., 2017; Naceur et al., 2022), Artificial Neural Networks (ANN) (Ermini et al., 2005; Arnone et al., 2014; Gelisli et al., 2015), Deep Learning algorithms (Azarafza et al., 2021; Habumugisha et al., 2022; Ji et al., 2023), Neuro-fuzzy system (Tien Bui et al., 2012; Dehnavi et al., 2015), Decision Tree (DT) (Pradhan, 2013;Wang et al., 2016) can effectively process and analyze multiple layers of spatial data simultaneously, discovering hidden patterns and interactions that might be missed by traditional statistical methods (Li et al., 2024).

Several research studies (Table 14.1) have analyzed landslide susceptibility in the Darjiling–Sikkim Himalayan region, but no LSMs currently exist for the NH-10 corridor, hindering effective landslide management policy implementation. This corridor holds significant socio-economic and political importance, yet frequent landslides repeatedly disrupt it, resulting in substantial economic losses. To address this gap, this study develops LSMs using an integrated approach that combines multiple methods: knowledge-based AHP, data driven FR bivariate statistical and advanced machine learning algorithms RF method. This comprehensive methodology aims to provide a reliable assessment of landslide hazards along the NH10 road corridor. AHP ability to handle both qualitative and quantitative criteria, weight assignment process, verify consistency through ratio checks, and produce reliable results that can be easily integrated with GIS for spatial analysis (Kumar & Anbalagan, 2016). Frequency ratio serves as an uncomplicated geospatial tool that calculates probability-based relationships between landslide occurrences and various causative factors across different map classes (Zhang et al., 2020). According to Ozdemir and Altural (2013) FR is a reliable method for landslide susceptibility assessment, consistently producing satisfactory results through its straightforward statistical calculations. Tree-based ensemble algorithms, especially Random Forest (RF), have proven to be highly effective for landslide susceptibility mapping compared to traditional machine learning methods (Merghadi et al., 2020; Akinci et al., 2020). Multiple studies confirm that RF consistently delivers superior accuracy and reliability, making it a top choice among researchers for predicting landslide-prone areas (Achour & Pourghasemi, 2020; Chen et al., 2017; Dou et al., 2019; Sevgen et al., 2019). A distinctive aspect of this research is the unprecedented implementation of knowledge-driven, data-driven, and machine learning approaches for spatial landslide susceptibility mapping. By employing these complementary models, the research seeks to provide a more robust and multifaceted analysis of landslide susceptibility, offering insights that go beyond the limitations of individual methodological techniques and addressing the critical need for comprehensive risk assessment in infra-structurally vulnerable regions. By synthesizing multiple methodological approaches, this study offers decision-makers a more comprehensive and reliable landslide susceptibility map, enabling more informed and strategic land use planning beyond the limitations of single-method assessments.

Table 14.1 Overview of literature on landslide susceptibility

Year	Authors	Study area	Paper title	Methodology	Key findings
2019	Roy et al.	Darjiling & Kalimpong, West Bengal	A Novel Ensemble Approach for Landslide Susceptibility Mapping (LSM) in Darjiling and Kalimpong Districts, West Bengal, India	Weight-of-Evidence & Support Vector Machine (SVM)	Classified LSMs into four susceptibility categories; very high susceptibility areas ranged from 15 to 20% across different models
2019	Roy & Saha	Darjiling District, West Bengal	Landslide Susceptibility Mapping Using Knowledge-Driven Statistical Models in Darjiling District, West Bengal, India	Fuzzy-LNRF & Fuzzy-AHP	34% and 22% of the study area were highly susceptible to landslides using LNRF and AHP-based fuzzy logic, respectively
2021	Islam et al.	Sikkim Himalaya	Landslide susceptibility modeling in a complex mountainous region of Sikkim Himalaya using new hybrid data mining approach	Decision Tree, Quantum-Particle Swamp Optimization and Random Forest	Identifies landslides as a major threat in the Sikkim Himalaya's mountainous regions. Concludes that the QPSO-ADTree ensemble model outperformed other approaches and suggests this novel ensemble approach could be valuable for LSM in mountainous regions globally

(continued)

Table 14.1 (continued)

Year	Authors	Study area	Paper title	Methodology	Key findings
2021	Nath et al.	Darjiling–Sikkim Himalaya	Remote Sensing GIS-Based Landslide Susceptibility & Risk Modeling in Darjiling–Sikkim Himalaya Together with FEM-Based Slope Stability Analysis of the Terrain	AHP, Logistic Regression (LR), Random Forest (RF), Artificial Neural Network (ANN)	Integrated 28 causative factors; classified terrain slopes into 'High to Severe' landslide susceptibility zones
2022	Sonker et al.	Sikkim Himalaya	Remote Sensing and GIS-Based Landslide Susceptibility Mapping Using Frequency Ratio Method in Sikkim Himalaya	Frequency Ratio (FR)	Used 28 causative factors; classified terrain into five susceptibility zones; 27% of the area falls in higher susceptibility zones
2023	Saha et al.	Darjiling District in North-East Himalaya	A Multi-Criteria Decision Analysis (MCDA) Approach for Landslide Susceptibility Mapping of a Part of Darjiling District in North-East Himalaya, India	MCDA-based model (analytical hierarchy process)	The result obtained is classified into four classes, i.e., very high (11.68%), high (26.18%), moderate (48.87%), and low (13.27%) landslide susceptibility. It is observed that an entire 37.86% of the area is in a high to very high susceptibility zone

(continued)

Table 14.1 (continued)

Year	Authors	Study area	Paper title	Methodology	Key findings
2023	Biswakarma et al.	West Sikkim, Indian Himalaya	An Integrated Quantitative and Qualitative Approach for Landslide Susceptibility Mapping in West Sikkim District, Indian Himalaya	FR, Information Value (IV), AHP	Classified study area into high-risk (3%), medium-risk (14%), and low-risk (83%) zones, emphasizing mitigation strategies
2024	Alam et al.	NH31A & East Sikkim Himalaya	Geospatial Intelligence for Landslide Susceptibility and Risk Analysis: Insights from NH31A and East Sikkim Himalaya Settlements	FR, Information Value (IV), Certainty Factor (CF)	FR (20.75%), IV (40.91%), and CF (18.78%) models identified high landslide risk on NH31A; similar trends in densely populated regions

2 Objectives

The present study has set forth the following objectives:

- Implement statistical (FR), heuristic (AHP), and machine learning (RF) techniques for landslide susceptibility mapping along the NH-10 corridor.
- To compare and validate the performance of AHP, FR, and RF methods in predicting landslide susceptibility zones while integrating multiple approaches to enhance reliability and overcome individual limitations.
- To identify and evaluate critical landslide conditioning factors specific to the NH-10 corridor through integrated multi-method approach.
- To establish a geospatial database of landslide-prone areas along NH-10 that can directly support evidence-based disaster management policy implementation and mitigation planning.

3 Study Area

The study area (Fig. 14.1) is identified by the 500mts buffer of the National Highway 10. The National Highway 10 (NH-10) formerly known as NH 31A represents one of India's most strategically significant transportation corridors, serving as the vital lifeline connecting the plains of West Bengal to the mountainous state of Sikkim

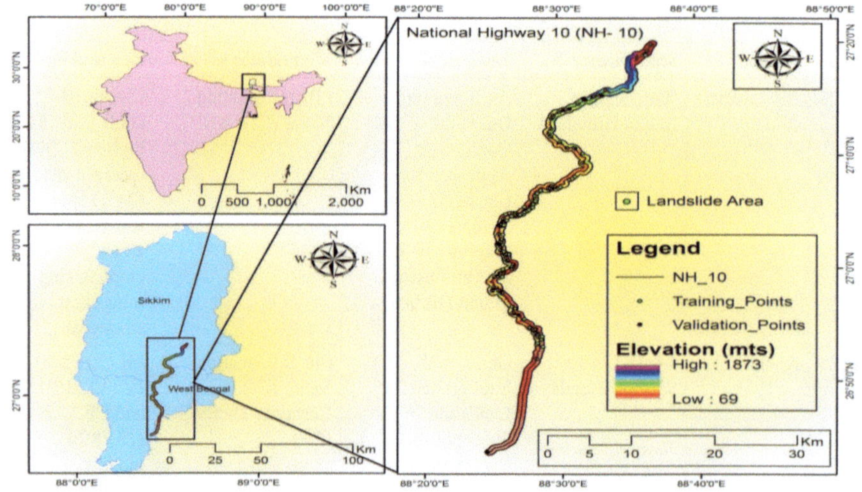

Fig. 14.1 Location map of the study area

(MoRTH, 2019). The highway begins 118 km its journey from Siliguri at an elevation of approximately 69 m above mean sea level (AMSL) and extends northeastward through the tectonically active Lesser Himalayan region before reaching Gangtok at 1,873 m AMSL. The route spans 52 km in Sikkim and 66 km in West Bengal (MoRTH, 2019). The study segment traverses through the tectonically active Lesser Himalayan terrain, characterized by rugged topography with slope angles ranging from 0° to 70° (Roy et al., 2022). This vital transportation corridor link Sikkim to the rest of India is the NH10, experiencing an average daily traffic volume of approximately 4,968 vehicles (Ghose et al., 2016).

Geologically this area is under the high-risk earthquake zone (falls under the seismic zone IV and V), because lies in a zone where the convergence vector between the Indian and Eurasian plates is almost perpendicular to the Himalayan deformation front and Main Central Thrust (MCT) is passing through this area (Bhattacharyya et al., 2015) (Fig. 14.2). The study area experiences a complex climatic regime influenced by both altitude and monsoon patterns, with the lower elevations exhibiting subtropical characteristics and the higher reaches demonstrating temperate conditions (Gupta et al., 2016). Annual rainfall varies significantly from 2,000 mm in the lower valleys to over 4,000 mm in the higher elevations, with the monsoon season (June to September) accounting for approximately 75% of the annual precipitation with mean temperature ranges from 1 °C to 30 °C (Gupta et al., 2016). The interplay of these climatic conditions and local geology creates a highly landslide-prone environment and the vulnerability is further exacerbated by human activities, including road cutting and widening, urbanization along highways, deforestation, inadequate drainage systems, and vibrations from traffic. NH-10 widely recognized as the "lifeline of Sikkim," serves as the sole arterial transportation corridor, facilitating the movement of essential commodities, supporting regional commerce,

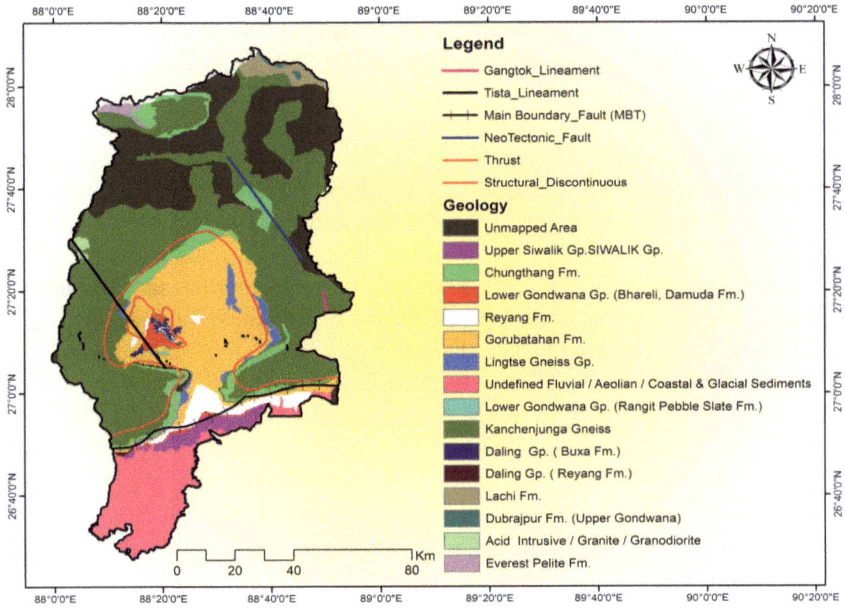

Fig. 14.2 Geological structure of the Darjiling–Sikkim Himalaya

enabling tourist mobility, and ensuring strategic connectivity to this geopolitically significant Himalayan state (DoT *2024*). It is worth noting that, NH-10 traverses the North-Eastern Hills of the Himalaya biodiversity hotspot, which harbors numerous endemic species (Arrawatia & Tambe, 2011; Ghose et al., 2016). Recurrent landslides along NH-10 critically disrupt the mountainous population by causing frequent highway closures, which impede regional economic activities and obstruct access to essential healthcare services, while simultaneously threatening the region's wildlife and ecosystem integrity.

4 Data Collection and Methodology

4.1 Data Sources for the Preparation of Thematic Map

No standardized guidelines exist for selecting landslide susceptibility parameters so the selection of appropriate instability factors is prior essential for mapping landslide susceptibility and understanding the slope failures (Guzzetti et al., 1999; Ayalew and Yamagishi, 2005). For this study, causative factors were identified based on previous landslide research, government reports and availability of data for the study area to ensure consistency with the findings. After reviewing relevant literature 11 critical factors were selected for landslide susceptibility mapping: slope, elevation, rainfall, distance to roads, distance to river, distance to lineaments, NDVI, TWI, LULC, aspect, and geology. High resolution ASTER GDEM (30 m spatial resolution) was used to generate slope, elevation, aspect and TWI and for annual rainfall map, grid-based (0.25° × 0.25°) long-term rainfall data (from 2014 to 2023) were used which is collected from the Indian Metrology Department, Pune (IMD). Landsat-8 series satellite image for the year of 2024 (01/09/2023–31/10/2024) was used for the creation of NDVI map, which was collected from https://earthexplorer.usgs.gov and prepared using Google Earth Engine platform. River, geology and lineament shapefiles are collected from the https://bhukosh.gsi.gov.in in vector format. To determine the Land use/cover in the study area Global 10 m resolution land use land cover datasets (2023) published by ESRI (https://livingatlas.arcgis.com/landcover). The road corridor (NH-10) shapefile has digitized from Roads and Bridges Department Government of Sikkim map. After all, for geospatial analysis and uniformity of the database, all the thematic layers converted into raster format same pixel size (30 m × 30 m) with the standard WGS 1984 UTM 45N zone with the help of resample technique in ArcGIS 10.8. Further, all the data sources are used in this study, their details are mentioned briefly in Table 14.2. Above all, the overall methodology and conceptual framework of the discussion study is presented Fig. 14.3 through a flow chart.

Table 14.2 Dataset used in the study and their sources

Parameters	Description of data	Acquisition date	Data characteristics	Data sources
NH-10 Road	Shapefile	2015	Vector data	Roads and bridges department government of Sikkim
Slope, Elevation, Aspect, TWI	ASTER GDEM	01/03/2000	Version 3, spatial resolution 30 m	https://search.earthdata.nasa.gov
Rainfall	Indian Metrology Department (IMD, Pune)	2014–2023	High-resolution a gridded time-series dataset, 0.25° × 0.25°	https://www.imdpune.gov.in
NDVI	Landsat-8 OLI/TIRS C2 L1	01/09/2023–31/10/2024	High-resolution 30 m × 30 m	https://earthexplorer.usgs.gov
Distance to river, Lineaments and Geology	Geological survey of India (GSI) data	2019	Vector layer	https://bhukosh.gsi.gov.in
LULC	ESRI land use/cover	2023	Sentinel-2 10 m Land Use/Land Cover Time Series map	https://livingatlas.arcgis.com/landcover

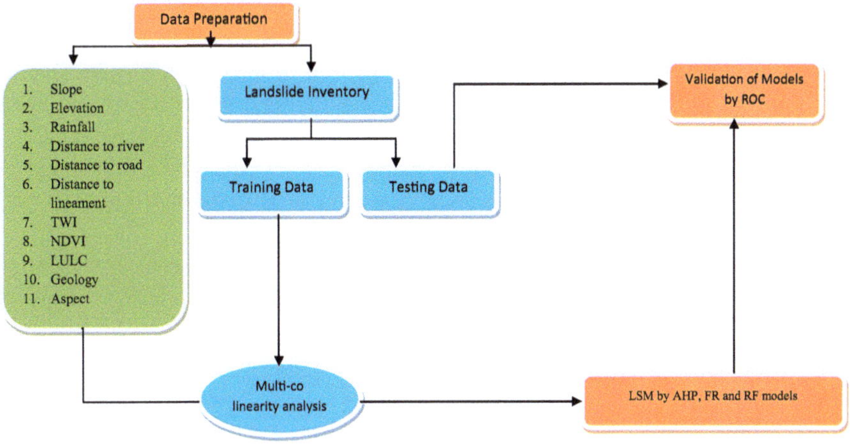

Fig. 14.3 Flowchart of the research work

4.2 Preparation of Landslide Inventory

Landslide inventory data plays a vital role in predicting where future landslides may occur (Pham et al., 2017). Past landslide locations must be compared with environmental conditions to understand their relationship, as similar conditions are expected to trigger future events (Youssef & Pourghasemi, 2020). So, create accurate landslide susceptibility maps, it's essential to understand both where previous landslides occurred and which conditions triggered them. In the present study, landslide inventory which was extracted from multiple data sources i.e. "BHUKOSH" (a web-based geospatial platform developed by the Geological Survey of India) and The Global Landslide Catalog (NASA open data portal). At first 304 landslide points (Fig. 1) extracted from these platforms and extracted data randomly divided into training (70%) and validation (30%) classes with help of subset feature tools available on arcGIS platform. 213 training points used for model building and 91for accuracy assessment. In this study the landslides were mapped as individual points because their areas were too small to be effectively shown as polygons at this scale (Yilmaz, 2009; Veerappan and Anbazhagan, 2015). This point-based inventory shows landslide events scattered throughout the NH-10 road's length. After all landslides polygons are converted into raster format using conversion tool polygon to raster in ArcGIS 10.8.

4.3 Causative Factors

Landslide susceptibility modeling represents a complex interplay of multiple environmental, geological, and anthropogenic factors that work in concert to determine an area's vulnerability to slope failures (Wang et al., 2020; Zhou et al., 2021). The underlying factors contributing to landslide occurrence vary significantly across different geographical and geological contexts, making the selection of input variables a critical step in developing effective prediction models. In this work, the inputs variables are categorized as nominal (geology, aspect, land use/cover) do not have a natural order or ranking, but represent distinct classes or categories and ordinal (elevation, slope, rainfall, NDVI, TWI, distance from lineament, distance from roads, and distance from river) are continuous data types where the order and magnitude of the values hold significance (Table 14.3).

4.4 Landslide Susceptibility Methods (LSM)

4.4.1 Analytic Hierarchy Process (AHP)

The Analytic Hierarchy Process (AHP) is a powerful Multiple-criteria Decision-Making (MCDM) methodology developed by Thomas L. Saaty in 1987. As a comprehensive measurement theory, AHP enables the conversion of both discrete and continuous paired comparisons into meaningful ratio scales (Saaty, 1987). AHP has found extensive applications across multiple domains, particularly in site selection, suitability analysis and landslide susceptibility mapping (Veerappan and Anbazhagan, 2015; Ayalew et al., 2005). AHP is breaking down complex decisions into a manageable hierarchical structure, organizing elements in a way that resembles a family tree (Zaim et al., 2012; Elvis et al., 2022). Determining the weights of the selected factors as well as their corresponding classes using AHP method determined by the following steps:

Step 1: Pair-wise comparison matrix:

Pair-wise comparison is a core component of the AHP model. In this approach, different factors or criteria are compared two at a time to establish their relative importance in relation to the overall goal and these comparisons use numerical values from Saaty's fundamental scale (Table 14.4), which ranges from 1 to 9 (Saaty, 1987). A score of 1 indicates that two factors are equally important, while a score of 9 indicates that one factor is extremely important compared to the other (Elvis et al., 2022; Saaty, 1980). The judgments are then organized into a comparison matrix to determine the priority of each parameter (Table 14.5).

Step 2: Normalized Pair wise comparison matrix (NPCM).

To create a normalized pair-wise comparison matrix (NPCM), each value in the original comparison matrix is divided by the sum of its respective column (Table 14.6). This normalization process helps reduce subjective bias in the assigned weights of thematic maps and their features (Elvis et al., 2022). The resulting normalized weights fall between 0 and 1, and their sum always equals 1 (Doke et al., 2021; Malczewski, 1999).

Table 14.3 The relationship between impact factors and landslide occurrence

Landslide triggering factors	Procedure of preparation of each factor
Slope	• The slope failure is closely linked to its steepness with increasing slope angle, the likelihood of landslides occurring also rises. The stability of a slope against failure can be determined by the relationship between the shear stresses and the shear strength (safety factor), where the driving force of mass movement increases with increasing the slope angle (Guillard & Zezere, 2012). Additionally, slope gradient plays a crucial role in subsurface flow and soil moisture concentration, which are directly related to the occurrence of landslides (Magliulo et al., 2008). For this reason, slope is a frequently used and well-documented parameter in the field of landslide hazard assessment • In the current study, the slope angle map was derived from a DEM with a resolution of 30 m, and the slope angles for the study area ranged from 0° to 69° (Fig. 14.4a)
Elevation	• Elevation is a key factor in landslide susceptibility assessment, widely used in risk analyses. In mountainous regions, landslide risk typically rises with elevation, as greater heights correspond to increased rainfall, changes in vegetation, and distinct patterns of human activity indicated that altitude is one of the topographic factors that influence slope instability (Youssef & Pourghasemi, 2020; Zhou et al., 2021) • The elevation map of study area was prepared from DEM and it's ranged between 69 to 1873 m above sea level (Fig. 14.4b)
Rainfall	• Landslides in mountainous terrain are primarily triggered by heavy, sustained rainfall as these mass movements typically in the aftermath of intense precipitation events (Wang et al., 2020). Heavy rainfall events can provoke landslides by elevating pore water pressure and compromising slope stability (Zhou et al., 2021). The study area experiences a subtropical monsoon climate, with heavy rainfall between June and October, when the majority of landslide events are recorded (Roy et al., 2022) • Therefore, this study used the multi-year average rainfall data of the study area from 2014 to 2023 for the assessment of landslide susceptibility and it's ranged from 2628 to 3112 mm (Fig. 14.4c)
Distance to river	• Rivers or water bodies play a crucial role in landslide occurrence. Water flowing along rivers erodes and undermines slopes, compromising their stability and potentially triggering landslides (Abedini et al., 2019). Slopes situated near rivers are particularly vulnerable, as they face both erosional forces and increased water pressure conditions that can weaken soil structure and lead to slope failure (Tran et al., 2024) • For this study area river shapefile extracted from 'Bhukosh' geo-spatial platform and distance to river has a range (0–17,490 m), was developed using the Euclidean distance tool in ArcGIS 10.8 (Fig. 14.5d)

(continued)

Table 14.3 (continued)

Landslide triggering factors	Procedure of preparation of each factor
Distance to road	• The construction and presence of roads can significantly impact the stability of surrounding terrain and increase landslide risks through several mechanisms. Initial construction impacts, the constant human activities and development near roads—from traffic vibrations to drainage modifications—further compromise slope stability, making these areas particularly vulnerable to landslides (Ha et al., 2022; Tran et al., 2024) • The road network dataset was extracted through digitization of 1:25,000 'Roads and Bridges Department Government of Sikkim'. The digitized roads were then converted into raster format, and successive 100-m buffer zones were created around the road network for analysis (Fig. 14.5e)
Distance to lineaments	• Lineaments generally weaken geological structure of the surface such as a fault, fracture zone, or shear zone, plays a role in influencing slope stability (Silalahi et al., 2019). The stability of a slope can be further compromised when external factors like heavy rainfall or seismic activity trigger failures along zones of weakness associated with these landscape features (Kanungo et al., 2009) • For this study area the Euclidean distance tool in ArcGIS 10.8 was used to produce distance from faults map and distance to lineaments has a range of 0–14810 m (Fig. 14.5f)
NDVI	• NDVI indicates vegetation health and density, which significantly influences slope stability by controlling water infiltration, surface runoff patterns, and soil weathering processes (Zhang et al., 2019). Due to its significance in understanding vegetation's role in slope stability, NDVI has become a standard factor in landslide susceptibility mapping studies (Zhou et al., 2021). The NDVI value ranging from -1 to $+1$, the negative values indicating the water body, values close to zero (-0.1–0.1) show the barren land, low positive values (0.2–0.4) represent the grasslands and shrub, and high values show dense vegetation cover (Youssef & Pourghasemi, 2020) • NDVI map of the study area prepared from Landsat 8 OLI satellite image was used, and the value was calculated using the following formula: $\text{NDVI} = \frac{\text{NIR}-\text{Red}}{\text{NIR}+\text{Red}}$ (1) Where the NIR denoted near-infrared light and the Red is the visible light. The NDVI map of the study area was categorized into five sub-categories ranging from -0.048 to 0.68 using the natural break method (Fig. 14.6g)

(continued)

Table 14.3 (continued)

Landslide triggering factors	Procedure of preparation of each factor
TWI	• TWI is a secondary topographical attribute that presents the spatial distribution of wetness conditions and soil moisture (Beven & Kirkby, 1989). TWI measures how topography influences water flow and accumulation patterns in the landscape. This index reveals areas where soil moisture tends to concentrate—higher TWI values indicate locations with greater soil water content, making these zones more susceptible to saturation and potential landslides (Zhou et al., 2021) • The TWI map was calculated from the DEM using following formula (Eq. (2)) $TWI = \text{Ln}\left(\frac{a}{\tan B}\right)$ (2) Where, a denoted the cumulative upslope area draining through a point (per unit contour length) and tan b is the slope angle at the point (Rahmati et al., 2016). The final TWI map was divided into five classes ranging from 2.63 to 21.60 (Fig. 14.6h)
LULC	• Land use and cover strongly impact slope stability, as factors like vegetation, infrastructure, and bare ground alter processes like infiltration, soil cohesion and runoff that affect landslide risk (Tran et al., 2024). Therefore, LULC has become a standard and essential component in landslide susceptibility mapping and has been widely used in previous studies (Zhou et al., 2021) • For this study, a land cover map was produced by classifying Sentinel-2 satellite imagery (2023) with 10-m spatial resolution into seven distinct land cover subclasses including water body, vegetation cover, crop land, build-up area, bare land, and rangeland respectively (Fig. 14.6i)
Aspect	• Slope-aspect plays a crucial role in influencing landslide occurrence through both direct and indirect processes by affecting environmental conditions like wind, rain, sunlight, soil moisture, and vegetation growth (Youssef & Pourghasemi, 2020). Studies have shown it correlates with other landslide-related factors including rock type, slope steepness, and proximity to water bodies and tectonic features (Regmi et al., 2010) • Slope aspect in the study area was constructed using DEM and classified into ten categories including flat (−1), North (0°–22.5°), Northeast (22.5°–67.5°), East (67.5°–112.5°), Southeast (112.5°–157.5°), South (157.5°–202.5°), Southwest (202.5°–247.5°),West (247.5°–292.5°), Northwest (292.5°–337.5°) and North (337.5°–360°) (Fig. 14.7j)
Geology	• Geology is an important factor affecting landslides, especially in mountainous areas (Shirzadi et al., 2018). Different geological units, with their varying physical and chemical properties like strength, weathering, durability, density, and permeability, significantly influence the occurrence of geo-hazards such as landslides and land subsidence (Rahmati et al., 2016) • In the current study, the geology was extracted from the (1:250,000-scale) acquired from the Geological Survey of India data base and six geological units were identified including, Lingtse gneiss, Gorubathan formation, Siwalik group, Daling formation, Bhareli/Damuda formation and Fluvial/ Aeolian, Coastal & Glacial sediments (Fig. 14.7k)

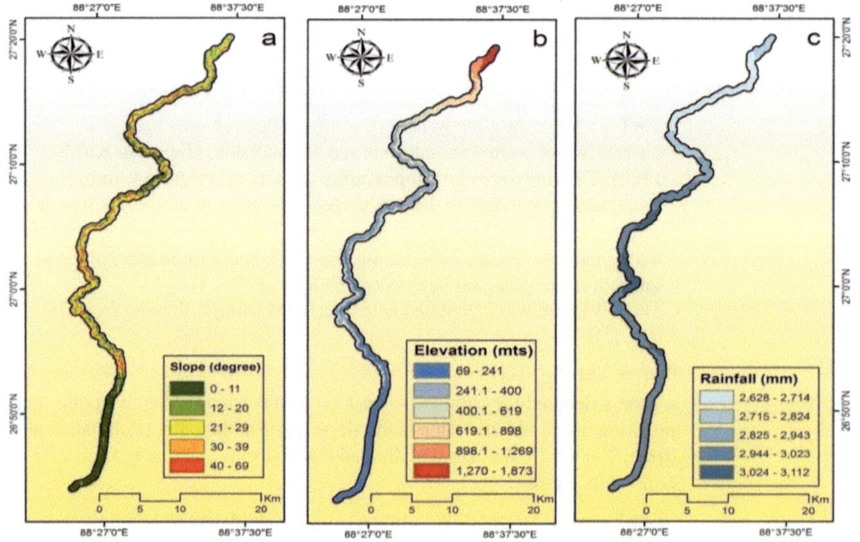

Fig. 14.4 Landslide causative factor maps **a** Slope, **b** Elevation, and **c** Rainfall

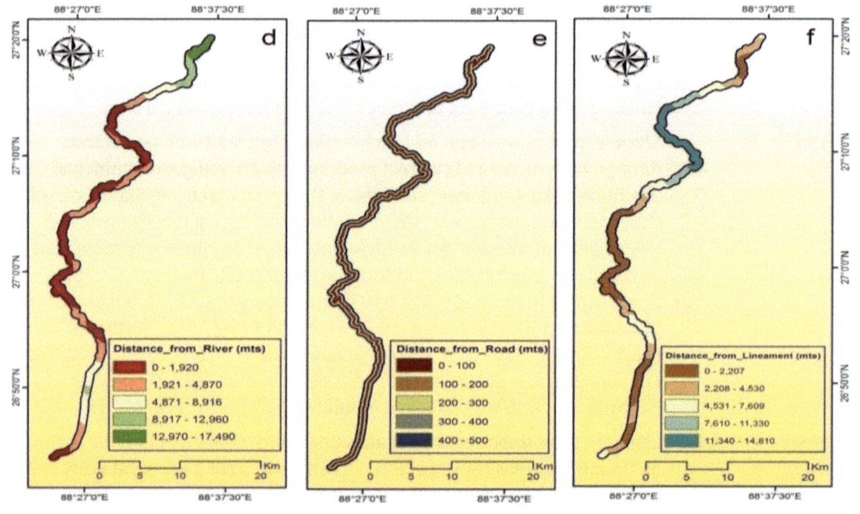

Fig. 14.5 Landslide causative factor maps **d** Distance from river, **e** Distance from roads, **f** Distance from lineaments

Step 3: the Consistency ratio (CR).

The final step in the Analytic Hierarchy Process (AHP) is to verify the consistency of the normalized weights assigned to different criteria. This verification is essential to ensure that the pair-wise comparisons are logically sound. The Consistency Ratio

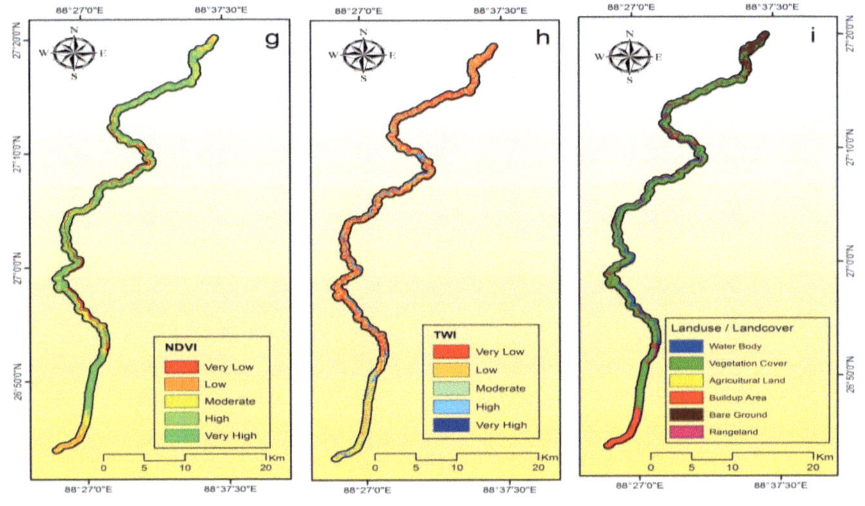

Fig. 14.6 Landslide causative factor maps **g** NDVI, **h** TWI, **i** LULC

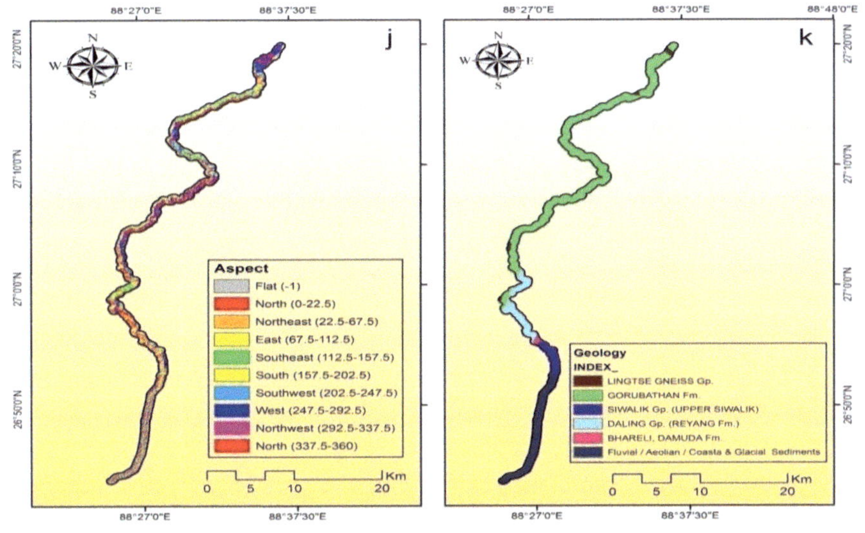

Fig. 14.7 Landslide causative factor maps **j** Aspect, **k** Geology

(CR), introduced by Saaty (1980), measures how consistent the weights are across all conditioning factors. Specifically, the CR value helps evaluate the reliability of weights derived from the normalized pair-wise comparison matrix (NPCM) using Eq. (3) (Ifediegwu, 2022; Liu et al., 2024).

Table 14.4 The fundamental scale (*Source* Saaty, 1987)

Intensity of importance on an absolute scale	Definition	Explanation
1	Equal importance	Two activities contribute equally to the objective
3	Moderate importance of one over another	Experience and judgment strongly favor one activity over another
5	Essential or strong importance	Experience and judgment strongly favor one activity over another
7	Very strong importance	An activity is strongly favored and its dominance demonstrated in practice
9	Extreme importance	The evidence favoring one activity over another is of the highest possible order of affirmation
2, 4, 6, 8	Intermediate values between the two adjacent judgments	When compromise is needed

$$CR = \frac{CI}{RI} \tag{3}$$

where, CR represent consistency ratio, CI represents Consistency Index derived from the Eq. (4) (motioned below), and RI represents Random Consistency Index (Table 14.7 adopted from Saaty, 1980).

$$CI = \frac{\lambda_{\max} - n}{n - 1} \tag{4}$$

where, λ max represents the largest Eigen value of the pair-wise comparison matrix and n represents the number of conditioning factors. Saaty (1980) established that a CR value must be less than 0.1 (10%) for the analysis to be considered valid. If the CR exceeds this threshold, the entire analysis must be redone. The study's calculations yielded a Consistency Ratio (CR) of 0.05, which is below the 0.1 threshold. With λ max at 11.69, a sample size of 11, a Consistency Index of 0.07, and a Random Index of 1.51 (Table 14.7), the analysis demonstrates acceptable consistency and reliable judgment in the hierarchical evaluation (Table 14.6).

Step 4: Score determination (Based on Saaty's scale).

In evaluating landslide susceptibility zoning, each parameter's sub-classes are weighted on a scale of 1 to 9 based on their influence on landslide occurrence (Abedini and Tulabi, 2018). Sub-classes with the strongest influence on landslide occurrence receive the highest score of 9, while those with minimal influence receive the lowest score of 1. Intermediate scores are assigned to sub-classes with moderate influence.

Table 14.5 Pair-wise comparison matrix of selected parameters (PCM)

Factors	Elevation	Slope	Rainfall	Distance to river	Distance to Road	Distance to Lineament	NDVI	TWI	LULC	Aspect	Geology
Elevation	1	1/2	1/3	1/4	1/5	1/5	3	1/3	2	1	1/3
Slope	2	1	1	1/3	1/4	1/3	5	1	3	1	1
Rainfall	3	2	1	1/3	1	1	7	1	5	2	1
Distance to river	4	3	3	1	2	2	5	2	4	3	2
Distance to road	5	4	2	1/2	1	1	6	3	4	2	2
Distance to lineament	5	3	2	1/2	1	1	6	2	4	2	2
NDVI	1/3	1/5	1/7	1/5	1/6	1/6	1	1/3	2	1/5	1/6
TWI	3	2	1	1/2	1/3	1/2	3	1	1	2	1
LULC	1/2	1/3	1/5	1/4	1/4	1/4	1/2	1	1	1/3	1/4
Aspect	2	1	1/2	1/3	1/2	1/2	5	1/2	3	1	1
Geology	3	2	1	1/2	1/2	1/2	6	2	4	2	1

Table 14.6 Normalized Pair-wise comparison matrix and criteria weight for each selected parameter based on AHP method

Factors	Elevation	Slope	Rainfall	Distance to river	Distance to road	Distance to lineament	NDVI	TWI	LULC	Aspect	Geology
Elevation	0.03	0.03	0.03	0.05	0.03	0.03	0.06	0.02	0.06	0.03	0.03
Slope	0.07	0.05	0.04	0.07	0.04	0.05	0.11	0.04	0.09	0.06	0.05
Rainfall	0.10	0.11	0.09	0.07	0.07	0.07	0.15	0.07	0.15	0.12	0.10
Distance to river	0.14	0.16	0.26	0.21	0.30	0.29	0.11	0.15	0.12	0.19	0.20
Distance to road	0.17	0.21	0.17	0.11	0.15	0.14	0.13	0.22	0.12	0.12	0.20
Distance to lineament	0.17	0.16	0.17	0.11	0.15	0.14	0.13	0.15	0.12	0.12	0.20
NDVI	0.01	0.01	0.01	0.04	0.02	0.02	0.02	0.02	0.06	0.01	0.02
TWI	0.10	0.11	0.09	0.11	0.05	0.07	0.06	0.07	0.03	0.12	0.05
LULC	0.02	0.02	0.02	0.05	0.04	0.04	0.01	0.07	0.03	0.02	0.02
Aspect	0.07	0.05	0.04	0.07	0.07	0.07	0.11	0.04	0.09	0.06	0.05
Geology	0.10	0.11	0.09	0.11	0.07	0.07	0.13	0.15	0.12	0.12	0.10
Result	Maximum Eigen value (L. Max) = 11.69, CI = 0.07, CR = 0.05, RI = 1.51, n = 11										

Table 14.7 Ratio index for different values of '*n*' based on Saaty's (1980)

No of selected parameters	1	2	3	4	5	6	7	8	9	10	11
Random consistency index (R.I.)	0	0	0.52	0.89	1.11	1.25	1.35	1.41	1.45	1.49	1.51

This weighting system was applied to all key parameters: slope, elevation, rainfall, distance to roads, distance to river, distance to lineaments, NDVI, TWI, LULC, aspect, and geology.

Step-5: Overlay analysis for the delineation of landslide susceptibility map.

Next step was calculation of landslide susceptibility index (LSI). It was computed using weighted arithmetic sum method which can be formulated as given below (Eq. (5)).

$$\text{LSM} = \sum_{\text{i}=1}^{\text{n}=11} \text{W}_i \tag{5}$$

where, LSM is the groundwater potential zone, Wi is the criteria weight, Ri is the ranking of parameter factors, and i denotes each of the eight influencing factors.

4.4.2 Frequency Ratio (FR) Methods

The Frequency Ratio (FR) is a statistical approach that quantifies the relationship between existing landslides and various environmental factors that influence their occurrence (Nicu, 2018; Shahabi et al., 2014). It works by analyzing the spatial relationship between historical landslide locations and the various factors that may trigger those (Roian et al., 2016). The ratio is calculated by dividing the percentage of landslide occurrences in a particular class by the percentage of the total area covered by that class (Anbazhagan and Ramesh, 2014). A ratio of 1 represents average conditions, with values above 1 indicating stronger landslide associations and values below 1 showing weaker associations (Youssef et al., 2015). To calculate the FR value, the following formula (6), (7) and (8) was used (Lee & Pradhan, 2007; Lee & Sambath, 2006):

$$\text{Frequency ratio} = \frac{\text{Landslide ratio}}{\text{Class ratio}} \tag{6}$$

where,

$$\text{Landslides ratio} = \frac{\text{Number of landslide in class}}{\text{Total number of landslides}} \tag{7}$$

$$\text{Class ratio} = \frac{\text{Number of grids in individual class}}{\text{Total number of grids in whole class}} \tag{8}$$

After calculating the Frequency Ratio (FR) value for each class within the thematic layers, the Landslide Susceptibility Index (LSI) was computed by summing the FR values of all layers using Eq. 9:

$$\text{LSI} = \text{FR}_1 + \text{FR}_2 + \text{FR}_3 \ldots\ldots \text{FR}_n \tag{9}$$

where, LSI is landslide susceptibility index; FR_1, FR_2, FR_3.....FR_n are the weight values for different parameter considered in the analysis and each FR value represents the relative contribution of that particular layer to landslide occurrence (Youssef and Hegab, 2005; Samanta et al., 2018).

4.4.3 Random Forest (RF)

Machine learning techniques (MLT) have become instrumental in spatial modeling for natural hazards like landslide susceptibility assessment and experts recommend using MLT as a supportive tool rather than replacing human expertise (Shirzadi et al., 2018; Youssef & Pourghasemi, 2020; Zhou et al., 2018). MLT combined with Geographic Information Systems (GIS) and field datasets (such as historical landslide inventories and geo-environmental factors) can produce accurate models for decision-makers and planners (Goetz et al., 2015). Random Forest (RF) methodology has emerged as a powerful machine learning approach, introduced by Breiman in 2001 for landslide susceptibility mapping, offering enhanced prediction capabilities compared to traditional statistical method (Wang et al., 2020). This ensemble learning technique combines multiple decision trees to generate a more accurate and stable prediction model while minimizing over fitting issues commonly associated with single decision trees (Breiman, 2001; Zhou et al., 2021). This method's robustness stems from its ability to handle complex, non-linear relationships between various natural hazards conditioning factors, including topographic, geological, hydrological, and land use parameters (Sarfraz et al., 2022). In the implementation of RF for landslide susceptibility assessment, the algorithm constructs numerous decision trees through bootstrap aggregating (bagging) and random feature selection processes (Zhou et al., 2021). Each tree in the forest is developed using a random subset of both the training data and predictor variables, with the final susceptibility prediction determined through a majority voting system among all trees (Zhou et al., 2021). The operating principle of random forest algorithm is shown in Fig. 14.8. Based on the above advantages, in this study, RF model was adopted to evaluate and the landslide susceptibility map of the study area has been created using Google Earth engine software.

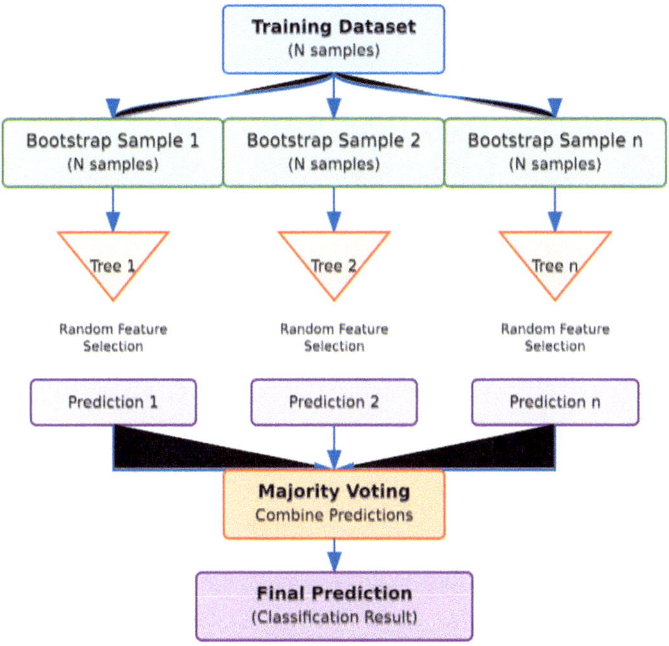

Fig. 14.8 Schematic diagram of the random forest process

4.4.4 Modeling Prediction and Performance

Model evaluation is the key to assessing performance and ROC (Receiver Operating Characteristic) curve analysis provides a robust graphical approach to evaluate classification accuracy (Liu et al., 2010; Wang et al., 2020). The AUC (Area under the Curve) value serves as the basis for determining model effectiveness, with higher AUC indicating better discriminatory power. This method plots sensitivity (the true positive rate) on the vertical axis against 1-specificity (the false positive rate) on the horizontal axis. The AUC values, which are in the range of 0–1, were used to classify the accuracy of the landslide predictive models as follows: 0.9–1: excellent accuracy; 0.8–0.9: very good accuracy; 0.7–0.8: good accuracy; 0.6–0.7: moderate accuracy; and 0.5–0.6: poor accuracy (Hussain et al., 2019). Therefore, this paper applied this indicator to evaluate and compare the performances of the thrice models in the present research.

5 Results and Discussion

5.1 Correlation Between Landslides Occurrence and Causative Factors Using AHP Model

Correlation is a phenomenon where multiple factors in a model are interconnected and jointly affect the landslide susceptibility predictions (Tran et al., 2024). In this research, experts with extensive field experience assigned ranks to the selected Landslide conditioning factors and determined appropriate weighting values for each parameter class.

Using the Analytical Hierarchy Process (AHP), the study evaluated the relative importance of 11 Landslide Conditioning Factors (LCFs) throughout the road corridor, achieving a Consistency Ratio of 0.05. The analysis revealed that five factors were most influential in triggering landslides along NH-10 Road: rainfall (0.10), distance to roads (0.17), distance to river (0.19), distance to lineaments (0.15), and geology (0.11). Moderate influence was attributed to slope (0.06), aspect (0.07), and TWI (0.08). The least significant factors were determined to be elevation (0.04), NDVI (0.02), and LULC (0.03) (Table 14.8) as indicated by their lower weight values.

5.2 Correlation Between Landslides Occurrence and Causative Factors Using FR Model

Landslide prediction relies on understanding the factors that have contributed to past landslide events, in order to anticipate and forecast the factors that may trigger future landslides (Demir et al., 2014). Linked to this approach, FR method is used to determine the correlation between previous landslide locations and the factors affecting landslide occurrence. If the ratio is less than 1, then the relationship between a landslide and the factor's range or type is weak and vice versa (Lee & Sambath, 2006).

Table 14.9 illustrates the correlation between landslide occurrence and the different classes of the landslide inventory, in relation to each causal factor. For the relationship between landslide occurrence and slope angle, the data shows that slopes below 20° have a frequency ratio less than 1, indicating a very low probability of landslide events in those areas. Conversely, slopes exceeding 20° exhibit a frequency ratio greater than 1, suggesting a high probability of landslide occurrence on steeper slopes. The frequency value shows that elevation class between 256 and 717 m the FR values are greater than 1, which indicates that the area is favorable for landslides occurrence. The areas with elevations between 717 and 1870 m are not contributed to landslide occurrence. For rainfall FR value in the 3,024–3,112 class had an apparent maximum weight (2.56) over other classes and more than half of

Table 14.8 Allocated and normalized weights of the conditioning factors for the landslide causative in the study area

Parameters	Class	Normalized weights	Influence (%)	Assigned Weight based on Saaty's scale
Elevation (mts)	69–255	0.04	4	2
	256–439			3
	440–717			5
	718–1,170			8
	1,180–1,870			9
Slope in degree	0–11	0.06	6	2
	12–20			3
	21–29			5
	30–39			8
	40–69			9
Rainfall	2,628–2,714	0.10	10	2
	2,715–2,824			3
	2,825–2,943			5
	2,944–3,023			8
	3,024–3,112			9
Distance to river (mts)	0–1,920	0.19	19	2
	1,921–4,870			3
	4,871–8,916			5
	8,917–12,960			8
	12,970–17,490			9
Distance to road (mts)	0–100	0.16	16	2
	100–200			3
	200–300			5
	300–400			8
	400–500			9
Distance to Lineament (mts)	0–2,207	0.15	15	2
	2,208–4,530			3
	4,531–7,609			5
	7,610–11,330			8
	11,340–14,810			9
NDVI	−0.048–0.17	0.02	2	9
	0.18–0.32			8
	0.33–0.43			6
	0.44–0.52			3
	0.53–0.68			2

(continued)

Table 14.8 (continued)

Parameters	Class	Normalized weights	Influence (%)	Assigned Weight based on Saaty's scale
TWI	2.63–5.46	0.08	7	2
	5.47–7.54			3
	7.55–10.7			5
	10.8–14.7			8
	14.8–21.6			9
LULC	Water body	0.03	3	9
	Vegetation cover			2
	Agricultural land			2
	Buildup area			3
	Bare ground			9
	Rangeland			7
Aspect	Flat (−1)	0.07	6	1
	North (0–22.5)			1
	Northeast (22.5–67.5)			6
	East (67.5–112.5)			6
	Southeast (112.5–157.5)			7
	South (157.5–202.5)			5
	Southwest (202.5–247.5)			5
	West (247.5–292.5)			7
	Northwest (292.5–337.5)			9
	North (337.5–360)			5
Geology	Fluvial/Aeolian/ Coastal & Glacial Sediments	0.11	11	1
	Siwalik Gp. (Upper Siwalik)			2
	Lower Gondwana Gp. (Bhareli, Damuda Fm.)			2
	Lingtse Gneiss Gp			5

(continued)

Table 14.8 (continued)

Parameters	Class	Normalized weights	Influence (%)	Assigned Weight based on Saaty's scale
	Daling Gp. (Reyang Fm.)			7
	Daling Gp. (Gorubathan Fm.)			9

the landslides occur in this class. The analysis revealed a strong correlation between landslide occurrences and their proximity to both drainage systems and roadways (table 8). Landslide events were predominantly concentrated in areas close to rivers and roads. Specifically, locations within 8000 m of rivers and within 100 m of roads showed a frequency ratio greater than 1, indicating an elevated probability of landslide occurrence. Conversely, areas beyond these threshold distances demonstrated a frequency ratio less than 1, suggesting a significantly lower likelihood of landslides taking place. Areas within 11,000 m of a lineament exhibited a frequency ratio greater than 1, signifying a heightened probability of landslide occurrence. This relationship can be explained by the fact that as proximity to a lineament increases, rock formations experience more extensive fracturing and undergo increased weathering processes (Lee & Sambath, 2006). A negative correlation exists between landslide events and both the NDVI and TWI. Areas with NDVI values exceeding 0.33 and TWI values below 5.46 showed frequency ratios greater than 1, indicating an increased probability of landslide occurrence. These findings suggest that landslide susceptibility increases in regions with higher vegetation index values and lower soil moisture conditions. The analysis of landslide occurrence in relation to land use patterns revealed that vegetated areas experienced higher frequencies of landslides, while barren land, built-up areas, agricultural lands, and rangelands showed lower occurrence rates. This distribution pattern can be attributed to the fact that landslides predominantly occurred in mountainous and steeply inclined terrain, which typically features more extensive vegetation cover. Analysis of geological formations revealed that three units demonstrated elevated landslide susceptibility, as indicated by their high Frequency Ratio (FR) values: the Gorubathan formation (1.28), Upper Siwalik formation (1.19), and Reyang formation (1.24). These higher FR values suggest these geological units have an increased probability of experiencing landslide events. Analysis of slope aspect revealed that several directional classes showed increased landslide susceptibility, with Frequency Ratio values exceeding 1. These high-risk aspects included northeast-facing slopes (22.5°–67.5°), southeast-facing slopes (112.5°–157.5°), west-facing slopes (247.5°–292.5°), northwest-facing slopes (292.5°–337.5°), and north-facing slopes (337.5°–360°), all of which demonstrated a higher probability of landslide occurrence (Table 14.9).

Table 14.9 statistical analysis results of landslide occurrence and landslide factors using the FR method

Tirggering factors	Class	Class Pixel	Landslide Pixel	FR	RF	RF (INT)	Min RF	Max RF	Max–Min RF	PR
Slope in degree	0–11	29,566	4	0.070	0.011	1	0.011	0.418	0.407	2.69
	12–20	21,403	32	0.771	0.127	12				
	21–29	26,462	69	1.344	0.221	22				
	30–39	22,443	59	1.355	0.223	22				
	40–69	9919	49	2.546	0.418	41				
Elevation (mts)	69–255	37,160	45	0.624	0.180	18	0.00	0.474	0.474	3.14
	256–439	37,953	121	1.643	0.474	47				
	440–717	18,919	44	1.199	0.346	34				
	718–1,170	10,631	0	0.000	0.000	0				
	1,180–1,870	5130	3	0.000	0.000	0				
Rainfall (mm)	2,628–2,714	21,732	21	0.498	0.108	10	0.021	0.556	0.536	3.55
	2,715–2,824	12,840	13	0.522	0.113	11				
	2,825–2,943	6047	3	0.095	0.021	2				
	2,944–3,023	52,874	95	0.926	0.201	20				
	3,024–3,112	16,300	81	2.561	0.556	55				
Distance to river (mts)	0–1,920	40,924	108	1.360	0.431	43	0.014	0.453	0.439	2.91
	1,921–4,870	34,953	97	1.430	0.453	45				
	4,871–8,916	12,179	4	0.169	0.054	5				
	8,917–12,960	11,383	1	0.045	0.014	1				
	12,970–17,490	10,354	3	0.149	0.047	4				
Distance to roads (mts)	0–100	26,059	113	2.235	0.483	48	0.076	0.483	0.407	2.70
	100–200	22,601	23	0.525	0.113	11				

(continued)

Table 14.9 (continued)

Tirggering factors	Class	Class Pixel	Landslide Pixel	FR	RF	RF (INT)	Min RF	Max RF	Max–Min RF	PR
	200–300	22,779	44	0.996	0.215	21				
	300–400	20,656	21	0.524	0.113	11				
	400–500	17,698	12	0.350	0.076	7				
Distance to lineaments (mts)	0–2,207	40,601	80	1.016	0.202	20	0.112	0.263	0.151	1.00
	2,208–4,530	24,043	46	1.000	0.196	19				
	4,531–7,609	18,310	47	1.323	0.263	26				
	7,610–11,330	9425	21	1.149	0.228	22				
	11,340–14,810	17,414	19	0.562	0.112	11				
TWI	2.63–5.46	43,026	116	1.390	0.362	36	0.039	0.362	0.323	2.14
	5.47–7.54	38,254	58	0.782	0.204	20				
	7.55–10.7	14,096	26	0.951	0.248	24				
	10.8–14.7	10,986	12	0.563	0.147	14				
	14.8–21.6	3431	1	0.150	0.039	3				
NDVI	−0.048–0.17	9197	10	0.560	0.130	12	0.080	0.300	0.219	1.45
	0.18–0.32	11,882	8	0.347	0.080	8				
	0.33–0.43	15,579	34	1.125	0.260	26				
	0.44–0.52	34,248	86	1.294	0.300	29				
	0.53–0.68	38,887	75	0.994	0.230	23				
LULC	Water Body	9983	14	0.723	0.262	26	0.000	0.492	0.492	3.26
	Vegetation Cover	71,709	189	1.359	0.492	49				
	Agricultural Land	39	0	0.000	0.000	0				
	Buildup Area	12,744	0	0.000	0.000	0				

(continued)

Table 14.9 (continued)

Tirggering factors	Class	Class Pixel	Landslide Pixel	FR	RF	RF (INT)	Min RF	Max RF	Max–Min RF	PR
	Bare Ground	9473	6	0.326	0.118	11				
	Rangeland	5845	4	0.353	0.128	12				
Aspect	Flat (−1)	33	0	0.000	0.000	0	0.000	0.179	0.179	1.18
	North (0–22.5)	5318	3	0.291	0.034	3				
	Northeast (22.5–67.5)	11,565	25	1.114	0.131	13				
	East (67.5–112.5)	15,047	28	0.959	0.113	11				
	Southeast (112.5–157.5)	14,920	31	1.071	0.126	12				
	South (157.5–202.5)	11,910	17	0.736	0.087	8				
	Southwest (202.5–247.5)	12,798	14	0.564	0.066	6				
	West (247.5–292.5)	15,736	33	1.081	0.127	12				
	Northwest (292.5–337.5)	16,660	49	1.516	0.179	17				
	North (337.5–360)	5806	13	1.154	0.136	13				
Geology	Lingtse Gneiss Gp	2440	1	0.211	0.048	3.427	0.011	0.290	0.279	1.85
	Daling Gp. (Gorubathan Fm.)	63,939	159	1.282	0.290	4				
	Siwalik Gp. (Upper Siwalik)	6470	15	1.195	0.270	28				
	Daling Gp. (Reyang Fm.)	14,493	35	1.245	0.281	26				

(continued)

Table 14.9 (continued)

Tirggering factors	Class	Class Pixel	Landslide Pixel	FR	RF	RF (INT)	Min RF	Max RF	Max–Min RF	PR
	Lower Gondwana Gp. (Bhareli, Damuda Fm.)	1159	1	0.445	0.100	28				
	Undefined Fluvial/ Aeolian/ Coastal & Glacial Sediments	21,292	2	0.048	0.011	10				

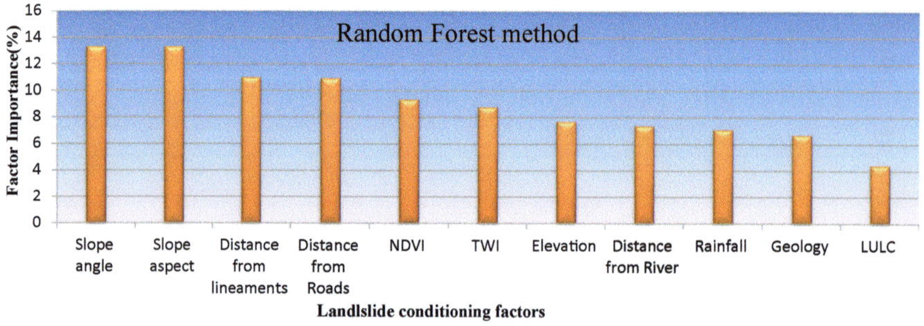

Fig. 14.9 Relative importance's of landslide conditioning factors using random forest model

5.3 *Correlation Between Landslides Occurrence and Causative Factors Using RF Model*

Accurate identification of influential environmental factors and evaluating there (each factor) contribution to landslide occurrence is fundamental for reliable spatial modeling and hazard management (Arabameri et al., 2021). Therefore, Random Forest (RF) model provides a robust method to rank these environmental factors based on their predictive importance, using metrics like means decrease accuracy (Zhou et al., 2021). Google earth engine platform (https://earthengine.google.com) was used to calculate the mean decrease accuracy of each factor and the factors with higher index values indicate greater importance (Wang et al., 2020; Arabameri et al., 2021) (Fig. 14.9).

The results from the RF analysis (mean decrease accuracy) showed that slope angle (13.34) is the most important contributing factor, followed by slope aspect (13.33), distance from lineaments (11.00), distance from roads (10.94), NDVI (9.35), TWI (8.79), elevation (7.71), distance from river (7.39), rainfall (7.10), geology (6.71) and LULC (4.43). However, geology (6.71) and LULC (4.43) appeared to have relatively less importance for landslide susceptibility modeling in this study area.

5.4 *Landslide Susceptibility Models*

The study aimed to develop and compare landslide susceptibility maps generated through three different methods: AHP, FR, and RF. To facilitate comparison between the output maps, the natural breaks classification method was employed, which determined to be the most appropriate approach for categorizing the probability results. Using this method, the probability maps were classified into five landslide susceptibility categories: very low, low, moderate, high, and very high (Figs. 14.10 and 14.11).

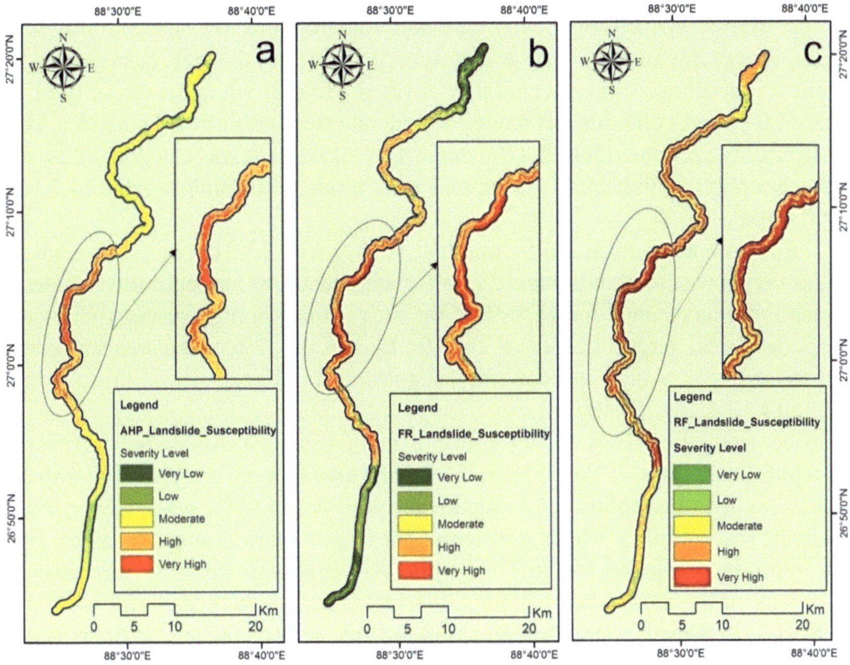

Fig. 14.10 Landslide susceptibility maps: **a** AHP model; **b** FR model; and **c** RF model

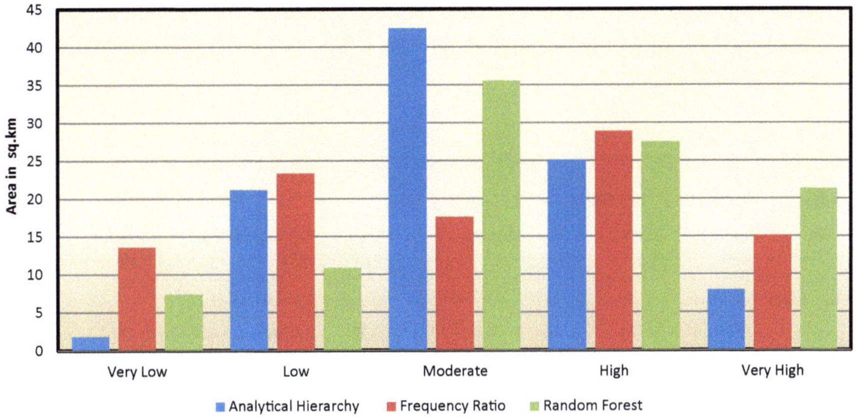

Fig. 14.11 Quantitative comparison of landslide susceptibility classes of different models (based on area in sq. km)

The AHP model systematically classified the area into five susceptibility categories. A very low susceptibility was assigned to 1.91% of the area, representing the smallest proportion. Low susceptibility covered 21.46%, while the majority of the area, 43.04%, was classified as having a moderate susceptibility to landslides. High susceptibility accounted for 25.47%, and 8.10% of the area was categorized as very high susceptibility, indicating the region's most prone to landslides (Table 14.10 and Fig. 14.10a).

The FR model showed a different distribution, with 13.84% of the area identified as very low susceptibility and 23.66% classified as low susceptibility. Moderate susceptibility accounted for 17.86% of the area, while the high susceptibility category covered the largest portion at 29.27%. Finally, 15.37% of the area was classified as having very high susceptibility, suggesting significant risk in those regions (Table 14.10 & Fig. 14.10b).

In contrast, the RF model exhibited a more balanced distribution across the susceptibility classes. It classified 7.25% of the area as very low susceptibility and 10.62% as low susceptibility. A significant portion, 34.62%, was categorized as moderate susceptibility, which represented the largest share among all classes. High susceptibility accounted for 26.77%, while very high susceptibility was assigned to 20.74% of the area, reflecting a notable spread across the higher risk categories (Table 14.10). Despite these differences in the precise percentages, all three models consistently highlighted that the majority of the study area falls within the moderate-to-high susceptibility range (Fig. 14.10c), indicating a significant landslide hazard that requires careful monitoring and mitigation planning. The variations between the models can be attributed to their distinct underlying assumptions and algorithms, but the overall alignment in the spatial patterns lends credibility to the landslide risk assessment for this region.

Table 14.10 Spatial distribution of landslide susceptibility class in the study area

Level	Analytical hierarchy			Frequency ratio			Random forest		
	Range	Area (km^2)	Area (%)	Range	Area (km^2)	Area (%)	Range	Area (km^2)	Area (%)
Very Low	3–4	1.89	1.91	172.7–415.7	13.68	13.84	0–0.11	7.45	7.25
Low	4.01–5	21.21	21.46	415.8–551.4	23.38	23.66	0.11–0.27	10.92	10.62
Moderate	5.01–6	42.54	43.04	551.5–690.8	17.65	17.86	0.27–0.64	35.60	34.62
High	6.01–7	25.17	25.47	690.9–830.1	28.92	29.27	0.64–0.80	27.53	26.77
Very High	7.01–9	8.01	8.10	830.2–1,084	15.18	15.37	0.80–1.00	21.33	20.74

5.5 Accuracy Assessment and Comparison

The performance and reliability of three distinct landslide susceptibility modeling approaches were rigorously evaluated using success rate and prediction rate Area under the Curve (AUC) analyses (Fig. 14.12). The success rate AUC metric critically assesses the retrospective classification accuracy of known landslide locations within the study area. The AHP model exhibited a robust performance with an AUC of 0.807, which falls within the "very good" classification of model fit. FR method marginally improved upon this, achieving a success rate AUC of 0.823, also categorized as "very good". However, RF model emerged as the most outstanding performer, securing the highest success rate AUC of 0.866 and solidifying its position as the most reliable approach for landslide susceptibility mapping in this research context. These comparative results underscore the potential of advanced machine learning techniques, particularly Random Forest, in developing more precise and reliable landslide susceptibility models compared to traditional statistical and heuristic methods.

The prediction rate Area under the Curve (AUC) provided crucial insights into the models' capability to forecast future landslide occurrences beyond the initial training dataset (Youssef & Pourghasemi, 2020). This metric offers a more forward-looking assessment of model reliability and predictive power. AHP model demonstrated strong predictive performance with an AUC of 0.803, firmly establishing itself in the "very good" performance category. FR model marginally enhanced the predictive accuracy, achieving an AUC of 0.812, which similarly falls within the "very good" range. Notably, the RF model once again distinguished itself by securing the highest prediction rate AUC of 0.851, consistently reinforcing its superior predictive capabilities across the analysis. These results underscore the potential of machine learning approaches, particularly Random Forest, in developing more robust and forward-looking landslide susceptibility models. The superior performance of the RF model,

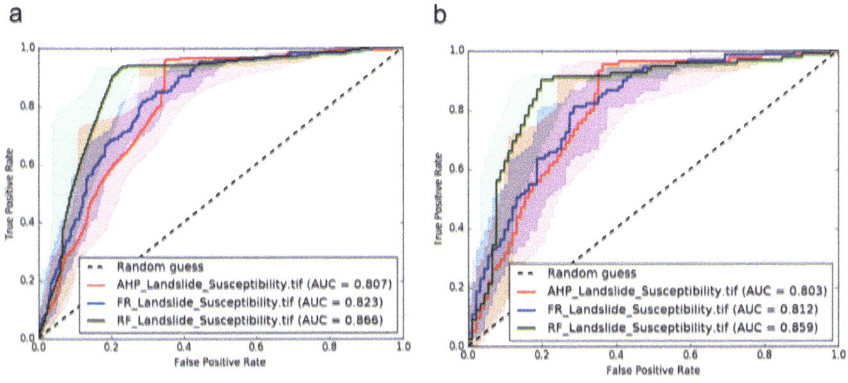

Fig. 14.12 Success rate (a) and Prediction rate (b) curves for the susceptibility maps produced in this study for different models

particularly in terms of prediction capability, suggests it may be the most reliable and transferable approach for assessing future landslide hazards in similar geological and climatic settings (Wang et al., 2020). The variations in AUC values provide a robust quantitative assessment of the models' efficacy, strengthening confidence in the landslide susceptibility maps produced and supporting their use in guiding risk mitigation strategies and resource allocation within the region.

The findings of this study underscore the complexity of landslide susceptibility in the NH-10 corridor, where both natural and anthropogenic factors play critical roles in influencing slope stability. The comparative analysis of AHP, FR, and RF models has provided a comprehensive perspective on the primary drivers of landslides in the region and their implications for disaster mitigation. The results indicate that slope angle, distance to rivers, distance to roads, rainfall, and geology is the most influential factors contributing to landslide susceptibility. The AHP model highlighted the significant impact of distance to roads (0.17), distance to rivers (0.19), and rainfall (0.10), reinforcing the role of hydrological and geo-morphological conditions in triggering landslides. Similarly, the FR analysis demonstrated a strong correlation between landslide occurrence and steep slopes exceeding 20°, high rainfall zones, and proximity to drainage networks and roads. The RF model further validated these findings, with slope angle (13.34), aspect (13.33), and distance from lineaments (11.00) emerging as the most critical determinants of landslide probability. These results suggest that landslides are predominantly controlled by terrain morphology and hydrological factors, with excessive rainfall acting as a primary triggering mechanism. The interplay between these factors creates high-risk zones, particularly in areas with steep slopes, unstable geological formations, and frequent water infiltration. The presence of highly susceptible geological units such as the Gorubathan, Upper Siwalik, and Reyang formations further exacerbates instability, highlighting the need for geotechnical interventions in these vulnerable zones.

Anthropogenic activities have significantly amplified landslide susceptibility along the NH-10 corridor. Road construction, deforestation, and unplanned urbanization have altered the natural slope stability, leading to increased erosion and slope failures. The study findings indicate that areas within 100 m of roads exhibit a higher frequency ratio (>1), suggesting that road-cut excavations and improper drainage management contribute to slope destabilization. Furthermore, changes in land use and vegetation cover play a critical role in influencing landslide occurrences. The FR model showed a negative correlation between NDVI and landslide susceptibility, indicating that deforested and barren areas are more susceptible to mass movements. As vegetation loss reduces soil cohesion and increases surface runoff, it accelerates erosion and slope failures, particularly in steep terrains. Landslide susceptibility maps are essential tools for planning, assessment, and landslide management, playing a crucial role in hazard evaluation and risk analysis (Youssef & Pourghasemi, 2020). The integration of AHP, FR, and RF models has provided valuable insights into the spatial distribution and contributing factors of landslide susceptibility along the NH-10 corridor. The findings reinforce the need for a multi-pronged mitigation approach, combining engineering solutions, environmental conservation, and policy-driven interventions to reduce landslide risks. As climate variability and human activities continue to shape landscape dynamics, ongoing monitoring and adaptive management strategies will be crucial in mitigating future hazards and ensuring sustainable infrastructure development in landslide-prone regions.

The present study identified key variables contributing to landslides and demonstrated methods applicable for landslide susceptibility mapping beyond the study area. The novelty of this research can be summarized as follows: [1] the application of AHP, FR, and RF models in the Darjiling–Sikkim Himalaya is innovative, [2] integrating spatial analytical methods for landslide susceptibility modeling is a valuable approach, and [3] providing a comparative analysis of the performance of these methods enhances the understanding of their effectiveness.

6 Limitations and Future Suggestions

In this study, the AHP, FR, and RF models were employed to assess landslide susceptibility along National Highway 10 (NH-10). Among these methods, the RF model demonstrated superior performance in predicting landslide susceptibility. However, despite its higher accuracy, RF and other machine learning approaches face several critical limitations. They are highly data-dependent, requiring extensive and high-quality datasets for accurate predictions. The most significant constraint is the temporal incompleteness of the landslide inventory data, with the last update by the Geological Survey of India in 2021. This outdated inventory potentially introduces substantial bias in the AHP, FR, and RF models, as recent landslide events were not incorporated into the model training, which may compromise the accuracy and representativeness of the susceptibility predictions. Each method—AHP with its pair-wise comparison approach, FR with its statistical frequency-based analysis, and RF with its ensemble learning technique- relies on comprehensive, up-to-date, and accurate training data, and any missing or outdated information can significantly impair model performance along the NH-10 corridor. Future research should prioritize conducting detailed field surveys and utilizing high-resolution satellite imagery to develop a more current and comprehensive landslide inventory for NH-10. Moreover, to enhance predictive capabilities, subsequent studies could explore advanced deep learning algorithms such as Convolution Neural Networks (CNN) and Recurrent Neural Networks (RNN), comparing their performance with the current AHP, FR, and RF approaches (Zhou et al., 2021). Various studies highlight issues in landslide susceptibility analysis, these models for site-specific planning due to scale sensitivity, while model performance depends on input variables, with RF performing well in some areas but poorly in others (Lee et al., 2017; Hong et al., 2016). By addressing these data limitations and exploring more sophisticated machine learning techniques, researchers can potentially improve the accuracy and reliability of landslide susceptibility mapping along the challenging terrain of NH-10.

7 Conclusion

The comprehensive landslide susceptibility analysis undertaken along the strategically important NH 10 transportation corridor has yielded invaluable insights to support risk-informed decision-making and effective hazard management strategies in this geologically complex region. By employing a robust multi-model approach, incorporating well-established techniques such as AHP, FR, and RF, the study has delivered a thorough assessment of the key factors influencing slope instability and their relative contributions to landslide occurrence. The results of the landslide susceptibility mapping showed a high degree of consistency across the three modeling methods, with the majority of the study area classified as having moderate to high susceptibility. Specifically, the AHP model designated 43.04% of the region as moderately susceptible, 25.47% as highly susceptible, and a relatively small 1.91% as very low susceptibility. This aligns with the FR model's identification of the highest proportions as high (29.27%) and low (23.66%) susceptibility, while the RF model struck a balance, with 34.62% classified as moderate, 26.77% as high, and 7.25% as very low susceptibility. The validation of the models through success rate and prediction rate AUC analyses lent further credibility and confidence to the results. The RF approach demonstrated the strongest performance, with excellent AUC values of 0.871 for success rate and 0.861 for prediction rate, indicating its superior ability to both retrospectively classify known landslide locations and prospectively predict future slope failures. The AHP and FR models also showed very good to excellent AUC scores, confirming their reliability in capturing the key drivers of landslide occurrence in this complex terrain.

These findings underscore the critical importance of comprehensive landslide risk assessment along major transportation corridors like NH 10, where slope instability poses a significant threat to infrastructure integrity, public safety, and economic activity (Fig. 14.13). The consistency in high-risk areas identified by the multiple modeling techniques provides a solid foundation for targeted mitigation strategies, resource allocation, and hazard management planning. By proactively addressing the landslide susceptibility hotspots, stakeholders can enhance the safety and resilience of this vital transportation link, safeguarding the flow of goods, services, and people essential for the region's economic and social well-being. Moving forward, the landslide susceptibility maps generated through this study should be regularly updated and integrated with other relevant datasets, such as rainfall patterns, seismic activity, and land use changes, to maintain a dynamic understanding of the evolving hazard landscape. Continued monitoring, coupled with robust emergency response protocols and community engagement initiatives, will be essential to effectively mitigate the risks posed by landslides along the NH 10 corridor and similar geologically challenging regions. Only through a holistic, data-driven approach can stakeholders ensure the long-term safety, sustainability, and resilience of critical infrastructure networks in the face of these persistent natural hazards.

Fig. 14.13 Field photographs of landslide events along the National High 10 (NH-10)

References

Abedini, M. B., Ghasemian, A., Shirzadi, D. T., & Bui, A. (2019). Comparative study of support vector machine and logistic model tree classifiers for shallow landslide susceptibility modeling. *Environment and Earth Science, 78*, 560. https://doi.org/10.1007/s12665-019-8562-z

Abedini, M., & Tulabi, S. (2018). Assessing LNRF, FR, and AHP models in landslide susceptibility mapping index: A comparative study of Nojian watershed in Lorestan province, Iran. *Environment and Earth Science, 77*, 405. https://doi.org/10.1007/s12665-018-7524-1

Abu, I., Asish, S., Bonosri, G., Subodh, P., Indrajit, C., Javed, M. (2021). Landslide susceptibility modeling in a complex mountainous region of Sikkim Himalaya using new hybrid data mining approach. *Geocarto International, 37*. https://doi.org/10.1080/10106049.2021.2009920

Achour, Y., & Pourghasemi, H. R. (2020). How do machine learning techniques help in increasing accuracy of landslide susceptibility maps? *Geoscience Frontiers, 11*, 871–883.

Akinci, H., Kilicoglu, C., & Dogan, S. (2020). Random forest-based landslide susceptibility mapping in coastal regions of Artvin, Turkey. *ISPRS International Journal of Geo-Information, 9*(9), 553. https://doi.org/10.3390/ijgi9090553

Alam, S. K., Sujit, M., & Ramkrishna, M. (2024). Geospatial intelligence for landslide susceptibility and risk analysis: Insights from NH31A and East Sikkim Himalaya Settlements. *Natural Hazards Research*. https://doi.org/10.1016/j.nhres.2024.10.001

Anbazhagan, S., & Ramesh, V. (2014). Landslide hazard zonation mapping in ghat road section of Kolli hills, India. *Journal of Mountain Science, 11*, 1308–1325.

Arabameri, A., Biswajeet, P., Khalil, R., & Chang-Wook, L. (2019). Assessment of landslide susceptibility using statistical- and artificial intelligence-based FR–RF integrated model and multiresolution DEMs. *Remote Sensing, 11*, 999. https://doi.org/10.3390/rs11090999

Arabameri, A., Santosh, S. S., Omid, G., Jagabandhu, R., John, T., Hossein, M., & Romulus-Dumitru, C. (2021). Spatial prediction of shallow landslide: Application of novel rotational forest-based reduced error pruning tree. *Natural Hazards and Risk, 12*, 1343–1370. https://doi.org/10.1080/19475705.2021.1914753

Arnone, E., Francipane, A., Noto, L. V., Scarbaci, A., & La Loggia, G. (2014). La Loggia G (2014) Strategies investigation in using artificial neural network for landslide susceptibility mapping: Application to a Sicilian catchment. *Journal of Hydroinformatics, 16*, 502–515.

Arrawatia, M.L., Tambe, S. (2011). Biodiversity of Sikkim exploring and conserving a global hotspot. Department of Information and Public Relations, Government of Sikkim, Gangtok. ISBN: 978-81-920437-0-9

Ayalew, L., & Yamagishi, H. (2005). The application of GIS-based logistic regression for landslide susceptibility mapping in the Kakuda–Yahiko Mountains, Central Japan. *Geomorphology, 65*, 15–31.

Ayalew, L., Yamagishi, H., Marui, H., & Kanno, T. (2005). Landslides in Sado Island of Japan: Part II. GIS-based susceptibility mapping with comparisons of results from two methods and verifications. *Engineering Geology, 81*, 432–445.

Azarafza, M., Azarafza, M., Akgün, H., et al. (2021). Deep learning-based landslide susceptibility mapping. *Science and Reports, 11*, 24112. https://doi.org/10.1038/s41598-021-03585-1

Ballabio, C., Sterlacchini, S. (2012). Support vector machines for landslide susceptibility mapping: The Staffora River basin case study, Italy. *Math Geosciences, 44*. https://doi.org/10.1007/s11004-011-9379-9

Batar, A., & Watanabe, T. (2021). Landslide susceptibility mapping and assessment using geospatial platforms and weights of evidence (WoE) method in the Indian Himalayan region: recent developments, gaps, and future directions. *International Journal of Geo-Information, 10*, 28. https://doi.org/10.3390/ijgi10030114

Beven, K., & Kirkby, M. (1989). A physically based, variable contributing area model of basin hydrology. *Hydro Science Bullet, 24*(1).

Bhattacharyya, K., Mtra, G, & Kwon, S. (2015). Geometry and kinematics of the Darjeeling–Sikkim Himalaya, India: Implications for the evolution of the Himalayan fold-thrust belt. *Journal of Asian Earth Sciences*. 101016/jjseaes201509008

Biswakarma, P., Joshi, V., Abdo, H.G., Almohamad, H., Abdullah Al Dughairi, A., Al-Mutiry, M. (2023). An integrated quantitative and qualitative approach for landslide susceptibility mapping in West Sikkim district, Indian Himalaya. *Geomatics Natural Hazards and Risk,14*(1). https://doi.org/10.1080/19475705.2023.2273781

Breiman, L. (2001). *Machine Learning, 45*(1), 5–32. https://doi.org/10.1023/a:1010933404324

Chen, E., Archana, R., & Tapas, M. (2013). Landslide susceptibility assessment using information value method in parts of the Darjeeling Himalayas. *Journal of Geological Society of India, 82*, 351–362. https://doi.org/10.1007/s12594-013-0162-z

Chen, W., Xie, X., Wang, J., Pradhan, B., Hong, H., Tien Bui, D., Duan, Z., & Ma, J. (2017). A comparative study of logistic model tree, random forest, and classification and regression tree models for spatial prediction of landslide susceptibility. *CATENA, 151*, 147–160.

Dai, F., & Lee, C. (2002). Landslide characteristics and slope instability modeling using GIS, Lantau Island, Hong Kong. *Geomorphology, 42*(3–4), 213–228. https://doi.org/10.1016/s0169-555x(01)00087-3

Das, S., Shantanu, S., & Kanungo, D. (2022). GIS-based landslide susceptibility zonation mapping using the analytic hierarchy process (AHP) method in parts of Kalimpong Region of Darjeeling Himalaya. *Environmental Monitoring and Assessment, 194*, 234. https://doi.org/10.1007/s10661-022-09851-7

Dehnavi, A., Aghdam, I. N., Pradhan, B., & Morshed, V. M. H. (2015). A new hybrid model using step-wise weight assessment ratio analysis (SWARA) technique and adaptive neuro-fuzzy inference system (ANFIS) for regional landslide hazard assessment in Iran. *CATENA, 135*, 122–148.

Demir, G., Aytekin, M., & Akgun, A. (2014). Landslide susceptibility mapping by frequency ratio and logistic regression methods: An example from Niksar-Resadiye (Tokat, Turkey). *Arabian Journal of Geosciences, 8*(3), 1801–1812. https://doi.org/10.1007/s12517-014-1332-z

Doke, A. B., Zolekar, R. B., Patel, H., & Das, S. (2021). Geospatial mapping of groundwater potential zones using multi-criteria decision-making AHP approach in a hardrock basaltic terrain in India. *Ecological Indicators, 127*, Article 107685. https://doi.org/10.1016/j.ecolind.2021.107685

DoT—Department of *Transport-Government of Sikkim (2024) News and Announcements* Sikkim CM Extends Deep Gratitude and Lauds Centre's Transformative NH-10 Decision. Gangtok. Available 06 Nov 2024

Dou, J., Yunus, A. P., Tien Bui, D., Merghadi, A., Sahana, M., Zh, Z., Chen, C. W., Khosravi, K., Yang, Y., & Pham, B. T. (2019). Assessment of advanced random forest and decision tree algorithms for modeling rainfall-induced landslide susceptibility in the Izu-Oshima Volcanic Island, Japan. *Science of the Total Environment, 662*, 332–346.

El Jazouli, A., Barakat, A., & Khellouk, R. (2019). GIS-multicriteria evaluation using AHP for landslide susceptibility mapping in Oum Er Rbia high basin (Morocco). *Geoenviron Disasters, 6*, 3. https://doi.org/10.1186/s40677-019-0119-7

Elvis, B. W. W., Arsene, M., Théophile, N. M., Bruno, K. M. E., & Olivier, O. A. (2022). Integration of shannon entropy (SE), frequency ratio (FR) and analytical hierarchy process (AHP) in GIS for suitable groundwater potential zones targeting in the Yoyo river basin, Méiganga area, Adamawa Cameroon. *Journal of Hydrology: Regional Studies, 39*, Article 100997.

EM-DAT International Disaster Database. (2020). *Université Catholique de Louvain.* Belgium.

Ermini, L., Filippo, C., & Nicola, C. (2005). Artificial Neural Networks applied to landslide susceptibility assessment. *Geomorphol., 66*, 327–343. https://doi.org/10.1016/j.geomorph.2004.09.025

Froude, M. J., & Petley, D. N. (2018b). Global fatal landslide.

Froude, M. J., & Petley, D. N. (2018a). Global fatal landslide occurrence from 2004 to 2016. *Natural Hazards and Earth Systems Sciences, 18*, 2161–2181. https://doi.org/10.5194/nhess-18-2161-2018

Gelisli, K., Kaya, T., & Babacan, A. E. (2015). Assessing the factor of safety using an artificial neural network: Case studies on landslides in Giresun, Turkey. *Environmental Earth Sciences, 73*, 8639–8646.

Ghose, M., Ratika, P., & Polash, B. (2016). GIS based spatial noise impact analysis (SNIA) of the broadening of national highway in Sikkim Himalayas: A case study. *AIMS Environmental Science, 3*, 714–738. https://doi.org/10.3934/environsci.2016.4.714

Goetz, J. N., Brenning, H., & Petschko, P. (2015). Leopold evaluating machine learning and statistical prediction techniques Forland slide susceptibility modeling. *Computers & Geosciences, 81*(2015), 1–11.

Guillard, C., & Zezere, J. (2012). Landslide susceptibility assessment and validation in the framework of municipal planning in Portugal: The case of Loures Municipality. *Environmental Management, 50*(2012), 721–735.

Gupta, S., Nayek, S., Chakraborty, D. (2016). Hydrochemical evaluation of Rangit River, Sikkim, India: using Water Quality Index and multivariate statistics. *Environment and Earth Sciences, 75*(7). 101007/s12665-015-5223-8

Guzzetti, F., Carrarra, A., Cardinali, M., & Reichenbach, P. (1999). Landslide hazard evaluation: A review of current techniques and their application in a multi scale study Central Italy. *Geomorphology, 31*, 181–216.

Guzzetti, F., Peruccacci, S., Rossi, M., et al. (2007). Rainfall thresholds for the initiation of landslides in central and southern Europe. *Meteorology and Atmospheric Physics, 98*, 239–267. https://doi.org/10.1007/s00703-007-0262-7

Ha, H., Bui, Q. D., Khuc, T. D., Tran, D. T., Pham, B. T., Mai, S. H., Nguyen, L. P., & Luu, C. (2022). A machine learning approach in spatial predicting of landslides and flashflood susceptible zones for a road network Model. *Earth Systems and Environment, 8*(2022), 4341–4357. https://doi.org/10.1007/s40808-022-01384-9

Habumugisha, J.M., Chen, N., Rahman, M., Islam, M.M., Ahmad, H., Elbeltagi, A., Sharma, G., Liza, S.N., Dewan, A. (2022). Landslide susceptibility mapping with deep learning algorithms. *Sustainability, 14*(3), 1734. https://doi.org/10.3390/su14031734

Hindustan Times (2024) NH10 shut down as heavy rains hit north Bengal triggering fresh landslides. Jun 29, 2024. https://www.hindustantimes.com.

Hong, H., Pourghasemi, H. R., & Pourtaghi, Z. S. (2016). Landslide susceptibility assessment in Lianhua County (China): A comparison between a random forest data mining technique and bivariate and multivariate statistical models. *Geomorphology, 259*(2016), 105–118.

Hussain, G., Singh, Y., Singh, K., et al. (2019). Landslide susceptibility mapping along national highway-1 in Jammu and Kashmir State (India). *Innovative Infrastructure Solutions, 4*, 59. https://doi.org/10.1007/s41062-019-0245-9

Ifediegwu SI (2022) Assessment of groundwater potential zones using GIS and AHP techniques: a case study of the Lafia district, Nasarawa State, Nigeria. *Applied Water Science 12*, 10 (2022). https://doi.org/10.1007/s13201-021-01556-5

Irjesh, S., Jayant, T., & Swarnim, M. (2022). Remote sensing and GIS-based landslide susceptibility mapping using frequency ratio method in Sikkim Himalaya. *Quaterrary Science Advances, 8*, Article 100067. https://doi.org/10.1016/j.qsa.2022.100067

Ji, J., Zhou, Y., Cheng, Q., Jiang, S., & Liu, S. (2023). Landslide susceptibility mapping based on deep learning algorithms using information value analysis optimization. *Land, 12*(6), 1125. https://doi.org/10.3390/land12061125

Kanungo, D., Manoj, A., Sarkar Shantanu, S., & Gupta, R. P. (2009). Landslide susceptibility zonation (LSZ) mapping—A review. *Journal of South Asia Disaster Studies, 2*, 81–105.

Kumar, K., Annadurai, R. (2015). Comparison of frequency ratio model and analytic hierarchy process methods upon landslide susceptibility mapping using geospatial techniques. *Disaster Advances, 8*(5).

Kumar, R., & Anbalagan, R. (2015). Landslide susceptibility zonation in part of Tehri reservoir region using frequency ratio, fuzzy logic and GIS. *Journal of Earth System Science, 124*, 431–448.

Kumar, R., & Anbalagan, R. (2016). Landslide susceptibility mapping using analytical hierarchy process (AHP) in Tehri reservoir rim region, Uttarakhand. *Journal of the Geological Society of India, 87*, 271–286. https://doi.org/10.1007/s12594-016-0395-8

Kumar, V., & Singh, K. E. (2019). Activeness of remote sensing and GIS-based landslide susceptibility zonation mapping using information value method. *Sustain Engineering, 17*, 225–234.

Lee, S., Hong, S.-M., & Jung, H.-S. (2017). A support vector machine for landslide susceptibility mapping in Gangwon Province, Korea. *Sustainability, 9*(1), 48. https://doi.org/10.3390/su9010048
Lee, S., & Pradhan, B. (2007). Landslide hazard mapping at Selangor, Malaysia using frequency ratio and logistic regression models. *Landslides, 4*, 33–41. https://doi.org/10.1007/s10346-006-0047-y
Lee, S., & Sambath, T. (2006). Landslide susceptibility mapping in the Damrei Romel area, Cambodia using frequency ratio and logistic regression models. *Environmental Geology, 50*(6), 847–855. https://doi.org/10.1007/s00254-006-0256-7
Leonardi, G., Palamara, R., Manti, F., Tufano, A. (2022). GIS-multicriteria analysis using AHP to evaluate the landslide risk in road lifelines. *Applied Science 12*, no. 9: 4707. https://doi.org/10.3390/app12094707.
Li, M., Wang, H., Chen, J., & Zheng, K. (2024). Assessing landslide susceptibility based on the random forest model and multi-source heterogeneous data. *Ecological Indicators.* https://doi.org/10.1016/j.ecolind.2024.111600
Liu, X., Shao, S., & Shao, S. (2024). Landslide susceptibility zonation using the analytical hierarchy process (AHP) in the Great Xi'an Region, China. *Scientific Reports, 14*, 2941. https://doi.org/10.1038/s41598-024-53630-y
Liu, Y. L., Yin, K. L., & Liu, B. (2010). Application of logical regression and artificial neural network model in spatial prediction of landslide disaster. *Hydrogeology and Engineering Geology, 37*, 92–96.
Magliulo, P., DiLisio, A., Russo, F., Zelano, A. (2008). Geomorphology and landslide susceptibility assessment using GIS and bivariate statistics: a case study in southern Italy. *Natural Hazards 47*(2008), 411–435
Malczewski, J. A. (1999). *GIS and Multi Criteria Decision Analysis* (p. 392). John Wiley & Sons.
Merghadi, A., Yunus, A. P., Dou, J., Whiteley, J., ThaiPham, B., Tien Bui, D., Avtar, R., et al. (2020). Machine learning methods for landslide susceptibility studies: A comparative overview of algorithm performance. *Earth Science Reviews, 207*, Article 103225.
MoRTH. (2019). The List of National Highways in the Country is as under; by Ministry of Road Transport and Highways, Govt. of India (https://morth.nic.in/sites/default/files/Details-of-National-Highways-as-on-31.03_1.pdf)
Naceur, H. A., Abdo, H. G., Igmoullan, B., et al. (2022). Performance assessment of the landslide susceptibility modelling using the support vector machine, radial basis function network, and weight of evidence models in the N'fis river basin, Morocco. *Geoscience Letters, 9*, 39. https://doi.org/10.1186/s40562-022-00249-4
Nath, S., Arnab, S., Anand, S. (2021). Remote sensing GIS-based landslide susceptibility & risk modeling in Darjeeling–Sikkim Himalaya together with FEM-based slope stability analysis of the terrain. *Natural Hazards, 108*. https://doi.org/10.1007/s11069-021-04823-5.
NDMA. (2019). Compendium of task force sub group reports on National Landslide Risk Management Strategy.
Nicu, L. C. (2018). Application of analytic hierarchy process, frequency ratio, and statistical index to landslide susceptibility: An approach to endangered cultural heritage. *Environment and Earth Science, 2018*(77), 79.
Nirmala, J., Roy, P., Ranjan Martha, T., Jalan, P., Nanda, A. (2023). Landslide Atlas of India (Mapping, monitoring and advance techniques using space-based inputs). NRSC special publication. NRSC/ISRO. Document number: NRSC-RSA-GSG-GMED-FEB 2023-TR-0002167-V1.0.
Ozdemir, A., & Altural, T. (2013). A comparative study of frequency ratio, weights of evidence and logistic regression methods for landslide susceptibility mapping: Sultan Mountains, SW Turkey. *Journal of Asian Earth Sciences, 64*, 180–197.
Pham, B.T., Pradhan, B., Tien Bui, D., Prakash, I., Dholakia, M.B. (2017). A comparative study of different machine learning methods for landslide susceptibility assessment: A case

study of Uttarakhand area (India). *Environmental Model Software, 84*, 240–250. 101016/jenvsoft201607005 101016/jenvsoft201607005

Pham, B. T., Tien Bui, D., Indra, P., & Dholakia, M. (2015). Landslide susceptibility assessment at a part of Uttarakhand Himalaya, India using GIS-based statistical approach of frequency ratio method. *International Journal of Engineering and Research Technology, 4*, 338–344.

Pourghasemi, H. R., Moradi, H. R., & Aghda, S. F. (2013). Landslide susceptibility mapping by binary logistic regression, analytical hierarchy process, and statistical index models and assessment of their performances. *Natural Hazards, 69*, 749–779. https://doi.org/10.1007/s11069-013-0728-5

Pradhan, B. A. (2013). (2013) Comparative study on the predictive ability of the decision tree, support vector machine and neuro-fuzzy models in landslide susceptibility mapping using GIS. *Computers & Geosciences, 51*, 350–365.

Rahaman, A., Venkatesan, M. S., & Ayyamperumal, R. (2021). GIS-based landslide susceptibility mapping method and Shannon entropy model: A case study on Sakaleshapur Taluk, Western Ghats, Karnataka, India. *Arabian Journal of Geosciences, 14*, 2154. https://doi.org/10.1007/s12517-021-08422-3

Rahmati, O., Pourghasemi, H.R., Zeinivand, H. (2016). Flood susceptibility mapping using frequency ratio and weights-of-evidence models in the Golastan Province, Iran. Geocarto International 2016. https://doi.org/10.1080/10106049.2015.1041559

Razavizadeh, S., Solaimani, K., Massironi, M., & Kavian, A. (2017). Mapping landslide susceptibility with frequency ratio, statistical index, and weights of evidence models: A case study in northern Iran. *Environment and Earth Science, 76*, 499.

Regmi, N. R., Giardino, J. R., & Vitek, J. D. (2010). Modeling susceptibility to landslides using the weight of evidence approach: Western Colorado, USA. *Geomorphology, 115*(2010), 172–187.

Reichenbach, P., Rossi, M., Malamud, B. D., Mihir, M., & Guzzetti, F. (2018). A review of machine learning techniques for landslide susceptibility mapping. *Earth-Science Reviews, 180*, 60–91.

Roian, G., Csaba, H., Kinga-Olga, R., Bo an, C.N., Gavrila, I.G. (2016). Assessing landslide vulnerability using statistical analysis and the frequency ratio model. Case study: Transylvanian Plain (Romania). *Geomorphology, 60*, 359–371.

Roy, L., Somasis, S., Sayantan, D., Arindam, S. (2022). Evaluating the relative tectonic response of the fluvial systems using multicriteria entropy method: A case study of the Rangit catchment, Eastern Himalayas, India. https://doi.org/10.1007/978-3-030-95435-2_12

Roy, J., & Saha, S. (2019). Landslide susceptibility mapping using knowledge driven statistical models in Darjeeling District, West Bengal, India. *Geoenvironmental Disasters, 6*, 11. https://doi.org/10.1186/s40677-019-0126-8

Roy, J., Saha, S., Arabameri, A., Blaschke, T., & Bui, D. T. (2019). A novel ensemble approach for landslide susceptibility mapping (LSM) in Darjeeling and Kalimpong Districts, West Bengal, India. *Remote Sensing, 11*(23), 2866. https://doi.org/10.3390/rs11232866

Saaty, R. W. (1987). The analytic hierarchy process—What it is and how it is used. *Mathematical Modeling, 9*(3–5), 161–176. https://doi.org/10.1016/0270-0255(87)90473-8

Saaty, T. L. (1980). *The Analytic Hierarchy Process*. McGraw-Hill.

Saha, A., Villuri, V. G. K., Bhardwaj, A., & Kumar, S. (2023). A multi-criteria decision analysis (MCDA) approach for landslide susceptibility mapping of a part of Darjeeling district in North-East Himalaya, India. *Applied Science, 13*(8), 5062. https://doi.org/10.3390/app13085062

Samanta, R. K., Bhunia, G. S., Shit, P. K., & Pourghasemi, H. R. (2018). Flood susceptibility mapping using geospatial frequency ratio technique: A case study of Subarnarekha River Basin, India. *Modeling Earth Systems and Environment, 4*, 395–408. https://doi.org/10.1007/s40808-018-0427-z

Sarfraz, Y., Basharat, M., Riaz, M., Akram, M., Xu, C., Ahmed, K., Shahzad, A., Al-Ansari, N., & Thuy Linh, N. (2022). Application of statistical and machine learning techniques for landslide susceptibility mapping in the Himalayan road corridors. *Open Geosciences, 14*(1), 1606–1635. https://doi.org/10.1515/geo-2022-0424

Sevgen, E., Kocaman, S., Nefeslioglu, H. A., & Gokceoglu, C. A. (2019). Novel performance assessment approach using photogrammetric techniques for landslide susceptibility mapping with logistic regression ANN and random forest. *Sensors, 19*, 3940.

Shadman, R., Majid, J.A., Himan, S., Safarrad, T. (2016). Fuzzy shannon entropy: a hybrid GIS-based landslide susceptibility mapping method. *Entropy 18*(10), 343. https://doi.org/10.3390/e18100343

Shahabi, H., Khezri, S., Ahmad, B. B., & Hasim, M. (2014). Landslide susceptibility mapping at central Zab basin, Iran: A comparison between analytical hierarchy process, frequency ratio and logistic regression models. *CATENA, 155*, 55–70.

Shirzadi, A. K., Soliamani, M., Habibnejhad, A., Kavian, K., Chapi, H., Shahabi, W., et al. (2018). Novel GIS based machine learning algorithms for shallow landslide susceptibility mapping. *Sensors, 18*, 3777.

Silalahi, F. E. S., Pamela Arifianti, Y., et al. (2019). Landslide susceptibility assessment using frequency ratio model in Bogor, West Java, Indonesia. *Geoscience Letters, 6*, 10. https://doi.org/10.1186/s40562-019-0140-4

Sreelakshmi, S., SS, V.C., Shaji, E. (2022). Landslide identification using machine learning techniques: Review, motivation, and future prospects. *Earth Science Informatics,15*, 2063–2090. https://doi.org/10.1007/s12145-022-00889-2

Sun, D., Wen, H., Wang, D., & Xu, J. (2020). A random forest model of landslide susceptibility mapping based on hyperparameter optimization using Bayes algorithm. *Geomorphology*. https://doi.org/10.1016/j.geomorph.2020.107201

Taalab, K., Cheng, T., & Zhang, Y. (2018). Mapping landslide susceptibility and types using Random Forest. *Big Earth Data, 2*(2), 159–178. https://doi.org/10.1080/20964471.2018.1472392

Tien Bui, D., Pradhan, B., Lofman, O., Revhaug, I., & Dick, O. B. (2012). Landslide susceptibility mapping at Hoa Binh province (Vietnam) using an adaptive neurofuzzy inference system and GIS. *Computers & Geosciences, 45*, 199–211.

Tran, V. A., Thanh, D. K., Xuan, Q. T., An, B. N., Truong, T. P. (2024). Application of potential machine learning models in landslide susceptibility assessment: A case study of Van Yen District, Yen Bai province, Vietnam. *Quaternary Science Advances, 14*, 2024. https://doi.org/10.1016/j.qsa.2024.100181

Varnes, D.J. (1978). Slope movement types and processes. In: Schuster, R.L. and Krizek, R.J., Eds., *Landslides, Analysis and Control, Transportation Research Board*, Special Report No. 176, Nat Aca of Sci, pp. 11–33.

Veerappan, R., & Anbazhagan, A. (2015). Landslide susceptibility mapping along Kolli hills Ghat road section (India) using frequency ratio, relative effect and fuzzy logic models. *Environment and Earth Science, 73*, 8009–8021. https://doi.org/10.1007/s12665-014-3954-6

Wang, L. J., Guo, M., Sawada, K., Lin, J., & Zhang, J. A. (2016). Comparative study of landslide susceptibility maps using logistic regression, frequency ratio, decision tree, weights of evidence and artificial neural network. *Geosciences Journal, 20*, 117–136.

Wang, L. J., Kazuhide, S., & Shuji, M. (2013). Landslide susceptibility analysis with logistic regression model based On FCM sampling strategy. *Computers & Geosciences, 57*, 81–92.

Wang, Y., Deliang, S., Haijia, W., Hong, Z., & Fengtai, Z. (2020). Comparison of random forest model and frequency ratio model for landslide susceptibility mapping (LSM) in Yunyang County (Chongqing, China). *International Journal of Environmental Research and Public Health, 17*, 4206. https://doi.org/10.3390/ijerph17124206

Wubalem, A., & Meten, M. (2020). Landslide susceptibility mapping using information value and logistic regression models in Goncha Siso Eneses area, northwestern Ethiopia. *SN Appl Sci, 2*, 807. https://doi.org/10.1007/s42452-020-2563-0

Yilmaz, I. (2009). Landslide susceptibility mapping using frequency ratio, logistic regression, artificial neural networks and their comparison: A case study from Kat landslides (Tokat-Turkey). *Computers & Geosciences, 35*, 1125–1138.

Youssef, A.M., Hegab, M.A. (2005). Using geographic information systems and statistics for developing a database management system of the flood hazard for Ras Gharib area, Eastern Desert, Egypt. In: *The fourth international conference on the geology of Africa*, vol 2.

Youssef, A.M., Pourghasemi, H.R. (2020). Landslide susceptibility mapping using machine learning algorithms and comparison of their performance at Abha Basin, Asir Region. Saudi Arabia Geoscience Frontiers. 101016/jgsf202005010

Youssef, A. M., Al-Kathery, M., & Pradhan, B. (2015). Landslide susceptibility mapping at Al-Hasher Area, Jizan (Saudi Arabia) using GIS-based frequency ratio and index of entropy models. *Geosciences Journal, 19*, 113–134.

Zaim, S., Ali, T., Mehmet, A., Al-Turki, U., & Omer, D. (2012). Maintenance strategy selection using AHP and ANP algorithms: A case study. *Journal of Quality in Maintenance Engineering, 18*, 16–29. https://doi.org/10.1108/13552511211226166

Zhang, Y.X., Lan, H.X., Li, L.p., et al. (2020). Optimizing the frequency ratio method for landslide susceptibility assessment: A case study of the Caiyuan Basin in the southeast mountainous area of China. *Journal of Mountain Science 17*, 340–357 (2020). https://doi.org/10.1007/s11629-019-5702-6

Zhang, G., Cai, Y., Zheng, Z., Zhen, J., Liu, Y., & Huang, K. (2016). Integration of the statistical index method and the analytic hierarchy process technique for the assessment of landslide susceptibility in Huizhou, China. *Catena, 142*, 233–244.

Zhang, S. H., Li, Z. H., Hou, X. N., & Yi, Y. J. (2019). Impacts on watershed-scale runoff and sediment yield resulting from synergetic changes in climate and vegetation. *Catena, 179*, 129–138.

Zhou, C., Yin, K., Cao, Y., Ahmed, B., Li, Y., Catani, F., & Pourghasemi, H. R. (2018). Landslide susceptibility modeling applying machine learning methods: A case study from Longju in the Three Gorges Reservoir area, China. *Computers and Geosciences, 112*(2018), 23–37. https://doi.org/10.1016/j.cageo.2017.11.019

Zhou, X., Wen, H., Zhang, Y., Xu, J., & Zhang, W. (2021). Landslide susceptibility mapping using hybrid random forest with GeoDetector and RFE for factor optimization. *Geoscience Frontiers, 12*(5), Article 101211. https://doi.org/10.1016/j.gsf.2021.10121

Anbazhagan, S., Ramesh, V. (2014). Landslide hazard zonation mapping in ghat road section of Kolli hills, India. *Journal of Mountain Science, 11*, 1308–1325. https://doi.org/10.1007/s11629-012-2618-9

Climate Change and Fragile Environment

Variability in Temperature and Precipitation and Its Impact on the Snow Cover Dynamics in the Eastern Himalayan Cryosphere

Vimha Ritse, Taposhi Sarmah, Nabajit Hazarika, and Latonglila Jamir

Abstract The Eastern Himalayan cryosphere plays a crucial role in regional hydrology which influences water availability and climate patterns. The present study examines climate-induced changes in snow cover, temperature and precipitation trends in Arunachal Pradesh from 1988 to 2020 using remote sensing and statistical techniques. Results indicate a significant decline in snow cover, with more pronounced observations during the pre-monsoon and monsoon season. Mean temperature revealed a steady increase during the monsoon season, while trends in precipitation showed variability. The study also showed significant warming trends and a strong correlation between temperature rise and loss of snow cover. The Eastern Himalayan region is experiencing a growing vulnerability towards climate change which could impact water resources, disaster risk and stability of the ecosystem. This study could provide information to policymakers and support conservation efforts, aiming to mitigate the climate change impacts in the region.

Keywords Snow cover · Temperature · Precipitation · Arunachal Pradesh · Climate change

1 Introduction

The Himalayas are also known as the 'Asian Water Tower—AWT' as it plays an important role in the region's climate and hydrological systems, which provides water to over a billion people downstream (Bolch et al., 2012; Yao et al., 2022). Snow is considered to be one of the most important variables which plays an important role in the regional climate, glacier and hydrology of the region (Kulkarni et al., 2011;

V. Ritse · L. Jamir
Department of Environmental Science, Nagaland University, Nagaland, India

T. Sarmah · N. Hazarika (✉)
Department of Environmental Biology and Wildlife Sciences, Cotton University, Guwahati, Assam, India
e-mail: nabajit4u@gmail.com; nabajit.hazarika@cottonuniversity.ac.in

© The Author(s), under exclusive license to Springer Nature Switzerland AG 2025

S. Ghosh (ed.), *The Himalaya Dilemma*, Springer Natural Hazards,
https://doi.org/10.1007/978-3-031-95083-4_15

Stigter et al., 2017; Singh et al., 2019). It acts as a natural reservoir, essential in determining the spring snowmelt run-off (Devkota & Shakya, 2023; Shrestha et al., 2012).

Snow cover change is spatially and temporally varied at different seasons of the year. Shrinkage or expansion in snow-covered areas serves as an indicator for climate change within the environment (Singh et al., 2018). The Himalayan snow cover is very sensitive to climatic changes where studies indicate a decreasing trend in snow cover over the recent decades (Mishra et al., 2014; Shekhar et al., 2010).Global warming indirectly causes reductions in snow and glaciers. These can contribute to rise in sea levels and change weather patterns which could further impact agricultural productivity and safety in downstream regions (Pathak et al., 2014; Zarenistanak et al., 2014). Snow cover loss can also alter the albedo effect that can exacerbate the climate change through positive feedback (He et al., 2014).

The Himalayas have complex terrains which influences precipitation patterns with changing altitudes and seasonal variation (Ouyang et al., 2020).The region experiences two types of precipitation patterns. In the Eastern Himalayas, precipitation occurs during the months of May to October which is called the Indian Summer Monsoon (ISM). Precipitation in Karakoram and Western Himalayas are called the Westerly Disturbances (WDs), which is dominated by cool season cycles (Riley et al., 2021). Altered precipitation patterns, rising temperatures and accelerated glacial melt could potentially trigger mass-wasting events such as Glacial Lake Outburst Floods (GLOFs) and landslides (Krusic et al., 2015). Recent findings also show an increased trend of precipitation extremes in Nepal, where frequent floods and landslides were observed (Karki et al., 2017).

The Himalayan region is experiencing a notable increase in temperature of 5–6 °C (Salick et al., 2014). With the significant rise in Earth's average surface temperature due to anthropogenic activities, global warming is considered one of the main causes of climate change (Li, 2021). The Intergovernmental Panel on Climate Change (IPCC) has established a strong consensus that the observed warming trends were caused by human-induced factors with a probability of about 90–100%. Certain weather extremes were observed as a result of these anthropogenic activities (Cody et al., 2015). Increased weather extremes could cause more risk in flooding and heatwaves, which could become more frequent, intense and challenging (Kundzewicz, 2016).

Remote Sensing (RS) and Geographical Information System (GIS) provides an effective tool to monitor environmental changes in large, rugged and inaccessible terrains. With a wide repository of data and a capacity to process large datasets, Google Earth Engine (GEE) offers an effective tool to analyse snow cover, precipitation and temperature. The present study aims to address a key gap in understanding the impacts of climate change on snow cover, precipitation and temperature patterns in Arunachal Pradesh during 1988 to 2020. The study will play a crucial role for policy makers, researchers and scientists in improving the climate adaptation strategies, water resource management and conservation of biodiversity in the region.

2 Materials and Methods

2.1 Study Area

Arunachal Pradesh, also known as 'The Land of Rising Sun', is the largest state in North-East India. Bounded by China to the north, Bhutan to the west, Myanmar to the east and the states of Assam and Nagaland to the south, the state covers an area of approximately 82,300 sq.km. Geographically, it extends between 26°38′59.349″N to 29°26′44.469″N latitudes and 91°33′43.967″E to 97°25′39.304″E longitudes (Fig. 1).

Currently, there are 28 districts in Arunachal Pradesh, some of which includes Tawang, West Kameng, East Kameng, Kurung Kumey, Upper Subansiri, Shi-Yomi,

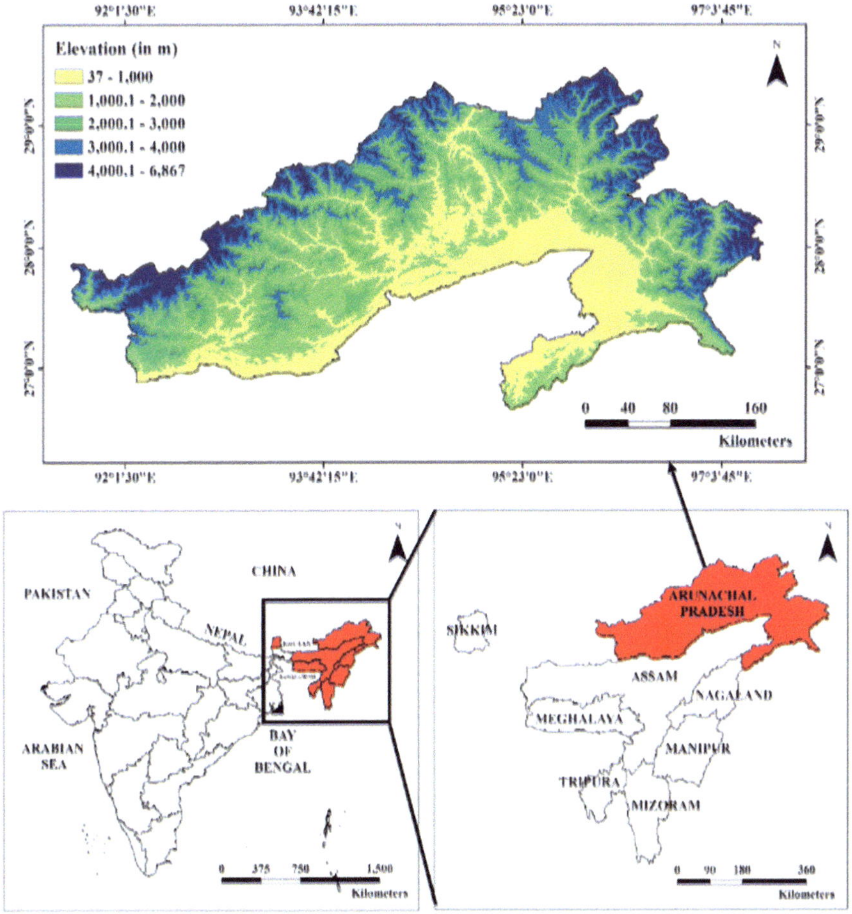

Fig. 1 Location of the study area with the elevation gradient

West Siang, Upper Siang, Dibang Valley and Anjaw. The climate of the region is generally subtropical which is characterized by significant temperature and precipitation variations across different altitudinal zones. Influenced by the monsoon, it receives heavy rainfall from June to September and contributes to lush vegetation (Narah et al., 2023; Sharma et al., 2015). The state is rich in biodiversity and constitutes part of the Eastern Himalayas hotspot which supports diverse flora and fauna (Sharma et al., 2015; & Parashuram 2019). The forests include subtropical evergreen, temperate and alpine types. Subtropical evergreen forests are found at lower elevations and are rich in species like hardwoods and bamboos (Sharma et al., 2015). With increasing altitude, the forests transition into temperate forests, dominated by conifers and a home to unique wildlife and many endemic species (Sharma et al., 2015; Sheth et al., 2019).

3 Data Collection

ERA5-Land monthly dataset (ECMWF/ERA5_LAND/MONTHLY_AGGR) was utilized in analysing the snow cover percentage. The dataset provides detailed land surface information such as snow depth and snow water equivalent (Sabater, 2019). This dataset has a monthly temporal resolution and a spatial resolution of about 9 km.

The Climate Hazards Group InfraRed Precipitation with Station data (CHIRPS) daily dataset (UCSB-CHG/CHIRPS/DAILY) was utilized in analysing high-resolution precipitation data (Funk et al., 2015). CHIRPS integrates satellite imagery with in-situ station data, which provides global precipitation values. This dataset has a daily temporal resolution and a spatial resolution of approximately 0.05°.

For temperature analysis, the ERA5 monthly reanalysis dataset from the was utilized (Copernicus Climate Change Service, 2017). This dataset provides monthly averages for multiple atmospheric, land surface and oceanic-wave variables. The dataset has a monthly temporal resolution with a spatial resolution of 0.25°. The minimum, mean and maximum temperatures were derived using this dataset.

4 Data Handling and Statistical Applications

The data processing steps included importing and pre-processing datasets in GEE where the datasets were clipped to the region of interest. The GEE platform was used to derive the monthly, seasonal and annual data of snow cover percentage, temperature and precipitation. The vector file of Arunachal Pradesh was imported to GEE and the algorithm "ee.ImageCollection" was applied to select the dataset. Filter command "ee.Filter.calendarRange" was applied to cover the study period (1988–2020) (Banerjee et al., 2020). Basic statistics were computed on a monthly and seasonal basis. Four seasons were defined based on months: January to March (winter), April to June (pre-monsoon), July to September (monsoon), and October to December (post-monsoon). Landsat data was processed in GEE to generate daily Normalized Difference Snow Index (NDSI) composites for the entire study area. The maximum snow cover during each identified season was extracted using a maximum value function. A threshold of 0.2 for NDSI was applied to distinguish snow-covered areas. While open regions generally exhibit NDSI values above 0.4, dense forest canopies may have values below 0.2. Using this threshold enhances snow detection accuracy for forested landscapes (Zhang et al., 2010). Finally, the seasonal snow cover data derived for each year was downloaded for further analysis.

5 Climatic Variables Trend Analysis

5.1 Mann-Kendell's Test and Sen's Slope Method

Non-parametric statistics are less influenced by outliers and non-normality and measure monotonic linear dependence (Davis, 1986; Lanzante, 1996; Rossi et al., 1992; Yadav et al., 2014). The Mann–Kendall (MK) test is the most commonly used non-parametric method for detecting trends in hydrologic variables. The MK test counts pairs of observations where a later value is greater than an earlier one. Sen's slope estimates the trend's direction and magnitude by calculating the slope between all pairs of points in the dataset (Eq. (1)). The combination of MK test and Sen's slope provides a robust method in identifying significant trends and quantifying their magnitudes in environmental data (Yadav et al., 2014).

The equation for the Sen's Slope (S) is given as:

$$S = \sum_{i=1}^{n} \left\{ \sum_{j=i+1}^{n} sgn(\boldsymbol{R}_i - \boldsymbol{R}_j) \right\} \tag{1}$$

where,

$sgn(x) = 1$ for $x > 0$.

$sgn(x) = 0$ for $x = 0$.

$sgn(x) = -1$ for $x < 0$.

R_i and R_j = Values at time points i and j respectively and j > i.

When the null hypothesis H_o is true, S follows an approximate normal distribution with $\mu = 0$ (Eq. (2)).

$$\delta = n\frac{(\boldsymbol{n} - 1)(2\boldsymbol{n} + 5)}{18} \tag{2}$$

where,

n = Total number of time series observations.

The MK z-statistic (Eq. (3)) is written as

$$Z = \frac{|S|}{\sigma^{0.5}} \tag{3}$$

where,

σ = Standard deviation of Sen's slope.

A positive S signifies an upward trend whereas negative S indicates a downward trend. These methods are applied to precipitation, temperature and snow cover percentage data to identify and measure changes spatially and temporally.

5.2 *Linear Regression*

Regression analysis is a powerful statistical method used for modelling the relationship between a dependent variable and one or more independent variables (Oguntunde et al., 2017). The trends in the climatic time series data were analysed by using linear regression method. The slope derived from this regression shows the average rate of change of the variable over time (Gadedjisso-Tossou et al., 2021). Additionally, correlation matrix analysis between the variables was performed.

6 Results

6.1 *Trend in Climatic Variables*

6.1.1 Trend in Annual Mean Precipitation

Linear regression analysis for the annual mean precipitation indicates a significant trend only for January–March with an R^2 value of 0.1209 and a p-value of 0.0474, pointing to a statistically significant but weak relationship. The other quarters (April–June, July–September and October–December) do not show significant trends with p-values of 0.6344, 0.0546 and 0.0951, respectively (Table 1). Figure 2 shows the trend in annual mean precipitation (in mm) of Arunachal Pradesh during 1988–2020.The MK test further assesses trends which showed a weak but non-significant decreasing trend for January–March, with Kendall's τ of −0.222, p-value of 0.077 and Sen's slope of −0.341 °C per year. The other seasons show no significant trends, as indicated by Kendall's τ values of −0.060, −0.202 and −0.117 with p-values of 0.638, 0.108 and 0.355 respectively. The other seasons: April-June, July–September and October–December has a Sen's slope of −0.159, −1.093 and −0.18 respectively.

Table 1 Summary of the linear regression and MK trend test in mean precipitation

	Observations	Obs. with missing data	Obs. without missing data	Minimum (in mm)	Maximum (in mm)	Mean (in mm)	Std. deviation
Jan–Mar	32	0	32	22.83	66.17	41.03	10.23
Apr–June	32	0	32	172.37	282.25	226.06	29.61
July–Sept	32	0	32	222.91	333.91	271.76	30.05
Oct–Dec	32	0	32	20.56	55.82	39.05	8.70

	Linear regression			MK test			
	R^2	p-Value	Significance	Kendall's τ	p-Value	Sen's slope	Significance
Jan–Mar	0.1209	0.0474	Significant	−0.222	0.077	−0.341	n.s
Apr–June	0.0074	0.6344	n.s	−0.060	0.638	−0.159	n.s
July–Sept	0.114	0.0546	n.s	−0.202	0.108	−1.093	n.s
Oct–Dec	0.0873	0.0951	n.s	−0.117	0.355	−0.18	n.s

Note n.s.: not significant

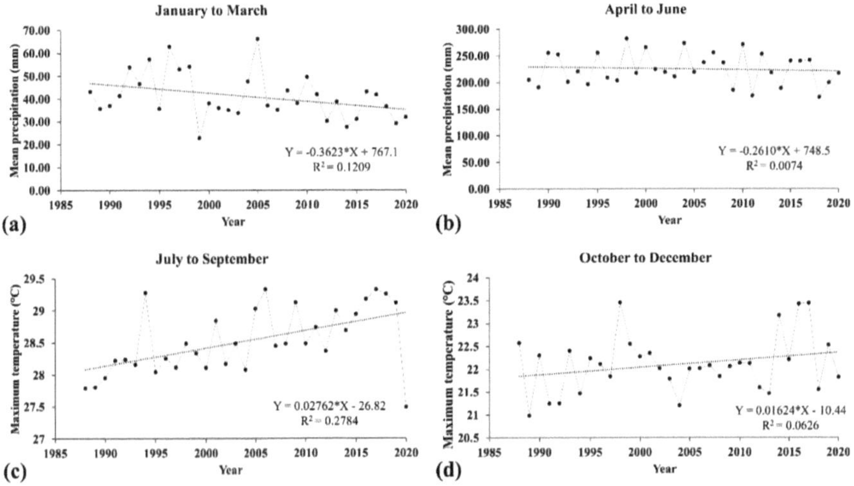

Fig. 2 Annual precipitation (in mm) averaged over Arunachal Pradesh during 1988–2020. **a–d** shows the mean precipitation patterns during January to March, April to June, July to September and October to December respectively

Table 2 Summary of the linear regression and MK trend test in total precipitation

	Observations	Obs. with missing data	Obs. without missing data	Minimum (in mm)	Maximum (in mm)	Mean (in mm)	Std. deviation
Jan–Mar	32	0	32	68.48	198.50	123.10	30.68
Apr–June	32	0	32	517.12	846.75	678.17	88.82
July–Sept	32	0	32	668.72	1001.74	815.29	90.15
Oct–Dec	32	0	32	61.67	167.46	117.14	26.11

	Linear regression			MK test			
	R^2	p-Value	Significance	Kendall's τ	p-Value	Sen's slope	Significance
Jan–Mar	0.1209	0.0474	Significant	−0.222	0.077	−1.023	n.s
Apr–June	0.007381	0.6345	n.s	−0.060	0.638	−0.476	n.s
July–Sept	0.1141	0.0546	n.s	−0.202	0.108	−3.28	n.s
Oct–Dec	0.08725	0.0951	n.s	−0.117	0.355	−0.539	n.s

Note n.s.: Not Significant

6.2 Trend in Annual Total Precipitation

Linear regression analysis for the annual total precipitation reveals a significant decreasing trend only for January–March, with an R^2 value of 0.1209 and p-value of 0.0474. The remaining seasons: April–June, July–September and October–December do not exhibit any significant trends with p-values of 0.6345, 0.0546 and 0.0951 respectively.

MK test further supports these findings (Table 2). During January–March, a weak but non-significant decreasing trend is observed with a Kendall's τ of −0.222, a p-value of 0.077 and a Sen's slope of −1.023 °C per year. The other seasons: April–June, July–September and October–December show no significant trends, as reflected by Kendall's τ values between −0.060 and −0.202, p-values above 0.05 and minimal Sen's slope values which indicates little to no consistent change in total precipitation patterns over the study period. Figure 3 shows the trend in annual total precipitation (in mm) of Arunachal Pradesh during 1988–2020.

6.3 Trend in Maximum Temperature

The average maximum temperature during January to March is 19.63 °C, with a standard deviation of 0.82. Linear regression analysis shows an insignificant trend with an R^2 value of 0.02768 and a p-value of 0.3548. Similarly, MK test shows no significant trend, with Kendall's τ of 0.1935 and a p-value of 0.123. Sen's slope estimate is 0.017 °C per year, indicating a slight but statistically insignificant increase.

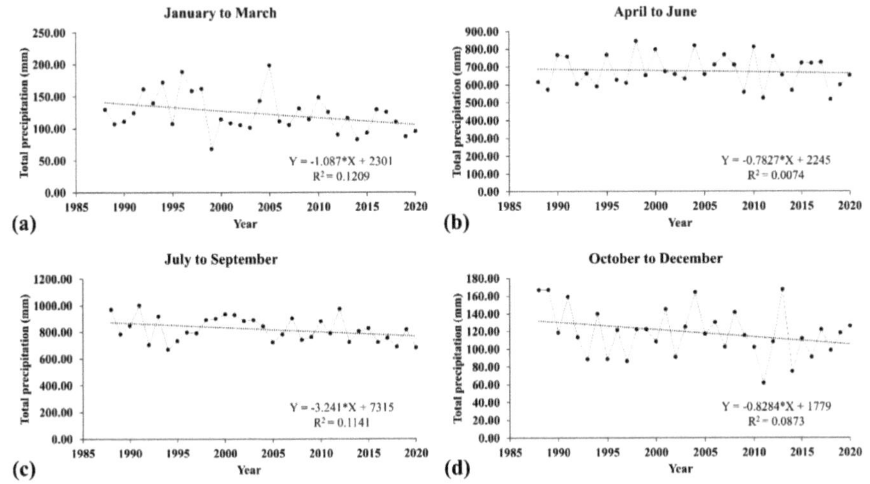

Fig. 3 Annual total precipitation (in mm) averaged over Arunachal Pradesh during 1988–2020. **a**–**d** shows the total precipitation patterns during January to March, April to June, July to September and October to December respectively

April to May season has a higher average maximum temperature of 26.42 °C with a standard deviation of 0.54. The linear regression analysis indicates a weak but non-significant trend ($R^2 = 0.09335, p = 0.0838$). However, MK test shows a statistically significant upward trend with Kendall's τ of 0.258 and a p-value of 0.039. Sen's slope of 0.021 °C per year suggests a positive trend in maximum temperatures during this period (Table 3). Figure 4a–d shows the trend in annual maximum temperature (in °C) of Arunachal Pradesh during 1988–2020.July to September season shows the most pronounced trend with an average maximum temperature of 28.55 °C and a low standard deviation of 0.50. Linear regression analysis reveals a significant trend ($R^2 = 0.2784, p = 0.0016$). MK supports this with a strong positive trend (Kendall's $\tau = 0.419$, $p = 0.001$). Sen's slope is 0.034 °C per year which shows a steady and statistically significant increase in maximum temperatures. During October to December, the average maximum temperature showed 22.09 °C with a standard deviation of 0.63. The linear regression analysis shows no significant trend ($R^2 = 0.06256$, p = 0.1604). Similarly, MK test shows an insignificant trend (Kendall's τ = 0.145, p = 0.250). Sen's slope of 0.016 °C per year suggests only a minor and non-significant increase.

6.4 Trend in Mean Temperature

The mean temperature during this season is 4.97 °C, with a standard deviation of 0.71. Linear regression reveals a significant increasing trend with an R^2 value of 0.1319 and a p-value of 0.0378. MK test confirms this trend as significant, with Kendall's

Table 3 Summary of the linear regression and MK trend test in maximum temperature

	Observations	Obs. with missing data	Obs. without missing data	Minimum (in °C)	Maximum (in °C)	Mean (in °C)	Std. deviation
Jan–Mar	32	0	32	17.87	22.35	19.63	0.82
Apr–June	32	0	32	25.43	27.58	26.42	0.54
July–Sept	32	0	32	27.49	29.34	28.55	0.50
Oct–Dec	32	0	32	20.98	23.46	22.09	0.63
	Linear regression			MK test			
	R^2	p-Value	Significance	Kendall's τ	p-Value	Sen's slope	Significance
Jan–Mar	0.02768	0.3548	n.s	0.1935	0.123	0.017	n.s
Apr–June	0.09335	0.0838	n.s	0.2581	0.039	0.021	Significant
July–Sept	0.2784	0.0016	Significant	0.4194	0.001	0.034	Significant
Oct–Dec	0.06256	0.1604	n.s	0.1452	0.250	0.016	n.s

Note n.s.: not significant

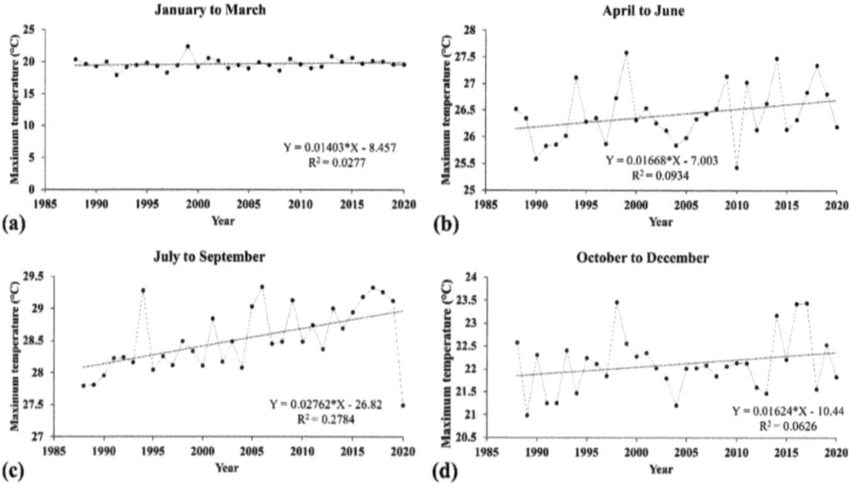

Fig. 4 Annual maximum temperature (in °C) averaged over Arunachal Pradesh during 1988–2020. **a–d** shows the trends in maximum temperature during January to March, April to June, July to September and October to December respectively

τ of 0.3024 ($p = 0.016$) and a Sen's slope of 0.032 °C per year. This suggests a steady rise in mean temperatures during the winter months. April to June season has an average mean temperature of 13.71 °C and a low standard deviation of 0.44. Although the linear regression analysis shows a non-significant trend ($R^2 = 0.08827$, $p = 0.0931$), MK test identifies a statistically significant upward trend (Kendall's τ

Table 4 Summary of the linear regression and MK trend test in mean temperature

	Observations	Obs. with missing data	Obs. without missing data	Minimum (in °C)	Maximum (in °C)	Mean (in °C)	Std. deviation
Jan–Mar	32	0	32	3.66	6.65	4.97	0.71
Apr–June	32	0	32	12.61	14.63	13.71	0.44
July–Sept	32	0	32	16.62	18.07	17.13	0.41
Oct–Dec	32	0	32	6.99	10.58	8.33	0.88

	Linear regression			MK test			
	R^2	p-Value	Significance	Kendall's τ	p-Value	Sen's slope	Significance
Jan–Mar	0.1319	0.0378	Significant	0.3024	0.016	0.032	Significant
Apr–June	0.08827	0.0931	n.s	0.2621	0.036	0.019	Significant
July–Sept	0.5407	< 0.0001	Significant	0.5565	<0.0001	0.033	Significant
Oct–Dec	0.1843	0.0127	Significant	0.2782	0.026	0.039	Significant

Note n.s.: not significant

$= 0.2621, p = 0.036$), with a Sen's slope of 0.019 °C per year which could indicate a subtle but consistent warming during this period (Table 4). Figure 5 shows the trend in annual mean temperature (in °C) of Arunachal Pradesh during 1988–2020.July to September records the highest mean temperature of 17.13 °C with a low standard deviation of 0.41. Linear regression shows a highly significant warming trend ($R^2 = 0.5407, p < 0.0001$). MK test strongly supports this result with Kendall's τ of 0.5565 ($p < 0.0001$) and a Sen's slope of 0.033 °C per year, showing the most pronounced and statistically significant warming trend. October to December has a mean temperature of 8.33 °C with a standard deviation of 0.88. Linear regression reveals a significant warming trend ($R^2 = 0.1843$, $p = 0.0127$). MK test also shows a significant trend (Kendall's $\tau = 0.2782, p = 0.026$) with a Sen's slope of 0.039 °C per year. This could indicate a steady rise in temperatures during the post-monsoon months.

6.5 *Trend in Minimum Temperature*

Both linear regression and MK tests reveal significant warming trends in most of the seasons. The mean minimum temperature during January to March is −12.88 °C, with a standard deviation of 1.03. Linear regression did not show statistically significant trend ($R^2 = 0.06845$, $p = 0.1414$). However, MK test identifies a significant upward trend, with Kendall's τ of 0.2500 ($p = 0.046$) and a Sen's slope of 0.039 °C per year which suggests a subtle but significant warming during the winter months (Table 5). Figure 6 shows the trend in annual minimum temperature (in °C) of Arunachal Pradesh during 1988–2020.

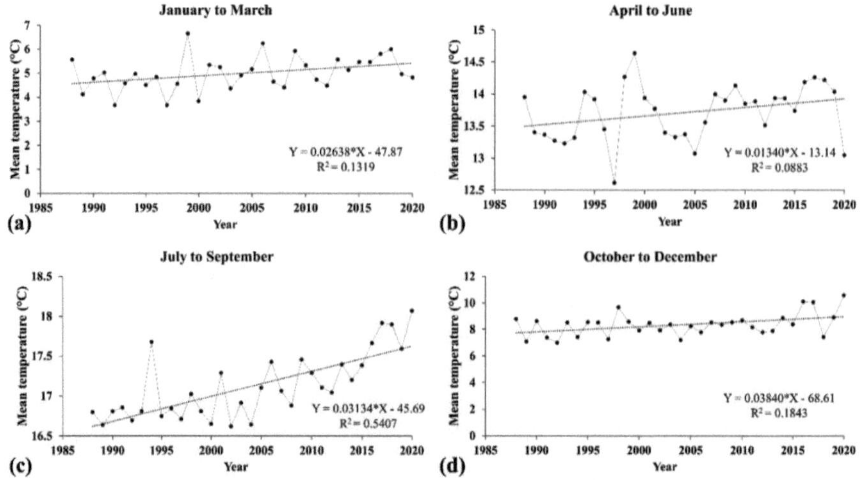

Fig. 5 Annual temperature (in °C) averaged over Arunachal Pradesh during 1988–2020. **a–d** shows the trends in mean temperature during January to March, April to June, July to September and October to December respectively

Table 5 Summary of the linear regression and MK trend test in minimum temperature

	Observations	Obs. with missing data	Obs. without missing data	Minimum (in °C)	Maximum (in °C)	Mean (in °C)	Std. deviation
Jan–Mar	32	0	32	−14.90	−10.65	−12.88	1.03
Apr–June	32	0	32	−4.05	−0.52	−2.03	0.75
July–Sept	32	0	32	1.64	3.79	2.59	0.56
Oct–Dec	32	0	32	−11.75	−6.46	−9.42	1.13

	Linear regression			MK test			
	R^2	p-Value	Significance	Kendall's τ	p-Value	Sen's slope	Significance
Jan–Mar	0.06845	0.1414	n.s	0.2500	0.046	0.039	Significant
Apr–June	0.09148	0.0871	n.s	0.2702	0.031	0.029	Significant
July–Sept	0.4612	<0.0001	Significant	0.4919	<0.0001	0.044	Significant
Oct–Dec	0.2352	0.0042	Significant	0.3347	0.007	0.059	Significant

Note n.s.: not significant

The mean minimum temperature for April to June season is −2.03 °C, with a standard deviation of 0.75. Linear regression shows a non-significant trend (R^2 = 0.09148, p = 0.0871). MK test reveals a significant warming trend with Kendall's τ of 0.2702 (p = 0.031) and a Sen's slope of 0.029 °C per year which could indicate a noticeable increase in minimum temperatures during the pre-monsoon period. July to

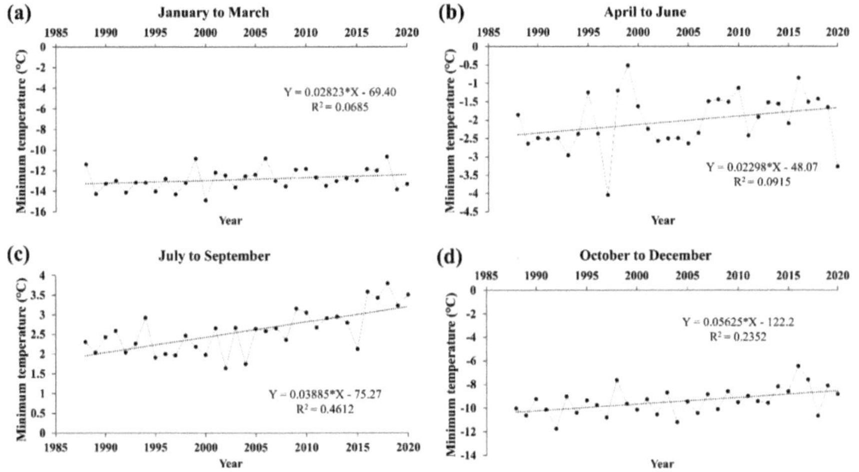

Fig. 6 Annual minimum temperature (in °C) averaged over Arunachal Pradesh during 1988–2020. **a–d** shows the trends in minimum temperature during January to March, April to June, July to September and October to December respectively

September shows a mean minimum temperature of 2.59 °C, with a standard deviation of 0.56. Linear regression shows a highly significant upward trend ($R^2 = 0.4612$, $p < 0.0001$). MK test strongly supports this result, with Kendall's τ of 0.4919 ($p < 0.0001$) and a Sen's slope of 0.044 °C per year. This period shows the most pronounced and statistically significant warming trend. The mean minimum temperature during October to December is −9.42 °C, with a standard deviation of 1.13. Linear regression reveals a significant trend ($R^2 = 0.2352$, $p = 0.0042$), indicating an upward trend in temperatures. MK test further confirms this with Kendall's τ of 0.3347 ($p = 0.007$) and a Sen's slope of 0.059 °C per year.

7 Seasonal Maximum Snow Cover and Its Changes

Table 6 provides data on the area of snow cover in Arunachal Pradesh during four decades: 1988, 2000, 2010 and 2020 which were categorized into four seasons: January–March, April–June, July–September and October–December. In the January–March period, snow cover remained relatively stable between 1988 (12,571.46 sq.km) and 2010 (12,837.61 sq.km), but it significantly decreased to 11,434.08 sq.km by 2020 (Fig. 7). During April–June, snow cover increased substantially from 5840.95 sq.km to 9746.72 sq.km during 1988–2010 respectively declined to 4396.16 sq.km in 2020 (Fig. 8). During July–September, snow cover consistently declined from 646.68 sq.km in 1988 to 275.45 sq.km in 2020 (Fig. 9). Similarly, the October–December period saw an initial increase 5723.07 sq.km in 1988 to 8008.09 sq.km in 2000 and a decrease to 6327.54 sq.km by 2020 (Fig. 10).

Table 6 Area (in sq.km) of the seasonal snow cover and decadal changes in Arunachal Pradesh (1988–2020)

	1988	2000	2010	2020
January–March	12,571.46	12,367.38	12,837.61	11,434.08
April–June	5840.95	8718.28	9746.72	4396.16
July–September	646.68	382.37	338.58	275.45
October–December	5723.07	8008.09	6967.71	6327.54
	Inter-periodic changes			
	1988–2000	2000–2010	2010–2020	1988–2020
January–March	−204.08	+470.23	−1403.53	−1137.38
April–June	+2877.33	+1028.44	−5350.57	−1444.80
July–September	−264.32	−43.79	−63.12	−371.23
October–December	+2285.02	−1040.38	−640.17	+604.46

Note Positive values indicate an expansion of the snow-covered area, while negative values signify a reduction

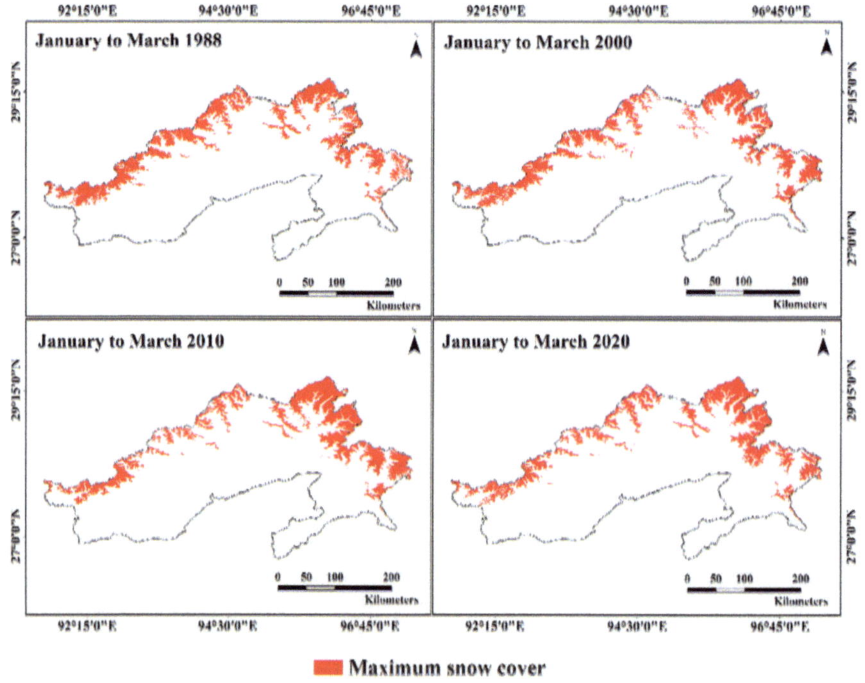

Fig. 7 Spatial distribution of seasonal snow cover extent in Arunachal Pradesh during January to March 1988–2020

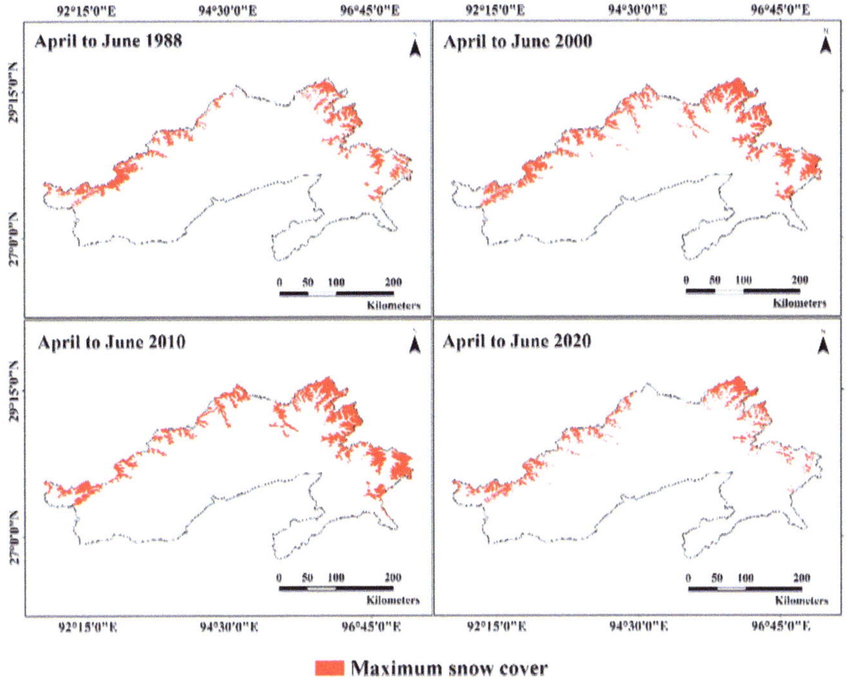

Fig. 8 Spatial distribution of seasonal snow cover extent in Arunachal Pradesh during April to June 1988–2020

The inter-decadal changes highlights both increase and decrease in snow cover. During 1988–2000, snow cover reduced marginally in January–March (−204.08 sq.km) but rose significantly in April–June (+2877.33 sq.km) and October–December (+2285.02 sq.km). During 2000–2010, there was an overall increase in snow cover for January–March (+470.23 sq.km) and April–June (+1028.44 sq.km), while October–December showed a decline (−1040.38 sq.km). However, the 2010–2020 period saw widespread reductions across all seasons, with the most notable decreases in January–March (−1403.53 sq.km) and April–June (−5350.57 sq.km).Overall, declining total snow cover was observed in most seasons. January–March period saw a reduction of 1137.38 sq.km, while the April–June period experienced the largest decrease of 1444.80 sq.km. The July–September season followed a consistent downward trend with an overall decline of 371.23 sq.km. In contrast, October–December period showed a slight overall increase of 604.46 sq.km.

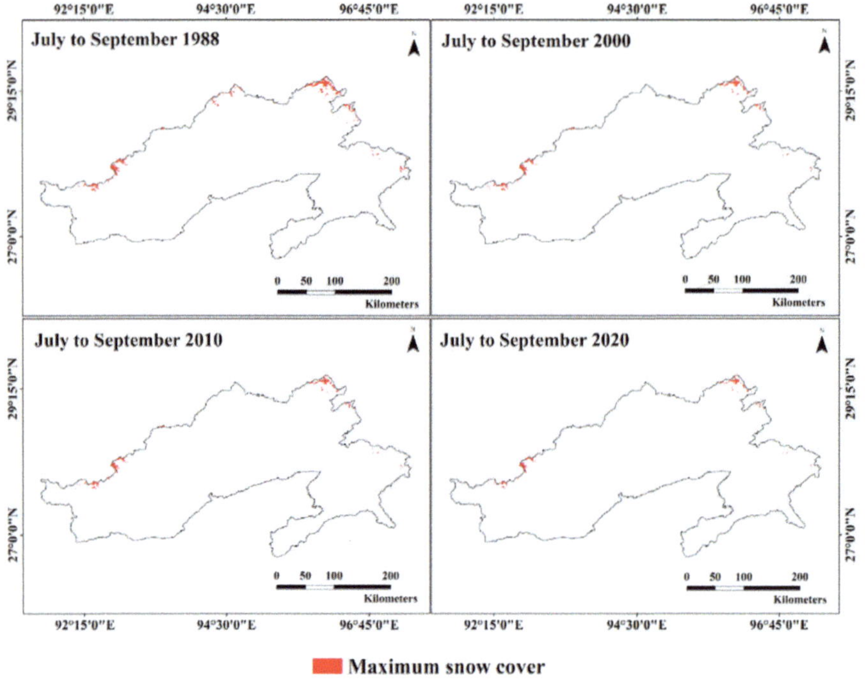

Fig. 9 Spatial distribution of seasonal snow cover extent in Arunachal Pradesh during July to September 1988–2020

8 Analysis of Annual Snow Cover Percentage

The snow cover percentage observed during January to March is 32.48%, with a standard deviation of 5.22. Linear regression analysis reveals a significant decreasing trend ($R^2 = 0.1708$, $p = 0.0168$). MK test supports this by showing a significant negative trend with Kendall's τ of −0.371 ($p = 0.003$) and a Sen's slope of −0.255% per year. These findings show a noticeable reduction in snow cover during this season. The snow cover percentage for April to June is 13.93%, with a standard deviation of 2.76. Linear regression reveals a statistically significant decline ($R^2 = 0.2654$, $p = 0.0022$). MK test further confirms this with a significant Kendall's τ of −0.407 ($p = 0.001$) and a Sen's slope of −0.185% per year, which show a consistent decrease in snow cover during this season (Table 7). Figure 11 shows the trend in annual minimum temperature (in °C) of Arunachal Pradesh during 1988–2020.During July to September, snow cover percentage is minimal at 1.45%, with a standard deviation of 0.61. Linear regression shows a significant decreasing trend ($R^2 = 0.304$, $p = 0.0009$). MK test confirms this with Kendall's τ of −0.363 ($p = 0.004$) and a Sen's slope of −0.034% per year. Although the snow cover is naturally limited in the monsoon period, the trend indicates further decline. The average snow cover percentage during October to December is 11.84%, with a standard deviation

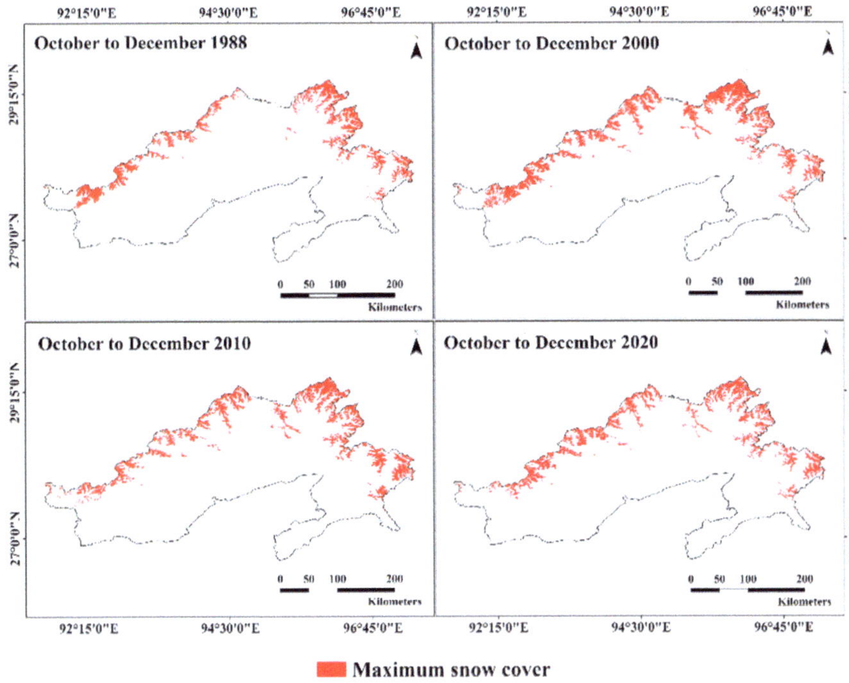

Fig. 10 Spatial distribution of seasonal snow cover extent in Arunachal Pradesh during October to December 1988–2020

of 2.90. Linear regression shows a non-significant trend ($R^2 = 0.0272$, $p = 0.359$). MK test also detects no significant trend (Kendall's $\tau = -0.105$, $p = 0.408$), with a Sen's slope of −0.042% per year. This suggests relatively stable snow cover during the post-monsoon season.

9 Correlation Analysis

The correlation analysis of all months consistently shows that temperature is the main driver of snow cover variation, with mean temperature negatively correlated with snow cover percentage (Fig. 12). With rise in temperature, there is a decrease in snow cover, which verifies the seasonal pattern of snowmelt. Precipitation has a weak or secondary influence, with minor positive correlations suggesting that while increased precipitation may contribute to snow accumulation, it does not significantly offset temperature-driven melting. Strong positive correlations exist among temperature variables (maximum, minimum and mean), which indicates redundancy where mean temperature alone can represent overall temperature trends. Similarly, mean and total precipitation are highly correlated. Only one can therefore be used to describe

Table 7 Summary of the linear regression and MK trend test in snow cover percentage

	Observations	Obs. with missing data	Obs. without missing data	Minimum (in %)	Maximum (in %)	Mean (in %)	Std. deviation
Jan–Mar	32	0	32	19.15	41.37	32.48	5.22
Apr–June	32	0	32	7.30	19.11	13.93	2.76
July–Sept	32	0	32	0.68	2.82	1.45	0.61
Oct–Dec	32	0	32	5.23	17.26	11.84	2.90

	Linear regression			MK test			
	R^2	p-Value	Significance	Kendall's τ	p-Value	Sen's slope	Significance
Jan–Mar	0.1708	0.0168	Significant	−0.371	0.003	−0.255	Significant
Apr–June	0.2654	0.0022	Significant	−0.407	0.001	−0.185	Significant
July–Sept	0.304	0.0009	Significant	−0.363	0.004	−0.034	Significant
Oct–Dec	0.0272	0.359	n.s	−0.105	0.408	−0.042	n.s

Note n.s.: not significant

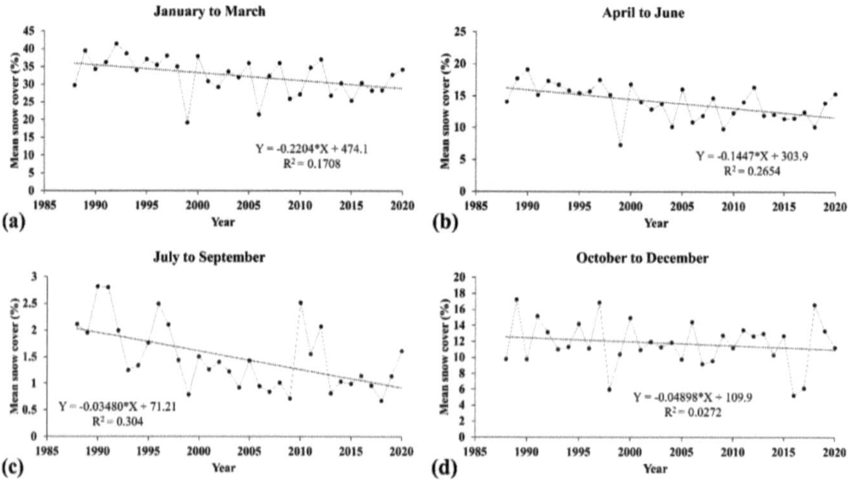

Fig. 11 Annual snow cover (in %) averaged over Arunachal Pradesh during 1988–2020. **a–d** shows the distribution pattern during January to March, April to June, July to September and October to December respectively

precipitation trends. Seasonal variations slightly affect the intensity of correlations but the overall pattern remains consistent, temperature governs snow cover changes while precipitation plays a limited role.

Fig. 12 Correlation analysis between snow cover percentage and different variables. (a–d) shows the correlation heatmap during January to March, April to June, July to September and October to December respectively

10 Discussion

Seasonal variability is vital to understand the hydrological cycles and agricultural productivity as it directly influences water availability and crop growth. The present study highlights significant changes in climatic variables and snow cover of Arunachal Pradesh during 1988–2020. Precipitation trends show variability across seasons, with the highest rainfall during July–September and the lowest during January–March. A weak but significant declining trend in precipitation was observed only for January–March, whereas an insignificant declining trend was observed for the other seasons. A study conducted by Kashyap et al. (2021) demonstrates that the Eastern Himalayas receives more precipitation in comparison to the Central and Western Himalayas. The Eastern Himalayas is largely governed by monsoon-controlled dynamics, with precipitation occurring during the summer season (Palazzi et al., 2013). Singh and Kumar (2022) analysed trend and variability of precipitation, demonstrating a decreasing trend in rainfall during monsoon and increasing trend during post-monsoon season over the Northeast India.

Over the past century, the Himalayan region is undergoing a notable increase in temperature of approximately 0.74 °C (Chhetri et al., 2018).The warming trend is particularly evident during summer, with significant temperature increases which makes them the most threatened outside of the polar regions (Negi et al., 2021; Salick et al., 2014). The present study indicated that temperature trends are more pronounced with steady increases in mean and minimum temperatures across all seasons. The warming trend during monsoon and pre-monsoon periods is particularly significant. A study by Mishra and Chaudhuri (2015) highlights about the average temperature in sub-tropical mountain systems, including the Himalayan system, has been rising in line with global warming patterns. Rapid reduction of glaciers in the region poses significant risks to downstream regions which can further be attributed to these temperature increases (Xu et al., 2009).

Snow cover dynamics reveal a substantial reduction during the study period. The largest decline is observed in April–June (−1444.80 sq.km), followed by January–March (−1137.38 sq.km). Snow cover during July–September consistently decreased by 371.23 sq.km and October–December showed a slight increase (+604.46 sq.km). Snow cover percentages also indicate significant decline, particularly during April–June and January–March. Singh et al. (2016) reported a steady decline in seasonal snow cover depth and duration over the Western Himalayas. Variations in temperature and precipitation typically cause snow to accumulate and melt faster, which reduces the snow cover duration (Gulahmadov et al., 2023). Rikiishi and Nakasato (2006) also highlighted the impact of climate change, documenting a declining length of snow cover in the Himalayas and Tibetan Plateau. Lau et al. (2010) also highlighted that anthropogenic aerosols accelerate snowmelt in the region. The strong correlations between the snow cover and temperature helps in explaining the trends in variations in snow cover (Maskey et al., 2011). Change in precipitation patterns causes extreme weather events such as floods, landslides and drought to occur regularly. This poses significant risks to the fragile ecosystems and local communities in various parts of the Himalayas (Das & Mishra, 2022).

11 Conclusion

The present study provides vital information about the climatic trends in Arunachal Pradesh during 1988 to 2020. Results indicated a significant decline in snow cover, especially during the pre-monsoon (April–June) and monsoon (July–September) periods. The steady increase in temperatures is closely linked to this loss of snow cover, which could indicate the impact of global warming on the region's cryosphere. The overall warming trend and its correlation with decreasing snow cover indicate significant changes in the region's hydrological processes, despite the variability in the precipitation trend.

The increase in temperatures and decline in snow cover could pose significant risk to regional water resources. The decrease in snow cover could severely affect the water availability for the downstream regions. The vulnerability of ecosystems and human populations to climate-induced disasters like GLOFs and landslides could become more frequent. These challenges are intensified by the unpredictability of precipitation which makes it difficult to forecast future water availability and disaster risks.

The present study highlights the urgency of developing adaptive strategies to mitigate the impacts of climate change in the region. In order to ensure sustainable water resources in the face of a changing climate, combination of scientific research and traditional knowledge with regional cooperation is essential. Therefore, studying the ongoing changes in the Eastern Himalayan cryosphere could serve as a wake-up call for urgent action to address the challenges posed by climate change. Collective, up-to-date and proactive measures are essential to understand and to minimize the adverse effects which would secure the future of this region.

Acknowledgements VR is thankful to the Ministry of Tribal Affairs (MoTA), Government of India for the NFST fellowship provided to him for pursuing Ph.D. course vide award no- 201920-NFST-NAG-02696.

References

Banerjee, A., Chen, R., E. Meadows, M., Singh, R. B., Mal, S., & Sengupta, D. (2020). An analysis of long-term rainfall trends and variability in the Uttarakhand Himalaya using google earth engine. *Remote Sensing*, *12*(4), 709. https://doi.org/10.3390/rs12040709

Bolch, T., Kulkarni, A., Kaab, A., Huggel, C., Paul, F., Cogley, J. G., Frey, H., Kargel, J. S., Fujita, K., Scheel, M., Bajracharya, S., & Stoffel, M. (2012). The state and fate of Himalayan glaciers. *Science, 336*(6079), 310–314. https://doi.org/10.1126/science.1215828

Chhetri, B., Badola, H. K., & Barat, S. (2018). Predicting climate-driven habitat shifting of the near threatened Satyr Tragopan (Tragopan Satyra; Galliformes) in the Himalayas. *Avian Biology Research*, *11*(4), 221–230. https://doi.org/10.3184/175815618x15316676114070

Cody, E. M., Reagan, A. J., Mitchell, L., Dodds, P. S., & Danforth, C. M. (2015). Climate change sentiment on Twitter: an unsolicited public opinion poll. *PLOS ONE*, *10*(8), e0136092. https://doi.org/10.1371/journal.pone.0136092

Copernicus Climate Change Service (C3S). (2017). *ERA5: Fifth generation of ECMWF atmospheric reanalyses of the global climate*. Copernicus Climate Change Service Climate Data Store (CDS). https://cds.climate.copernicus.eu/cdsapp

Das, S., & Mishra, A. J. (2022). Climate change and the Western Himalayan community: Exploring the local perspective through food choices. *Ambio, 52*(3), 534–545. https://doi.org/10.1007/s13280-022-01810-3

Davis, J. C. (1986). *Statistics and data analysis in geology* (2nd ed.). New York Wiley.

Devkota, N., & Shakya, N. M. (2023). A study on temporal variation of snow and glacier in Langtang Basin. *Kathford Journal of Engineering and Management*, *2*(01), 36–45. https://doi.org/10.3126/kjem.v2i01.61763

Funk, C., Peterson, P., Landsfeld, M., Pedreros, D., Verdin, J., Shukla, S., Husak, G., Rowland, J., Harrison, L., Hoell, A., & Michaelsen, J. (2015). The climate hazards infrared precipitation with stations—a new environmental record for monitoring extremes. *Scientific Data*, *2*(1). https://doi.org/10.1038/sdata.2015.66

Gadedjisso-Tossou, A., Adjegan, K. I., & Kablan, A. K. M. (2021). Rainfall and temperature trend analysis by Mann–Kendall test and significance for Rainfed Cereal Yields in Northern Togo. *Sci*, *3*(1), 17. https://doi.org/10.3390/sci3010017

Gulahmadov, N., Chen, Y., Gulakhmadov, M., Satti, Z., Naveed, M., Davlyatov, R., Ali, S., &Gulakhmadov, A. (2023). Assessment of temperature, precipitation, and snow cover at different altitudes of the Varzob River basin in Tajikistan. *Applied Sciences*, *13*(9), 5583–5583. https://doi.org/10.3390/app13095583

He, C., Li, Q., Liou, K.-N., Takano, Y., Gu, Y., Qi, L., Mao, Y., & Leung, L. R. (2014). Black carbon radiative forcing over the Tibetan Plateau. *Geophysical Research Letters*, *41*(22), 7806–7813. https://doi.org/10.1002/2014gl062191

Karki, R., Hasson, S. ul, Schickhoff, U., Scholten, T., & Böhner, J. (2017). Rising precipitation extremes across Nepal. *Climate, 5*(1), 4. https://doi.org/10.3390/cli5010004

Kashyap, R., Pandey, A. C., & Parida, B. R. (2021). Spatio-temporal variability of monsoon precipitation and their effect on precipitation triggered landslides in relation to relief in Himalayas. *Spatial Information Research, 29*(6), 857–869. https://doi.org/10.1007/s41324-021-00392-8

Krusic, P. J., Cook, E. R., Dukpa, D., Putnam, A. E., Rupper, S., & Schaefer, J. (2015). Six hundred thirty-eight years of summer temperature variability over the Bhutanese Himalaya. *Geophysical Research Letters, 42*(8), 2988–2994. https://doi.org/10.1002/2015gl063566

Kulkarni, A. V., Rathore, B. P., Singh, S. K., & Bahuguna, I. M. (2011). Understanding changes in the Himalayan cryosphere using remote sensing techniques. *International Journal of Remote Sensing, 32*(3), 601–615. https://doi.org/10.1080/01431161.2010.517802

Kundzewicz, Z. W. (2016). Extreme weather events and their consequences. *Papers on Global Change IGBP, 23*(1), 59–69. https://doi.org/10.1515/igbp-2016-0005

Lanzante, J. R. (1996). Resistant, robust and non-parametric techniques for the analysis of climate data: Theory and examples, including applications to historical radiosonde station data. *International Journal of Climatology, 16*(11), 1197–1226. https://doi.org/10.1002/(SICI)1097-0088(199611)16:11%3C1197::AID-JOC89%3E3.0.CO;2-L

Lau, W. K. M., Kim, M.-K., Kim, K.-M., & Lee, W.-S. (2010). Enhanced surface warming and accelerated snow melt in the Himalayas and Tibetan Plateau induced by absorbing aerosols. *Environmental Research Letters, 5*(2), 025204. https://doi.org/10.1088/1748-9326/5/2/025204

Li, K. (2021). Climate change and aerosol sciences. *Journal of Earth Sciences and Geotechnical Engineering*, 1–13. https://doi.org/10.47260/jesge/1131

Maskey, S., Uhlenbrook, S., & Ojha, S. (2011). An analysis of snow cover changes in the Himalayan region using MODIS snow products and in-situ temperature data. *Climatic Change, 108*(1–2), 391–400. https://doi.org/10.1007/s10584-011-0181-y

Mishra, B., Tripathi, N. K., & Babel, M. S. (2014). An artificial neural network-based snow cover predictive modeling in the higher Himalayas. *Journal of Mountain Science, 11*(4), 825–837. https://doi.org/10.1007/s11629-014-2985-5

Mishra, N. B., & Chaudhuri, G. (2015). Spatio-temporal analysis of trends in seasonal vegetation productivity across Uttarakhand, Indian Himalayas, 2000–2014. *Applied Geography, 56*, 29–41. https://doi.org/10.1016/j.apgeog.2014.10.007

Narah, J., Streinzer, M., Chakravorty, J., Megu, K., Spaethe, J., Brockmann, A., & Schmitt, T. (2023). Cuticular hydrocarbon profiles of Himalayan bumble bees (Hymenoptera:*Bombus*Latreille) are species-specific and show local and elevation related variation. *BioRxiv (Cold Spring Harbor Laboratory)*. https://doi.org/10.1101/2023.08.20.554003

Negi, V. S., Tiwari, D. C., Singh, L., Thakur, S., & Bhatt, I. D. (2021). Review and synthesis of climate change studies in the Himalayan region. *Environment, Development and Sustainability, 24*(9), 10471–10502. https://doi.org/10.1007/s10668-021-01880-5

Oguntunde, P. G., Lischeid, G., & Dietrich, O. (2017). Relationship between rice yield and climate variables in southwest Nigeria using multiple linear regression and support vector machine analysis. *International Journal of Biometeorology, 62*(3), 459–469. https://doi.org/10.1007/s00484-017-1454-6

Ouyang, L., Yang, K., Lu, H., Chen, Y., Lazhu, N., Zhou, X., & Wang, Y. (2020). Ground-Based Observations Reveal Unique Valley Precipitation Patterns in the Central Himalaya. *Journal of Geophysical Research Atmospheres, 125*(5). https://doi.org/10.1029/2019jd031502

Palazzi, E., von Hardenberg, J., & Provenzale, A. (2013). Precipitation in the Hindu-Kush Karakoram Himalaya: Observations and future scenarios. *Journal of Geophysical Research: Atmospheres, 118*(1), 85–100. https://doi.org/10.1029/2012jd018697

Pathak, H., Pramanik, P., Khanna, M., & Kumar, A. (2014). *Climate change and water availability in Indian agriculture: Impacts and adaptation. 84*(6).

Rikiishi, K., & Nakasato, H. (2006). Height dependence of the tendency for reduction in seasonal snow cover in the Himalaya and the Tibetan Plateau region, 1966–2001. *Annals of Glaciology, 43*, 369–377. https://doi.org/10.3189/172756406781811989

Riley, C., Rupper, S., Steenburgh, J. W., Strong, C., Kochanski, A. K., & Wolvin, S. (2021). Characteristics of historical precipitation in High Mountain Asia based on a 15-Year high resolution dynamical downscaling. *Atmosphere*, *12*(3), 355. https://doi.org/10.3390/atmos12030355

Rossi, R. E., Mulla, D. J., Journel, A. G., & Franz, E. H. (1992). Geostatistical tools for modeling and interpreting ecological spatial dependence. *Ecological Monographs*, *62*(2), 277–314. https://doi.org/10.2307/2937096

Sabater, M. J. (2019). *ERA5-Land monthly averaged data from 1950 to present*. Copernicus Climate Change Service (C3S) Climate Data Store (CDS). https://doi.org/10.24381/cds.68d2bb30

Salick, J., Ghimire, S. K., Fang, Z., Dema, S., & Konchar, K. M. (2014). Himalayan alpine vegetation, climate change and mitigation. *Journal of Ethnobiology*, *34*(3), 276. https://doi.org/10.2993/0278-0771-34.3.276

Sharma, D., Tiwari, B. K., Chaturvedi, S. S., & Diengdoh, E. (2015). Status, utilization and economic valuation of non-timber forest products of Arunachal Pradesh, India. *Journal of Forest and Environmental Science*, *31*(1), 24–37. https://doi.org/10.7747/jfes.2015.31.1.24

Shekhar, M. S., Chand, H., Kumar, S., Srinivasan, K., &Ganju, A. (2010). Climate-change studies in the western Himalaya. *Annals of Glaciology*, *51*(54), 105–112. https://doi.org/10.3189/172756410791386508

Sheth, C., Datta, A., & Parashuram, D. (2019). Persistent loss of biologically-rich tropical forests in the Indian Eastern Himalaya. *Preprint Retrieved from*. https://doi.org/10.1101/827360

Shrestha, M., Wang, L., Koike, T., Xue, Y., & Hirabayashi, Y. (2012). Modeling the spatial distribution of snow cover in the Dudhkoshi Region of the Nepal Himalayas. *Journal of Hydrometeorology*, *13*(1), 204–222. https://doi.org/10.1175/jhm-d-10-05027.1

Singh, D., Juyal, V., & Sharma, V. (2016). Consistent seasonal snow cover depth and duration variability over the Western Himalayas (WH). *Journal of Earth System Science*, *125*(7), 1451–1461. https://doi.org/10.1007/s12040-016-0737-3

Singh, D. K., Gusain, H. S., Mishra, V., & Gupta, N. (2018). Snow cover variability in North-West Himalaya during last decade. *Arabian Journal of Geosciences*, *11*(19). https://doi.org/10.1007/s12517-018-3926-3

Singh, M. K., Thayyen, R. J., & Jain, S. K. (2019). Snow cover change assessment in the upper Bhagirathi basin using an enhanced cloud removal algorithm. *Geocarto International*, *36*(20), 2279–2302. https://doi.org/10.1080/10106049.2019.1704069

Singh, P., & Kumar, N. (2022). Analysis of trend and temporal variability in rainfall over Northeast India. *MAUSAM*, *73*(2), 307–314. https://doi.org/10.54302/mausam.v73i2.5479

Stigter, E. E., Wanders, N., Saloranta, T. M., Shea, J. M., Bierkens, M. F. P., & Immerzeel, W. W. (2017). Assimilation of snow cover and snow depth into a snow model to estimate snow water equivalent and snowmelt runoff in a Himalayan catchment. *The Cryosphere*, *11*(4), 1647–1664. https://doi.org/10.5194/tc-11-1647-2017

Xu, J., Grumbine, R. E., Shrestha, A., Eriksson, M., Yang, X., Wang, Y., & Wilkes, A. (2009). The melting Himalayas: cascading effects of climate change on water, biodiversity, and livelihoods. *Conservation Biology*, *23*(3), 520–530. https://doi.org/10.1111/j.1523-1739.2009.01237.x

Yadav, R., Tripathi, S. K., Pranuthi, G., & Dubey, S. K. (2014). Trend analysis by Mann-Kendall test for precipitation and temperature for thirteen districts of Uttarakhand. *Journal of Agrometeorology*, *16*(2), 164–171. https://doi.org/10.54386/jam.v16i2.1507

Yao, T., Bolch, T., Chen, D., Gao, J., Immerzeel, W., Piao, S., Su, F., Thompson, L., Wada, Y., Wang, L., Wang, T., Wu, G., Xu, B., Yang, W., Zhang, G., & Zhao, P. (2022). The imbalance of the Asian water tower. *Nature Reviews Earth & Environment*. https://doi.org/10.1038/s43017-022-00299-4

Zarenistanak, M., Dhorde, A. G., Kripalani, R. H., & Dhorde, A. A. (2014). Trends and projections of temperature, precipitation, and snow cover during snow cover-observed period over southwestern Iran. *Theoretical and Applied Climatology*, *122*(3–4), 421–440. https://doi.org/10.1007/s00704-014-1287-8

Zhang, Y., Yan, S., & Lu, Y. (2010). Snow cover monitoring using MODIS data in Liaoning Province, Northeastern China. *Remote Sensing*, *2*(3), 777–793. https://doi.org/10.3390/rs2030777

Seasonal Geo-Spatial Assessment of Vegetational Responses to Changing Climate in the Brahmaputra River Basin of the Eastern Himalaya

Shukla Acharjee and Biswajit Nath

Abstract Understanding temperature–precipitation–vegetation interaction is very important for designing adaptation and mitigation measures for terrestrial ecosystems under climate change (CC) scenarios in the eastern Himalaya. Many studies have been carried out in the past on the spatial pattern of precipitation–temperature–vegetation correlations. In this genre, a comprehensive investigation of spatio-seasonal patterns of vegetation response to temperature–precipitation anomalies in the Brahmaputra River Basin (BRB) was performed. To support this, multi-sensor satellite-derived processed datasets were retrieved from the NASA-Earthdata Giovanni web portal, which provided a vegetation index and climatic information for the years 2000–2019. Therefore, this chapter has specific focuses with the aim is to examine in detail what specific factors, climatic or biogeographic are responsible for the spatio-temporal patterns of precipitation–temperature–vegetation changes. The results show that BRB's are strongly affected by Skin Surface temperature (ST_{skin}) and precipitation anomalies. More importantly, at the basin level, the spatio-temporal distribution of Normalized Difference Vegetation Index (NDVI)-precipitation correlations is closely related to the vegetation changes. To justify this phenomena, Mann–Kendall test was applied to get the trend of the data and observed NDVI has an increasing trend. The outputs from this study is found effective for the sustainable practices of the terrestrial ecosystem, hydrological and water resources of the BRB.

Keywords Normalized difference vegetation index · Precipitation · Skin surface temperature · Multi-sensors · Spatio-seasonal changes · Brahmaputra river basin

S. Acharjee (✉)
Centre for Studies in Geography, Dibrugarh University, Dibrugarh 786004, Assam, India
e-mail: sacharjee@dibru.ac.in

B. Nath
Lab of Geoinformatics and Earth Observation Research (LGEOR), Department of Geography and Environmental Studies, Faculty of Biological Sciences, University of Chittagong, Chittagong 4331, Bangladesh
e-mail: nath.gis79@cu.ac.bd

© The Author(s), under exclusive license to Springer Nature Switzerland AG 2025

S. Ghosh (ed.), *The Himalaya Dilemma*, Springer Natural Hazards,
https://doi.org/10.1007/978-3-031-95083-4_16

1 Introduction

Analysis of changes in vegetation cover resulting from climate change (CC) is a growing scientific field mirrored by public concern. Therefore, many researchers are focusing on the assessment of vegetation response to CC. A significant role is played by vegetation in the exchange of energy, water, and carbon between the land surface and the atmosphere (Jong et al., 2011), changes in air temperature (Braswell et al., 1997; Root et al., 2003), precipitation (Fay et al., 2008; Nemani et al., 2002), solar radiation (Kong et al., 2017), and sea surface temperature (SST) (Forzieri et al., 2013). Several studies suggest that global climate change (CC) has impacted the distribution, structure, composition, and diversity of vegetation communities (Alo & Wang, 2008; Weltzin et al., 2003; Zhao et al., 2019). Among the above-mentioned climatic variables involve in vegetation–atmosphere interactions, variability in precipitation is thought to be largely responsible for vegetation change and ecosystem structure by its effect on water availability driving the biological processes in over 40% of Earth's vegetated surfaces (Nemani et al., 2003; Notaro et al., 2010). As a result, in many areas of the world (Africa, Central Asia, and North America), extensive studies have been conducted on the relationship between precipitation and vegetation (Yang et al., 1998; Zeng et al., 2002; Martiny et al., 2005; Propastin et al., 2008; Jiang et al., 2017).

The remote sensing (RS) and geographic information systems (GIS) are essential for assessing seasonal vegetation responses to CC in the Brahmaputra River Basin (BRB) due to their ability to monitor large, remote, and diverse landscapes effectively. RS provides multi-temporal, multi-spectral data to track vegetation health, phenology, and climate-driven changes over time. GIS integrates this data with environmental variables like topography, hydrology, and climate for spatial analysis and modeling. These tools help identify patterns, detect vulnerable areas, and analyze ecological processes influenced by climate variability. In this aspect, RS and GIS offer a cost-effective, scalable, and comprehensive approach found effective for informed decision-making and sustainable resource management in this ecologically sensitive region.

RS and GIS technology is increasing at a tremendous pace, used globally for the study and identification of patterns in spatial data and vegetation changes based on the multi-temporal satellite images (Anderson et al., 2010; Xie et al., 2015). Monitoring of vegetation changes in large areas mainly based on field surveys is extremely difficult and not cost-effective. Therefore, geospatial technology is considered as the advantageous over traditional field surveys, as it provides better means of studying vegetation dynamics in large areas (Xie et al., 2008). The recently launched NASA-Giovanni portal Goddard Earth Sciences Data and Information Services Center (GES-DISC) (http://disc.gsfc.nasa.gov) facilitates the use of long term processed multi-sensor satellite data, and this study is found suitable to consider this data source as it takes less processing time, compared to the other freely available earth observation satellite image based analysis. The multi-sensors processed temporal satellite data are available as the Network Common Data Format (NetCDF) for storing multi-dimensional data and a comma-separated values (CSV) file format, with the option of accessing seasonal spatio-temporal data for user-specific requirement by drawing area of interest (AOI) along with specific time frame.

In this chapter, the present study has a specific research question about at what specific factors, climatic or biogeographic are responsible for the spatio-temporal patterns of precipitation–temperature–vegetation changes in the BRB? The aim of this study is to analyze the spatio-temporal patterns of NDVI-precipitation correlations in relation to vegetation changes at the basin scale by applying NASA-Giovanni web portal ((http://disc.gsfc.nasa.gov) based multi-sensors processed data, is considered as a novel approach which helps to identify the causes behind the ongoing Himalayan crisis. By achieving this aim and to solve the raised question, the study has move with the specific objectives which are: (i) to analyze spatio-temporal trends of vegetation dynamics in the last 19 years (2000 December–2019 November) in the BRB based on climate and multi-sensors remote sensing processed data; (ii) to identify the main causes of vegetation change in relation to climatic factors (i.e., temperature and precipitation); and (iii) to compare correlations between seasonal NDVI versus STskin and seasonal NDVI versus Precipitation. Moreover, if the climatic variables can be shown to provide a robust explanation of vegetation change, soil must be considered to be of secondary importance but not irrelevant.

2 Material and Methods

2.1 Study Area

The BRB is of irregular shape and the maximum east–west length is 1,540 km and the maximum north–south width is 682 km. The BRB lies between 23°00′N to 32°00′N latitude and 82°00′E to 97°50′E longitude (Fig. 1). According to Immerzeel (2008), the BRB has three different physiographic zones: (a) the Tibetan Plateau (TP), (b) Himalayan Belt (HB), and (c) the floodplain (FP). These three zones respond differently to the anticipated CC. The part of the TP has 44.4% coverage of the BRB has an elevation varying from 3,000 to 5,000 m a.s.l and has numerous glaciers (Singh et al., 2004), whereas HB covers 28.6% of the basin with elevations varying from 100 m to 3,500 m a.s.l. The area with an elevation of less than 100 m a.s.l. is considered as FP which comprises about 27% of the entire BRB.

The basins receive over 70% of their annual precipitation during the monsoon season (June–September), with significant contributions from snow and glacier melt occurring between May and October. During the monsoon, rain runoff dominates due to the high intensity of precipitation, while the contribution from snow and glacier melt to water yield remains relatively low. In contrast, rainfall is much lower in other seasons, including post-monsoon (October–November), winter and dry season (December–March), and pre-monsoon (April–May). The Southern Himalayan slope

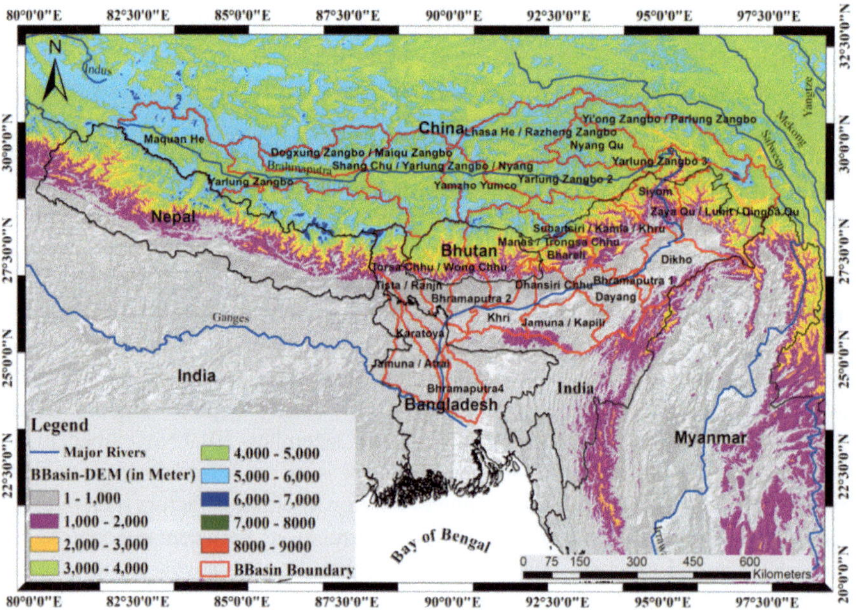

Fig. 1 Location of the study area; (base imagery DEM data acquisition from USGS earth explorer, and administrative boundaries of all countries downloaded from spatial data section of Diva-GIS web portal (https://www.diva-gis.org/gdata), and river GIS shapefile accessed from the Data catalog of the World Bank Group available as major rivers of the world (https://datacatalog.worldbank.org/search/dataset/0042032). Map prepared by authors using ArcGIS 10.8 software

of the basin, influenced by the South Asian monsoon, receives an average annual rainfall of 1400–6000 mm (average 2300 mm). Meanwhile, the upper basin in the TP (Northern Himalayan slope) experiences significantly lower annual rainfall, ranging from 300–1200 mm (average 750 mm). Approximately 70–80% of the annual rainfall occurs during the monsoon (June–September), 15–20% during the pre-monsoon period (April–May), and the remaining 5–10% during the post-monsoon (October–November) and winter and dry season (December–March) (Bajracharya et al., 2015).

The Brahmaputra River drains diverse environments from the cold dry plateau of Tibet, the rain-drenched Himalayan slopes, alluvial plains of Assam, and the vast deltaic lowlands of Bangladesh. In the BRB area, the specific single year can be divided into four seasons: the relatively dry, cool winter from December–February (DJF); the dry, hot summer from March–May (MAM); the southwest monsoon from June–August (JJA) when the predominating southwest maritime winds bring rains; and the retreating monsoon of September–November (SON).

Frigid winds from the Himalayas can depress temperatures near the BRB. The two Himalayan states in the east India, Sikkim and Arunachal Pradesh, receive substantial snowfall. The extreme north of West Bengal, centered on Darjiling, also experiences snowfall, but only rarely. Winter rainfall and occasional snowfall are associated with the large storm systems known as '*Nor'westers*' and '*Westerly disturbances*'; the latter coming from the west in Summer in the BB, lasting from March to May. The southwest summer monsoon, when massive convective thunderstorms dominate the weather in the BRB, originates from a high-pressure mass, centered over the southern Indian Ocean; attracted by a low-pressure region centered over South Asia giving rise to surface winds that ferry humid air into the BRB from the southwest. These inflows ultimately result from a Northward shift of the local jet stream which itself results from rising summer temperatures over Tibet and the Indian subcontinent. The gap left by the jet stream, which switches from a route just south of the Himalayas to one tracking north of Tibet, attracts warm humid air from the southwest. The main factor behind this shift is the high summer temperature difference between Central Asia and the Indian Ocean. The hottest month for most of the BRB is May. In cooler regions of North India, immense pre-monsoon squall-line thunderstorms, locally known as '*Nor'westers*', commonly drop large hailstones and rain. Most summer rainfall occurs during powerful thunderstorms associated with the southwest summer monsoon; occasional tropical cyclones also contribute to this. The Bay of Bengal (BoB) monsoon, moves northward and spreads over most of Assam (Brahmaputra and Meghna Basin) by the first week of June. On encountering the barrier of the Great Himalayan Range, it is deflected westward along the Indo-Gangetic Plain (i.e. over the Ganges basin) towards New Delhi (North-west part of the Ganges basin).

2.2 Major Geographic Features of the BRB in the Eastern Himalaya

The details of major geographic features such as climate, elevation, landform, hydrology, biotic importance and hazard information of the BRB in the Eastern Himalaya is presented in Table 1.

2.3 Datasets Used

In this study, multi-sensors satellite datasets were selected and retrieved from the NASA's Goddard Earth Sciences Data and Information Services Center (GES-DISC) web portal domain (https://disc.gsfc.nasa.gov). The application available in this archive is a coupling of all types of geophysical parameters that allow users to visualize data that can be saved in a local environment for further processing of the individual datasets. The details of multi-sensors satellite datasets used in this study are summarized in Table 2.

Table 1 Details of major geographic features of the BRB

Climate	The annual precipitation is about 300 mm in the Brahmaputra basin in Tibet. Annual rainfall in the southern slopes of the Himalayan drainage of the Brahmaputra is between 1000 to 3000 mm, and between 2000 and 4000 mm over the Brahmaputra Valley of Assam as well as the Naga Patkai and Indo-Myanmar Ranges
Elevation	The Brahmaputra River is a large international river flowing through Tibet (China), India and Bangladesh. Rising near the Chema Yundung glacier in southwest Tibet (elevation 5300 m, 82°10′E, 30°30′N)
Landform description	The BRB features diverse landforms, including high-altitude glaciers and steep valleys in the Tibetan Plateau, rugged mountains along the Eastern Himalayas, and vast alluvial plains in Assam and Bangladesh. It is characterized by braided river channels, fertile floodplains, wetlands, and deltaic formations near its confluence with the Bay of Bengal
Hydrology	The Brahmaputra traverses a total distance of 2880 km, which comprises an easterly course of 1625 km in Tibet, a southerly and westerly course of 918 km in India and a southerly course of 337 km in Bangladesh. Total basin area of the river is 580,000 km^2 It is the fourth largest river in the world in terms of average flow discharge at its mouth with a flow of 19,830 m^3/s^{-1}, whereas the river ranks 22nd in terms of drainage area. Hence, discharge per unit drainage area in the Brahmaputra is amongst the highest in the world. At Pandu, the Brahmaputra yields 0.0306 $m^3 s^{-1} km^{-2}$, and the mean annual flood discharge is 51,156 $m^3 s^{-1}$. The rainy season (May through October) accounts for up to 82% of the mean annual flow at Pandu with high fluctuations in daily discharge
Biotic importance	The BRB supports rich biodiversity, hosting dense tropical forests, wetlands, and grasslands critical for numerous species, including the one-horned rhinoceros and Bengal tiger. Its fertile floodplains sustain agriculture, while its aquatic ecosystems provide habitats for fish and freshwater species, supporting livelihoods and ecological balance in the region
Hazards	The BRB faces severe hazards, including frequent flooding, riverbank erosion, and landslides, driven by intense monsoon rains and glacial melt. CC exacerbates these issues with increased rainfall intensity and rising temperatures. Additionally, earthquakes and sedimentation contribute to hydrological instability, posing significant risks to communities and infrastructure of the BRB located in the Eastern Himalaya

2.4 Methods

The study followed several steps to fulfill the research objectives is shown in the flowchart of the methodology (Fig. 2). The details of each step is shown in the following several sub-sections.

Table 2 Details of multi-sensors processed satellite datasets used

Multi-sensors satellite data	Data acquisition period	Temporal resolution	Model	Spatial resolution	Version of datasets	Source
MODIS NDVI	Dec 2000–Nov 2019	monthly	MODIS-Terra	0.05°	MOD13C2 v006	NASA GES-DISC
MODIS ST_{Skin}	Dec 2000–Nov 2019	monthly	MERRA-2	0.5*0.625°	(M2TMN*SLV v5.12.4)	NASA GES-DISC
Precipitation	Dec 2000–Nov 2019	monthly	MERRA-2	0.5*0.625°	M2TMN*FLX v5.12.4	NASA GES-DISC

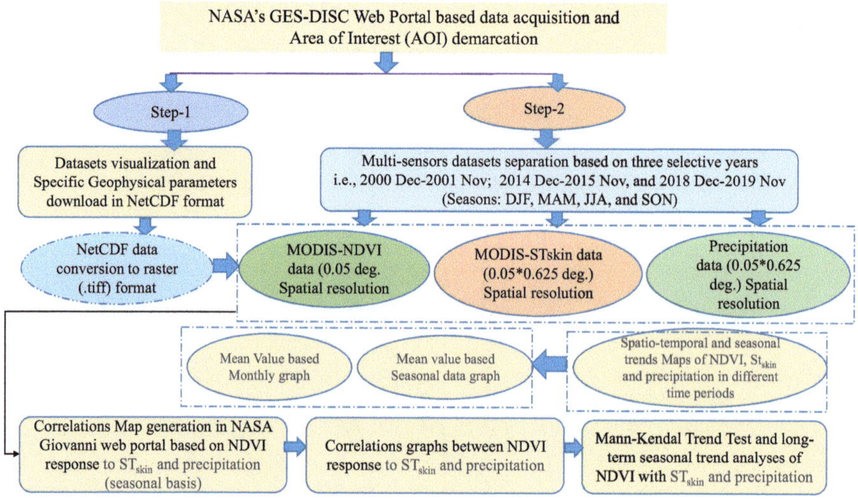

Fig. 2 Flowchart of the methodology used in this study

2.4.1 Multi-Sensors Satellite Datasets Retrieval and Mean Value Calculations for NDVI, ST$_{skin}$ and Precipitation

In this study, at the initial stage, multi-sensors processed datasets retrieved from NASA's GES-DISC archive were considered to observe spaio-seasonal variations of multi-temporal parameters such as NDVI, surface skin temperature (ST$_{Skin}$), and precipitation and it's mapping in four different seasons since 2000 December to 2019 November. Monthly observations of Normalized difference vegetation index (NDVI), ST$_{Skin}$ data measured by NDVI MODIS-Terra (MOD13C2 v006) (Friedl et al., 2002; Gao et al., 2005; Schaaf et al., 2002; Wan & Dozier, 1996).

From December 2000–November 2019 with monthly temporal resolution on a 0.05° latitude/longitude grid. Moreover, ST$_{skin}$ datasets in the same time period are used to analyze the ST$_{skin}$ including to observe how it relates to surface properties. The MODIS ST$_{skin}$ data is only retrieved on clear days on a monthly basis having a sensor of the MERRA-2 model with spatial resolution of 0.5*0.625°. The ST$_{skin}$ values are averaged from the daytime and nighttime of Land surface temperature (LST) values, which is the product of the monthly ST$_{skin}$ (M2TMN*SLV v5.12.4).

Besides these, for the assessment of total surface precipitation, monthly 0.5*0.6250 spatial resolution data retrieved from the MERRA-2 model surface precipitation product (M2TMN*FLX v5.12.4; unit: kg $m^{-2}s^{-1}$) from the same platform. All these processed images were saved in NetCDF format including time-series averaged monthly data for the selected time period, with further processing to produce an average of the specific season.

However, in the statistical analyses part, mean values of the different variables in specific seasons are presented for a seasonal comparison, illustrated by bar diagrams with error bars. The error bars are the standard errors for different seasonal changes in three selected years for NDVI, STskin, and precipitation.

2.4.2 Map Correlations and Statistical Analyses for Seasonal NDVI, ST_{Skin} and Precipitation

In this stage of the present study, correlations map among the variables, NDVI in relation with ST_{skin} and precipitation, were performed within the NASA Giovanni web portal (https://disc.gsfc.nasa.gov) considering two variables at each grid cell in a specific season, by adopting Pearson's correlation technique. For example, NDVI (MOD13C2 v006)-MODIS-Terra, monthly, spatial resolution 0.05° as the 1st variable and ST_{skin} (MERRA-2 model with spatial resolution 0.5*0.625°), monthly data as the 2nd variable was used to develop a spatial correlation map for specific months in each season. Similar to the time frame in each season, we have performed the same analysis for the other seasons. In addition to this, NDVI and precipitation correlation maps were also similarly retrieved for all four seasons based on three different years. All the correlation maps were saved in NetCDF format and further processed in ArcGIS 10.8 software based on high and low values of each correlation map. This has been done by adopting the natural breaks (Jenks) classification method or optimization system is the data classification method designed to optimize the arrangement of a set of values into "natural" classes. The natural class is the most optimal class range found naturally in a data set.

Moreover, a scatter plot diagram for each season i.e., 2000 December–2001 February (DJF) in three different years (2000 December–2019 November) was created in the MS Excel 2019 environment by considering Mean ST_{skin} on the X-axis and Mean NDVI on the Y-axis. Similarly, all other seasons (MAM, JJA, SON) have been analyzed in the same manner for Mean Precipitation versus Mean NDVI. The corresponding p values and equations are illustrated in each seasonal correlation graph. In this case, the linear method was adopted to draw trend lines that represent either positive or negative correlation.

2.4.3 Trend Analysis Method

In the final stage, this study moves forward by analyzing the statistical technique (Mann–Kendall) (Mann, 1945; Kendall, 1957), which is often utilized to understand the quantification of the significance of long-term trends in temporal studies of hydro-meteorological datasets (Da Silva et al., 2015). The Mann–Kendall test (Mann, 1945; Kendall, 1957) is a non-parametric statistical method used to detect trends in time series data, commonly applied in environmental and climate studies. It evaluates whether data values consistently increase or decrease over time without requiring the data to follow a specific distribution. The test calculates the Kendall's tau statistics, which measures the strength and direction of a monotonic trend, where a positive tau indicates an increasing trend, and a negative tau shows a decreasing trend. The test's significance is determined using *p-values* or *Z-scores*. It is robust to missing values, outliers, and non-normal data, making it suitable for diverse datasets. The statistical method measures the degree of consistently increasing or decreasing within a range from −1 to +1. The value with +1 demonstrates a consistent increasing and never decreases; and −1 refers as decreasing, and 0 value indicates no consistent trend in the data. This particular test has been calculated in the XLSTAT add-in software (https://www.xlstat.com/en/solutions/features/mann-kendall-trend-tests) worked in the MS Excel 2019 software environment based on time series values.

3 Results and Discussion

3.1 Spatio-Temporal and Seasonal Variations in Vegetation Cover (NDVI)

A 19-year long time series of monthly NDVI data at a spatial resolution of 0.05° from Moderate Resolution Imaging Spectro-radiometer (MODIS) Terra (MODIS13C2 v 006) was exploited in our study for quantifying NDVI and its trend. The products provide detailed measurements of pixel values and can detect the reflected light in areas of high biomass and corrects for distortions of background scattering, which is applicable in vegetation health-related studies (Didan, 2018). The processed NDVI netcdf data, converted in raster format in ArcGIS 10.8, was used further for classification based on seasons for a particular year. Table 3 and Figure 3(a–l) demonstrated the spatial patterns of the trends of NDVI over the period of 2000 Dec–2019 Nov. In order to evaluate the spatial and seasonal heterogeneity of NDVI trends, time series seasonal variations (2000 Dec–2019 Nov) based on NDVI maximum, minimum, mean, and Standard Deviation (SD) values were derived from classification statistics for each classified image under the classify tabs in ArcGIS 10.8 software. From Table 3, the NDVI seasonal mean values over the period of 2000–2019 (in three different time periods of 2000–2001, 2014–2015, and 2018–2019) were further considered for seasonal comparison which is demonstrated in following section.

Among all the above seasonal datasets of NDVI from 2000 Dec–2019 Nov, the largest increase in NDVI is observed in SON (0.82–0.93) in 2019, compared to other seasons. On the other hand, the lowest NDVI values were observed in MAM (−0.07–0.09) in 2015 compared to all others seasons in the respective years. However, in comparison to all four seasons from 2000 to 2019, NDVI values sharply increased in and SON, compared to the other two seasons.

Table 3 Seasonal time series NDVI, surface skin temperature (ST_{skin}) (°C) and Total Surface Precipitation (kg m^{-2} s^{-1}) Monthly Mean, Maximum and SD variations

NDVI	Mean				Maximum				SD			
	DJF	**MAM**	**JJA**	**SON**	**DJF**	**MAM**	**JJA**	**SON**	**DJF**	**MAM**	**JJA**	**SON**
2000–2001	0.33	0.32	0.45	0.47	0.88	0.82	0.90	0.92	0.26	0.27	0.26	0.28
2014–2015	0.35	0.36	0.52	0.48	0.9	0.86	0.92	0.93	0.28	0.28	0.27	0.29
2018–2019	0.5	0.52	0.36	0.35	0.9	0.85	0.92	0.93	0.29	0.26	0.28	0.28
ST_{Skin} (°C)												
2000–2001	0.36	10.10	10.47	9.94	18.06	30.10	30.62	25.62	10.82	10.85	10.88	9.93
2014–2015	1.25	10.55	16.64	10.54	20.17	30.20	29.41	26.27	12.59	11.23	7.96	9.85
2018–2019	0.11	10.21	16.99	10.34	19.55	31.18	29.72	25.62	12.37	11.12	8.08	10.03
Precipitation (kg m^{-2} s^{-1})												
2000–2001	0.02	0.14	0.32	0.15	0.13	0.47	1.05	0.42	0.02	0.09	0.2	0.12
2014–2015	0.02	0.15	0.38	0.1	0.35	0.45	1.48	0.35	0.02	0.12	0.29	0.07
2018–2019	0.03	0.13	0.32	0.14	0.25	0.51	1.17	0.42	0.02	0.11	0.22	0.1

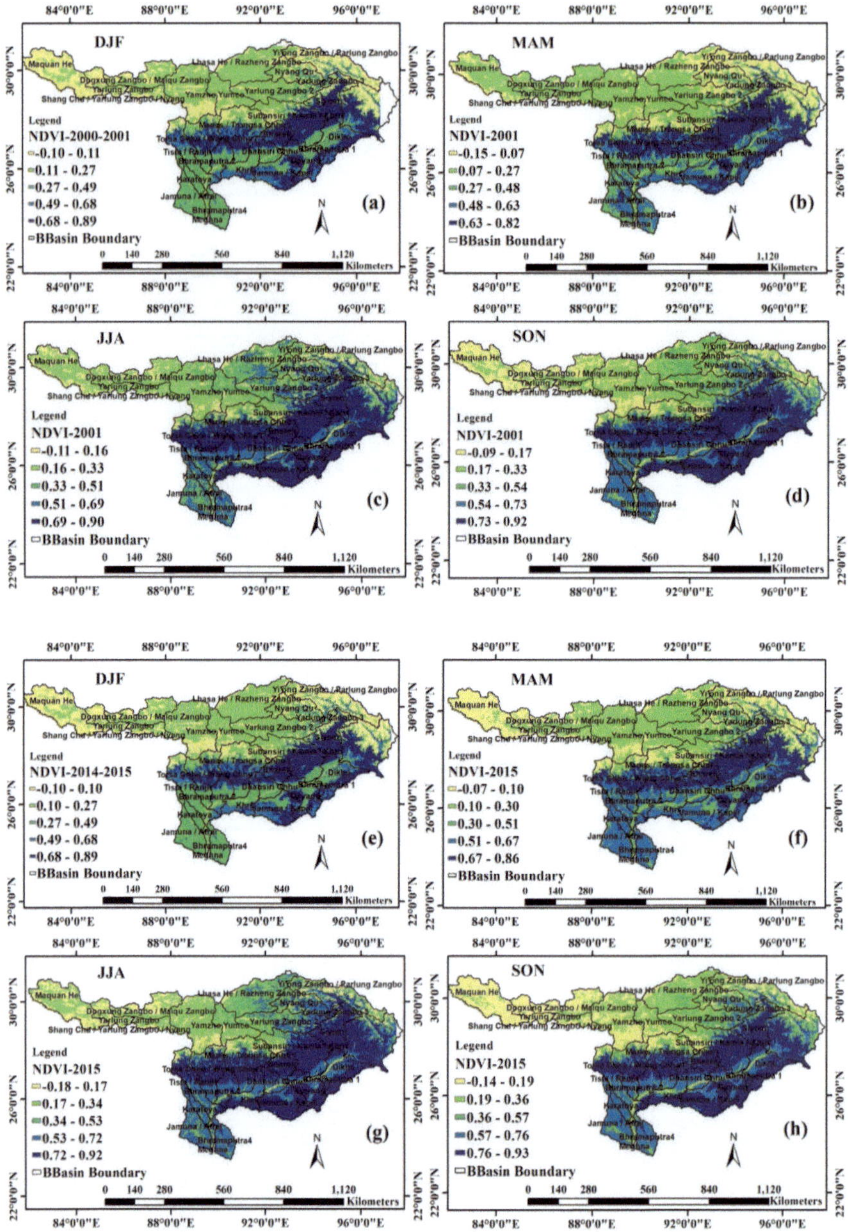

Fig. 3 Spatio-temporal and seasonal trends of vegetation cover in three different time frames and four seasons from 2000 December–2019 November; Fig. 3(**a–d**) for NDVI 2000 December–2001 November, (**e–h**) same as (**a–d**) but for NDVI 2014 December–2015 November and (**i–l**) same as **a–d** but for NDVI 2018 December–2019 November

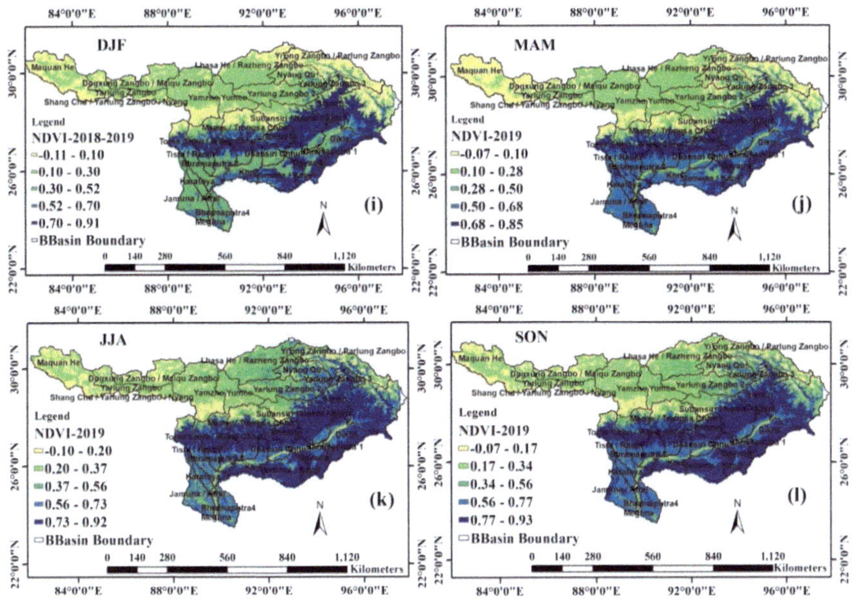

Fig. 3 (continued)

3.2 *Spatio-Temporal and Seasonal Variations in Surface Skin Temperature (ST_{skin})*

The term 'skin temperature' has been used for 'radiometric surface temperature' (Jin et al., 1997). It can be measured by either a hand-held or aircraft-mounted radiation thermometer, a derived from upward longwave radiation based on the Stefan–Boltzmann law (Holmes, 1969; Oke, 1987), or retrieved from satellite observations and mapped over large areas, after removing the effects of atmospheric attenuation on the satellite measured.

The reason for considering ST_{skin} observations is because they provide more coverage than T_{air}. High spatial resolution is one of the key strengths of satellite data, and it can reveal valuable details over the globe. Since the T_{air} observations over the globe are not uniformly distributed (one of the limits of T_{air} observations), many areas do not have observations of T_{air}. Therefore, solely relying on T_{air} data may be misleading for accurate land surface change characterization over the globe. In general, the major patterns of T_{air} are consistent with those of ST_{skin}, although details may differ. Therefore, in this study, processed satellite sensors datasets were considered to observe the ST_{skin} over the BRB. The winter (DJF), spring (MAM), summer (JJA) and autumn (SON) (2000–2001, 2014–2015, 2018–2019) ST_{skin} (°C) has been used for the present study (Figs. 4a–l).

The result of the ST_{skin} suggests that the spring season (MAM) in both the years (i.e., 2001, 2019) represented the highest ST_{skin} of 31.19 °C and lowest of −3.14

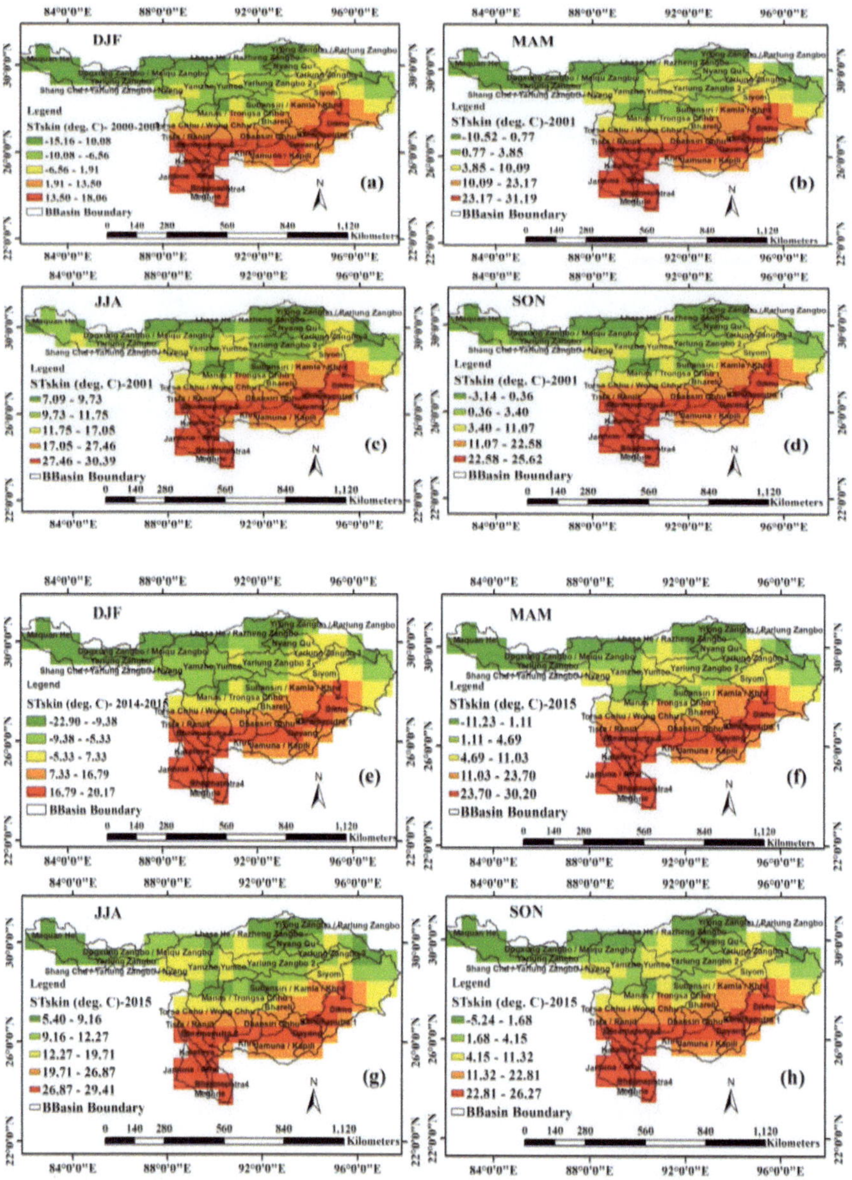

Fig. 4 Spatio-temporal and seasonal variation of surface skin temperature (ST_{skin}) based on different time period from 2000 December–2019 November; **a** ST_{skin} from 2000 Dec–2001 February (DJF), **b** ST_{skin} from 2001 March–2001 May (MAM), **c** ST_{skin} from 2001 June–2001 August (JJA), **d** ST_{skin} from 2001 September–2001 November (SON), **e–h** same as (**a–d**), but for 2014 December–2015 November, and (**i–l**) same as previous but for 2018 December–2019 November

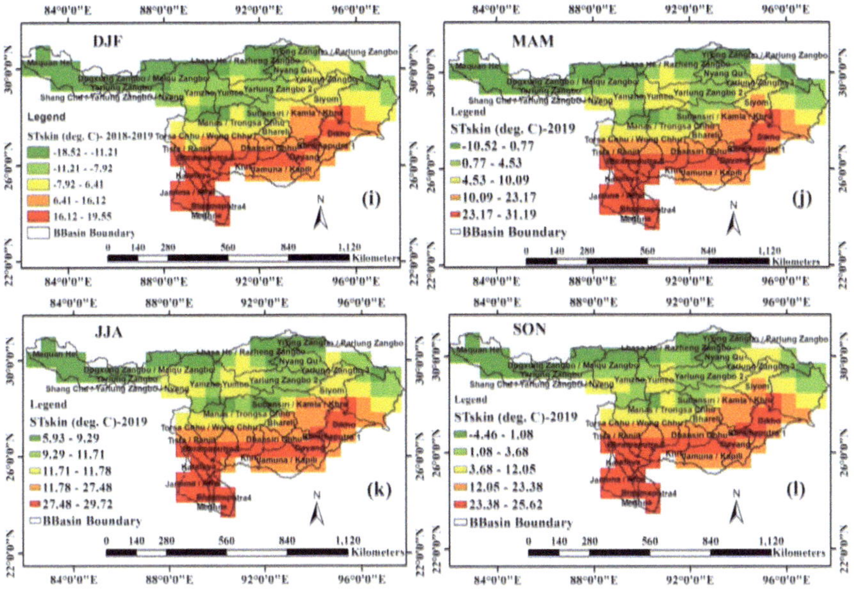

Fig. 4 (continued)

°C was observed in autumn (SON) in the year 2001 compare to the seasons of other year's. The ST_{skin} values have substantial seasonality, with their maximum values occurring in the Brahmaputra valley during spring (MAM) and summer months (JJA) and minima in the winter months (DJF) as a result of solar insolation variations.

3.3 *Spatio-Temporal and Seasonal Distribution of Precipitation*

Figure 5a–l shows the spatial distributions of seasonal total surface precipitation (unit: kg m^{-2} s^{-1}) over the BB from 2000 to 2019 derived from MERRA-2 model monthly data sets with spatial resolution of 0.5*0.625° in each grid cell. The timing of the data sets (the four seasons) is the same as the other two variables as mentioned earlier.

Heavy precipitation occurs as a result of the migration of the Intertropical Convergence Zone (ITCZ) over the eastern equatorial Pacific Ocean and the Atlantic Ocean and the Asia summer monsoon. In the south-southeastern, part of the BB, covering Arunachal Pradesh in the India Himalayas, receives a large amount of precipitation. Among all three years, the highest rainfall of 1.48 kg m^{-2} s^{-1} was observed during summer (JJA) and the lowest of 0.00 kg m^{-2} s^{-1} was estimated by satellite sensor during winter (DJF) in 2014 Dec–2015 Feb.

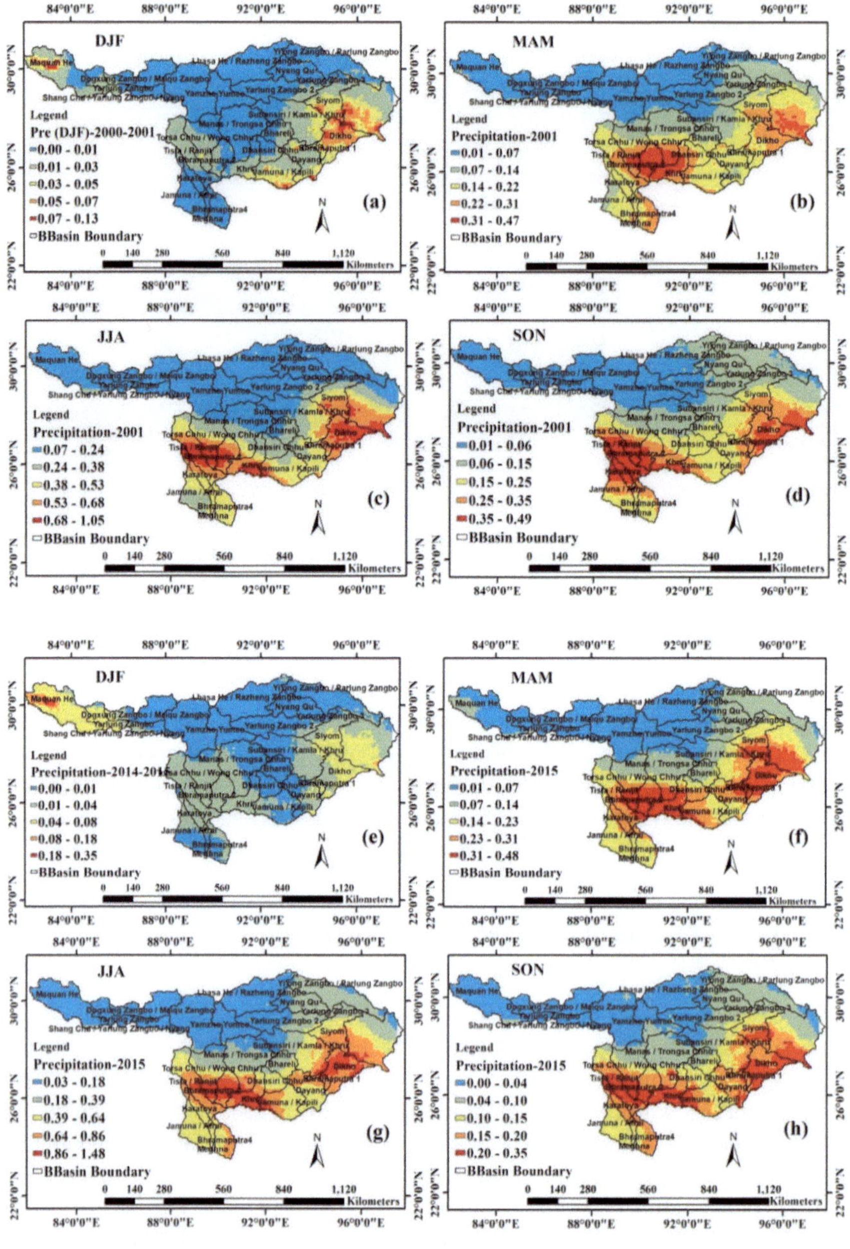

Fig. 5 Spatio-temporal and seasonal trends of precipitation in different time periods from 2000 December–2019 November; **a** Precipitation from 2000 December–2001 February (DJF), **b** Precipitation from 2001 Mar to 2001 May (MAM), **c** Precipitation from 2001 June to 2001 August (JJA), **d** Precipitation from 2001 September–2001 November (SON), **e–h** same as (**a–d**), but for 2014 December–2015 November, and (**i–l**) same as previous but for 2018 December–2019 November

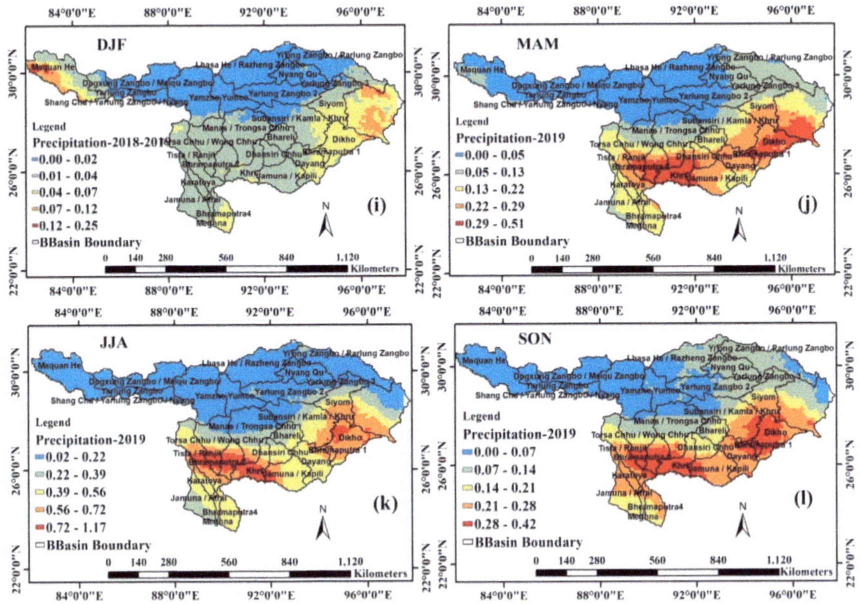

Fig. 5 (continued)

3.4 *Seasonal Effects of Climatic Variations on NDVI Trends*

In this particular section, we discuss the seasonal effects of climatic variables on NDVI trends by four selected seasons: such as DJF, MAM, JJA, and SON. The NDVI mean in all seasons increased in most areas over time with an overall increase in NDVI of 10%. Seasonal NDVI variation for different seasons can give us a better understanding of the patterns of NDVI changes in the BB (Fig. 6). Table 4 and Fig. 6b show the SD annual increasing trend based on different seasons over the period from 2000 to 2019.

The NDVI peaks in the BB are well observed and it is all above 10 °C in MAM, JJA, and SON in 2000–2001, 2014–2015, and 2018–2019, except for SON in 2000–2001 (Fig. 6c) and in DJF in 2018–2019, the lowest minima observed as 0.11 °C (Fig. 6c). Furthermore, inter-annual variations are also evident among all three years in four different seasons; SD values in the DJF months show higher values compared to the other three seasons (Fig. 6d). The season JJA of the year 2019 was the warmest among all seasons from 2001 to 2019.

The mean of all seasonal values of ST_{skin} increased in most areas over time with an overall increase in ST_{skin} by 2.00 °C. In seasonal ST_{skin}, SD can give us a better understanding of the patterns of ST_{skin} change. Table 2 and Fig. 6d showed the SD annual and seasonal STskin increasing rates over the period from 2000 to 2019. The Total Surface Precipitation (TSP) showed spatially and seasonally heterogeneous patterns in the trends with increasing rates of mean TSP during summer (JJA) over

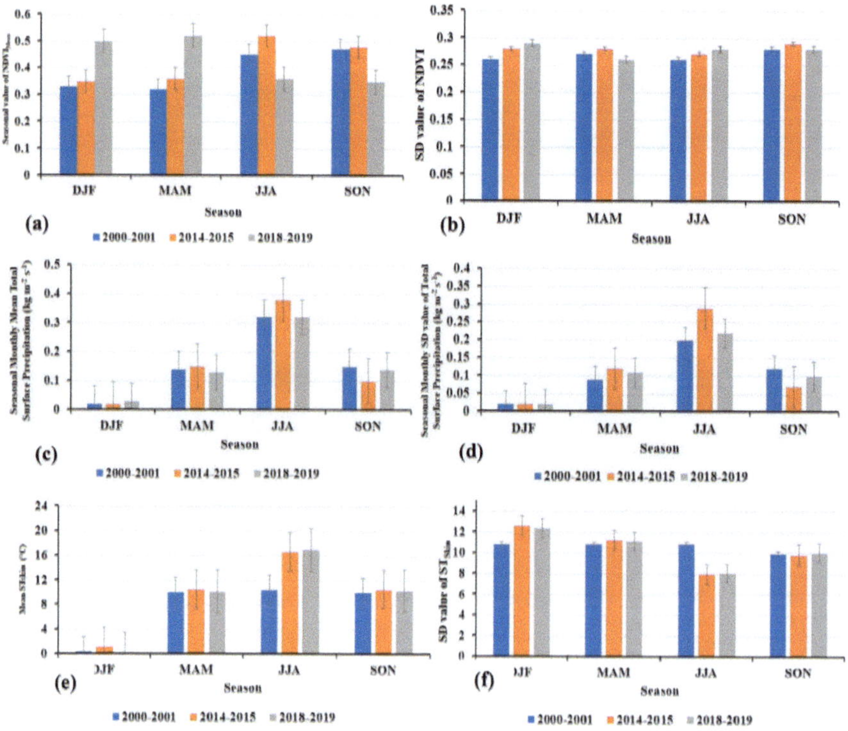

Fig. 6 Seasonal time series of Monthly Mean NDVI and SD variations in response to ST_{skin} and total surface precipitation from 2000 Dec–2019 Nov; **a** seasonal Monthly Mean NDVI and **b** SD variations of NDVI; **c** seasonal Monthly mean surface skin temperature (ST_{skin}) (°C) and **d** SD variations of ST_{skin}; **e** Monthly Mean Total Surface Precipitation (kg m^{-2} s^{-1}), and **f** SD variations of Total Surface Precipitation (kg m^{-2} s^{-1})

Table 4 Correlation between NDVI and ST_{skin}, NDVI and Precipitation based on different time periods and seasonal data from 2000 December–2019 November

ST_{skin} & NDVI	DJF	MAM	JJA	SON	Precipitation & NDVI	DJF	MAM	JJA	SON
	–	–	–			–			
r	0.585	0.102	0.119	−0.12	r	0.994	0.755	0.827	0.391
R^2	0.343	0.01	0.014	0.015	R^2	0.988	0.571	0.68	0.153
P	0.414	0.897	0.88	0.879	P	0.006	0.244	0.172	0.608

the period of 2000–2019 (Table 3 and Fig. 6e). The seasonal increase in maximum TSP values was observed during MAM over the same time period (see Table 3). The mean of all seasons TSP increased in most areas over time. Overall, the study area indicated an increase in ST_{skin} by 0.01 kg m^{-2} s^{-1} in the two time periods except in 2014–2015. During that period there was a 0.02 kg m^{-2} s^{-1}, 0.06 kg m^{-2} s^{-1}

higher in MAM and JJA seasons and 0.05 kg m^{-2} s^{-1} lower in SON, respectively compared to the time period of 2018–2019 (Table 3). The actual cause of the increase and decrease in that year is unknown. Moreover, the SD of Seasonal TSP can give us a better understanding of the patterns of TSP change. Table 3, and Fig. 6f showed the SD annual and seasonal TSP increasing trend over the period from 2000 to 2019.

3.5 Correlation Between Climatic Variables with NDVI in Different Seasons

For exploring the effects of climate variations on vegetation growth, maps depicting Pearson correlation coefficients between NDVI and the two climatic variables (i.e., ST_{skin} and precipitation) (Figs. 7 and 8) were calculated for each pixel automatically based on the method discussed in Section 2.2.2, which has been produced within the NASA Giovanni portal and was further classified using ArcGIS 10.8 software. Even as climate factors change through multiple periods, the vegetation growth takes time to respond, causing a time lag (Piao et al., 2003; Wang et al., 2003). Therefore, it can be said that the variations in climatic factors often lead to variations in vegetation cover for a specific site, which is mostly represented through NDVI temporal data sets but later than the climatic variation. Previous research has been carried out to document and understand the correlation between seasonal NDVI and climatic variables such as, temperature and precipitation considering a possible time lag of vegetation growth response to seasonal climate conditions (Hussein et al. 2017; Potter & Brooks, 1998; Hao et al., 2012; Los et al. 2001; Piao et al., 2003, 2006; Chu et al. 2007; Hu et al., 2011; Zhao et al., 2011; Wang et al., 2003; Chuai et al., 2013; Daham et al., 2018; Huang et al., 2019).

But in this study, we have derived the processed correlations maps between NDVI with ST_{skin} and NDVI with total surface precipitation derived from two variables in each season in a particular year. The spatial distribution and correlation maps between NDVI and ST_{skin}, and NDVI versus Precipitation in four different seasons of each year are demonstrated in Figs. 7a–l and 8a–l, respectively. Moreover, the mean values of two variables, namely mean ST_{skin} and mean precipitation in each season for the respective years (from 2000 Dec to 2019 Nov) are used in this study based on the method (scatter diagram) discussed at the end of Section 2.4.2.

For the entire study area, the spring NDVI (MAM) negatively correlated significantly with spring temperature (MAM) ($r = -0.102$, $R^2 = 0.010$, $P = 0.897$) (Fig. 9b, and Table 4), and also correlated negatively with precipitation (MAM) ($r = -0.755$, $R^2 = 0.571$, $P = 0.244$) (Fig. 9f, and Table 4). On the other hand, Summer NDVI (JJA) is negatively correlated with summer temperature (JJA) ($r = -0.119$, $R^2 = 0.014$, $P = 0.880$) (Fig. 9c, and Table 4); whereas, Summer NDVI (JJA) is positively correlated with summer precipitation (JJA) ($r = -0.827$, $R^2 = 0.68$, $P = 0.172$) (Table 4). Autumn NDVI (SON) has a negative correlation with autumn temperature (SON) ($r = -0.120$, $R^2 = 0.015$, $P = 0.879$) and precipitation in autumn (SON) ($r = -0.391$,

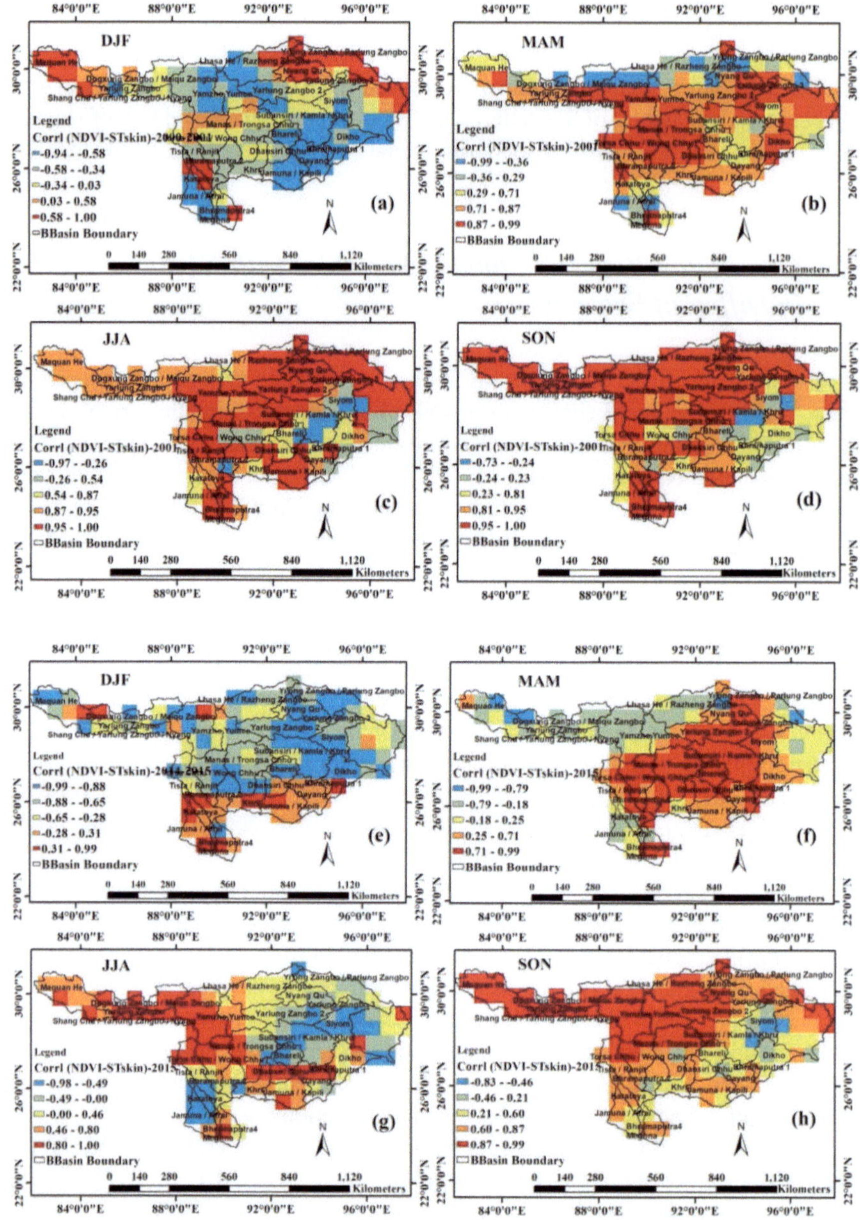

Fig. 7 Correlation map between NDVI and Surface Skin Temperature (ST_{skin}) based on different time periods and seasonal change data from 2000 Dec to 2019 Nov; Fig. 7 **a–d** for correlation between NDVI and ST_{skin} from 2000 Dec–2001 Nov, **e–h** same as **a–d** but for NDVI and ST_{skin} from 2014 Dec–2015 Nov, and (**i–l**) same as **a–d** but for NDVI and ST_{skin} from 2018 Dec to 2019 Nov

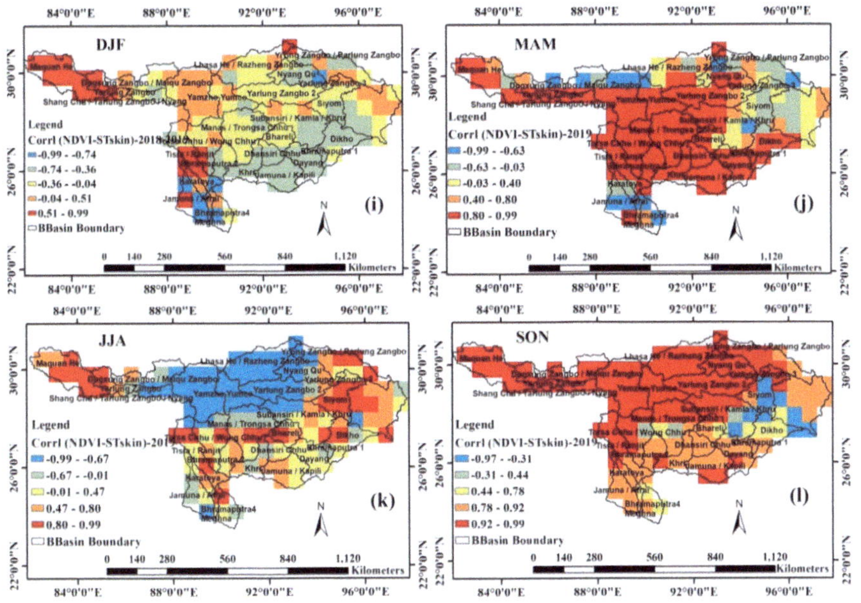

Fig. 7 (continued)

R^2=0.153, P = 0.608) (Table 4). Winter NDVI (DJF) has a negative correlation with winter temperature (DJF) (r = −0.585, R^2 = 0.343, P = 0.414) and precipitation in winter is positively correlated with winter NDVI (r = 0.994, R^2 = 0.988, P = 0.006) (Table 4). From the results of these correlations among all seasons, a strong positive correlation between winter precipitation and NDVI is observed, because in monsoon regions the annual onset of precipitation occurs quickly and the vegetation requires time to respond which happens in winter. Photosynthesis is probably decreased during the monsoonal wet and cloudy periods due to less solar energy reaching the surface and NDVI is subdued at this time. Another possible explanation is that cloudy sky may decrease actual estimates of the NDVI. Further work should be conducted to ascertain the actual cause of the negative correlations.

3.6 Overall Change Effect Analysis of NDVI in Responses to ST_{skin} and Precipitation in the BRB

Vegetation information is represented by the NDVI from MODIS remote sensing. NDVI less than 0 is related to water or snow surfaces and NDVI from 0 to 0.9 indicates vegetation-cover. These phenomena have been clearly reflected in the presence of seasonal variability in the BRB with consistent results achieved based on the processed MODIS data sets used in this study from 2000 Dec to 2019 Nov. The

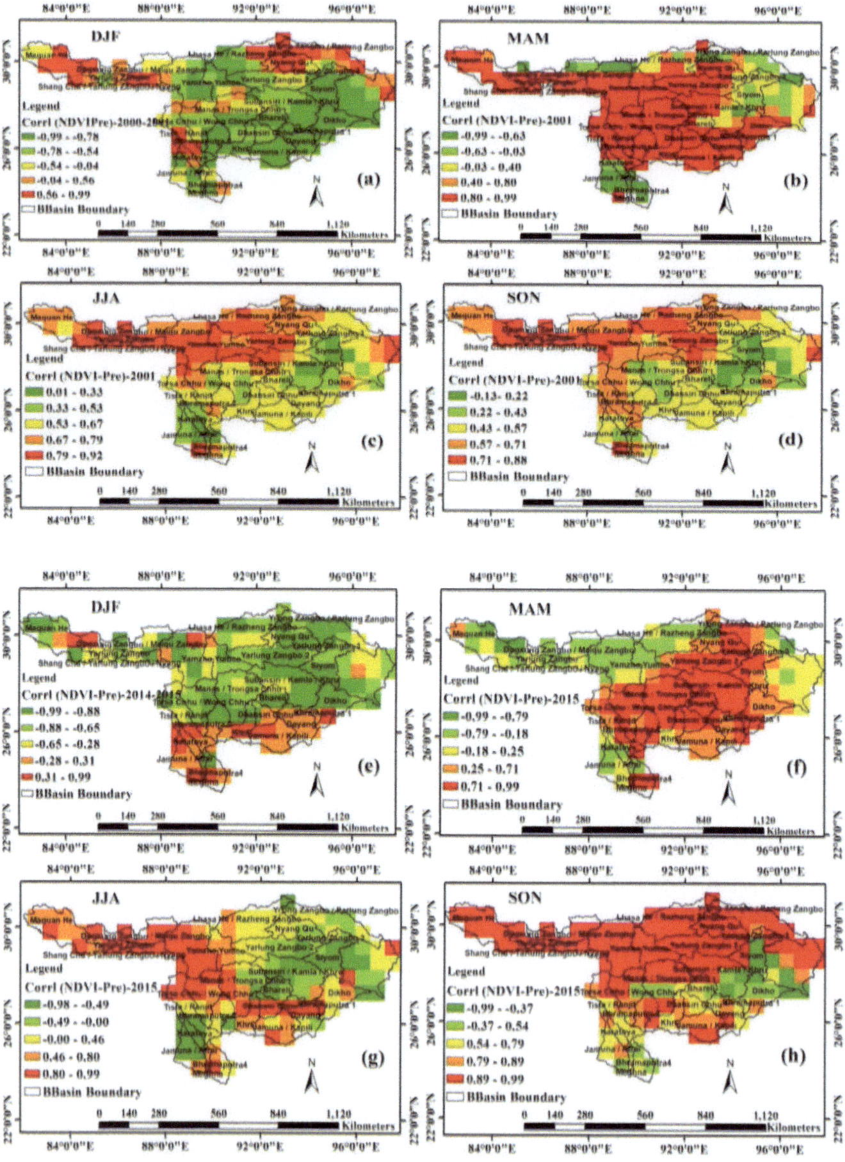

Fig. 8 Correlations between NDVI and Precipitation based on different time periods and seasonal data from 2000 Dec–2019 Nov; Fig. 8a–d for correlation between NDVI and Precipitation from 2000 Dec–2001 Nov, **e–h** same as **a–d** but for NDVI and Precipitation from 2014 Dec–2015 Nov and (**i–l**) same as **a–d** but for NDVI and Precipitation from 2018 Dec to 2019 Nov

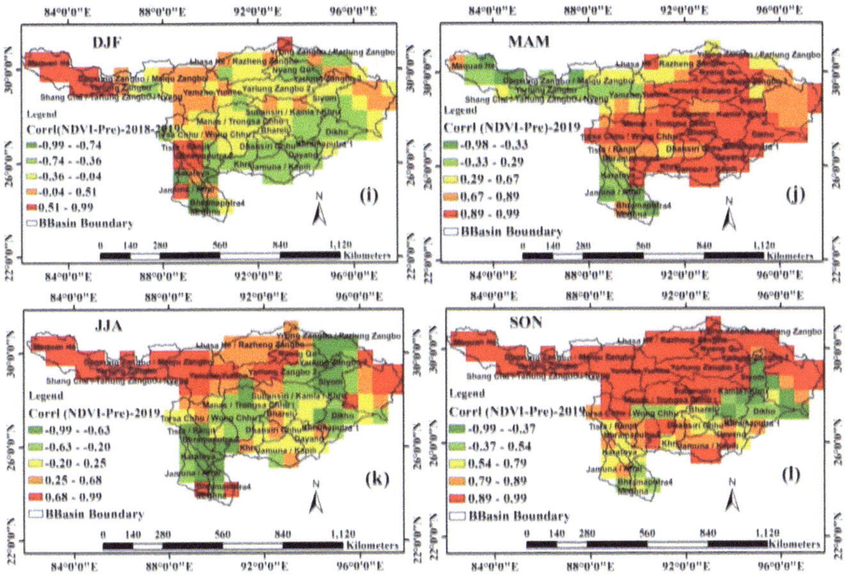

Fig. 8 (continued)

scenario is different for climatic parameters such as ST_{skin} and precipitation influences on NDVI. ST_{skin} almost linearly decreases with the increase of NDVI values, suggesting that more vegetation reduces ST_{skin}. Such a decrease is partly because there is more vegetation, more evapotranspiration, and therefore, less heat is used to warm the ground surface. On the other hand, since ST_{skin} is observed from space, the higher NDVI value represents dense forest, where ST_{skin} mainly represents the canopy top temperature. In addition, for the studied region, ST_{skin} is the highest, up to 31.19 °C in MAM-2001 and MAM-2019, when NDVI is 0.82 and 0.85 in the same month but in two different years, respectively. Further, this study shows that ST_{skin} influences NDVI, which is consistent with previous studies, all of which reported a major increase in NDVI witnessed in the early growing season due to a gradual rise of temperature as the growing season advances (Myneni et al., 1997; Los et al., 2001; Zhou et al., 2001; Tucker et al., 2001; Hicke et al.,, 2002; Slayback et al., 2003; Piao et al., 2003). On the other hand, in our study, variation in mean seasonal precipitation across the BB from 2000 Dec to 2019 Nov increased in MAM, and JJA, compared to the other two seasons. But in the 2015 summer (JJA), the precipitation increased by 0.24 kg m^{-2} s^{-1} compared to spring (MAM).

However, in recent time, precipitation has suddenly decreased in both MAM, and JJA in 2019; more specifically in summertime (JJA), a sharp decrease of 0.07 kg $m^{-2}s^{-1}$ is observed in 2019 which is the same as in JJA in 2001 (19 years before). In comparison to all four seasons in each corresponding year, the least precipitation is observed in winter (DJF) based on data from 2000 Dec to 2019 Nov.

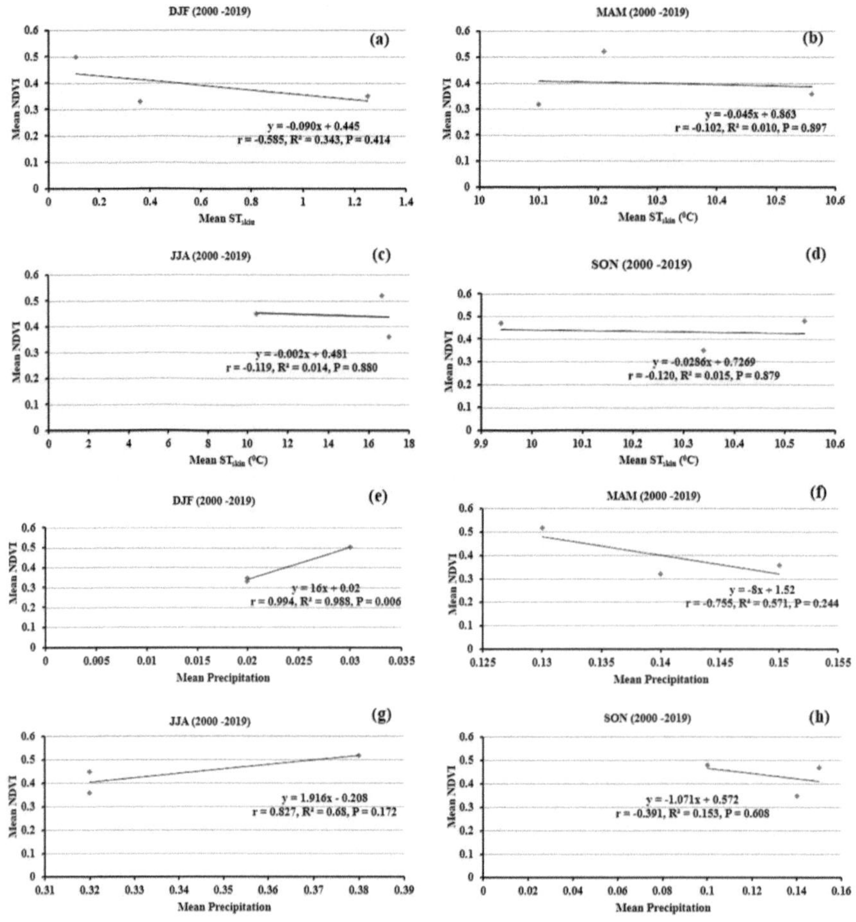

Fig. 9 Correlations between NDVI and ST_{skin} and NDVI and Precipitation based on different time periods and seasonal data from 2000 Dec tov 2019 Nov; Fig. 9a–d for correlation between NDVI and ST_{skin} from 2000 Dec to 2019 Nov, **e–h** same as a–d but for NDVI and Precipitation from 2000 Dec to 2019 Nov

Moreover, a slight increase of precipitation was observed in DJF and SON in 2018–2019, compared to 2014–2015. The autumn (SON) precipitation in 2019 shows a marked decrease compared to summer (JJA) in the same year, with a clear effect on vegetation growth in autumn (SON) in 2019, completely different from the monthly and seasonal study reported by Piao et al. (2003) for China from 1982 to 1999 which, with an autumn precipitation decrease, showed that a large effect on plant growth in autumn may not exist. However, no effect on vegetation growth was observed in summer (JJA) in 2001 and 2015, when sufficient precipitation occurred in the BB. This phenomenon has been clearly tracked and presented by the mean values derived from the processed satellite datasets.

3.7 *Trend Analysis of NDVI with Seasonal ST_{skin} and Precipitation*

Over the 19-year period (2000 December to 2019 November), the details of long-term trend analysis of seasonal ST_{skin} with NDVI and seasonal precipitation with NDVI is shown in Table 5. In this long-term seasonal trend, NDVI, ST_{skin}, and precipitation are defined spatio-temporal variations of the mean data, which varying and behave differently in different seasons of the specific years. Among three different years, mean NDVI value was observed lowest in 0.32 in the MAM of the year 2001, and maximum in the season of MAM in 2019, and JJA in 2015 time periods (Table 5). In the MAM, the mean NDVI value has gradually increased as the time progresses, and in recent time 2019 (MAM) it was observed as 0.50 compared to 0.33 in 2000–2001 (DJF) time period, a significant increase with a mean NDVI value of 0.17 was noticed in the BRB. On the other hand, mean ST_{skin} was observed maximum as 16.64 °C in JJA in 2015, which was observed 10.47 °C in the same season of the year 2001, a 6.17 °C sharp increase was noticed in the BRB.

The results confirm the gradual increasing phenomena of warming temperature in the BRB, thus reflecting a climate warming scenario in the entire BRB. On the other hand, in SON, the mean temperature increased was observed 0.40 °C in between 2000 and 2019 time period. However, with the changing trend of NDVI, climatic warming was clearly noticed in the BRB, whereas, the mean precipitation was also observed increasing trend with the value of 1.48 kg m^{-2} s^{-1} in JJA of the year 2015, compared to 1.05 kg m^{-2} s^{-1} in the same season of 2001 (Table 5).

Conversely, both this trend analyses showed small variability in the magnitude and direction of trends (Fig. 10). In the case of significance of trends, only in 2015 (JJA) shows strong correlation between ST_{skin} and NDVI.

Table 5 Long term Seasonal trends of NDVI with ST_{skin} and precipitation from 2000 December to 2019 November

Time period	NDVI	ST_{Skin}	Precipitation
2000–2001 (DJF)	0.33	0.36	0.13
2001 (MAM)	0.32	10.10	0.47
2001 (JJA)	0.45	10.47	1.05
2001 (SON)	0.47	9.94	0.42
2014–2015 (DJF)	0.35	1.25	0.35
2015 (MAM)	0.36	10.55	0.45
2015 (JJA)	0.52	16.64	1.48
2015 (SON)	0.48	10.54	0.35
2018–2019 (DJF)	0.50	0.11	0.25
2019 (MAM)	0.52	10.21	0.51
2019 (JJA)	0.36	16.99	1.17
2019 (SON)	0.35	10.34	0.42

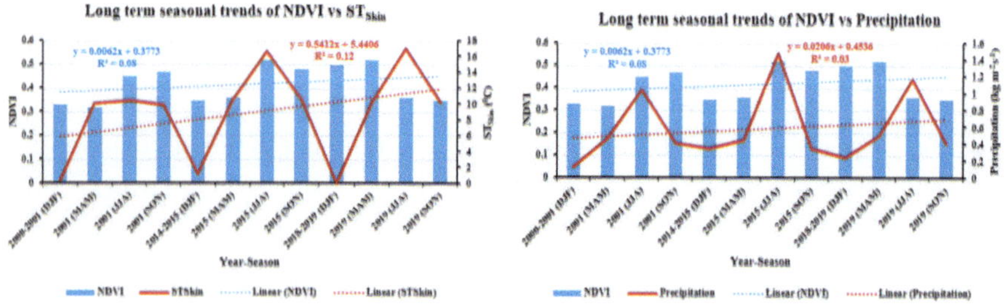

Fig. 10 Long term Seasonal trends of NDVI with ST_{skin} and Precipitation from 2000 December to 2019 November; **a** Seasonal trends of NDVI with ST_{skin} from 2000 December to 2019 November; **b** Seasonal trends of NDVI with Precipitation from 2000 December to 2019 November

On the other hand, the strong correlation was observed between precipitation and NDVI only in 2001 (JJA), 2015 (JJA) and 2019 (JJA) compared to precipitation trend line, where the ST_{skin} trend line was identified as more pronounced and sharper increasing trend. However, in most of the study area, the Mann–Kendall test result revealed that precipitation and ST_{skin} shows no confidence in trend, but NDVI confidence trend shows it is probably increasing.

4 Conclusions and Future Directions

The results indicated that seasonal changes of NDVI, ST_{skin}, and precipitation were all single-peak curves with the peak values ranging from July to August. NDVI, ST_{skin}, and precipitation all roughly had a downtrend from the southeast to the northwest. Moreover, NDVI was more susceptible to the variations of precipitation than ST_{skin} as a whole. The relationships had confirmed the presence of strong seasonal dynamics and distinct regional characteristics that were related to snow melting, soil condition, climate change, and vegetation changes, etc. This research study overall discussed the spatial–temporal patterns and its relationships of NDVI with ST_{skin} and grided satellite precipitation throughout the study period.

The study finally highlights that the above analyses were basically consistent with actual situations in Tibet, which demonstrated the potential of using the processed data for regional research. The present research considered climate data from 2000 December to 2019 November which is not sufficient for climate model as it requires more than 30 years of data. There is a future scope of this study by considering more than 30 years of data which will finally correlating with NDVI. However, in addition to the above, this study will further focus on this basin related to stress-related vegetation indices, green chlorophyll vegetation index and enhanced vegetation index, and so on to find out the actual causes, climate variability and its relation with extreme event including the climate induced vegetation changes with the study of coupling effects of natural and anthropogenic impacts.

References

Alo, C. A., & Wang, G. (2008). Potential future changes of the terrestrial ecosystem based on climate projections by eight general circulation models. *Journal of Geophysical Research, 113*, G01004.

Anderson, L.O., Malhi, Y., Arag, OLEOC, Ladle, R., Arai, E., Barbier, N., Phillips, O. (2010) Remote sensing detection of droughts in Amazonian forest canopies. *New Phytology,187*, 733–750

Anding, D., & Kauth, R. (1970). Estimation of sea surface temperature from space. *Remote Sensor Environment, 1*, 217–220.

Bajracharya, S. R., Palash, W., Shrestha, M. S., Khadgi, V. R., Duo, C., Das, P. J., & Dorji, C. (2015). Systematic evaluation of satellite-based rainfall products over the Brahmaputra basin for hydrological applications. *Advances in Meteorology, 2015*(9), 398687(1–17).

Becker, F., & Li, Z. (1995). Surface temperature and emissivity at various scales: Definition, measurement problems. *Remote Sensing Reviews, 12*, 225–253.

Braswell, B. H., Schimel, D. S., Linder, E., & MooreIII, B. (1997). The response of global terrestrial ecosystems to interannual temperature variability. *Science, 278*(5339), 870–873.

Chu, D., Lixin, L., Tingjun, Z. (2007). Sensitivity of normalized difference vegetation index (ndvi) to seasonal and interannual climate conditions in the Lhasa Area, Tibetan Plateau, China, Arctic, Antarctic, and *Alpine Research, 39*, 4: 635–641, https://doi.org/10.1657/1523-0430(07-501)[CHU]2.0.CO;2

Chuai, X. W., Huang, X. J., Wang, W. J., & Bao, G. (2013). NDVI, temperature and precipitation changes and their relationships with different vegetation types during 1998–2007 in Inner Mongolia, China. *International Journal of Climatology, 33*(7), 1696–1706.

Coll, C., Caselles, V., & Schmugge, T. J. (1994). Estimation of land surface emissivity differences in the split-window channels of AVHRR. *Remote Sensing of Environment, 48*, 127–134.

Da Silva, R. M., Santos, C., Moreira, A. G., Corte-Real, M. J., Silva, V. C. L., & Medeiros, I. C. (2015). Rainfall and river rainfall and river fow trends using Mann–Kendall and Sen's slope estimator statistical tests in the cobres river basin. *Natural Hazards, 77*, 1205–1221.

Daham A, Han D, Rico-Ramirez M, Marsh A (2018) Analysis of NDVI variability in response to precipitation and air temperature indifferent regions of Iraq, using MODIS vegetation indices. *Environmental Earth Science*, 77–389. https://doi.org/10.1007/s12665-018-7560-x.

Didan, KMOD13Q1 V006, LP DAAC: NASA Land Data Products and Services. Available online: https://lpdaac.usgs.gov/node/844 (accessed on 17 December 2020)

Fay, P. A., Kaufman, D. M., Nippert, J. B., Carlisle, J. D., & Harper, C. D. (2008). Changes in grassland ecosystem function due to extreme rainfall events: Implications for responses to climate change. *Global Change Biology, 14*, 1600–1608.

Forzieri, G., Vivoni, E. R., & Feyen, L. (2013). Ecosystem biophysical memory in the southwestern North America climate system. *Environmental Research Letters, 8*(4), 4016.

Friedl, M. A., McIver, D. K., Hodges, J. C. F., Zhang, X. Y., Muchoney, D., Strahler, A. H., Woodcock, C. E., Gopal, S., Schneider, A., Cooper, A., Baccini, A., Gao, F., & Schaaf, C. (2002). Global land cover mapping from MODIS: Algorithms and early results. *Remote Sensing of Environment, 83*(1–2), 287–302.

Gao, F., Schaaf, C., Strahler, A., Roesch, A., Lucht, W., & Dickinson, R. (2005). The MODIS BRDF/Albedo climate modeling grid products and the variability of albedo for major global vegetation types. *Journal of Geophysical Research, 110*, D01104. https://doi.org/10.1029/2004JD005190

Hao, F., Zhng, X., Ouyang, W., Skidmore, A. K., & Toxopeus, A. G. (2012). Vegetation NDVI linked to temperature and precipitation in the upper catchments of yellow river. *Environmental Modeling and Assessment, 17*, 389–398.

Hicke, J.A., Asner, G.P., Randerson, J.T. et al. (2002) Trends in North American net primary productivity derived from satellite observations, 1982–1998. *Global Biogeo-Chemical Cycles 16*, 2–1:2–14

Holmes RM (1969) Airborne measurements of thermal discontinuities in the lowest layers of the atmosphere Paper presented at 9th Conf. Agriculture Meteorology (Seattle, WA), p 18

Hu, M. Q., Mao, F., Sun, H., & Hou, Y. Y. (2011). Study of normalized difference vegetation index variation and its correlation with climate factors in the three-river-source region. *International Journal of Applied Earth Observation and Geoinformation, 13*, 24–33.

Huang, X. L., Zhang, T., Yi, G., He, D., Zhou, X., Li, J., et al. (2019). Dynamic changes of NDVI in the growing season of the Tibetan Plate auduring the past 17 years and its response to climate change. *International Journal of Envionmental Research and Public Health, 16*, 3452.

Hussein, S. O., Kovacs, F., & Tobak, Z. (2017). Spatiotemporal assessement of vegetation indices and land cover Erbil city and its sur-rounding using MODIS imageries. *Journal of Environmental Geography, 10*(1–2), 31–39.

Immerzeel, W. (2008). Historical trends and future predictions of climate variability in the Brahmaputra basin. *International Journal of Climatology, 28*(2), 243–254.

Jiang, L., Jiapaer, G., Bao, A., Guo, H., & Ndayisaba, F. (2017). Vegetation dynamics and responses to climate change and human activities in Central Asia. *Science of the Total Environment, 599–600*, 967–980.

Jin, M., Dickinson, R. E., & Vogelmann, A. M. (1997). A comparison of CCM2/BATS skin temperature and surface-air temperature with satellite and surface observations. *Journal of Climate, 10*, 1505–1524.

Jong, R. D., Bruin, S. D., Wit, A. J. W. D., Schaepman, M. E., & Dent, D. L. (2011). Analysis of monotonic greening and browning trends from global NDVI time-series. *Remote Sensing of Environment, 115*(2), 692–702.

Kendall, M. G. (1957). Rank correlation methods. *Biometrika, 44*(1/2), 298. https://doi.org/10.2307/2333282

Kong, D., Zhang, Q., Singh, V. P., & Shi, P. (2017). Seasonal vegetation response to climate change in the Northern Hemisphere (1982–2013). *Global and Planetary Change, 148*, 1–7.

Los, S.O., Collatz, G.J., Bounoua, L., Sellers, P.J., Tucker, C.J. (2001). Global interannual variations in sea surface temperature and landsurface vegetation, air temperature, and precipitation. *Journal of Climatology, 14*:1535–1549

Mann, HB. (1945). Nonparametric tests against trend. *Econometrica, 13*(3), 245. https://doi.org/10.2307/190.7

Martiny, N., Richard, Y., & Camberlin, P. (2005). Interannual persistence efects in vegetation dynamics of semi-arid Africa. *Geophysical Research Letters, 32*, L24403.

Myneni, R. B., Keeling, C. D., Tucker, C. J., Asrar, G., & Nemani, R. R. (1997). Increased plant growth in the northern high latitudes from 1981 to 1991. *Nature, 386*, 698–702. https://doi.org/10.1038/386698a0

Nemani, R. R., Keeling, C. D., Hashimoto, H., Jolly, W. M., Piper, S. C., Tucker, C. J., Myneni, R. B., & Running, S. W. (2003). Climate-driven increases in global terrestrial net primary production from 1982 to 1999. *Science, 300*(5625), 1560–1563.

Nemani, R., White, M., Thornton, P., Nishida, K., Reddy, S., Jenkins, J., & Running, S. (2002). Recent trends in hydrologic balance have enhanced the terrestrial carbon sink in the United States. *Geophysical Research Letters, 29*(10), 1468.

Notaro, M., Liu, Z., Gallimore, R. G., Williams, J. W., Gutzler, D. S., & Collins, S. (2010). Complex seasonal cycle of ecohydrology in the Southwest United States. *Journal of Geophysical Research, 115*, G04034.

Oke, T. R. (1987). *Boundary Layer Climates* (2nd ed.). Methuen.

Palas, W., Bajracharya, S. R., Shrestha, A. B., Wahid, S., Hossain, M. S., Mogumder, T. K., & Mazumder, L. C. (2023). Climate change impacts on the hydrology of the Brahmaputra River Basin. *Climate, 11*(18), 1–30. https://doi.org/10.3390/cli11010018

Piao, S. L., Fang, J. Y., Zhou, L. M., Guo, Q. H., Henderson, M., Ji, W., Li, Y., & Tao, S. (2003). Interannual variations of monthly and seasonal NDVI in China from 1982 to 1999. *Journal of Geophysical Research, 108*(D14), 4401.

Piao, S. L., Mohammat, A., Fang, J. Y., Cai, Q., & Feng, J. M. (2006). NDVI-based increase in growth of temperate grasslands and its responses to climate changes in China. *Global Environmental Change, 16*, 340–348.

Potter, C. S., & Brooks, V. (1998). Global analysis of empirical relations between annual climate and seasonality of NDVI. *International Journal of Remote Sensing, 15*, 2921–2948.

Prata, A. J. (1993). Land surface temperatures derived from the advanced very high resolution radiometer and the along-track scanning radiometer: 1 Theory. *Journal of Geophysical Research, 98*, 16689–16702.

Price, J. C. (1984). Land surface temperature from measurements from the split window channels of the NOAA 7 advanced very high resolution radiometer. *Journal of Geophysical Research, 89*, 7231–7237.

Propastin, P. A., Kappas, M., & Muratova, N. R. (2008). Inter-annual changes in vegetation activities and their relationship to temperature and precipitation in Central Asia from 1982 to 2003. *Journal of Environmental Informatics, 12*, 75–87.

Root, T. L., Price, J. T., Hall, K. R., Schneider, S. H., Rosenzweig, C., & Pounds, J. A. (2003). Fingerprints of global warming on wild animals and plants. *Nature, 421*, 57–60.

Sarma, J.N. (2004). An overview of the Brahmaputra River system. In: *The Brahmaputra Basin Water Resources*. Kluwer Academic Publishers: London, UK: 72–87.

Sarma JN, Acharjee S (2018) A study on variation in channel width and braiding intensity of the Brahmaputra River in Assam, India. *Geosciences, 8*, 343: 19. https://doi.org/10.3390/geosciences8090343

Saunders, P. M. (1967). Aerial measurements of sea surface temperature in the infrared. *Journal of Geophysical Research, 72*, 4109–4117.

Schaaf, C. B., Gao, F., Strahler, A. H., Lucht, W., Li, X., Tsang, T., Strugnell, N. C., Zhang, X., Jin, Y., Muller, J. P., Lewis, P., Barnsley, M., Hobson, P., Disney, M., Roberts, G., Dunderdale, M., Doll, C., d'Entremont, R. P., Hu, B., ... Roy, D. (2002). First operational BRDF, Albedo and Nadir reflectance products from MODIS. *Remote Sensing of Environment, 83*, 135–148.

Singh, V.P., Sharma, B., Shekhar, C., Ojha, P. (2004). The Brahmaputra Basin water resources. Kluwer Academic Publishers, ISBN 1-4020-1737-5.

Slayback, D. A., Pinzon, J. E., Los, S. O., & Tucker, C. J. (2003). Northern hemisphere photosynthetic trends 1982–99. *Global Change Biology, 9*, 1–15.

Sobrino, J. A., Li, Z. L., Stoll, M. P., & Becker, F. (1994). Improvements in the split-window technique for land surface temperature determination. *IEEE Transactions on Geoscience Remote Sensing Letters, 32*, 243–253.

Stephens, G. L. (1994). *Remote Sensing of the Lower Atmosphere: An Introduction*. Oxford University Press.

Tucker, C. J., Slayback, D. A., Pinzon, J. E., Los, S. O., Muneni, R. B., & Taylor, M. G. (2001). Higher northern latitude normalized difference vegettion index and growing season trends from 1982 to 1999. *International Journal of Biometeorology, 45*, 184–190.

Ulivieri, C., Castronuovo, M. M., Francioni, R., & Cardillo, A. (1994). A split window algorithm for estimating land surface temperature from satellites. *Advances in Space Research, 14*, 59–65.

Wan, Z., & Dozier, J. (1996). A generalized split-window algorithm for retrieving land-surface temperature from space. *IEEE Transactions on Geoscience Remote Sensing Letters, 34*, 892–904.

Wang, Q., Shi, J., Zhang, Z., & Meng, Z. (2003). Current situation of the environment in Shiyang river basin and its evolutionary trends. *Journal of Desert Research, 23*(1), 46–52.

Wasson, R., Acharjee, S., & Rakshit, R. (2022). Towards identification of sediment sources, and processes of sediment production, in the Yarlung-Tsangpo-Brahmaputra River catchment for reduction of fluvial sediment loads. *Earth-Science Reviews, 226*, 1–15. https://doi.org/10.1016/j.earscirev.2022.103932

Water and Power Consultancy Services (India). Morphological Studies of the River Brahmaputra. Available online: http://www.wapcos.gov.in/ (accessed on 11 September 2018).

Weltzin, J. F., Loik, M. E., Schwinning, S., Williams, D. G., Fay, P. A., Haddad, B. M., Harte, J., Huxman, T. E., Knapp, A. K., Lin, G., Pockman, W. T., Shaw, R. M., Small, E. E., Smith, M. D.,

Smith, S. D., Tissue, D. T., & Zak, J. C. (2003). Assessing the response of terrestrial ecosystems to potential changes in precipitation. *BioScience, 53*(10), 941–952.

Xie, B., Jia, X., Qin, Z., Shen, J., & Chang, Q. (2015). Vegetation dynamics and climate change on the Loess Plateau, China: 1982–2011. *Regional Environmental Change, 16*, 1583–1594.

Xie, Y., Sha, Z., & Yu, M. (2008). Remote sensing imagery in vegetation mapping: A review. *Journal of Plant Ecology, 1*, 9–23.

Yang, L., Wylie, B. K., Tieszen, L. L., & Reed, B. C. (1998). An analysis of relationships among climate forcing and time-integrated NDVI of grasslands over the U.S. Northern and Central Great Plains. *Remote Sensing of Environment, 65*, 25–37.

Zeng, N., Hales, K., & Neelin, J. D. (2002). Nonlinear dynamics in a coupled vegetation atmosphere system and implications for desert–forest gradient. *Journal of Climate, 15*(23), 3474–3487.

Zhao, F., Wu, Y., Sivakumar, B., Long, A., Qiu, L., Chen, J., Wang, L., Liu, S., & Hu, H. (2019). Climatic and hydrologic controls on net primary production in a semiarid loess watershed. *Journal of Hydrology, 568*, 803–815.

Zhao, X., Tan, K., Zhao, S., & Fang, J. (2011). Changing climate affects vegetation growth in the arid region of the northwestern China. *Journal of Arid Environments, 75*, 946–952.

Zhou, L., Tucker, C. J., Kaufmann, R., Slayback, D., Shabanov, N., & Myneni, R. B. (2001). Variations in northern vegetation activity inferred from satellite data of vegetation index during 1981 to 1999. *Journal of Geophysical Research, 106*, 20069–20083.

Hydro-meteorological Hazards and Associated Risk

Historical Glacial Lake Outburst Floods in the Indus, Ganga, and Brahmaputra River Basins of the Hindu Kush Himalaya

Atul Kumar, Udo Schikchoff, and Suraj Mal

Abstract Under ongoing climate change, rapid melting of glaciers, and continuous expansion of glacial lakes, the risk of Glacial Lake Outburst Floods (GLOFs) is increasing in the mountainous regions of the Indus, Ganga, and Brahmaputra (IGB) river basins of the Hindu Kush Himalaya (HKH). These river basins are some of the most populated river basins in the world and have the highest exposure to future GLOF events. Therefore, continuous monitoring of GLOFs is crucially important for better preparing for the future GLOF events by better understanding the GLOF mechanisms, developing mitigation strategies, and better community preparedness. This study tries to figure out the patterns of past GLOFs that have been reported in the IGB river basins by looking at the GLOF dataset created at the sub-basin level by the International Centre for Integrated Mountain Development (ICIMOD) in 2022. The study reports 423 historical GLOF events in different subbasins of the IGB between 1533 and 2024. Among the major basins, the Indus River basin had the highest number of reported GLOF events (203), followed by the Brahmaputra River basin (122) and the Ganga River basin (98). The mean elevation of reported GLOFs in the IGB river basins is 4432.13 m asl, which is expected to increase in the upcoming future due to elevation-dependent warming, rapid melting of glaciers at higher elevations, and formation of new glacial lakes at higher elevations. Furthermore, the study found that reported GLOFs are highly influenced by regional topographic, cryospheric, hydrological, and climatic characteristics, which makes the prediction and modelling of these events difficult. Given these challenges, the present study will be of importance to researchers and local and international authorities as an input for better understanding and preparing for future GLOF events. This study investigates the impact of elevation-dependent warming on the rapid melting of glaciers at higher altitudes, alongside the formation of new glacial lakes in these regions. Our findings reveal

A. Kumar (✉) · U. Schikchoff
CEN Center for Earth System Research and Sustainability, Institute of Geography, University of Hamburg, 20146 Hamburg, Germany
e-mail: atul.sbsc@gmail.com

S. Mal
Centre for the Study of Regional Development, School of Social Sciences, Jawaharlal Nehru University, New Delhi 110067, India

© The Author(s), under exclusive license to Springer Nature Switzerland AG 2025

S. Ghosh (ed.), *The Himalaya Dilemma*, Springer Natural Hazards,
https://doi.org/10.1007/978-3-031-95083-4_17

that reported Glacial Lake Outburst Floods (GLOFs) are significantly influenced by a combination of regional topographic, cryospheric, hydrological, and climatic characteristics, complicating predictions and modelling efforts for such events. Given these complexities, the present research is poised to provide valuable insights for researchers and local as well as international authorities, serving as a crucial resource for enhancing understanding and preparedness for future GLOF occurrences.

Keywords Glacial lakes · Hindu Kush Himalaya (HKH) · Glacial lake outburst flood (GLOF)

1 Introduction

Glacial Lake Outburst Floods (GLOFs) can be defined as sudden releases of water from glacial lakes, mainly triggered either by the mechanism of dam breach or overtopping waves (Emmer et al., 2022; Taylor et al., 2023). Historically, these events have caused serious devastation in the downstream regions (Veh et al., 2019; ICIMOD, 2022; Shrestha et al., 2023; Zhang et al., 2025). Trigger events primarily divide the mechanism of GLOF failure into two categories: (i) dam breach and (ii) overtopping waves. Dam breach can be caused by degradation of the dam due to piping, displacement of dam material, seismic activity or excessive hydrostatic pressure, whereas overtopping waves can be caused by excessive precipitation, ice and rock avalanches, or sudden release of excessive water from upstream smaller glacial lakes (Nie et al., 2018; Veh et al., 2020; Emmer et al., 2022; Taylor et al., 2023; Zhang et al., 2024). Therefore, it is of immense importance to continuously monitor, predict, and develop mitigation strategies to limit the damage caused by GLOF events. Furthermore, with rapid melting of glaciers, expansion of glacial lakes and ongoing climate change in the high mountainous regions, better understanding of GLOFs is of urgent importance for mitigating hazards and developing effective warning systems.

GLOFs have been observed and studied across the cryospheric environment of the world (Taylor et al., 2023; Zhang et al., 2024). However, prior to 2000, the majority of the studies were limited to the regional aspect of the GLOF: origin, mechanisms, impact, exposure, and policy framework (Mool, 1995; Teiji and Rothacher, 1996; Reynolds, 1999; Yamada, 2000). This is primarily due to the limited data availability, the isolated location of GLOF events, and the capital-, labour-, and time-intensive nature of the studies (Nie et al., 2018). Prior to the 2000s, the majority of the studies focused on isolated events, which laid the foundation for GLOF studies by documenting events and attempting to understand the process behind the origin of GLOFs (Teiji and Rothacher, 1996; Reynolds, 1998). Furthermore, it is important to mention here that due to the isolated location of GLOF events, especially in the high mountainous regions Indus, Ganga and Brahmaputra (IGB) river basins of the Hindu Kush Himalaya (HKH), there is still no comprehensive study available for the historical GLOF events. Based on the available information, several studies have attempted to generate the GLOF inventories for the macro scale (ICIMOD, 2022; Nie et al., 2018; Shrestha et al., 2023; Veh et al., 2019; Zhang et al., 2024). All of these studies were published after 2000. The advancements in big data technologies, remote sensing and geoinformatics have dramatically improved our capabilities to not only map the historical GLOF events but also monitor GLOF events, identify future GLOF threats, map GLOF exposures and help in building mitigation strategies (Wilson et al., 2019; Mal et al., 2021; Shrestha et al., 2023; Taylor et al., 2023; Zhang et al., 2024).

Nie et al. (2018) attempted to develop the first comprehensive GLOF inventory for the Himalayan region of the HKH. They mapped 62 GLOF events between 1930 and 2018 based on existing literature, reports, news articles and assessment of glacial lake area changes based on remote sensing observations. Veh et al. (2018) mapped 32 moraine-dammed GLOFs in the Himalaya between 1988 and 2016 based on Landsat satellite images. ICIMOD (2022) mapped 736 historical GLOF events in the Hindu Kush Himalaya (HKH) between 1533 and 2024; Lützow et al. (2023) reported 569 GLOF events in High Mountain Asia (HMA) between 1900 and 2022; and Shrestha et al. (2023) identified 697 GLOF events in the HMA between 1833 and 2022. The discrepancies in the number of reported GLOF events in different studies are mainly due to the different data sources and methodologies used to identify the historical GLOF events.

However, there is still a research gap in understanding the evolution of historical GLOFs at sub-basin level in the case of the IGB river basins of the HKH. A comprehensive understanding of historical GLOF events at the sub-basin level is of urgent need, especially when more than 1 million people live within the proximity of potential GLOF in the IGB river basins (Taylor et al., 2023). Therefore, the aims of the present study are to (i) understand the spatial distribution of GLOFs in the IGB river basins at the sub-basin level in the IGB river basins; (ii) analyse the spatio-temporal change in the frequency of the GLOFs at the sub-basin level in the IGB river basins; and (iii) understand the changes in the characteristics of GLOFs at the sub-basin level in the IGB river basins.

2 Study Area

The IGB river basins are the most important river basins of the Indian subcontinent, extending from 21°40′N to 36°58′N and 66°14′E to 97°14′E, latitude and longitude, respectively (Fig. 1). IGB river basins are trans-political boundary river basins, which are spread over 8 countries (Afghanistan, Bangladesh, Bhutan, China, India, Myanmar, Nepal, and Pakistan) with a total area of ~2.65 million km^2. The physiography of the IGB river basins is of complex origins. The elevations range from 0 m asl at the southwestern and southeastern margins, whereas the elevation reaches up to the elevation of 8,848 m asl (Mt Everest) in the central north part of the IGB river basins. All three major rivers, Indus, Ganga and Brahmaputra, originate from the glaciers and lakes of the HKH in the northern part of the IGB river basins. The

IGB river basin contains ~32,000 glaciers with a total area of 71,182 km^2, which is one of the highest outside the polar regions (Sharma et al., 2013). And is home to 19,284 glacial lakes with an area of 1191.81 ± 209.21 km^2 (Kumar et al., 2025). Recent studies found that due to ongoing climate change, glaciers are melting at a faster rate, leading to the rapid expansion of glacial lakes in the region (Bolch et al., 2019; Kumar et al., 2025; Shugar et al., 2020). The climate of the region is monsoon dominated, where the majority of the precipitation (~80%) takes place in summer months, whereas ~10% of the precipitation is received in winter months, which is brought by Western Disturbances (Dimri & Chevuturi, 2016). In summer months, the temperature can reach up to ~50 °C on the southwestern margins, whereas in winters the temperature can fall up to ~−40 °C in the northwestern margins of the IGB river basins (Krishnan et al., 2019). Furthermore, the IGB rivers support the lives of ~1.4 billion people living in the basin area (ICIMOD, 2023). Among these river basins, the Ganga River basin supports the largest number of people, with ~650 million people living in the Ganga River basin, followed by the Indus River basin with ~268 million people and the Brahmaputra River basin with ~130 million people (ICIMOD, 2023).

3 Data Sources and Methodology

3.1 Data Sources

To understand the evolution of historical GLOFs in the IGB river basins, we used a GLOF database developed by ICIMOD (2022) for High Mountain Asia (HMA). The dataset provides a comprehensive inventory of historical GLOFs for the HMA from 1533. The dataset is based on existing literature, media reports, and thorough monitoring of satellite images. The dataset was initially published in the year 2022, with the last update done in the year 2024. The dataset provides vital information about GLOF characteristics, such as the exact year of the event, dam type, lake area, elevation, damage caused, country of origin, river basin, repeatability, and others. By 2024, the dataset had comprehensively recorded 736 GLOF events in the HMA. In addition to the GLOF dataset, we also used major basin and sub-basin shapefiles developed by ICIMOD (https://rds.icimod.org/Home/DataDetail?metadataId=7952).

3.2 Methodology

The first step for understanding the distribution, spatio-temporal changes, and changes in the GLOF characteristics at the sub-basin level in the IGB basin was to convert the CSV format data into the vector format point-based data using the XY

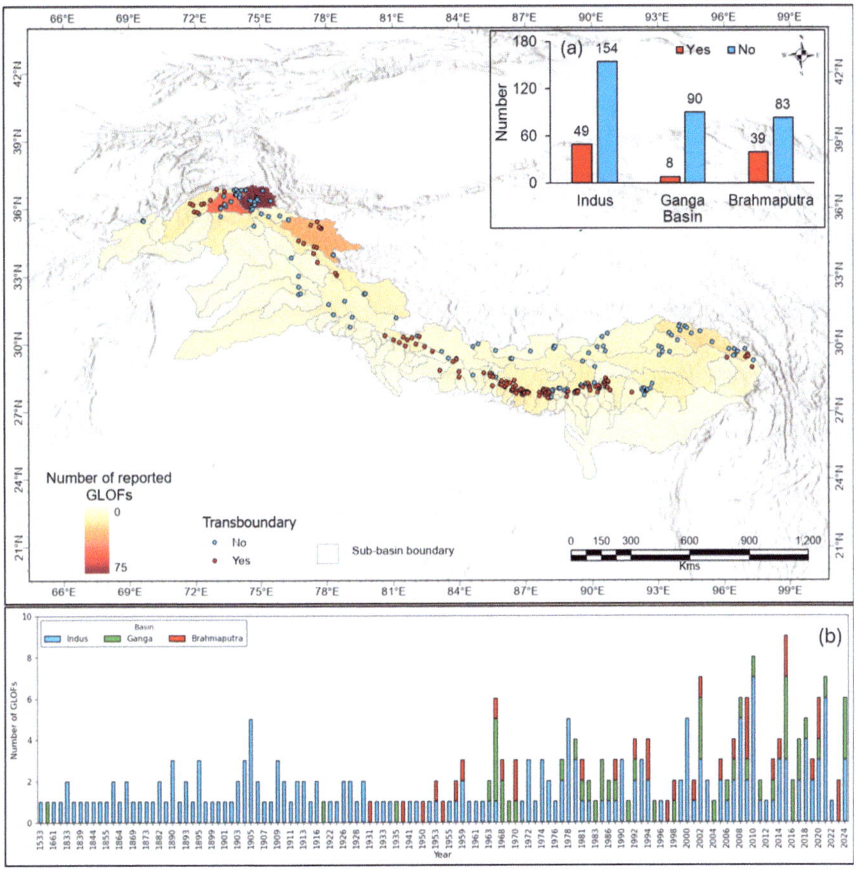

Fig. 1 Map showing the distribution of GLOFs in the IGB basins; the red and blue points show whether the GLOF events were transboundary (red) or not (blue). **a** shows the distribution of transboundary GLOF events at the major basin levels. And **b** shows the temporal distribution of glacial lakes between 1533 and 2024 at the major basin level

converter tool in the ArcGIS Pro environment. Once the data was converted, the next step was to overlay the XY point data over the sub-basin shapefile. After overlaying the two layers, the next step was to use the spatial join tool to generate a new XY point layer with updated information about major basins and sub-basins. We also used a heat map tool in the ArcGIS Pro environment to show the density of repeated GLOFs.

4 Results

4.1 Distribution of Historical GLOFs in the IGB River Basins

In total there were 423 GLOF events which occurred in the IGB river basins between 1533 and 2024. Out of these, 423 GLOF events, which occurred in the IGB river basins, the highest number of GLOFs occurred in the Indus River basin (203), followed by the Brahmaputra River basin (122) and the Ganga River basin (98) (Fig. 1). The first recorded GLOF in the Indus River basin was recorded in the year 1533, whereas the first recorded GLOF in the Ganga River basin was recorded in 1560, and in the Brahmaputra River basin, it was recorded in 1931 (Fig. 1). At the sub-basin level, GLOFs were reported in the 53 sub-basins. Out of the 53 sub-basins with reported GLOFs, the Brahmaputra River basin reported GLOFs in 21 sub-basins, followed by the Ganga basin, which reported GLOFs in 20 sub-basins, and the Indus River basin, which observed reported GLOFs in only 12 sub-basins. Among the 53 sub-basins, the Hunza sub-basin (Indus River basin) had the highest number of reported GLOFs (75), whereas 14 sub-basins only reported 1 GLOF event between 1533 and 2022 (Fig. 1).

Furthermore, out of the 423 reported GLOF events, 96 were of transboundary nature, meaning that 96 of these events impacted more than one country. At the major basin level, the Indus River basin had the highest number of transboundary GLOF events (49), followed by the Brahmaputra River basin (39) and the Ganga River basin (8).

4.2 Temporal and Seasonal Changes in the GLOF Frequencies

In total, 423 GLOF events were reported between 1533 and 2024. Between 1533 and 2024, there is a significant shift in the occurrence of reported GLOFs (Fig. 2). Out of the 423 reported GLOF events, information regarding the exact year of the GLOF event is only available for the 273 GLOF events (Fig. 1b). At the major basin level, out of these 273 GLOF events, the Indus River basin had 189, followed by 50 in the Ganga River basin and 34 in the Brahmaputra River basin. Interestingly, although the Brahmaputra River basin (122) had a higher number of total reported GLOFs than the Ganga River basin (98), however the Ganga River basin (50) had a higher number of reported GLOFs with information about the year in which they occurred as compared to the Brahmaputra River basin (34) (Fig. 1b). Furthermore, out of the 273 reported GLOF events with available information regarding the exact year, we found that GLOFs were scarcely reported between 1533 and 1950. Between 1533 and 1950, only 85 GLOF events were reported, whereas since 1950, 188 GLOFs have been reported. This shows a sudden rise in the number of observed GLOF events in the past 70 years. The year 2015 reported the highest number of GLOF events (9).

In fact, between 1533 and 2024, 13 years observed more than 4 GLOF events, and 12 out of these 13 years were between 1950 and 2024 (Fig. 1b). At the major basin level, the Indus River basin recorded the highest GLOF events (10) in 2010, whereas the Ganga River basin recorded the highest number of GLOF events (4) in 2015, and the Brahmaputra River basin recorded the highest number of GLOF events (3) in 2009 (Fig. 1b).

Additionally, the dataset provides information about the month for 148 GLOF events between 1860 and 2024. Information about the month of the GLOF events is of immense importance as it helps with identifying the seasonal patterns, high-risk periods, development of early warning systems and enhancing community preparedness. Out of these 148 reported GLOF events, the highest number of them was reported in July (41) (Fig. 2). At the major basin level, the Indus river basin reported

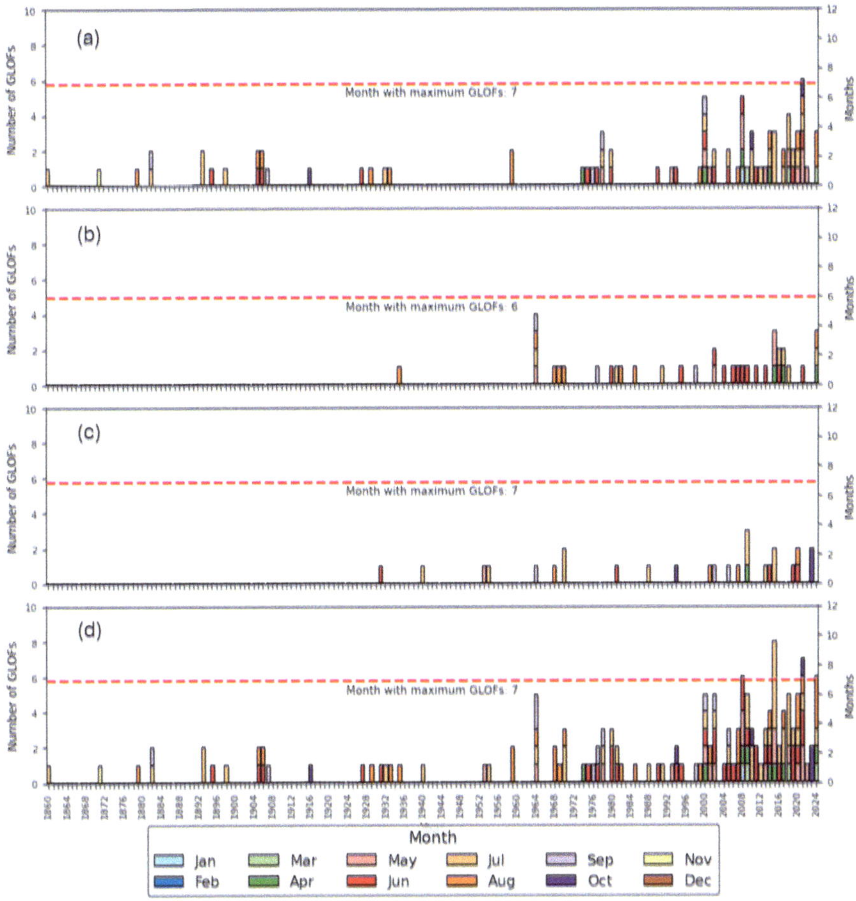

Fig. 2 Stacked bar diagram showing the temporal distribution of historical GLOF events in the IGB river basins, where **a** shows the temporal distribution of reported GLOFs in the Indus River basin, **b** in the Ganga River basin, **c** in the Brahmaputra River basin, and **d** overall

Table 1 Seasonal distribution of GLOFs in the IGB river basins

Season	Months	Number of GLOFs
Winter	December to February	1
Pre-monsoon	March to May	28
Monsoon	June to September	112
Post-monsoon	October to November	7

the highest number of GLOF events (24) in the month of July, the Ganga river basin reported the highest number of GLOF events (12) in the month of June, and the Brahmaputra river basin reported the highest number of GLOF events (10) in the month of July (Fig. 2). Furthermore, to understand the seasonal distribution of reported GLOF events, we distributed the available monthly reported GLOF data (Table 1) into four seasons (winter, pre-monsoon, monsoon, and post-monsoon). We found that the monsoon season (June to September) recorded the highest number of GLOF events (112) between 1860 and 2024, followed by the pre-monsoon season (28), the post-monsoon season (7) and the winter season, which only recorded 1 GLOF event.

4.3 Altitudinal Distribution of Reported GLOFs in the IGB River Basins

The altitude of reported GLOFs varies between 2,562 m asl (Hunza sub-basin of Indus river basin) and 5,982 m asl (Kali Gandaki sub-basin of Ganga river basin). There is a west-to-east gradient in the mean elevation of reported GLOFs (Fig. 3). The overall mean elevation of reported GLOFs is 4432.13 m asl, whereas at the major basin level, the Brahmaputra river basin had the highest mean elevation of GLOFs (5013.71 m asl), followed by the Ganga river basin (4964.26 m asl), and the Indus river basin (3825 m asl). We further divided reported GLOFs into seven elevation zones (Fig. 3a). Among these elevation zones, the maximum of reported GLOFs is observed in the elevation zone of 5000–5500 m asl (107), followed by the elevation zone of 4500–5000 m asl (99), 3000–3500 m asl (60), 4000–4500 m asl (58), 2500–3000 m asl (40), 5500–6000 m asl (31), and 3500–4000 m asl (28) (Fig. 3a). Furthermore, the southern sub-basins had a lower mean elevation of reported GLOFs as compared to the northern sub -basins in the IGB river basin (Fig. 3). At the sub-basin level, the Astor sub-basin (2940 m asl) in the Indus river basin had the lowest mean elevation of reported GLOFs, whereas the Kali Gandaki sub-basin (5717.66) in the Ganga river basin had the highest mean elevation of reported GLOFs (Fig. 3b). Additionally, out of 53 sub-basins, 9 sub-basins had a lower mean elevation than the average elevation of the reported GLOFs, and 44 sub-basins had a higher mean elevation than the average elevation of the reported GLOFs (Fig. 3b). Out of these 9 sub-basins with lower mean elevation than the mean elevation of all the GLOFs, 6 were in the Indus river basin and 3 were in the Ganga river basin (Fig. 3b).

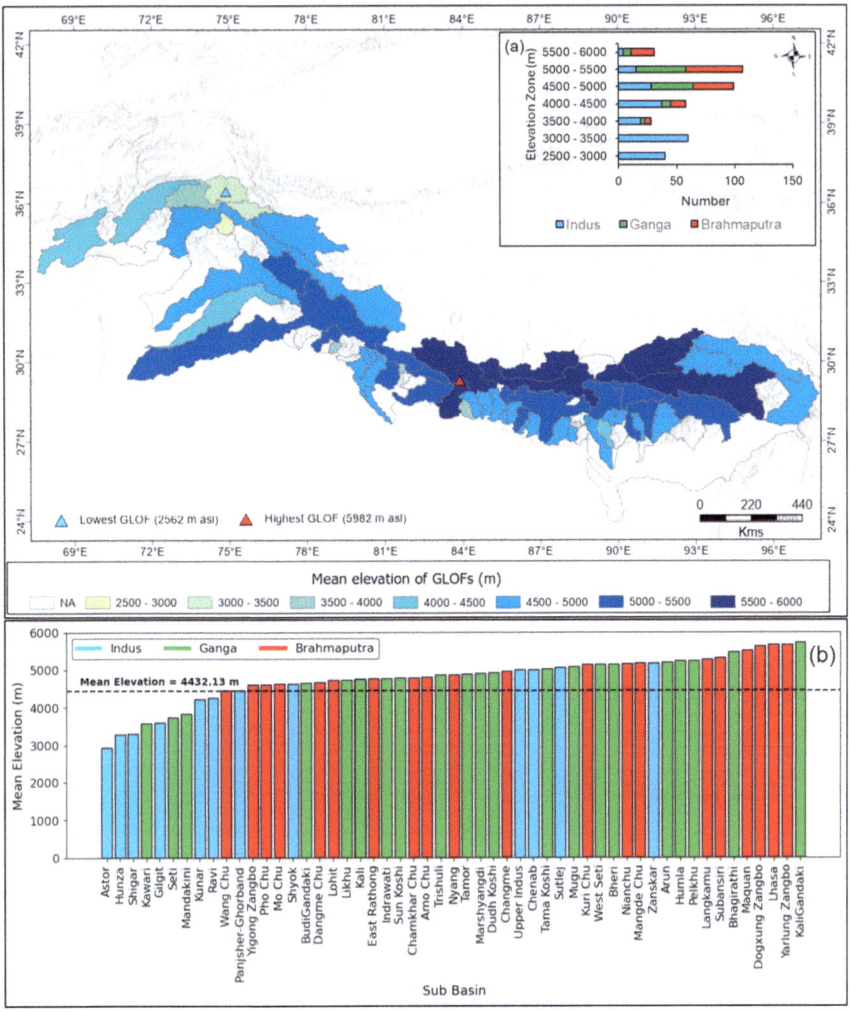

Fig. 3 Distribution of mean elevation of reported GLOFs at the sub-basin level in the IGB river basins. The blue/red triangle is showing the GLOF with the lowest and highest elevation, respectively. Additionally, **a** shows the distribution of reported GLOFs in different elevation zones in the IGB river basins, and **b** shows the distribution of mean elevation using the bar diagrams for the sub-basins

4.4 Type-Wise Distribution of Reported GLOFs in the IGB River Basins

ICIMOD categories reported GLOFs into seven categories based on the dam material (ICIMOD, 2022). These seven categories are moraine dammed, ice dammed, landslide dammed, supraglacial, bedrock, water pockets, and unknown. Out of the 423 reported GLOFs, moraine dammed GLOFs constituted the largest proportion with 220 (~52%), followed by ice dammed GLOFs (n: 106 & p: 25.05%), supraglacial GLOFs (n: 55 & p: ~13%), unknown GLOFs (n: 28 & p: 6.61%), water pockets GLOFs (n: 8 & p: 1.89%), bedrock dammed GLOFs (n: 3 & p: 0.70%), and landslide GLOFs (n: 3 & p: 0.70%) (Fig. 4).

4.4.1 Moraine Dammed GLOFs

About 52% of the reported GLOFs between 1533 and 2024 were from moraine dammed lakes in the IGB river basins. There is a west-to-east gradient in the distribution of reported moraine dammed GLOFs (Fig. 4). At the major basin level, interestingly, the Brahmaputra river basin had the highest number of reported moraine dammed GLOFs (109), followed by the Ganga river basin (77) and the Indus river basin (34). At the sub-basin level, out of 53 sub-basins which reported GLOFs, 49 of them reported moraine dammed GLOFs (Fig. 4). Out of these 49 sub-basins, Yigong Zangbo a sub-basin of the Brahmaputra river basin reported the highest number of moraine dammed GLOFs (14) (Fig. 4). Furthermore, the average elevation of moraine dammed GLOFs is 4935.15 m asl.

4.4.2 Ice Dammed GLOFs

Ice dammed GLOFs were the second most common type of reported GLOF in the IGB river basin. They account for 25.05% of total reported GLOFs. All 106 of the reported ice dammed GLOFs took place in the Indus river basin; there were no reported ice dammed GLOFs in the Ganga and Brahmaputra river basins, respectively. At the sub-basin level, out of 53 sub-basins, only 4 sub-basins reported ice dammed GLOFs, and all of them were in the Indus river basin (Fig. 4). Among these 4 sub-basins, Hunza sub -basin reported the highest number of ice dammed GLOFs (52), followed by the Shyok sub-basin (30), the Gilgit sub-basin (23) and the upper Indus sub-basin (1). The mean elevation of ice dammed GLOFs is 3724.34 m asl.

4.4.3 Supraglacial GLOFs

Supraglacial GLOFs are the third most common type of reported GLOF in the IGB river basin. They account for ~13% of the total reported GLOFs in the IGB river

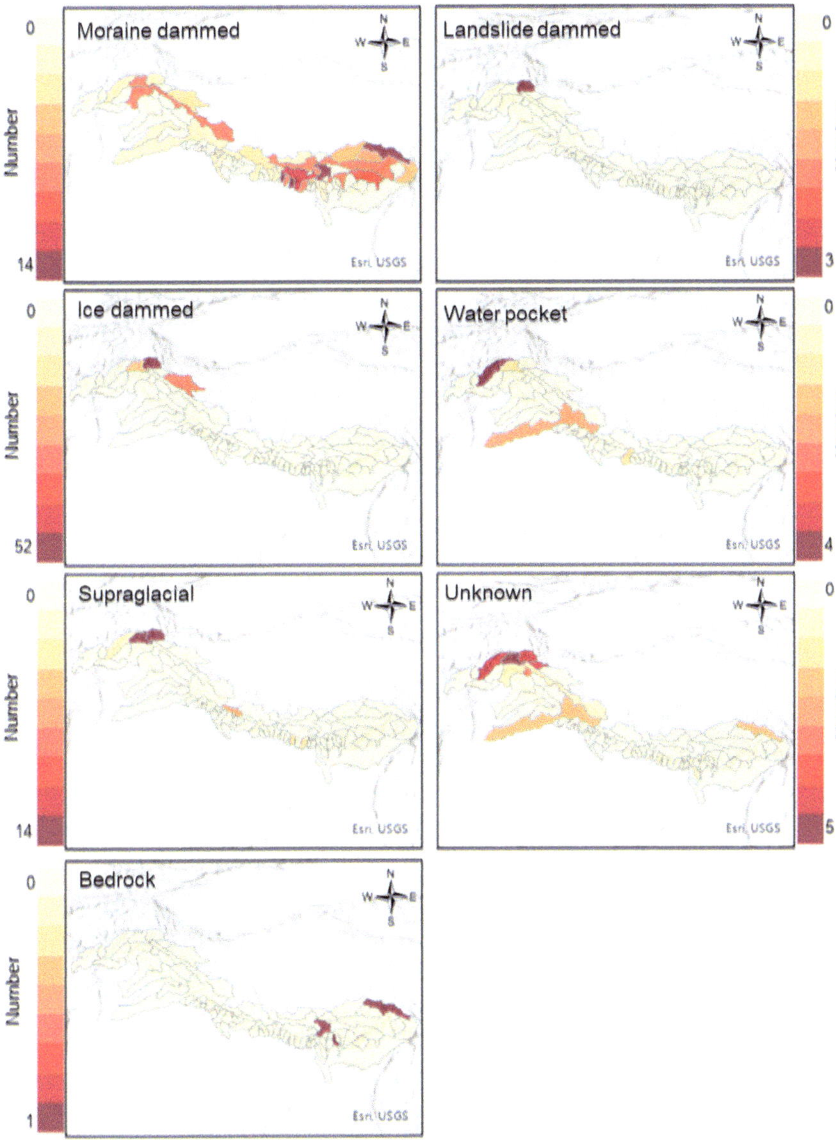

Fig. 4 The subplots show distribution of different types of reported GLOFs at the sub basin level in the IGB river basins

basins (Fig. 4). At the major basin level, out of 55 supraglacial GLOFs, the Indus river basin reported the highest number (30), followed by the Ganga river basin (19) and the Brahmaputra river basin (6). At the sub-basin level, 17 sub-basins reported the supraglacial GLOF events. Out of these 17 sub-basins, the Ganga river basin had the highest number of sub-basins (7) with supraglacial GLOFs, followed by the

Brahmaputra river basin (6) and the Indus river basin (4). Furthermore, among the sub-basins, the Hunza sub-basin in the Indus river basin had the highest number of supraglacial GLOF events (14) (Fig. 4). And the mean elevation of supraglacial GLOF events is 4002.72 m asl.

4.4.4 Unknown GLOFs

Unknown GLOFs were those GLOF events for which there was no information available about the lake type to ICIMOD. ICIMOD reported 28 (6.61%) of unknown GLOFs, putting them forth on the list (Fig. 4). At the major basin level, 23 of these events took place in the Indus river basin, followed by 4 in the Brahmaputra river basin, and 1 in the Ganga river basin. At the sub-basin level, 11 sub-basins reported unknown GLOFs, out of which 7 sub-basins were in the Indus river basin, 3 in the Brahmaputra river basin and 1 in the Ganga river basin. Among the sub-basins, the Gilgit sub-basin (5) in the Indus river basin reported the highest number of unknown GLOFs (Fig. 4). The mean elevation of the unknown GLOF events is 3972 m asl.

4.4.5 Water Pocket GLOFs

Water pocket GLOFs were defined by ICIMOD as GLOF events caused by stored water within the glacier (ICIMOD, 2022; Shrestha et al., 2023). There were only 8 (1.89%) reported events of water pockets GLOFs in IGB river basins between 1533 and 2024 (Fig. 4). Out of these 8 events, at the major basin level, 7 took place in the Indus river basin and 1 took place in the Ganga river basin. The Brahmaputra river basin did not report any water pocket GLOF event. At the sub-basin level, only 4 sub-basins reported the water pocket GLOFs (Fig. 4). Among these 4 sub-basins, the Kunar sub-basin (4) in the Indus river basin reported the highest number of water pocket GLOFs, followed by the Sutlej sub-basin (2), the Gilgit sub-basin (1) and the Kali Gandaki sub-basin (1). The mean elevation of water pocket GLOFs is 4356.62 m asl.

4.4.6 Bedrock GLOFs

The IGB river basin combined only reported 3 bedrock GLOFs. All three of these GLOFs took place in the Brahmaputra river basin (Fig. 4). The Indus and Ganga river basins did not report any occurrence of bedrock GLOFs. Furthermore, at the sub-basin level, 3 sub-basins (the Yigong Zangbo, Nianchu and the Mangde Chu) in the Brahmaputra river basin reported 1 bedrock event each between 1533 and 2024. The mean elevation of bedrock GLOFs is 5109.66 m asl.

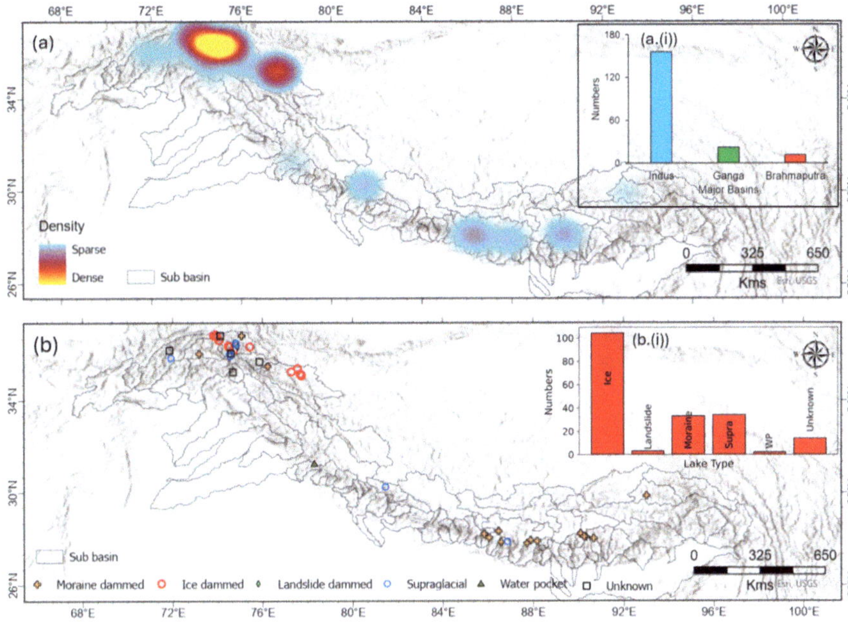

Fig. 5 **a** shows the density distribution of repeated GLOFs at the sub basin level in the IGB river basins and the inset diagram **a**.(**i**) shows the distribution of repeated GLOFs at the major basin level. **b** shows the type wise distribution of repeated GLOFs at the sub basin level in the IGB river basins and the inset diagram **b**(**i**) shows the distribution of different types of repeated GLOFs events using bars

4.4.7 Landslide Dammed GLOFs

The landslide dammed GLOFs also reported only 3 events between 1533 and 2024 in the IGB river basins (Fig. 4). All three of these events took place in the Indus river basin, whereas the Ganga and Brahmaputra river basins did not report any landslide dammed GLOFs. At the sub-basin level, the Hunza sub-basin in the Indus river basin reported all three of the landslide dammed GLOFs. The mean elevation of the landslide dammed GLOFs is 4244 m asl (Fig. 3).

4.5 Repeated GLOFs and Their Characteristics in the IGB River Basins

Repeated GLOF events are those GLOFs which originated from the same lake, however within a specific time interval. Out of the 423 reported GLOF events, 190 were of repeated nature. At the major basin level, the Indus river basin reported the highest number of repeated GLOFs (156), followed by the Ganga river basin (22) and the Brahmaputra river basin (12) (Fig. 6a). Furthermore, only 16 sub-basins

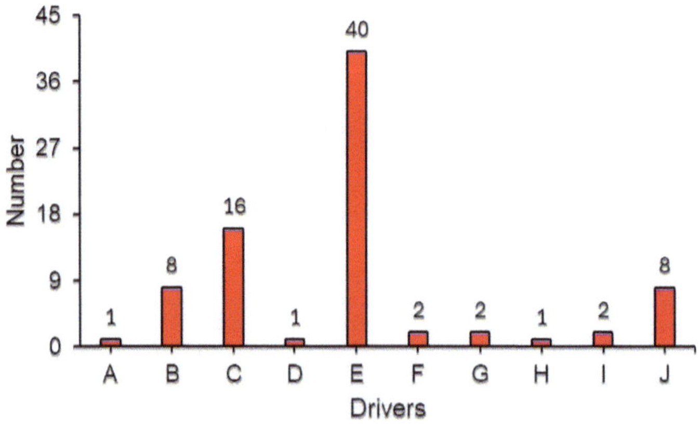

Fig. 6 Bar diagram showing the different drivers of reported GLOFs (n: 81), where A represents debris flow, B represents high temperature, C represents intense rainfall, D represents landslide, E represents ice avalanche, F represents water level rise, G represents ice calving, H represents overflow water from upper lake, I represent rockfall and J represents more than one driver causing the reported GLOF

reported repeated GLOFs in the IGB river basins. Out of these 16 sub-basins, 7 lie in the Indus river basin, 5 in the Ganga river basin, and 4 in the Brahmaputra river basin. Among these 16 sub-basins, the Hunza sub-basin in the Indus river basin reported the highest number of repeated GLOF events (72) (Fig. 6a). The mean elevation of repeated GLOFs is 3887.20 m asl, which is relatively lower than the average elevation of GLOFs in the IGB river basins (4432.13 m asl). At the major basin level, the Ganga river basin had the highest mean elevation of repeated GLOFs (5011.04 m asl), followed by the Brahmaputra river basin (4921.66 m asl) and the Indus river basin (3649.14 m asl).

Additionally, in terms of different types of repeated GLOF events, ice dammed had the highest number of repeated GLOFs (n: 104 & p: 54.73%), followed by supraglacial (n: 34 & p: 17.89%), moraine dammed (n: 33 & p: 17.36%), unknown (n: 14 & p: 7.36%), landslide (n: 3 & p: 1.57%), water pocket (n: 2 & p: 1.05%) and bedrock dammed did not experience any repeated GLOFs (Fig. 6b). In terms of the mean elevation of the different types of repeated GLOFs, moraine dammed repeated GLOFs had highest mean elevation (4744.24 m asl), followed by water pockets (4578 m asl), landslide dammed (4244 m asl), ice dammed (3733.23 m asl), supraglacial (3635.79 m asl) and unknown (3446.28 m asl).

5 Discussion

5.1 *Causes of Historical GLOF Events in the IGB River Basins*

Predominant causes of GLOF events can be divided into two main categories: (i) dam failure and (ii) overtopping waves (ICIMOD, 2022; Nie et al., 2018; Shrestha et al., 2023). However, there can be several drivers of these causes (ICIMOD, 2022; Shrestha et al., 2023). Furthermore, overtopping waves can further lead to dam failures in extreme cases (ICIMOD, 2022; Shrestha et al., 2023). Primarily dam failure is driven by the inability of glacial lake dams to hold back the water within the lake, which can be caused by several factors such as an increase in the hydrostatic pressure on the natural dam, piping within the natural dam, seismic activity, degradation of the natural dam due to the melting of ice cores within the dam, movement in the natural dam due to movement in glaciers, excessive precipitation, and the sudden release of water from upstream smaller lakes (ICIMOD, 2022; Shrestha et al., 2023). Furthermore, overtopping waves primarily caused by ice and rock avalanches and calving of glaciers can also cause damage to the dam structure, resulting in the GLOF events (Shrestha et al., 2023; Zhang et al., 2024). Several researchers have attempted to identify the causes and mechanisms of GLOFs; however, due to the nearly inaccessible locations and high cost of conducting field-based studies, our knowledge about the causes and mechanisms of the GLOFs is still limited.

The present study, based on the ICIMOD dataset, we reported 423 GLOF events between 1533 and 2024; however, information about the possible driver for the GLOFs is only available for 81 GLOF events. The possible drivers for the reported GLOFs for which GLOF drivers were identified were debris flow, high temperature, intense rainfall, landslides, ice avalanches, water level rise, ice calving, overflowing water from upper lakes, rockfalls and unknown factors in the IGB river basins. Out of the 423 reported GLOFs, the possible cause of the reported GLOF was not identified for 342 glacial lakes. This highlights the huge research gap in understanding the mechanisms of GLOFs in the IGB river basins, which further influences the development of early warning systems and mitigation strategies. Out of the 81 reported GLOFs, for which information about possible GLOF drivers was available, 40 were caused by ice avalanches, followed by 16 by intense rainfall, 8 by high temperature, 2 each were caused by water level rise, ice calving, and rockfall, and 1 each was caused by debris flow, landslide, and overflow water from the upper lake (Fig. 6). Furthermore, 8 reported GLOFs were driven by more than one driver (Fig. 6).

5.2 *Factors Affecting Rise in the Reported GLOFs in the Recent Decades*

It has been established that the number of reported GLOFs has increased in the recent past (Harrison et al., 2018; ICIMOD, 2022). ICIMOD (2022) provides a dataset of 423 reported GLOFs between 1533 and 2024; out of these 423 reported GLOFs, only 114 GLOFs were between 1533 and 1972 (the year in which the Landsat satellite mission was launched), whereas 309 GLOFs were reported between 1972 and 2024 (Fig. 7). This highlights that prior to the early 1970s we had limited information about the reported GLOFs, especially in the IGB river basins of the HKH region, mainly because of limited monitoring capabilities (Shrestha et al., 2023). The advent of remote sensing technologies has allowed researchers to closely monitor the GLOFs in the inaccessible regions like the IGB river basins of the HKH region. Furthermore, between 1970 and 2024, there is no significant trend regarding the growth of the number of reported GLOFs (Fig. 7) (Harrison et al., 2018; Shrestha et al., 2023; Veh et al., 2019). One of the possible reasons for the increase in the numbers of reported GLOF events in the IGB river basins can be the continuous monitoring of the water bodies through the remotely sensed data provided by several satellite missions such as Landsat, IRS, Sentinel, etc. (Shrestha et al., 2023).

However, it is important to highlight that although there is no significant trend regarding the growth of reported GLOFs in the IGB river basins, the threat has increased in the past decades due to the ongoing climate change, rapid melting of glaciers, continued expansion of glacial lakes, high population pressure and increase in the extreme climatic events (Bolch et al., 2019; Kumar et al., 2025; Shrestha et al., 2023; Taylor et al., 2023).

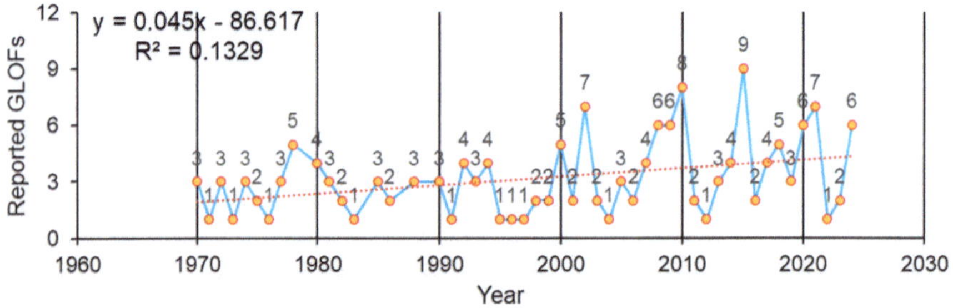

Fig. 7 Trend of reported GLOFs between 1970 and 2024 in the IGB river basins based on ICIMOD (2022)

6 Conclusion

The present study attempted to understand the spatial and temporal distribution of reported GLOFs in the IGB river basins of the HKH region based on the dataset generated by ICIMOD (2022) at the sub-basin level. The analyses found that there is uneven spatial and temporal distribution of reported GLOFs in the IGB river basins. In total, 423 reported GLOF events occurred between 1533 and 2024, out of which the highest were in the Indus river basin (203), followed by the Brahmaputra river basin (122) and the Ganga river basin (98). We found that the number of reported GLOF events has increased in the recent past; however, there is no significant trend. Advancements in the Earth Observation (EO) satellites have greatly helped the reporting of the GLOFs in inaccessible areas of the IGB river basins. Furthermore, ~45% of the reported GLOFs are of repeated nature. These GLOF events have caused huge damage to the human population, downstream infrastructure, and surrounding environment. Under the ongoing climate change, rapid melting of glaciers and continuous expansion of glacial lakes, the number of GLOFs are expected to be increased in the IGB river basins.

Therefore, the present study can help in providing valuable insights in identifying local GLOF hotspots at the sub-basin level, understanding the potential drivers and mechanisms of GLOFs, developing better mitigation strategies, and better community preparedness at the sub-basin level in the IGB river basins.

Acknowledgements The work is supported by doctoral research grant provided by German Academic Exchange Service (DAAD) (funding ID: 57507871). The authors are also thankful towards International Centre for Integrated Mountain Development (ICIMOD) for providing dataset on the reported GLOFs.

References

Bolch, T., Shea, J. M., Liu, S., Azam, F. M., Gao, Y., Gruber, S., et al. (2019). Status and change of the cryosphere in the extended Hindu Kush Himalaya region. In: *The Hindu Kush Himalaya assessment: Mountains, climate change, sustainability and people*, pp. 209–255. https://doi.org/10.1007/978-3-319-92288-1_7

Dimri, A. P., & Chevuturi, A. (2016). Western disturbances-an Indian meteorological perspective. https://doi.org/10.1007/978-3-319-26737-1

Emmer, A., Wood, J. L., Cook, S. J., Harrison, S., Wilson, R., Diaz-Moreno, A., et al. (2022). 160 glacial lake outburst floods (GLOFs) across the Tropical Andes since the Little Ice Age. *Global and Planetary Change, 208*, Article 103722. https://doi.org/10.1016/j.gloplacha.2021.103722

Harrison, S., Kargel, J. S., Huggel, C., Reynolds, J., Shugar, D. H., Betts, R. A., et al. (2018). Climate change and the global pattern of moraine-dammed glacial lake outburst floods. *The Cryosphere, 12*(4), 1195–1209. https://doi.org/10.5194/tc-12-1195-2018

ICIMOD. (2022). GLOF database of High Mountain Asia. ICIMOD. https://doi.org/10.26066/RDS.1973283

ICIMOD. (2023). *Water, ice, society, and ecosystems in the Hindu Kush Himalaya: An outlook.* In: Wester, P., Chaudhary, S., Chettri, N., Jackson, M., Maharjan, A., Nepal, S., Steiner, J. F. [Eds.]. *ICIMOD*. https://doi.org/10.53055/ICIMOD.1028

Krishnan, R., Shrestha, A. B., Ren, G., Rajbhandari, R., Saeed, S., Sanjay, J., et al. (2019). Unravelling climate change in the Hindu Kush Himalaya: rapid warming in the mountains and increasing extremes. In: *The Hindu Kush Himalaya assessment: Mountains, climate change, sustainability and people*, pp. 57–97. https://doi.org/10.1007/978-3-319-92288-1_3

Kumar, A., Mal, S., Schickhoff, U., & Dimri, A. P. (2025). Basin-scale spatio-temporal development of glacial lakes in the Hindukush-Karakoram-Himalayas. *Global and Planetary Change, 245*, Article 104656. https://doi.org/10.1016/j.gloplacha.2024.104656

Lützow, N., Veh, G., & Korup, O. (2023). A global database of historic glacier lake outburst floods. *Earth System Science Data Discussions, 2023*, 1–27. https://doi.org/10.5194/essd-15-2983-2023

Mal, S., Allen, S. K., Frey, H., Huggel, C., & Dimri, A. P. (2021). Sectorwise assessment of glacial lake outburst flood danger in the Indian Himalayan region. *Mountain Research and Development, 41*(1), R1. https://doi.org/10.1659/MRD-JOURNAL-D-20-00043.1

Mool, P. K. (1995). Glacier lake outburst floods in Nepal. *Journal of Nepal Geological Society, 11*, 273–280. https://doi.org/10.3126/jngs.v11i0.32802

Nie, Y., Liu, Q., Wang, J., Zhang, Y., Sheng, Y., & Liu, S. (2018). An inventory of historical glacial lake outburst floods in the Himalayas based on remote sensing observations and geomorphological analysis. *Geomorphology, 308*, 91–106. https://doi.org/10.1016/j.geomorph.2018.02.002

Reynolds, J. M. (1998). High-altitude glacial lake hazard assessment and mitigation: A Himalayan perspective. *Geological Society, London, Engineering Geology Special Publications, 15*(1), 25–34. https://doi.org/10.1144/GSL.ENG.1998.015.01.03

Reynolds, J. M. (1999). Glacial hazard assessment at Tsho Rolpa, Rolwaling, Central Nepal. *Quarterly Journal of Engineering Geology, 32*(3), 209–214. https://doi.org/10.1144/GSL.QJEG.1999.032.P3.01

Sharma, A. K., Singh, S. K., Kulkarni, A. V., et al. (2013). Glacier Inventory in Indus, Ganga and Brahmaputra Basins of the Himalaya. *National Academy Science Letters, 36*, 497–505. https://doi.org/10.1007/s40009-013-0167-6

Shrestha, F., Steiner, J. F., Shrestha, R., Dhungel, Y., Joshi, S. P., Inglis, S., et al. (2023). A comprehensive and version-controlled database of glacial lake outburst floods in High Mountain Asia. *Earth System Science Data, 15*(9), 3941–3961. https://doi.org/10.5194/essd-15-3941-2023

Shugar, D. H., Burr, A., Haritashya, U. K., Kargel, J. S., Watson, C. S., Kennedy, M. C., et al. (2020). Rapid worldwide growth of glacial lakes since 1990. *Nature Climate Change, 10*(10), 939–945. https://doi.org/10.1038/s41558-020-0855-4

Taylor, C., Robinson, T. R., Dunning, S., Rachel Carr, J., & Westoby, M. (2023). Glacial lake outburst floods threaten millions globally. *Nature Communications, 14*(1), 487. https://doi.org/10.1038/s41467-023-36033-x

Teiji, W., & Rothacher, D. (1996). The 1994 Lugge Tsho glacial lake outburst flood, Bhutan Himalaya. *Mountain Research and Development, 16*(1), 77–81. https://doi.org/10.2307/3673897

Veh, G., Korup, O., Roessner, S., & Walz, A. (2018). Detecting Himalayan glacial lake outburst floods from Landsat time series. *Remote Sensing of Environment, 207*, 84–97. https://doi.org/10.1016/j.rse.2017.12.025

Veh, G., Korup, O., von Specht, S., Roessner, S., & Walz, A. (2019). Unchanged frequency of moraine-dammed glacial lake outburst floods in the Himalaya. *Nature Climate Change, 9*(5), 379–383. https://doi.org/10.1038/s41558-019-0437-5

Veh, G., Korup, O., & Walz, A. (2020). Hazard from Himalayan glacier lake outburst floods. *Proceedings of the National Academy of Sciences, 117*(2), 907–912. https://doi.org/10.1073/pnas.1914898117

Wilson, R., Harrison, S., Reynolds, J., Hubbard, A., Glasser, N. F., Wündrich, O., et al. (2019). The 2015 Chileno Valley glacial lake outburst flood, Patagonia. *Geomorphology, 332*, 51–65. https://doi.org/10.1016/j.geomorph.2019.01.015

YAMADA, T. (2000). Glacier lake outburst floods in Nepal. *Journal of the Japanese Society of Snow and Ice*, *62*(2), 137–147. https://doi.org/10.5331/seppyo.62.137

Zhang, G., Carrivick, J. L., Emmer, A., Shugar, D. H., Veh, G., Wang, X., et al. (2024). Characteristics and changes of glacial lakes and outburst floods. *Nature Reviews Earth & Environment, 5*(6), 447–462. https://doi.org/10.1038/s43017-024-00554-w

Zhang, T., Wang, W., & An, B. (2025). A massive lateral moraine collapse triggered the 2023 South Lhonak Lake outburst flood, Sikkim Himalayas. *Landslides, 22*(2), 299–311. https://doi.org/10.1007/s10346-024-02358-x

Regional Warming and Its Impact on Precipitation Patterns in the Indus–Ganges–Brahmaputra Watersheds: Implications for Glacial Lake Outburst Floods

Monisha Chetia and Siddhartha Kumar Lahiri

Abstract The Himalayas are experiencing considerable warming due to global climate change, significantly affecting precipitation patterns. Our analysis of precipitation and temperature trends over the last four decades (1981–2023) and the estimation of climatic extreme indices (Z-statistics of trend analysis derived from Mann–Kendall's S-statistics and cross verified by Sen's slope analysis) in the Indus–Ganges–Brahmaputra (IGB) watersheds reveals important findings. Precipitation is increasing in the Indus basin, with both pre-monsoon and monsoon strengthening while decreasing in the Ganges and Brahmaputra basins. The entire Himalayan belt shows a remarkable positive increment in the daily maximum temperature; however, the western Himalayas show a more prominent temperature rise during the pre-monsoon period, whereas the middle and the eastern Himalayas experience temperature rise during the post-monsoon period. These changes are likely linked to various interconnected climatic and atmospheric factors, such as orographic effects, shifting temperature gradients due to climate change, strengthening Western Disturbances, interactions between the monsoon and mid-latitude circulation, and rising cyclonic activity in the Arabian Sea. The decrease in precipitation in the Ganges and Brahmaputra regions could be attributed to atmospheric circulation changes, including the Quasi-Biennial Oscillation (QBO) and the El Niño Southern Oscillation (ENSO), the uplift of the Shillong Plateau causing a rain shadow effect, and reduced land–ocean temperature gradients. Warming is particularly pronounced in the Eastern Himalayas compared to the central and western regions, as evidenced by an increased maximum of daily maximum temperature (TXx), a minimum of daily maximum temperature (TXn), and a maximum of daily minimum temperature extremes (TNx), alongside a decrease in the Ice Days (ID). These trends may indicate a higher risk of Glacial Lake Outburst Floods (GLOFs) in the future. Compared to the Hindu Kush, Western, and Central Himalayas, the Eastern Himalayas face a greater frequency and intensity of

M. Chetia · S. K. Lahiri (✉)
Department of Applied Geology, Dibrugarh University, Dibrugarh, Assam 786004, India
e-mail: siddharthalahiri@dibru.ac.in

M. Chetia
Department of Physics, Nanda Nath Saikia College, Titabar, Assam 785630, India

© The Author(s), under exclusive license to Springer Nature Switzerland AG 2025

S. Ghosh (ed.), *The Himalaya Dilemma*, Springer Natural Hazards,
https://doi.org/10.1007/978-3-031-95083-4_18

GLOFs. The significant regional warming accelerates glacier melting, contributing to the formation and expansion of glacial lakes. Ongoing monitoring and effective risk management strategies are crucial to mitigating the impacts of these natural hazards.

Keywords Himalayas · Indus–Ganges–Brahmaputra · Climatic extremes · GLOF

1 Introduction

Himalayas, a vast and intricate mountain range, stretches approximately 2500 km from Nanga Parbat in the west to Namche Barwa in the east. This gigantic mountain range is home to 10 of the world's 14 peaks that exceed 8000 m in altitude and 31 peaks rising above 7600 m (Nandargi & Dhar, 2011). This loftiest mountain range consists of a series of narrow mountain chains intersected by several trans-Himalayan rivers and their tributaries, shaping actively this high mountain range.

The Himalayas play a very significant role in hydrological budgeting in the Indus, Ganges, and Brahmaputra watersheds (Fig. 1). The Himalayan–Tibetan region hosts the world's third largest snow and glacier resources after Antarctica and the Arctic and is often referred to as the Third Pole. The climate is significantly varying throughout this unique folded orogen. The western Himalayan regions are colder than the eastern part and, hence, contain a larger proportion of glaciers, while the eastern regions receive more rainfall than snowfall. Many major transboundary rivers in South Asia have been originating from this mountain range. The IGB watershed covers seven nations: Tibet (China), India, Nepal, Bhutan, Afghanistan, Pakistan, and Bangladesh. The Indus watershed covers Tibet (China), Pakistan, Afghanistan and India. The Ganges watershed is distributed across Tibet (China), India, Nepal, and Bangladesh, while the Brahmaputra watershed occupies Tibet (China), India, Bhutan and Bangladesh. The highest proportion of Himalayan glaciers lies in the Indus watershed (~18,495 glaciers covering 21,000 km^2), followed by the Brahmaputra (~11,495 glaciers covering 14,000 km^2) and the Ganges (~7963 glaciers covering 9000 km^2) (Bajracharya & Shrestha, 2011; Qazi et al., 2020).

Two major circulation systems, the Indian Summer Monsoon (ISM) and the Western Disturbances (WD,) control the hydrology of the Himalayas (Bookhagen & Burbank, 2010). The ISM loses its strength as it moves from the eastern to the western Himalayas, while the variation of WD is just the reverse. The WD weakens as it moves from west to east. There are some transition zones in the Indus watershed where precipitation from both the ISM and WD is received (Azam et al., 2016). Due to the orographic barrier, the foothills of the Himalayas receive a larger proportion of precipitation, while the Tibetan plateau receives lesser rainfall because of the rain-shadow effect (Chandel & Ghosh, 2021).

Global warming and climate change are of top significance to the scientific community today. As per the IPCC's (Intergovernmental Panel on Climate Change) sixth report in 2022 (Kikstra et al., 2022), global temperatures are rising due to

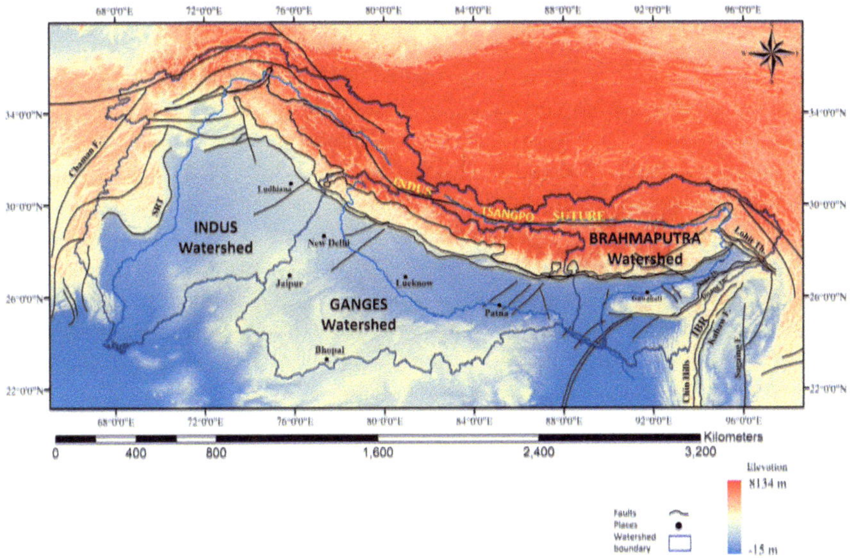

Fig. 1 Shaded relief map of the Himalayas and foreland region prepared using Shuttle Radar Topography Mission (SRTM) DEM (30 m resolution) data (https://earthexplorer.usgs.gov) showing the major tectonic elements. The abbreviations used in the figure are as follows: IBR, Indo Burma Ranges; SRT, Salt Range Thrust; Th., Thrust; F., Fault.

increased human-induced greenhouse gas emissions with a projected increase of 1.5 °C over the next two decades. This temperature rise is expected to bring about changes in temperature extremes across large regions, intensifying heavy precipitation, tropical cyclones, and droughts. Climatic patterns have already been altered as a result of the global temperature (Trenberth, 2011). Globally, the number of cool days and nights is decreasing in contrast to the increasing number of warm days and nights (Sharma & Goyal, 2020). The rising temperature and extreme precipitation events increase the likelihood of Glacial Lake Outburst Floods (Ahmed et al., 2021; Harrison et al., 2018). Global warming and changing climate have a significant influence on water resources, agricultural productivity and overall economic growth (Dimri et al., 2021; Mishra, 2019; Saini et al., 2023).

The Indus–Ganges–Brahmaputra alluvial plain lies south of the Himalayas and is one of the most densely populated regions of the world. Since the hydrological budgeting of the rivers that have been draining the region is entirely dependent on the climate, climate change may have a pronounced effect on the upstream and downstream regions as well. Recent studies reported the effect of climate change in the Himalayas. These studies were focused on rainfall trend variation and changes in ISM (Naidu et al., 2009; Smith & Bookhagen, 2020), strengthening westerly disturbances (Cannon et al., 2014, 2015), intensified storms (Bookhagen & Burbank, 2010; Yao et al., 2012; Singh et al., 2014), rise in average temperature (Barman & Bhattacharjya, 2015; Vaughan et al., 2013), variation in snow-covered area (Barman &

Bhattacharjya, 2015; Bolch et al., 2012; Guo et al., 2022; Singh et al., 2018) and change in snow-water storage resulting in overall change in hydrological budget (Immerzeel et al., 2010; Smith & Bookhagen, 2018; Smith et al., 2017). It has been frequently reported that due to regional climatic changes, the Himalayan glacier is thinning and retreating, and the behaviour of monsoonal precipitation is showing erratic changes (Scherler et al., 2011; Bolch et al., 2012; Patel et al., 2021). Although these studies recognise the impact of climate change on precipitation, temperature, and snowmelt patterns, they provide limited insight into how these effects differ across various regions of the IGB watershed. Moreover, glacier retreats driven by varying climate conditions may not only significantly affect water availability but also pose serious climatic hazards. Therefore, a quantitative analysis of climatic extremes such as extreme temperature and precipitation is essential to assess how these extremes have been changing in recent years.

The hydrology of a region is significantly influenced by two key components: precipitation and snowmelt. Both factors play a crucial role in determining water availability, flow patterns, and the overall water cycle in a watershed. Climate change can lead to alterations in these components, resulting in variations in precipitation patterns and snowmelt timing, which, in turn, can affect the hydrological behaviour of the region. Understanding these changes is essential for assessing the impact of climate change on water resources and managing water systems effectively.

The primary objective of this study is to evaluate the changes in precipitation and temperature within the Indus–Ganges–Brahmaputra (IGB) watershed over the past forty years. This period (1981–2023) has been chosen to analyse long-term trends and identify significant shifts that may have occurred due to climate change. Additionally, the study aims to estimate various climatic indices, such as extreme temperature and precipitation values, to quantify the changes in climatic extremes.

To achieve this, the study analysed two specific time windows (1981–2000 & 2001–2023), allowing for a comparison of climatic conditions at different points in time. This temporal analysis of climatic extremes—such as maximum and minimum temperatures, extreme precipitation events, and other relevant indices helps to provide a clearer picture of how the climate has evolved in this region over the last several decades.

By focusing on these objectives, the study seeks to offer insights into the long-term trends of precipitation and temperature changes and their potential impact on hydrological processes in the IGB watershed.

2 Data and Methodology

Trend analysis of precipitation and temperature from 1981 to 2023 is analysed using remote sensing data utilising the Google Earth Engine platform. The CHIRPS (Climate Hazards Center Infra-Red Precipitation with Station Data) precipitation data was used for trend analysis on a seasonal and annual basis and also for estimating climatic indices (https://developers.google.com/earth-engine/datasets/catalog/UCSB-CHG_CHIRPS_DAILY). This dataset contains daily precipitation estimates with spatial resolution 0.05° × 0.05° (5566 X 5566 square meters) covering latitude 50°N to 50°S and all longitude. It provides precipitation in mm/day for each pixel.

The ERA5-Land Daily Aggregated—ECMWF Climate Reanalysis daily maximum and minimum temperature data from 1981 to 2023 was accessed from https://developers.google.com/earth-engine/datasets/catalog/ECMWF_ERA5_LAND_DAILY_AGGR#description. This global climatic dataset provides daily climatic parameter aggregates with spatial resolution 0.1° × 0.1° (11,132 × 11,132 square meters).

The Mann–Kendall test and Sen's slope estimator have been used as complementary tools for trend analysis and slope determination of changes in precipitation and temperature. These two methods are commonly used in climatic analysis because they are effective at detecting trends and estimating the magnitude of changes in climatic data over time. As non-parametric techniques, they are particularly well-suited for analysing climatic data that does not follow a normal distribution (Gocic & Trajkovic, 2013; Kamal & Pachauri, 2019; Ray et al., 2021). These methods are widely used in the analysis of climatic variables globally (Mir et al., 2015; Dawood, 2017; Patel et al., 2021; Sahu & Gupta, 2020; Choudhury et al., 2021; Venkatesh et al., 2021; Guo et al., 2022; Samanth et al., 2022; Garba and Udokpoh, 2023; Hossen & Evan, 2023).

2.1 Mann–Kendall (MK) Test

The MK test is a monotonic, non-parametric rank-based procedure test used to detect trends in climatic variables such as temperature, precipitation, snow-covered areas, etc. It is a robust test making no assumption of the underlying probability distribution of the data (Kendall, 1975; Mann, 1945). This method tests whether a monotonic trend is present in the data series or not. If there is no trend, then the null hypothesis H_0 is accepted, and if there is a monotonic increasing or decreasing trend, the null hypothesis H_0 is rejected, and the alternative hypothesis H_1 is accepted.

The Mann–Kendall test statistics (S) is estimated by,

$$S = \sum_{i=1}^{n-1} \sum_{j=i+1}^{n} sign(X_j - X_i)$$

$$sign(X_j - X_i) = \left\{ \begin{array}{l} +1,\ if\ (X_j - X_i) > 0 \\ 0,\ if\ (X_j - X_i) = 0 \\ -1,\ if\ (X_j - X_i) < 0 \end{array} \right\}$$

where, X_i and X_j are sequential data values of i-th and j-th position ($j = i + 1$) and n is the data series length. If the sample size $n \geq 10$, then the Z statistics is computed from the variance of the S statistics which gives significance to the trend present in the data series.

$$var(S) = \frac{n(n-1)(2n+5)}{18}$$

$$Z = \left\{ \begin{array}{c} \frac{S-1}{\sqrt{var(S)}}, ifS > 0 \\ 0, ifS = 0 \\ \frac{S+1}{\sqrt{var(S)}}, ifS < 0 \end{array} \right\}$$

The positive value of Z statistics indicates an increasing trend, the negative value indicates a decreasing trend, and the zero value indicates no trend in the time series data. At 99% confidence limit, if $|Z| > 2.576$, at 95% confidence limit, if $|Z| > 1.96$, and at 90% confidence limit, if $|Z| > 1.645$, the null hypothesis is rejected (Patel et al., 2021).

2.2 *Sen's Slope Estimator*

Sen's slope estimator is also a non-parametric method proposed by Sen (1968) used for the climatic data set (Venkatesh et al., 2021). In this method, slopes are computed for all pairs of points, and the median of those slopes is considered as the overall slope. This method can also be used for trend analysis.

2.3 *Climatic Indices*

The Expert Team on Climate Change Detection and Indices (ETCCDI) has suggested 27 core climatic indices (Detailed descriptions are available at https://etccdi.pacificclimate.org/list_27_indices.shtml) to monitor the climatic change. Amongst them, 13 climatic indices (Precipitation-related indices: 5; Temperature-related indices: 8) have been estimated in this work. These are,

i. Consecutive Dry Days (CDD): This index is calculated in days as the maximum number of consecutive days with precipitation less than 1 mm.
ii. Consecutive Wet Days (CWD): It is also calculated in days as the maximum number of consecutive days with precipitation greater than or equal to 1 mm.
iii. R10 (Number of heavy precipitation days): This index gives the annual count of days when precipitation is greater than or equal to 10 mm.
iv. R20 (Number of very heavy precipitation days): This index gives the annual count of days when precipitation is greater than or equal to 20 mm.

v. PRCPTOT (Annual total wet-day precipitation): This index is estimated in mm as the annual total precipitation on wet days when precipitation is greater than or equal to 1 mm.
vi. FD (Frost Days): It is estimated as the annual number of days when daily minimum (TN) temperature is less than 0 °C.
vii. Ice Days (ID): It is estimated as the annual number of days when the daily maximum (TX) temperature is less than 0 °C.
viii. Summer Days (SU): It is estimated as the annual number of days when the daily maximum (TX) temperature is greater than 25 °C.
ix. Tropical Nights (TR): It is estimated as the annual number of days when daily minimum (TN) temperature is greater than 20 °C.
x. TXx (Max Tmax): It is the annual maximum value of daily maximum temperature in degrees Celsius.
xi. TXn (Min Tmax): It is the annual minimum value of daily maximum temperature in degrees Celsius.
xii. TNx (Max Tmin): It is the annual maximum value of daily minimum temperature in degrees Celsius.
xiii. TNn ((Min Tmin): It is the annual minimum value of daily minimum temperature in degrees Celsius.

The climatic indices have been estimated in two-time intervals, W1 (1981–2000) and W2 (2001–2023), and calculated using the Google Earth Engine platform. Changes during the time windows have also been analysed.

3 Results

3.1 *Trend Analysis of Precipitation and Temperature*

3.1.1 Annual Trend of Precipitation and Temperature

The results from the MK test have been presented in Fig. 2. The annual precipitation is showing an increasing trend across most of the Indus watershed with significant positive Z values at 99–95% confidence level in large areas of the watershed, including the western Himalayas. In the Ganges watershed, the valley region shows a similar increasing trend, significant in the southern parts. However, the central Himalayan region exhibits a significant decreasing trend, marked by highly negative Z values. In the Brahmaputra watershed, a significant decrease is observed in the Eastern Himalayan syntaxis and Indo Burma Ranges, while other areas show an insignificant decrease, and some higher Himalayan regions exhibit a significant increase.

Regarding temperature trends, daily maximum temperatures are rising across the Himalayas, with some areas showing significant increases, while parts of the Indus and Ganges valley regions show a non-significant decrease. The daily minimum

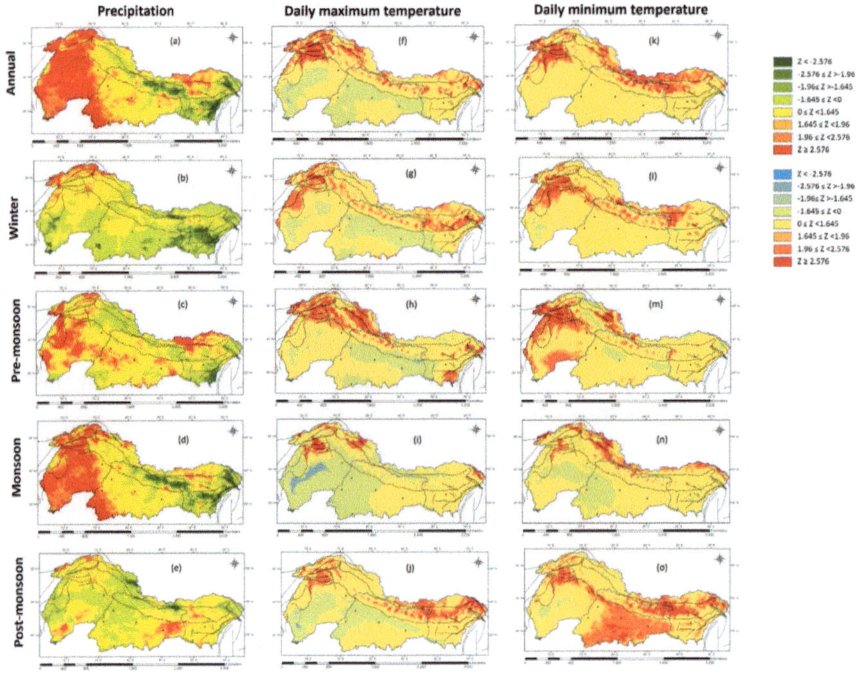

Fig. 2 Spatial distribution of trends in precipitation (**a–e**) and temperature (**f–o**) annually and seasonally computed and plotted from Z-statistics derived from the Mann–Kendall test-based S-statistics

temperatures also show an overall increase, with significant trends in most parts of the Himalayas.

3.1.2 Seasonal Trend of Precipitation and Temperature

For seasonal trend analysis, each year is divided into four seasons based on the climatic features of the months given by the Indian Meteorological Department: Winter season (January and February), Pre-monsoon season (March, April, and May), Monsoon (June, July, August and September) and Post monsoon (October, November and December). The seasonal trends for the spatial distribution of precipitation and temperature are shown in Fig. 2.

Winter Scenario

In winter, annual precipitation decreases in the Ganges and Brahmaputra watersheds, with a significant decline in parts of the Brahmaputra basin. The Indus watershed, however, shows an increasing trend, particularly significant in the higher Himalayan zone beyond the ITS. Both maximum and minimum temperatures are significantly rising in the Himalayan region.

Pre-Monsoon Variability

During the pre-monsoon season, precipitation increases across the Indus and Ganges watersheds, with a significant rise in the Indus basin. The northern Brahmaputra watershed near the ITS shows a significant increase in precipitation, while other areas experience a decline. Maximum temperatures significantly increase in the western and central Himalayas, though the central Ganges basin shows a non-significant decrease. The Brahmaputra watershed observes a significant increase in both maximum and minimum temperatures in some areas.

Monsoonal Changes

In the monsoon season, the Indus watershed shows a significant increase in precipitation, except for the eastern part, which remains non-significantly increasing. The Himalayan region of the Ganges watershed shows a significant decrease in precipitation, while the valley areas show a non-significant increase. The southwestern part of the Ganges basin experiences a significant increase in precipitation. The Brahmaputra watershed shows a significant decline in monsoonal precipitation, except for some higher Himalayan regions where precipitation increases non-significantly. Daily maximum temperatures generally rise across the Himalayas, with significant trends in the western and eastern parts. Minimum temperatures also increase significantly, particularly in the northern Himalayan arc. Daily maximum temperatures decrease in the valley regions of the Indus and western Ganges basins, while the eastern Ganges basin and Brahmaputra watershed show non-significant increases.

Post-Monsoonal Precipitation-Temperature Variability

Post-monsoonal precipitation shows a slight decreasing trend in the eastern Indus and western Ganges watersheds, with the eastern Ganges region experiencing a significant increase. The Brahmaputra watershed overall shows an increasing trend, except for the eastern syntaxis, where a notable decline is observed. Daily maximum temperatures are rising across the Himalayan region, with significant increases in the eastern, central, and western syntaxis areas, while temperatures are decreasing in some regions of the Ganges and Indus watersheds. Daily minimum temperatures are rising throughout the region, with significant increases in the foothills of the Indus and Ganges watersheds, as well as the central and eastern Himalayas.

Figure 3 depicts the results from the Sen's slope estimator. It gives the rate of change of precipitation and temperature per year. A negative value of slope indicates

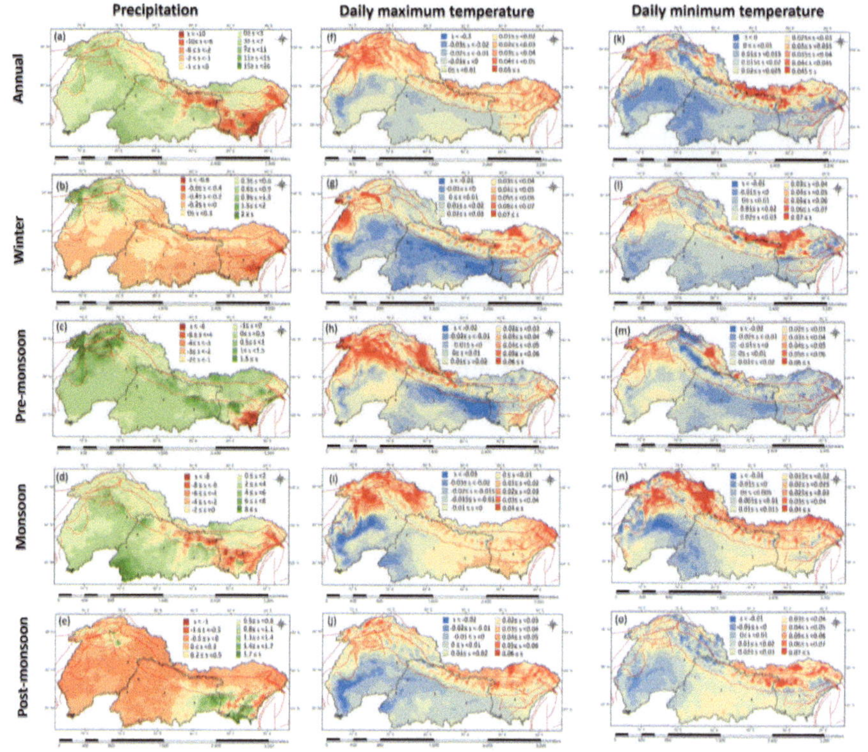

Fig. 3 Trend of precipitation (**a**–**e**) and temperature (**f**–**o**) annually and seasonally based on Sen's slope analysis

a decrease in parameters, and a positive value indicates an increase in parameters. The Himalayas are experiencing a notable rise in annual as well as seasonal temperatures.

3.2 *Variation in Climatic Indices*

Climatic indices were analysed for two time periods: W1 (1981–2000) and W2 (2001–2023). The spatial variations and changes of five precipitation indices across these periods are shown in Fig. 4. The CDD (Consecutive Dry Days) and CWD (Consecutive Wet Days) indices reflect dry and wet spells, respectively. In W1, CDD values are highest in the southern Indus watershed and lower in the northern Himalayan region. By W2, CDD decreases in the valley of the Indus basin, while the Himalayan region experiences an increase. The Ganges watershed also sees a decrease in CDD, especially in the south. The Brahmaputra watershed, however, shows a general increase in CDD. CWD increases in the southern Indus and Ganges basins but decreases in the Himalayan regions. The R10 and R20 indices, which track heavy precipitation days, show a marked decrease in most regions, except for the eastern Indus basin and western Ganges basin, where increases are observed, especially in the Himalayan foothills. The R20 index increases in the foothills of the Himalayas. The PRCPTOT index, indicating total precipitation on wet days, shows mixed trends. The Indus and far western part of the Ganges watersheds show an overall increase in PRCPTOT. The Brahmaputra basin sees a decline, particularly in the south.

Figure 5 illustrates the spatial variation of extreme temperature indices. In the Brahmaputra watershed, TXx (maximum daily maximum temperatures) increase overall, except in the far west and trans-Himalayan region. In the Ganges watershed, TXx decreases, with exceptions in the Himalayan zone and southern regions. In the Indus watershed, TXx decreases in the central and Himalayan regions between the foothills and Karakoram thrust, while other areas show increases. Similar patterns are seen in TXn (minimum daily maximum temperatures). TNx (maximum daily minimum temperatures) increases in Brahmaputra, except in the far south, while in Ganges, the foothills show a decrease and the higher Himalayan regions an increase.

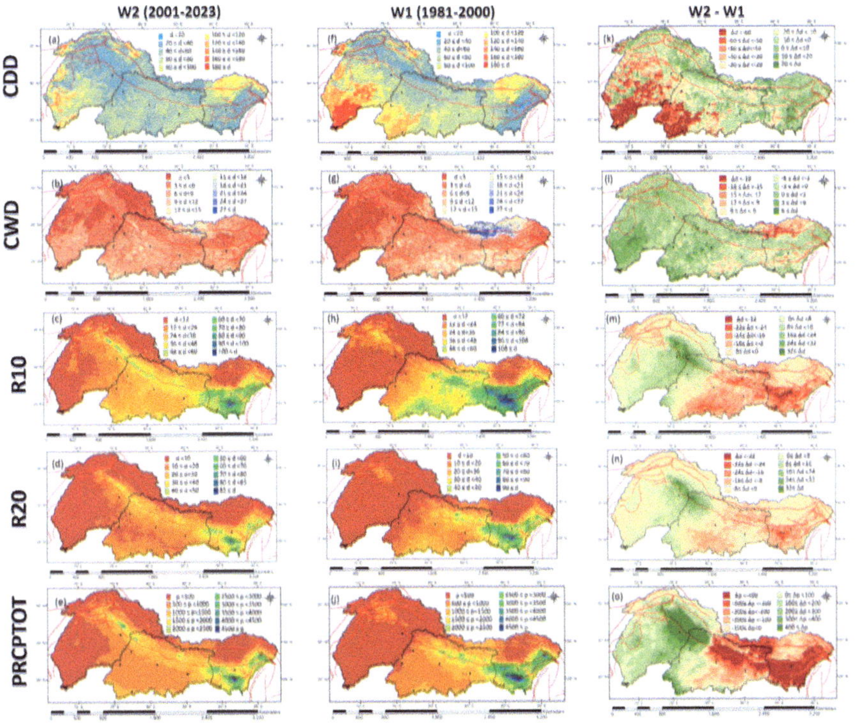

Fig. 4 Spatial distribution of five precipitation extremes during W1 (1981–2000) and W2 (2001–2023). Mean annual Consecutive Dry Days (CDD), Consecutive Wet Days (CWD), Heavy precipitation Days (R10), very heavy precipitation Days (R20) and total precipitation on wet days (PRCPTOTOT) are shown. The difference in indices W2-W1 is shown in figures (**k**–**o**))

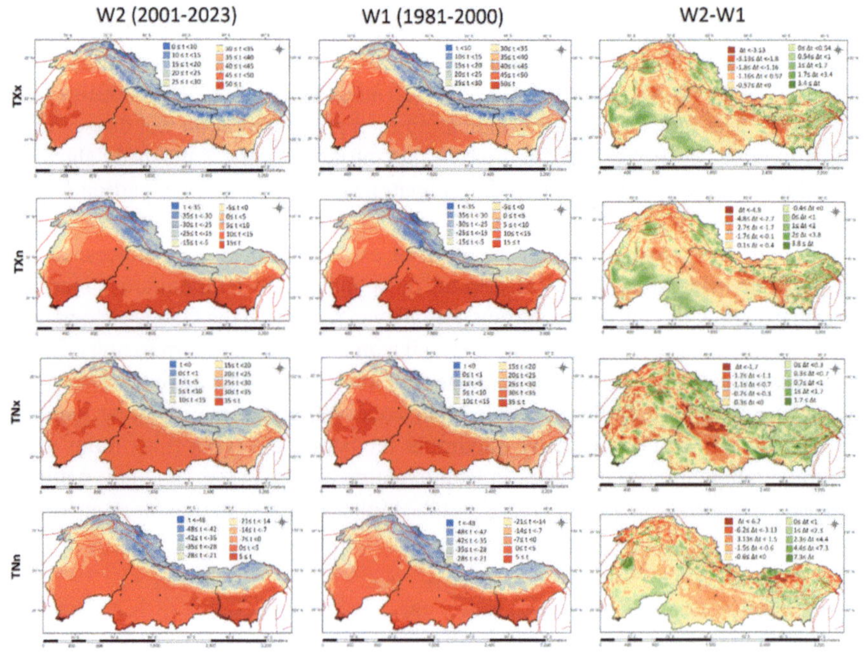

Fig. 5 Spatial distribution of eight temperature extremes during W1 (1981–2000) and W2 (2001–2023). The difference in indices W2–W1 is also shown

The western syntaxis shows a decline in TNx, while the higher Himalayan regions of Indus show an increase. TNn (minimum daily minimum temperatures) generally decrease across the region, except in some Himalayan areas. Frost days (FD) decrease along the southern rim of the Himalayas but increase in the higher Himalayan regions. Ice days (ID) show similar trends, with a decline in the western Brahmaputra's higher Himalayan region. Summer days (SU) increase in the foothills and northern Ganges and Brahmaputra basins but decrease in the foothills of the Indus. Tropical nights (TR) are increasing along the Himalayan foothills.

4 Discussion

The Himalayas are experiencing a notable shift in climatic conditions. The annual and seasonal temperature trend is increasing significantly in the Himalayas. The precipitation trend shows a significant variation from west to east. In the Indus watershed covering the western Himalayas, the precipitation is increasing in contrast to the Ganges and Brahmaputra watersheds covering the central and eastern Himalayas, respectively. A combination of different climatic and atmospheric factors might be affecting these drastic variations.

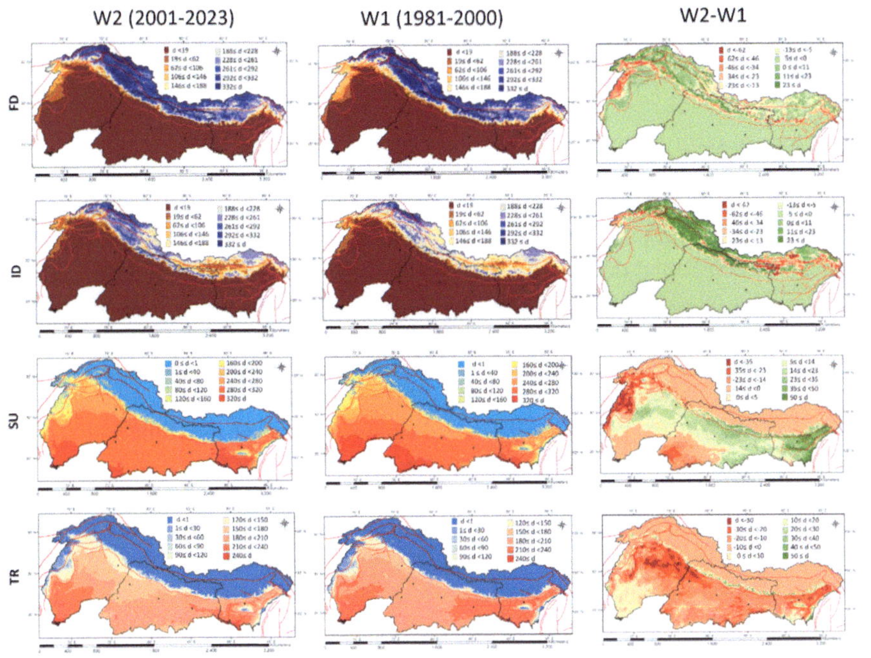

Fig. 5 (continued)

4.1 Changing Characteristics of Precipitation Extremes

The orographic effect is a primary factor responsible for the spatial variation of precipitation. The western Himalayas' complex topography influences precipitation patterns and other factors (Singh & Kumar, 1997). Change in temperature gradient induced by climate change is another reason for increased precipitation. Due to global warming, the enhanced surface warming over the eastern Tibetan plateau compared to the western side enhances the east–west temperature gradient, which is increasing the precipitation extremes (Krishnan et al., 2019). Western disturbances are significant contributors to non-monsoonal precipitation. Due to increased human-induced GHG emissions, the temperature gradient is changing across the Tibetan plateau, which further led to the enhanced intensity of WD and orographic precipitation (Krishnan et al., 2019). The weakening of the southwest monsoon circulation and increased activity of transient upper-air westerly troughs have enhanced moisture supply by southerly winds and increased extreme precipitation events in the western Himalayas (Vellore et al., 2016; Priya et al., 2017).

4.2 Uneven Sea Surface Temperature Variation, Monsoonal Intensity and Cyclones

The Tropical cyclonic activity is generally higher in the Bay of Bengal than in the Arabian Sea because the shallow and wider Bay of Bengal favours the increased sea surface temperature (SST) and thus the cyclone generation. But recently (2001–2019), the frequency of occurrence of cyclonic storms in the Arabian Sea has increased by 52% and decreased by 8% in the Bay of Bengal (Tiwari et al., 2022). This is again related to the abnormal warming of the Arabian Sea. The SST in the Arabian Sea has increased by 1.4 °C, while in the Bay of Bengal, it has increased by 0.8 °C recently. As the temperature rises, hot air moves upward, creating pressure depressions and generating cyclones under favourable atmospheric conditions.

In contrast to increased precipitation in the Indus watershed, the Himalayan region of the Ganges watershed and Brahmaputra watershed experienced a decrease in precipitation. Changes in atmospheric circulations, such as Quasi-Biennial Oscillations (QBO) and El Nino Southern Oscillation (ENSO), might be affecting the monsoon circulations (Shrestha et al. 2018). Also, the upliftment of the Shillong plateau has created a rain shadow effect, which further led to decreased precipitation in eastern Bhutan (Grujic et al., 2018).

Land ocean thermal gradient significantly affects the Indian Summer Monsoon circulation. A high gradient attracts more moisture-laden air from the sea, while if the gradient is low, the circulation will be weaker. The western Indian Ocean has had a warming of 1.2 °C in the last century (Roxy et al., 2014, 2015). A significant decreasing trend for monsoonal rainfall in the Ganges and Brahmaputra watersheds is also reported by Roxy et al. (2015) due to the increase in Indian Ocean warming, resulting in a decrease in land–ocean thermal gradient. In the Arctic region, the enhanced land-sea temperature gradient is responsible for the intensified Arctic cyclones (Day & Hodges, 2018).

Moreover, the Sea Surface Temperature (SST) plays a significant role in monsoon intensity. During the last decade, a significant temperature difference has been occurring between the Arabian Sea and the Bay of Bengal (Roxy et al., 2015). This anomalous warming of the Arabian Sea reduces the land-sea thermal contrast, thus suppressing the monsoon circulation. Due to this, less moisture circulates to the Indian subcontinent, leading to decreased rainfall (Roxy et al., 2015; Mishra et al., 2020). This rapid warming of the northern Arabian Sea is responsible for the threefold rise of extreme rainfall events in Central India and Pakistan (Roxy et al., 2017).

4.3 Impacts Due to Changing Temperature Extremes and GLOFs

The analysis of temperature extremes revealed that the Himalayas are supposed to experience serious impacts due to global warming, particularly in the eastern regions. Compared to the western and central Himalayas, the Eastern Himalayas are experiencing significant warming, as reflected by the increased TXx, TXn, and TNx while ID is decreasing. A decrease in Snow-covered areas in the Brahmaputra watershed was reported in Barman and Bhattacharjya (2015) in the period from 2002 to 2012 due to global warming. These signs might be alarming signals for future Glacial Lake Outburst Floods (GLOF). Compared to the Hindu-Kush, Western and central Himalayas, the Eastern Himalayas experience a higher frequency and magnitude of GLOFs. This region is identified as a hotspot for GLOF hazards, with a higher rate of historical lake outbursts compared to other Himalayan regions (Veh et al., 2020). The significant regional warming accelerates glacier melting, leading to the formation and expansion of glacial lakes. The geomorphological factors, such as the presence of unstable moraines and steep topography, contribute significantly to the enhanced frequency of GLOFs (Dubey & Goyal, 2020; Ahmed et al., 2021; Islam and Patel, 2022).

5 Conclusion

The temporal and spatial variations in precipitation and temperature over the last 40 years in the Indus, Ganges, and Brahmaputra watersheds reveal significant climatic impacts on the Himalayan region, particularly in terms of water resources, climate patterns, and natural hazards. These watersheds are not only critical for millions of people dependent on their water but also sensitive indicators of broader climate change trends. The analysis highlights distinct trends in precipitation and temperature changes that have important implications for the hydrology and natural hazards in these regions.

In the Indus watershed, annual average precipitation has shown an increasing trend, which suggests that the region is experiencing more rainfall over time. This increase in precipitation is linked to changes in atmospheric conditions, such as the warming of the Arabian Sea, which is believed to enhance cyclonic activity in the region. As a result, the monsoon season in the Indus watershed has been receiving more rainfall, possibly due to stronger moisture-laden winds from the Arabian Sea and changes in atmospheric circulation patterns.

Conversely, the Ganges and Brahmaputra watersheds have experienced a decrease in precipitation, particularly during the monsoon season. This reduction in monsoonal rainfall can be attributed to a weakening of the Indian Summer Monsoon, which has been linked to a decrease in the land–ocean thermal gradient. The weakening thermal gradient disrupts the normal circulation patterns that drive the monsoon, resulting in less intense or shortened monsoon rains in these watersheds. Such changes in precipitation patterns could have significant implications for water availability, agriculture, and ecosystem health in these regions.

The Eastern Himalayas, in particular, are especially vulnerable to the effects of climate change, with increasing temperatures contributing to accelerated glacial melting. This warming, combined with the significant expansion of glacial lakes, is heightening the risk of Glacial Lake Outburst Floods (GLOFs) in the region. GLOFs occur when a glacial lake, often formed by melting glaciers, breaches its natural dam, releasing large amounts of water downstream. In the Eastern Himalayas, the frequency and magnitude of GLOFs are higher compared to other parts of the Himalayas, largely due to the combination of glacial melting, rapid glacial lake expansion, and the region's unique geomorphological characteristics, which create conditions conducive to such catastrophic events.

These changes not only pose direct threats to local populations but also disrupt the hydrological systems, leading to risks such as flooding, landslides, and changes in river flow patterns. Continuous monitoring of climatic trends, glacial behaviour, and lake dynamics is crucial to understanding the evolving risks and enabling timely responses. Effective risk management strategies, including early warning systems and climate adaptation plans, are essential to mitigate the potential impacts of GLOFs and other climate-induced hazards in the region. By improving understanding and preparedness, the vulnerability of communities and infrastructure in the Himalayas can be reduced while preserving the vital ecosystem services that these watersheds provide.

Acknowledgements The first author thanks the Department of Science and Technology, Ministry of Science and Technology, Government of India (DST, India), for granting the INSPIRE fellowship that supported this research. Gratitude is extended to various web sources for providing free access to essential data. Further, the authors would like to thank the Department of Applied Geology, Dibrugarh University, for their invaluable assistance and unwavering support throughout this study.

Data Availability Data used in the present work from various sources are sincerely acknowledged. The list is given below: For precipitation data: The CHIRPS (Climate Hazards Center Infra-Red Precipitation with Station Data), Climate Hazards Center, University of California, Santa Barbara, CA 93,106. Source: https://developers.google.com/earth-engine/datasets/catalog/UCSB-CHG_CHIRPS_DAILY. For climatic parameter aggregates: Source: https://developers.google.com/earth-engine/datasets/catalog/ECMWF_ERA5_LAND_DAILY_AGGR

Nomenclature

Temperature terms

TX	Daily maximum temperature
TN	Daily minimum temperature
TXx	The maximum of daily maximum temperature in a year
TXn	The minimum of daily maximum temperature in a year
TNx	A maximum of daily minimum temperature in a year
TNn	A minimum of daily minimum temperature in a year
FD	Frost days (number of days in a year when TN < 0 °C)
ID	Ice days (number of days in a year when TX < 0 °C)
SU	Summer days (number of days in a year when TX > 25 °C)

SST	Sea Surface Temperature
TR	Tropical nights (number of days when TN > 20 °C)

Precipitation terms

CDD	Consecutive dry days (The maximum number of consecutive days with precipitation <1 mm)
CWD	Consecutive wet days (The maximum number of wet days with precipitation ≥ 1 mm)
R10	Number of rainy days in a year with precipitation ≥ 10 mm
R20	Number of rainy days in a year with precipitation ≥ 20 mm
PRCPTOT	Total number of annual wet-day precipitation ≥ 1 mm

Geoscientific terms

DEM	Digital Elevation Map
ENSO	El Niño Southern Oscillation
GHG	nGreen House Gases
GLOF	Glacial Lake Outburst Flood
IBR	Indo-Burman Range
IGB	Indus–Ganges–Brahmaputra
ISM	Indian Summer Monsoon
ITS	Indus Tsangpo Suture
SRT	Salt Range Thrust
SRTM	Shuttle Radar Topography Mission
QBO	Quasi-Biennial Oscillation
WD	Western Disturbance

Institutional terms

CHG	Climate Hazard Group
CHIRPS	Climate Hazards Center Infra-Red Precipitation with Station Data
ECMWF	European Centre for Medium-Range Weather Forecasts
ERA5	Fifth generation ECMWF atmospheric reanalysis of the global climate covering the period from January 1940 to present
ETCCDI	Expert Team on Climate Change Detection and Indices
IPCC	Intergovernmental Panel on Climate Change
UCSB	University of California, Santa Barbara

References

Ahmed, R., Wani, G. F., Ahmad, S. T., Sahana, M., Singh, H., & Ahmed, P. (2021). A review of glacial lake expansion and associated glacial lake outburst floods in the Himalayan region. *Earth Systems and Environment, 5*(3), 695–708.

Azam, M. F., Ramanathan, A. L., Wagnon, P., Vincent, C., Linda, A., Berthier, E., Sharma, P., Mandal, A., Angchuk, T., Singh, V. B., & Pottakkal, J. G. (2016). Meteorological conditions, seasonal and annual mass balances of Chhota Shigri glacier, western Himalaya, India. *Annals Glaciology, 57*(71), 328–338.

Bajracharya, S.R., Shrestha, B.R. (2011). The status of glaciers in the Hindu Kush-Himalayan region. In: *International Centre for Integrated Mountain Development (ICIMOD)*

Barman, S., & Bhattacharjya, R. K. (2015). Change in snow cover area of Brahmaputra River basin and its sensitivity to temperature. *Environmental Systems Research, 4*(16), 1–10. https://doi.org/10.1186/s40068-015-0043-0

Bolch, T., Kulkarni, A., Kääb, A., Huggel, C., Paul, F., Cogley, J. G., Frey, H., Kargel, J. S., Fujita, K., Scheel, M., Bajracharya, S., & Stoffel, M. (2012). The state and fate of Himalayan glaciers. *Science, 336*(6079), 310–314.

Bookhagen, B, Burbank, D. W. (2010). Toward a complete Himalayan hydrological budget: spatiotemporal distribution of snowmelt and rainfall and their impact on river discharge. *Journal of Geophysical Research: Earth Surface, 115*(F03019). https://doi.org/10.1029/2009JF001426

Cannon, F., Carvalho, L., Jones, C., Bookhagen, B. (2014). Multi-annual variations in winter westerly disturbance activity affecting the Himalaya. *Climate Dynamics*, 1–15

Cannon, F., Carvalho, L.M., Jones, C., Norris, J. (2015). Winter westerly disturbance dynamics and precipitation in the western Himalaya and Karakoram: a wave-tracking approach. *Theoretical Applied Climatology*, 1–18

Chandel, V.S., Ghosh, S. (2021). Components of Himalayan River flows in a changing climate. *Water Resources Research,57*: e2020WR027589. https://doi.org/10.1029/2020WR027589

Choudhury, A., Yadav, A. C., & Bonafoni, S. (2021). A response of snow cover to the climate in the northwest Himalaya (NWH) using satellite products. *Remote Sensors, 13*, 655.

Dawood, M. (2017). Spatio-statistical analysis of temperature fluctuation using Mann-Kendall and Sen's slope approach. *Climate Dynamics, 48*(3), 783–797.

Day, J. J., & Hodges, K. I. (2018). Growing land-sea temperature contrast and the intensification of Arctic cyclones. *Geophysical Research Letters, 45*, 3673–3681. https://doi.org/10.1029/2018GL077587

Dimri, A. P., Allen, S., Huggel, C., Mal, S., Ballesteros-Cánovas, J. A., Rohrer, M., et al. (2021). Climate change, cryosphere and impacts in the Indian Himalayan region. *Current Science, 120*, 1012–1012.

Dubey, S., Goyal, K. (2020). Glacial lake outburst flood hazard, downstream impact, and risk over the Indian Himalayas. *Water Resources Research, 56*(4): p.e2019WR026533

Garba, H., & Udokpoh, U. U. (2023). Analysis of trend in meteorological and hydrological time-series using Mann–Kendall and Sen's slope estimator statistical test in Akwa Ibom state, Nigeria. *International Journal of Environment and Climate Change, 13*(10), 1017–1035.

Gocic, M., & Trajkovic, S. (2013). Analysis of changes in meteorological variables using Mann–Kendall and Sen's slope estimator statistical tests in Serbia. *Global and Planetary Change, 100*, 172–182.

Grujic, D., Govin, G., Barrier, L., Bookhagen, B., Coutand, I., Cowan, B., Hren, M. T., & Najman, Y. (2018). Formation of a rain shadow: O and H stable isotope records in authigenic clays from the Siwalik Group in eastern Bhutan. *Geochemistry, Geophysics, Geosystems, 19*(9), 3430–3447.

Guo, R. Y., Ji, X., Liu, C. Y., Liu, C., Jiang, W., & Yang, L. Y. (2022). Spatiotemporal variation of snow cover and its relationship with temperature and precipitation in the Yarlung Tsangpo-Brahmaputra River Basin. *Journal of Mountain Science, 19*(7), 1901–1918.

Harrison, S., Kargel, J. S., Huggel, C., Reynolds, J., Shugar, D. H., Betts, R. A., Emmer, A., Glasser, N., Haritashya, U. K., Klimeš, J., & Reinhardt, L. (2018). Climate change and the global pattern of moraine-dammed glacial lake outburst floods. *The Cryosphere, 12*(4), 1195–1209.

Hossen, M. S., & Evan, R. I. (2023). Assessing historical climate trends in Dhaka City: A multivariate analysis using Mann-Kendall and Sen's slope method. *International Journal of Climate Research, 7*(1), 24–45.

Immerzeel, W. W., Van Beek, L. P., & Bierkens, M. F. (2010). Climate change will affect the Asian water towers. *Science, 328*, 1382–1385.

Islam, N., & Patel, P. P. (2022). Inventory and GLOF hazard assessment of glacial lakes in the Sikkim Himalayas, India. *Geocarto International, 37*(13), 3840–3876.

Kamal, N., & Pachauri, S. (2019). Mann-Kendall, and Sen's slope estimators for precipitation trend analysis in north-eastern states of India. *International Journal of Computer Applications., 177*(11), 7–16.

Kendall, M.G. (1975). Rank correlation methods. 4th ed. Charles Griffin London

Kikstra, J. S., Nicholls, Z. R., Smith, C. J., Lewis, J., Lamboll, R. D., Byers, E., Sandstad, M., Meinshausen, M., Gidden, M. J., Rogelj, J., & Kriegler, E. (2022). The IPCC Sixth Assessment Report WGIII climate assessment of mitigation pathways: From emissions to global temperatures. *Geoscientific Model Development, 15*(24), 9075–9109.

Krishnan, R., Sabin, T. P., Madhura, R. K., Vellore, R. K., Mujumdar, M., Sanjay, J., Nayak, S., & Rajeevan, M. (2019). Non-monsoonal precipitation response over the Western Himalayas to climate change. *Climate Dynamics, 52*, 4091–4109.

Mann HB (1945) Nonparametric tests against trend. *Econometrica: Journal of the Econometric Society*, 245–259

Mir, R. A., Jain, S. K., Saraf, A. K., & Goswami, A. (2015). Decline in snowfall in response to temperature in Satluj basin, western Himalaya. *Journal of Earth System Science, 124*, 365–382.

Mishra, A.K. (2019). Quantifying the impact of global warming on precipitation patterns in India. *Meteorological Applications, 26*, 153–160. https://doi.org/10.1002/met.1749

Mishra, A. K., Dwivedi, S., & Das, S. (2020). Role of Arabian Sea warming on the Indian summer monsoon rainfall in a regional climate model. *International Journal of Climatology, 40*(4), 2226–2238.

Naidu, C.V., Durga Lakshmi, K., Muni Krishna, K., Ramalingeswara Rao, S., Satyanarayana, G.C., Lakshminarayana, P., Malleswara Rao, L. (2009). Is summer monsoon rainfall decreasing over India in the global warming era? *Journal of Geophysical Research: Atmospheres 114*(D24)

Nandargi, S., & Dhar, O. N. (2011). Extreme rainfall events over the Himalayas between 1871 and 2007. *Hydrological Sciences Journal, 56*(6), 930–945.

Patel, A., Goswami, A., Dharpure, J. K., & Thamban, M. (2021). Rainfall variability over the Indus, Ganga, and Brahmaputra river basins: A spatio-temporal characterisation. *Quaternary International, 575*, 280–294.

Priya, P., Krishnan, R., Mujumdar, M., & Houze, R. A. (2017). Changing monsoon and midlatitude circulation interactions over the Western Himalayas and possible links to occurrences of extreme precipitation. *Climate Dynamics, 49*, 2351–2364.

Qazi, N.Q., Jain, S.K., Thayyen, R.J., Patil, P.R., Singh, M.K. (2020). Hydrology of the Himalayas. *Himalayan Weather and Climate and their Impact on the Environment*, 419–450

Ray, S., Das, S. S., Mishra, P., & Al-Khatib, A. M. G. (2021). Time series SARIMA modelling and forecasting of monthly rainfall and temperature in the South Asian countries. *Earth Systems and Environment, 5*, 531–546.

Roxy, M. K., Ghosh, S., Pathak, A., Athulya, R., Mujumdar, M., Murtugudde, R., Terray, P., Krishnan, R., & Rajeevan, M. (2017). A threefold rise in widespread extreme rain events over central India. *Nature Communications, 8*(1), 708.

Roxy, M. K., Ritika, K., Terray, P., & Masson, S. (2014). The curious case of Indian Ocean warming. *Journal of Climate, 27*(22), 8501–8509.

Roxy, M. K., Ritika, K., Terray, P., Murtugudde, R., Ashok, K., & Goswami, B. N. (2015). Drying of Indian subcontinent by rapid Indian Ocean warming and a weakening land-sea thermal gradient. *Nature Communications, 6*(1), 7423.

Sahu, R., & Gupta, R. D. (2020). Snow cover area analysis and its relation with climate variability in Chandra basin, Western Himalaya, during 2001–2017 using MODIS and ERA5 data. *Environmental Monitoring and Assessment, 192*, 1–26.

Saini R, Sharma N, Attada R (2023) Delving into Recent Changes in Precipitation Patterns over the Western Himalayas in a Global Warming Era. In *Global Warming-A Concerning Component of Climate Change.* IntechOpen.

Samanth, A., Rakesh, V., Purwar, S., Gavaskar, S. M., Jagadeesha Pai, B., & Mohapatra, G. N. (2022). Analysis of spatio-temporal variability in observed rainfall over Karnataka using different data analytical techniques. *Journal of Earth System Science, 131*(2), 66.

Scherler, D., Bookhagen, B., & Strecker, M. R. (2011). Spatially variable response of Himalayan glaciers to climate change affected by debris cover. *Nature Geoscience, 4*, 156–159.

Sen PK (1968) Estimates of the regression coefficient based on Kendall's tau. *Journal of the American statistical association, 63*(324), 1379–1389

Sharma, A., & Goyal, M. K. (2020). Assessment of the changes in precipitation and temperature in Teesta River basin in Indian Himalayan Region under climate change. *Atmospheric Research, 231*, Article 104670.

Shrestha, S., Yao, T., & Adhikari, T. R. (2019). Analysis of rainfall trends of two complex mountain river basins on the southern slopes of the Central Himalayas. *Atmospheric Research, 215*, 99–115.

Singh, D. K., Gusain, H. S., Mishra, V., & Gupta, N. (2018). Snow cover variability in North-West Himalaya during last decade. *Arabian Journal of Geosciences, 11*(579), 1–12. https://doi.org/10.1007/s12517-018-3926-3

Singh, D., Tsiang, M., Rajaratnam, B., & Diffenbaugh, N. S. (2014). Observed changes in extreme wet and dry spells during the South Asian summer monsoon season. *Nature Clinical Practice Endocrinology & Metabolism, 4*, 456–461.

Singh, P., & Kumar, N. (1997). Effect of orography on precipitation in the western Himalayan region. *Journal of Hydrology, 199*(1–2), 183–206.

Smith, T., Bookhagen, B., Rheinwalt, A. (2017). Spatiotemporal patterns of High Mountain Asia's snowmelt season identified with an automated snowmelt detection algorithm, 1987–2016. *Cryosphere*

Smith, T., Bookhagen, B. (2018). Changes in seasonal snow-water equivalent distribution in High Mountain Asia (1987 to 2009). *Science Advances, 4*(1)

Smith, T., Bookhagen, B. (2020). Remotely sensed rain and snowfall in the Himalayas. In: *Himalayan Weather and Climate and Their Impact on the Environment*, 119–139

Tiwari, G., Kumar, P., & Tiwari, P. (2022). The appraisal of tropical cyclones in the North Indian Ocean: An overview of different approaches and the involvement of Earth's components. *Frontiers in Earth Science, 10*, Article 823090.

Trenberth, K. E. (2011). Changes in precipitation with climate change. *Climate Research, 47*, 123–138.

Vaughan, D., Comiso, J., Allison, I., Carrasco, J., Kaser, G., Kwok, R., Mote, P., Murray, T., Paul, F., Ren, J., Rignot, E., Solomina, O., Steffen, K., Zhang, T. (2013). Observations: cryosphere. In: *Climate change 2013: the physical science basis. Contribution of Working Group I to the Fifth Assessment Report of the IPCC*

Veh, G., Oliver, K., & Ariane, W. (2020). Hazard from Himalayan glacier lake outburst floods. *Proceedings of the National Academy of Sciences, 117*(2), 907–912.

Vellore, R. K., Kaplan, M. L., Krishnan, R., Lewis, J. M., Sabade, S., Deshpande, N., Singh, B. B., Madhura, R. K., & Rama Rao, M. V. S. (2016). Monsoon-extratropical circulation interactions in Himalayan extreme rainfall. *Climate Dynamics, 46*, 3517–3546.

Venkatesh, B., Nayak, P. C., Thomas, T., Jain, S. K., & Tyagi, J. V. (2021). Spatio-temporal analysis of rainfall pattern in the Western Ghats region of India. *Meteorology and Atmospheric Physics, 133*, 1089–1109.

Yao, T., Thompson, L., Yang, W., Yu, W., Gao, Y., Guo, X., Yang, X., Duan, K., Zhao, H., & Xu, B. (2012). Different glacier status with atmospheric circulations in Tibetan Plateau and surroundings. *Nature Clinical Practice Endocrinology & Metabolism, 2*, 663–667.

Glacial Lake Outburst Floods in Indian Hilly Regions and Geospatial Applications

Anup Das and Manisha Giri

Abstract Glacial Lake Outburst Floods (GLOFs) pose significant risks to life and property in the Indian Himalayan region. This chapter delves into the increasing issue of GLOFs, exploring their causes, consequences, and the essential function of geospatial technologies in reducing their impacts. We examine how remote sensing, GIS, and GPS are utilized for Monitoring glacial lake dynamics: Observing changes in lake size, volume, and water levels; assessing risk: Identifying potentially hazardous lakes and mapping areas at risk of flooding. Developing early warning systems: Facilitating real-time monitoring and prompt alerts for affected communities and Informing mitigation strategies: Assisting in designing and implementing drainage systems, early warning measures, and community-based disaster risk reduction initiatives. This chapter underscores the vital importance of geospatial technologies in deepening our understanding of GLOFs and aiding informed decision-making for their mitigation and adaptation in the Indian Himalayas.

1 Introduction

The Himalayas, recognized as the planet's youngest and largest mountain range, contain the third-largest snow and ice reserves, surpassed only by the Polar Regions. The Hindu Kush–Karakoram–Himalaya (HKH) region is home to approximately 54,252 glaciers and is aptly designated as the "Water Tower of Asia," supplying an annual volume of 86,000,000 m^3 of water. These glaciers sustain some of the world's principal rivers—including the Ganga, Indus, Brahmaputra, Salween, Mekong, Yangtze, and Huang Ho—providing critical water resources for nearly one billion downstream individuals. The freshwater sourced from these high-altitude glaciers underpins various sectors, such as agriculture, navigation, fishing, hydropower, and tourism. Nevertheless, the glaciers within the Himalayas are particularly vulnerable to the impacts of climate change and global warming, with many experiencing significant mass loss as temperatures continue to escalate. The accelerated melting of these

A. Das (✉) · M. Giri
North Eastern Space Applications Centre (NESAC), Umiam, Meghalaya, India
e-mail: 786anupdas@gmail.com

© The Author(s), under exclusive license to Springer Nature Switzerland AG 2025
S. Ghosh (ed.), *The Himalaya Dilemma*, Springer Natural Hazards,
https://doi.org/10.1007/978-3-031-95083-4_19

glaciers also leads to the formation and enlargement of glacial lakes. While these glaciers can present societal benefits, they simultaneously pose substantial risks to the lives and properties of those inhabiting the lower elevations. The phenomena of glacier retreat and thinning, attributable to climate change, are associated with various natural hazards, including snow, ice, and rock avalanches, landslides or debris flows triggered by rainfall, glacial lake outburst floods (GLOFs), and landslide lake outburst floods (LLOFs), all of which possess the potential to inflict considerable damage. This atlas endeavours to compile information regarding the glacial lakes in the Uttarakhand Himalayas, illustrating the distribution of the various glacial lakes predominantly found in the Central Himalaya (Indian Institute of Remote Sensing et al. 2020).

Glaciers are made of fallen snow that, over many years, compresses into large, thickened ice masses. A glacier is a large, long-lasting mass of ice formed on land that moves downhill under its weight. It is part of the Earth's hydrosphere. About 75% of the world's freshwater supply is in glacial ice. These lakes are visible in aerial photos of landforms in glaciated regions during the last ice age. Glaciers form when snow remains in one location long enough to transform into ice—form where snow and ice accumulation exceed snow and ice melt. Under the pressure of the layers of ice and snow above it, this granular ice fuses into denser and denser firn; Firn is partially compacted neve. Neve is a young, granular snow that has been partially melted, refrozen and compacted. Glacial Lakes are found in the Himalayan Region of India. The water of such lakes is dammed by snow. In winter, the water of those lakes is formed as ice or snow. However, in the summer, the ice or snows melt, and the water becomes visible and looks like a lake.

The creation and features of glacial lakes differ by location. They can be categorized into six types (Table 19.1): glacial erosion lakes, ice-blocked lakes, moraine-dammed lakes, other glacial lakes, supraglacial lakes, and subglacial lakes. Mostly, those types of lakes are facing this natural disaster. Moraine-dammed lakes and Ice-dammed lakes are susceptible to Glacial Lake Outburst Floods (GLOFs). GLOFs are a serious threat to local communities in the Himalayan regions. It is essential to detect the lakes to monitor them for potential flooding. Nowadays, the biggest threat to glaciers is global warming; global warming is melting our glaciers, causing sea levels to rise dramatically, and melting glaciers means losing homes or shelter and way of life, creating tragedy in ecosystems worldwide. Many regions will lose fresh water sources and be submerged in the ocean or sea.

2 Overview

The Indian Himalaya, a treasure trove of natural beauty and biodiversity, are also home to a growing threat: glacial lake outburst floods (GLOFs). These sudden and devastating events, triggered by the rapid release of water from moraine-dammed lakes, pose a significant risk to the region's life, property, and infrastructure (Fig. 19.1). A glacial lake outburst flood (GLOF) Fig. 19.1 in a hilly area of India

Table 19.1 Glacial lake types and their identification keys

Sl. no	Lake type	Lake subtype	Identification keys
1	Moraine dammed lake	End-moraine dammed lake	Lake dammed by end (terminal) moraines
2		Lateral moraine dammed lake	Lake dammed by lateral moraines(s), not in contact with glacial ice
0.3		Lateral moraine dammed lake (with Ice)	Lake dammed by lateral moraines(s) in contact with glacial ice
4		Other moraine dammed lake	Lake dammed by other moraines
5	Ice-dammed lake	Supra-glacial lake	Pond or lake on the surface of a glacier
6		Glacial ice-dammed lake	Lake dammed by glacier ice with no lateral moraines
7	Glacier erosion lake	Cirque erosion lake	A small pond occupying a cirque
8		Glacier trough valley erosion lake	Lakes formed in the glacier trough as a result of the glacier erosion process
9		Other glacial erosion lake	Bodies of water occupying depressions formed by the glacial erosion process
10	Other glacier lake	Other glacial lake	Lakes form in a glaciated valley and fed by glacial melt, but damming material not directly part of the glacial process

Source https://www.nrsc.gov.in/sites/default/files/pdf/Announcements/Glacial_Lake_Atlas_Ganga_Basin_NRSC.pdf (modified)

would most likely occur in the Himalayas, specifically in states like Uttarakhand and Sikkim, where retreating glaciers form glacial lakes that can potentially burst their natural dams, causing devastating floods downstream, a prominent example is the 2013 Kedarnath flood in Uttarakhand, triggered by a GLOF from the Chorabari Tal lake. Since the Little Ice Age, Earth has lost over 50% of its glaciers. The current rise in glacier retreat due to climate change has led to a transition from frozen to liquid water, resulting in global glacial lake growth. Most existing glacial lakes are found in Asia, Europe, and North America. The Southern Tibetan Plateau region, particularly where debris-covered glaciers exist, is expected to experience the most significant increase in lake formation. This rise in glacial lakes also indicates a greater likelihood of glacial lake outburst floods, which occur when moraine and ice dams break.

Remote Sensing and Satellite Technology: This technology uses satellite imagery and remote sensing methods to monitor changes in glacial lakes and identify potential threats. It systematically tracks variations in lake size, water levels, and the formation of new lakes, thereby providing critical data for risk assessment. Ground-Based Monitoring Stations: This initiative sets up ground-based monitoring stations

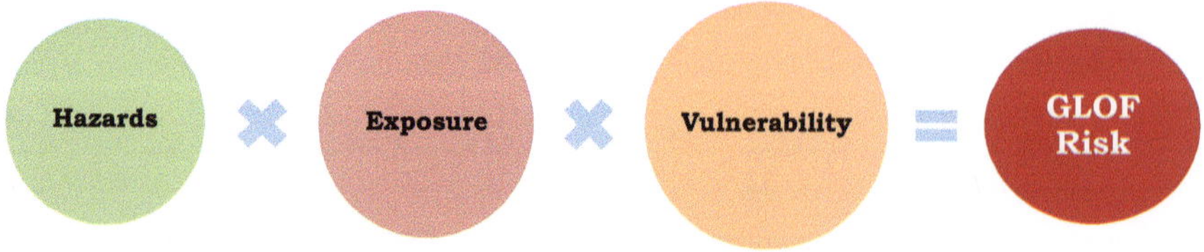

Fig. 19.1 GLOF risk is a function of hazard, exposure and vulnerability

strategically located in high-risk areas. These stations measure various parameters, including water levels, temperature, and seismic activity, improving the overall surveillance and early warning system. Real-Time Data Collection and Analysis: This approach implements real-time data collection and analysis systems to process information from remote sensors and monitoring stations. Advanced algorithms identify anomalies and potential indicators of lake instability or landslide activity.

Risk Assessment Models: This initiative develops sophisticated risk assessment models that integrate data from diverse sources, including geological surveys, climate data, and historical records. These models predict the likelihood of glacial lake outbursts or landslide-induced lake outbursts, facilitating proactive measures.

Community Engagement and Early Warning Systems: This initiative actively engages local communities in surveillance activities, fostering awareness and understanding of associated risks. Furthermore, it establishes early warning systems capable of efficiently communicating alerts to downstream communities, enabling timely evacuations and emergency responses. (National Institute of Disaster Management, Ministry of Home Affairs, Govt. of India 2023).

This chapter will delve into the causes, consequences, and mitigation strategies of GLOFs in the Indian Himalayas (Fig. 19.2).

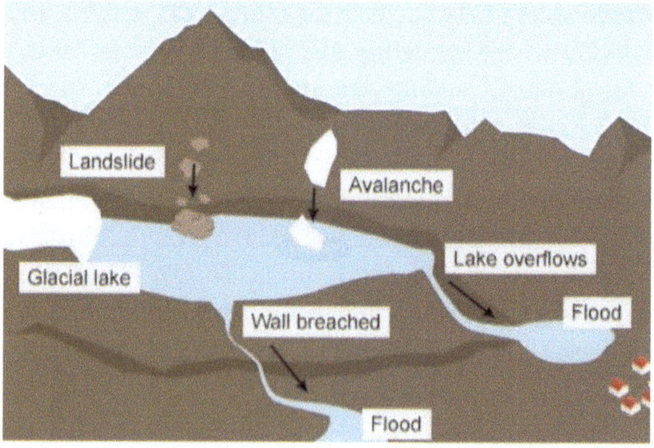

Fig. 19.2 Glacial lake outburst flood (GLOF). *Source* https://www.google.com/url?sa=i&url=https%3A%2F%2Fstaging.iasgyan.in%2Fdaily-current-affairs%2Fglaciallakes&psig=AOvVaw2rgRI3g1Q0YjZJ82LwKOh9&ust=1738825103403000&source=images&cd=vfe&opi=89978449&ved=0CB

3 Causes of GLOFs

3.1 Causal Factors of GLOFs

- **Climate Change:** Rising global temperatures are accelerating the melting of glaciers, leading to the formation and expansion of glacial lakes. As these lakes grow, they exert increasing pressure on the moraine dams that contain them, making them more susceptible to failure. When glaciers melt, they can form lakes; if these lakes fill quickly, their natural dams may break. Additionally, as glaciers recede, they can create new glacial lakes or expand existing ones, raising the risk of glacial lake outburst floods (GLOFs) in certain areas. Therefore, the retreat of glaciers and GLOFs are interconnected.
- **Moraine Instability:** Glaciers often deposit rock and sediment, called moraines, along their edges. These moraines can create natural dams that hold back glacial lakes. A failure of these moraines due to erosion, overtopping, or other factors can lead to GLOFs; Moraines, the debris deposited by glaciers, are inherently unstable. Earthquakes, heavy rainfall, or even the weight of the impounded water can trigger landslides or erosion, leading to dam failure and a subsequent GLOF.
- **Ice Avalanches:** The collapse of ice cliffs or the sudden movement of glaciers can generate large waves in glacial lakes, potentially overtopping or breaching the moraine dams.
- **Geothermal activity:** Volcanic or geothermal activity can cause the rapid melting of glaciers and the formation of glacial lakes that can burst their natural dams.
- **Avalanches:** Avalanches can cause the formation of a natural dam, which can be destroyed by the pressure of the water accumulating behind it.
- **Earthquakes:** Seismic activity can cause the rupture of natural dams holding back glacial lakes, resulting in floods downstream.
- **Landslides:** Landslides can also create natural dams, which can be breached by water accumulation or seismic activity.
- **Human activities:** Human activities such as mining, construction and other forms of development in mountainous areas can cause destabilization of slopes and increase the likelihood of landslides or avalanches, leading to GLOFs.
- **Glacier lake expansion:** Glacier lakes can expand due to increased meltwater runoff, rainfall, or snowmelt. The resulting increase in water pressure can cause the natural dam to burst.
- **Glacier surging:** A surge is a sudden acceleration of a glacier's flow, creating new glacial lakes or expanding existing ones. The resulting pressure can cause the dam to fail, leading to GLOFs.
- **Glacial lake overtopping:** When a glacial lake's water level rises too high, it can overtop the natural dam, leading to its erosion and eventual failure.
- **Glacier calving:** Glacier calving is the breaking off ice chunks from the front of a glacier. These ice chunks can fall into a glacial lake, creating waves damaging the natural dam.

- **Extreme weather events:** Intense rainfall or snowfall can increase the volume of water in a glacial lake, leading to an increased risk of GLOFs.

4 Consequences of GLOFs

- **Loss of Life and Property:** GLOFs can cause widespread devastation, including loss of human life and destruction of homes, infrastructure, and agricultural land. According to a recent study, up to 15 million people worldwide are at risk from GLOFs. The substantial risk to human settlements is demonstrated by the fact that almost one million people reside in Asia alone, just 10 kms from a glacial lake. Because of their intense and swift nature, GLOFs can swiftly sweep away or damage roads, bridges, and structures. Their force can destroy essential infrastructure, upsetting daily life and negatively affecting the economy. The abrupt inflow of massive amounts of water, muck, and debris can impact aquatic life and the populations that rely on these systems, changing river ecosystems. GLOFs can potentially flood agricultural areas, resulting in crop loss and damaging the local economy and food security.
- **Environmental Damage:** The sudden release of large volumes of water can erode riverbeds, damage ecosystems, and pollute water sources. Destabilization of river ecosystems, Destruction of natural habitats and erosion, deforestation, also land degradation.
- **Economic Disruption:** GLOFs can disrupt transportation, communication, and economic activities in affected regions, leading to significant financial losses.

5 Mitigation Strategies

- **Early Warning Systems**: The development and implementation of robust early warning systems, including remote sensing and ground-based monitoring, are crucial for predicting and mitigating GLOFs. As the study suggests, early warning systems require better design. Enhanced EWS can deliver timely information, facilitating quicker evacuations and protection of assets. Modern communication technology, such as smartphones, can quickly alert those at risk populations.
- **Controlled Drainage:** Controlled drainage channels and tunnels can be constructed to gradually release water from potentially dangerous lakes, reducing the risk of sudden outbursts.
- **Community Awareness and Preparedness:** Educating local communities about the risks of GLOFs and developing evacuation plans can help minimise casualties and damage. Local communities are key in the initial response. Preparing local personnel and implementing evacuation drills can protect lives during a GLOF.

- **Sustainable Development:** Promoting sustainable development practices in the Himalayan region, such as reducing deforestation and promoting eco-friendly tourism, can help mitigate the impacts of climate change and reduce the risk of GLOFs.
- **Utilization of RS technology:** Using technologies such as Synthetic-Aperture Radar imagery enables the detection of alterations in water bodies, providing a proactive method for monitoring glacial lakes.
- **Construction Codes:** Establishing and following consistent construction codes in GLOF-prone areas can enhance infrastructure resilience to such events.

6 Data and Software Used in This Study

Monitoring glacier-associated lakes is crucial in the IHR for identifying critical glacial lakes, necessitating a detailed inventory of these lakes and their types. Glacial lakes can be categorized into several types based on their location relative to the glacier and the damming mechanism. Inventorying glacial lakes in these remote mountain areas with rugged terrain and harsh weather conditions is challenging and labor-intensive; therefore, Remote Sensing (RS) data is essential in gathering information on these lakes (Table 19.2). Satellites equipped with high spatial, spectral, and temporal resolution sensors effectively provide lake information with improved accuracy and consistency. GIS tools and software can be used for data analysis, evaluation, and monitoring. By utilizing both technologies, researchers and experts can enhance their assessments, enabling early warning systems and preparedness to mitigate the vulnerabilities posed by natural events. Data collected between September and December were predominantly utilized due to the minimal snow or cloud cover. United States Geological Survey (USGS) satellite data from Landsat 5 and 7 (Thematic Mapper and Enhanced Thematic Mapper Plus) has been extensively employed for mapping glacial lakes, owing to its free accessibility. In contrast, Indian Remote Sensing (IRS) satellite data from Advanced Wide Field Sensor (AWiFS), Linear Imaging Self-Scanning Sensor LISS-III and LISS-IV has also been utilized for such inventory. The Digital Elevation Model (DEM) has been used for topographic information and watershed boundary generation. Other pieces of information, like names of lakes and rivers, have been gathered from digital toposheets and open-source platforms rather than all of those data like LULC (pre- and Post-Disaster), River cross-sections (pre- and post-event) for river modelling and precipitation data (INSAT/ Kalpana, CHIRPS, etc.) are the most important.

6.1 Data Interpretation Using Satellite Images

The spectral Reflectance curve of water in the visible spectrum starts with a low in the Blue region (0.4–0.5 μm), reaches a peak in the Green region (0.5–0.6 μm), decreases

Table 19.2 Data used for this study

Types	Data	Resolution (m)	Source
DEM	ALOS DEM	12	JAXA, Bhoonidhi, USGS etc
	CartoDEM	10	
	SRTM DEM	90	
Satellite Image	SPOT-6 & SPOT-7	1.5	USGS, Bhoonidhi, ESA Copernicus, etc
	Cartosat-1	2.5	
	Cartosat-2	1	
	LISS-IV (Resourcesat-1,2)	5.8	
	LISS-III	23.5	
	Landsat 4-5 (TM)	30	
	Landsat 8-9 (OLI)	30	
	Sentinel 1 (SAR) & 2	10	
UAV	UAV Images	–	Collected by authorised UAV service providers
	DEM, DTM	–	

in the Red region (0.6–0.7 μm) and probably the most distinctive characteristic is the energy absorption at Near InfraRed (NIR) wavelengths. Identifying and delineating water bodies with remote sensing data are carried out efficiently in near-infrared wavelengths because of this absorption property in the IR region. However, various physical conditions of water bodies (water depth, turbidity, chlorophyll content, etc.) manifest spectral changes. As a result of various conditions of lakes, the water in satellite images in False Colour Composite (FCC) ranges in appearance from light to dark blue to black. In the case of frozen lakes, it appears white.

Glacial lake sizes are generally small, having circular, semi-circular, or elongated shapes with wonderful textures and are generally associated with glaciers in high-altitude areas. Glacial lakes, like erosion and cirque lakes, are not necessarily associated with glaciers. Knowledge of the glacial lake's physical characteristics and associated features is essential for interpreting the images. Satellite data interpretation can be done using visual image interpretation keys such as colour, size, tone, texture, pattern, association, shape, shadow, and orientation. Some remote sensing methods have been developed for glacial lake detection and mapping or development of inventory. Manual or automated lake mapping methods have difficulties identifying the lakes described in the following section. The visual interpretation method was the most accurate.

6.2 *Identification of Lakes*

Glacial lakes are identified via visual analysis or automated mapping, but the latter faces challenges from varied terrains, including mountain shadows, snow cover, and partial freezing. Both methods struggle in snow-covered glacier tongues or ablation zones, where many supraglacial lakes may exist.

Lake water absorbs radiation, appearing darker in false colour composite (FCC) satellite imagery. Darkening effects in cloud or mountain-shaded areas can lead to mapping inaccuracies. Mountain shadows over lakes complicate boundary detection, causing many lakes to go undetected, especially in high-altitude shadows. Frozen lakes may appear white in cold climates, hindering identification in automated mapping, though they can be detected through visual assessment.

Automatic mapping can speed up glacial lake detection but is not applicable across the entire Himalayan region due to significant variations in satellite scenes over time. For example, frozen lakes, snow-covered areas, or clouded regions in shadows are undetectable through these methods. Thus, manual interpretation becomes essential for accurate mapping. Therefore, until a certain point, glacial lake mapping can only be partially automated, with visual interpretation techniques ensuring greater accuracy and reducing misclassification. This study has manually identified and mapped glacial lakes and their types across the Indian Himalayan River Basins using high-resolution satellite data from Bhoonidhi and Google Earth to refine the glacial lake database.

6.3 Software

By exploiting the unique spectral properties of surfaces, it is possible to distinguish the properties of surfaces; it is possible to distinguish open water from snow and rock. This software is applicable for detecting and monitoring glacier lakes and GLOFs analysis and different types of research: Google Earth, Google Earth Pro (for collecting attribute information, etc.) Erdas Imagine, ENVI (Image processing and image classification), Google Earth Engine (cloud-computing-based image analysis), ArcMap & QGIS (for analysis, modelling and map preparation) and for Hydrological modelling HEC-RAS, HEC-HMS, and river parameters estimation and watershed analysis required ArcGIS based GeoHMS, GeoRAS etc.

7 Some of the GLOFs Events in the Indian Himalayan Region

7.1 Sikkim GLOFs (October, 2023)

South Lhonak Lake (SLL) in Sikkim, India, is located 5,200 m above sea level in the Upper Teesta basin. It is one of the largest, fastest-growing, and most hazardous lakes in Sikkim, posing a significant risk of downstream damage in the event of a glacial lake outburst flood (GLOF). On October 3, 2023, SLL experienced an outburst that triggered a devastating flood cascade, resulting in the deaths of 55 people, leaving 74 missing, and destroying the 1,200-megawatt (MW) Teesta-III hydropower dam (Fig. 19.3). This flood cascade occurred from October 3 to 4 and affected Sikkim and West Bengal, with transboundary implications for Bangladesh.

The flood cascade caused damage to approximately 25,900 buildings, 59% of which were built in the last decade. Most of the affected structures are located below Chungthang, within a distance of 200 to 385 km from the lake, with the most severely inundated area situated between 290 and 385 km downstream in Bangladesh. Additionally, an estimated 276 square kilometers of agricultural land was flooded. A total of 31 significant bridges, constructed from Reinforced Cement Concrete (RCC) or steel, along the Teesta River, were damaged, including 14 located

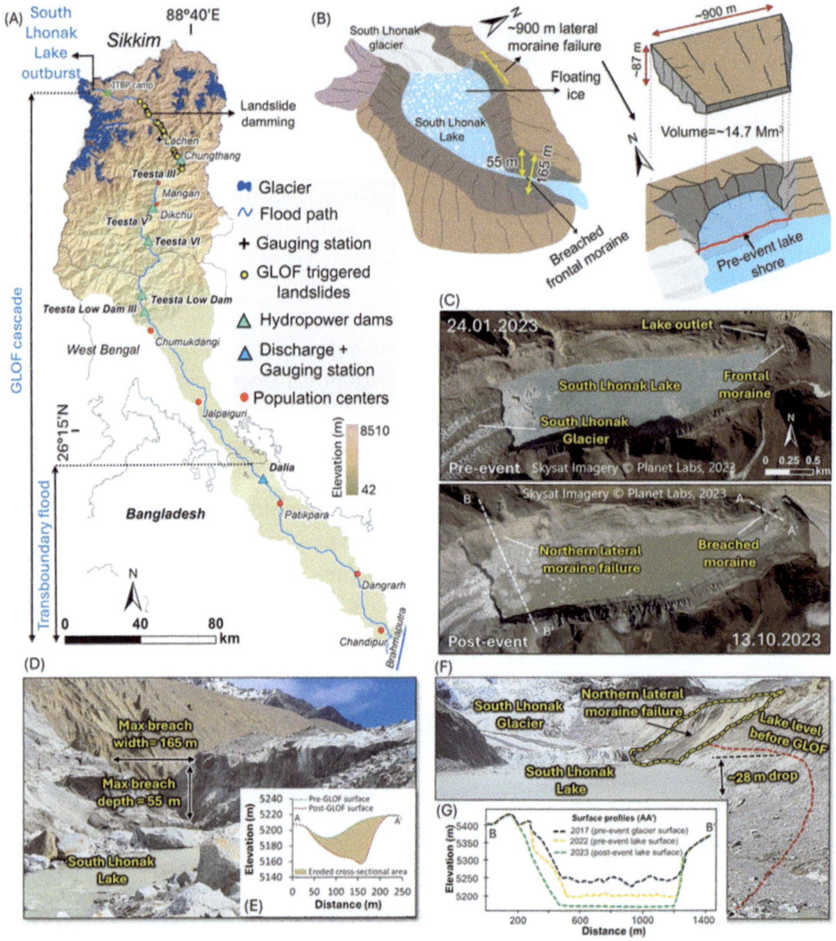

Fig. 19.3 Overview of the 3–4 October 2023 GLOF cascade from South Lhonak Lake. *Source* Sattar et al. (2025)

upstream of Chungthang. Furthermore, around 20 small pedestrian bridges in Sikkim were also affected (Fig. 19.4). About 18.5 km of roadway sustained damage, with roughly 6.4 km attributed to secondary landslides. Approximately 200 buildings were impacted by these triggered landslides, with 90% of the damage resulting from the two largest nearby landslides, L33 and L35, located 60 km downstream of SLL. In total, 10 landslides affected the road network, leading to the maximum extent of road length damage.

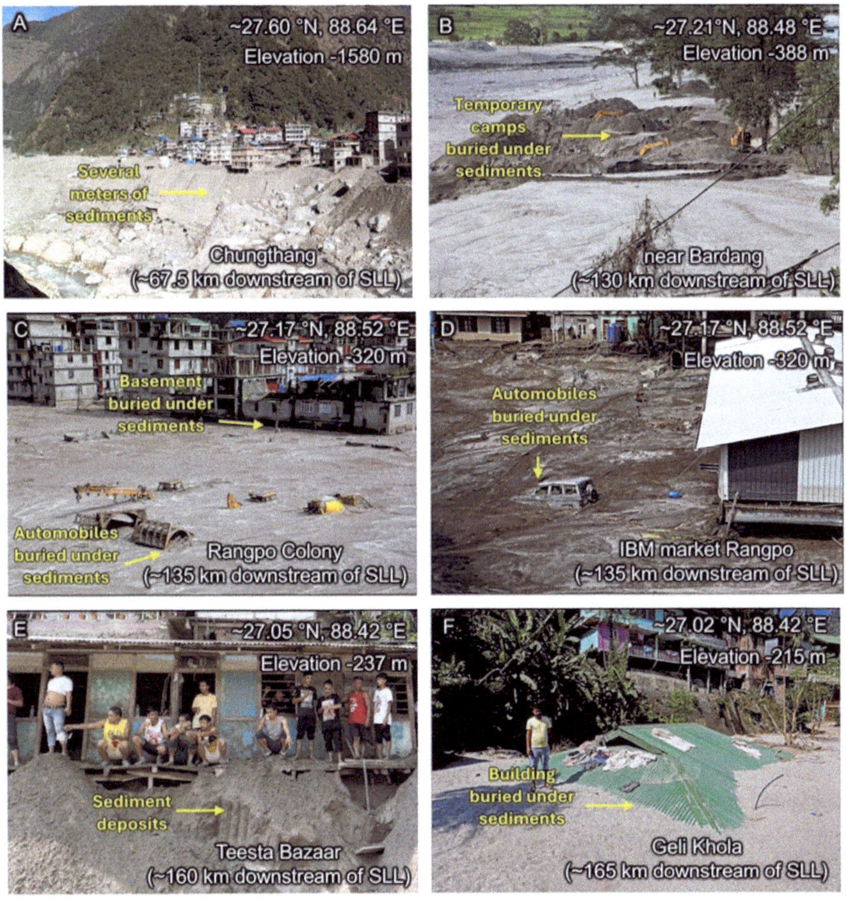

Fig. 19.4 Field evidence of sediment aggradation. *Source* Sattar et al. (2025)

7.2 Uttarakhand (Kedarnath) GLOFs (June 2013)

The 2013 Kedarnath disaster was caused by several factors, notably a glacial outburst from Chorabari Tal, a glacial lake. Intense rainfall and melting glaciers led to flash floods, resulting in widespread devastation throughout the Kedarnath region (Fig. 19.5). This disastrous event claimed thousands of lives, destroyed infrastructure, and severely impacted the area around the Kedarnath temple. It raised alarms regarding the vulnerability of glacial lakes in the area and their potential to trigger floods. Across the region, flash floods and landslides resulted in over 6,000 fatalities, the majority of which were associated with the Kedarnath event. Numerous roads and bridges were damaged, and at least 30 hydropower plants were destroyed or severely impaired (Sati and Gahalaut, 2013). The devastation of roads and trekking paths left approximately 100,000 pilgrims and tourists stranded until military and civic rescue operations could complete evacuation efforts (Martha et al., 2014).

Day-June 16 debris flow—

Local reports show that by the afternoon of June 16, surface flooding, erosion, and damage were already visible in the villages of Kedarnath and Rambara after two days of intense rainfall (Sengupta, 2013). Around 7 pm that evening, a significant debris flow hit Kedarnath, likely caused by a shallow landslide on the steep (25°–35°) slopes above the Companion Glacier's terminus (Martha et al., 2014; Petley, 2013). As the flow descended, it collected a large volume of debris while passing through a snow and debris avalanche channel that was actively eroding along steep moraine and outwash gravels at the glacier's eastern margin and proglacial area.

Day-June 17 Lake Outburst flood and debris flow—

The second and most disastrous debris flow occurred at approximately 6:45 am the following day (Dobhal et al., 2013a), triggered by the failure of a small seasonal lake located in a depression at the outer edge of the Chorabari western lateral moraine. Known as Chorabari or Gandhi Sarovar Lake, this body of water was around 400 m long and 200 m wide, with depths estimated between 11 and 15 m. It had no direct connection to the glacier and relied entirely on runoff from rainfall and snowmelt. The lake's maximum volume was estimated at 0.4 million cubic meters (Das et al., 2015; Durga Rao et al., 2014), and it featured no outlet channel. Evidence suggests that debris was not only washed into Chorabari Lake from runoff but also directly transported from the surrounding moraine and upstream, where significant new erosion is visible in post-disaster images. The lake's accumulation of water and debris eventually caused a catastrophic breach at a low point in the moraine wall.

7.3 Uttarakhand GLOFs (February, 2021)

On February 7, 2021, a massive rockfall, accompanied by a hanging glacier, occurred in the catchment area of Raunthi Gadhera, a tributary of Rishi Ganga, located in

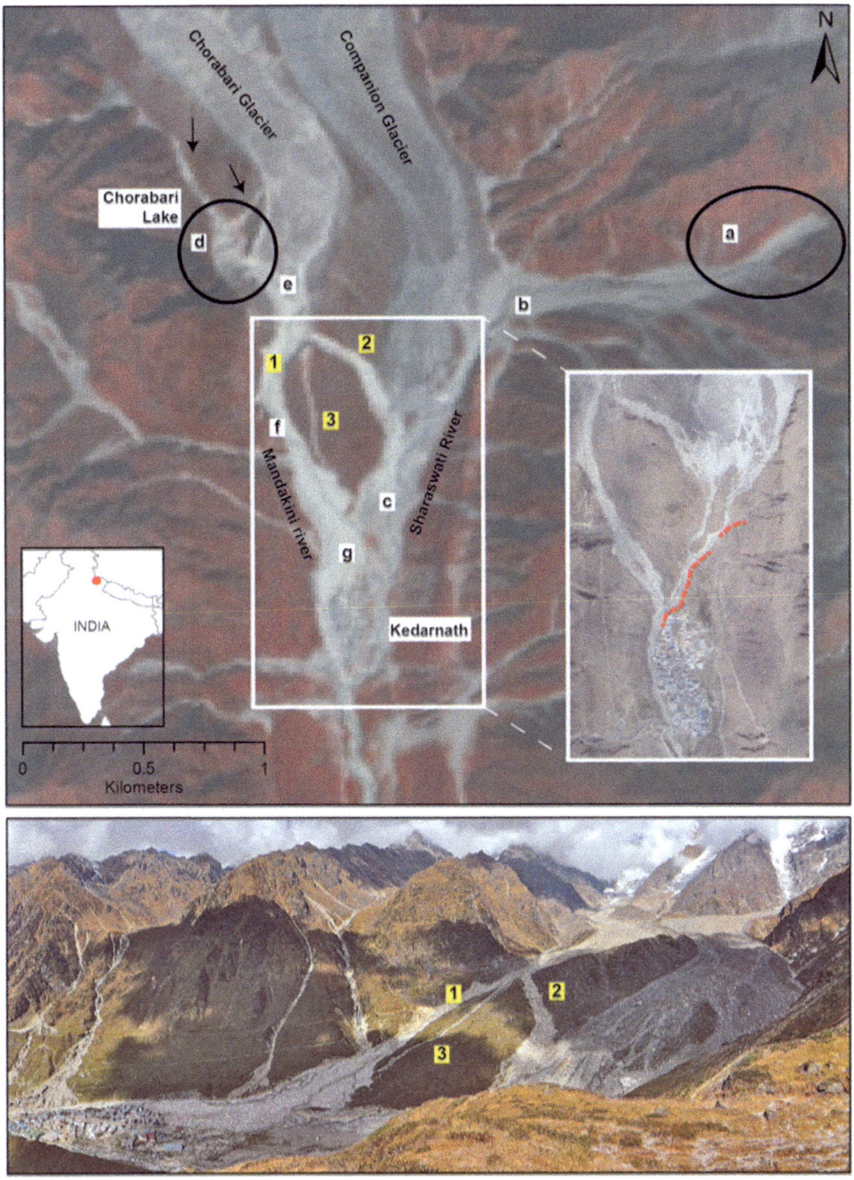

Fig. 19.5 Overview of the Kedarnath disaster (Up one), showing the origin of debris flows on June 16 and June 17, 2013

Fig. 19.6 Significant damming activities were observed at the confluence of Raunthi Gadhera and Rishi Ganga

the Chamoli district of Uttarakhand, India (Figs. 19.6 and 19.7). This led to the blockage of the river's flow and the sudden breach of the landslide dam, along with successive damming at the confluence of Raunthi Gadhera with Rishi Ganga, north of Murunna and at the junction of Rishi Ganga with the Dhauli Ganga River at Rini. The resulting impacts caused significant damage and destruction to the Rishi Ganga, Dhauli Ganga, and Alaknanda rivers, especially around the areas of Rini and Tapoban. Between February 9 and 11, 2021, the flow of the Rishiganga stream became permanently obstructed by debris from Raunthi Gadhera after the breach of the landslide dam. Floodwaters saturated the debris, which included boulders and ice blocks from Raunthi Gadhera that collided with the valley wall, leading to backflow upstream along the Rishi Ganga. This resulted in a significant accumulation of sediments and the formation of an artificial lake.

In Uttarakhand, another two major glacial lake outburst floods (GLOFs) occurred in the last decade. The first took place in Kedarnath Valley in June 2013, leading to 6,000 fatalities. The latest GLOF occurred in the Rishighanga valley of Chamoli in February 2021, resulting in 72 deaths. With the monsoon season looming, the Uttarakhand State Disaster Management Department (USDMA) is set to initiate a vulnerability study of 13 glacial lakes next month, five of which are identified in a "high-risk zone." This study seeks to gather data to help avert disasters like lake outbursts. Among the 13 lakes, the high-risk ones include those in the Darma, Lasaryanghati, and Kutiyangti valleys of Pithoragarh district, as well as Vasundhara Tal Lake in the Dhauli Ganga basin of Chamoli district. These lakes vary in size from 0.02 to 0.50 square kilometers and are located at elevations exceeding 4,000 m above sea level (Mishra, 2024).

Fig. 19.7 An aerial perspective reveals the submerged Tapovan hydel power project following a glacier burst in Chamoli district, Uttarakhand, on February 12, 2021

7.4 Ladakh GLOFs (August, 2014)

A GLOF impacted Gya Village, Ladakh, during the night of August 6–7, 2014, triggered by the collapse of a moraine-dammed lake linked to the Gya Glacier, situated 14 km upstream. This flood damaged multiple agricultural terraces, a concrete bridge, and several homes in the village. The Gya Glacier and the nearby Gya Village are in the Trans-Himalayan region of India's Ladakh (Fig. 19.8). Gya Village (33°39′12″ N; 77°44′07″ E; altitude: approx. 4100 m asl) lies approximately 73 km from Leh along the Leh-Manali highway. Currently, the glacier covers an area of 0.50 km^2 (50 ha) and stretches about 1150 m in length, with its snout positioned at an elevation of 5450 m asl. It is associated with a proglacial lake spanning 0.11 km^2. The meltwater from the glacier satisfies various agricultural and domestic requirements. (Majeed et al., 2020) According to the 2011 Census of India, Gya Village had 140 households and a population of 658 people.

8 Geospatial Applications on GLOFs

Geospatial technologies are essential for understanding, monitoring, and mitigating the risks related to GLOFs. By utilizing these tools, scientists, policymakers, and communities can more effectively prepare for and respond to these catastrophic events, ultimately saving lives and safeguarding livelihoods (Fig. 19.9).

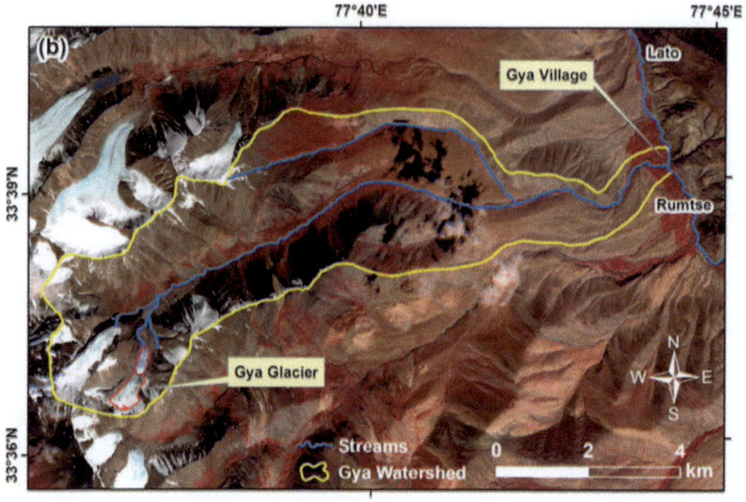

Fig. 19.8 Gya watershed featuring Gya Glacier, a stream, and Gya village. Planet satellite image of 15 Sept, 2019

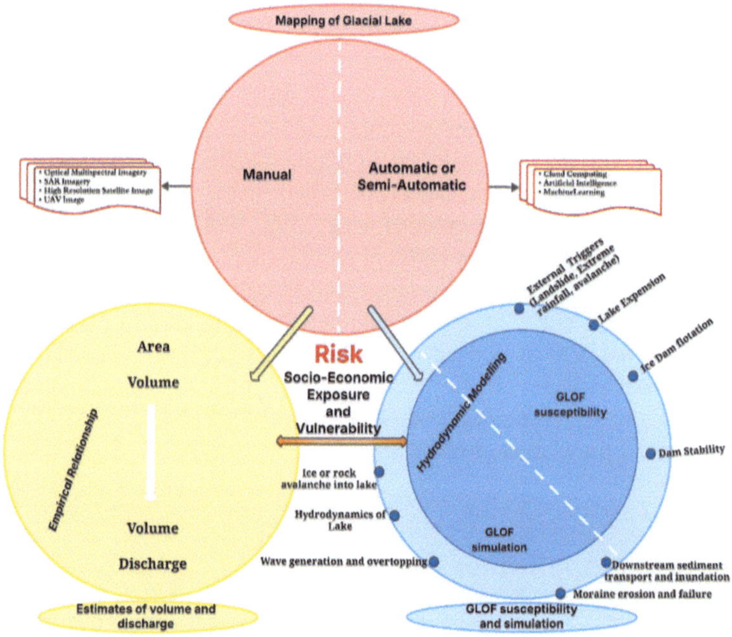

Fig. 19.9 Dimensions of geospatial applications to study GLOFs

8.1 Mapping Snow and Glacier

The Himalayan regions of India, notably Sikkim, Uttarakhand, and Arunachal Pradesh, are characterised by snow and glacial covers. Snow's runoff is a dependable water source for agricultural activities and drinking purposes, particularly in semi-arid climates. However, substantial snow cover accumulation followed by abrupt melting can precipitate severe flooding, resulting in loss of life and the destruction of property within downstream areas. Consequently, it is imperative to accurately map and ascertain the geographic extent of snow cover and water equivalent in critical glacial areas on a seasonal basis. In instances where cloud cover is absent, identifying snow cover distribution using visible and/or near-infrared images is relatively straightforward, as snow appears brighter than the surrounding vegetation, soil, and water. When confusion arises between snow and cloud cover, multiple images of the study area can be procured quickly using an Advanced Very High-Resolution Radiometer (AVHRR) or Geostationary Operational Environmental Satellites (GOES). The snow cover will remain stable during this short interval. Typically located in high-altitude terrains, snow cover poses challenges for accurate mapping due to mountain shadows. A Normalized Difference Snow Index method can delineate snow cover extent in such cases. Snow exhibits a higher reflectance in the optical region than other land features, such as vegetation, rocks, and water; however, it displays lower reflectance in the Short-Wave Infrared (SWIR) region than in rocks and vegetation. Consequently, snow appears white in the visible spectrum and black in the SWIR region in satellite imagery. This characteristic underpins the use of the Normalized Difference Snow Index (NDSI) for mapping snow cover and proves to be a valuable technique in the context of Himalayan snow, where mountain shadows are particularly pronounced (Kulkarni et al., 2002; Hall et al., 1995). The Normalized Difference Snow Index (NDSI), calculated through the ratio of differences between visible and SWIR channels, is presented as follows:

$$NDSI = \frac{Green\,Band\,(B2) - SWIR\,Band\,(B5)}{Green\,Band\,(B2) + SWIR\,Band\,(B5)}$$

8.2 Geospatial Modelling for GLOFs Susceptibility Analysis

Glacial lakes have increased significantly throughout the Himalayan region. Thus, the present study focuses on a geospatial modelling approach to analyse the susceptibility to glacial lake outburst floods (GLOF) in the central Himalayan region (Fig. 19.10). Multitemporal satellite imagery, including Landsat-TM and Sentinel-2, was utilised. The hydrodynamic compatibility tool HEC-RAS 5.0 was employed for GLOF prediction, while the lake volume and surface area were calculated using established empirical equations. The breach analysis indicates that the flood flow is approximated,

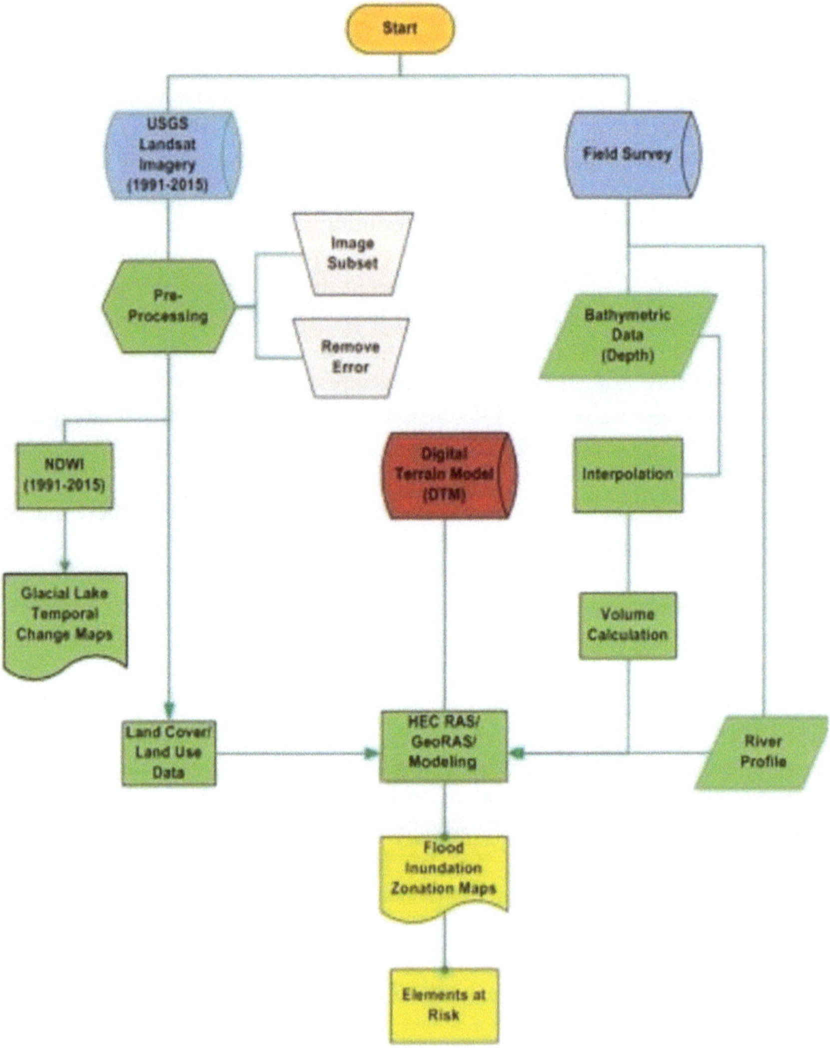

Fig. 19.10 Methodology for hydrological modelling. *Source* Amin et al. (2020)

and the associated uncertainties are examined through three outburst scenarios. The study's findings reveal that the number of lakes and their sizes have increased significantly. In this research, Vasundhara Lake (VL), located at 4,676 m above sea level, has been classified as hazardous for further examination. It has been observed that the lake's area grew over the past three decades (i.e., 1994–2020), rising from 0.15 to 0.29 km^2. The study indicates that, in the worst-case scenario, infrastructure along riverbanks may be vulnerable to damage. Additionally, it offers insights into the potential impacts of GLOF events in the region, which could benefit early warning systems and water resource management plans (Thapliyal & Kimothi, 2024).

8.3 *Change Detection Mapping of Glacier Lakes*

A time series of Landsat imagery has been utilized to identify prospective lake sites and to monitor the spatial and temporal variations in lake areas. This series illustrates the change detection in the glacial lake area from 1990 to 2020. After analyzing temporal variations in satellite imagery, we found that a glacial lake forms in the lower ablation zones of the mother glacier, exhibiting evidence of breaching, and the lake's area continues to grow as the glacier retreats. Therefore, South Lhonak glacial lake has been identified as the most hazardous lake, with a maximum area of 1.39 km^2 (Rawat et al., 2022) (Fig. 19.11).

To analysis this research, use two different indices, NDWI and MNDWI. The MNDWI was introduced by Han-Qiu in 2005 as a modification of the NDWI. While the NDWI uses the near-infrared (NIR) band, the MNDWI replaces it with the short-wave infrared (SWIR) band (Fig. 19.12). This substitution helps suppress information from built-up areas and other background features, allowing for a more accurate extraction of water body information in the target area. The MNDWI leverages the higher reflectance in the green and NIR bands to effectively highlight water features.

$$NDWI = \frac{Green - NIR}{Green + NIR}$$

$$MNDWI = \frac{Green - SWIR}{Green + SWIR}$$

Fig. 19.11 Changes in the area of the glacial lake (1990–2020)

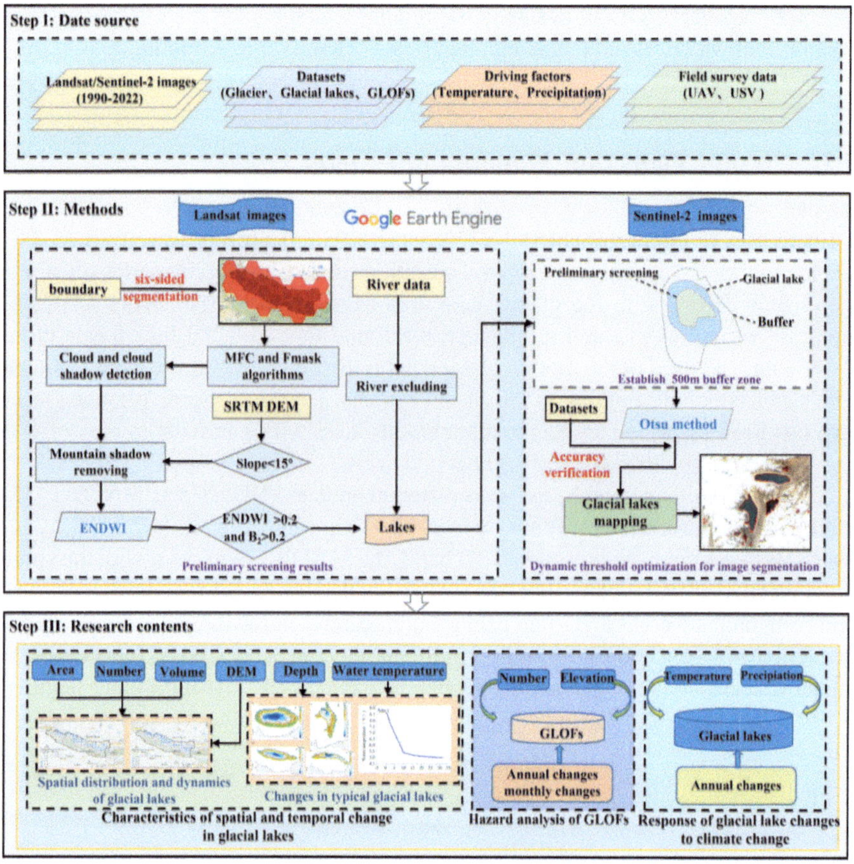

Fig. 19.12 Methodology of temporal changes of glacial lakes. *Source*Cheng et al. (2024)

8.4 GLOFs Risk Assessment and GLOF Inundation Modelling

Preparing GLOF risk management plans demands comprehensive data to assess the lake's potential for outbursts, its physical characteristics, potential triggers for failure, anticipated discharge magnitude, downstream flood wave behaviour, and its effects on communities along the route. Given these lakes' problematic terrains, collecting in-situ data is often laborious and, at times, unfeasible. In such cases, Remote Sensing (RS) and Geographical Information Systems (GIS) are vital for obtaining crucial insights into glacial lakes. Satellites with high spatio-temporal and spectral resolution provide essential data regarding glacial lakes' location, shape, water coverage, and other attributes. Integrating RS and GIS techniques with hydrologic and hydraulic modelling enables swift analysis of these challenging glaciated regions. These tools are crucial for identifying priority glacial lakes and monitoring real-time GLOF occurrences. There is abundant research regarding RS and GIS applications in studying glacial lakes and GLOFs. This research encompasses glacial lake inventories, distribution assessments, vulnerability analyses, modelling, and risk evaluations. In the Indian Himalayas, inventoried glacial lakes have been subject to risk assessments, while ICIMOD has similarly documented glacial lakes in the Hindu Kush Himalayas. Many regional studies on glacial lakes have also taken place, relying heavily on remotely sensed data from sources like Landsat, Sentinel, and RS-2 imagery, among others, processed through GIS software and tools. (Rao et al., 2024).

8.5 *GLOFs Susceptibility Analysis*

The MCDM-based AHP method assigns weights to each GLOF susceptibility factor. This semi-quantitative decision-making technique employs weights through pairwise comparisons among various factors to avoid inconsistencies. Several studies have utilized the AHP method to assess GLOF susceptibility (Fig. 19.11) in glacial lakes within the Himalayan region.

To determine the susceptibility of individual lakes to potential GLOF events (Fig. 19.13) in the future, weights were assigned to each parameter. These weights were primarily based on the authors' and users' expertise and understanding of how the different parameters interact and impact potential GLOF events for each lake. The AHP technique succinctly summarizes the authors' perceptions and judgments into a multi-level hierarchical configuration that reflects their significance. It clarifies the complex relationships among various parameters through the comparison matrix and minimises biases by ensuring the consistency of the analysis. To assess the relative significance of the parameters, a pairwise comparison matrix was calculated using a degree of importance scale ranging from 1 to 9, where 1 indicates equal importance, and 9 signifies extreme importance of one parameter relative to another. This matrix was then normalized to determine the weights and ranks of each parameter. The final weights for each parameter were calculated by multiplying the AHP factor weights with the class index according to the given formula (Das et al., 2024):

$$\text{GLOF index} = \sum_{i=0}^{n} F_w * C_i$$

where Fw is the factor weight, and C_i is the class index value.

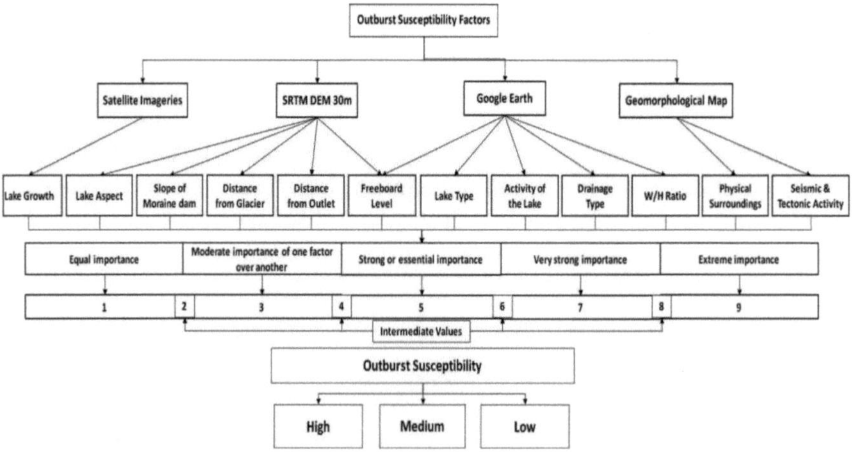

Fig. 19.13 Flowchart for GLOF susceptibility. *Source* Aggarwal et al. (2017)

9 Conclusion

Glacial lake outburst floods (GLOFs) pose a considerable threat to mountainous regions, chiefly attributable to the cascading effects of climate change. With millions of individuals exposed to risk, adopting a multifaceted approach that integrates technological solutions with community-centred strategies is imperative. Emphasizing early warning systems, utilizing contemporary technology, and enhancing community resilience is crucial for protecting against the catastrophic consequences of GLOFs.

Glacial Lake Outburst Floods (GLOFs) present a growing and serious threat to the Indian Himalayan region. The rapid melting of glaciers due to climate change is causing the creation and enlargement of many glacial lakes, raising the chances of disastrous incidents. Geospatial technologies such as remote sensing, GIS, and GPS provide essential resources for tracking the changes in glacial lakes, evaluating risks, and crafting efficient mitigation strategies. By utilizing these technologies, scientists and policymakers can- Accurately chart and observe glacial lakes, tracking shifts in size, volume, and water levels over time. Identify lakes that may be dangerous. Assess the stability of moraine dams and locate lakes at significant risk of outbursts. Establish early warning systems to utilize real-time data for forecasting and alerting communities about imminent GLOF events. Develop and implement mitigation strategies, directing the creation of drainage channels, early warning systems, and other infrastructure to lessen the impacts of GLOFs.

Ongoing research and advancements in geospatial technologies are essential for comprehending GLOF dynamics and creating successful strategies to lessen their severe impacts in the Indian Himalayas.

References

Aggarwal, S., Rai, S., Thakur, P., & Emmer, A. (2017). Inventory and recently increasing GLOF susceptibility of glacial lakes in Sikkim, Eastern Himalaya. *Geomorphology, 295*, 39–54. https://doi.org/10.1016/j.geomorph.2017.06.014

Allen, S. K., Linsbauer, A., Randhawa, S. S., Huggel, C., Rana, P., & Kumari, A. (2016). Glacial lake outburst flood risk in Himachal Pradesh, India: An integrative and anticipatory approach considering current and future threats. *Natural Hazards, 84*(3), 1741–1763. https://doi.org/10.1007/s11069-016-2511-x

Allen, S. K., Rastner, P., Arora, M., Huggel, C., & Stoffel, M. (2015). Lake outburst and debris flow disaster at Kedarnath, June 2013: Hydrometeorological triggering and topographic predisposition. *Landslides, 13*(6), 1479–1491. https://doi.org/10.1007/s10346-015-0584-3

Amin, M., Bano, D., Hassan, S. S., Goheer, M. A., Khan, A. A., Khan, M. R., & Hina, S. M. (2020). Mapping and monitoring of glacier lake outburst floods using geospatial modelling approach for Darkut valley, Pakistan. *Meteorological Applications, 27*(1). https://doi.org/10.1002/met.1877

Cheng, X., Shangguan, D., Yang, C., Li, W., Zhou, Z., Liu, X., Li, D., Zhang, X., Ding, H., Liu, Z., Yu, Y., Wang, X., He, B., Yang, Q., Li, Y., Wang, R., Liu, Y., Deng, L., & Shi, Y. (2024). Temporal and spatial changes of glacial lakes in the central Himalayas and their response to climate change based on multi-source remote sensing data. *Global and Planetary Change*, 104675. https://doi.org/10.1016/j.gloplacha.2024.104675

Chouhan, S., & Mukherjee, M. (2023). Design and application of a multi-hazard risk rapid assessment questionnaire for hill communities in the Indian Himalayan region. *Natural Hazards and Earth System Sciences, 23*(4), 1267–1286. https://doi.org/10.5194/nhess-23-1267-2023

Das, S., Das, S., Mandal, S. T., Sharma, M. C., & Ramsankaran, R. (2024). Inventory and GLOF susceptibility of glacial lakes in Chenab basin, Western Himalaya. *Geomatics Natural Hazards and Risk, 15*(1). https://doi.org/10.1080/19475705.2024.2356216

Das, S., Kar, N. S., & Bandyopadhyay, S. (2015). Glacial lake outburst flood at Kedarnath, Indian Himalaya: A study using digital elevation models and satellite images. *Natural Hazards, 77*(2), 769–786. https://doi.org/10.1007/s11069-015-1629-6

Dobhal, D. P., Gupta, A. K., Mehta, M., & Khandelwal, D. D. (2013). Kedarnath disaster: facts and plausible causes. *Current Science, 105*(2), 171–174. https://www.cabdirect.org/cabdirect/abstract/20133312000

Examine the causes and consequences of glacial lake outburst floods (GLOFs). How do these events pose a threat to mountainous regions, and what measures can be taken to mitigate the impact of GLOFs? (250 words, 15 marks)—PWOnlyIAS. (n.d.). PWOnlyIAS. https://pwonlyias.com/mains-answer-writing/examine-the-causes-and-consequences-of-glacial-lake-outburst-floods-glofs-how-do-these-events-pose-a-threat-to-mountainous-regions-and-what-measures-can-be-taken-to-mitigate-the-impact-of-glofs/

Guru, N. (2022). Evaluation of potential lakes susceptible to GLOF using multicriteria assessment in Jhelum sub-basin of Indus Basin. In *Disaster Resilience and Green Growth* (pp. 607–624). https://doi.org/10.1007/978-981-19-7100-6_33

Hall, D. K., Riggs, G. A., & Salomonson, V. V. (1995). Development of methods for mapping global snow cover using moderate resolution image spectroradiometer data. *Remote Sensing of Environment, 54*, 127–140.

Indian Institute of Remote Sensing, Sivan, K., Chauhan, P., Pandey, P., Champati Ray, P. K., Sharma, P., Chattoraj, S. L., Sharma, R., Nainwal, H. C., Nawaz Ali, S., & Singh, R. (2020). Glacial Lakes of Uttarakhand. In Indian Institute of Remote Sensing. *Indian Institute of Remote Sensing*. https://www.iirs.gov.in/iirs/sites/default/files/pdf/2021/Glacial_Lake_Atlas_HR_Uttarakhand.pdf

Khanduri, S. (2021). Flash Flood struck Dhauliganga valley on February 7, 2021: A Case study of Chamoli district of Uttarakhand Himalaya in India. *Academic Platform Journal of Natural Hazards and Disaster Management, 2*(1), 1–15. https://doi.org/10.52114/apjhad.903387

Kulkarni, A. V., Srinivasulu, J., Manjul, S. S., & Mathur, P. (2002). Field based spectral reflectance to develop ndsi method for snow cover monitoring. *J ISRS, 30*, 221–228.

Majeed, U., Rashid, I., Sattar, A., Allen, S., Stoffel, M., Nüsser, M., & Schmidt, S. (2020). Recession of Gya Glacier and the 2014 glacial lake outburst flood in the Trans-Himalayan region of Ladakh, India. *The Science of the Total Environment, 756*, Article 144008. https://doi.org/10.1016/j.scitotenv.2020.144008

Mandal, S. P., & Chakrabarty, A. (2016). Flash flood risk assessment for upper Teesta river basin: using the hydrological modeling system (HEC-HMS) software. *Modeling Earth Systems and Environment, 2*(2). https://doi.org/10.1007/s40808-016-0110-1

Martha, T. R., Roy, P., Govindharaj, K. B., Kumar, K. V., Diwakar, P. G., & Dadhwal, V. K. (2014). Landslides triggered by the June 2013 extreme rainfall event in parts of Uttarakhand state, India. *Landslides, 12*(1), 135–146. https://doi.org/10.1007/s10346-014-0540-7

Mishra, I. (2024). Uttarakhand to study risk posed by 13 glacial lakes during monsoon. The Hindu. https://www.thehindu.com/sci-tech/energy-and-environment/uttarakhand-will-study-glacial-lakes-to-avoid-outburst-scenarios/article68340516.ece.%20Accessed%2028%20June%202024.

National Institute of Disaster Management, Ministry of Home Affairs, Govt. of India. (2023). How Glacial Lake Outburst Flood (GLOF) & Landslide Lake Outburst surveillance system works? In *National Institute of Disaster Management, Ministry of Home Affairs, Govt. Of India* [Report]. https://www.gofundme.com/f/sikkim-glacial-lake-outburst-flood-recovery-fund

Petley, D. (2013). Reconstructing the events at Kedarnath using data, images and eye-witness reports. *The Landslide Blog*, 4 July 2013.Accessed 20 Nov 2014. http://blogs.agu.org/landslideblog/

Rao, B. S., Rao, Joshi, S., R, A., Raju, P. V., & National Remote Sensing Centre, ISRO. (2024). GLOF risk assessment of Suru Glacial Lake in Indus River Basin [Technical report]. National Remote Sensing Centre, ISRO. https://www.nrsc.gov.in

Rawat, M., Pandey, A., & Gupta, P. K. (2022). Geospatial analysis of glacial lake outburst flood (GLOF). In *Water science and technology library* (pp. 141–160). https://doi.org/10.1007/978-3-030-98981-1_6

Sarwar, M., & Mahmood, S. (2023). Exploring potential glacial lakes using geo-spatial techniques in Eastern Hindu Kush Region, Pakistan. *Natural Hazards Research, 4*(1), 56–61. https://doi.org/10.1016/j.nhres.2023.07.003

Sarwar, M., & Mahmood, S. (2023). Exploring potential glacial lakes using geo-spatial techniques in Eastern Hindu Kush Region, Pakistan. *Natural Hazards Research, 4*(1), 56–61. https://doi.org/10.1016/j.nhres.2023.07.003

Sati, S. P., & Gahalaut, V. K. (2013). The fury of the floods in the north-west Himalayan region: The Kedarnath tragedy. *Geomatics Natural Hazards and Risk, 4*(3), 193–201. https://doi.org/10.1080/19475705.2013.827135

Sattar, A., Cook, K. L., Rai, S. K., Berthier, E., Allen, S., Rinzin, S., et al. (2025). The Sikkim flood of October 2023: Drivers, causes and impacts of a multihazard cascade. *Science.* https://doi.org/10.1126/science.ads2659

Schmidt, S., Dame, J., & Nüsser, M. (2016). Glacier-induced hazards in the Trans-Himalaya of Ladakh (NW-India). EGUGA. http://ui.adsabs.harvard.edu/abs/2016EGUGA..1818019S/abstract

Schmidt, S., Nüsser, M., Baghel, R., & Dame, J. (2020). Cryosphere hazards in Ladakh: The 2014 Gya glacial lake outburst flood and its implications for risk assessment. *Natural Hazards, 104*(3), 2071–2095. https://doi.org/10.1007/s11069-020-04262-8

Schmidt, S., Nüsser, M., Baghel, R., & Dame, J. (2020). Cryosphere hazards in Ladakh: The 2014 Gya glacial lake outburst flood and its implications for risk assessment. *Natural Hazards, 104*(3), 2071–2095. https://doi.org/10.1007/s11069-020-04262-8

Shugar, D. H., A. Burr, U. K. Haritashya, et al. 2020. Rapid worldwide growth of glacial lakes since 1990. *Nature Climate Change, 10*, 939–945. https://https://doi.org/10.1038/s41558-020-0855-4

Thapliyal, A., & Kimothi, S. (2024). A geospatial modelling for glacial lake outburst floods susceptibility analysis in the Central Himalayas. *Journal of Water and Climate Change, 15*(9), 4293–4306. https://doi.org/10.2166/wcc.2024.024

Thapliyal, A., & Kimothi, S. (2024). A geospatial modelling for glacial lake outburst floods susceptibility analysis in the Central Himalayas. *Journal of Water and Climate Change, 15*(9), 4293–4306. https://doi.org/10.2166/wcc.2024.024

The Landslide blog. (n.d.). The Landslide Blog. https://blogs.agu.org/landslideblog/

Westoby, M. J., N. F. Glasser, M. J. Hambrey, J. Brasington, J. M. Reynolds, and M. A. A. Hassan. 2014. Reconstructing historic Glacial Lake Outburst Floods through numerical modelling and geomorphological assessment: Extreme events in the Himalaya. *Earth Surface Processes and Landforms, 39*, 1675–1692, https://doi.org/10.1002/esp.3617

Zheng, G., Allen, S. K., Bao, A., et al. (2021). Increasing risk of glacial lake outburst floods from future Third Pole deglaciation. *Nature Climate Change, 11*, 411–417. https://doi.org/10.1038/s41558-021-01028-3

Evaluation and Assessment of the Geomorphic Changes in the Upper Catchment of the Teesta Basin Due to the Extreme Flash Flood (South Lhonak Lake, 2021)

Olivia Halder and Arindam Sarkar

Abstract Due to the numerous and varied processes in the catchment, flash floods can cause significant harm to societies. Globally, it is evident that extreme rainfall events have increased in recent years due to climate change, and flash flooding is quickly becoming one of the most common types of natural disasters worldwide. Managing flash flooding is extremely challenging due to its suddenness, complex patterns, and difficulty in forecasting. According to popular belief, flash floods are highly concentrated, short-lived phenomena that peak at a high altitude less than six hours after rainfall. Nonetheless, these floods significantly affected most of the downstream portion of the river, where the valley gradient is minimal and is dramatically impacted by floods. The year 2021 witnessed a significant flash flood event that led to substantial geomorphic alterations in the upper watershed of the Teesta Basin, particularly in the South Lhonak Lake region in Sikkim, India. This event, due to its scale and impact, serves as a crucial case study for understanding the geomorphic changes caused by flash floods. This study will primarily examine the geomorphic modifications caused by the flash flood in the upper catchment area, focusing on the amount of precipitation, sediment dynamics, and landform deformation. It will also analyse the pre- and post-flood scenarios of the basin using data from remote sensing and GIS-based analysis. The probability of glacial lake outburst floods (GLOFs) has increased due to the rapid expansion of the proglacial South Lhonak Lake caused by accelerated glacial melting in recent years. Understanding the process is essential, as extreme events present additional challenges for risk management. This study will also emphasize the assessment of land use and land cover change after the flood scenario and analyze the peak discharge during the flood event using the hypsometric curve. Evaluate hazard management and minimize damage caused

O. Halder (✉)
Department of Geography, University of Calcutta, Calcutta, West Bengal, India
e-mail: oliviahalder2501@gmail.com

A. Sarkar
Department of Geography, Purash Kanpur Haridas Nandi College (University of Calcutta), Calcutta, West Bengal, India

© The Author(s), under exclusive license to Springer Nature Switzerland AG 2025
S. Ghosh (ed.), *The Himalaya Dilemma*, Springer Natural Hazards,
https://doi.org/10.1007/978-3-031-95083-4_20

by extreme flood events. Therefore, to comprehend exposure, people's socioeconomic vulnerability, and their ability to respond to floods, a flood risk assessment is required. The findings indicated that the midstream sections and upstream portions of the right bank of the Teesta River posed the greatest danger of flooding. There were notable differences in the level of flood risk among the upper watershed of the Teesta basin. A noteworthy inverse relationship was observed between vulnerability and capacity. The primary objectives of flood risk assessment are to enhance preparedness for potential future flood hazards, reduce flood risk through mitigation activities, and to compare pre- and post-flood scenarios.

Keywords South Lohnak Basin · Teesta basin · Geomorphic changes · Glacial Lake Outburst Flood (GLOF) · Remote sensing · Flash flood

1 Introduction

One of the most destructive natural disasters in India's hilly regions is flash floods. The Teesta Watershed in the Sikkim-Darjiling Himalayas is a likely area for flash floods. With a direct connection to catastrophic flood events, the current project aims to create a simulation model of surface runoff in the upper Teesta basin. The first step involves predicting the streamflow response at the watershed's outlet to create a primary unit hydrograph with time for an excess rainfall event. In particular, the methodology relied on data editing and the processing of morphological and meteorological data in a geospatial setting. The India Meteorological Department provided the rainfall time-series data, which was then analysed to determine water flow, assess peak discharge over time, and estimate runoff volume. To enhance flash flood forecasting, numerous studies have been carried out in the past 20 years (Zanchetta & Coulibaly, 2020; Nanditha & Mishra, 2021; Yordanova et al., 2022). Damages haven't been considerably lessened, though (Montz & Gruntfest, 2002). For proactive risk management of flash flood tactics, it is essential to understand the characteristics of flash floods and their potential impacts (Hapuarachchi et al., 2011). Many studies have been carried out globally to describe risk management (Xia et al., 2011; Ma et al., 2021) and flash floods (Bonacci et al., 2006; Rusjan et al., 2009).

Extreme flash flood events in the Sikkim Himalayan region have not been well studied (Marchi et al., 2010; Kumar et al., 2024). Due to extraordinary circumstances presenting new challenges for risk management, a clear understanding of the process is essential. The transboundary braided river Teesta experienced an intense flash flood in October 2021, which has been examined in this study. Since the majority of the population in the Teesta basin relies on agriculture as their primary source of income, it is a vital source of freshwater for both Bangladesh and India, particularly in West Bengal (Sarker et al., 2011). Originating in the Paohunri Mountain range in the Eastern Himalayas, the Teesta is a transboundary river that flows through Bangladesh and the Indian states of Sikkim and West Bengal before joining the mighty Brahmaputra–Jamuna River as a tributary. With a catchment area of 12,160

km^2, the river is approximately 414 km long overall, with only 19% of its length located in Bangladesh (Das et al.,2024; Talukdar et al., 2020). The river's braiding is a well-known feature. The Teesta basin is also heavily impacted by human activity, as seen by the three barrages that run along its course—two in India and one in Dalia, Bangladesh. A frequent hydrologic occurrence in the basin's Bangladeshi and Indian portions is flooding. The years 2015, 2000, 1996, 1993, 1978, 1976, 1975, 1973, 1968, and 1950 were the years with the most devastating floods (Das et al., 2024). The Teesta floods cannot be stopped even with three barrages. The July 2015 flood is an example. An unprecedented level of surplus discharge. The flash flood that affected a large area of the Teesta basin in India, resulting from an abrupt cloud burst in the Sikkim and Darjiling Himalayas (West Bengal), was 5,500 m^3/s. According to Das et al., 2024, the majority of the Jalpaiguri and Maynaguri blocks were under water for four days.

2 Overview of the Study Area

At approximately 5,200 m above sea level, South Lhonak Lake is situated southeast of the Indian state of Sikkim. The research region encompasses the downstream drainage area of the Chungthang Dam site, which is the confluence of the Lachen and Lachung rivers, as well as the South Lhonak glacier lake, located at coordinates 27.913° N and 88.199° E. The south Lhonak glacier provides the majority of the meltwater that enters the lake, in addition to the outflow from the northern Lhonak glacial lake. In 2019, the glacier's total extent was mapped as 12.5 km^2. The lake has expanded significantly over time, increasing from 0.42 km^2 in 1990 to 1.35 km^2 in 2019, in tandem with glacial retreat (Sattara et al.,, 2021). The lake terminal moraine, which forms the lake's downstream boundary, is composed of glacial deposits that are not well-cemented. On the other hand, the lateral moraines on the lake's margins are composed of loose material. The lake's outflow drainage channel has a rather steep slope of 30:1 V, making it ideal for high flood wave velocities. Due to the significant precipitation in the local environment, flash floods can occur in areas downstream. The lake's only exit is the Goma channel, which drains into the *Zemu* River 36 km downstream. While the area around the lake is sparsely populated, downstream of the lake are several communities and significant infrastructure, including dams and barrages (Sharma et al., 2018) (Fig. 20.1).

3 Data and Methodology

An extensive literature review was done to prepare a data inventory for the study. The primary sources of secondary data include published articles in various journals, book chapters, magazine articles, personal weblogs, and various social media groups. Data have been compiled from published papers, maps, theses, and other

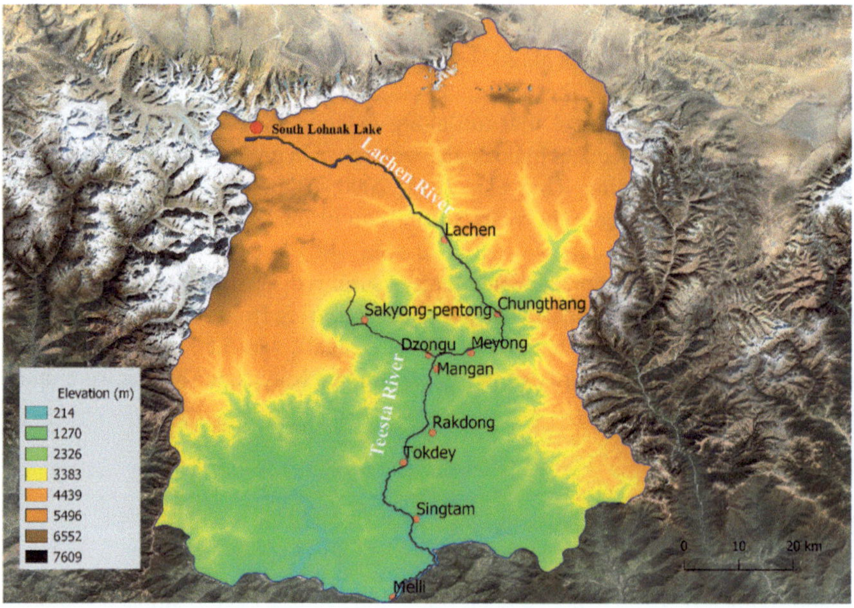

Fig. 20.1 Physiography of Sikkim. It shows, a small Himalayan state in northeastern India, is characterized by rugged mountainous terrain, deep valleys, and steep slopes. It lies with elevations ranging from about 300 m in the south to over 8,586 m at Kanchenjunga, the third-highest peak in the world. The state is divided into three physiographic zones: the Lesser Himalayas, the Greater Himalayas, and the Teesta River Valley, which serves as the main drainage system. The diverse topography influences the region's climate, biodiversity, and settlement patterns

sources. A comprehensive interpretation and evaluation have been conducted using topographical maps from the Survey of India, geological and geotectonic maps from the Geological Survey of India, and the 2011 Census report from the Census of India. Using GIS software, satellite photos have been used to create landscape and human habitation maps. Mapping has been conducted using geospatial techniques, incorporating data from multiple sources for a comprehensive view. However, Sentinel-2 data has been used to show the change in the lake volume of the months of August and October.

4 Geology and Topography of the Study Area

The terrain and geology of Sikkim's South Lhonak Lake are significant factors in the lake's susceptibility to eruptions. The lateral and terminal borders of the lake, located in the eastern Himalayas at an elevation of approximately 5,200 m, are formed by weakly consolidated glacier moraines. These moraines are naturally unstable and prone to collapsing under stress because they are composed of loose sediment and unconsolidated elements that the retreating glaciers deposited (Sattara et al., 2021). High-velocity floodwaters during a breach are made possible by the lake's outflow drainage channel's steep slope, which has a gradient of 30:1. The lake's location increases the potential of debris-filled outbursts in a glacial valley created by previous glacier movements, which supplies an ample supply of silt (Das et al., 2024). Furthermore, the region's geological instability is exacerbated by the nearby Higher Himalayan Crystalline rocks, which are characterised by faults and fractures. This makes the area vulnerable to landslides and seismic activity, which can cause moraine failure. The moraine walls are further destabilized by the tremendous hydrological strain caused by the lake's rapid expansion, which is fueled by seasonal snowmelt and accelerated glacier melt. South Lhonak Lake's catastrophic outburst in October 2023 was the result of a fragile situation exacerbated by geological and topographical features, as well as environmental stressors such as heavy rainfall and warming temperatures.

5 Slope Factor

The destabilization of the lateral moraine, resulting from permafrost thawing, was a leading cause. Rising temperatures associated with climate change have caused permafrost, which serves as a natural binder for rock and silt, to deteriorate. The slopes surrounding the lake become less structurally sound as a result of this melting, making abrupt collapses more likely. Due to constant pressure from the growing lake waters, the lateral moraine, composed of loose rock and glacier debris, became extremely unstable (Maurer et al., 2020). Accelerated glacial melting over the past few decades has caused South Lhonak Lake to expand quickly, increasing the volume and strain on its natural dam. The system was highly vulnerable to failure because the already unstable moraine slopes were further destabilized by the rising water level and its weight. A disastrous glacial lake outburst flood (GLOF) was ultimately triggered by a massive displacement wave generated by the moraine's collapse into the lake, which breached the natural dam (Wieczorek et al., 2022). The steep topography of the surrounding region, combined with unstable glacier deposits and increased water pressure, significantly contributed to exacerbating the disaster.

6 Results and Discussion

6.1 Historical Events of GOLF Events in India

Highlighting the main GOLF events which has been taking place in the Himalayan Region in the last 30 years in order to interpret the vulnerability and the disasters it caused.

The Table 20.1 highlights major Glacial Lake Outburst Flood (GLOF) events in India from 1990 to 2024, showcasing their locations, triggering glacial lakes, event descriptions, and associated disasters. Notable incidents include the 1990 Pangong Lake flood in Ladakh, one of the earliest documented GLOF-like events, and the catastrophic 2013 Chorabari Lake breach in Uttarakhand, which caused over 5,700 deaths and extensive damage. Sikkim's South Lhonak Lake, a hotspot for recent GLOF events, experienced destructive floods in both 2020 and 2023, with the latter resulting in over 30 fatalities and widespread infrastructure damage. The data reveals an alarming increase in the frequency and intensity of GLOF events, primarily driven by climate change and accelerated glacial melting, underscoring the need for proactive monitoring, risk mitigation strategies, and infrastructure adaptation in vulnerable Himalayan regions.

6.2 *Changes in Overall Geomorphology of Landscape and River Channel Migration and Widening of Channel*

The geomorphological impacts of glacial lake outburst floods (GLOFs) are profound and multifaceted, particularly in regions experiencing significant glacial retreat and climate change (Fig. 20.2). The South Lohnak Lake, like many glacial lakes, is susceptible to such catastrophic events, which can lead to substantial alterations in landscape morphology and river channel dynamics. GLOFs are characterized by sudden and often massive releases of water from glacial lakes, which can result in rapid channel migration, widening, and the formation of new geomorphological features. Research indicates that GLOFs can dramatically reshape river channels and surrounding landscapes through processes such as erosion, sediment transport, and deposition.

For instance, Wells et al. emphasize that understanding the geomorphological impacts of jökulhlaups (a type of GLOF) is crucial for reconstructing paleofloods and predicting future events, highlighting the role of extreme hydrological events in landscape evolution (Wells et al., 2022). Similarly, Thorndycraft et al. note that the geomorphic evidence from glacial lake outburst floods reveals unique landscape signatures that reflect the variability in geological and topographical settings, which influence the extent of geomorphological changes (Thorndycraft et al., 2015). The dynamics of GLOFs can lead to significant channel reconfiguration. Westoby et al. discuss how these floods can perform extensive geomorphological reworking of river channels, including the creation of new fluvial pathways and the alteration of existing ones (Westoby et al., 2014). This is echoed by Maurer et al., who provide insights into the geomorphic changes resulting from a GLOF in Bhutan, illustrating how such events can lead to the mobilization of large sediment loads and the formation of new landforms (Maurer et al., 2020).

Furthermore, Cook et al. emphasize that GLOFs can induce fluvial erosion, significantly affecting river morphology and sediment dynamics in the Himalayas (Cook

Table 20.1 Year and location of golf events in India

Year	Location	Glacial lake	Event description	Disasters occurred
1990	Jammu & Kashmir, Ladakh	Pangong Lake	GLOF-like event, triggered by rapid glacier melt	Flash floods led to property damage, erosion, and displacement of communities
2000	Sikkim, Eastern Himalayas	Lhonak Lake	Lake outburst caused by rapid melting	Flooding damaged infrastructure, including roads and agricultural land
2004	Himachal Pradesh, Kullu Valley	Bara Shigri Glacier	GLOF-like event after melting	Damaged property, disrupted agriculture, and affected local communities
2010	Uttarakhand, Garhwal Region	Chorabari Lake	GLOF-like event following monsoon rains	Triggered flash floods, landslides, and the loss of life in the Kedarnath valley
2013	Uttarakhand, Garhwal Region	Chorabari Lake	Major GLOF event post-cloudburst	Over 5,700 deaths, extensive damage to villages, infrastructure, and roads
2016	Jammu & Kashmir, Ladakh	Tso Moriri Lake	GLOF event triggered by warm weather	Flash floods led to loss of crops and damage to infrastructure
2017	Himachal Pradesh, Kullu Valley	Bara Shigri Glacier	GLOF-like water surge from melting glaciers	Water surge caused destruction of homes, crops, and displaced villagers
2020	Sikkim, Eastern Himalayas	South Lhonak Lake	Breach of the glacial lake	Flash floods, property damage, infrastructure destruction, and community displacement
2023	Sikkim, North Sikkim District	South Lhonak Lake	Major GLOF event triggered by lake breach	Over 30 fatalities, widespread destruction of bridges, hydroelectric dams, homes, and agricultural lands

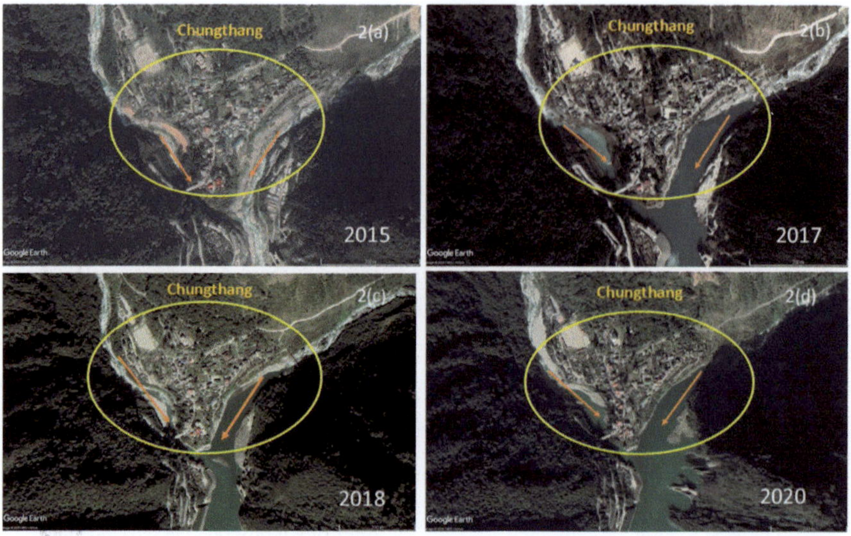

Fig. 20.2 These historical satellite images show an overall geomorphological change of the Teesta River over the past 5 years. **a** shows that the river exhibits a braided and slightly meandering pattern, with visible sediment deposits indicating active erosion and deposition. Steep banks and deforested slopes indicate ongoing erosion, with gully formations suggesting slope instability. Human activities, including infrastructure development, embankments, and possible hydroelectric projects, have likely altered the river's natural flow. **b** shows that the river appears more stabilized, with reduced visible sediment deposition, indicating a decreased sediment load or better erosion control measures. Additionally, human settlements have expanded, and infrastructure development has intensified, suggesting growing anthropogenic influence. **c** shows increased sediment deposition along the riverbanks, particularly on the right bank, forming a mid-channel sandbar indicative of altered sediment transport and deposition patterns. **d** highlights ongoing geomorphological changes in the Teesta River, particularly in sediment deposition and water dynamics. The formation and expansion of mid-channel sandbars are evident, indicating continued sediment transport and deposition influenced by fluctuating water levels, likely due to dam regulation. The water appears more stabilized, suggesting controlled flow patterns. There is a noticeable increase in exposed sediment patches, possibly due to seasonal variations or reduced discharge. Human interventions, such as infrastructure expansion, remain visible, reinforcing the anthropogenic impact on the river system

et al., 2018). The widening of river channels due to GLOFs is a critical concern, particularly in the context of increasing glacial melt and the frequency of outburst events. Gao points out that factors such as temperature and precipitation are key drivers of GLOF hazards, which can exacerbate the risk of channel migration and widening (Gao, 2024). This is particularly relevant in the context of climate change, which is expected to increase the frequency of glacial lake outbursts, thereby intensifying their geomorphological impacts (Wieczorek et al., 2022).

6.3 *Landslide-Sediment-Entrainment*

The upper catchment of the Teesta Basin is particularly vulnerable to landslides and erosion, exacerbated by hydrological changes resulting from glacial lake outburst floods (GLOFs), such as those potentially originating from South Lohnak Lake. The geomorphological dynamics in this region are influenced by both natural processes and anthropogenic factors, resulting in significant alterations to landscape stability and river morphology. Landslides in mountainous regions, such as the Teesta Basin, are often triggered by intense rainfall and rapid snowmelt, which can exacerbate GLOFs. Chakraborty and Biswas demonstrate the effectiveness of hydrological models in simulating flow conditions in ungauged river sites, indicating that accurate modelling can help predict flood events and their potential impacts on erosion and landslide occurrences (Chakraborty & Biswas, 2021). The study emphasizes the significance of understanding surface runoff dynamics in mountainous river basins, which is essential for evaluating the risks associated with GLOFs. The role of geological and hydrological processes in the Teesta River, emphasizing that the basin experiences significant CO_2 consumption rates due to weathering processes, which are closely linked to erosion dynamics (Das et al., 2021).

The interaction between chemical weathering and physical erosion can lead to increased sediment loads in river systems, further destabilizing slopes and enhancing the likelihood of landslides. The unstable behaviour of the Teesta River, characterized by frequent channel migration and braiding patterns, is a direct consequence of these geomorphic processes. Das and Santosh note that variations in bank erosion severity can lead to significant changes in channel morphology, which in turn impact both the riverine ecosystem and the livelihoods of communities dependent on the river (Das et al., 2022). This instability is exacerbated by the effects of climate change, which can increase the frequency and intensity of extreme weather events, thereby contributing to increased erosion and landslide risks (Arfanuzzaman et al., 2020). In addition to natural factors, human activities, such as deforestation and changes in land use, exacerbate erosion in the upper catchment. The work of Khan and Islam highlights significant rates of bank erosion observed in the Brahmaputra–Jamuna River system, which can be indicative of similar patterns in the Teesta Basin (Khan & Islam, 2003). Anthropogenic modifications to riverbanks, including the construction of infrastructure, can destabilize slopes and increase the susceptibility to landslides. Furthermore, the impact of GLOFs on sediment transport and river morphology cannot be understated. Liu et al. describe how GLOFs can evolve into debris flows, which significantly increase peak discharge and sediment loads downstream, leading to destructive erosion and landslides in the affected areas (Liu et al., 2020). This is particularly relevant for the Teesta Basin, where the interplay between glacial dynamics and hydrological events poses a serious risk to both the environment and local communities.

6.4 Flood Quantification and Detection of Flood Extent

South Lhonak Lake's high elevation over 5000 m above sea level in the Higher Himalayan Crystalline zone, which forms the hanging wall of the South Tibetan Detachment System, is the leading contributing cause. Significant vegetation cover, which would typically act as a barrier to prevent erosion from intense rainfall, is conspicuously lacking in this area. It is essential to recall that the valley itself was formed by previous glacial activity. Therefore, it is not limited by a lack of available sediment. The Lhonak glacier melted to form the portion of South Lhonak Lake, which is surrounded by terminal and lateral moraines (Yu et al., 2024). These moraines are prone to falling, sliding, and subsiding due to their loose, unconsolidated material, a result of their recent development. The satellite imagery available on websites like Google Earth clearly reveals this vulnerability (Fig. 20.3). The lake's water level also increased significantly as a result of environmental changes, including more excellent summertime snow and ice melting and some precipitation at this high elevation, which is likely related to global warming. The surrounding moraine may have experienced significant compressional stresses as a result of the lake's water volume increasing due to freezing or the transition from a liquid to a solid state. A massive impulse wave would have also been produced by the calving of the NNW margin and the South Lhonak Lake glacier snout. Ultimately, this pressure led to the collapse of a 500-m-long section of the left lateral moraine. The lake's outlet eroded and widened as a result of the force created by the collapse of the left lateral moraine, which caused a rapid discharge of water (Brodsky et al., 2024). The slope on the NNW side remained active, which is another significant finding from the time-series data, particularly evident in the Sentinel-2B image of October 9, 2023. Fresh mass movements triggered by this continuous action have significantly altered the lake's shoreline in that area. Although the lake remains in its original location, its current condition makes it more hazardous for potential future occurrences. Analysis of satellite images reveals that the massive surge of water and silt caused the river's flow to alter near the exit, where the break occurred. Following the catastrophe, the lake's exit widened from 18 m to about 35 m. As a result, the river began to fill with too much water and silt. The condition was made worse by river deposits and the loose, unconsolidated moraine. As a result, massive amounts of water carrying loose debris moved quickly downstream.

A comparative study before and after the event revealed an increase in sediments and river flow. Carrying debris from the surrounding areas and glacial moraines, the massive water flowed toward the Sikkim Urja Ltd dam (1200 MW) at Chungthang, causing the most devastating destruction ever seen in the area. At an elevation of 2481 m above sea level, a significant landslip event was also noted at Lachen, leading to a partial obstruction of the Teesta River. This blockage would have momentarily slowed the flow of further water as it proceeded to emerge from the glacier. The Sikkim Urja Ltd dam might have been breached as a result of a large amount of water being released downstream when the dam formed by the landslide gave way. Another element that amplifies the destructive power of the gradient-laden, ruptured

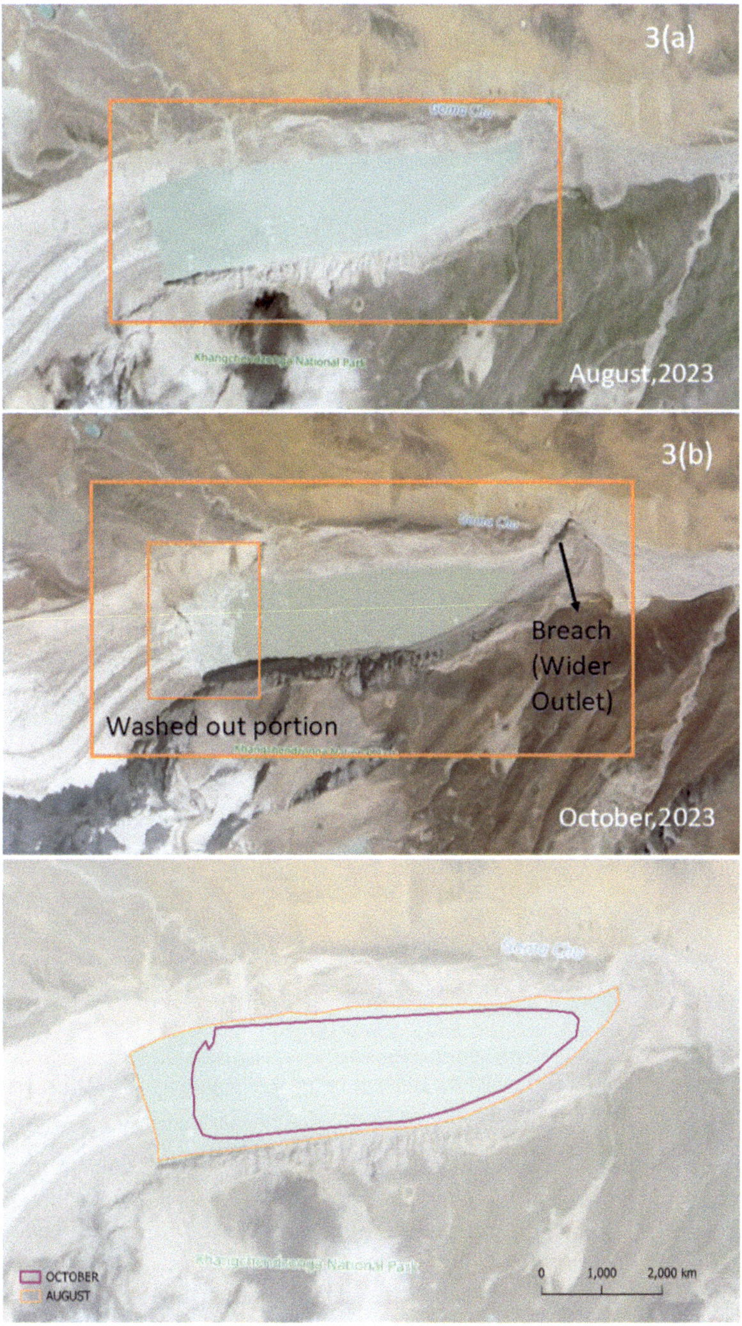

Fig. 20.3 Glacial Lake and Surrounding Terrain Morphological Development Before and After Outburst. **a** Glacial Lake condition prior to outburst. **b** Glacial Lake condition, following **c,** showing the change that has taken place due to the breakdown of the lateral moraine

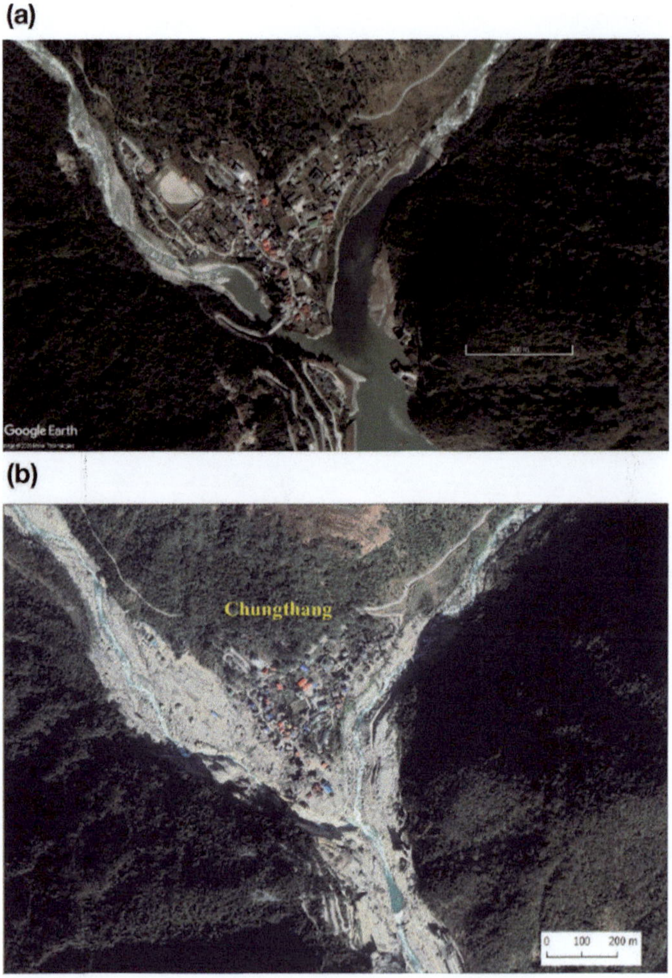

Fig. 20.3 The October 2023 GLOF event has completely washed out a permanent concrete bridge in Chungthang, as shown in (**a**). This bridge was destroyed in 2023, as depicted in (**b**). A large amount of sediment and debris accumulation can be observed in recent pictures. A bamboo bridge was constructed at Chungthang through the combined efforts of the army, police, and local residents

lake water is the Teesta River's straight path from the junction of Lhonak Chu and the Teesta River to Chungthang. At Chungthang, where the Teesta and Lachung Rivers converge, the upper catchment area of the Lachung River experienced heavy rainfall, which may have contributed to the break of the Sikkim Urja Ltd. dam. Lately, both summer and winter disasters in the Himalayas have become more frequent and intense. The rise in anthropogenic activity has led to an increase in the frequency of disasters (Dubey & Goyal, 2020). Given that development activities pose a direct threat to the environment, this tendency is likely to worsen in the future. The construction of man-made structures has disrupted the rivers' natural flow patterns, causing them to deviate from their intended course. Choosing safe land use locations would be a challenging task to accomplish, given the area's increasing development activities and trend towards urbanization. More than 10,000 people resided in over 1,900 houses in Chungthang, and the state's rate of urbanization increased by around 25% between 1971 and 2011, according to the 2011 census report. Therefore, when rebuilding the damaged area in the future, careful consideration of these factors is essential.

Long-term glacial melting and the partial collapse of the left-lateral moraine were two key contributing factors to the October 2023 glacial lake outburst flood (GLOF) from South Lhonak Lake. Increased water levels and the potential for ice calving near the glacier snout posed weighty compressive loads on the lake, which had already reached a critical threshold for stability. Moraine destabilization was a result of this pressure building, which was exacerbated by probable impulse waves caused by ice block separation. Approximately 38.49 million cubic meters of water were released rapidly downstream as a result of the sudden collapse of a substantial portion of the moraine (Joshi et al., 2024). Floodwaters moved quickly due to the high gradient and steep outflow channel, carrying large volumes of silt and debris that eroded the riverbanks and altered the downstream morphology.

6.5 *Impact Assessment on Infrastructure, Livestock, and Agriculture*

The entire length of South Lhonak Lake, which first flows through the Lhonak River before joining the Zeemu River and finally meeting the Teesta River, was severely damaged and destroyed by the GLOF in October 2023. There have been several kinds of disasters and harm. There was significant damage to the upper riparian area, particularly in terms of cattle washing away, infrastructure destruction, loss of connectivity, and erosion of land. Five significant bridges that served as the primary means of communication between the area and the rest of the nation were destroyed. Many minor bridges were also washed away (Zhang et al., 2024). In addition to enhancing national security and connecting the area to the rest of the country, these bridges facilitated the transportation of animals. For instance, the summer grazing fields were connected to the winter pasture lands by the Zeema 1 bridge, situated at the confluence of the Teesta and Zeemu rivers. The livestock's ability to move was eventually hampered and restricted by the erosion of the bridges. Numerous roads were entirely washed away, and the GLOF triggered numerous landslides. The route between Chungthang (Fig. 20.3) and Lachen has been washed away entirely in several locations. There were certain areas where the conditions were so bad that people could not even walk on them. Livestock became immobile due to the roads' destruction, which was exacerbated by the absence of bridges. The livestock should have been moved to the winter grazing regions by the first week of November, but instead, they

Fig. 20.4 Shows the Settlement of Munshithang along the side of the river Teesta

were stranded in the summer pasture area. Smaller settlements, including Daray, Munshithang (Fig. 20.4) Latong, Denya, Rabom, Phem Cheka, and Zeema, were destroyed or washed out. The herder used to spend an average of five months of the year in these communities, which served as winter grazing areas for sheep. Since the herders now have nowhere to go, the loss of these villages has seriously disrupted the movement of the animals. Additionally, the herders face considerable difficulty and trouble as they are unable to relocate their herds from summer pasture areas to winter pasture areas due to winter snowfall.

6.6 Sustainability and Assessment of Hazards in the Lower Basin of the Teesta Basin

The growing risk of Glacial Lake Outburst Floods (GLOFs) poses significant sustainability issues and hazards for the lower Teesta basin, which encompasses parts of Bangladesh, West Bengal, and India. These floods are triggered by abrupt glacial lake breaches, which are often caused by landslides, heavy precipitation, or rapid glacier melting resulting from climate change. The source of the Teesta River is the Himalayan glaciers, where unstable glacial lakes, such as South Lhonak Lake in Sikkim, have formed and expanded due to warming temperatures. The magnitude of this threat was brought to light by the terrible GLOF event in October 2023, which resulted in devastating floods that destroyed significant infrastructure, such as bridges and a hydroelectric power plant, while uprooting thousands of people and causing damage to huge agricultural regions (Allen et al., 2022). In downstream areas, floodwaters reportedly reached 15–20 feet, upsetting livelihoods, ecosystems, and financial stability. These recurrent disasters seriously threaten both the ecology and human populations in the lower basin. A multifaceted strategy is necessary to address this threat, which includes strengthening climate-resilient infrastructure, implementing advanced early warning systems, and conducting regular glacial monitoring (Lachenpa, 2024). Furthermore, encouraging transboundary collaboration between Bangladesh and India for the management of water resources is crucial to mitigating the adverse effects of GLOFs and ensuring the basin's long-term sustainability.

The situation was swiftly addressed by local communities, activists, academics, scientists, and members of civil society. There is a noticeable lack of mountain-specific legislation that effectively manages the Himalayan region's vulnerable ecosystems, and anti-dam sentiments have gained significant momentum. Power, water, and infrastructure designers, as well as government agencies, must acknowledge that the little state of Sikkim (Kaushik et al., 2024), which uses only about 400 MW of electricity but was producing 2,300 MW, has far outgrown its power generation capability. The relief efforts and rescue missions moved swiftly. To deliver supplies and food to the isolated villages, the locals constructed zip lines, showcasing the mountain communities' enduring tenacity. The October 2023 GLOF and downstream flood on the Teesta has become a wake-up call for Sikkim as well as all Himalayan states in India and beyond, given the comparable dangers that have already affected the Western Himalayas earlier this year. When making decisions about development, national and international organisations must use scientific knowledge. Natural hazards in the Himalayas are expected to become more frequent due to the combined effects of changing socioeconomic and climatic conditions, increasing the risk of harm to local infrastructure and populations.

6.7 *Future Insights of the Golf Events in Sikkim*

Significant lake area expansions were identified during the examination of glacial lake changes between 1990 and 2023, particularly for lakes classified as high to very high risk using both the AHP and FAHP methodologies. For instance, the North and South Lohnak lakes, as well as the Gurudongmar lakes, have grown significantly, most likely due to climate variables such as rising temperatures and increased glacial melt (Vilímek et al., 2014). The majority of Sikkim's glacial lakes appear to be at a low risk of outburst floods, as indicated by the AHP method's classification of 34 out of 47 glacial lakes as being in very low to low susceptibility zones. Seven lakes were identified as being in the high to very high susceptibility zones; nevertheless, this highlights the necessity of careful observation and preventive measures. This implies that, although most lakes are currently regarded as safe, the high-risk lakes pose potential risks that should be addressed through efficient planning and mitigation techniques to prevent future catastrophes. The integration of critical factors, such as lake volume, area, seismic activity, rainfall, elevation, slope, and proximity to river systems, highlights the complex relationships that influence GLOF susceptibility. While the FAHP approach offers a more nuanced evaluation, focusing on moderate risk categories and showcasing its potential for detailed risk assessments, the AHP method classifies most glacial lakes as low-risk. However, it identifies specific lakes in high-risk zones, highlighting the need for vigilant monitoring and proactive mitigation strategies (Zeller et al., 2024). Both methods demonstrated impressive predictive accuracy, with AHP achieving a higher AUC of 0.92 compared to FAHP's 0.88, confirming their reliability in identifying high-risk lakes and guiding targeted risk reduction initiatives (Das et al., 2024). The expansion of high-risk lakes over the past three decades is a concerning trend that underscores the urgent need for adaptive management techniques to address the increasing threats posed by glacial dynamics and climate change. Although the majority of Sikkim's glacial lakes fall into the low to moderate risk range, a sizable portion still resides in the high to very high susceptibility zones, underscoring the need for targeted management approaches. Furthermore, given the continued effects of climate change, proactive monitoring is necessary due to the developing correlation between increasing glacial lake volumes and the severity of outburst floods. This study enhances our understanding of Sikkim's glacier lake dynamics and provides an invaluable resource for decision-makers and disaster relief agencies. Policymakers can effectively mitigate potential GLOF risks and safeguard infrastructure and human life in sensitive areas by integrating these findings into comprehensive risk management systems. To improve forecast accuracy, add more parameters, and refine susceptibility models, more research and interdisciplinary cooperation will be necessary. Stakeholders can increase resilience against upcoming environmental issues (Schmidt & Nusser, 2009) and ensure the safety and sustainability of communities in Sikkim and other comparable locations worldwide by prioritizing proactive measures such as enhanced monitoring systems and infrastructure fortification.

The rapid development of glacial lakes, driven by climate change, has raised concerns about the future of glacial lake outburst floods (GLOFs) in Sikkim. According to studies, the region's glacial lakes have expanded significantly in size over the last three decades, thereby increasing the likelihood of GLOFs. The South Lhonak Lake, for example, has experienced significant growth, and its most recent eruption in October 2023 resulted in devastating flooding in the Teesta River basin, causing substantial infrastructure damage and fatalities. To guard against future GLOFs, researchers have emphasized the importance of critical and operational monitoring of such lakes. The growing occurrence of extreme weather phenomena, such as abrupt cloudbursts, further increases the risk. Implementing thorough hazard assessments, establishing early warning systems, and developing adaptable infrastructure planning are crucial to reducing the likelihood of future disasters. To increase resilience against possible GLOF events in Sikkim, proactive steps and ongoing monitoring are crucial.

7 Conclusion

Significant new information about the intricate dynamics of glacial lake outburst floods (GLOFs) and their geomorphic, environmental, and socioeconomic impacts can be gained from studying the geomorphic changes in the upper catchment of the Teesta Basin, especially following the devastating flash flood triggered by the South Lhonak Lake outburst. A crucial case study for evaluating the environmental vulnerabilities of the Himalayan region and the increasing threats posed by glacial melting due to climate change is the October 2023 flood. The results show that the South Lhonak Lake has become a potential hotspot for hazards due to extended glacier retreat and unstable moraine structures. A tremendous amount of water was unleashed by its abrupt eruption, causing flash floods that destroyed infrastructure, agricultural lands, and towns downstream.

To detect changes in the lake's size and topography, the research emphasizes the importance of using sophisticated remote sensing techniques and GIS-based analysis. This provides an efficient tool for proactive flood risk assessment. Comparisons of satellite imagery taken before and after the flood revealed notable changes to the landscape, including increased erosion, enlargement of river channels, and sediment displacement. The catastrophic scope of the geomorphic changes brought on by the GLOF was demonstrated through a thorough analysis of sediment dynamics and flood discharge. Human communities that depend on the Teesta River, as well as natural ecosystems, were impacted by these long-lasting changes in river morphology. The study's focus on the complex link between environmental vulnerabilities and socioeconomic risks is one of its main contributions. According to the study, the Teesta River's right-bank upstream and midstream segments were most vulnerable to flooding, resulting in significant losses to infrastructure, property, and agricultural output. Homes were washed away, and important bridges were destroyed, severely disrupting communication and transportation networks in small communities, including Zeema, Latong, and Phem Cheka. The situation was exacerbated by the loss of agricultural areas and grazing fields, which endangered the stability of the agropastoral populations' means of subsistence in the area.

To summarize, the results of this investigation underscore the importance of implementing a multifaceted strategy for mitigating GLOF hazards. Incorporating sophisticated remote sensing technology, thorough geomorphic evaluations, and robust socioeconomic studies can yield valuable information for informed decision-making. To ensure sustainable management of the Teesta Basin and protect its populations from the changing threat of glacial hazards, it will be crucial to strengthen climate-resilient infrastructure, raise community awareness, and promote cross-border collaboration. Policymakers, researchers, and local stakeholders must prioritize proactive actions that increase resilience and reduce disaster risks in this ecologically sensitive and geologically unstable region, in light of the lessons learned from the 2023 South Lhonak Lake outburst.

References

Allen, S. K., Sattar, A., King, O., Zhang, G., Bhattacharya, A., Yao, T., & Bolch, T. (2022). Glacial lake outburst flood hazard under current and future conditions: Worst-case scenarios in a transboundary Himalayan basin. *Natural Hazards and Earth Systems Sciences, 22*, 3765–3785. https://doi.org/10.5194/nhess-22-3765-2022

Arfanuzzaman, M., Hassan, S. M. T., & Syed, M. A. (2020). Cost-benefit analysis of promising adaptations for resilient development in climate hotspots: Evidence from the lower Teesta basin in Bangladesh. *Journal of Water and Climate Change, 12*(1), 44–59. https://doi.org/10.2166/wcc.2020.130

Bonacci, O., Ljubenkov, I., & Roje-Bonacci, T. (2006). Karst flash floods: An example from the Dinaric karst (Croatia). *Natural Hazards and Earth System, Sciences 6*(2), 195–203. https://doi.org/10.5194/nhess-6-195-2006

Brodský, L., Rusnák, S., Schmidt, S., Vilímek, V., Azzoni, R., Nüsser, M., Tronti, G., Kropáček, J., Pandey, A. (2024). Interannual spatiotemporal evolution of the supraglacial lakes on the Belvedere Glacier between 2000 and 2023. *AUC GEOGRAPHICA, 59*: 1–16. https://doi.org/10.14712/23361980.2024.24

Chakraborty, S. and Biswas, S. (2021). Simulation of flow at an ungauged river site using the HEC-HMS model for a mountainous river basin. *Arabian Journal of Geosciences, 14*(20). https://doi.org/10.1007/s12517-021-08385-5

Cook, K., Andermann, C., Gimbert, F., Adhikari, B. R., & Hovius, N. (2018). Glacial lake outburst floods as drivers of fluvial erosion in the Himalayas. *Science, 362*(6410), 53–57. https://doi.org/10.1126/science.aat4981

Das, S., Tripathy, G. R., Santosh, K., Danish, M., Thakur, D., Dutt, Sarangi, S. (2021). The role of sulfuric acid in continental weathering: Insights from dissolved major ions and inorganic carbon isotopes of the Teesta River, Lower Brahmaputra System. *Geochemistry, Geophysics, Geosystems, 22*(4). https://doi.org/10.1029/2020gc009324

Das, M., Parveen, M. T., Ghosh, D., Sarkar, A., Alam, J., & Saha, S. (2022). Unstable behavioural pattern of the Teesta River and its impact on riverine dwellers: A case study of the confluence area of the Teesta and Dharala rivers, India. https://doi.org/10.20944/preprints202212.0510.v1

Das, A., Singh, S. K., Kanga, S., Sajan, B., Meraj, G., & Kumar, P. (2024). Glacial lake outburst flood susceptibility mapping in Sikkim: A comparison of AHP and fuzzy AHP models. *Climate, 12*(11), 173. https://doi.org/10.3390/cli12110173

Dubey, S., Goyal, M. (2020). Glacial Lake Outburst Flood (GLOF) hazard, downstream impact, and risk over the Indian Himalayas. *Water Resources Research, 56*, e2019WR026533. https://doi.org/10.1029/2019WR026533

https://www.icimod.org/

Hapuarachchi, H. A. P., Wang, Q. J., & Pagano, T. C. (2011). A review of advances in flash flood forecasting. *Hydrological Processes, 25*(18), 2771–2784. https://doi.org/10.1002/hyp.v25.1810.1002/hyp.8040

Joshi, M., Kumar, K., Rajeevan, K., & Barfal, S. S. (2024). Tragedies due to human folly: A learnt lesson from the year 2023 Sikkim avalanche. *Quaternary Science Advances, 13*. https://doi.org/10.1016/j.qsa.2023.100162

Kaushik, S., Rafiq, M., Dharpure, J. K., Howat, I., Moortgat, J., Joshi, P. K., Singh, T., & Dietz, A. J. (2024). Increasing risk of glacial lake outburst flood in Sikkim, Eastern Himalaya under climate warming. *Remote Sensing Applications: Society and Environment, 36*. https://doi.org/10.1016/j.rsase.2024.101286

Khan, N. I., & Islam, A. (2003). Quantification of erosion patterns in the Brahmaputra–Jamuna river using geographical information systems and remote sensing techniques. *Hydrological Processes, 17*(5), 959–966. https://doi.org/10.1002/hyp.1173

Lachenpa, S. (2024). *The South Lhonak Lake, Sikkim Glacial Lake outburst flood: Impact on agro-pastoral practices and local responses*. https://chs.snu.edu.in

Liu, M., Chen, N., Yong, Z., & Deng, M. (2020). Glacial lake inventory and lake outburst flood/debris flow hazard assessment after the Gorkha Earthquake in the Bhote Koshi Basin. *Water, 12*(2), 464. https://doi.org/10.3390/w12020464
Marchi, L., Borga, M., Preciso, E., & Gaume, E. (2010). Characterisation of selected extreme flash floods in Europe and implications for flood risk management. *Journal of Hydrology, 394*(1–2), 118–133. https://doi.org/10.1016/j.jhydrol.2010.07.017
Ma, M., Zhao, G., He, B., Li, Q., Dong, H., Wang, S., & Wang, Z. (2021). XGBoost-based method for flash flood risk assessment. *Journal of Hydrology, 598*, 126382. https://doi.org/10.1016/j.jhydrol.2021.126382
Maurer, J. A., Schaefer, J. M., Russell, J., Rupper, S., Wangdi, N., Putnam, A. E., … & Young, N. E. (2020). Seismic observations, numerical modeling, and geomorphic analysis of a glacier lake outburst flood in the Himalayas. *Science Advances, 6*(38). https://doi.org/10.1126/sciadv.aba3645
Montz, B. E., & Gruntfest, E. (2002). Flash flood mitigation: Recommendations for research and applications. *Environmental Hazards, 4*(1), 15–22. https://doi.org/10.3763/ehaz.2002.0402
Nanditha, J. S., & Mishra, V. (2021). On the need of ensemble flood forecast in India. *Water Security, 12*, 100086. https://doi.org/10.1016/j.wasec.2021.100086
Rusjan, S., Kobold, M., & Mikoš, M. (2009). Characteristics of the extreme rainfall event and consequent flash floods in W Slovenia in September 2007. *Natural Hazards and Earth System Sciences, 9*(3), 947–956. https://doi.org/10.5194/nhess-9-947-2009
Sarker, D. C., Pramanik, B. K., Zerin, A. I. & Ara, I. (2011). Climatic impact assessment: a case study of Teesta barrage irrigation project in Bangladesh. *International Journal of Civil & Environmental Engineering, 11*(1), 99–110. https://doi.org/10.12691/jgg-3-3-1
Sattar, A., Goswami, A., Kulkarni, A. V., Emmer, A., Haritashya, U. K., Allen, S., Frey, H., & Huggel, C. (2021). Future Glacial Lake Outburst Flood (GLOF) hazard of the South Lhonak Lake, Sikkim Himalayas. *Geomorphology, 388*. https://doi.org/10.1016/j.geomorph.2021.107783
Sattara, A., Ajanta Goswami, A., Kulkarni, V., Emmer, A., Haritashya, U. K., Allen, S., Frey, H., & Huggel, C. (2021). Future glacial lake outburst flood (GLOF) hazard of the South Lhonak Lake, Sikkim Himalayas. *Geomorphology, 388*, Article 107783.
Schmidt, S., Nüsser, M. (2009). Fluctuations of Raikot Glacier during the past 70 years: a case study from the Nanga Parbat massif, northern Pakistan. *Journal of Glaciology—J GLACIOLOGY, 55*. https://doi.org/10.3189/002214309790794878.
Sharma, R. K., Pradhan, P., Sharma, N. P., & Shrestha, D. G. (2018). Remote sensing and in situ-based assessment of rapidly growing South Lhonak glacial lake in eastern Himalaya, India. *National Hazards, 93*, 393–409. https://doi.org/10.1007/s11069-018-3305-0
Talukdar, S., Ghose, B., & Shahfahad, et al. (2020) Flood susceptibility modeling in Teesta River basin, Bangladesh using novel ensembles of bagging algorithms. *Stoch Environ Res Risk Assess, 34*, 2277–2300. https://doi.org/10.1007/s00477-020-01862-5
Thorndycraft, V. R., Cripps, J. E., & Eades, G. L. (2015). Digital landscapes of deglaciation: Identifying late Quaternary glacial lake outburst floods using LiDAR. *Earth Surface Processes and Landforms, 41*(3), 291–307. https://doi.org/10.1002/esp.3780
Vilímek, V., Klimeš, J., Emmer, A., Benešová, M. (2014). Geomorphologic impacts of the glacial lake outburst flood from Lake No. 513 (Peru). *Environmental Earth Science, 73*, 5233–5244
Wells, G., Dugmore, A., Beach, T., Baines, E. R. C., Saemundsson, T., & Luzzadder-Beach, S. (2022). Reconstructing glacial outburst floods (jökulhlaups) from geomorphology: Challenges, solutions, and an enhanced interpretive framework. *Progress in Physical Geography: Earth and Environment, 46*(3), 398–421. https://doi.org/10.1177/03091333211065001
Westoby, M., Glasser, N. F., Hambrey, M. J., Brasington, J., Reynolds, J. M., & Hassan, M. (2014). Reconstructing historic glacial lake outburst floods through numerical modelling and geomorphological assessment: Extreme events in the Himalayas. *Earth Surface Processes and Landforms, 39*(12), 1675–1692. https://doi.org/10.1002/esp.3617

Wieczorek, I., Stachnik, Ł, & Yde, J. C. (2022). *Inventory and Classification of the Post Little Ice Age Glacial Lakes in Svalbard*. https://doi.org/10.5194/tc-2021-364

Xia, J., Falconer, R. A., Lin, B., & Tan, G. (2011). Modelling flash flood risk in urban areas. In *Proceedings of the institution of civil engineers-water management* (Vol. 164, No. 6, pp. 267–282). https://doi.org/10.1680/wama.2011.164.6.267

Yu, Y., Li, B., Li, Y., & Jiang, W. (2024). A retrospective analysis of glacial lake outburst floods (GLOFs) using AI Earth InSAR and optical images: A case study of south Lhonak Lake, Sikkim. *Remote Sensing, 16*(13), 2307. https://doi.org/10.3390/rs16132307

Zanchetta, A., & Coulibaly, P. (2020) Recent advances in real-time pluvial flash flood forecasting. *Water, 12*(2), 570. https://doi.org/10.3390/w12020570

Zeller, L., Mcgrath, D., Mccoy, S., & Jacquet, J. (2024). Seasonal to decadal dynamics of supraglacial lakes on debris-covered glaciers in the Khumbu region, Nepal. *The Cryosphere, 18*, 525–541. https://doi.org/10.5194/tc-18-525-2024

Zhang, T., Wang, W., An, B. (2024). A massive lateral moraine collapse triggered the 2023 South Lhonak Lake outburst flood, Sikkim Himalayas. *Landslides, 22*. https://doi.org/10.1007/s10346-024-02358-x.

Deep Learning-Enhanced Flood Risk Mapping for Urban Planning and Disaster Mitigation in the Morang District, Nepal Himalaya

Pawan Thapa, Narayan Thapa, and Mandil Thapa

Abstract The impact of flooding is exacerbated by unplanned settlements, particularly in developing countries. Addressing these issues involves mapping flood-prone areas and assessing their impact on populations and households. This study utilized datasets from Google Earth Engine (GEE), Food and Agriculture Organization (FAO), Central Bureau Statistics (CBS), and Earth Data to prepare maps of slope, drainage density, digital elevation model, rainfall, land use, and soil. These maps were created using Google Earth Engine (GEE) and QGIS through overlay analysis, considering factors such as influence and slope. Deep learning techniques were also integrated to enhance flood risk modeling, leveraging neural networks to analyze spatial and temporal data patterns. The risk assessment indicates that approximately twenty-four percent of the population is at high risk, with over three thousand settlements prone to flooding. The deep learning model revealed a significantly increasing trend of floods in the Morang district, providing more accurate and detailed risk predictions. The settlement risk map, enhanced by deep learning methods, can more effectively identify flood-safe and high-risk areas in the Morang district. This tool will aid local governments, urban planners, and communities planning residential areas to mitigate flooding risks.

Keywords Safe area · Deep learning · Flooding · Risk analysis · Disaster mitigation

P. Thapa (✉)
Department of Geography and the Environment, University of Alabama, Tuscaloosa, AL, USA
e-mail: pthapa2@crimson.ua.edu

N. Thapa
Department of Geomatics Engineering, Kathmandu University, Dhulikhel, Nepal
e-mail: nt02249417@student.ku.edu.np

M. Thapa
Department of Humanities and Social Sciences, Tribhuvan University, Kirtipur, Nepal
e-mail: mandil_thapa@swastikcollege.edu.np

© The Author(s), under exclusive license to Springer Nature Switzerland AG 2025

S. Ghosh (ed.), *The Himalaya Dilemma*, Springer Natural Hazards,
https://doi.org/10.1007/978-3-031-95083-4_21

1 Introduction

Flooding poses a significant global threat, impacting over two billion people between 1998 and 2017 and causing widespread property damage, infrastructure (including essential services like schools and hospitals), and tragically, resulting in loss of life (UNISDR, 2018). Flood vulnerability is often exacerbated by a combination of factors, including inadequate residential planning (Satterthwaite et al., 2009), insufficient early warning systems (Birkmann et al., 2010), and a lack of public awareness (Armaş et al., 2017). A growing body of literature indicates a global increase in flood events, a trend widely attributed to the impacts of climate change and the expansion of unplanned settlements in flood-prone areas (Meresa et al., 2022; Mishra et al., 2024). This issue disproportionately affects developing countries, where impoverished communities often reside in high-risk zones, facing severe consequences (PreventionWeb, 2024). Key issues surrounding flood risk include the complex interplay between climate change, human activity (such as urbanization and land use change), and the resulting increase in both the frequency and intensity of flood events. For example, changes in precipitation patterns due to climate change can intensify rainfall and increase runoff, while deforestation and unplanned urbanization can reduce natural drainage capacity and further exacerbate flood risk.

In Nepal, floods represent a major recurring natural disaster, causing billions of dollars in damage and claiming numerous lives annually. The Morang district, a vital industrial hub in eastern Nepal, is particularly susceptible to flooding during the monsoon season. Rapid and often unplanned urbanization along riverbanks has intensified flood exposure in this region. Each year, the monsoon rains trigger flooding in low-lying areas and landslides in higher terrains, resulting in significant casualties (ICIMOD, 2023; Petley, 2012). While the risks are well-documented, a significant gap remains in the availability of comprehensive flood risk maps and vulnerability assessments for populations and households in Nepal (Aksha et al., 2020; Walker et al., 2019). Although existing studies have contributed to mapping flood risks and their impacts (De Risi et al., 2020; Mioc et al., 2015; Rufat et al., 2015), there is a critical need for more detailed, actionable data to inform effective mitigation and adaptation strategies. Specifically, existing research often lacks the granularity needed to identify highly vulnerable communities and inform targeted interventions. Furthermore, the integration of advanced techniques like deep learning into flood risk assessment has not been fully explored, limiting the potential for improved accuracy and predictive capability.

This study aims to address these critical gaps by developing a detailed flood risk map for the Morang district. We utilize six key criteria known to influence flood risk: land use, drainage density, slope, rainfall, digital elevation model (DEM), and soil type. Critically, this study integrates deep learning techniques to enhance the accuracy and detail of the flood risk assessment, offering a novel methodological approach. The resulting flood risk maps will provide crucial spatial information, delineating high-risk and safe areas for settlement, thereby empowering government and policymakers to make informed decisions aimed at minimizing flood risks.

This information can also support efforts to relocate vulnerable populations through targeted awareness campaigns and relief packages. This study contributes to the existing body of knowledge by providing a more granular and accurate assessment of flood risk in the Morang district, leveraging the power of deep learning to improve flood prediction and inform effective risk management strategies.

2 Materials and Methods

2.1 *Study Area*

The Morang district, located in the southeastern Terai region of Nepal (Fig. 1), is characterized by a complex interplay of geological, geomorphological, climatic, and land use/land cover characteristics that contribute to its high susceptibility to flooding. Geologically, the district is primarily composed of alluvial deposits from the Koshi River system, which have created a flat, low-lying terrain (Kafle et al., 2015). This flat topography, coupled with a dense network of rivers and streams, including the Ratuwa, Bakraha, Lohandra, Chisang, Keshliya, and Singiya rivers, significantly impedes drainage and increases the risk of inundation (Flood Forecasting Division (FFD), 2024; Nepal Red Cross Society Morang Chapter, 2018).

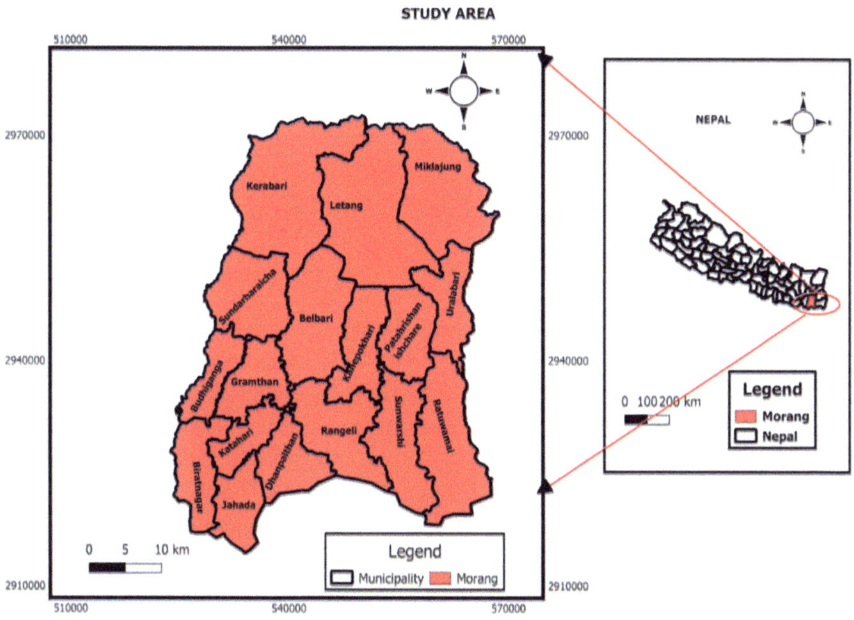

Fig. 1 Study area used for flood risk mapping

The geomorphology of the district is dominated by floodplains, which are inherently prone to flooding. The dynamic nature of these river systems, with frequent changes in course and sediment deposition, further exacerbates the flood hazard (ICIMOD, 2019). The climate of Morang is characterized by a subtropical monsoon pattern, with a distinct wet season from June to September. During this period, intense and prolonged rainfall events are common, often exceeding the drainage capacity of the rivers and leading to widespread flooding (DoHM, 2020). Furthermore, climate change projections suggest an increase in the intensity and frequency of these extreme rainfall events, further compounding the flood risk (IPCC, 2021).

Land use and land cover in the Morang district have undergone significant changes in recent decades. Rapid urbanization and agricultural expansion, often occurring in an unplanned manner, have led to deforestation, soil degradation, and altered drainage patterns (MoE, 2014). The conversion of natural wetlands and floodplains to built-up areas and agricultural land has reduced the natural buffering capacity of the landscape, increasing runoff and flood vulnerability (MEA, 2005). These anthropogenic factors, combined with the natural hazards, have created a complex and dynamic flood risk environment in the Morang district.

The increasing frequency and severity of floods in the region have prompted its inclusion in Flagship 4 by the Nepal Risk Reduction Consortium (Nepal Risk Reduction Consortium, 2014). Over the past five years, these floods have displaced more than two thousand families and resulted in the deaths of hundreds of individuals, highlighting the devastating impacts on communities and livelihoods. This study aims to identify high-risk and low-risk areas within the Morang district to aid in the planning of safer settlements. Previous research has highlighted the urgent need for detailed flood risk assessments to mitigate the impacts of such disasters (Aksha et al., 2020; De Risi et al., 2020). By providing spatial information on flood-prone areas, this study will support local governments, communities, and urban planners in making informed decisions to reduce the vulnerability of populations to flooding.

2.2 Data Collection

This study utilized various datasets to model flood risk in the Morang district. Historical rainfall data from 2000 to 2019 was used to analyze precipitation patterns. Additional datasets were acquired from multiple sources, including land use data from Google Earth Engine (GEE) and soil data from the Food and Agriculture Organization (FAO). Drainage Density, Slope, Population, and Household data were obtained from the Central Bureau of Statistics (CBS, 2011), and Digital Elevation Model (DEM) data was sourced from Earth Data, and provides a summary of the datasets used (Table 1).

For historical flood data in the Morang district from the Department of Hydrology and Meteorology (DoHM) of Nepal. The DoHM provides comprehensive data on hydrological and meteorological activities, including flood events, rainfall, and river

Table 1 Datasets used for modelling

Datasets	Data source	Description
Population, household	Central Bureau of Statistics (CBS)	Provides demographic information crucial for assessing the impact of floods on communities
DEM, drainage density, slope	Earth data	Essential for understanding the topography and hydrology of the area, which influence flood behavior
Soil data	Food and Agriculture Organization (FAO)	Soil characteristics affect water absorption and runoff, which are critical for flood modeling
Land use, rainfall	Google Earth Engine (GEE)	Land use patterns and rainfall data are key factors in determining flood risk
Historical flood data	Local and national sources	Includes records of past flood events, affected areas, and damage extent, providing a basis for model validation

discharge levels. This data is crucial for understanding flood patterns and assessing flood risks.

2.3 Methods

This study utilized datasets from Earth Data, using Google Earth Engine (GEE), and the Food and Agriculture Organization (FAO) data repository, as detailed in Table 1, to prepare maps of land use, soil, digital elevation, rainfall, drainage density, and slope (Fig. 2). These maps were generated using Google Earth Engine to assess the risk of residential areas in the Morang district. In addition to traditional Geographic Information System (GIS) methods, deep learning techniques were employed to enhance the accuracy of flood risk modeling. Convolutional Neural Networks (CNNs) were used to analyze spatial data, while Recurrent Neural Networks (RNNs) were applied to temporal data to predict flood occurrences based on historical patterns. Integrating these deep learning models allowed for more precise identification of high-risk and low-risk areas. The convolution operation involves a filter (kernel) sliding over the input data to produce a feature map. Mathematically, the convolution operation can be expressed as:

$$(I * K)(x, y) = (I * K)(x, y) = \sum_{i=0}^{F-1} \cdot \sum_{j=0}^{F-1} I(x + i, y + j) \cdot K(i, j) \tag{1}$$

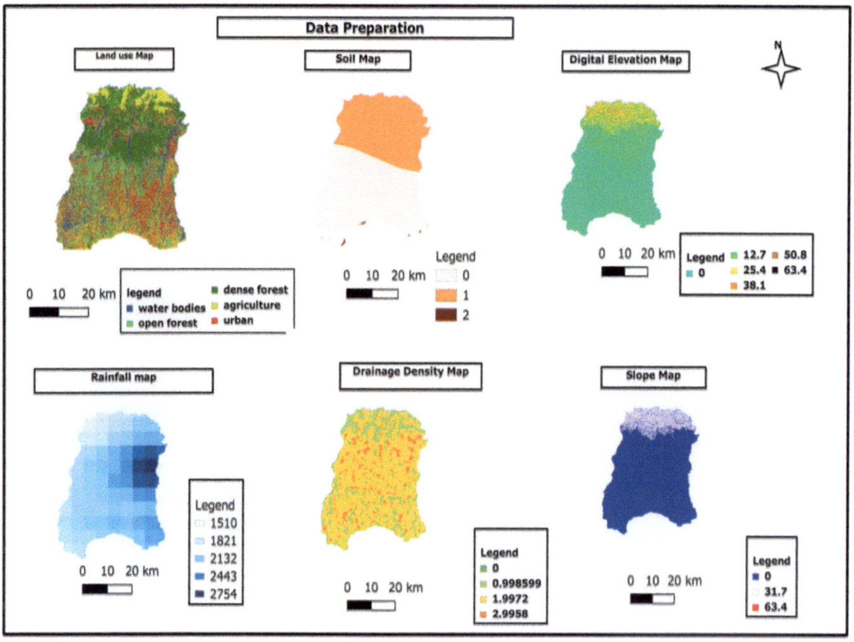

Fig. 2 Six maps to estimate the risk of flood in Morang district

where (I) is the input image, (K) is the filter, and (F) is the filter size. After convolution, an activation function (e.g., ReLU) is applied to introduce non-linearity:

$$A(x, y) = \max(0, (I * K)(x, y)) \tag{2}$$

Pooling reduces the spatial dimensions of the feature map, typically using max pooling:

$$P(x, y) = \max_{i=0}^{P-1} \max_{j=0}^{P-1} A(x + i, y + j) \tag{3}$$

where (P) is the pooling size. RNNs are well-suited for temporal data analysis as they can capture dependencies over time. The key mathematical operations in an RNN include the hidden state update and the output calculation. The hidden state at time (t) is updated based on the input at time (t) and the previous hidden state:

$$h_t = \tan h(W_h \cdot x_t + U_h \cdot h_{t-1} + b_h) \tag{4}$$

where (x_t) is the input at time (t), (h_{t-1}) is the previous hidden state, (W_h) and (U_h) are weight matrices, and (b_h) is the bias. The output at time (t) is computed from the hidden state:

Table 2 The data layers are used for flood risk analysis

Layers	Influence (100%)
Land use	24
Slope	10
Soil	8
Digital elevation model	18
Rainfall	20
Drainage density	20

$$y_t = f\left(W_y \cdot h_t + b_y\right) \tag{5}$$

where (W_y) is the weight matrix, (b_y) is the bias, and (f) is an activation function (e.g., softmax for classification). These mathematical equations form the basis of CNNs and RNNs, enabling them to analyze spatial and temporal data. The deep learning models were trained using historical flood data and environmental variables. The models' performance was evaluated using accuracy, precision, recall, and the F1 score. This approach provided a robust framework for assessing flood risk and informing settlement planning in the Morang district.

2.4 *Overlay Analysis*

The overlay analysis has two basic terms influence and scale. Influence is fundamental for analysis from six layers (Table 2); land use is the highest influencer as proper land use planning in this study area can reduce flood impacts. It is critical overall of the layer. In contrast, the soil is less significant compared to the other five parameters.

2.5 *Evaluation Criteria*

The attributes of the data are classified into five categories, with weightage ranging from 1 to 5. In this analysis, a scale from 1 to 5 is used to evaluate the importance of each attribute. Here, 1 indicates less significance, and 5 indicates the highest importance. This scale is essential for flood risk mapping and assessment (Table 3).

This structured approach ensures a comprehensive flood risk assessment, aiding in effective planning and mitigation strategies.

Table 3 The attribute used for flood risk analysis

Attribute	Scale
Rainfall	Higher rainfall higher the scale. That is lower rainfall value is 1 followed by highest elevation 5
Digital elevation model	Higher value is less important as compare to lower elevation
Slope	Lower slope more chance of flood so more important
Soil	Soil based on the fertility of soil
Land use	Urban are taken more important Agriculture land with important Waterbodies with moderate important Openforest with less important Dense forest with least important
Driange density	Higher the drainage density more the important

3 Results

The combination of drainage network and elevation data allows for a preliminary assessment of flood risk zones. Areas with a dense drainage network and low elevation are likely to be more vulnerable to flooding. The red boundary line of the Morang district provides a clear demarcation of the study area, enabling focused analysis within this region (Fig. 3). This figure serves as a crucial baseline for subsequent flood risk modeling and analysis, as it visually represents the key geographical factors influencing flood susceptibility. The satellite imagery background adds contextual information about land cover and surrounding areas, providing a holistic view of the region.

It provides a fundamental understanding of the Morang district's geographical and hydrological characteristics (Fig. 4). The spatial distribution of drainage lines, as depicted in blue, reveals the density and pattern of river systems, which are critical for assessing flood vulnerability. Areas with a high density of drainage lines are likely to experience increased surface runoff and potential flooding during heavy rainfall events. The elevation profile, represented by the color gradient, further enhances our understanding of the district's topography. Higher elevations, shown in yellow/orange, indicate mountainous or hilly regions, while lower elevations, shown in green/blue, represent plains or low-lying areas. Lower elevations are inherently more susceptible to flooding due to their proximity to water bodies and reduced capacity for water drainage.

The spatial extent and intensity of floodplain zones within the Morang district are critical components for understanding flood vulnerability (Fig. 5). The identification of "High" and "Very High" risk zones is crucial for targeted flood risk management and mitigation efforts. The light blue areas, designated as "High" risk, indicate regions where flooding is likely to occur during significant rainfall events, potentially causing damage to infrastructure and affecting livelihoods. The darker blue areas, classified as "Very High" risk, represent zones with a significantly elevated

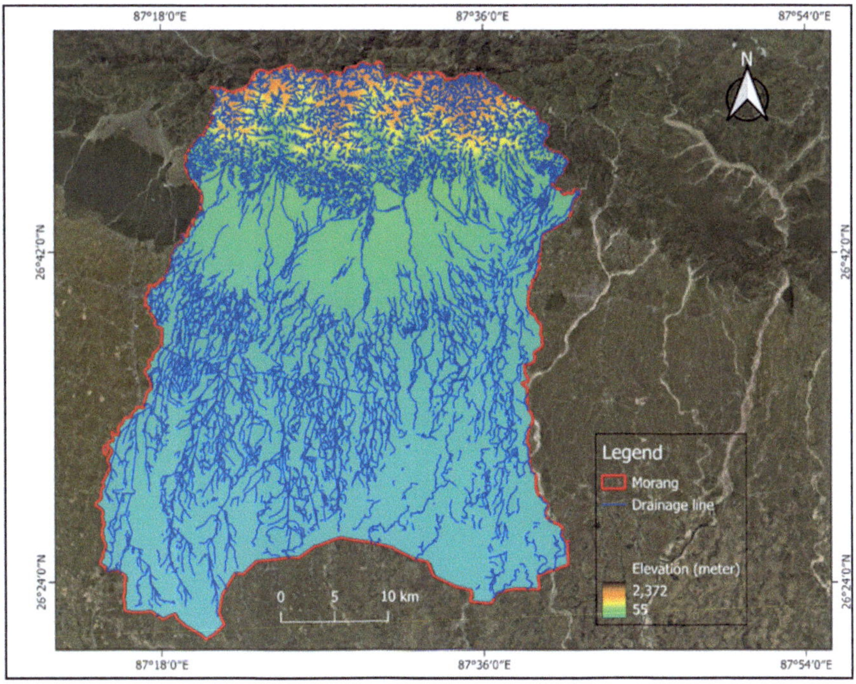

Fig. 3 Spatial distribution of drainage network and elevation in Morang district, Nepal

probability of inundation, where severe flooding is expected during even moderate rainfall events. These areas are particularly vulnerable and require urgent attention.

This study generated a comprehensive flood risk map for the Morang district, Nepal, classifying the region into five distinct risk levels: very high, high, medium, low, and flood safe (Fig. 6). These classifications are visually represented on the map using a color gradient, with darker blues indicating areas of very high flood risk and lighter blues representing medium risk. Areas classified as "flood safe" are represented by a distinct color (green). This visualization allows for a clear and immediate understanding of the spatial distribution of flood risk across the district. The analysis encompassed thirteen municipalities and rural municipalities, revealing a wide range of flood vulnerability. Notably, four municipalities were identified as having low flood risk or being flood safe, suggesting that these areas may be suitable for future development, while other areas may be suitable for agriculture or industry.

The analysis revealed that approximately 24% of the population in the Morang district resides in areas classified as high risk for flooding. This translates to over three thousand settlements directly exposed to the threat of inundation. The spatial distribution of risk indicates a strong correlation between proximity to rivers and flood vulnerability. Areas located closer to the major rivers, such as the Ratuwa, Bakraha, and others, are significantly more likely to experience flooding, while areas at higher elevations generally exhibit lower risk. The study also found that over 50% of the

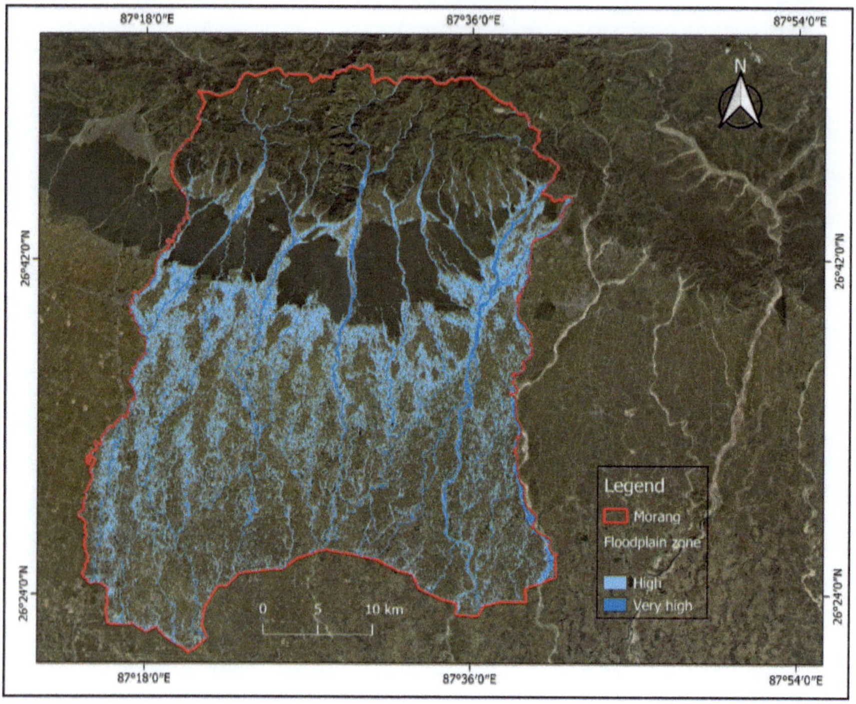

Fig. 4 Identified floodplain zones in Morang district, Nepal

district is prone to some level of flooding (combining very high, high and medium risk categories).

The detailed spatial analysis of floodplain risk zones (Fig. 7) within the Morang district offers a comprehensive overview of flood susceptibility across the region. The classification of risk into five categories—"Very Low," "Low," "Medium," "High," and "Very High"—allows for a nuanced understanding of flood vulnerability. "Very Low," and "Low", represented by the lightest shades of blue, indicate regions with minimal flood risk. They are likely located at higher elevations or in areas with effective drainage systems. "High" and "Very High" Risk Zones, represented by the darkest shades of blue, indicate areas with a high probability of flooding. They are typically located along riverbanks, low-lying areas, or regions with poor drainage systems. "Very High" risk zones represent areas where severe flooding is expected even during moderate rainfall.

The spatial distribution of these risk zones reveals the extent of flood-prone areas within the district. This highlights the widespread nature of the flood hazard in Morang and underscores the urgent need for effective mitigation and adaptation strategies. Observed patterns reveal a clear concentration of high and very high flood risk zones along the floodplains of the Ratuwa, Bakraha, and other major rivers. These zones extend significantly into the low-lying areas adjacent to the river channels, demonstrating the direct influence of river proximity and topography on

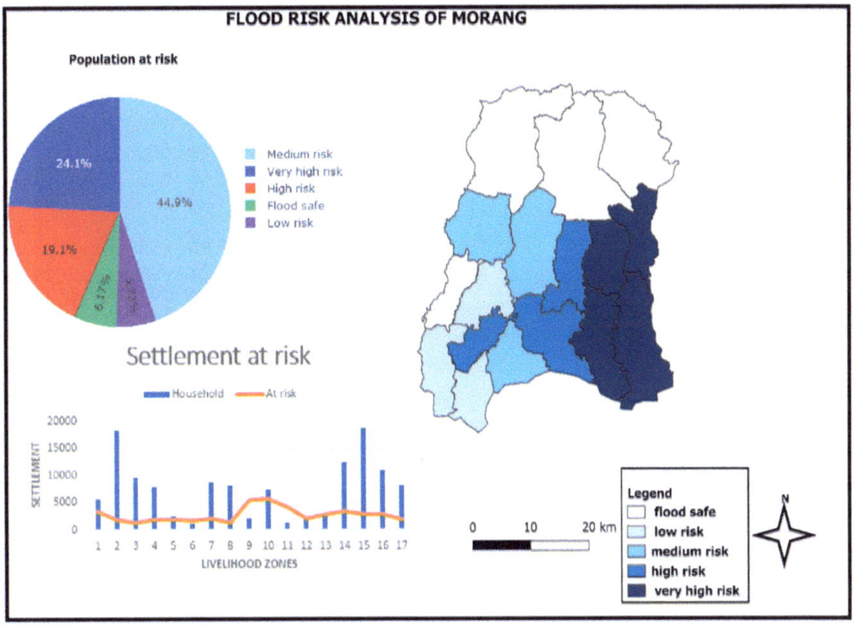

Fig. 5 Flood risk analysis map and associated pie charts showing the proportion of the population and settlements at each risk level

flood susceptibility. Furthermore, the higher elevation areas in the northern part of the district exhibit lower flood risk, confirming the protective effect of elevation. The floodplain zones were delineated using a combination of remote sensing analysis using Landsat and DEM data, and GIS-based spatial analysis. Specifically, the hydrological modeling incorporated rainfall data, river discharge, and topographic information to simulate flood inundation extents for various return periods. Remote sensing data, including digital elevation models (DEMs), were used to map terrain characteristics and identify low-lying areas prone to flooding. These datasets were integrated within a GIS environment to generate the final floodplain risk map.

This multi-faceted methodology was chosen to ensure accuracy and reliability in floodplain mapping. Hydrological modeling provides a physically based approach to simulating flood dynamics, capturing the complex interactions between rainfall, river flow, and topography. Remote sensing data offers a synoptic view of the terrain and land cover, enabling the identification of flood-prone areas at a regional scale. GIS-based spatial analysis facilitates the integration and analysis of diverse datasets, allowing for the generation of comprehensive floodplain risk maps. By combining these techniques, the study overcomes the limitations of individual methods and provides a robust assessment of flood risk in the Morang district.

A detailed breakdown of flood risk at the municipality and rural municipality levels (Fig. 8). It shows that Sunawarshi, Ratuwamai, Pathari-Sanischare, and Urlabari municipalities are classified as having a very high risk of flooding. These four

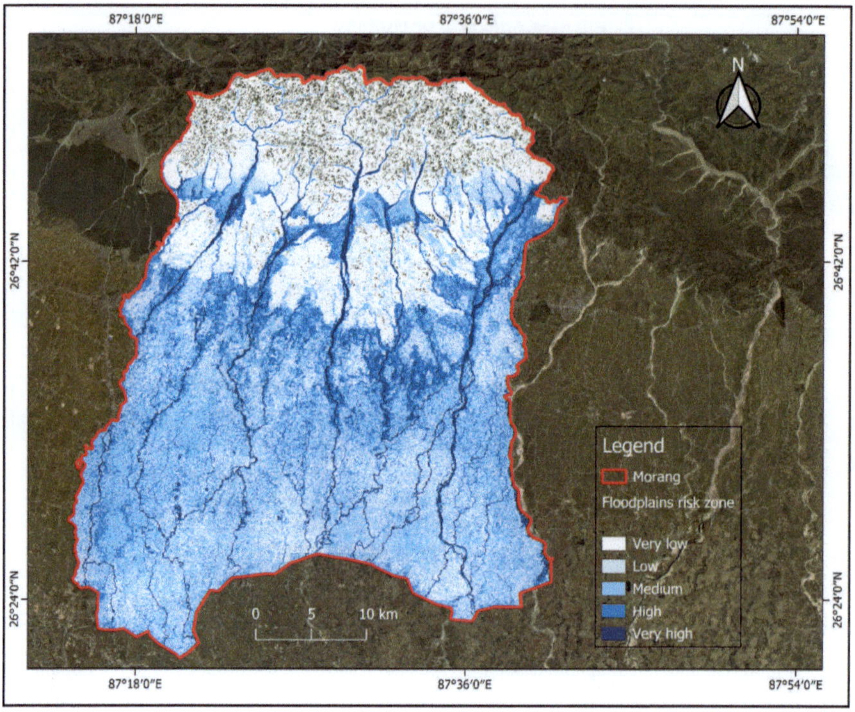

Fig. 6 Floodplain risk assessment in Morang district, Nepal

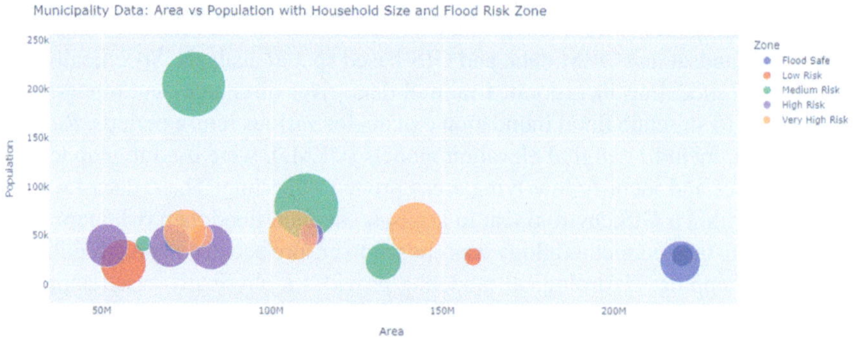

Fig. 7 Chart showing the flood risk classification, population, area, and number of households affected for each municipality/rural municipality

municipalities alone have a combined population of 212,632 and contain 33,821 households, demonstrating the significant concentration of vulnerable populations in these high-risk zones. In contrast, Miklajung and Budhiganga rural municipalities were identified as having low flood risk, with a total population of 50,385 and 10,978 households. Importantly, only the Letang and Kerabari municipalities were

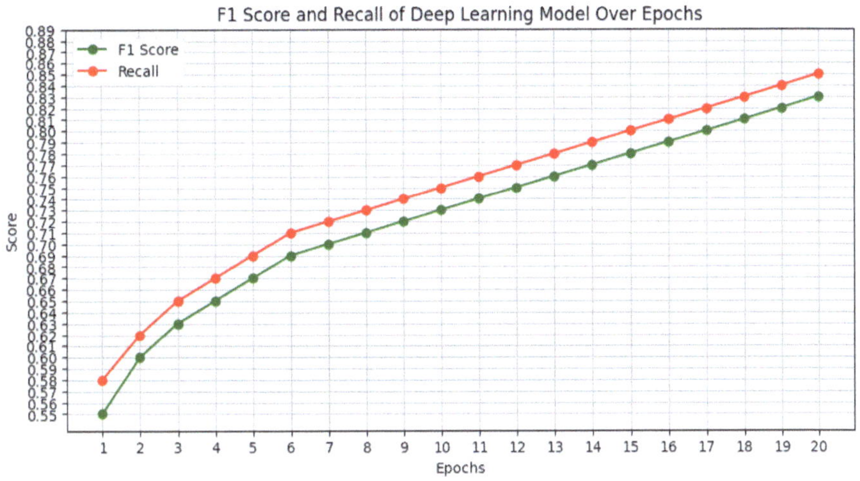

Fig. 8 Graph showing the accuracy of the deep learning model over the training epochs

classified as flood-safe areas, suggesting that these areas could be prioritized for future settlement planning. The information presented offers valuable insights for targeted interventions and resource allocation, enabling local governments and other stakeholders to prioritize the most vulnerable communities.

The integration of deep learning techniques, specifically Convolutional Neural Networks (CNNs) for spatial data analysis and Recurrent Neural Networks (RNNs) for temporal data analysis, significantly enhanced the flood risk assessment. These advanced methods allowed for the identification of more nuanced patterns and correlations in the data that traditional GIS methods might have missed. For example, the deep learning models were able to identify the combined influence of drainage density, elevation, and land use patterns on flood risk with greater precision.

The performance of the deep learning model during the training process (Fig. 8). The model's accuracy steadily improved over the epochs, starting at approximately 60% and reaching 88% by the 20th epoch. This consistent improvement demonstrates that the model effectively learned from the training data and progressively refined its ability to predict flood risk. The high accuracy achieved by the deep learning model underscores its potential as a powerful tool for flood risk assessment and prediction.

The deep learning models confirmed that areas with higher drainage density and lower elevation are particularly susceptible to flooding. Furthermore, the models highlighted the critical role of land use planning in mitigating flood risk. Specifically, urban areas characterized by poor drainage systems were identified as high-risk zones, emphasizing the need for improved infrastructure, better urban planning, and potentially, the implementation of sustainable drainage systems (SUDS). The deep learning-enhanced flood risk map provides a valuable resource for local governments, urban planners, and communities to make informed decisions regarding land use, settlement planning, and infrastructure development, ultimately contributing to a reduction in flood vulnerability and enhanced community resilience.

4 Discussion

This research will provide information about the district's flood risk area, which will locate a place that will be impacted by flood in the long-term using six factors. It creates a potential flood risk analysis map for the settlement and population of Morang district. Morang is one of the most vulnerable communities to floods that affected thousands of households. Flood impact rises yearly due to land-use changes, unplanned residential areas, and climate change. This increasing trend of flooding is worldwide, causing the lives of millions of people. It is a rapid growth of urbanization without proper study of its terrain and risk of disaster in developing countries (Marsh, 2020). Few studies try to understand the most common natural disasters like floods, landslides, and earthquakes locally, regionally, and nationally in Nepal (Thapa & Thapa, 2021; Tuladhar et al., 2015). Even proper study information is unable to capture the uncertainties that exist in these hazards and vulnerabilities. Flood risk assessments in data-limited environments face challenges due to assumptions, incomplete datasets, and imperfect models, which can lead to errors (Rogelis et al., 2014). Furthermore, the accuracy of these assessments depends on considering site-specific factors, such as slope and the influence of surrounding areas (Papaioannou et al., 2015). A spatial approach plays a vital role in risk analysis and assessing residential vulnerability, particularly in flood-prone regions (Lindley et al., 2007; Thapa, 2021). For instance, studies integrating GIS and remote sensing have demonstrated how terrain, land use, and hydrological connectivity influence flood exposure (Merz et al., 2014; Pradhan, 2010). In the context of Nepal, results indicate that municipalities such as Uralabari, Ratuwamau, Sunwarshi, and Patahri are highly prone to flooding due to their proximity to river systems and low-lying topography. In contrast, Letang and Kerbari exhibit lower flood risk, likely attributable to elevated terrain and effective drainage systems (World Bank, 2020). It will provide information about the location of very high risk, high-risk areas to support disaster management programs. Therefore, remotely sensed data with geospatial technologies such as google earth engine and QGIS prepare maps and assess the risk of flood on settlement areas. This study introduces a spatial technique to identify the risk in the Morang district. Flood risk mapping and its vulnerability assessment provide basic policy formulation guidelines and its implementation of land use planning. The first time an approach to assess flood risk mapping on the region and method contain limitations and potential for factors that drive flood in the Morang. Its spatial flood analysis and prediction are feasible for the given period of 2000–2019. Similarly, this can be a guideline for further prediction of flood analysis. In the absence of a national guide, this risk mapping and assessment determines the safer places for a settlement based on everyday practice.

Water is an essential part of human life and needs agriculture, fishing, drinking water, electricity production, and transportation. Hence, water is the source of income for many people to sustain their livelihood. But on the other hand, excessive flow of water such as floods. People in one part of the Morang district suffer from a lack of water for agriculture and flooding, whereas in another part of the same district, people are migrating to flood. Nepal government has not done surveying to know the actual cause of the migration of the people. People migrate from the hilly region to eastern Nepal's Terai region seeking jobs, education, and many more. But in these districts, it seems that people are migrating due to the problem of a flood. The most common problem of these people is water. So, the planner should plan in such a way that both issues are solved. The inhabitants are at high risk and should be helped by building the dam, diverting the river, and planning for the storage house of water so that it becomes the problem solver. The challenge for the people living in this district, planners, politicians, and the economists to allocate the budget for hydropower, use this water for agriculture, and housing. Controlling the flood can be the source of income for the Morang. As people of this part mostly depend on agriculture and industry. Agriculture and industry both need water. Suppose there is water storage in a high-risk area. In that case, it can supply to the other municipality of this district since Morang is the district that touches Nepal's Indian boundary. There has always been a debate between Nepal and India when India clams the Nepal water cause flood in India. On the other hand, Nepal claims that India's water carries all the nutrients, minerals, and crude protein. The country that can clam with India can build the dam and make the people living in this part problem free from the flood.

5 Conclusion

The flood risk map and assessment determine safer places from floods such as Letang, Kerbari municipality, where people can settle. Uralabari, Ratuwamau, Sunwarshi, Patahri are the municipalities prone to upcoming flood. From this response, local government, residence, disaster management teams, and security forces should be prepared during every summer and monsoon in these municipalities to reduce the loss of lives and property damage. Similar studies need the whole country for better preparedness, response, recovery, and natural disaster reduction. However, the government and exposed people to these prone are unaware of this due to a lack of economic and human resources, and people have no place to shift their settlement. Despite significant input from nonprofit organizations, and international organizations warning people about a flood and its impact through awareness, there is an increasing number of deaths and property damage every year. This study suggests that more people live in a very high, high-risk area, increasing the risk every summer and monsoon. The solution might be living in safer places and performing other low and medium-risk activities such as farming and forests in high-flood risk places to reduce its risk.

References

Aksha, S. K., Resler, L. M., Juran, L., & Carstensen, L. W. (2020). A geospatial analysis of multi-hazard risk in Dharan, Nepal. *Geomatics, Natural Hazards and Risk, 11*(1), 88–111. https://doi.org/10.1080/19475705.2019.1710580.

Armaş, et al. (2017). Public perception of flood risk: A case study from Romania. *Natural Hazards, 85*(1), 169–187.

Birkmann, et al. (2010). Measuring vulnerability to promote resilience. In *Measuring vulnerability to natural hazards* (pp. 9–53). Routledge.

CBS. (2011). *National Population and Housing Census 2011*. Central Bureau of Statistics (CBS). https://cbs.gov.np/wp-content/upLoads/2018/12/27-Kathmandu_VDCLevelReport.pdf.

De Risi, R., Jalayer, F., De Paola, F., Carozza, S., Yonas, N., Giugni, M., & Gasparini, P. (2020). From flood risk mapping toward reducing vulnerability: The case of Addis Ababa. *Natural Hazards, 100*(1), 387–415. https://doi.org/10.1007/s11069-019-03817-8.

DoHM. (2020, March 31). *Oxfam in Nepal and ICIMOD join hands with communities to manage flood risk in Nepal—ICIMOD* (Department of Hydrology and Meteorology). https://www.icimod.org/success-stories/oxfam-in-nepal-and-icimod-join-hands-with-communities-to-manage-flood-risk-in-nepal/.

Flood Forecasting Division (FFD). (2024). *Over dozen villages prone to flooding in Morang*. https://myrepublica.nagariknetwork.com/index.php/news/over-dozen-villages-prone-to-flooding-in-morang.

ICIMOD. (2019, July 24). *Partnering with private enterprise and communities to manage flood risk—ICIMOD*. https://www.icimod.org/partnering-with-private-enterprise-and-communities-to-manage-flood-risk/.

ICIMOD. (2023). *Flood risk rising despite below-average monsoon—ICIMOD*. https://www.icimod.org/flood-risk-rising-despite-below-average-monsoon/.

IPCC. (2021). *Climate change 2021: The physical science basis*. Intergovernmental Panel on Climate Change. https://www.slvwd.com/sites/g/files/vyhlif1176/f/uploads/item_10b_4.pdf.

Kafle, K. R., Khanal, S. N., & Dahal, R. K. (2015). Dynamics of the Koshi River on the perspective of morphology and sedimentation with emphasis on post disaster impact of the 2008 Koshi flood. *Kathmandu University Journal of Science Engineering and Technology, 11*(1). https://journals.ku.edu.np/kuset/article/view/422.

Lindley, S. J., Handley, J. F., McEvoy, D., Peet, E., & Theuray, N. (2007). The role of spatial risk assessment in the context of planning for adaptation in UK urban areas. *Built Environment, 33*(1), 46–69.

Marsh, S. (2020). *Nepal: Priority river basins flood risk management project*. https://policycommons.net/artifacts/403350/nepal/1372356/.

MEA. (2005). *Ecosystems and human well-being* (Vol. 5). Island Press Washington, DC. https://www.unioviedo.es/ranadon/Ricardo_Anadon/docencia/DoctoradoEconomia/Millenium%20Eco%20Assesment%2005%20Oppor%20Business%20Industry.pdf.

Meresa, H., Tischbein, B., & Mekonnen, T. (2022). Climate change impact on extreme precipitation and peak flood magnitude and frequency: Observations from CMIP6 and hydrological models. *Natural Hazards, 111*(3), 2649–2679. https://doi.org/10.1007/s11069-021-05152-3.

Merz, B., et al. (2014). Flood risk curves and uncertainty bounds. *Natural Hazards, 67*(2), 459–478.

Mioc, D., Nkhwanana, J. N., Moreiri, K. K., Nickerson, B., Santos, M., McGillivray, E., Morton, A., Anton, F., Ahmad, A., & Mezouaghi, M. (2015). Natural and man-made flood risk mapping and warning for socially vulnerable populations. *International Journal of Safety and Security Engineering, 5*(3), 183–202.

Mishra, A., Maraun, D., Schiemann, R., Hodges, K., & Zappa, G. (2024). *Long-lasting intense cut-off lows to become more frequent in the Northern Hemisphere*. https://www.researchsquare.com/article/rs-3982488/latest.

MoE. (2014). *Final report of participatory land use planning and implementation project* [SDG Help Desk]. Ministry of Environment (MoE). https://sdghelpdesk.unescap.org/e-library/final-report-participatory-land-use-planning-and-implementation-project.

Nepal Red Cross Society Morang Chapter. (2018). *Nine districts in eastern Nepal face flood risks*. https://myrepublica.nagariknetwork.com/index.php/news/nine-districts-in-eastern-nepal-face-flood-risks.

Nepal Risk Reduction Consortium. (2014). NRRC Flagship 4: Community based disaster risk reduction. *PrepareCenter*. https://preparecenter.org/resource/nrrc-flagship-4-community-based-disaster-risk-reduction/.

Papaioannou, G., Vasiliades, L., & Loukas, A. (2015). Multi-criteria analysis framework for potential flood prone areas mapping. *Water Resources Management, 29*, 399–418.

Petley, D. (2012). Global patterns of loss of life from landslides. *Geology, 40*(10), 927–930.

Pradhan, B. (2010). Flood susceptible mapping using GIS and remote sensing in Nepal. *Environmental Earth Sciences, 61*(3), 539–550.

PreventionWeb. (2024). *Least polluting populations face highest flood risks*. PreventionWeb. https://www.preventionweb.net/news/pioneering-research-reveals-some-worlds-least-polluting-populations-are-much-greater-risk.

Rogelis, M. C., Lam, J. C., & Ramirez Cortes, F. (2014). Advances in flood risk assessments for data-limited environments. In *Proceeding of the 6th International Conference on Flood Management (ICFM6)*, PAP014415.

Rufat, S., Tate, E., Burton, C. G., & Maroof, A. S. (2015). Social vulnerability to floods: Review of case studies and implications for measurement. *International Journal of Disaster Risk Reduction, 14*, 470–486.

Satterthwaite, et al. (2009). Building resilience to climate change in cities. *Annual Review of Environment and Resources,34*, 1–28.

Thapa, P. (2021). Using geospatial technologies for disaster management in developing countries. *Journal of Remote Sensing & GIS, 9*(4), 1–5.

Thapa, P., & Thapa, N. (2021). *Mapping floods risk and assessing flood vulnerability for settlement areas*. https://www.academia.edu/download/91966865/fee8ef80-8ca5-4811-8519-85d14d1b83ab.pdf.

Tuladhar, G., Yatabe, R., Dahal, R. K., & Bhandary, N. P. (2015). Disaster risk reduction knowledge of local people in Nepal. *Geoenvironmental Disasters, 2*(1), 5. https://doi.org/10.1186/s40677-014-0011-4.

UNISDR. (2018). *Economic losses, poverty and disasters 1998–2017*. UNISDR.

Walker, T., Kawasoe, Y., & Shrestha, J. (2019). Risk and vulnerability in Nepal. *Findings from the household risk and*. https://documents1.worldbank.org/curated/en/318281581716869564/pdf/Risk-and-Vulnerability-in-Nepal-Findings-from-the-Household-Risk-and-Vulnerability-Survey.pdf.

World Bank. (2020). *Nepal: Building resilience to floods and erosion*.

Anthropogenic Perspective of Hazards

River Sand Mining: A Developing Geohazard in the Himalaya

Rajiv Sinha, Moumita Akuria, Kanchan Mishra, and Dipro Sarkar

Abstract Indiscriminate in-channel sand mining leads to serious off-site and onsite impacts on channel morphology and sediment dynamics of rivers. Owing to high sediment flux and the ready availability of gravels and sand, the Himalayan rivers have been exploited for decades. The most observed impacts of in-channel sand mining are channel incision and riverbed coarsening. The former disturbs the equilibrium state of the river, causing erosion downstream, and the latter reduces sediment supply and availability for transport, disrupting longitudinal connectivity. Such effects are now widespread in several Himalayan rivers and considered a major geohazard in this region. Unregulated Sand mining is not just impacting the river channels in terms of channel modifications in the mining stretches but also resulting in cascading effects such as bank erosion, channel incision and flooding in downstream reaches. We provide here a compilation of geomorphic impacts of unregulated sand mining on riverine landscapes that have been documented globally with a special focus on the Himalayan rivers. We have then presented a case study from the Upper Ganga River between Raiwala and Bhogpur in northern India, which has been exploited for gravel and sand for several decades. The Ganga River in this stretch has high sediment flux potential from the tectonically active hinterland, making it suitable for sand mining. We present a hydrogeomorphic approach to study the impact of sand mining on channel morphology and riverine processes of the Upper Ganga River. We use channel belt width, active channel width, and centreline shifts derived from Landsat-based interdecadal planform maps to analyse the impacts on the river channel. We finally present a critical review of the existing sand mining policy in India and some recommendations.

Keywords Geomorphic impacts · Sand mining · Himalayan rivers · Geohazards · River health

R. Sinha (✉) · M. Akuria · D. Sarkar
Department of Earth Sciences, Indian Institute of Technology Kanpur, Kanpur, India
e-mail: rsinha@iitk.ac.in

K. Mishra
Department of Geosciences, University of Tübingen, Tübingen, Germany

© The Author(s), under exclusive license to Springer Nature Switzerland AG 2025

S. Ghosh (ed.), *The Himalaya Dilemma*, Springer Natural Hazards,
https://doi.org/10.1007/978-3-031-95083-4_22

1 Introduction

Rivers are not just the conduits of water, but they also carry sediments, and this is one of their natural functions. Many of the Himalayan rivers are indeed large sediment dispersal systems. The Ganga being the axial river system of northern India, together with the Brahmaputra, discharge close to 1 billion tonnes of sediments into the Bay of Bengal annually. Similarly, the cratonic rivers of India also carry a significant sediment load. A large fraction of this sediment, as a part of the natural process, does not reach the sink but is accommodated within the channels and the floodplains. The process fosters the geomorphological development of the river and supports riverine habitat with nutrients. However, excessive sedimentation in channels also poses serious threats in terms of channel instability, bank erosion and frequent flooding, often prompting local authorities to consider regulated sand mining as a mitigation measure. Conversely, sand and gravel are high-value commodities, and therefore lucrative resources for mining operations. This eventually results in unregulated sand mining which disrupts the natural equilibrium of the river. Therefore, unregulated sand mining is fast evolving as a major geohazard in Himalayan River basins and this needs to be assessed carefully to design sustainable mining policies.

Unregulated sand mining has emerged as a major global issue, and the indiscriminate extraction of riverine sand and gravel is manifold higher than natural replenishments in several river basins across the world. This has serious offsite and onsite impacts, leading to changes in channel form, physical habitats, food webs and engineering structures associated with river channels and their watersheds. As these adverse effects gain recognition, instream sand mining/aggregate extraction has invited increasing scientific scrutiny.

The Himalayan rivers are known for exceptionally high sediment flux (Hovius, 1998; Sinha & Friend, 1994; Sinha et al., 2019; Tandon & Sinha, 2007), and several reaches are influenced by anthropogenic modifications, primarily interventions, such as barrages and bridges, influencing the natural flow regime and, hence, sediment transport. This prompts us to scrutinise and identify the reaches of excessive sedimentation which, if left unmonitored, may lead to serious problems in river management, including river-related hazards (Sinha, 2018). A balance of sediment supply, transport, and deposition is essential to prevent significant changes in river morphology and to guide sustainable sand mining policies for the Ganga basin.

Although more focused research is needed for restoration of rivers impacted by sand mining, it is amply clear that large-scale unregulated mining of sand, gravel and stones from riverbeds and banks alters the natural flow of a river and may severely affect the entire river system. As reflected from the published literature, excessive in-channel sediment mining from a river can cause the following changes:

(a) It may disrupt the longitudinal continuity of the river system (Calle et al., 2017; Kondolf, 1997).
(b) An altered equilibrium profile of the rivers can initiate rapid aggradation or degradation of the channel bed of formerly stable rivers. This instability may

further modify the river planform from initially braided to a single-thread channel or vice-versa (Arróspide et al., 2018; Brousse et al., 2021; Comiti et al., 2011; Ghosh & Maiti, 2021; Surian et al., 2009).

(c) It may cause significant channel instability leading to widespread loss of riparian vegetation and direct loss of wildlife habitat and other indirect ecological impacts (Gumiero et al., 2015; Zou et al., 2019).

(d) Sediment mining in riverbeds locally initiates channel incision and results in a steeper longitudinal gradient, which further locally increases the flow shear stress and results in rapid bed and bank erosion (Bhattacharya et al., 2019a; Hackney et al., 2020; Martín-Vide et al., 2010).

(e) Excessive riverbed incision can cause several negative impacts on infrastructures, such as undermining of bridge piers and other structures (Kondolf, 1994; Padmalal & Maya, 2014; Rinaldi et al., 2005).

(f) Incision of the riverbed typically causes the alluvial aquifer to drain to a lower level, resulting in a loss of aquifer storage and lowering of the groundwater table in the surrounding floodplain (Padmalal et al., 2008; Park, 2024).

These major impacts of excessive sediment mining in the rivers have been illustrated in Fig. 1. All these impacts may or may not be visible in all rivers. Rather, some or all of the above-listed impacts may be more severe than others in rivers of different geographical locations. Several rivers draining the Ganga basin are severely impacted by sand mining, and unsustainable sand mining has been rampant despite the 2016 guidelines issued by the Ministry of Environment and Forests. To strengthen the existing system and to curb illegal and unsustainable mining, the Ministry of Environment and Forests has introduced 'Enforcement and Monitoring Guidelines for Sand Mining 2020'. It is a detailed document, that incorporates the use of technologies like drones with night vision for site surveillance, procedures for replenishment of sand, post-environmental clearance monitoring of sites, environmental audit of such areas and other strategies to combat illegal mining.

2 Impact of Sand Mining on Riverine Ecosystems: A Global Overview

2.1 Geomorphic Impacts of Sand Mining

Sand plays an important role in fluvial geomorphology, sustaining the dynamics and ecology of riverine networks (Hauer et al., 2018). It functions as a reservoir by storing water at shallow depths and increasing residence time (Hohensinner et al., 2004). Further, sand provides essential breeding and hiding grounds for fish and microorganisms, thereby fostering and supporting diverse riverine ecosystems and biotas (Hauer et al., 2018; Meadows & Anderson, 1966). The increasing demand and supply of sand due to rapid urbanization and industrialization has severely affected river channels (Bendixen et al., 2021; Peduzzi, 2014). Unsustainable sediment extraction has

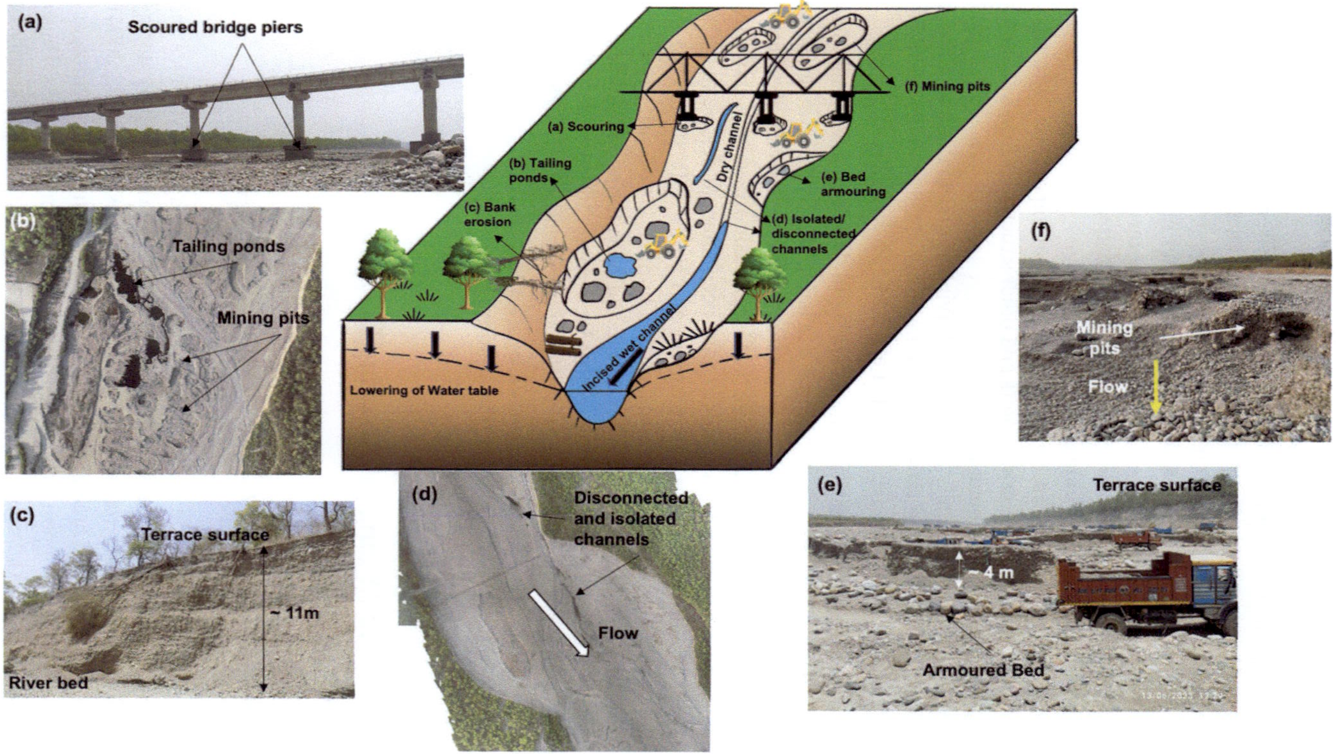

Fig. 1 A riverine landscape impacted by in-channel sand mining. Field/drone photos illustrate the various impacts **a** Scoured bridge piers, **b** Tailing ponds and mining pits as seen in drone images, **c** Eroding banks, **d** Disconnected and isolated channels, **e** river bed excavation and bed armouring, **f** mining pits

caused extensive damage to the entire river ecosystem, altering morphology, ecology, and hydrology (Koehnken et al., 2020; Rentier & Cammeraat, 2022).

In-channel sediment mining involves the extraction of sand and gravel directly from the river bed, point bars and braided bars, altering channel geometry and bed elevation (Dufour et al., 2015; Rinaldi et al., 2005; Surian et al., 2009). In-channel sediment mining comprises two main types: pit mining, excavating pits below the thalweg, and bar skimming, which entails scraping the upper layer of a gravel bar (Padmalal & Maya, 2014). In-channel pit mining is further categorised into dry-pit mining, where the groundwater table is unaffected or remains at greater depths as seen in ephemeral rivers (Calle et al., 2017; Akuria & Sinha, 2025), and wet-pit mining, which involves excavating pits below the surface water of perennial rivers or the alluvial groundwater table (Hackney et al., 2020, 2021; Ng & Park, 2021). Wet-pit mining often requires equipment such as draglines or hydraulic excavators to dredge sand and gravel from beneath the water surface. These practices significantly disrupt riverine processes, threatening the ecological and hydrological balance of river systems.

The geomorphic processes and resulting forms in rivers are driven by a balance between the driving forces such as water discharge (Q_W), channel slope (S), and resisting forces such as sediment discharge or sediment load (Q_S), and sediment bed grain size or particle diameter (D_S) (Lane, 1955). The imbalance of the driving and resisting forces causes the river to adjust in three dimensions: vertically, laterally and longitudinally which can further cause a wholesale adjustment in rivers (Fryirs & Brierley, 2005). Koehnken et al. (2020) reported that channel incision is the most widely reported geomorphic impact of riverine sand mining, followed by channel narrowing across the globe. Several Italian rivers have experienced bed incisions ranging from 2 m in the Piave River (Comiti et al., 2011), up to 4–5 m in the Lower Orba River from 1950 to 1970s (Mandarino, 2022) and as much as 10 m in certain reaches of the Panaro River, which occurred during or shortly after a period of extensive sediment mining (Gumiero et al., 2015). Isik et al. (2008) observed channel bed incision up to 7 m in the lower Sakarya River in Turkey during the 41-year period (1965–2006), which was mostly related to sand and gravel mining. In Spain, gravel extraction resulted in incision of over 5 m in Gállego River (Martín-Vide et al., 2010), at least 3.5 m in Rambla de la Viuda (Calle et al., 2017), and an average of 1.5 m over the whole reach in lower Tordera, leading to damage to infrastructures and severely affecting the river ecosystem and groundwater (Rovira et al., 2005). Several sections of the Maipo River in Chile experienced very large incision up to 20 m, with an increase in the area affected by gravel mining from 86.62 to 368.13 ha from 1954 to 2015 (Arróspide et al., 2018). Sand mining is likely generating a net deficit in the Mekong and the Bassac channels in Cambodia, which showed significant bed material losses amounting to approximately 90 and 110 Million cubic meters, respectively, and an increase of mean depth by an excess of 1.3 m over a period of 10 years (1998 and 2008) (Brunier et al., 2014). Binh et al. (2021) reported an increase in riverbed incision rate from 0.16 m/yr in 1998–2014 to 0.5 m/yr in 2014–2017, resulting in lowering of water levels in the Mekong River.

In planform view, the river changes often from a wide, braided channel to a straight, single thread and/or wandering narrow channel, causing a wholesale change in rivers (Fryirs, 2017; Fryirs & Brierley, 2005). This implies disruption of the pre-existing channel morphology and sediment deficiency. Such morphological changes, caused by excessive sand mining, have been documented in several rivers, including the Tagliamento, Brenta, Magra, Piave and Panaro rivers in Italy (Comiti et al., 2011; Dufour et al., 2015; Gumiero et al., 2015; Rinaldi et al., 2005, 2009; Surian & Cisotto, 2007; Ziliani & Surian, 2016), in Rambla de la Viuda and the Palancia River in Spain (Calle et al., 2017; Sanchis-Ibor et al., 2017) and the Maipo River, Chile (Arróspide et al., 2018). Some studies indicate a strong correlation between channel incision and channel narrowing/degradation induced by sand mining (Calle et al., 2017; Gumiero et al., 2015; Rinaldi et al., 2005; Surian & Cisotto, 2007; Surian et al., 2009), though some suggested a weaker but still evident correlation (Comiti et al., 2011; Mandarino, 2022).

Channel incision induced by riverbed mining has also disrupted the lateral hydrological connectivity between river and wetland in the Vietnamese Mekong Delta as the connected days of the river and wetland decreased from 360 to 270 within about 15 years (after 2000) (Park et al., 2020). This decline in hydrological connectivity directly reduces available water resources for irrigation and local consumption. Comiti et al. (2011) revealed that bed incision due to gravel mining promotes vegetation encroachment leading to channel narrowing and only intense, infrequent flood events of return period of 10–15 years can inundate the floodplains, thereby reducing the lateral connectivity. Removal of fine grained sediments from riverbeds can lead to bed armouring, resulting in a layer of coarser sediment that is more resistant to, erosion (Kondolf, 1994; Rinaldi et al., 2005) eventually disrupting the longitudinal connectivity of the river.

The processes of sediment excavation and bed armouring decrease the sediment load within the river system. Channel excavation creates localised steep gradients, generating over-steepened knickpoints. The reduced bedload supply along with the localised steep gradient increases stream power because of the 'hungry water effect' (Kondolf, 1994, 1997) exacerbating bed and bank erosion (Hackney et al., 2020). This causes head-cutting that can propagate several kilometres upstream. A topographic survey of Cache Creek, California, conducted in 1992, documented a 3-m-deep knickpoint resulting from excavation that created a 4-m-high headwall, situated 1400 m downstream of Capay Bridge. By 1993, the knickpoint had migrated 960 m upstream. Subsequent flooding in 1995 exacerbated its progression, leading it beneath Capay Bridge and nearly causing its structural failure (Kondolf, 1997). Similarly, instream pit mining on the Russian River near Healdsburg, California in the 1950s and 1960s led to channel incision of 3–6 m over an 11 km length of the river (Kondolf, 1997). The incision is so severe that some studies have documented bedrock exposure in certain rivers. In the Drac River, France, incision intensified as the channel rapidly eroded ancient fluvial-lacustrine clay deposits, increasing scouring rates (Brousse et al., 2021). Similarly, the impact of bedload particles eroding soft bedrock in the Bernesga River, Spain, led to the concentration of incisions in narrower segments creating a positive feedback loop of further incisions (Ferrer-Boix et al., 2023).

In Austria's Johnsbach Valley, channel incision led to the exposure of talus-covered bedrock (Rascher et al., 2018). Channel incision can increase bank height and induce instability, threatening key infrastructure and communities along the river's banks (Hackney et al., 2020). On the other hand, bar skimming and undercutting can also cause bank failure leading to the widening of the river channel showcasing channel instability (Kondolf, 1994). Apart from in-channel mining, sand mining also occurs on floodplains and older terrace deposits of perennial rivers, as observed in the Amite River in the USA (Mossa & McLean, 1997) and the Cibin River floodplain in Romania (Costea, 2018). In the event of an avulsion or gradual migration, floodplain pits may integrate into the active channel, termed 'pit capturing', posing potential instream or in-channel implications over a time scale of decades (Padmalal & Maya, 2014).

The sand mining industry, crucial for construction and urbanization, is particularly prevalent in developing countries like India, where 61.6% of surveyed river reaches showed evidence of mining (Dujardin et al., 2024). Due to reduced sediment supply, the Dwarkeswar and Mayurakshi Rivers in West Bengal have undergone significant narrowing and incision, transforming from multi-threaded to single-thread channels (Ghosh, 2024; Ghosh & Maiti, 2021). Channel incision has often triggered shifts in the thalweg, leading to substantial channel migration, instability and pool-riffle alterations in the Damodar and Kangsabati rivers (Ghosh et al., 2016; Bhattacharya et al., 2019a, b; Ghosh & Jana, 2021). Channel incision has resulted in the destabilisation of engineering structures, increased bank erosion, and a lowering in the water table, as observed in the rivers of Kerala, South India (Padmalal et al., 2008; Sreebha & Padmalal, 2011). Research on the impacts of sand mining in the Himalayan rivers has emerged in the past decade, highlighting similar trends, including severe channel narrowing and incision exposing bedrock, reduced channel bar areas and a shift from braided to single-thread sinuous channels (Mitra et al., 2023; Tamang & Mandal, 2015; Wiejaczka et al., 2018). Mitra et al. (2023) further emphasised that sediment extraction is the primary driver of channel degradation surpassing other anthropogenic factors like urbanization, vegetation removal, and channel embankments. Additionally, riverbed mining in the Himalayas exacerbates flood hazards by intensifying erosion, making floodplains more vulnerable to flooding, particularly in the monsoon season due to increased downstream transport of loose landmass (Singh & Sidhu, 2016). Vercruysse and Grabowski (2021) associated with the impacts of sand mining with a timescale-dependent evolutionary trajectory, highlighting a pulse response in the Sutlej River in the Himalayas. In a recent study in the Gaula River in the northwest Himalaya, Akuria and Sinha (2025) have shown that in-channel sand mining creates a positive feedback system by increasing the local gradient and stream power, which exacerbates channel narrowing and incision and concentrates flow, leading to thalweg fixing and a severe decline in geomorphic diversity. The Gaula River has undergone four evolutionary phases: barren conditions (1976–1995), mobilisation and incision (1995–2009), channel belt widening (2009–2014), and the development of an underfit channel (2014–2021). These evolutionary phases were caused primarily by extensive sand mining activities in the river for the last several decades.

2.2 Ecological Impacts

Geomorphic impacts of sand mining have often led to disrupting the ecological functioning of the rivers as the habitats of the aquatic life get severely damaged. Gravel mining from active channel beds and adjacent floodplains contributes to the fragmentation of riparian forests such as in the lower Eygues River in France (Kondolf et al., 2007). Human-induced channel bed incision disconnects the adjacent floodplains from the main channel causing a lowering of the groundwater table and impacting the surrounding ecology and riparian vegetation. Gumiero et al. (2015) reported that riverbed incision in some reaches of the Panaro River in Italy caused a decline of hygrophilous trees such as willows and poplars and an inclination towards xerophilous conditions. Some incised reaches also showed recolonization of the abandoned floodplains by vegetation. Skalski et al. (2016) reported that river incision reduced the species richness and diversity of ground beetles on the lowest bench, which was frequently inundated, but increased them on the highest bench, which was rarely flooded. The abundance of small, predatory, macropterous, and spring-breeding species, which are adapted to dynamic habitats, was negatively impacted by river incision, especially on the lowest bench. Conversely, the abundance of large, herbivorous, brachypterous, and autumn-breeding species, which are typical of stable habitats, was positively impacted by river incision, especially on the highest bench (Skalski et al., 2016). Mingist and Gebremedhin (2016) revealed that sand mining interfered with the migratory routes of fish and resulted in the loss of their spawning grounds in the inflowing rivers to Lake Tana, Ethiopia. Gravel dredging in large rivers diminishes fish abundance, diversity, and trophic positions by altering habitats and nutrient pathways, reducing benthic primary production, and simplifying food webs. This poses a threat, particularly to native and specialist fish species, dependent on gravel and rock substrata for spawning, leading to potential biodiversity loss and shifts in community composition in the river (Freedman et al., 2013).

In India, sand mining and the laying of roads up to the riverbank have led to a reduction in the nesting sites of crocodiles, turtles, and birds in the Chambal River (Singh & Rao, 2017). A study of in-channel sand mining in the Ganga River near Haridwar has significantly altered the channel morphology over the past two decades leading to the degradation of water quality and a decline in aquatic biodiversity (Kamboj & Kamboj, 2019). Meanwhile, floodplain mining has adversely affected the riparian vegetation along adjacent riverbanks. Prabhakar et al. (2019) further observed that sand mining activities increased turbidity and decreased the transparency of water, which negatively impacted the primary production and feeding efficiency of zooplankton in the River Ganga in Bihar.

3 Case Study: Upper Ganga River

We present a case study based on a pilot project on rapid assessment of sand mining in the upper reaches of the Ganga River between Raiwala and Bhogpur (Fig. 2) to highlight the geomorphic impacts of sand mining in a large river. This pilot project utilized historical remote sensing data as well as drone technology which have provided important insights about the geomorphic processes operating in this region. We identified several 'hotspots' of aggradation/degradation based on the historical remote sensing data and validated by drone surveys. This formed the basis for the identification of the reaches impacted by excessive sand extraction and vulnerable to channel morphodynamics and flooding. This case study has been designed to illustrate a comprehensive protocol for the assessment of geomorphic impacts of sand mining in a Himalayan River.

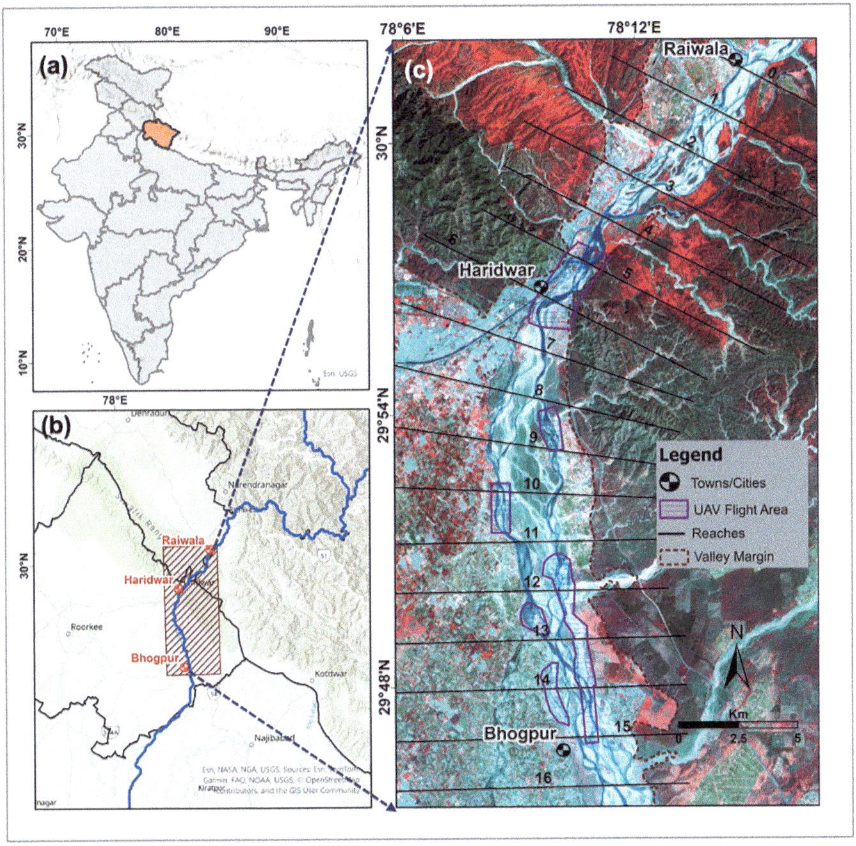

Fig. 2 Study area in the Upper Ganga River between Raiwala and Bhogpur; the reach length is 2 km approx. covering ~32 km of the Ganga River

3.1 Data, Approach and Methodology

We conducted a remote sensing-based study for the stretch between Raiwala and Bhogpur, covering an area of approximately 32 km (Fig. 2), to assess the impact of sand mining on river morphology and the adjoining floodplains. A series of satellite images spanning from 1992 to 2020 were analyzed to (a) systematically document the temporal changes in river morphology and (b) assess the long-term impacts, if any, of sand mining. In addition to Landsat data, high-resolution CORONA images from 1973 were analyzed after appropriate rectification and processing to establish the 'reference' condition of the river before the onset of mining activities. We performed detailed planform mapping of the Ganga River from Raiwala to Bhogpur at a decadal scale for the years 1992, 2000, 2010, 2013, 2015, and 2020. Pre-monsoon satellite data, as presented in Table 1, were used due to their low flow conditions and high consistency in mapping planform features on an annual basis. Figure 3 presents the complete flowchart of the methodology followed in this study.

3.1.1 Planform Mapping

For planform mapping, the primary satellite data used for this study are listed in Table 1. For the temporal study, we utilized pre-monsoon satellite data because of the higher likelihood of obtaining cloud-free images. To map the planform features, we spectrally enhanced the satellite images using the histogram equalization method, along with different band combinations, during the mapping process. Additionally, we employed two indices i.e., the Normalized Difference Water Index (NDWI) and the Normalized Difference Vegetation Index (NDVI) to map and verify planform

Table 1 Data used for mapping of planform dynamics

Data type			
Satellites/sensors	Spectral resolution (μm)	Date of acquisition	Ground resolution/scale
Hexagon	Visible (9″ × 18″)	1973–03-18	3.6 m
Landsat 5 TM	Visible and infrared (bands 1–5, 7)	1992–03-16 2010–04-03	30 m
Landsat 7 ETM +	Visible and infrared (bands 1–5, 7) Panchromatic (PAN) (band 8)	2000–03-14	Bands 1–5 & 7: 30 m, band 8: 15-m
Landsat 8 OLI	Visible and infrared (bands 1–5, 7) Panchromatic (PAN) (band 8)	2013–04-11 2015–04-07	Bands 1–7 & 9: 30 m, band 8: 15-m
Sentinel 2 MSI	Visible and infrared (bands 1–8)	2020–04-24	Bands 2–4 & 8: 10 m, bands 5–7 and B8A: 20 m

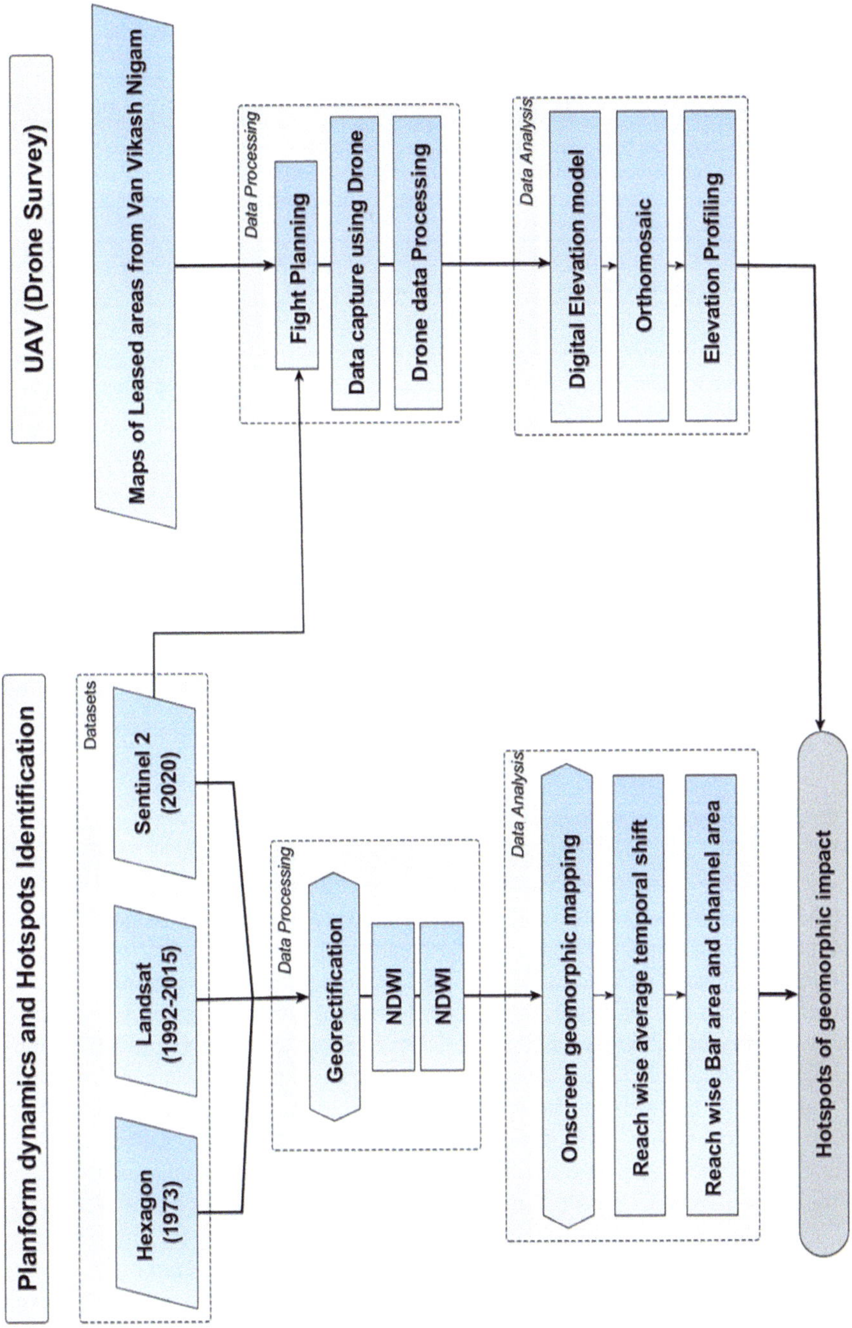

Fig. 3 Various data sets used, and methodology followed in the present study

Table 2 Definition of the different geomorphic features mapped in the present study

Feature	Definition and characteristics
Wet channels	Channel-like features carved by the active flow of water. It includes the major as well as minor active channels within the channel belt
Dry channels	There are two types of Dry channels mapped in the present study i. Large-scale sediment deposits along the vegetated bars which lie within the channel belts. These are mainly large sand bars having channel flow patterns but remain dry most of the time ii. The contributing streams to the main river system/channel belt which remain dry most of the year
Fresh bars	Fresh sediment deposition in and around the river channel. It includes mid-channel bars, point bars, lateral bars etc
Vegetated bars	Elongated and elevated sand deposits mostly aligned along the flow direction of the river. They are mostly stable, due to vegetation cover
Active flood plain	Areaon either side of a river which gets flooded regularly mapped using distinctive geomorphic signatures and temporal satellite images

features such as wet channels, channel bars, and floodplains. To minimize errors due to misclassification, all features were manually digitized. The mapping was conducted at a resolution scale ranging from 1:6,000 to 1:8,000.

Our planform mapping was aimed at understanding the impacts of sand mining (or the absence of it) on river morphology, as well as the natural processes of erosion and deposition. Accordingly, we focused on four major morphological elements within the channel belt, viz. wet channels, dry channels, fresh bars, and vegetated bars. Table 2 provides a list of the morphological features mapped in this study.

3.1.2 Morphometric Analysis

To observe subtle morphological changes in the river planform, we divided the entire stretch between Raiwala and Bhogpur into several reaches of approximately 2 km in length (Fig. 2). Several parameters were measured during the mapping process, including (a) channel dynamics, (b) thalweg dynamics, (c) dry channel and fresh bar area (DC + FB), (d) vegetated bar area (VA), and (e) channel area (CA). The wet channel, representing the current flow of the river, was extracted from pre-monsoon data. We also assessed channel dynamics by analyzing changes in channel form and the migration of the wet channel within the channel belt. The thalweg line, which defines the deepest flow path of the river and its lateral migration, was derived by analyzing both pre- and post-monsoon NDWI maps generated from the satellite data for the respective years. In this study, DC + FB included all dry channels (DC) and fresh bars (FB), which signify sediment deposition within the channel belt. The vegetated bar area (VA) represented stabilized bar forms covered by trees and grassland. The channel area (CA) included only wet channels that remained active during the study period, as defined in Table 2.

3.1.3 UAV Surveys

UAV (Unmanned Aerial Vehicle) based survey ensures precision, reliability and repeatability for evaluating mining impacts at high resolution partciularly in inaccesible areas. Between Haridwar and Bhogpur, multiple mining sites were identified from the data provided by the Uttarakhand Geology and Mining Department and Van Vikas Nigam. The locations were mapped onto the GIS platform to plan UAV flight paths, ensuring comprehensive coverage. In total, we conducted 18 UAV flights, capturing 2880 high-resolution images with a front image overlap of 85 and 70% side overlap to generate a detailed digital topographic model.

We employed a quadcopter (DJI Phantom IV Pro) for image data collection due to its flexibility, stability, and high-resolution imaging capabilities. The UAV was operated at a flight altitude of 200 m above ground level, The camera with an FOV (Field of View) of 84°, provided a Ground Sampling Distance (GSD) of approximately 5 cm. Flight planning was conducted using DroneDeploy software, integrating AutoCAD drawings from the mining department with Sentinel satellite imagery to delineate sand mining boundaries. The study area was segmented into six parts, with each flight covering approximately 1 km^2 to optimize data acquisition. UAV deployment was executed in multiple sessions, with take-off and landing sites selected in situ. The rapid data collection process minimized temporal discrepancies, enabling near-instantaneous morphometric analysis compared to conventional long-duration surveys. Data processing steps for UAV images involved the generation of dense point clouds, orthomosaics, and digital elevation model (DEM) following Sarkar et al. (2023).

4 Results and Analysis

4.1 Planform Dynamics Reconstructed from Historical Satellite Images

Figure 4 presents the various satellite imageries and the corresponding planform maps covering the period from 1973 to 2020, for the stretch between Raiwala and Haridwar. The channel belt is occupied by numerous fresh bars and vegetated bars, while active floodplains are prominently visible on both sides of the channel belt. Downstream of the Bhimgoda barrage, the river has undergone considerable alterations in its flow path and channel form. In 1973, the river exhibited a single-thread channel with minimal braiding (Fig. 4a), flowing westward within the channel belt, with many vegetated and fresh bars present on both sides. However, after 2010, the river's main channel began migrating from west to east, as observed in the planform maps for 2013, 2015, and 2020 (Fig. 4b). The downstream reaches towards Bhogpur have experienced widespread braiding between 2000 and 2020. The central sections of the reaches between Haridwar and Bhogpur are predominantly occupied by vegetated bars, which are surrounded by numerous dry channels consisting mostly of fresh sediment deposits with no flow.

The temporal distribution of planform parameters related to major geomorphic features between Raiwala and Bhogpur is illustrated in Fig. 4c. Between 1973 and 1992, the cumulative area of dry channels did not show any major change, measuring 18.79 km^2 in 1973 and 19.80 km^2 in 1992. However, between 2000 and 2010, the area covered by dry channels increased from 9.85 km^2 to 13.35 km^2 but subsequently decreased to 4.79 km^2 by 2020. Fresh bars remained relatively consistent throughout the study period, except in the years 2000 (21.06 km^2) and 2013 (20.27 km^2).

Regarding vegetated bars, the year 2020 recorded the maximum area (22.08 km^2), whereas the year 2000 had the least area (13.37 km^2). The wet channel, representing the active flow of the river, showed a significant increase in area from 6.29 km^2 in 1973 to 14.54 km^2 in 2020, indicating notable channel expansion. Significant fresh sediments are deposited in dry channels and fresh bars; thus, their combined area is used as a measure of in-channel sedimentation. Figure 4c demonstrates that the combined area of dry channels and fresh bars (DC + FB) has considerably decreased from 37.75 km^2 in 1973 to 20.69 km^2 in 2020. This reduction is attributed partly to channel expansion and partly because some of these areas have transitioned into vegetated bars, which have shown a significant increase over the years.

Figure 5a presents the reach-scale temporal variation of the combined area of dry channels and fresh bars (DC + FB) relative to the wet channel area between Raiwala and Bhogpur. Reaches 5 and 6 include the Bhimgoda barrage and Chandi bridge, respectively. Upstream of the Bhimgoda barrage, reaches 1–3 show an increase in the (DC + FB) area, with the highest recorded value in 2015 (~3.00 km^2) and the lowest in 2020 (~1.5 km^2). A sudden drop in areal distribution is observed at reach 5, where it falls to less than 0.5 km^2, reflecting the inundation effect caused by the Bhimgoda barrage. Downstream of the barrage, reaches 6–9 exhibit a sharp rise in the (DC + FB) area, reaching its peak in reach 8 and remaining relatively high in the downstream reaches. In the downstream reaches, the year 1973 shows the highest values, followed by 1992, 2000, 2013, and 2015, with 2020 recording the lowest values in the series. In contrast to the (DC + FB) area (Fig. 5a), the wet channel area increases at the barrage, followed by a decline between reaches 6–8. From reach 9 to 16, the channel area shows a rising trend. It is important to note that mapping dry channels and bars from the panchromatic images of 1973 posed certain challenges; therefore, all further analyses were restricted to the period from 1992 to 2020.

We further analysed two ratios to understand the processes of reach-scale aggradation and degradation: (a) VA/CA, representing the ratio of vegetated bars to wet channel area (Fig. 5b), and (b) (DC + FB)/CA, representing the ratio of dry channels and fresh bars to wet channel area (Fig. 5c). The VA/CA plot (Fig. 5b) indicates that two reaches (2 and 3), located just upstream of the Bhimgoda barrage, have consistently high values. Reaches 8–10 exhibit an increasing trend in the VA/CA ratio, while the remaining reaches, except for reach 14, show high values. Reaches with a high VA/CA ratio are interpreted as regions where the bars have stabilized, and fresh sedimentation no longer occurs; however, they serve as nuclei that encourage sediment deposition in the surrounding areas. Since most of the fresh sediment deposition

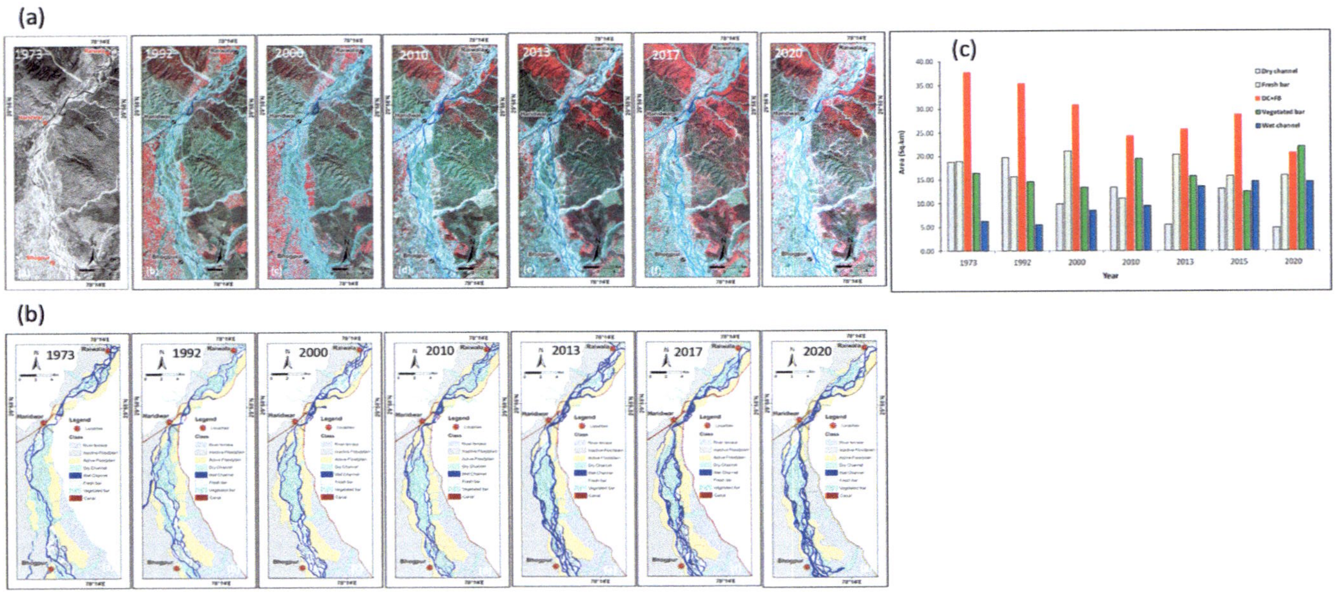

Fig. 4 **a** Satellite images and planform maps of the Ganga river between Raiwala and Bhogpur covering the period 1973–2020. **b** Planforms maps for the corresponding satellite images. Significant fresh sediments are deposited in the dry channel and fresh bars. So, the combined area serves as an indicator of in-channel sedimentation. **c** Plot showing temporal changes in major geomorphic units. The total area of the dry channel and fresh bars (DC + FB) has significantly declined from 37.75 km^2 in 1973 to 20.69 km^2 in 2020 due to (i) channel expansion and (ii) the conversion of the dry channel and fresh bars into vegetated bars

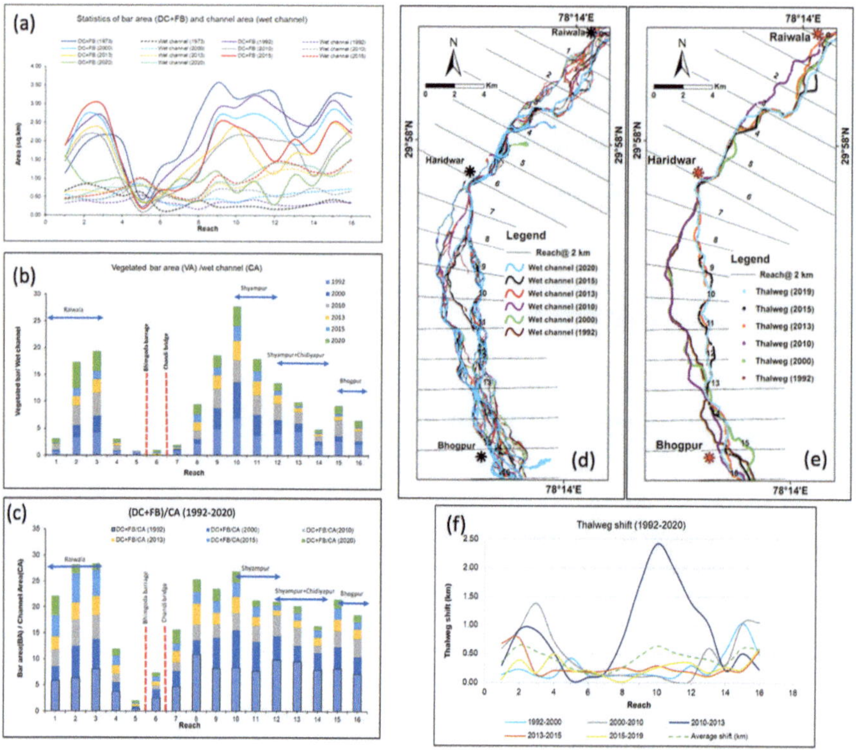

Fig. 5 **a** Reach-wise comparison of bar area (dry channel and fresh bar) and channel area (wet channel) during the period of 1973–2020; **b** Reach-wise comparison of vegetated bar area (VA) vs channel area (CA) during the period of 1992–2020; **c** Reach wise variation in bar area to channel area ((DC + FB)/CA) ratio during the period of 1973–2020; **d** Wet channel dynamics between 1992 and 2020; **e** Channel shift of the Ganga river between Raiwala and Bhogpur between 1992–2020; **f** Thalweg shift between Raiwala and Bhogpur during 1992–2020

occurs in the form of dry channels and fresh bars, the (DC + FB)/CA ratio (Fig. 5c) offers a better representation of channel aggradation relative to channel expansion. Upstream of the Bhimgoda barrage, reaches 1 and 3 have been identified as aggradation hotspots in almost all the years from 1992 to 2015. However, in 2020, the (DC + FB)/CA ratio decreased significantly in reaches 2 to 4, suggesting the inundation of sand bars in these reaches in recent years. Downstream of the Bhimgoda barrage, most reaches exhibited high values of the (DC + FB)/CA ratio between 1992 and 2010, after which the ratio decreased considerably.

To further analyze the channel pattern and migration, we extracted the wet channels from satellite images spanning the period from 1992 to 2020 (Fig. 5e and f). Additionally, the Thalweg line was derived from the NDWI maps (Fig. 5e and f) to gain insights into channel dynamics and lateral migration. The study shows that the river has migrated between right and left banks in several reaches during the study period (1992–2020). Between Raiwala and Haridwar, the river is highly braided in the reaches 1 and 2 but shows an anabranching pattern in reach 2, 3 and 4. Between Haridwar and Bhogpur, the river initially shows a large anastomosing pattern confined to reach 8–13 and then shows a highly braided pattern. In total, the channel is characterized by low sinuosity and high multiplicity except those close to the Bhimgoda bridge.

4.2 Integrating UAV Imaging for Geomorphic Analysis

Six selected windows were analyzed using UAV surveys (Fig. 6a). Among these, one representative study window is presented here. Figure 6b presents the orthoimage draped on digital elevation model (DEM) for one of the selected study windows 2 generated from UAV data acquired in 2022 based on which several cross-sections were constructed (Fig. 6c). The calculated mean elevation of the islands is 323 ± 0.50 m. The elevation of the east bank is approximately 325.80 ± 0.40 m, while the west bank is lower at 318.20 ± 0.30 m. The average elevation of the embankment is 323.60 ± 0.20 m, making it 5.5 m higher than the surrounding areas (Fig. 6c).

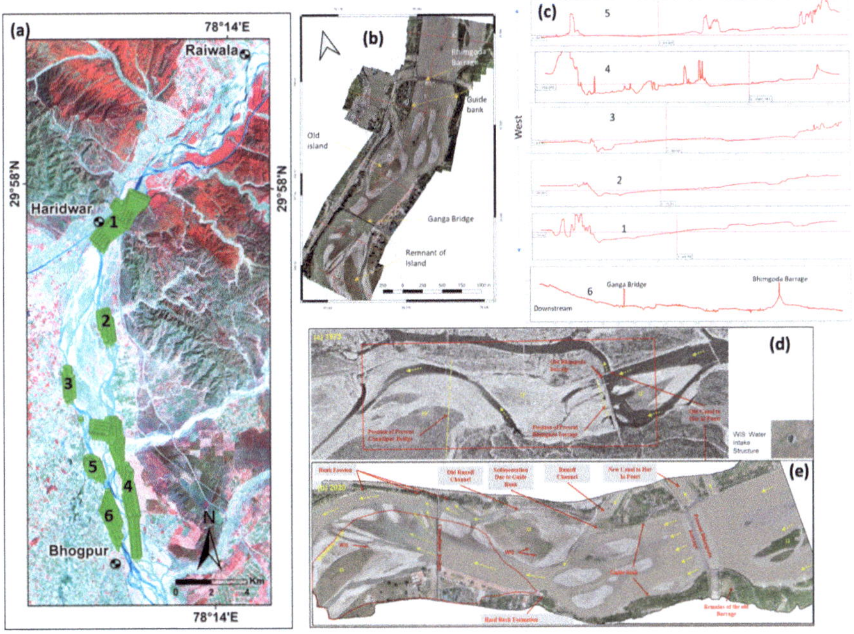

Fig. 6 **a** Satellite image showing the coverage by drone flights during field survey; **b** Drone image of window 2; **c** Cross sections and one longitudinal section of window 2; **d** The old Bhimgoda barrage on Ganga in 1973 as seen on the Corona image. The red boundary represents approximate coverage by (**e**); **e** the present-day situation of the reaches at barrage and downstream as seen on the drone image

Comparing UAV imagery with 1973 imagery indicates that Island I2 (denoted as I2' in the 1973 images) originally covered a much larger area, but a part of this has since eroded. Another island, I3, with an elevation of 338.5 ± 0.4 m, was once a part of the floodplain (Fig. 6d and e). It currently stands 5–8 m above the riverbed, as evidenced by scouring around a water intake structure in the UAV images. We estimated that the river shifted westward by 100–250 m between 1973 and 2020, leading to the erosion of the eastern banks. The west bank has an average elevation of 328 m, while the east bank is slightly higher at 330 m.

Figure 6e shows an UAV image of the reach of the Ganga River between Bhimgoda barrage and Chandipur bridge at Haridwar highlighting the impact of human interventions on channel sedimentation. Our detailed analysis, based on UAV urveys and field visits, suggests that the presence of the Chandipur Bridge has reduced the expected protective effect of the guide bank. High turbulence downstream of the bridge, caused by the piers, is evident in aerial imagery (Fig. 6d and e). This turbulence has bifurcated the river flow, destabilizing and degrading Island I3. It may also be responsible for the erosion of a nearly 15-m-high island on the east bank. Furthermore, the east bank has been reinforced with concrete, which has increased its stability. This stabilization has led the primary flow to shift westward, ultimately causing erosion of the west bank in the reach between the Bhimgoda Barrage and the Chandipur Bridge.

4.3 Identification of 'Hotspots' of Aggradation/Degradation and Assessment of Impacts of Sand Mining (or No Mining)

We have integrated the results from historical remote sensing data analysis and drone surveys to generate reach-scale hotspot map of the stretch between Raiwala and Bhogpur (Fig. 7). We have used three parameters namely (a) average (DC + FB)/CA, (b) average VA/CA, and (c) thalweg shift of each reach and have integrated them using a relative weightage scheme to generate a hotspot index (see Table 4). This hotspot index was calculated using the equation: Hotspot Index $= 0.6 \times [(DC + FB)/CA] + 0.3 \times [VA/CA] + 0.1 \times [\text{Thalweg shift}]$. For each of these parameters, we determined thresholds for different classes and then the final classes of the hotspot index were computed. The thresholding was based on our understanding of the geomorphic processes, field observations and drone surveys (see Table 4 and Fig. 7 for details).

The reaches with high values of hotspot index represent highly aggrading reaches with significant thalweg dynamics. The underlying geomorphic process is that channel aggradation leads to the development of large bars, makes the river shallow and therefore induces channel shifting. Reaches 2 and 3 upstream of Bhimgoda barrage are mapped as highly aggrading reaches. This is probably driven by a natural break in slope although some influence of the barrage cannot be ruled out. These reaches are a part of the Reserved Forests; hence, no further drone investigations

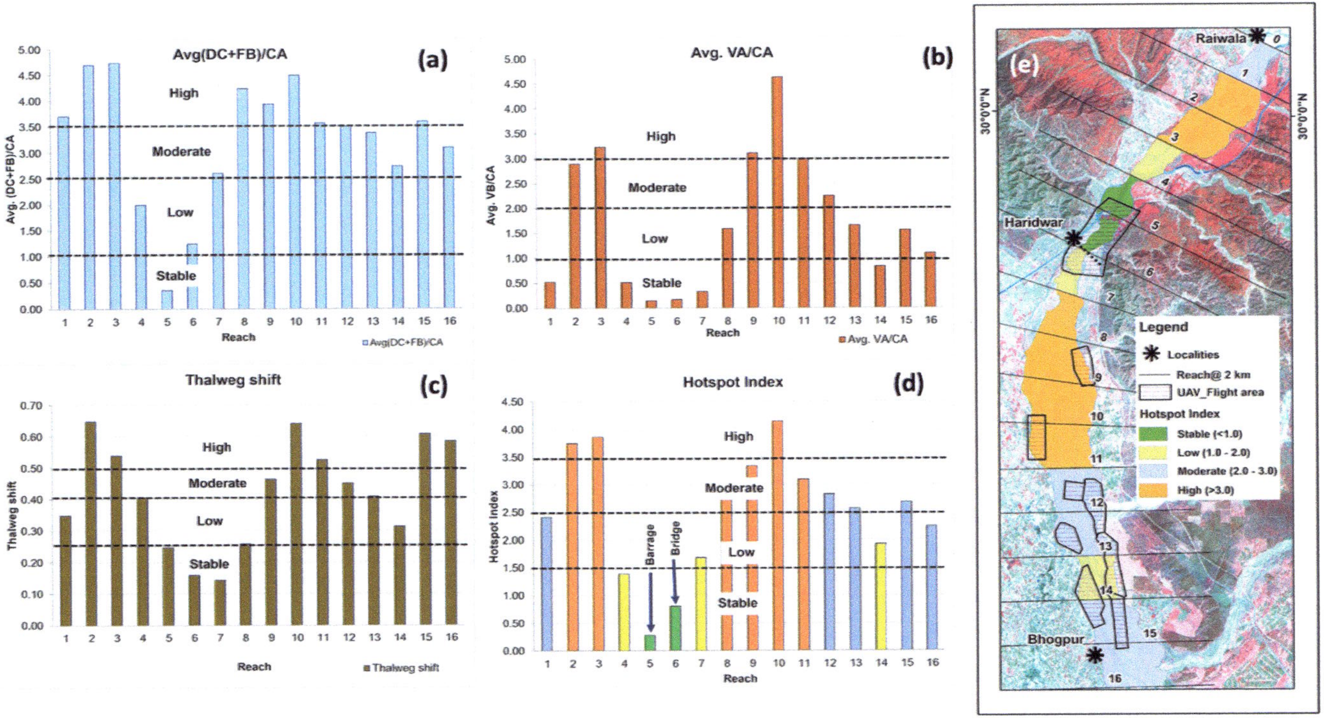

Fig. 7 **a** Average DC + FB/CA ratio map; **b** Average VA/CA ratio map; **c** Average thalweg shift map; **d** Final hotspot index map based on the combined analysis of maps (**a**), (**b**) and (**c**) between Raiwala and Bhogpur during 1992–2020 **e** Identification of hotspots of reach-scale aggradation and channel dynamics

Table 4 Computation of hotspot index

Reach	Average (DC + FB)/CA (1992–2020)	Average VA/CA (1992–2020)	Average thalweg shift (km)	Hotspot index	Inferences from UAV surveys	Recommendations
1	3.70	0.53	0.35	2.42	No UAV survey could be done as these reaches fall in Reserved Forests	Several reaches are highly aggraded and there are known incidences of severe floods. River training work and channelization might be necessary after detailed investigations
2	4.70	2.90	0.65	3.75		
3	4.74	3.23	0.54	3.87		
4	2.00	0.52	0.41	1.39		
5	0.35	0.15	0.25	0.28	Window 1 of UAV survey, moderately aggrading reaches in the lower part, downstream of the Bhimgoda barrage; marked by siltation and channel expansion/ instability	River training work in reaches downstream of the Bhimgoda barrage for channelization; might require partial removal of some bars
6	1.24	0.17	0.16	0.81		
7	2.61	0.33	0.14	1.68		
8	4.23	1.59	0.26	3.04	Window 2 of UAV surveys; highly aggrading reach	Urgent river training works such as channelization should be taken up; strategic sand mining may also be feasible and helpful
9	3.94	3.11	0.46	3.34		
10	4.49	4.63	0.64	4.15		

(continued)

Table 4 (continued)

Reach	Average (DC + FB)/CA (1992–2020)	Average VA/CA (1992–2020)	Average thalweg shift (km)	Hotspot index	Inferences from UAV surveys	Recommendations
11	3.57	2.99	0.53	3.09	Window 3 of UAV survey, Aggrading tendency but the replenishment may be slow;	Some strategic mining integrated with river training works may be feasible in this window, but the rate of extraction should match the replenishment
12	3.51	2.25	0.45	2.82	Window 4: Left bank of reaches 12–15 around Shyampur; Uppermost part falls in the aggradational regime; sandy area, not used for mining. Window 5: Right bank of reach 13, severely affected by past mining; Window 6: Right bank of reach 14, 15 around Bhogpur; moderate aggradation in the lower parts	Window 4: Not suitable for mining, some river training work might be needed to alleviate the flood risk
13	3.38	1.65	0.41	2.56		Window 5: No mining is advisable at present as the window is degrading
14	2.74	0.83	0.31	1.92		Window 6: No mining advisable at present
15	3.60	1.55	0.61	2.68		
16	3.10	1.09	0.58	2.24		

were possible. Reaches 8–11 also record high values of hotspot index and parts of this stretch were surveyed by drone (windows 2 and 3). Our assessment suggests that reach 9 (window 2) is a highly aggrading reach where urgent river training works such as channelization is required, and strategic sand mining may also help. Reach 11 (window 3) also shows an aggrading tendency but the replenishment may be slow. Therefore, some strategic mining integrated with river training works may be feasible in this window, but the rate of extraction should match the replenishment.

We have identified several reaches with moderate values of hotspot index, and these are characterized by high aggradation, but slightly lower thalweg shifts. The average (DC + FB)/CA ratios are generally high to moderate, but VA/CA ratios are slightly lower. Reach 1, upstream of the Bhimgoda barrage, falls into this category and we observed several stretches of high aggradation with a history of severe floods. However, this area also falls in the Reserved Forests and hence no further investigations using drones were possible. Several reaches in downstream such as reaches 12, 13, 15, and 16 (around Shyampur and Bhogpur) also fall in the moderate hotspot zone. Drone surveys in parts of these reaches along the left bank (window 4) map moderately aggradational zones that could be used for mining and more importantly for river training works to alleviate this region from flood risk.

4.4 Sand Mining Policy and Regulations in India: A Critical Review and Further Recommendations

Rivers are a major source of sand and gravel for numerous reasons: (a) cities tend to be located near rivers so transport costs are low; (b) river energy grinds rocks into gravels and sands, thus eliminating the cost of mining, grinding, and sorting rocks; and (c) the material produced by rivers tends to consist of resilient minerals of angular shape that are preferred for construction (whereas wind-blown deposits in deserts are rounder and less suitable) (Koehnken et al., 2020). The term "sand mining" is used as a generic term to embrace extraction of riverine aggregates regardless of particle size. Sand mining activities are one of many recognized pressures affecting riverine ecosystems, where biodiversity is already in rapid decline (World Wildlife Fund, 2018). Increasingly, there are media reports about the negative environmental and social impacts of sand mining, and as calls grow for stronger regulation of mining (Schandl et al., 2016), there is a need to understand the scientific evidence of mining impacts to underpin management.

In recent times, the concern about the impacts of sand mining has arisen globally due to several factors, and the most important of these include the lack of reliable data on sand mining and extraction rates as much of the sand mining activity is unregulated and unmonitored (Hackney et al., 2020). There is also a lack of sediment transport data which makes the estimates of replenishment unreliable and uncertain. However, the availability of modern technology such as high-resolution remote sensing from drones can overcome some of these problems and offer reliable solutions. Further, it is also important to understand the impacts of sand mining which could be direct as well as indirect (Koehnken et al., 2020). Direct impacts are those in which the extraction of material is directly responsible for the ecosystem impact, such as due to the removal of floodplain habitat. Indirect impacts are related to ecosystem changes that are propagated through the system due to physical changes in the river system resulting from sand extraction. For example, the removal of material from a river can alter the channel, river hydraulics, or sediment budget which in turn can alter the distribution of habitats and ecosystem functioning. These types of impacts can be difficult to attribute to sand mining, as they may require long time frames to emerge, and other interventions can result in similar changes. The situation is further complicated by the existence of geomorphic thresholds in river systems (Schumm, 1979). Alterations linked to the removal of sand from rivers may not be gradual and/or linear, and only limited changes may be observed for an extended period, but once a threshold is reached, change may become rapid and irreversible. Whether the impacts of sand mining are positive, neutral, or negative depends on the situation and perceptions of different stakeholders. For example, river incision could be perceived as positive by stakeholders if it reduces flood risk; however, it can also be considered a change from natural or a decline in geomorphic or ecosystem characteristics as negative.

In India, unsustainable sand mining has been rampant despite the 2016 guidelines issued by the Ministry of Environment, Forests and Climate Change (MoEFCC), Government of India. To strengthen the existing system and to curb illegal and unsustainable mining, the MoEFCC released the 'Enforcement and Monitoring Guidelines for Sand Mining 2020' with the following objectives:

- Identification and quantification of and efficient use of Mineral Asset.
- To control sand and gravel mining in the Country from its recognition by consumers and the general public to its final end-use.
- Use of IT-enabled systems and the latest sand mining monitoring technologies at every phase.
- Reduced demand and shortages in supply.
- Development of a sand replenishment process.
- Monitoring post-Environmental Clearance.
- Development of Environmental Assessment Protocol
- Monitor illegal mining instances.

The MoEFCC sand mining policy 2020 is a very comprehensive document, and among other things, it suggests the use of technologies like drones with night vision for surveillance of sand mining sites, steps to identify sources of sand, procedures for replenishment of sand, post-environmental clearance monitoring of sand mining sites, a procedure for environmental audit of such areas and steps to control the instances of illegal mining. The salient points of this document are as follows:

- A comprehensive District Survey Report and Mining Plan must be developed in which all potential river mining zones/areas should be identified. Table 5 shows the typical structure and content of the DSR and Mining Report
- The latitude and longitude of each mining lease shall be mentioned in the Letter of Intent issued to the potential mining lease.
- Environmental Clearance must be obtained for all mining sites.
- No riverbed mining in monsoon, as defined by IMD (typically after 30th June), should be permitted.
- Replenishment studies should be conducted regularly for which the following protocols are prescribed:

Table 5 Structure of the district survey report and mining plan

District survey report (DSR)	Mining plan
• Inventory of RBM • Defining the sources of sand • Channelization of river bed on a map showing areas of aggradation and degtadation, zones of bank protection and 'no mining zone' (use of UAVs encouraged for surveys) • Permanent boundary pillars • Mining lease to be selected in the depositional areas • Mining outside the river bed only if there is enough replenishment • Potential impacts on forest, protected area, habitation, bridges to be avoided (necessary certificates required)	• Original ground level recorded at an interval not more than 10 × 10 M along and across the length of the river • Mapping of the deposition or aggradation reaches in the river • Sediment rating curve for the potential sites (to be checked against the extracted RBM) • Extraction across the active channel in dry season—dry and abandoned channels to be preferred (no diversion allowed) • No extraction in erosional reaches • Extent of lateral migration to be identified (extraction within the zone) • Mining depth <3 m and distance from bank 1/4th of river width • Piedmont zone—mining away from channel banks

- 1st Survey: To be done in April to estimate the level of mining lease before monsoon.
- 2nd Survey: Closing of the mines for monsoon season for quantification of the materials excavated before the offset of monsoon.
- 3rd Survey in the post-monsoon period to estimate the material deposited/ replenished.
- 4th Survey at the end of March of the subsequent year to quantify how much material is excavated.
- The areas of deposition and erosion for each cross-section should be determined.
- Estimation of the bed load transport should be done with analytical models for replenishment estimation
- Iso-pluvial maps should be acquired from the Meteorological Department for estimation of rainfall.
- Runoff coefficient measurement should be carried out.
- Peak flood discharge estimation should be done every year.

This case study on the Upper Ganga River presented above utilized historical remote sensing data as well as UAV technology which have provided important insights about the geomorphic processes operating in this region. Based on this, we would like to offer a few additional guidelines with specific reference to the Himalayan rivers:

(i) The decision about sand mining in a river should not be governed by its present state or morphology but by defining a '**reference state**' i.e., a minimum acceptable or agreed upon morphological and ecological state of the river. The reference state does not have to be the pristine state of the river but should be based on the dynamic behavior of the river and an understanding of river processes and forms sensitivity.

(ii) A particular stretch of the river is characterized by aggradation or degradation (sediment deposition or erosion) depending upon (a) stream power and (b) sediment supply. While the stream power is a function of water discharge and slope, the sediment supply is influenced by upstream/hinterland characteristics. Therefore, one of the first criteria to identify the stretches for riverbed mining is to mark the zones of aggradation or degradation along the river. All mining should be located around the aggradational stretches.

(iii) Identification of aggradational stretches must be done through historical analysis of satellite images and available maps, and the focus should be on the long-term trend rather than one-time mapping. In addition to mapping the hotspots of aggradation, the associated processes such as channel dynamics and bank erosion should also be investigated and the causal factors (natural vs anthropogenic) should be established.

(iv) While the spatial distribution of aggradational areas may be crucial, it is also important to understand the depth of permissible mining at these locations. Depth of mining should be determined based on the rates of accumulations (sand budget) at these locations, and it should not alter the general slope of the stretch in a major way. This requires site-specific topographic, hydrologic and hydraulic information and some measurements. Such data are used to determine the amount and depth of sand that can be removed from a site. In general, the extractable volume of sand should never exceed the volume that can be replenished by normal hydraulic transport. Therefore, bar skimming or scraping is considered the most sustainable way of sand mining.

(v) Several river systems or several reaches in a river are naturally aggrading due to exceptionally large sediment flux from the source areas. These rivers or specific reaches pose serious threats to the existing infrastructure and also enhance the flood risk due to large-scale channel aggradation. At the same time, they offer potential areas for sand mining or dredging. It is therefore crucial to have a sediment management strategy for these rivers. It may be extremely beneficial to integrate river training works with sand mining activities and the concerned departments should work in tandem. However, precautions should be taken not to change the form and slope of the river in a major way.

(vi) Several rivers are extremely dynamic, and they keep changing their position at very short intervals. Thalweg shifting might occur due to the capture of pits during water level rise. An in-channel pit is formed from an off-channel pit. Mid channel bars are eroded and the side channel bars come under threat. If the wet channel stays on one bank, it starts eroding the bank and depositing along the opposite bank creating channel instability. Therefore, attempts should be made to keep the thalweg line in the centre of the channel belt as far as possible.

(vii) Limited mining of the filled paleochannels away from the active channels is a good option but again some caution is required as they also serve as important areas for groundwater recharge. Lowering of the riverbed results in a corresponding lowering of the water table, causing the wells in the vicinity of the river to dry out, thereby increasing the chances of the local communities losing their sources of drinking water. Excessive mining from these areas may, therefore, impact the groundwater potential of the region, and therefore, a minimum of 3–4 m sand layer (undisturbed unsaturated zone) should be left for the protection of the aquifer.

(viii) Riverbed mining in selected areas should always be taken up in patches to maintain the riffle-pool structure of the river longitudinally. This is an important ecological criterion and must be followed. A sudden change in slope due to mining can create a 'knick point' and this can induce downstream degradation thereby damaging bridge piers and buried pipelines, if any. Uniform mining all along the river also alters the slope and increases the runoff in high flow period leading to increased flood risk. Therefore, the maintenance of a desired slope in a river is the key to maintaining the river form, and this must be established through field/modelling studies before commencing the mining.

(ix) Cutting of river banks and widening of river channels must be avoided during sand mining. This will disturb the hydrologic regime of the river significantly and will also harm the riparian vegetation and soil. Therefore, management measures must be taken up in areas earmarked for sand mining such as river stabilization, revegetation of buffer strips and impacts of linked floodplain.

(x) In areas where mining has been done before or is being carried out, continuous monitoring of the river form and position both upstream and downstream of the sand mining areas must be carried out to avoid any undesirable impacts. Alluvial rivers often transmit the disturbance upstream and downstream and such responses are governed by geomorphic thresholds, feedback and lags. Continuous monitoring can help detect such responses and take necessary action.

(xi) All ecological hotspots along the river should be avoided for sand mining irrespective of their regime. The first level zoning of all such habitats must be carried out before planning the sand mining activities and these areas should be excluded from the plans.

(xii) The most fundamental point about sand mining is that if undertaken at all, the extraction of riverbed sand should be conducted at rates that are less than the rate at which sand is redeposited from either bedload or suspended load transported from upstream. Therefore, enough time should be provided to the river to replenish its sediments in a stretch, and therefore, mining areas should shift regularly along the river.

5 Concluding Remarks

Sediment transport and dynamics control the nutrient supply in large river systems and eventually influence the morphological and biological balance. Waterways collect the precipitation, erode the mountains, and distribute the sediments in the plains. Changes in this sensitive system by natural or anthropogenic factors will likely lead to long-lasting major changes in natural environments, where recovery will be difficult. These aspects might be exacerbated in these areas through multiple socio-economic trends: the effects of population increase and expansion of settlements into high-risk areas, the extraction of natural resources and hydropower generation, or the construction of critical infrastructure (power lines, pipelines) and recreational facilities. Sand mining in rivers is one such problem that has attracted enormous attention in several parts of the world. However, there is very little information available in terms of total sand extraction in India, its usage and the geomorphic impacts of sand mining on rivers from India.

The case study from the Upper Ganga River presented in this chapter was focused on a small stretch of the main Ganga River between Raiwala and Bhogpur in Uttarakhand where sand mining impacted the river severely, after which it has been restricted. Consequently, the river has shown signs of recovery except for a few reaches, and therefore, very limited impacts of sand mining are identifiable at present. However, the situation is very different in other Himalayan rivers e.g., Gaula River in Uttarakhand where severe degradation of the river has been documented (Akuria & Sinha, 2025). It is important to take a wholesome stock of the status of mining activity in the entire Himalayan region and assess the impact on river morphology and ecology to evolve a long-term strategy. It is important to note that several Himalayan rivers are highly sediment-charged, and therefore, a balance must be maintained between excessive sediment aggradation and channel stability through river training and strategic sand mining. This calls for a more comprehensive study and development of a sustainable sediment management plan.

Sand mining has emerged as a critical environmental and socio-economic issue driven by unsustainable extraction rates exceeding 50 billion tonnes annually. The impacts are multifaceted, spanning ecological degradation, infrastructure risks, and conflicts with global sustainability goals, while technological advancements offer new monitoring solutions. Technological integration is pivotal to balancing these trade-offs. For instance, satellite-derived sediment budgets could inform "sand credits" to regulate extraction, while AI-powered monitoring can enhance transparency in high-risk regions. However, long-term viability requires coupling technology with policy reforms to address root causes like urbanization demands and illegal trade networks.

Acknowledgements We are grateful to the National Mission for Clean Ganga (NMCG), Ministry of Jal Shakti for funding the pilot project in upper Ganga followed by another major project in Uttarakhand related to the geomorphic impacts of sand mining. We acknowledge the support of the Haridwar administration for supporting the field campaign and drone survey during a very difficult time when the entire country was struggling with the COVID-19 pandemic.

References

Arróspide, F., Mao, L., & Escauriaza, C. (2018). Morphological evolution of the Maipo River in central Chile: Influence of instream gravel mining. *Geomorphology, 306*, 182–197. https://doi.org/10.1016/j.geomorph.2018.01.019.

Akuria, M., Sinha, R., 2025. Spatiotemporal morphodynamics of an ephemeral Himalayan River impacted by sand mining: A process-response framework. *Science of the Total Environment, 964*, 178526. https://doi.org/10.1016/j.scitotenv.2025.178526.

Bendixen, M., Iversen, L. L., Best, J., Franks, D. M., Hackney, C. R., Latrubesse, E. M., & Tusting, L. S. (2021). Sand, gravel, and UN sustainable development goals: Conflicts, synergies, and pathways forward. *One Earth, 4*, 1095–1111. https://doi.org/10.1016/j.oneear.2021.07.008.

Bhattacharya, R., Dolui, G., & Das Chatterjee, N. (2019a). Effect of instream sand mining on hydraulic variables of bedload transport and channel planform: An alluvial stream in South Bengal basin, India. *Environmental Earth Sciences, 78*, 303. https://doi.org/10.1007/s12665-019-8267-3.

Bhattacharya, R. K., Das Chatterjee, N., & Das, K. (2019b). Geomorphic response to riverine land cover dynamics in a quarried alluvial river Kangsabati, South Bengal, India. *Environmental Earth Sciences, 78*, 633. https://doi.org/10.1007/s12665-019-8652-y.

Binh, D. V., Kantoush, S. A., Sumi, T., Mai, N. P., Ngoc, T. A., Trung, L. V., & An, T. D. (2021). Effects of riverbed incision on the hydrology of the Vietnamese Mekong Delta. *Hydrological Processes, 35*, Article e14030. https://doi.org/10.1002/hyp.14030.

Brousse, G., Liébault, F., Arnaud-Fassetta, G., Breilh, B., & Tacon, S. (2021). Gravel replenishment and active-channel widening for braided-river restoration: The case of the Upper Drac River (France). *Science of the Total Environment, 766*, Article 142517. https://doi.org/10.1016/j.scitotenv.2020.142517.

Brunier, G., Anthony, E. J., Goichot, M., Provansal, M., & Dussouillez, P. (2014). Recent morphological changes in the Mekong and Bassac river channels, Mekong delta: The marked impact of river-bed mining and implications for delta destabilisation. *Geomorphology, 224*, 177–191. https://doi.org/10.1016/j.geomorph.2014.07.009.

Calle, M., Alho, P., & Benito, G. (2017). Channel dynamics and geomorphic resilience in an ephemeral Mediterranean river affected by gravel mining. *Geomorphology, 285*, 333–346. https://doi.org/10.1016/j.geomorph.2017.02.026.

Comiti, F., Da Canal, M., Surian, N., Mao, L., Picco, L., & Lenzi, M. A. (2011). Channel adjustments and vegetation cover dynamics in a large gravel bed river over the last 200 years. *Geomorphology, 125*, 147–159. https://doi.org/10.1016/j.geomorph.2010.09.011.

Costea, M. (2018). Impact of floodplain gravel mining on landforms and processes: A study case in Orlat gravel pit (Romania). *Environmental Earth Sciences, 77*, 119. https://doi.org/10.1007/s12665-018-7320-y.

Dufour, S., Rinaldi, M., Piégay, H., & Michalon, A. (2015). How do river dynamics and human influences affect the landscape pattern of fluvial corridors? Lessons from the Magra River, Central-Northern Italy. *Landscape and Urban Planning, 134*, 107–118. https://doi.org/10.1016/j.landurbplan.2014.10.007.

Dujardin, E., Vercruysse, K., Cohen, S., Poesen, J., & Vanmaercke, M. (2024). Mapping and modeling riverine sand and gravel mining at the sub-continental scale: A case study for India. *Science of the Total Environment, 912*, Article 169200. https://doi.org/10.1016/j.scitotenv.2023.169200.

Ferrer-Boix, C., Scorpio, V., Martín-Vide, J. P., Núñez-González, F., & Mora, D. (2023). Massive incision and outcropping of bedrock in a former braided river attributed to mining and training. *Geomorphology, 436*, Article 108774. https://doi.org/10.1016/j.geomorph.2023.108774.

Freedman, J. A., Carline, R. F., & Stauffer, J. R. (2013). Gravel dredging alters diversity and structure of riverine fish assemblages. *Freshwater Biology, 58*, 261–274. https://doi.org/10.1111/fwb.12056.

Fryirs, K. A. (2017). River sensitivity: A lost foundation concept in fluvial geomorphology: River sensitivity: A lost foundation concept in fluvial geomorphology. *Earth Surface Processes and Landforms, 42*, 55–70. https://doi.org/10.1002/esp.3940.

Fryirs, K. A., & Brierley, G. J. (2005). *Geomorphic analysis of river systems: An approach to reading the landscape*. Wiley & Sons.

Ghosh, B. (2024). Assessing the effects of human interventions on the morphodynamics and health of a lowland tropical river: A case study of the river Dwarkeswar, India. *Geosystems and Geoenvironment, 3*, Article 100234. https://doi.org/10.1016/j.geogeo.2023.100234.

Ghosh, A., & Maiti, R. (2021). Development of new Ecological Susceptibility Index (ESI) for monitoring ecological risk of river corridor using F-AHP and AHP and its application on the Mayurakshi river of Eastern India. *Ecological Informatics, 63*, Article 101318. https://doi.org/10.1016/j.ecoinf.2021.101318.

Ghosh, P. K., Bandyopadhyay, S., Jana, N. C., & Mukhopadhyay, R. (2016). Sand quarrying activities in an alluvial reach of Damodar River, Eastern India: Towards a geomorphic assessment. *International Journal of River Basin Management, 14*, 477–489. https://doi.org/10.1080/15715124.2016.1209509.

Ghosh, P. K., & Jana, N. C. (2021). Study of river sensitivity for sustainable management of sand quarrying activities in Damodar river, West Bengal, India. *Current Science, 121*, 810. https://doi.org/10.18520/cs/v121/i6/810-822.

Gumiero, B., Rinaldi, M., Belletti, B., Lenzi, D., & Puppi, G. (2015). Riparian vegetation as indicator of channel adjustments and environmental conditions: The case of the Panaro River (Northern Italy). *Aquatic Sciences, 77*, 563–582. https://doi.org/10.1007/s00027-015-0403-x.

Hackney, C. R., Darby, S. E., Parsons, D. R., Leyland, J., Best, J. L., Aalto, R., Nicholas, A. P., & Houseago, R. C. (2020). River bank instability from unsustainable sand mining in the lower Mekong River. *Nature Sustainability, 3*, 217–225. https://doi.org/10.1038/s41893-019-0455-3.

Hackney, C. R., Vasilopoulos, G., Heng, S., Darbari, V., Walker, S., & Parsons, D. R. (2021). Sand mining far outpaces natural supply in a large alluvial river. *Earth Surface Dynamics Discussions, 9*, 1323–1334. https://doi.org/10.5194/esurf-9-1323-2021.

Hauer, C., Leitner, P., Unfer, G., Pulg, U., Habersack, H., & Graf, W. (2018). The role of sediment and sediment dynamics in the aquatic environment. In S. Schmutz, & J. Sendzimir (Eds.), *Riverine ecosystem management* (pp. 151–169). Cham: Springer International Publishing. https://doi.org/10.1007/978-3-319-73250-3_8.

Hohensinner, S., Habersack, H., Jungwirth, M., & Zauner, G. (2004). Reconstruction of the characteristics of a natural alluvial river–floodplain system and hydromorphological changes following human modifications: The Danube River (1812–1991). *River Research & Apps, 20*, 25–41. https://doi.org/10.1002/rra.719.

Hovius, N. (1998). Controls on sediment supply by large rivers. In G. Kocurek (Ed.), *Relative role of eustacy, climate, and tectonism in continental rocks* (p. 0). SEPM Society for Sedimentary Geology. https://doi.org/10.2110/pec.98.59.0002.

Isik, S., Dogan, E., Kalin, L., Sasal, M., & Agiralioglu, N. (2008). Effects of anthropogenic activities on the Lower Sakarya River. *CATENA, 75*, 172–181. https://doi.org/10.1016/j.catena.2008.06.001.

Kamboj, N., & Kamboj, V. (2019). Riverbed mining as a threat to in-stream agricultural floodplain and biodiversity of Ganges River, India. In *Contaminants in agriculture and environment: Health risks and remediation* (pp. 250–263). Haridwar, India: Agro Environ Media-Agriculture and Ennvironmental Science Academy. https://doi.org/10.26832/AESA-2019-CAE-0151-019.

Koehnken, L., Rintoul, M. S., Goichot, M., Tickner, D., Loftus, A., & Acreman, M. C. (2020). Impacts of riverine sand mining on freshwater ecosystems: A review of the scientific evidence and guidance for future research. *River Research and Applications, 36*, 362–370. https://doi.org/10.1002/rra.3586.

Kondolf, G. M. (1994). Geomorphic and environmental effects of instream gravel mining. *Landscape and Urban Planning, 28*, 225–243. https://doi.org/10.1016/0169-2046(94)90010-8.

Kondolf, G. M. (1997). PROFILE: Hungry water: Effects of dams and gravel mining on river channels. *Environmental Management, 21*, 533–551. https://doi.org/10.1007/s002679900048.

Kondolf, G. M., Piégay, H., & Landon, N. (2007). Changes in the riparian zone of the lower Eygues River, France, since 1830. *Landscape Ecology, 22*, 367–384. https://doi.org/10.1007/s10980-006-9033-y.

Lane, E. W. (1955). Design of stable channels. *Transactions of the American Society of Civil Engineers, 120*, 1234–1260.

Mandarino, A. (2022). Morphological adjustments of the lower Orba River (NW Italy) since the mid-nineteenth century. *Geomorphology, 410*, Article 108280. https://doi.org/10.1016/j.geomorph.2022.108280.

Martín-Vide, J. P., Ferrer-Boix, C., & Ollero, A. (2010). Incision due to gravel mining: Modeling a case study from the Gállego River, Spain. *Geomorphology, 117*, 261–271. https://doi.org/10.1016/j.geomorph.2009.01.019.

Meadows, P. S., & Anderson, J. G. (1966). Micro-organisms attached to marine and freshwater sand grains. *Nature, 212*, 1059–1060.

Mingist, M., & Gebremedhin, S. (2016). Could sand mining be a major threat for the declining endemic Labeobarbus species of Lake Tana, Ethiopia? *Singapore Journal of Tropical Geography, 37*, 195–208. https://doi.org/10.1111/sjtg.12150.

Mitra, S., Mondal, M., Datta, D., & Tamang, L. (2023). Planform adjustments of sub-Himalayan Mahananda-Balason River over five decades in response to anthropogenic interventions. *Arabian Journal of Geosciences, 16*, 676. https://doi.org/10.1007/s12517-023-11809-z.

Mossa, J., & McLean, M. (1997). Channel planform and land cover changes on a mined river floodplain. *Applied Geography, 17*, 43–54. https://doi.org/10.1016/S0143-6228(96)00026-4

Ng, W. X., & Park, E. (2021). Shrinking Tonlé Sap and the recent intensification of sand mining in the Cambodian Mekong River. *Science of the Total Environment, 777*, Article 146180. https://doi.org/10.1016/j.scitotenv.2021.146180.

Padmalal, D., Maya, K., Sreebha, S., & Sreeja, R. (2008). Environmental effects of river sand mining: A case from the river catchments of Vembanad lake, Southwest coast of India. *Environmental Geology, 54*, 879–889. https://doi.org/10.1007/s00254-007-0870-z.

Padmalal, D., & Maya, K. (2014). *Sand mining: Environmental impacts and selected case studies, environmental science and engineering*. Dordrecht: Springer Netherlands. https://doi.org/10.1007/978-94-017-9144-1.

Park, E. (2024). Sand mining in the Mekong Delta: Extent and compounded impacts. *Science of the Total Environment, 924*, Article 171620. https://doi.org/10.1016/j.scitotenv.2024.171620.

Park, E., Ho, H. L., Tran, D. D., Yang, X., Alcantara, E., Merino, E., & Son, V. H. (2020). Dramatic decrease of flood frequency in the Mekong Delta due to river-bed mining and dyke construction. *Science of the Total Environment, 723*, Article 138066. https://doi.org/10.1016/j.scitotenv.2020.138066.

Peduzzi, P. (2014). Sand, rarer than one thinks. *Environmental Development, 11*, 682.

Prabhakar, R., Kumari, A., & Sinha, R. K. (2019). *Impact of sand mining on Zooplankton of River Ganga in and around Patna, Bihar, India.*

Rascher, E., Rindler, R., Habersack, H., & Sass, O. (2018). Impacts of gravel mining and renaturation measures on the sediment flux and budget in an alpine catchment (Johnsbach Valley, Austria). *Geomorphology, 318*, 404–420. https://doi.org/10.1016/j.geomorph.2018.07.009.

Rentier, E. S., & Cammeraat, L. H. (2022). The environmental impacts of river sand mining. *Science of the Total Environment, 838*, Article 155877. https://doi.org/10.1016/j.scitotenv.2022.155877.

Rinaldi, M., Wyżga, B., & Surian, N. (2005). Sediment mining in alluvial channels: Physical effects and management perspectives. *River Research and Applications, 21*, 805–828. https://doi.org/10.1002/rra.884.

Rinaldi, M., Simoncini, C., & Piégay, H. (2009). Scientific design strategy for promoting sustainable sediment management: The case of the Magra River (Central-Northern Italy). *River Research and Applications, 25*, 607–625. https://doi.org/10.1002/rra.1243.

Rovira, A., Batalla, R. J., & Sala, M. (2005). Response of a river sediment budget after historical gravel mining (the lower Tordera, NE Spain). *River Research and Applications, 21*, 829–847. https://doi.org/10.1002/rra.885.

Sanchis-Ibor, C., Segura-Beltrán, F., & Almonacid-Caballer, J. (2017). Channel forms recovery in an ephemeral river after gravel mining (Palancia River, Eastern Spain). *CATENA, 158*, 357–370. https://doi.org/10.1016/j.catena.2017.07.012.

Sarkar, D., Sinha, R., & Bookhagen, B. (2023). Towards a guideline for UAV-based data acquisition for geomorphic applications. *Remote Sensing, 15*(14), 3692. https://doi.org/10.3390/rs15143692

Schandl, H., Hatfield-Dodds, S., Wiedmann, T., Geschke, A., Cai, Y., West, J., Newth, D., Baynes, T., Lenzen, M., & Owen, A. (2016). Decoupling global environmental pressure and economic growth: Scenarios for energy use, materials use and carbon emissions. *Journal of Cleaner Production, 132*, 45–56. https://doi.org/10.1016/j.jclepro.2015.06.100.

Schumm, S. A. (1979). Geomorphic thresholds: The concept and its applications. *Transactions of the Institute of British Geographers, 4*, 485. https://doi.org/10.2307/622211.

Singh, H., & Rao, R. J. (2017). Status, threats and conservation challenges to key aquatic fauna (Crocodile and Dolphin) in National Chambal Sanctuary, India. *Aquatic Ecosystem Health & Management, 20*, 59–70. https://doi.org/10.1080/14634988.2017.1298964.

Singh, R., & Sidhu, M. S. (2016). An overview of environmental impacts of riverbed mining in Himalayan terrain of Himachal Pradesh. *Journal of Applied Geochemistry, 18*, 473–479.

Sinha, R., & Friend, P. F. (1994). River systems and their sediment flux, Indo-Gangetic plains, Northern Bihar, India. *Sedimentology, 41*, 825–845. https://doi.org/10.1111/j.1365-3091.1994.tb01426.x.

Sinha, R., Gupta, A., Mishra, K., Tripathi, S., Nepal, S., Wahid, S. M., & Swarnkar, S. (2019). Basin-scale hydrology and sediment dynamics of the Kosi river in the Himalayan foreland. *Journal of Hydrology, 570*, 156–166. https://doi.org/10.1016/j.jhydrol.2018.12.051.

Sinha, R. (2018). Sediment management of The Himalayan rivers: A challenge for river managers. *Current Science, 115*, 367. https://doi.org/10.18520/cs/v115/i3/367-368.

Skalski, T., Kędzior, R., Wyżga, B., Radecki-Pawlik, A., Plesiński, K., & Zawiejska, J. (2016). Impact of incision of gravel-bed rivers on ground beetle assemblages: Ground beetles of mountain river incised channel. *River Research and Applications, 32*, 1968–1977. https://doi.org/10.1002/rra.3027.

Sreebha, S., & Padmalal, D. (2011). Environmental impact assessment of sand mining from the small catchment rivers in the southwestern coast of India: A case study. *Environmental Management, 47*, 130–140. https://doi.org/10.1007/s00267-010-9571-6.

Surian, N., & Cisotto, A. (2007). Channel adjustments, bedload transport and sediment sources in a gravel-bed river, Brenta River, Italy. *Earth Surface Processes and Landforms, 32*, 1641–1656. https://doi.org/10.1002/esp.1591.

Surian, N., Rinaldi, M., Pellegrini, L., Audisio, C., Maraga, F., Teruggi, L., et al. (2009). Channel adjustments in northern and central Italy over the last 200 years. https://doi.org/10.1130/2009.2451(05).

Tamang, L., & Mandal, D. K. (2015). Bed material extraction and its effects on the forms and processes of the lower Balason River in the Darjeeling Himalayas, India. *Geographia Polonica, 88*. https://doi.org/10.7163/GPol.2015.3.

Tandon, S. K., & Sinha, R. (2007). Geology of large river systems. In *Large rivers* (pp. 7–28). Wiley & Sons, Ltd. https://doi.org/10.1002/9780470723722.ch2.

Vercruysse, K., & Grabowski, R. C. (2021). Human impact on river planform within the context of multi-timescale river channel dynamics in a Himalayan river system. *Geomorphology, 381*, Article 107659. https://doi.org/10.1016/j.geomorph.2021.107659.

Wiejaczka, Ł, Tamang, L., Piróg, D., & Prokop, P. (2018). Socioenvironmental issues of river bed material extraction in the Himalayan piedmont (India). *Environment and Earth Science, 77*, 718. https://doi.org/10.1007/s12665-018-7897-1.

World Wildlife Fund. (2018). Living planet report (2018) risk and resilience in a new era. Gland, Switzerland: WWF International.

Ziliani, L., & Surian, N. (2016). Reconstructing temporal changes and prediction of channel evolution in a large Alpine river: The Tagliamento river, Italy. *Aquatic Sciences, 78*, 83–94. https://doi.org/10.1007/s00027-015-0431-6.

Zou, W., Tolonen, K. T., Zhu, G., Qin, B., Zhang, Y., Cao, Z., Peng, K., Cai, Y., & Gong, Z. (2019). Catastrophic effects of sand mining on macroinvertebrates in a large shallow lake with implications for management. *Science of the Total Environment, 695*, Article 133706. https://doi.org/10.1016/j.scitotenv.2019.133706.

Navigating the Anthropogenic Hazards of Tourism in the Himalaya

Shreeansh Mishra

Abstract Himalayan tourism has become a key economic force, which has contributed extensively to regional growth. Yet the increased tourism has also exacerbated anthropogenic hazards, which have brought about extreme environmental and socio-cultural problems. The present research delves into the intricate relationship between tourism development, anthropogenic hazards, and measures of mitigation in the Himalayas. Through qualitative examination of secondary data, such as contextual literature and case studies, the study recognizes major environmental effects like deforestation, biodiversity loss, and pollution, as well as socio-cultural disturbances like cultural commodification and local community displacement. The results highlight the necessity for an inclusive mitigation framework that incorporates sustainable tourism practices, policy enforcement, and community-based approaches. Sustainable tourism efforts emphasize environmental footprints minimization, whereas robust policy systems facilitate compliance with regulation and conservation. Further, engaging the local communities in participatory systems enhances resilience and upholds cultural integrity. The study finds, regenerative tourism plays a key role in recovering ecological balance and strengthening the resilience of local communities. By offering a comprehensive insight into tourism-related risks and mitigation measures, this study hopes to guide policymakers, tourism operators, and local communities towards a balanced tourism development strategy in the sensitive Himalayan environment.

Keywords Anthropogenic hazards · Sustainable tourism · Regenerative tourism · Himalayan region · Community based solutions

S. Mishra (✉)
Department of Tourism and Hospitality Management, Central University of Tamil Nadu, Thiruvarur, India
e-mail: shreeansh.galaxy@gmail.com

© The Author(s), under exclusive license to Springer Nature Switzerland AG 2025
S. Ghosh (ed.), *The Himalaya Dilemma*, Springer Natural Hazards,
https://doi.org/10.1007/978-3-031-95083-4_23

1 Introduction

The Himalayas, commonly termed as the "Third Pole," are one of the most ecologically and culturally diverse areas in the world (Wambulwa et al., 2021). Spanning across several South Asian Nations such as India, Nepal, Bhutan, and China, the spectacular mountainous system is a pivotal environmental treasure trove consisting of varied fauna and flora along with a singular culture (Singh, 2006). In recent years, this Himalayan region has become increasingly popular as an international tourism destination, welcoming millions of tourists every year who come to experience its natural charm, spiritual values, and adventure activities (Chakraborty & Ghosal, 2022). The adoption of digital marketing strategies by companies and the proliferation of social media use has demonstrably increased tourism footfall in the Himalayas (Thakur et al., 2024), contributing to a significant upsurge in visitation.

While tourism has certainly helped the economy of Himalayan communities to flourish, its unregulated and unmanaged growth has brought a number of anthropogenic risks, undermining the sensitive ecosystem and socio-cultural life of the region (Anup & Fernandez, 2022). On the positive side, it provides income opportunities to the local communities, facilitate infrastructure development, and promotes cross-cultural interactions (Chakraborty & Ghosal, 2022). Conversely, a range of problems follows, such as deforestation, pollution, habitat loss, and socio-economic inequalities (Ziegler et al., 2021). The development of tourism infrastructure, including roads, hotels, and trekking trails, tends to interfere with natural landscapes, causing instability in slopes and making them more susceptible to landslides and soil erosion. The large amounts of waste produced by tourists also have serious environmental implications (Sharma & Kumar, 2023). Moreover, this unregulated development and human intervention hamper the aesthetic beauty of the region.

Further, the cultural and spiritual ethos of the Himalayas are being undermined by the masses rushing to visit them. Historic monasteries, pilgrimage centers, and indigenous cultural practices are under the twin threats of commercialization and cultural homogenization (Howard, 2016). Places such as Leh-Ladakh, Manali, and Char Dham circuit have been especially subject to quick changeovers, where the fine balance between development and conservation is increasingly shifting in favor of exploitation (Apollo et al., 2021). Ashraf et al. (2024) explained in their study that increased exposure to external influences through tourism can contribute to the gradual assimilation of Himalayan cultures into more globalized norms, potentially leading to a loss of unique cultural identities.

The once pristine slopes of the Himalayas are now increasingly marked by overflowing waste bins, eroding trails, and the scars of unplanned construction, highlighting the growing environmental footprint of tourism in the Himalayas (Wilson & Dashper, 2023). The problem of anthropogenic hazards is also compounded by climate change (Nyaupane et al., 2014). The Himalayas, being vulnerable to global warming, are already witnessing enhanced glacial melting, unpredictable weather patterns, and a rise in the occurrence of natural disasters like flash floods and avalanches (Tiwari, 2008). The cumulative effect of the stress caused by tourism activities multiplies these threats, threatening human and ecological systems significantly. Despite rising awareness of these issues, existing literature and policy interventions are still fragmented. While numerous studies on the environmental and economic effects of Himalayan tourism have been conducted, fewer attempts have been made in the case of analyzing anthropogenic risks that accompany tourism in the Himalayas in general. Moreover, issues of local stakeholders who are mostly influenced by such consequences are underrepresented in policy and academic discussion.

This study intends to fill in these gaps through a critical examination of the complexities of anthropogenic hazards of tourism in the Himalayas. The study aims to conduct an advanced analysis of ways in which tourism activities support environmental degradation, cultural loss, and socio-economic issues. By investigating the linkage between tourism actions, environmental dangers, and the resilience of individuals, this study provides practical inputs to enhance responsible tourism practices, ensuring the protection of the Himalayan ecosystem while protecting local livelihood. The unchecked unsustainable tourism practices can amplify and exacerbate anthropogenic hazards of tourism in the Himalayas.

In the process, the study emphasizes the imperative to shift the paradigm of tourism planning and management in the Himalayas. The study explores the multifaceted anthropogenic hazards arising from deforestation, unregulated infrastructure development, overexploitation of water resources, air and noise pollution, and biodiversity loss. Additionally, it examines the socio-cultural disruptions, including cultural erosion and displacement of local communities, that are often overlooked in discussions on sustainable tourism. The study also highlights the exacerbating role of climate change, particularly glacial retreat and glacial lake outburst floods, which amplify the vulnerabilities of the region. Drawing insights from existing literature and case studies, the chapter highlights the interplay between human activities and environmental degradation, offering a critical perspective on the contemporary challenges faced by the Himalayan region.

2 Background of Study

The Himalayas, characterized by their fragile ecosystem and rich biodiversity, have become a hub for global tourism (Dhyani, 2023; Geneletti & Dawa, 2009). However, the increasing pressure from tourism activities has introduced numerous anthropogenic hazards, posing significant environmental, cultural, and socio-economic challenges (Tiwari & Rani, 2024). This section reviews the existing literature to provide a comprehensive understanding of the impacts of tourism in the Himalayan region.

2.1 Environmental Impacts of Tourism

The expansion of tourism infrastructure in the Himalayas has led to deforestation, habitat fragmentation, and soil erosion. Studies show that road construction and resort development destabilize slopes, triggering landslides and increasing vulnerability to natural disasters (Nepal, 2005; Tiwari & Rani, 2024). Waste mismanagement is another critical issue. Research by Neupane et al. (2024) highlights that most of the waste generated by trekkers, garbage left by hikers in Nepal's Annapurna region remains uncollected, leading to pollution in pristine areas. Additionally, the high volume of tourist vehicles has contributed to rising air pollution in hill stations like Shimla and Manali (Balyan, 2021). Furthermore, the noise pollution from increased vehicular traffic and tourist activities disrupts the natural behavior of wildlife, affecting their breeding and feeding patterns (Saqib et al., 2019).

2.2 Socio-Cultural Consequences of Tourism

Tourism has significantly altered the cultural landscape of the Himalayas. Local traditions and spiritual practices often become commodified to cater to tourist expectations. For instance, Apollo et al. (2023) observed that pilgrimage sites like Kedarnath and Yamunotri have increasingly prioritized commercial activities over traditional rituals, eroding their sanctity. Moreover, rapid urbanization driven by tourism has led to gentrification, displacing local populations and altering community structures (Frolich & Schmidt, 2022). Increased interaction with tourists also leads to cultural diffusion, influencing local languages, attire, and lifestyle choices, potentially diluting indigenous identities (Carter & Beeton, 2004). It was further stated that cultural expressions are interconnected, and any alteration in one aspect can lead to a shift in others, influencing traditions, social practices, and collective identities within the communities.

2.3 Climate Change and Tourism Nexus

The Himalayan region is particularly vulnerable to climate change, with tourism acting as both a victim and a contributor to this phenomenon. Glacial retreat and erratic weather patterns have been exacerbated by unsustainable tourism practices, such as overuse of water resources and unregulated trekking (Bolch et al., 2012; Xu et al., 2009). The correlation between tourism activities and the acceleration of glacial melting in areas like the Gangotri Glacier has been highlighted in studies by (Bhattacharya et al., 2024). These findings highlight the need for integrating climate resilience into tourism policies. Rising temperatures and altered precipitation patterns also affect the seasonality of tourism, impacting local livelihoods dependent on specific climate conditions (Chandel & Sharma, 2024).

2.4 Policy and Governance Challenges

Existing governance frameworks in the Himalayan region often fail to address the cumulative impacts of tourism. While there are policies aimed at promoting ecotourism, their implementation remains inconsistent. Chakraborty and Ghosal (2022) critique the lack of coordination among stakeholders, which results in fragmented efforts that fail to achieve sustainable outcomes. Similarly, inadequate enforcement of regulations for waste disposal and land use has compounded the environmental degradation caused by tourism (Sharma & Kumar, 2023). The lack of comprehensive environmental impact assessments (EIAs) for tourism projects further exacerbates the issue, as potential negative consequences are often overlooked during the planning and approval stages and need to strengthened (Jha-Thakur et al., 2020).

2.5 Need for Sustainable Tourism Practices

The concept of sustainable tourism has gained immense popularity in recent years as a potential solution to mitigate the anthropogenic hazards of tourism in the Himalayas. Case studies from Bhutan's high-value, low-impact tourism model illustrate how limiting tourist numbers and focusing on community-based tourism can preserve cultural and ecological integrity (Dasgupta & Vogelaar, 2022). However, replicating such models in other Himalayan regions requires tailored approaches that account for local socio-economic and environmental conditions. Technological innovations, such as the use of renewable energy and implementation of smart waste management systems, could also play a crucial role in minimizing the environmental footprint of tourism.

While research has been done on the environmental consequences of tourism in the Himalayas, there is a lack of comprehensive models that interlink anthropogenic hazards, socio-economic factors, and climate vulnerabilities. Further, limited studies have been conducted on the micro-level effects on certain areas or communities most vulnerable to tourism. Policy analysis regarding the efficacy of prevailing regulations in lessening these hazards is lacking. This study bridges these shortcomings by providing an inclusive analysis of anthropogenic hazards within the context of the Himalayan tourism environment, incorporating environmental, socio-economic, and policy contexts while prioritizing community participation. Highlighting the role of regenerative tourism practices, this research throws light beyond sustainability and concentrates on rejuvenating and enriching the natural and cultural heritage of the Himalayas.

3 Methodology

This study employs a qualitative research methodology, utilizing secondary data sources to provide a comprehensive understanding of anthropogenic hazards associated with tourism in the Himalayas. Contextual literature, is used to form the foundation for examining existing theoretical frameworks and empirical evidence. To enhance and support the validity and reliability of findings, the research adopts insights from two case studies, focusing on specific Himalayan regions experiencing pronounced tourism impacts. Case studies are beneficial, and their units of analysis predominantly consist of a wide variety of factors; individuals, social groups, organizations, and institutions emphasized on detailed contextual analysis (Flick, 2009; Yin, 1994). This method allows for an in-depth exploration of localized dynamics, providing nuanced insights into environmental, socio-economic, and policy-related aspects. Data triangulation is achieved by cross-referencing contextual literature with case study observations, ensuring the accuracy and robustness of the analysis.

4 Findings and Discussion

The Himalayas, renowned for their natural beauty and ecological significance, face a mounting threat from anthropogenic hazards (Neupane, 2023). These hazards arise from human activities and exert tremendous pressure on the region's fragile ecosystem, posing risks to biodiversity, local communities, and the environment. This section explores the major anthropogenic hazards in the Himalayan region, emphasizing their origins, impacts, and potential solutions. Tourism at high altitudes has resulted in more resource use and waste disposal problems, especially in alpine environments.

4.1 Deforestation and Land Degradation

The increasing demand for agricultural land, timber, and tourism infrastructure has resulted in extensive deforestation in the Himalayas. The development of new lodges and infrastructure to support tourists has also stretched these sensitive environments (Stevens, 2003). Forests are cleared for roads, hotels, and other facilities, leading to habitat loss and soil erosion. This deforestation disrupts the delicate balance of the Himalayan ecosystem, affecting the hydrological cycle and reducing carbon sequestration capacity (Upadhyay et al., 2004). The destabilization of slopes due to deforestation often triggers landslides, endangering both human lives and ecosystems. Additionally, the loss of forest cover impacts local biodiversity, leading to the displacement of species and altering predator–prey dynamics (Manchado et al., 2022). The over-dependency of the tourism sector on the natural resources of mountains negatively impacts the mountain ecological balance.

4.2 Unregulated Tourism and Infrastructure Development

Tourism, while economically beneficial, has contributed to the degradation of the Himalayan environment. The construction of roads, resorts, and trekking routes has disrupted natural landscapes, leading to habitat fragmentation and soil erosion (Geneletti & Dawa, 2009; Nyaupane et al., 2014). Popular destinations like Manali, Ladakh, and the Annapurna Circuit have become hotspots for overcrowding, waste mismanagement, and pollution. Unregulated tourism also leads to overconsumption of natural resources, including water and energy, which strains local ecosystems (Brandt et al., 2019; Ziegler et al., 2021). According to Kuniyal (2002) unmanaged waste from trekking expeditions contributes to the contamination of rivers and soil, posing risks to both humans and wildlife. To address these challenges, sustainable tourism practices, strict waste management policies, and visitor capacity regulations are needed Kuniyal et al. (2003). Additionally, The widespread infrastructure development to facilitate trekking, mountaineering, and expedition tourism tends to cause habitat fragmentation, enhanced waste production, and disruption of sensitive ecosystems as explained in case study 1. It is essential to reduce the logistical elements involved in Himalayan adventure travel to reduce environmental disruption and conserve the ecological balance of the area (Apollo, 2016).

4.3 Overexploitation of Water Resources

Rapid urbanization supported by tourism have placed immense stress on the region's limited water resources. The construction of dams and hydropower projects has disrupted river ecosystems, affecting aquatic life and the livelihoods of downstream communities (Bartwal & Nautiyal, 2023). Overextraction of groundwater for urban centers and tourist hubs often exceeds replenishment rates, leading to water scarcity in regions such as Uttarakhand and Himachal Pradesh (Mahar et al., 2024). Rising population in the fragile Himalayan region accompanied by higher tourist footfall place increased pressure on water resources for the purpose of drinking, sanitation and for recreational activities. This often surpasses the natural replenishment rate specially during peak season leading to water scarcity for local resources. Moreover, trekkers and hiking activist irresponsible behavior and improper waste disposal pollute the water bodies directly and indirectly, and poses a serious threat to the fragile Himalayan ecosystem.

4.4 Climate Change and Glacier Retreat

Human-induced climate change is accelerating glacial melting in the Himalayas, leading to changes in river flow patterns and an increased risk of glacial lake outburst floods (Stuart-Smith et al., 2021). The transport service used by tourists in the Himalayan region cause damage in multiple ways, such as vibrations from the vehicle, the heat generated, and carbon emissions, worsening the climate change impacts and catalyzing the melting of glaciers (Misra et al., 2024). Additionally, the energy consumption by hotels, resorts, and other tourism-related infrastructure, reliant on fossil fuels in many Himalayan regions, further adds to carbon emissions intensifying the climate change impacts (Sundriyal et al., 2018). This glacier retreat, driven in part by tourism-related activities, threatens water resources, increases the risk of glacial lake outburst floods Research by Bolch et al. (2012) found that Himalayan glaciers are retreating at an average rate of 10–15 m annually, with anthropogenic factors significantly contributing to this phenomenon. Adaptation measures such as glacier monitoring, early warning systems for GLOFs, and community-based disaster preparedness programs are essential to minimize the risks associated with climate change in the Himalayas (Rasul & Molden, 2019).

4.5 Air and Noise Pollution

The flow of vehicles, such as cars, buses, and motorbikes carrying tourists, results in high concentrations of particulate matter, nitrogen oxides, and other air pollutants, particularly in highly populated tourist zones and along major trekking trails (Roy & Mitra, 2024; Sharma, 1997). The use of diesel generators to support unreliable power supplies, especially in rural areas, also contributes to poor air quality (Balyan, 2021). At the same time, rising noise pollution from traffic, tourist infrastructure construction activities (hotels, roads), and recreational activities disturb the peace of the Himalayan ecosystem, affecting wildlife and tourists' experiences who are looking for peace and solitude (Roy & Mitra, 2024). This double pollution negatively impacts the health of indigenous people and reduces the overall environmental quality of the area. Additionally, noise pollution from tourism activities, such as adventure sports and vehicular traffic, disrupts wildlife behavior and affects the tranquility of the region (Potvin et al., 2023). To combat these issues, green initiatives such as stricter emission regulations, promotion of electric vehicles, and noise control measures are necessary (Bhatia et al., 2024).

4.6 *Habitat Fragmentation and Biodiversity Loss*

Anthropogenic activities have led to the fragmentation of habitats, reducing biodiversity in the region. Tourism induced infrastructure development often cuts through critical wildlife corridors, increasing the incidence of human-wildlife conflicts (Dhyani, 2023; Sharief et al., 2023). The Himalayan musk deer, red panda, and snow leopard are among the species most affected by these changes (Kandel et al., 2014; Sharief et al., 2023). Additionally, invasive species introduced through human activities threaten native flora and fauna, leading to alterations in ecosystem dynamics (Gupta et al., 2021; Pathak et al., 2019). Invasive species introductions, by and large inadvertent through travelers or in construction material, may further endanger local vegetation and animals, resulting in overall biodiversity as well as ecosystem loss. The loss of biodiversity not only disrupts ecological functions but also impacts ecosystem services crucial for local communities.

4.6.1 Cultural Erosion and Social Displacement

Tourism-driven modernization and urbanization have disrupted traditional lifestyles and cultural practices. The commodification of cultural heritage often leads to the loss of authenticity and identity of local communities (Sharma, 2018). Even though all the destinations hope to make the hosts gain by developing the tourism sector, the reluctant voice has already been raised by some of the destinations, such as Manali (Nimble, 2023), Shimla (Gupta & Chomplay, 2021). Without the support of the hosts, making sustainable development in the tourism sector might be an uphill battle. Infrastructure projects such as dams and highways frequently displace communities, creating socio-economic challenges and weakening the social fabric (Rai & Srivastava, 2014). Social displacement also leads to conflicts over land rights and access to natural resources thereby provoking the community members to encroach land and construct unsustainable infrastructure leading to ecological imbalance.

Case Study 1: Anthropogenic Hazards in Ghatte Khola, Nepal Himalaya

The Ghatte Khola area in the Myagdi District of Western Central Nepal illustrates the challenges posed by anthropogenic hazards in the Himalayan region. This area experiences frequent debris flows triggered by heavy cloudbursts, which are exacerbated by human activities such as haphazard infrastructure development and increased population pressure. The construction of a motor bridge and an electric power station near the confluence of Ghatte Khola and Kali Gandaki rivers illustrates the vulnerability of man-made structures to natural hazards intensified by human interventions. On May 25, 2019, a flash flood in Ghatte Khola destroyed the newly built bridge, highlighting the consequences of inadequate hazard assessment and poor land-use planning.

The study emphasizes that the debris flows in Ghatte Khola are linked to slope instabilities in the upstream catchment area, worsened by deforestation and unregulated construction activities. The increased sediment fluxes due to these anthropogenic activities accelerate bank erosion, threatening downstream settlements like Dana village. This case demonstrates how natural hazards transform into disasters due to human-induced changes in the landscape. Local communities are aware of the stream's ephemeral yet destructive behavior, but insufficient disaster preparedness and mitigation measures increase their vulnerability.

This case study highlights the need for sustainable infrastructure development and integrated disaster risk management in the Himalayas. It highlights the importance of comprehensive environmental impact assessments, community-based disaster preparedness, and stricter regulations on land use and construction in ecologically sensitive areas. By adopting such measures, the region can minimize the risks associated with anthropogenic hazards while ensuring sustainable development.

Source Gurung (2019).

Case Study 2: Community-Driven Ecotourism in Ladakh Himalaya

Community-driven ecotourism (CDE) in Ladakh, a region in the trans-Himalayan area, serves as a successful model of sustainable tourism by balancing economic growth with cultural and environmental preservation. This approach involves local communities managing tourism activities such as homestays, wildlife tours, and heritage conservation, ensuring that tourism revenues directly benefit the local population. CDE empowers communities by involving them in decision-making processes and promoting local employment, thereby reducing economic leakage and fostering cultural pride (Wani et al., 2025). Additionally, this model supports environmental stewardship by minimizing resource exploitation and maintaining ecological resilience.

In Ladakh, community-managed homestays are pivotal, offering tourists an authentic cultural experience while preserving traditional lifestyles. These homestays reduce the environmental footprint by using locally sourced materials, traditional construction techniques, and sustainable waste management practices. Wildlife tours led by local guides enhance environmental education and promote wildlife conservation, reducing human-wildlife conflicts. The model aligns with Sustainable Development Goals (SDGs) 8, 12, and 15 by fostering economic growth, promoting sustainable consumption, and conserving ecosystems.

However, the initiative faces challenges, including infrastructure deficits, limited training in sustainable tourism practices, and inconsistent policy support. Despite these issues, CDE in Ladakh illustrates the potential for community-centric tourism models to promote cultural integrity, environmental sustainability, and economic empowerment in ecologically sensitive regions like the Himalayas. This case underscores the significance of integrating local knowledge systems into tourism development and suggests that replicating this model could enhance sustainable tourism practices across other Himalayan destinations.

Source Wani et al. (2025).

5 Strategies for Sustainable Tourism in the Himalaya

The escalating anthropogenic hazards resulting from tourism in the Himalayas necessitate a multifaceted approach grounded in sustainability. This requires a synergistic combination of ecotourism promotion, robust policy implementation, and community-based solutions, all working in concert to mitigate environmental damage, preserve cultural integrity, and foster socio-economic well-being. As shown in Fig. 1, tourism expansion has a direct impact on anthropogenic risks, and these latter result in environmental degradation and socio-cultural disturbances. To face these threats, three important mitigation approaches are identified by the study: encouraging sustainable tourism, enforcing proper policies, and promoting community-based approaches. Sustainable tourism methodologies aim at reducing environmental impacts, policy enforcement guarantees compliance with regulations and protection of the environment, whereas community-based approaches empower stakeholders at the local level and increase resilience. This holistic approach presents a combined strategy for addressing tourism-induced risks in the Himalayas.

5.1 Sustainable Tourism Practices

5.1.1 Ecotourism Promotion

Encouraging eco-friendly tourism practices, such as low-impact trekking and wildlife conservation tours, which minimize environmental degradation (Chaudhary et al., 2021). This can be further enhanced by certifying tour operators based on adherence to strict environmental standards and promoting educational tours that emphasize ecological awareness and conservation ethics (Choegyal, 2022). Moreover, investing in the development of trails using sustainable materials and promoting the use of

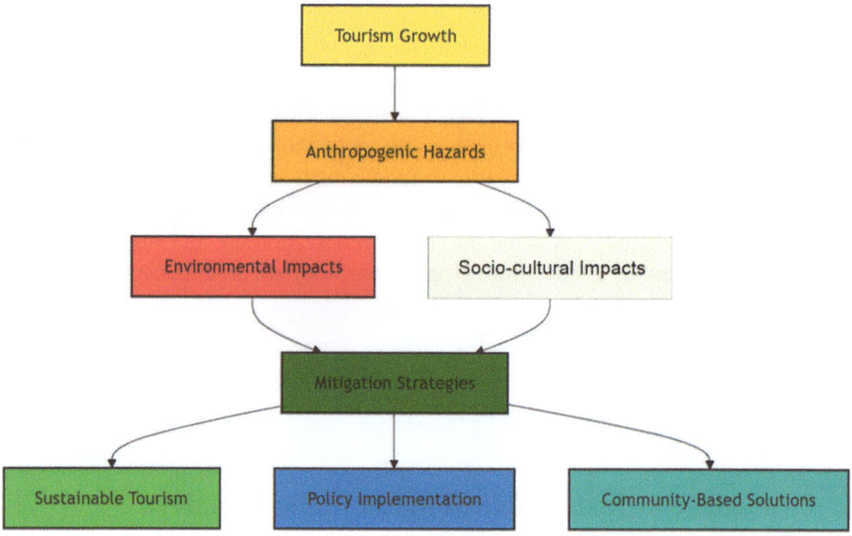

Fig. 1 Strategies for sustainable tourism in the Himalaya

local guides can further minimize the environmental impact while supporting local economies (Singh et al., 2021). Ecotourism must be cautiously marketed, and tourists and tour operators must see the importance of maintaining the balance of nature for generations to come. In addition, "controlled tourism" could be encouraged in order not to destroy the land, water, and air as a consequence of increased waste production, car utilization, and disposal of waste in rivers (Dłużewska & Giampiccoli, 2020).

5.1.2 Technological Integration

The technological integrated method such as Remote sensing (RS) and geographical information system (GIS) proves to be applicable in addressing universal and complex problems like biodiversity conservation, protected area management, and ecotourism development in the mountains, pointing towards the importance of accurate and effective planning to counteract the negative effects of tourism (Chaudhary et al., 2021). Protected areas that are rich in biodiversity requires patrolling and vigilance to reduce habitat destruction from transportation and deforestation, and to limit human-animal interaction which could be very well taken care by smart devices and early warning systems.

5.1.3 Carrying Capacity Regulation

Developing comprehensive tourism plans that incorporate carrying capacity assessments and zoning regulations is crucial. This involves identifying sensitive areas that require protection, limiting the number of visitors allowed in specific areas, and regulating the type and scale of tourism development (Nepal, 2000). Implementing visitor caps and seasonal restrictions to prevent overcrowding and overuse of natural resources. This requires a data-driven approach, utilizing ecological assessments and visitor surveys to determine optimal carrying capacities for specific regions and attractions (Dey et al., 2024). Differential pricing strategies, offering lower rates during off-peak seasons, could also help to distribute tourist flow more evenly throughout the year, reducing pressure on resources during peak periods (Lozano et al., 2020).

5.1.4 Green Initiatives

It is a strategic way of developing and managing green areas that is planned to provide ecosystem services to human populations, combining plants and other natural components to enhance the sustainability and environmental friendliness of urban development. Effective measures such as afforestation, community-based forest management, and sustainable land-use practices are essential to mitigate the impacts developing sustainable infrastructure, including energy-efficient hotels and waste management systems, to reduce ecological footprints (Gautam et al., 2021). This includes incentivizing the adoption of renewable energy sources, such as solar and wind power, in tourism facilities, and implementing comprehensive waste segregation and recycling programs (Chiwaridzo, 2024). Furthermore, promoting the construction of buildings using locally sourced, eco-friendly materials and implementing water conservation measures can significantly minimize the environmental impact of tourism infrastructure, as discussed in case study 2.

5.1.5 Regenerative Tourism

Regenerative tourism extends beyond the idea of sustainability in its commitment not merely to minimize harm, but also to restore and expand the cultural and natural endowments of the Himalayas. In contrast with sustainable tourism's emphasis on preservation of present levels of environment, regenerative tourism is actively promoting restoration of environments and resiliency within local communities (Dredge, 2022). It entails restoring over-degraded environments through rewilding, collaborating with initiatives promoting local conservation of biodiversity, and revitalizing local indigenous practices (Bellato & Pollock, 2023). In the context of the Himalayas, regenerative tourism might involve projects such as afforestation of degraded trekking trails, restoration of glacial water catchments, and low-impact building designs that blend with the local ecological and cultural contexts. In addition, regenerative tourism promotes more profound tourist involvement with local communities in the form of genuine cultural encounters and active contribution to conservation processes. By incorporating regenerative principles, tourism in the Himalayas can be a force for ecological and cultural regeneration, where the benefits of tourism are extended beyond economic benefit to environmental and social regeneration.

6 Policy Implementation

6.1 *Environmental Legislation*

Enforcing strict environmental regulations to control deforestation, waste disposal, and construction activities in sensitive areas is needed for the hour (Bhattacharya, 2019). This requires strengthening regulatory frameworks, increasing monitoring efforts, and imposing stricter penalties for environmental violations. Establishing protected areas and buffer zones with clearly defined regulations can also help safeguard vulnerable ecosystems from the adverse impacts of tourism. The enforcement of building codes must be instituted to ensure that buildings in areas susceptible to hazards are constructed to withstand natural disasters such as earthquakes and landslides (Arendt et al., 2017). As discussed in Case Study 1, implementing institutional frameworks for disaster preparedness, response, and mitigation, including steps to lessen the susceptibility of communities to natural hazards that are exacerbated by human activities, could help to improve the situation.

6.2 *Community Involvement in Policy Making*

Involving local communities in decision-making processes ensures that policies are culturally appropriate and sustainable. This necessitates the creation of participatory platforms where community members can voice their concerns, contribute their knowledge, and actively participate in the development and implementation of tourism policies (Barandiarán et al., 2019). Providing training and resources to enable local communities to effectively participate in policymaking can ensure that their perspectives are adequately represented (Mishra et al., 2024). Through the active involvement of local communities in designing and implementing environmental development plans and regulations, policies are better adapted to the region's specific ecological and socio-economic conditions. Through such participatory strategies, local expertise and traditional knowledge can be integrated into policies, leading to more effective means of preventing hazards such as deforestation, water contamination, and mass tourism (Mishra & Mishra, 2023). Empowering communities to control their own development pathways promotes resilience, enhances social cohesion, and results in more sustainable and equitable outcomes, reducing the effects of anthropogenic hazards on the environment and the local population.

6.3 Monitoring and Evaluation

Establishing monitoring systems to assess the effectiveness of environmental policies and adjust them based on ecological feedback (Bolch et al., 2012). This includes the establishment of key performance indicators (KPIs) for environmental sustainability and the regular collection and analysis of data on indicators such as water quality, air pollution, and forest cover (Morrison et al., 2024). Using remote sensing technologies and GIS mapping can provide valuable data for monitoring environmental changes and assessing the effectiveness of conservation efforts (Chaudhary et al., 2021). Moreover, use smart technologies should be part of destination management policies (Ivars-Baidal et al., 2017), so that emerging technologies could be made instrumental for identifying various malpractices like bonfire in sensitive zones, traffic parking congestion or illegal encroachment.

7 Community-Based Solutions

7.1 Local Stewardship Programs

Empowering local communities to manage natural resources and tourism activities to ensure sustainable use and conservation (Wani et al., 2025). This includes providing training and resources to local communities to manage tourism enterprises, conserve biodiversity, and promote sustainable agricultural practices, as discussed in case study 2. Establishing community-managed protected areas and ecotourism ventures can provide economic incentives for local communities to conserve natural resources; moreover, they will feel responsible and attached to the destination (Nyaupane et al., 2020). The Himalayan indigenous community possess intellectual knowledge which could be further developed and propagated when tourist and community members are engaged in mutual interaction (Singh, 2006). The targeted stewardship programs to involve such indigenous communities for managing tourists and related activities will conserve the pristine Himalayan ecosystem, facilitating coexistence with nature, as discussed in case study 2.

7.2 Education and Awareness

Conducting environmental awareness programs for locals and tourists to promote responsible tourism behavior (Singh et al., 2020). Such programs will cultivate positive behaviours resulting in villagers' engagement in public–private partnerships, their involvement in tourism-related development programmes and motivate them to contribute towards environmental activities This requires designing targeted awareness campaigns that educate tourists about responsible trekking practices, waste management, and cultural sensitivity, and educating local communities and tourist about the importance of conservation and sustainable tourism (Anjusha & Thomas, 2024). Using social media and online platforms to disseminate information about sustainable tourism practices can reach a wider audience and promote responsible tourism behavior (Han et al., 2017; Hysa et al., 2021).

7.3 Economic Diversification

Encouraging alternative livelihoods to reduce over-dependence on tourism, thereby minimizing environmental and cultural impacts (Milano et al., 2019). Diversifying the economic base through investments in sustainable agriculture, renewable energy, ecotourism (managed responsibly), small-scale manufacturing, and other sectors creates alternative income streams and reduces dependence on environmentally damaging activities. Providing access to micro-finance and entrepreneurship training can empower local communities to diversify their livelihoods, adopt sustainable practices and reduce their reliance on tourism (Prasad & Choubey, 2023).

8 Conclusion

Anthropogenic risks in the Himalayas, fueled primarily by unchecked tourism expansion, deforestation, infrastructural development, and climate change, have enormous implications for the region's sensitive ecosystem, biodiversity, and cultural heritage. Environmental effects such as land degradation, depletion of water resources, and air pollution are exacerbated by socio-cultural issues like cultural degradation and social displacement. Yet, the study points out that sustainable tourism methods, good policy enforcement, and community-driven interventions provide sound mechanisms for managing such risks. Encouraging ecotourism, controlling the number of tourists, implementing environmental laws, and empowering local populations are listed as key measures for sustainable development. The study highlights that incorporating regenerative tourism activities can further enhance the resilience of the Himalayan environment by actually rehabilitating natural and cultural heritage, guaranteeing long-term sustainability. Case studies of areas such as Ladakh and Bhutan reveal that managed tourism and the active involvement of the local population can effectively balance economic development with the preservation of the environment. An integrative approach involving environmental, social, and economic factors is vital for successful application. Joint initiatives among governments, host communities, and tourists are imperative to protecting the Himalayan environment and guaranteeing long-term sustainability.

8.1 Implications

This study on anthropogenic risks in the Himalayas has serious implications for policymakers, tourism developers, environmentalists, and the local population. For policymakers, the study highlights the importance of strict regulation of tourism growth and natural resource utilization. Conducting environmental impact assessments (EIAs) and imposing zoning regulations can reduce habitat loss and pollution. Tourism planners can utilize these findings to advance sustainable tourism frameworks, including ecotourism and community-based tourism, which focus on the conservation of the environment and empowering local communities. The research also emphasizes the need to build infrastructure that is consistent with environmental sustainability, thus mitigating negative impacts on sensitive ecosystems. Environmentalists can utilize the research to promote biodiversity conservation by raising awareness and engaging communities in the protection of threatened species and ecosystems. Local people, as major stakeholders, can gain through community-based resource management and sustainable livelihood as they are weaned from their socio-economic vulnerability. The study supports the idea that encouraging regenerative tourism can strengthen local communities and promote greater environmental rehabilitation, developing a more sustainable and resilient tourism pattern in the Himalayas. The study also contributes to scholarly knowledge by establishing a connection between tourism development and environmental and socio-cultural issues, offering an exhaustive framework for future research on sustainable mountain tourism. By combining climate change adaptation strategies with tourism and conservation policies, this research provides a blueprint for maintaining the long-term sustainability of the Himalayan ecosystem.

8.2 Limitations and Scope for Future Research

This study is constrained by the use of secondary data, which might not represent the dynamic changes in the environment and society of the discussed region. It concentrates on most visited sites, and as such, the effects in less-travelled regions can be overlooked. The research also generalizes mitigation without considering localized solutions that are specific to the variety of Himalayan sub-regions. In addition, it highlights tourism and infrastructure development and underplays other major factors such as agriculture, mining, and urbanization. The multifaceted interplay between climate change and human activities calls for a more interdisciplinary treatment than examined in this research.

Future studies should include empirical field studies for in situ data collection and longitudinal studies to assess long-term effects, especially with respect to climate change. Broadening the geographic scope to the lesser-known areas will give a richer picture. Studies should also assess localized mitigation measures, community-based conservation, and adaptive tourism management. An interdisciplinary study combining environmental science, social anthropology, and climatology is suggested for a holistic view. Furthermore, policy research is necessary to evaluate the existing regulations and create adaptive models of governance for sustainable development in the Himalayas.

References

Anjusha, P., & Thomas, T. K. (2024). Sustainable hiking through the green trails: Examples of exceptional responsible practices. *Tourism Cases*. https://doi.org/10.1079/tourism.2024.0011.

Anup, K. C., & Fernandez, M. (2022). Tourism and himalayan indigenous communities. In G. P. Nyaupane, & D. J. Timothy (Eds.), *Tourism and development in the himalaya social, environmental and economic forces* (pp. 149–169). Routledge. https://doi.org/10.4324/9781003030126-12.

Apollo, M. (2016). The true accessibility of mountaineering: The case of the High Himalaya. *Journal of Outdoor Recreation and Tourism, 17*, 29–43. https://doi.org/10.1016/j.jort.2016.12.001.

Apollo, M., Andreychouk, V., Mostowska, J., Rawat, K. (2021). Indo-Himalayan protected areas: Peak-hunters, pilgrims and mountain tourism. In: T. E. Jones, H. T. Bui, & M. Apollo (Eds.), *Nature-based tourism in Asia's mountainous protected areas. Geographies of tourism and global change*. Cham: Springer. https://doi.org/10.1007/978-3-030-76833-1_11.

Apollo, M., Andreychouk, V., & Wengel, Y. (2023). Pilgrimage tourism to sacred places of high Himalaya and its impact on residents across generations. The case of Yamunotri temple. In *Perspectives on Asian tourism* (pp. 89–105). https://doi.org/10.1007/978-981-19-9677-1_6.

Arendt, L., Hortacsu, A., Jaiswal, K., Bevington, J., Shrestha, S., Lanning, F., et al. (2017). Implementing Nepal's national building code: A case study in patience and persistence. *Earthquake Spectra, 33*(1_suppl), 167–183. https://doi.org/10.1193/121716eqs242m.

Ashraf, M. V., Misri, S., Pant, S., & Ahmad, S. (2024). Unorganized tourism practices and the allied challenges in the himalayas. In *Advances in Asian human-environmental research* (pp. 193–220). https://doi.org/10.1007/978-3-031-56858-9_9

Balyan, P. (2021). Impacts of air pollution on Himalayan region. In *Springer atmospheric sciences* (pp. 57–85). https://doi.org/10.1007/978-3-030-70509-1_5

Barandiarán, X., Restrepo, N., & Luna, Á. (2019). Collaborative governance in tourism: Lessons from Etorkizuna Eraikiz in the Basque Country, Spain. *Tourism Review, 74*(4), 902–914. https://doi.org/10.1108/tr-09-2018-0133.

Bartwal, T., & Nautiyal, P. (2023). Hydropower development in the Himalaya: Identifying critical river stretches. *Environmental Science and Pollution Research, 30*(16), 46741–46747. https://doi.org/10.1007/s11356-023-25401-1

Bellato, L., & Pollock, A. (2023). Regenerative tourism: A state-of-the-art review. *Tourism Geographies*, 1–10. https://doi.org/10.1080/14616688.2023.2294366.

Bhatia, A., Sharma, A., & Sharma, R. (2024). The transformative aspects of green tourism: A Himalayan perspective. In *Emerald Publishing Limited eBooks* (pp. 67–84). https://doi.org/10.1108/978-1-83608-326-920241004.

Bhattacharya, S. (2019). Environmental crisis in the eastern himalayan landscapes in India. *Consilience, 21*, 66–85. https://www.jstor.org/stable/26775084

Bhattacharya, U., Singh, A., & Jolly, R. (2024). Thirty years of climate and land-cover change and its effects on tourism resources in valley of flower national park. *IOP Conference Series Earth and Environmental Science, 1285*(1), Article 012021. https://doi.org/10.1088/1755-1315/1285/1/012021.

Bolch, T., Kulkarni, A., & Kääb, A. (2012). The state and fate of Himalayan glaciers. *Science, 336*(6079), 310–314. https://doi.org/10.1126/science.1215828.

Brandt, J. S., Radeloff, V., Allendorf, T., Butsic, V., & Roopsind, A. (2019). Effects of ecotourism on forest loss in the Himalayan biodiversity hotspot based on counterfactual analyses. *Conservation Biology, 33*(6), 1318–1328. https://doi.org/10.1111/cobi.13341.

Carter, R. W., & Beeton, R. J. S. (2004). A model of cultural change and tourism. *Asia Pacific Journal of Tourism Research, 9*(4), 423–442. https://doi.org/10.1080/1094166042000311282

Chakraborty, P., & Ghosal, S. (2022). Status of mountain-tourism and research in the Indian Himalayan Region: A systematic review. *Asia-Pacific Journal of Regional Science, 6*(3), 863–897. https://doi.org/10.1007/s41685-022-00243-w.

Chandel, V. B. S., & Sharma, M. R. (2024). Climate change and weather extremes in tourist towns: Challenges for livelihood in Himachal Pradesh, India. In *Climate change, vulnerabilities and adaptation* (pp. 61–78). https://doi.org/10.1007/978-3-031-49642-4_4.

Chaudhary, S., Kumar, A., Pramanik, M., & Negi, M. S. (2021). Land evaluation and sustainable development of ecotourism in the Garhwal Himalayan region using geospatial technology and analytical hierarchy process. *Environment Development and Sustainability, 24*(2), 2225–2266. https://doi.org/10.1007/s10668-021-01528-4.

Chiwaridzo, O. T. (2024). Fostering sustainable practices in Zimbabwean tourism for firewood energy sustainability and efficiency amid costly renewables, weak policies, scarce electricity, and extensive deforestation. *Sustainable Development, 32*(4), 3840–3858. https://doi.org/10.1002/sd.2871.

Choegyal, L. (2022). Tourism for development in the Himalaya: Using ecotourism to improve livelihoods. In *Tourism and development in the Himalaya* (pp. 129–148). Routledge.

Dasgupta, P., & Vogelaar, A. (2022). Brand Bhutan and the political economy of sustainable tourism development. In *Tourism, hospitality & event management* (pp. 205–224). https://doi.org/10.1007/978-3-030-92208-5_13.

Dey, S., Das, S., & Roy, S. K. (2024). Landslide susceptibility assessment in Eastern Himalayas, India: A comprehensive exploration of four novel hybrid ensemble data driven techniques integrating explainable artificial intelligence approach. *Environmental Earth Sciences, 83*(22). https://doi.org/10.1007/s12665-024-11945-z.

Dhyani, S. (2023). Are Himalayan ecosystems facing hidden collapse? Assessing the drivers and impacts of change to aid conservation, restoration and conflict resolution challenges. *Biodiversity and Conservation, 32*(12), 3731–3764. https://doi.org/10.1007/s10531-023-02692-x

Dłużewska, A., & Giampiccoli, A. (2020). Enhancing island tourism's local benefits: A proposed community-based tourism-oriented general model. *Sustainable Development, 29*(1), 272–283. https://doi.org/10.1002/sd.2141.

Dredge, D. (2022). Regenerative tourism: Transforming mindsets, systems and practices. *Journal of Tourism Futures, 8*(3), 269–281. https://doi.org/10.1108/jtf-01-2022-0015.

Flick, U. (2009). *An introduction to qualitative research* (4th ed.). London: SAGE Publications Ltd.

Frolich, L. M., & Schmidt, M. (2022). Population movements, colonization trends, and amenity Migrants in mountainscapes. In *Springer eBooks* (pp. 357–374). https://doi.org/10.1007/978-3-031-13298-8_20.

Gautam, A., Bhandari, R., Mishra, B. K., & Baral, B. (2021). Sustainable utilization of natural resources for socio-environmental resilience and transformation in the mountains of Nepal. In *Environmental resilience and transformation in times of COVID-19* (pp. 357–372). Elsevier. https://doi.org/10.1016/b978-0-323-85512-9.00021-8

Geneletti, D., & Dawa, D. (2009). Environmental impact assessment of mountain tourism in developing regions: A study in Ladakh, Indian Himalaya. *Environmental Impact Assessment Review, 29*(4), 229–242. https://doi.org/10.1016/j.eiar.2009.01.003.

Gupta, S., Kumar, N., Verma, D., Bachheti, A., Arya, A. K., Joshi, K. K., & Bachheti, R. K. (2021). Impacts of invasive alien plant species on biodiversity in the regions of Western Himalayas, India: An overview. *Biological Diversity: Current Status and Conservation Policies, 1*, 123–135.

Gupta, V., & Chomplay, P. (2021). Local residents' perceptions regarding the negative impacts of overtourism: A case of Shimla. In *Emerald Publishing Limited eBooks* (pp. 69–80). https://doi.org/10.1108/978-1-83909-706-520211006.

Gurung, N. (2019). Natural disasters versus anthropogenic activities in Nepal Himalayas: Case study from Ghatte Khola. *Myagdi, 1*(1), 65–70. https://doi.org/10.3126/TJ.V1I1.27593.

Han, W., McCabe, S., Wang, Y., & Chong, A. Y. L. (2017). Evaluating user-generated content in social media: An effective approach to encourage greater pro-environmental behavior in tourism? *Journal of Sustainable Tourism, 26*(4), 600–614. https://doi.org/10.1080/09669582.2017.1372442.

Howard, C. A. (2016). Touring the consumption of the other: Imaginaries of authenticity in the Himalayas and beyond. *Journal of Consumer Culture, 16*(2), 354–373. https://doi.org/10.1177/1469540516635400.

Hysa, B., Karasek, A., & Zdonek, I. (2021). Social media usage by different generations as a tool for Sustainable tourism marketing in Society 5.0 idea. *Sustainability, 13*(3), 1018. https://doi.org/10.3390/su13031018.

Ivars-Baidal, J. A., Celdrán-Bernabeu, M. A., Mazón, J., & Perles-Ivars, Á. F. (2017). Smart destinations and the evolution of ICTs: A new scenario for destination management? *Current Issues in Tourism, 22*(13), 1581–1600. https://doi.org/10.1080/13683500.2017.1388771.

Jha-Thakur, U., Khosravi, F., Quattrone, G., Bandyopadhyay, S., Magedera, I., & Garikipati, S. (2020). Exploring the role of strategic environmental assessment in cultural heritage tourism planning: A case study of the Srirangapatna-Mysore region in India. *Impact Assessment and Project Appraisal, 39*(2), 138–150. https://doi.org/10.1080/14615517.2020.1841595

Kandel, K., Huettmann, F., Suwal, M. K., Regmi, G. R., Nijman, V., Nekaris, K., Lama, S. T., Thapa, A., Sharma, H. P., & Subedi, T. R. (2014). Rapid multi-nation distribution assessment of a charismatic conservation species using open access ensemble model GIS predictions: Red panda (Ailurus fulgens) in the Hindu-Kush Himalaya region. *Biological Conservation, 181*, 150–161. https://doi.org/10.1016/j.biocon.2014.10.007.

Kumar, S., Kumar, P., Singh, A., Ashwani, N., & Kumar, M. (2022). Socio economic livelihood vulnerability to mountain hazards: A case of Uttarakhand Himalaya, India. In *Advances in geographical and environmental sciences* (pp. 169–190). https://doi.org/10.1007/978-981-16-7731-1_8

Kuniyal, J. C. (2002). Mountain expeditions: Minimising the impact. *Environmental Impact Assessment Review, 22*(6), 561–581. https://doi.org/10.1016/s0195-9255(02)00031-8.

Kuniyal, J. C., Jain, A. P., & Shannigrahi, A. S. (2003). Solid waste management in Indian Himalayan tourists' treks: A case study in and around the Valley of Flowers and Hemkund Sahib. *Waste Management, 23*(9), 807–816. https://doi.org/10.1016/s0956-053x(03)00027-8.

Lozano, J., Rey-Maquieira, J., & Sastre, F. (2020). An integrated analysis of tourism seasonality in prices and quantities, with an application to the Spanish hotel industry. *Journal of Travel Research, 60*(7), 1581–1597. https://doi.org/10.1177/0047287520947807

Mahar, N., Dobriyal, P., Badola, R., & Hussain, S. A. (2024). Tourism on the roof of the world: Socio-ecological impacts of tourism on the Indian Trans-Himalaya. *Land Use Policy, 138*, 107031. https://doi.org/10.1016/j.landusepol.2023.107031

Manchado, A. M., Ballesteros-Cánovas, J. A., Allen, S., & Stoffel, M. (2022). Deforestation controls landslide susceptibility in Far-Western Nepal. *CATENA, 219*, Article 106627. https://doi.org/10.1016/j.catena.2022.106627

Milano, C., Novelli, M., & Cheer, J. M. (2019). Overtourism and tourismphobia: A journey through four decades of tourism development, planning and local concerns. *Tourism Planning & Development, 16*(4), 353–357. https://doi.org/10.1080/21568316.2019.1599604

Mishra, S., Mishra, J. M., & Bhatt, V. (2024). Catalysing destination Sustainability: Role of Self-Help Groups. In *Emerald Publishing Limited eBooks* (pp. 251–263). https://doi.org/10.1108/978-1-83608-326-920241015.

Mishra, S., & Mishra, J. M. (2023). Self-help groups and community-based tourism: A study of interrelationship and convolutions. *Journal of Tourism, 24*(1), 71.

Misra, L. I., Samal, A., Chaini, S. R., Bisoi, S., & Gowda, D., V. (2024). Tourism development and environmental conservation in the Himalayas. In *Advances in hospitality, tourism and the services industry (AHTSI) book series* (pp. 194–219). https://doi.org/10.4018/979-8-3693-0823-3.ch017.

Morrison, A. M., Cheah, J., & Kumar, R. (2024). A destination performance measurement framework: Exploring the relationships among performance criteria and revisit intentions. *Current Issues in Tourism,* 1–19. https://doi.org/10.1080/13683500.2024.2309149

Nepal, S. K. (2000). Tourism in protected areas. *Annals of Tourism Research, 27*(3), 661–681. https://doi.org/10.1016/s0160-7383(99)00105-x.

Nepal, S. K. (2005). Tourism and remote mountain settlements: Spatial and temporal development of tourist infrastructure in the Mt Everest region, Nepal. *Tourism Geographies, 7*(2), 205–227. https://doi.org/10.1080/14616680500072471

Neupane, S. (2023). *Mountain tourism and sustainability: the dark story of Everest trash in Nepal.* http://hdl.handle.net/10400.1/20065.

Neupane, S., Pinto, H., & Pintassilgo, P. (2024). Mountain tourism stakeholders' perspectives on waste management: The case of Everest in Nepal. *Tourism Planning & Development,* 1–21. https://doi.org/10.1080/21568316.2024.2377565

Nimble, N. (2023). Community-based rural tourism development: An intersectional exploration of possibilities and challenges. In *Routledge handbook of trends and issues in tourism sustainability, planning and development, management, and technology* (pp. 171–185). Routledge.

Nyaupane, G. P., Lew, A. A., & Tatsugawa, K. (2014). Perceptions of trekking tourism and social and environmental change in Nepal's Himalayas. *Tourism Geographies, 16*(3), 415–437. https://doi.org/10.1080/14616688.2014.942233.

Nyaupane, G. P., Poudel, S., & York, A. (2020). Governance of protected areas: An institutional analysis of conservation, community livelihood, and tourism outcomes. *Journal of Sustainable Tourism, 30*(11), 2686–2705. https://doi.org/10.1080/09669582.2020.1858089

Pathak, R., Negi, V. S., Rawal, R. S., & Bhatt, I. D. (2019). Alien plant invasion in the Indian Himalayan Region: State of knowledge and research priorities. *Biodiversity and Conservation, 28*(12), 3073–3102. https://doi.org/10.1007/s10531-019-01829-1.

Potvin, D. A., Gilby, B. L., Anderson, M. K., & Watson, N. I. M. (2023). Effects of temporal variations in ecotourist noise on an avian community: A case study from a UNESCO world heritage site. *Journal of Ecotourism, 23*(3), 328–346. https://doi.org/10.1080/14724049.2023.2202361

Prasad, S., & Choubey, M. (2023). Factors influencing micro-entrepreneurship among women SHG members of Sikkim: A propensity score matching approach. *International Journal of Social Economics, 51*(6), 741–756. https://doi.org/10.1108/ijse-01-2023-0070

Rai, S. C., & Srivastava, A. (2014). Small hydro power projects and community participation. In *Advances in geographical and environmental sciences* (pp. 239–248). https://doi.org/10.1007/978-4-431-54868-3_17

Rasul, G., & Molden, D. (2019). The global social and economic consequences of mountain cryospheric change. *Frontiers in Environmental Science, 7,* 91. https://doi.org/10.3389/fenvs.2019.00091

Roy, S., & Mitra, S. (2024). Exploring the ramifications of unorganized tourism practices and associated challenges in the Sikkim Himalayas through machine learning approach. In *The Himalayas in the anthropocene* (pp. 133–150). https://doi.org/10.1007/978-3-031-50101-2_5

Saqib, N. U., Yaqub, A., Amin, G., Khan, I., Faridullah, N., Ajab, H., Zeb, I., & Ahmad, D. (2019). The impact of tourism on local communities and their environment in Gilgit Baltistan, Pakistan: A local community perspective. *Environmental & Socio-economic Studies, 7*(3), 24–37. https://doi.org/10.2478/environ-2019-0015

Sharief, A., Joshi, B. D., Kumar, V., Singh, H., Singh, V. K., Dar, S. A., Graham, C., Ramesh, C., Quyoom, I., Thakur, M., & Sharma, L. K. (2023). Empirical data suggest that the Kashmir Musk Deer (Moschus cupreus, Grubb 1982) is the one musk deer distributed in the Western Himalayas: An integration of ecology, genetics and geospatial modelling approaches. *Biology, 12*(6), 786. https://doi.org/10.3390/biology12060786.

Sharma, C. (1997). Urban air quality of kathmandu valley "Kingdom of Nepal". *Atmospheric Environment, 31*(17), 2877–2883. https://doi.org/10.1016/s1352-2310(96)00346-9

Sharma, B. K. (2018). English and discourses of commodification among tourism workers in the Himalayas. *Journal of Sociolinguistics, 22*(1), 77–99. https://doi.org/10.1111/josl.12274

Sharma, R., & Kumar, A. (2023). Determinants of quantification of tourism waste in the hilly terrain of Himalayas: A spatial approach. *Waste Management, 166*, 46–57. https://doi.org/10.1016/j.wasman.2023.04.039.

Singh, S. (2006). Tourism in the sacred Indian Himalayas: An incipient theology of tourism? *Asia Pacific Journal of Tourism Research, 11*(4), 375–389. https://doi.org/10.1080/10941660600931226.

Singh, P. B., Mainali, K., Jiang, Z., Thapa, A., Subedi, N., Awan, M. N., Ilyas, O., Luitel, H., Zhou, Z., & Hu, H. (2020). Projected distribution and climate refugia of endangered Kashmir musk deer *Moschus cupreus* in greater Himalaya, South Asia. *Scientific Reports, 10*(1), 1511. https://doi.org/10.1038/s41598-020-58111-6

Singh, R., Sharma, R. K., Bhutia, T. U., Bhutia, K., & Babu, S. (2021). Conservation policies, ecotourism, and end of pastoralism in Indian Himalaya? *Frontiers in Sustainable Food Systems, 5*. https://doi.org/10.3389/fsufs.2021.613998.

Stevens, S. (2003). Tourism and deforestation in the Mt Everest region of Nepal. *Geographical Journal, 169*(3), 255–277. https://doi.org/10.1111/1475-4959.00089

Stuart-Smith, R. F., Roe, G. H., Li, S., et al. (2021). Increased outburst flood hazard from Lake Palcacocha due to human-induced glacier retreat. *Nature Geoscience, 14*, 85–90. https://doi.org/10.1038/s41561-021-00686-4.

Sundriyal, S., Shridhar, V., Madhwal, S., Pandey, K., & Sharma, V. (2018). Impacts of tourism development on the physical environment of Mussoorie, a hill station in the lower Himalayan range of India. *Journal of Mountain Science, 15*(10), 2276–2291. https://doi.org/10.1007/s11629-017-4786-0

Thakur, A. S., Kumar, N., Chamoli, V., & Goyal, N. (2024). Social media marketing in the Himalayas: An examination of its influence on tourist decision-making in Himachal Pradesh. *International Journal of Business Competition and Growth, 9*(1), 43–60. https://doi.org/10.1504/ijbcg.2024.142220

Tiwari, P. (2008). Land use changes in Himalaya and their impacts on environment, society and economy: A study of the Lake Region in Kumaon Himalaya, India. *Advances in Atmospheric Sciences, 25*(6), 1029–1042. https://doi.org/10.1007/s00376-008-1029-x.

Tiwari, C., & Rani, M. (2024). Trouble in paradise: Evaluating the effects of unorganized tourism on the Himalayan ecology. In: A. Borthakur, & P. Singh (Eds.), The Himalayas in the anthropocene. Cham: Springer. https://doi.org/10.1007/978-3-031-50101-2_7.

Upadhyay, T., Sankhayan, P. L., & Solberg, B. (2004). A review of carbon sequestration dynamics in the Himalayan region as a function of land-use change and forest/soil degradation with special reference to Nepal. *Agriculture Ecosystems & Environment, 105*(3), 449–465. https://doi.org/10.1016/j.agee.2004.09.007.

Wambulwa, M. C., Milne, R., Wu, Z., Spicer, R. A., Provan, J., Luo, Y., Zhu, G., Wang, W., Wang, H., Gao, L., Li, D., & Liu, J. (2021). Spatiotemporal maintenance of flora in the Himalaya biodiversity hotspot: Current knowledge and future perspectives. *Ecology and Evolution, 11*(16), 10794–10812. https://doi.org/10.1002/ece3.7906.

Wani, M. D., Dada, Z. A., Bhat, W. A., & Shah, S. A. (2025). Community-driven ecotourism in the trans Himalayan region: A sustainable model for cultural and environmental preservation. *Community Development*, 1–23. https://doi.org/10.1080/15575330.2025.2451358.

Wilson, J., & Dashper, K. (2023). In the shadow of the mountain: The crisis of precarious livelihoods in high altitude mountaineering tourism. *Journal of Sustainable Tourism, 31*(10), 2270–2290. https://doi.org/10.1080/09669582.2022.2108038

Xu, J., Grumbine, R. E., Shrestha, A., Eriksson, M., Yang, X., Wang, Y., & Wilkes, A. (2009). The melting Himalayas: Cascading effects of climate change on water, biodiversity, and livelihoods. *Conservation Biology, 23*(3), 520–530. https://doi.org/10.1111/j.1523-1739.2009.01237.x.

Yin, R. K. (1994). *Case study research: Design and method*. Thousand Oaks, CA: SAGE Publications.

Ziegler, A. D., Wasson, R. J., Sundriyal, Y., Srivastava, P., Sasges, G., Ramchunder, S. J., Ong, C. E., Nepal, S. K., McAdoo, B. G., Gillen, J., Bishwokarma, D., Bhardwaj, A., & Apollo, M. (2021). A call for reducing tourism risk to environmental hazards in the Himalaya. *Environmental Hazards, 22*(1), 1–28. https://doi.org/10.1080/17477891.2021.1984196

Natural Resource and Environment Challenges

Securing Water Quality of the Asian Water Tower: A Myth or a Real Science?

Kshitiz Kandel, Rutger Willem Vervoort, Subodh Sharma, Chhatra Mani Sharma, Aastha Pandey, Rukumesh Paudyal, Sonam Rinzin, and Bakhat Rawat

Abstract The Hindu Kush-Himalaya (HKH)—the Tibetan Plateau and surrounding mountain region system, known as the "Third Pole" or the "Asian Water Tower," is vital for global environmental health, sustaining nearly 2 billion people through its water resources. Despite extensive research on water quantity, water quality remains underexplored, with rising threats from climate change, urbanization, and agriculture. Recent studies reveal increasing contamination, particularly downstream, marked by pollutants and trace metals. This paper proposes the Himalayan Mountain Water Health (HMWH) framework, calling for an integrated, science-driven approach to monitor and protect water quality in this critical region.

1 Introduction

The Himalayas, the Hindu Kush-Himalaya (HKH) region, often referred to as the "Third Pole" or "Asian Water Tower (AWT)" due to its extensive ice reserves, is critical for regional and global environmental health (Eriksson et al., 2009; Jianchu et al., 2009; Talukdar et al., 2021; Chettri et al., 2023). This mountain range supports vast ecosystems and human populations, as the rivers that originate from these Himalayas

K. Kandel (✉) · A. Pandey · R. Paudyal · B. Rawat
University of Chinese Academy of Sciences, Beijing 100049, China
e-mail: kshitizkandel310@gmail.com

K. Kandel · B. Rawat
Research Insight-Nepal, Bharatpur, Chitwan, Nepal

K. Kandel · S. Sharma · A. Pandey · B. Rawat
Kathmandu University, Dhulikhel, Nepal

R. W. Vervoort
The University of Sydney, Sydney, NSW 2006, Australia

C. M. Sharma
Central Department of Environmental Science, Tribhuvan University, Kirtipur, Kathmandu, Nepal

S. Rinzin
Newcastle University, Newcastle Upon Tyne, UK

© The Author(s), under exclusive license to Springer Nature Switzerland AG 2025

S. Ghosh (ed.), *The Himalaya Dilemma*, Springer Natural Hazards,
https://doi.org/10.1007/978-3-031-95083-4_24

supply water to almost 2 billion people (Qazi et al., 2020; Sharma et al., 2020; Liu et al., 2023; Suo et al., 2023) and are equally significant in hydrological and ecological functions (Bandyopadhyay and Gyawali, 1994; Haleem et al., 2022; Momblanch et al., 2020). The Himalayas are crucial in ensuring water, food, energy, and environmental security in much of Asia, significantly influencing the socio-economic aspects of a fifth of the world's population (Yao et al., 2012; Scott et al., 2019). However, the water quality of the Himalayas has been less studied than the water quantity and the disturbances it accounts for (Qu et al., 2019; Zhang et al., 2022). Nevertheless, there is reported evidence of natural and anthropogenic interferences in the mountain ecosystems and the water quality of the region under climatic change, hydroclimatic, and anthropogenic extremes (Gurung et al., 2018; Sharma et al., 2019; Prajapati et al., 2023; Van Vliet et al., 2023). Under the reported recent conditions, downstream are more vulnerable to anthropogenic interferences, such as urbanization, agriculture, traffic, and tourism (Zhang et al., 2022). This is supported by the latest findings on anthropogenic footprints, characterized by a dense population, agriculture, and highly enriched trace meals in the lower reaches of the Himalayan waters by (Kandel et al., 2024). In light of these water quality uncertainties, modern water science, combined with the latest technology, programs, and models, plays a crucial role. By leveraging advanced tools and methodologies, researchers and policymakers can better monitor, predict, and manage water resources to ensure safety and sustainability. On this crucial note, we propose an open invitation to the researchers, collaborators, and enthusiasts to secure the water quality of the region by introducing a holistic approach known as **Himalayan Mountain Water Health** termed "**HMWH**" as clearly illustrated in Fig. 1.

2 HMWH: An Interdisciplinary Approach for Securing the Himalayan Water Quality

Comprehensive research on the interplay between mountains, water, and humans is essential for sustainable development and environmental conservation in the fragile Himalayan region. We aim to achieve this objective by introducing a holistic approach, **Himalayan Mountain Water Health**, termed (**HMWH**). HMWH is an approach further connected to the subcategories under research packages (RPs) (Fig. 1). Each RP will be developed to scale up a comprehensive database of the entire Himalaya water, aid regional collaborations, and integrate hydrological science, and social science for human-nature harmony, as a holistic approach of HMWH.

RP1: Area Coverage: East Himalaya (He), West Himalaya (Hw), Central Himalaya (Hc). **BIG**

The Third Pole serves as the headwater region for major Asian rivers (Zhang et al., 2019), where Brahmaputra, Indus, and Ganges (**BIG**) form significant downstream reaches of the Third Pole waters and are particularly susceptible to reductions of

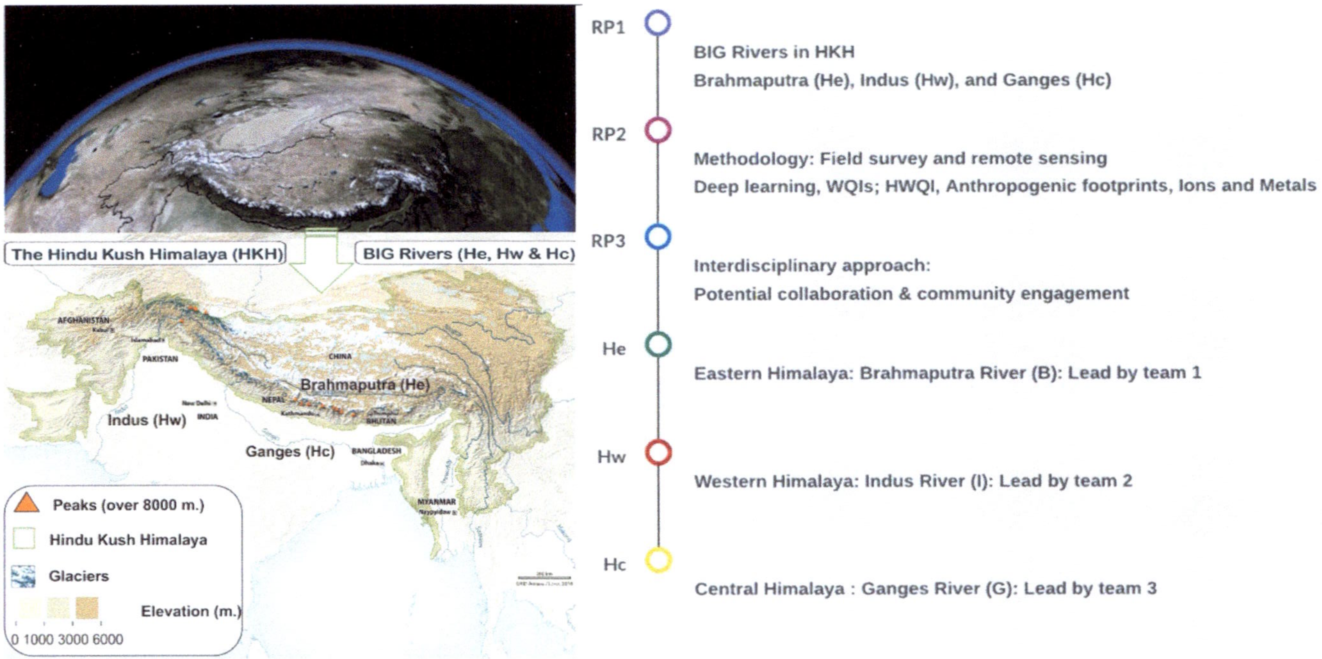

Fig. 1 HMWH in the BIG Rivers of HKH: An interdisciplinary approach for securing the Himalayan water quality (adopted and edited from GRID-Arendal resources library)

water flow associated with climate change impacts. This, subsequently threatens the food security, water availability, and livelihoods of an estimated 60 million people (Immerzeel et al., 2010). Based on these evidences, these three major rivers (**BIG**) are of significant attention in HMWH approach. Other small rivers/tributaries at the local scale that merge with these **BIG** rivers are of consideration as well which is separated under the following tasks.

Task 1: **B**: East Himalaya (**He**); Brahmaputra and its tributaries (China, India and Bangladesh).

Task 2: **I**: West Himalaya (**Hw**); Indus and its tributaries/ rivers (China, India and Pakistan).

Task 3: **G**: Central Himalaya (**Hc**): Ganges and its tributaries (most of Nepal rivers mix to the Ganges).

RP2: **Methodological Areas of Concern:** Deep learning for water quality, WQIs, Anthropogenic footprints, Ions, Metals, and Health. **Field Surveys and Remote Sensing**: Conduct extensive field surveys and utilize remote sensing technology to gather data on topography, vegetation, and water resources. **Community Engagement**: Collaborate with local communities to gather traditional ecological knowledge and understand their health and livelihood challenges. **Interdisciplinary Approach**: Integrate findings from environmental science, public health, and social sciences to develop a holistic understanding of the region.

RP3: **Potential Collaboration on Comprehensive Database:** Establish a detailed database on the Himalayan environment, including human activities, water resources, and health statistics. **Policy Recommendations**: Provide actionable insights for policymakers to promote sustainable development, conservation strategies, and public health initiatives. **Community Empowerment**: Develop educational programs to empower local communities with knowledge and tools for sustainable living and health improvement. **Public Health Issues**: Addressing the health challenges faced by mountain communities and enhancing their resilience to environmental changes.

3 Preliminary Findings in the Central Himalaya (Hc): Ganges and Its Tributaries, Incorporating Most of Nepal Rivers that Mix to the Ganges

Nepal and its surrounding regions encompass much of the Central Himalaya (HC) of the (HKH), including one of the greatest mountain systems in the world. The major rivers of Nepal, including the Koshi, Gandaki, and Karnali systems, act as significant tributaries to the Ganges River in India, contributing approximately 40% of its annual flow and playing a vital role in the hydrology of the Indo-Gangetic Plain (Immerzeel et al., 2010; Shrestha and Devkota, 2010). These rivers, originating in the Himalayas, provide crucial water resources for irrigation, hydropower, and ecosystem services in both Nepal and downstream regions in India (Eriksson et al., 2009; Shrestha & Devkota, 2010).

Over the years, diverse environmental matrices have been analyzed in understanding trace metal pollution and its broader implications for the ecosystem. While individual studies provide valuable insights, to truly safeguard the fragile health of the Third Pole environment, these environmental matrices must be assessed as an interconnected whole, offering a holistic and unified approach to sustainable management (Rawat et al., 2021). Among these matrices, glaciochemical records from the Central Himalayas stand out as reliable indicators of regional and global anthropogenic influences (Bolch et al., 2012). Trace metal dynamics in the Himalayas are shaped by the intricate interplay of natural processes and human activities, as evident in their spatial distribution and seasonal variations. For instance, elevated zinc concentrations are attributed to both natural phenomena, such as forest fires and dust storms, and anthropogenic sources, including industrial emissions and traffic pollution (Tripathee et al., 2014). Similarly, the seasonal behavior of metals such as lead highlights the dual pressures of local natural inputs during monsoons and the long-range transport of pollutants in non-monsoon periods, emphasizing the transboundary nature of atmospheric pollution (Kang et al., 2019; Rawat et al., 2021).

The observed spatial trends further support this understanding, with trace metal concentrations diminishing at higher elevations, suggesting a direct link between pollution sources and their proximity to glacier environments. The enrichment factor (EF), a ratio used to assess the degree of contamination by comparing the concentration of a metal to its background level, particularly for metals like zinc (Zn), copper (Cu), and cadmium (Cd), emphasizes the disproportionate susceptibility of certain metals to anthropogenic pressures (Rawat et al., 2021). This raises concerns about their potential long-term impacts on glacial health and downstream ecosystems. Additionally, trace metal concentrations are more concentrated during the non-monsoon season, while their sources and behavior shift during the monsoon due to varying weather conditions, highlighting the seasonal variations that complicate pollution management in the region.

A key pollutant that warrants focused attention in the Himalayan rivers is mercury (Hg). As a persistent and toxic pollutant, Hg is increasingly becoming a significant environmental concern in the region (Kocman et al., 2017). Its presence and movement in water systems are influenced by both natural processes and human activities, with growing evidence of Hg export from the Koshi River Basin (KRB), which collects water from several glacial-fed and rain-fed rivers and covers a sizable area of the Himalaya (Nepal et al., 2016; Sun et al., 2020). This trend is especially prominent during the post-monsoon season, when river runoff typically increases, facilitating greater Hg export. Such findings raise serious concerns about downstream water quality, as the mercury may be transported to lowland areas, exacerbating contamination risks. The seasonal fluctuation of Hg levels in the KRB, particularly with higher concentrations in the post-monsoon season, further reflects this pattern, driven by the seasonal changes in river runoff (Sun et al., 2020). During the monsoon, Hg export peaks, mirroring the increased river flow, while it significantly decreases during winter and pre-monsoon periods.

The increase in Hg export from a trans-Himalayan River such as the KRB has broader implications, contributing to transboundary pollution that can impact ecosystems downstream. As climate change is expected to alter seasonal runoff patterns, it is anticipated that Hg export from the KRB will rise, particularly in the post-monsoon season (Sun et al., 2020). This underscores the critical need for region-specific monitoring and pollution management strategies to mitigate the growing threat of Hg contamination in the Himalayan River systems.

In addition to Hg, the analysis of heavy metals in the sediments of the KRB reveals that most metals have a predominantly natural origin, with minimal human-induced pollution (Li et al., 2021). Spatial comparisons show no significant differences in heavy metal concentrations between upstream and downstream areas, although slightly higher levels of cadmium (Cd) and Pb were observed upstream. The average concentrations of heavy metals align closely with natural background levels and fall below sediment quality guidelines, indicating that, overall, the Koshi River maintains a relatively pristine environment with limited anthropogenic disturbances. These findings underscore the resilience of the river's sediment geochemistry, even amid varying environmental conditions and human activities across its upper and lower reaches. However, this consistency with natural backgrounds does not negate the potential for future risks. The geochemical composition of river sediments is a critical indicator of environmental change, and as climate change and human pressures intensify, long-term monitoring of trans-Himalayan River sediments will be essential to detect emerging trends and manage the region's environmental health effectively.

In light of these findings, studies underscore the urgent need for systematic monitoring of riverine trace metals exports to fully understand their environmental impacts. With the projected increase in Hg exports, especially during the post-monsoon period, it becomes even more crucial to adopt an integrated, science-driven approach to monitoring Hg levels in trans-Himalayan rivers. Such monitoring will play a pivotal role in the sustainable management of water resources in the Himalayas and in addressing the rising contamination of river systems, ensuring effective mitigation measures are in place for future environmental protection.

Moreover, there is growing attention to characterization of the "HKH" waters using different indices. One significant introduction is the Himalaya Water Quality Index (HWQI) in the latest finding by Kandel et al. (2024). In simpler terms, it is a new modified index, particularly subjected to the water quality of the Himalayan waters, termed "HWQI." It can be of fundamental significance in the region, with its categorization under the score obtained from the implication of its equation (H1).

$$\mathrm{H1} = \frac{\sum WnQn}{\sum Wn}$$

where, (Wn) represents the unit weight factor for each parameter, (Qn) is the sub-index value by using the formula:

$$Qn = \frac{[(Vn - Vo)]}{[(Sn - So)]} * 100$$

Here, Vn is the mean concentration of the nth parameter, and Vo is the Ideal value based on commonly accepted International Standard Values (Vo = Ideal values set by national or local regulatory authorities, particular International Standard, for instance, pH = 7.5 and DO = 7.00), Sn = standard desirable value of the nth parameter. HWQI is graded under the following category:

- H1 = (0–25): Excellent (Category A)
- H1 = (26–50): Good (Category B)
- H1 = (51–75): Poor (Category C)
- H1 = (76–100): Very Poor (Category D)
- H1 > 100): Unsuitable (Category E).

4 The Indrawati River Basin: A Representative Himalayan Glacial-Fed River

The Indrawati River Basin, a significant tributary of the Koshi River and a key component of Nepal's larger river system provides an illustrative example of how Himalayan glacial-fed rivers interact with both natural and anthropogenic influences (Bartlett et al., 2011). Located in the southern-central Himalayas, the Indrawati River Basin (IRB) is a critical water source for local communities and plays a vital role in the Melamchi Drinking Water Scheme, which serves approximately 2.5 million people in the Kathmandu Valley (Vibhāga, 2017). This dual importance emphasizes the basin's vulnerability to both natural and human-induced pollution, making it an important case study for understanding water quality dynamics in trans-Himalayan rivers.

In-depth chemical analyses of the IRB have revealed that rock weathering is the primary source of major ions in the river, with calcium and bicarbonate—derived from carbonate minerals such as calcite and dolomite—dominating the hydrochemical composition (Pandey et al., 2017). The hydrochemical facies, classified as Ca-HCO_3 type, confirm that natural geochemical processes largely influence the river's chemistry. While trace element concentrations in the IRB are below the WHO guidelines for drinking water, subtle seasonal variations between the monsoon and post-monsoon periods indicate dilution effects during the rainy season.

Interestingly, trace element analyses highlight a predominantly natural origin for the metals, stemming from lithogenic sources and weathering processes. However, there are localized increases in elements like zinc, cadmium, and tin downstream, likely linked to anthropogenic activities such as farming, brick production, and sand quarrying. Despite the generally low levels of trace metals, statistical analyses confirm that human activities play a minor yet notable role in shaping the distribution of these pollutants in the river basin. Ecological and health risk assessments, including the Hazard Quotient (HQ) and Hazard Index (HI), suggest that water poses minimal risk for both drinking and agricultural purposes, though localized erosion and particle release in downstream regions necessitate ongoing monitoring and management efforts.

The Indrawati River Basin exemplifies the growing importance of integrated monitoring frameworks that account for both natural and human-induced pressures on Himalayan River ecosystems. This case study underscores the need for a holistic approach to managing water quality and ensuring the sustainable use of resources in these sensitive, high-altitude regions.

In addition to the analysis of water quality and sediments, studies on the biota of the Indrawati River, specifically its fish species, provide crucial insights into mercury (Hg) dynamics within the river's ecosystem (Pandey et al., 2017). Among research conducted, one of the studies investigated the fish specimens, including a dominant species in the river: *Schizothorax richardsonii* (snow trout) and the *Barilius* species to understand the Hg concentrations and its biomagnification along the food web using total mercury (THg) levels and stable isotopes of nitrogen ($\delta^{15}N$) and carbon ($\delta^{13}C$) (Pandey et al., 2017).

The results indicated that Hg concentrations varied between species, with younger fish showing higher mercury concentrations. This pattern suggests that smaller fish accumulate Hg more readily, possibly due to differences in their dietary habits, growth rates, or habitat use. In addition, a positive correlation between THg and $\delta^{15}N$ values in *Schizothorax richardsonii* demonstrated biomagnification, with higher levels of Hg observed in fish higher up the food chain.

These findings emphasize the importance of species-specific differences in Hg accumulation, suggesting varying ecological roles and exposure risks for different fish species. The research highlights the need for further studies on mercury cycling in trans-Himalayan rivers, particularly with regard to potential risks for both aquatic wildlife and local communities that depend on fish for sustenance. Continuous monitoring of Hg levels in the biota is essential to assess ecological health and manage emerging threats to both wildlife and human populations in these fragile ecosystems.

5 Way Forward/Direction: Introduced System Dynamics

Addressing the challenges of heavy metal pollution in surface waters, particularly in ecologically sensitive and complex regions like the Himalayas, requires innovative and comprehensive strategies. Among these, system dynamics emerges as a promising tool for safeguarding Himalayan water health. By capturing the intricate interplay between natural processes, human activities, and policy interventions, system dynamics can help model and predict the outcomes of various management strategies, offering a pathway to informed decision-making.

Advanced monitoring techniques such as remote sensing and biomonitoring also hold immense potential to transform how pollution is tracked and understood. These tools enable the visualization of contamination patterns across time and space, offering a clearer picture of pollution dynamics and helping prioritize targeted mitigation efforts. Complementing these monitoring approaches, the development of remediation technologies tailored to the unique environmental and socio-economic contexts of the Himalayan region is equally critical. Sustainable, in situ solutions designed to function in remote and rugged terrains can provide viable pathways to restore polluted water bodies. To ensure effectiveness, these technologies must strike a balance between simplicity for widespread application and robustness to address significant contamination levels.

Integrated Water Resources Management (IWRM) offers a broader framework to address the interconnected challenges of water quality, quantity, and accessibility in the region. By considering the diverse needs of upstream and downstream communities, IWRM ensures that both natural and human factors influencing water quality are managed holistically. Engaging local communities through targeted education campaigns and skill development further enhances grassroots support for conservation efforts, empowering them to adopt sustainable practices that reduce pollution at the source.

At a governance level, legislation and policy serve as foundational pillars for sustainable water resource management. Developing and enforcing stringent water quality standards can drive better practices across sectors, while international collaborations foster shared understanding and harmonized approaches to monitoring, remediation, and regulation for transboundary rivers. Such partnerships are particularly vital for addressing the regional and global dimensions of pollution in the Himalayan River systems.

Finally, the urgency for research focused on scalable, efficient, and environmentally friendly remediation methods tailored to high-pollution scenarios cannot be overstated. The Himalayan rivers, lifelines for millions, demand innovative solutions to ensure their long-term health and viability. By integrating advanced monitoring, community engagement, policy support, and cutting-edge research, stakeholders can collectively safeguard these critical water resources for generations to come.

Early research progress in the central Himalaya of "HKH", (Hc) has provided valuable insights into hydrogeochemistry, surface water and atmospheric metals concentration, climatic impacts on river hydrology, and some heavy metals concentration on biotic life, which are essential for understanding regional water security. However, the variability in geography, climate, and socio-economic dependence across the HKH necessitates a comprehensive, harmonized research approach in all region, including western Himalaya (Hw), central Himalaya (Hc), and eastern Himalaya (He). Extending the methodologies employed in the (Hc) to other HKH rivers ensures comparability of data, enabling holistic policy-making and regional cooperation in mitigating climate change effects and water-related risks. Therefore, the necessity of conducting research on all major rivers in the Hindu Kush Himalaya (HKH) region using methods consistent with early studies in the Central Himalaya (Hc) is paramount due to the region's critical role in sustaining global water resources.

6 Conclusion and Perspectives

This study aims to bridge the gap between human needs and environmental conservation in the Himalayas. By understanding the intricate relationships between mountains, water, and health, we can develop strategies that ensure the long-term sustainability and well-being of this critical region.

We propose a research network to collect standardized basin-scale, cross-border data on water quality, pollutant release, and human health in the region. Crucially, we need to focus on identifying parameters and processes that influence water quality and health, for instance, the anthropogenic footprints resulting from human interferences such as agricultural practices, metal pollution, and pesticide uses. This will elucidate upstream–downstream connections in distinct climatic and lithological regions. To ensure future water resources of the Third Pole, advanced coupled atmosphere–cryosphere–hydrosphere–lithosphere (Earth System) models with collaboration between the upstream and downstream countries is crucial as proposed by Zhang and the team (Zhang et al., 2019). Likewise, integrating AI into water management ensures safer, cleaner water supplies, protecting public health and the environment. As a researcher, we must have a nexus between these advanced latest technologies and the general people in simplifying the issues and connecting them practically with the results to safeguard the water and health.

References

Bandyopadhyay, J., Gyawali, D. J. M. R., & Development (1994). *Himalayan water resources: ecological and political aspects of management* (pp. 1–24).

Bartlett, R., et al. (2011). *Freshwater ecosystem vulnerability assessment: The Indrawati sub-basin, Nepal.*

Bolch, T., et al. (2012). *The state and fate of Himalayan glaciers* (Vol. 336, issue 6079, pp. 310–314).

Chettri, N., et al. (2023). *Changing discourses in the third pole: A systematic review of climate change impact on biodiversity in the Hindu Kush Himalaya* (Vol. 155, p. 111046).

Eriksson, M., et al. (2009). *The changing Himalayas: Impact of climate change on water resources and livelihoods in the greater Himalayas.* International Centre for Integrated Mountain Development (ICIMOD).

Gurung, S., et al. (2018). *Hydrochemistry of Lake Rara: A high mountain lake in western Nepal* (Vol. 23, issue 2, pp. 87–97).

Haleem, K., et al. (2022). *Hydrological impacts of climate and land-use change on flow regime variations in upper Indus basin* (Vol. 13, issue 2, pp. 758–770).

Immerzeel, W. W., Van Beek, L. P., & Bierkens, M. F. J. S. (2010). *Climate change will affect the Asian water towers* (Vol. 328, issue 5984, pp. 1382–1385).

Jianchu, X., et al. (2009). *Climate change and its impacts on glaciers and water resource management in the Himalayan Region* (Vol. 44, p. 54).

Kandel, K., et al. (2024). *Synthesis analysis of hydrogeochemistry of nepal himalayan rivers: Perspective from major ions and trace elements* (Vol. 163, p. 112080).

Kang, S., et al. (2019). *Linking atmospheric pollution to cryospheric change in the Third Pole region: current progress and future prospects* (Vol. 6, issue 4, pp. 796–809).

Kocman, D., et al. (2017). *Toward an assessment of the global inventory of present-day mercury releases to freshwater environments* (Vol. 14, issue 2, p. 138).
Li, M., et al. (2020). Heavy metals in surface sediments in the trans-Himalayan Koshi River catchment: Distribution, source identification and pollution assessment. *Chemosphere, 244*, Article 125410.
Liu, J., et al. (2023). *Identification of dust aerosols, their sources, and the effect of soil moisture in Central Asia* (Vol. 868, p. 161575).
Momblanch, A., et al. (2020). *Enhancing production and flow of freshwater ecosystem services in a managed Himalayan river system under uncertain future climate* (Vol. 162, pp. 343–361).
Nepal, S. J. J. O. H. -E. R. (2016). *Impacts of climate change on the hydrological regime of the Koshi river basin in the Himalayan region* (Vol. 10, pp. 76–89).
Pandey, A., et al. (2017). *Mercury concentrations in the fish community from Indrawati River, Nepal* (Vol. 99, pp. 500–505).
Prajapati, R. N., Ibrahim, N., & Thapa, B. R. J. H. (2023). *Climate change impact on water availability in the Himalaya: Insights from Sunkoshi River basin, Nepal* (Vol. 6, pp. 279–292).
Qazi, N. Q., et al. (2020). *Hydrology of the Himalayas* (pp. 419–450).
Qu, B., et al. (2019). *Water quality in the Tibetan Plateau: Major ions and trace elements in rivers of the "Water Tower of Asia"* (Vol. 649, pp. 571–581).
Rawat, B., et al. (2021). *Glacial record of trace metal pollution over the Central Himalayas and its surroundings: Distribution, variation, and anthropogenic signals* (Vol. 251, p. 105428).
Scott, C. A., et al. (2019). *Water in The Hindu kush himalaya* (pp. 257–299).
Sharma, C. M., Kang, S., & Tripathee, L. (2019). *Water quality in the third pole: The roles of climate change and human activities*. Elsevier.
Sharma, C. M., et al. (2020). *Major ions and irrigation water quality assessment of the Nepalese Himalayan rivers*. Environment, Development and Sustainability.
Shrestha, A. B., & Devkota, L. P. (2010). *Climate change in the Eastern Himalayas: observed trends and model projections.*
Sun, X., et al. (2020). *Mercury variation and export in trans-Himalayan rivers: Insights from field observations in the Koshi River* (Vol. 738, p. 139836).
Suo, X., Yu, Z., & Cao, S. J. C. (2023). *Utilizing water resources from the Tibetan Plateau: Benefits, risks, and future research requirements* (Vol. 230, p. 107256).
Talukder, B., et al. (2021). *Melting of Himalayan glaciers and planetary health* (Vol. 50, pp. 98–108).
Tripathee, L., et al. (2014). *Concentrations of trace elements in wet deposition over the central Himalayas, Nepal* (Vol. 95, pp. 231–238).
Vibhāga, N. K. T. (2012). *National population and housing census 2011: National report* (Vol. 1). Government of Nepal, National Planning Commission Secretariat, Central.
Van Vliet, M. T., et al. (2023). *Global river water quality under climate change and hydroclimatic extremes* (Vol. 4, issue 10, pp. 687–702).
Yao, T., et al. (2012). *Third pole environment (TPE)* (Vol. 3, pp. 52–64).
Zhang, F., et al. (2019). *Water availability on the Third Pole: A revie.*(Vol. 7, p. 100033).
Zhang, F., et al. (2022). *Securing water quality of the Asian Water Tower* (Vol. 3, issue 10, pp. 611–612).

Preserving the Himalaya: Addressing the Devastating Impacts of Deforestation and Pathways to Sustainable Conservation

Sreedeep Dey

Abstract The aim of this chapter is to discuss the adverse consequences of deforestation on the Himalayan ecosystem and propose viable solutions promoting long-term conservation. It is worth noting that the Himalaya is an important natural resource for biodiversity, water supply, and climate regulation. On the other hand, environmental concerns are being raised due to widespread deforestation, climate change, and unplanned development programs. This chapter presents the theoretical background and motivation for investigating the emergence of ecological health maintenance in this region for minimizing the occurrence of natural disasters, protecting biodiversity, and supporting the livelihoods of millions of people. The methodology combines literature reviews of successful conservation programs and assessments of environmental policies. In particular, reforestation and sustainable forest management are the community-driven approaches that have been especially emphasized. Rearing practices like pastoralism and livestock management, belief in sacred sites, local knowledge of agrobiodiversity, nutrient cycling (e.g., composting and crop rotation), uses of ethnobotanical knowledge, and community involvement are the traditional ecological knowledges (TEKs) that help to restore the biodiversity, enhance soil fertility, and amplify fodder production. Building on this, this chapter discusses the potential application of biogas plants as a sustainable fuel source empowering local women (reduction of time spent gathering firewood: 15 h/month) and reducing loads on forests (greenhouse gas emissions reduction: 1938.6 kg/year CO_2e; firewood usage reduction by 5%). Not only does this chapter provide efficient access to the long-standing challenges for implementing comprehensive rules and carrying capacity monitoring, but it also introduces collaboration among government agencies, local communities, and non-governmental organizations (NGOs) for developing policies promoting sustainable development. In the quest for environmental sustainability, future studies should focus on climate resilience and devising measures that can reconcile human growth with ecological preservation.

S. Dey (✉)
Department of Chemical Engineering, University of Calcutta, Kolkata 700009, India
e-mail: sreedeepdey04@gmail.com

© The Author(s), under exclusive license to Springer Nature Switzerland AG 2025

S. Ghosh (ed.), *The Himalaya Dilemma*, Springer Natural Hazards,
https://doi.org/10.1007/978-3-031-95083-4_25

Keywords Himalayan conservation · Deforestation · Sustainable management · Biodiversity · Reforestation · Climate resilience

1 Introduction

The chapter originated from the emergence of addressing the environmental, societal, and economical issues in the Himalayan region as the results of detrimental effects of deforestation, climate change, and unsustainable development. This chapter is focused on maintenance of fragile ecosystems and reduction of climatic impacts, including melting of glaciers and other natural calamities. In addition, this chapter discusses that sustainable growth must be achieved according to the region's carrying capacity.

Mountains are important ecosystems that have an impact far beyond their geographical borders. More than 50 million people live in the Himalaya, and more than a billion rely on the ecosystem services they provide (Rasul, 2014). Anthropogenic activities have resulted in simultaneous deforestation and regeneration in the Himalaya, and actions to reduce forest loss and enhance forest recovery necessitate a comprehensive understanding of the complex and interconnected drivers of forest change. Verma et al. (2021) conducted a systematic review of 137 case studies from 1984 to 2020, combining a system dynamics approach with a causal network analysis to identify, map, and articulate the relationships between the drivers, actors, and mechanisms of forest change across the entire Himalayan Mountain range. In total, the analysis revealed five proximate drivers, 12 underlying drivers, two institutional factors and five 'other' factors connected by a total of 221 linkages. Forest change dynamics have been dominated by widespread smallholder agriculture, extensive non-timber forest product extraction, widespread commercial and non-commercial timber extraction, and high rates of agricultural abandonment. Underlying drivers include population growth, poor agricultural productivity, international support for development projects, and successful community forest management systems. Contradictory linkages emerge from a combination of contextual factors, which can have negative impacts on conservation goals (Kabeyi & Olanrewaju, 2022). Global processes such as shifts in governance, transnational infrastructure-development programs, economic slowdowns, labour migrations and climate change threaten to destabilize established dynamics and change forest trajectories. The underlying and proximate drivers interact via different paths, which can be used to achieve conservation objectives. Based on this data, researchers have identified five theme target areas to reduce forest loss and boost recovery: (1) reducing population pressure, (2) increasing agricultural production in a sustainable manner, (3) improving forest management institutions, (4) capitalizing on tourism growth and sustainable infrastructure expansion, and (5) transitioning to fuel and sowing firewood plants on degraded land. The wider adoption of systems thinking, particularly a system dynamics approach and causal network analysis, will greatly enhance the rigour of policy development, help design site-specific interventions at multiple spatial scales that can respond to local and global changes, and guide deeper inquiry to enhance our understanding of driver-forest dynamics (Jameel et al., 2024).

However, the previous reviews were primarily focused on mapping and analyzing the key drivers and their interactions. Their emphasis was on a basically system dynamics approach. The chapter is extremely relevant as it deals with issues that concern the ecological health of the Himalaya. This chapter implies that this region is critical to biodiversity, water security, climate regulation, and cultural legacy (Raman et al., 2024). This chapter emphasizes why deforestation has to be reduced; natural resources should be effectively managed for preventing climate change effects, including melting of glaciers and flash floods. It advocated creative reforestation, biogas uptake, and community-driven conservation methods which provide sustainable development in vulnerable environments. The importance of this review has been extended to global climate action as it has demonstrated that local actions in the Himalaya is helping with carbon sequestration, environmental resilience, and sustainable livelihoods.

2 Ecosystem of the Himalaya

2.1 The Eastern Himalaya

The Eastern Himalaya is the home to the world's tallest mountains (Kangtö 7,090 m; Nyegyi Kansang 7,047 m; Hkakabo Razi 5,881 m), the mightiest rivers (the Brahmaputra, along with its tributaries like the Teesta, Manas, and Subansiri), the world's largest delta (Ganges-Brahmaputra Delta), and the planet's largest mangrove forest (Sundarbans) (Bateman et al., 2023). Its lands and rivers sustain a billion people. It is home to one-eighth of the world's biodiversity, and its trees store carbon to help control the climate. In brief, it is the most significant region the world has ever heard of—until now (O'Neill, 2019).

2.2 The Uniqueness of Himalayan Ecosystem

The Himalayan ranges are the world's youngest (formed approximately 50 million years ago) and highest mountain systems, and they are extremely complex and diverse in terms of biological and physical characteristics. The dense forest in the Eastern Himalaya with trees like chir pine serves multiple ecological and economic roles, as they are drought-resistant, fast-growing, and utilized for timber production, resin extraction, and as a source of fuel (Fig. 1). The Himalayan environment is critical to the ecological security of the Indian landmass and occupies the vital position of the country's entire northern border (Basnet et al., 2019).

Fig. 1 Picture portraying dense pine woods of Himalayan region. These woods help prevent soil erosion, regulate water flow, and play a role in carbon sequestration, contributing to climate change mitigation. Additionally, they provide valuable resources like timber and resin, supporting local livelihoods.

Only 42% of the Himalayan region's total geographic area is covered by forests, with 43% under a dense canopy of trees and 50 percent under extremely good forest cover (O'Neill, 2019). The Himalayan woods, one of 36 global biodiversity hotspots, contain an incredible diversity of life organisms over longitudinal and altitudinal gradients. The Himalaya is home to up to 241 of the nation's 372 known animal species (Wani et al., 2023). The foothills are home to three key terrestrial species, the tiger, elephant, and rhinoceros, while the high-altitude habitat supports rare animal species such as the snow leopard, red panda, hangul, chiru, musk deer, and Himalayan tahr (Basnet et al., 2019).

2.3 The Hindukush Himalaya

The Hindukush Mountain Range is the world's youngest mountain range, with abundant natural resources. This mountain range, known as the *Third Pole*, contains more ice and snow than any other location outside of the polar region (Antarctica and the Arctic). These mountains' glaciers feed some of the world's most powerful rivers, upon which millions of people rely. As the rocky landscape progresses into the hills, it gives way to vast rangelands, various people villages, fertile agricultural areas, and lush forests (Premlata et al., 2024). Mountain forests play a vital role in maintaining a variety of ecosystem functions. Not only can they maintain steep slopes, but they also support a diverse range of flora and fauna, manage groundwater recharge, which is critical for springs, and provide an important source of energy. Forests naturally trap and store carbon, helping to maintain the global carbon balance and mitigate climate change (Mukherji et al., 2019).

2.4 Ecological Health Restoration of the Himalaya

Strategic infrastructure development, diversification of tourism products, and focused marketing efforts have attracted both domestic and international tourists to the Himalaya for last few decades. However, due to the emerging surge in disasters like landslides, flash floods, and cloudburst events in the Himachal Pradesh region, it becomes high time to rethink about the carrying capacity of the Himalayan ecological system (Premlata et al., 2024), identify the region's primary challenges (e.g., increased tourism, construction, and resource exploitation) and implement effective solutions (e.g., reforestation, wetland restoration, and land-use regulation). Additionally, this region protects India and includes Ladakh, which are now Union Territories. Recent studies claim that the primary concern is especially on West Himalaya, due of the ongoing environmental degradation which may have far-reaching consequences for downstream populations (Haq et al., 2022).

Thereafter, the question that emerges first is: what is a region's *carrying capacity?* The carrying capacity of an ecological system is the number of individuals or animals that an environment can support over a given time period without causing harm to natural resources or overall environmental health. There is a pattern which represents the population that can be sustained in an ecological system (Xu & Yang, 2024). States such as Himachal Pradesh are implementing corrective measures with direct involvement of the State Pollution Control Board (SPCB) along with strengthening their State Action Plan on Climate Change (SAPCC). They are making strategies to analyze the region's capacity, which poses issues in several Himalayan locations. Before moving further with any development in core areas of towns like Kasauli and Shimla, the region's carrying capacity should be analyzed and satellite towns are urged to develop to reduce pressure on existing urban centers (Premlata et al., 2024; Haq et al., 2022; Wani et al., 2023). Recently, Johnson et al. (2025) have introduced a flood hazard event-disaster database 'HiFlo-DAT' for the Kullu District, Himachal Pradesh, Indian Himalaya which contributes to the disaster risk reduction (DRR) agenda.

2.5 The Significance of the Himalaya's Ecological Health

(i) Biodiversity hotspot: Comprehending the significance of the Himalaya's ecological health is imperative as the Himalaya is essential to a large number of people in terms of climatic influence, water source, biodiversity hotspot and economic significance. Among many, one of the most important point of concern is biodiversity loss. The Himalaya includes diverse flora and fauna, with the majority of these species currently threatened. The snow leopard (inhabits alpine and subalpine zones at elevations of 3,000–4,500 m) is a species located exclusively in the Himalaya, and it, along with others (red panda, musk deer, Himalayan brown bear, Himalayan tahr, Asiatic elephant, wild water buffalo, Himalayan yew, *Aconitum heterophyllum*, *Podophyllum hexandrum*, *Nardostachys jatamansi*, Brahma Kamal), must be protected (O'Neill, 2019). The biodiversity hotspot of Himalaya has been demonstrated in Fig. 2. (ii) Water Resources: The Himalaya produce water for rivers, which serve 1.5 billion people for drinking, agriculture, energy, or other purposes. (iii) Climatic Regulation: The Himalaya is also crucial for climate control, helping to attenuate the most severe effects of climate change. The region also protects the Indians from severe weather events. It operates as a natural barrier, protecting India (Wani et al., 2023). (iv) Cultural and Spiritual Value: In addition, the Himalaya (often referred to as the "abode of the gods") has cultural and spiritual value. People in India adore the Himalaya spiritually, and they are a popular tourist destination in many locations (Arora et al., 2023).

2.6 Climate Change and Its Impact on Cryosphere and Water Resources of Himalayan Region

Himalayan mountain range, known for its soaring peaks and rugged terrain, runs from Afghanistan to Myanmar. This region includes the vertical, one of the most conspicuous physiographic features on Earth's surface. The region's climate is governed by a monsoon circulation pattern and westerly winds during the winter. In high altitudes,

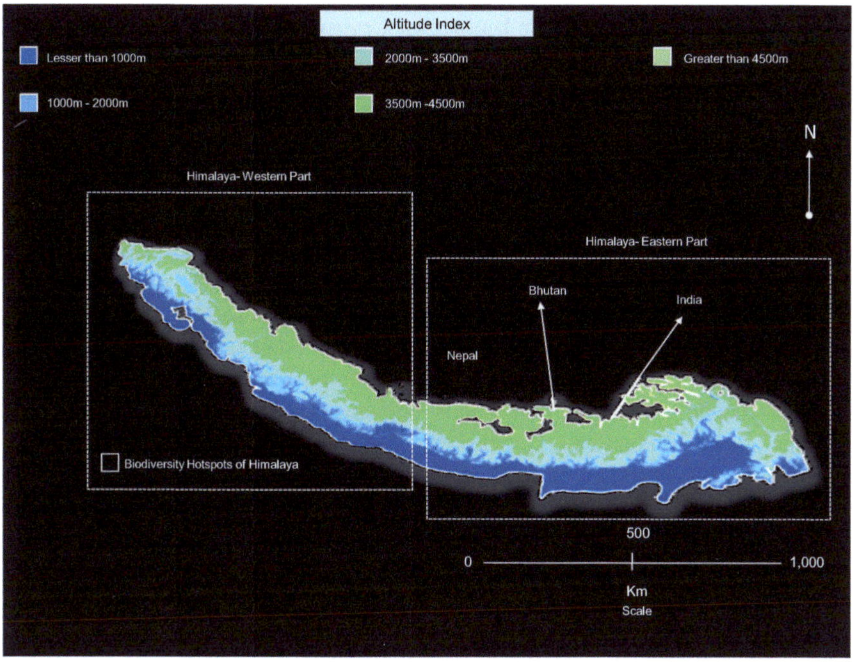

Fig. 2 Himalayan Biodivesrity Hotspot. The areas in the map were identified using the informations available on official website of Ministry of Environment, Forest and Climate Change of India (*Source* https://moef.gov.in/)

moisture is deposited as snow cover. Some of the recent incidences provide evidence for climate change. It is quite obvious that over time, a portion of the snow cover turns into glacier ice (Khan et al., 2017). In general, the region receives the most melt from the Hindukush Himalayan range's snow-covered glaciers and processes water for the region's ten largest basins including the Ganges, Brahmaputra, Yellow, and Yangtze. Needless to say, the water supplied by the region has a profound impact both locally and globally.

Some signs of climate change in the Hindukush Himalayan region have been outlined below. Temperature wise India can be divided into five physiographic zones (Dimri et al., 2020). Global averages indicate that high altitude sections of the country warm much faster than low elevation regions (Khan et al., 2017). The analysis of Du et al. (2017) conducted with 12-year field observations along two slopes with altitude differences about 1200 m (3250 m to 4400 m and 4280 m to 5538 m) of warming trends from Tibetan sites provides much clearer evidence of warming rates and shows that higher elevations warm faster than lower elevation places.

2.6.1 Climate Change and Its Impact on Cryosphere

An essential component of the Himalayan region is the cryosphere which provides material services like the supply of water for irrigation and agriculture, non-material services like the spirituality of landscapes and disservices such as disasters. Consequently, this schema is affected by climate change in terms of variations in temperature, precipitation, and other atmospheric parameters. These effects include snow, permafrost changes, and deglaciation (Dimri et al., 2020). (i) Glacial lakes: The first major effect is the development and expansion of glacial lakes, which is closely related to deglaciation. Glacial lakes may expand significantly before exploding, resulting in what are known as glacial lake outburst floods, or GLOFs (Sattar et al., 2021). The other impact results from alterations in precipitation patterns, wetlands, and the cryosphere. Drought, flash floods, or changes in hydrographs can be caused by variations in temperature and hydrological regimes. The region's human population is eventually impacted by these effects on irrigation and other elements (Khan et al., 2017). (ii) Glaciers: According to Ahmed et al. (2022) GLOFs and modifications to the hydrological system are the two mostly visible effects directly linked to glacier melt and the formation of glacial lakes. There are a few instances of deglaciation in the area and the catastrophic failure of potentially dangerous glacial lakes could be detrimental to human life and infrastructure in the adjacent low-lying areas. The Ras Samba glacier, also known as Rikha Samba, is located in Nepal's middle Himalaya (Dimri et al., 2020). The primary argument presented in the study of Bhambri et al. (2023) is that Gangotri glacier, which is the Ganges' headwater (the geographical point of its beginning) has been retreating and furthermore, from 2017 to 2022, the frontal retreat accelerated by about 1.5 times (33.8 $\pm$ 6.7 m a$-$1) compared to the period between 2006 and 2017 (Sattar et al., 2021; Bhambri et al., 2023). The Tibetan Plateau's glaciers are divided into three groups: *retreating, advancing,* and *stationary*, majority of which are retreating (Lee et al., 2021). The Hindukush Himalayan region is experiencing widespread deglaciation. Though they are small in number and confined to particular areas, some glaciers do exhibit increasing trends (Sattar et al., 2021; Bhambri et al., 2023). (iii) Glacial Lake Events: Extremely destructive floods can be caused by glacial lake outbursts. Based on Landsat 5 data of 1990, previously, as the glacier moved forward, the lakes in this area were formed (51,300 m^2 in 2021). However, it has since retreated to the area where glacier-deposited moraines border the lake water. Moreover, the lake's surrounds are extremely precarious and a strong correlation between lake area and glacier area was observed with r^2 of 0.93 (Lee et al., 2021; Sharma et al., 2023). The lake may release water at a very rapid rate and do significant damage if the moraine breaches for any uncertain reason. According to previous studies, there have been 35 GLOF episodes in the area, with 10 of them starting in Tibet and 25 in Nepal. In most of the cases, Nepal was significantly impacted by these events (Khan et al., 2017). More than 200 glacier lakes in the Hindukush Himalayan region have been identified by studies as potentially hazardous. The eruption of Tibet's Jang Jango glacier lake serves as a noteworthy illustration of the scope of the damage (Sattar et al., 2021). Jang Jango Lake erupted from 1964 until 1981 and both Tibet and

Nepal were affected by the GLOF in 1981. An additional illustration is the Dig Tsho GLOF, which nearly destroyed the Namche hydropower station in Nepal's Everest area (Dimri et al., 2020). Dig Tsho Lake has changed significantly since the GLOF (0.3 km^2) in 1985 by an ice avalanche of 0.15 mcm (Poudel et al., 2025). Byers et al. (2020) mentioned that a minor GLOF became prominent in Nepal when a large lake quickly dried up around June 23, 1980, in the Tamor valley. Villages may also sustain damage as a result of a chain of tiny supraglacial lakes. Dig Tsho Lake is one instance, where the GLOF event resulted in a massive overflow of the moraine dam (Dimri et al., 2020). From supraglacial ponds in the 1950s to a sizable glacial lake, the area grew to 1.76 km^2. More devastation than Dio Jang Jang is anticipated if the lake bursts. Because there are so many villages downstream, the lake's size and water storage capacity increase the risk (Khan et al., 2017). (iv) Mitigation Measures: A mitigation initiative was put into place in an effort to lower the probability of GLOFs. The project involved draining the water by creating an open route through the moraine dam. Due to the lack of road access to the project site, there were difficulties at the initial stage (Dimri et al., 2020). The site required the transportation and assembly of materials. There are still far too many potentially hazardous glacial lakes in the area as a result of the project's implementation at this magnitude. Nepal's surrounds are home to a number of potentially hazardous glacial lakes, as well as ongoing or planned hydroelectric projects. Nearly every project that has been suggested is at risk because of glacial lakes (Khan et al., 2017). A regional method, sometimes known as a *phased approach*, has been implemented in recognition of the urgency. This method integrates a variety of risk-based research, hydrological monitoring, desk analysis, and remote sensing (Sattar et al., 2021).

2.6.2 Climate Change and Its Impact on Water Resources and Hydrology

Glacier mass decreases as warming continues, with certain changes initially increasing and then reducing as the glacier shrinks. Although there are many unknowns, this phenomenon is portrayed in a very conceptual way. There is ambiguity about the behavior of the climate, the changes in the glacier mass, and, of course, the way the hydrological regime changes. Getting baseline data to address this problem in the region is quite difficult. The future of the region's water resources may be diminished as a result of these uncertainties (Khan et al., 2017).

Future conflicts and a rise in the demand for water could result from the shrinking of glaciers due to climate change. Problems may arise between far-reaching social, economic, and geopolitical sectors and between upstream and downstream ecosystems. Mostly areas that are inherently water-stressed face these effects. The issue may worsen in the eastern regions as opposed to the western ones (Dimri et al., 2020). Water from intricate and historically maintained irrigation systems provides around 80% of the flow in Pakistan's upper Indus. Areas may essentially be arid and parched without glacier-fed canalse (Khan et al., 2017). Numerous canals in this area get water straight from the glaciers, and even a small amount of glacial retreat might make it more difficult to get water. In past years, glacier retreat caused water shortages in a number of settlements in Pakistan's Upper Hunza region. As a result, people were forced to abandon the settlement. The Himalayan region is already experiencing climate change, albeit perhaps at a naturally accelerated rate (Dimri et al., 2020). Deglaciation, changes in the cryosphere, and a decrease in snow cover are the effects being felt in the area. These modifications may lead to serious environmental issues and ecosystem disruption (Dimri et al., 2020).

While changes in the hydrological regime occur slowly, they may have significant long-term effects on people and livelihoods. In contrast, as glacial lake outburst floods are quite common, numerous residents of the Hindukush Himalayan region, as well as their homes and means of subsistence, could be at risk (Khan et al., 2017).

3 Medicinal Plants of Himalayan Hills and Surrounding Areas of Himalaya

The variety of medicinal plants found in the Himalaya and their environs is introduced and examined in this section. In essence, these medicinal herbs have been used to treat a wide range of diseases, including COVID-19, cancer, cardiac issues, hypertension, and blood pressure. People experience allergy symptoms, such as sneezing, fever, coughing, and throat infections, when the seasons change. While some plants in the Himalayan foothills are used to treat allergies, other plants might trigger allergic reactions (Joshi et al., 2016). The flowers of *Adhatoda vasica* have a creamy, white hue. This flower's herbal tea is an extremely effective COVID-19 treatment. Additionally, bronchial disorders, asthma, and allergies can be treated with this flower herbal tea. The Himalayan foothills are home to a number of indigenous communities. These regions are home to cultural communities whose members use the variety of plants for fuelwood, food, and medicine. Numerous communities have lived in this area for years, and they have a wealth of information about the diversity of plants that has been passed down through the generations (Heinrich et al., 2021). In their daily lives, the locals harvest wood for fuelwood. *Rosa indica* is the botanical name for the rose plant. The rosehip refers to the base of this flower. The high vitamin C content of the herbal tea made from rosehips makes it a beneficial immunoboost. This helps with blood pressure. Thus, this plant is frequently utilized to treat cardiac issues. Wild mint is another significant plant. This herb is crucial for lowering allergic symptoms. Viral diseases are also treated with it. Additionally, this herb is used to strengthen the immune system. In the Himalayan region, *Carissa opaca* is referred to as granda (Joshi et al., 2016). Vitamin C is abundant in this plant. The fruits are edible and a good source of vitamin C, which helps to strengthen the immune system. This plant is used to treat a variety of illnesses, including heart disease, hepatitis, joint disease, and viral diseases (Naz et al., 2022). It must be kept in mind that nature desires love and care just as much as human do. Nature will provide human with health if they treat it with respect. Human get nourishment and energy from nature. Therefore, they must protect Mother Nature.

According to Ayurvedic founder *Charaka*, the Himalayan hills are a treasure trove of medicinal medicines. The subtropical to cold desert parts of the Western Himalaya, which include temperate and alpine pasturelands, lush green pine forests, and picturesque river valley landscapes, are home to a wide variety of colorful flowering plant resources, most of which are aromatic and medicinal. Many of these plants are unusual, uncommon, and possess healing qualities that are hard to get elsewhere (Joshi et al., 2016). A good example of this area is Himachal Pradesh, which is located in the Western Himalaya. In Vedic times as well, it was a source of pride. In the past, Indian rishis, munis, monks, and peers sought to alleviate human suffering by pursuing their academic and meditational endeavors in the sacred land, away from the bustling metropolis, in a serene and tranquil setting, and beneath the expansive paradise of nature. Generation after generation, from practitioners to apprentices, century after century, this enormous and priceless legacy has been passed down from the academic and rich experiences of the great Indian peers (Naz et al., 2022).

For almost 140 years, the Indians have had a family of Ayurvedic and Unani doctors. From generation to generation, this practice persists. There is a reference to Lord Hanuman transporting Sanjeevani from the Kailash Mountains to Lanka in Valmiki's Ramayana. Indra, Kashyapa's son, drank soma in the Chandra and Bhaga catchment areas in the Himachal Pradesh district of Lahaul-Spiti (Heinrich et al., 2021). The father of medicinal chemistry, Nagarjuna, studied in Yumthang in the Kangra district before moving to Trilokinath in Lahaul in the seventh century. Due to the absence of written documentation, the technique was either verbal or relied on the practitioners' exceptionally strong memory, and it was typically transmitted orally. With the practitioners' deaths, this important knowledge was permanently lost due to taboos or inhibitions (Joshi et al., 2016). In the medical field, traditional medicinal herbs are still highly regarded for their remarkable therapeutic qualities, which can be used to treat anything from common colds to the most complex and long-term illnesses. In the global market, herbal medications and other goods are in great demand (Naz et al., 2022).

Numerous sayings and poems concerning the use of plants including Berna, Bisuti, Tabu, and Ten Hua are still widely used in the state as a result of tradition. As long as one is aware of the ratio, proportion, mode, and technique of administration, "*where Berna, Bisuti, and Beret grow, one cannot die of diseases.*" Approximately 500 medicinal, 150 aromatic, and a sizable number of strong alternative and substitute medicine plant sources are currently available for use out of the more than 3,500 identified plant species. There may be more than 1,000 kinds of beneficial medicinal and aromatic plants if the plants used in home medicines are also counted (Joshi et al., 2016). In the lower regions of the Shivalik mountains, practically all of these plants grow wild in valleys, woodlands, and pastures, or they appear as weeds close to fields and villagers' homes. Mentha, *Centella asiatica*, *Withania somnifera*, *Mucuna pruriens*, *Costus speciosus*, Asclepias, *Plumbago zeylanica*, *Aloe barbadensis*, *Silybum marianum*, *Asparagus racemosus*, *Yucca gloriosa*, and *Typhonium trilobatum* are some of the species that are used as medications (Heinrich et al., 2021). There are 10 drug plants in the mullah group. It helps with digestive diseases, strengthens and energizes the body, helps avoid bronchial asthma, and helps women recover from childbirth. There are several commercially viable herbs in the temperate zone. They include: *Morchella esculenta, Taxus wallichiana, Prunella vulgaris, Angelica glauca, Scoparia dulcis, Atropa belladonna,* Heracleum candicans, Boerhaavia diffusa, Rheum, Salvia, Solanum, Aconitum, *Rhododendron arboreum,* and Berberis (Joshi et al., 2016). Eight herbs make up the Ashtavarga group's life-rejuvenating medications. These are a key component of common Ayurvedic formulas such as Chyavanprash. It is used as a tonic, helps with respiratory and bronchial issues, and is beneficial for urogenital and cardiac issues. The tree limit is approximately 3,700 m above mean sea level, which is the perennial snow line. This area is home to conifers such as oak, birch, chestnut, rhododendron, and silver fir (Naz et al., 2022).

The border separating the alpine and temperate zones is formed by peppers and dwarfs' alex, which reaches its highest point in the alpine grasslands. Around 3800 m are typically covered in snow, which is amazing given the diversity of exquisitely colored flowers that emanate both pleasant and unpleasant aromas and lead a brief but splendid existence despite the demands of the climate. They create a wealth of fragrant and therapeutic plants (Joshi et al., 2016). *Aconitum heterophyllum,* rhododendron, *Nardostachys grandiflora, Taxus wallichiana, Valeriana jatamansi,* and *polygonum* species are a few of the valuable medicinal herbs. *Ephedra gerardiana, Artemisia brevifolia, Capparis spinosa, Hyoscyamus niger,* and *Juniperus macropoda* are abundant in the chilly desert area. As a result, Himachal Pradesh is home to a wide range of fragrant and medicinal plants, from the Shivalik zone to the chilly desert regions. Numerous of these priceless herbs are native to the area. These meet the needs of the Indian medical system's Ayurvedic, homeopathic, Unani, allopathic, and even veterinary medications (Heinrich et al., 2021).

Tunics, avalehas, asavas, powders, and syrups are among the items. In addition to meeting national requirements, several of these items—such as guggal, the patradioscorea group, costus, valeriana, Artemisia, Kala zeera, Aconitum, Cichorium, berberis roots, hopa, valerian, and rhododendron roots—are also exported in their raw form. Many fragrant and medicinal plants, such as ephedra, which contains ephedrine, which is used to treat bronchial asthma, Artemisia brevifolia, which contains santonin, which is used as a vermicide, and Achillea, Capparis, and Picrorhiza, which protect the liver, are produced in greater quantities in Himachal Pradesh than any other state (Naz et al., 2022). Colic, diarrhea, and dysentery are treated using *Aconitum heterophyllum. Rheum australe* is a strong anti-inflammatory medication and a source of anthraquinone derivatives. Diosgenin, which is used in anti-fertility medications, comes from *Dioscorea deltoidea.* Furanocoumarins, which are utilized in leukoderma, are found in *Heracleum candicans* (Joshi et al., 2016). One source of taxol, which is helpful for breast and uterine cancer, is *Taxus wallichiana.* Berberine salts, which are found in berberis roots, are helpful for piles, cholera, menorrhea, and dysentery. Valeric acid, which is found in *Valeriana jatamansi,* has heart-stimulating properties. Bergenin, which is helpful in treating skin cancer, is found in Bergeniaciliata. Vitamin C-rich *hippophae rhamnoides* fruits can treat a variety of illnesses, including cancer, TB, and facial attractiveness issues (Naz et al., 2022). Cones of the *Humulus lupulus* plant are used to preserve drinks and add flavor. Tibetan medical professionals also utilize *Hyssopus officinalis* as a replacement since it helps with lung, cough, and cold conditions. While *Withania somnifera*, Digitalis, *Mucuna pruriens*, and Asparagus are aphrodisiacs, *Nardostachys grandiflora*, *Centella asiatica*, and *Acorus calamus* are brain tonics. Heart tonics include Tagar, Jatamansi, Inula, and Digitalis. Essential oils from plants including Artemisia, Ocimum, Salvia, Lavandula, and Matricaria are in high demand worldwide for use in cosmetics, perfumes, scented soaps, aerated water, and pesticides. *Pyracantha crenulata* and *Ginkgo biloba* are two possible 21st-century medications. Drugs valued at several crores of rupees are gathered and sold outside the state each year, totaling between 18,000 and 47,000 quintals (Joshi et al., 2016). The local shopkeepers and contractors purchase the material from the village people and then sell it to major metropolitan markets such as Delhi, Amritsar, Haridwar, and Mumbai after obtaining export permits from the State Forest Department and paying the required taxes. The quantity of active chemical ingredients found in plants and their diverse portions is marked and influenced by the time of collection (Heinrich et al., 2021).

Certain standards have to be observed for the collecting of good quality medications. Since it is a dormant phase and the full concentration of the plant's major contents is found in the underground section, it is best to gather underground parts such as bark, rhizomes, roots, tubers, etc., in late autumn or early spring. When the plants have finished their life cycle by bearing fruit, shedding seeds, and withering out, the goal is to gather the leaves and flowering tips. Leaves should be picked fresh or right before pollination, when flowers are fully formed and before fruits and seeds are mature (Naz et al., 2022). Seeds should be gathered when fully matured and, if at all possible, before the fruits have dehisced, and fruits should be picked when fully developed. If the entire plant is covered in dew, to minimize harm to the active ingredients, the materials are essentially dried in the shade after collecting. While some materials, such as matrica area, hyssop, lavender, and tussilago, are typically dried in the sun, others are dried in shade. Prior to marketing, the materials are graded, packed, stored, and garbled after drying. The utilization of new, authentic, and high-quality raw materials is essential to the legitimacy of any medical system. Degraded quality is frequently the result of wild collection, particularly when overexploitation occurs (Naz et al., 2022).

This issue can be resolved to help accomplish goals like reducing the strain on natural resources, creating high-quality raw materials, creating genetically modified plant material, preventing the use of adulterants and fake substitutes, standardizing post-harvest procedures, and encouraging the preservation of biological diversity and medicinal plants. In a variety of agroclimatic zones, Himachal Pradesh has enormous potential for the production of aromatic and medicinal plants. Medicinal and aromatic plants are the main crops grown in Lahaul, Spiti, Kinnaur, and Pangi. For several decades, this profitable activity has supported and sustained those involved in the trade (Joshi et al., 2016). Several decades of such operations have considerably boosted the state's economy. The state exports a huge quantity, including Ephedra, Berberis roots, and other therapeutic plants, worth Rs. 100 million each year. The demand for the material from industries dealing in pharmaceutical products, insecticides, and allied sectors is substantial, and the state can supply such demands if managed properly over time. Pharmaceuticals produced from plants used in traditional remedies have a global market worth of more than $20 billion, with India accounting for a sizable share.

However, rising demand places significant strain on natural resources due to overexploitation, development activities, floods, landslides, deforestation, agricultural expansion, and overgrazing (Heinrich et al., 2021). Thus, it is critical to identify and document the natural flora through substantial herbarium work, preserving medicinal and aromatic species that are as old as human civilization itself. The Government of Himachal Pradesh promotes and encourages the preservation of this rich legacy. It specializes in the development, cultivation, testing, and quality control of medicinal plants. The Department of Forests has upgraded its efforts to educate the public on the value and potential of therapeutic herbs. The government has decided to create a block in the state for commercializing medicinal resources in order to generate revenue. These initiatives will guide the country in maintaining its precious biological resources. Both in-situ and ex-situ conservation strategies are being utilized smoothly (Naz et al., 2022).

Recent advancements have brought Indian medicinal plants and their ecosystems to global attention. A slew of adverse effects induced by synthetic pharmaceuticals has tipped the scales in favor of herbal remedies. These offer treatments for a variety of ailments, from simple to complex (Heinrich et al., 2021). What is required is to make wise use of this herbal wealth, encourage commercial cultivation by farmers, establish a chain of cottage industries for value addition, raise awareness about the importance of this natural wealth through training and visitor programs, and make it a required component of school, college, and university curricula. Who knows, a bead of information today could be a gift to fortune tomorrow. A list of exotic plant species of Himalaya has been demonstrated in Table 1.

4 Threats to the Himalayan Ecosystem

To sum up, a complex mix of climatic and geological processes, economic marginalization, and destructive practices of resource exploitation has resulted in considerable resource degradation.

People in the Eastern Himalaya are at the forefront of climate change. As glaciers melt, water levels rise, and storms strengthen, the threat to lives and livelihoods grows. But people here understand what needs to be done. Communities are employing local knowledge to restore their landscapes, planting trees to safeguard their farms, preserving forests to preserve water supplies, and demonstrating that when human collaborate with nature, economies develop and communities benefit (Premlata et al., 2024). Global warming has had substantial effects on the Himalayan region, including fluctuation in the volumetric flow of water in rivers, biodiversity loss, unsustainable ecological changes, glacier retreat, deforestation, and degradation (Kattel,

Table 1 Exotic Plant Species of Himalaya. The data was obtained from the official website of *Botanical Survey of India* (https://bsi.gov.in/) on 27-Feb-2025

Type	Common name/local name	Scientific name	Family	Native region
Trees	Darwin black wattle	*Acacia auriculiformis* A.Cunn. ex Benth	Mimosaceae	Australia
	Sweet thorn	Acacia karroo Hayne		South Africa
	Splendid thorn	*Acacia robusta* Burch		South Africa
	Dundathu pine	Agathis robusta (Moore) F.M. *Bailey*	Araucariaceae	Australia
	Tung oil tree	*Aleurites fordii* Hemsl	Euphorbiaceae	China
	Candlenut tree	*Aleurites moluccana* Willd		Malaysia
	Bunya pine	*Araucaria bidwillii* Hk	Araucariaceae	Australia
	Neem tree	*Azadirachta indica* A.Juss	Meliaceae	Myanmar
Herbs	Okra	Abelmoschus esculentus (L.) Moench	Malvaceae	Africa
	Mexican prickly Poppy	*Argemone mexicana* L.	Papaveraceae	West Indies
	Dill	Anethum graveolens L.	Apiaceae	South & Southwest Asia
	Visadodi	*Ageratum conyzoides* L.	Asteraceae	America
	Shwanmukh	*Antirrhinum majus* L.	Scrophulariaceae	Europe, Syria, America
Climbers	Pilaghanti	*Allamanda cathartica* L.	Apocynaceae	Brazil
	Mexican creeper	Antigonon leptopus Hook. & Arn	Polygonaceae	America
Shrubs	Century plant	*Agave americana* L.	Agavaceae	America
	Red orchid tree	*Bauhinia galpinii* B.Pr	Caesalpiniaceae	South Africa
Bamboos	Hedge bamboo	*Bambusa glaucescens* (Willd.) Sieb	Poaceae	China
	Guadua bamboo	Bambusa guadua Hump.&bonds	Poaceae	USA
	Common bamboo	Bambusa vulgaris Schrad.ex Wendl	Poaceae	Bangladesh

2022). Percentage of remaining forest area over time in the Indian Himalaya has been graphically illustrated in Fig. 3. Impending natural disaster conditions, as well as the displacement of indigenous civilizations vulnerable to Himalayan ecosystem changes, are considered as some of the major issues in these regions (Isyaku, 2021). It is worth noting that natural resources from the Himalayan ecosystem are critical in the face of human activity, climate change, and natural disasters (Premlata et al.,

2024). The Himalayan environment is expected to suffer as a result of climate change, including increased temperatures, altered precipitation patterns, drought spells, and biotic factors. Climate change may also cause vector-borne diseases, such as malaria and yellow fever, which are generally transmitted by insects (Kattel, 2022).

4.1 Human-Nature Interactions

Humans and the environment have coexisted in the Hindukush Himalaya for generations, and in some distant areas of the rough mountain landscape, people continue to live without civilization and rely heavily on natural resources. Nepal can be considered as an example which pledged to increase forests cover 45 percent of the Himalayan nation by 2030 (Isyaku, 2021). However, with a population of 28 million and rising, land conversion for agriculture and urbanization is putting strain on the country's forests. This is exacerbated by a strong reliance on fuelwood for cooking, especially among poor and rural people. More than 75% of Nepal's household energy demand is satisfied by fuelwood, and this dependence will persist until other energy sources become more inexpensive and accessible to forest-dependent populations (Baral Gautam et al., 2013).

4.2 Impact of Tourism Activities on Biogeography of Himalaya

The country's northern region is characterized by a continuous sequence of mountain ranges. These mountains are referred to as the Himalayan Range. The Karakoram and Hindukush ranges border the Himalayan range to the northwest (Dandabathula et al., 2021). The Himalayan range is bounded by the Tibetan Plateau to the north and the Indo-Gangetic plain to the south. Himalaya is a Sanskrit term. '*Him*' signifies '*snow*,' while '*alaya*' means '*abode*'. Together, '*Himalaya*' signifies '*abode of snow*.' The Himalaya is divided into three parallel ranges: the Greater Himalaya, also known as the Himadri range, the Middle Himalaya, or the Himachal range, and the Lesser Himalaya, or the Shivalik range. Mount Everest, the world's highest peak, is located in the greater Himalaya (Mahar et al., 2024). Other high peaks in this range are Kanchenjunga, Nanga Parbat, and Annapurna. The Greater Himalayan range contains numerous mountain passes. A mountain pass is a route that leads through a mountain. Zoji La, Shipki La, and Nathu La are some of the most important Himalayan passes. The Greater Himalaya is covered in snow all year. There are many glaciers in this Himalayan range. Glaciers are massive bodies of flowing ice. Many Himalayan rivers are generated as glaciers melt. The rivers Indus, Ganga,

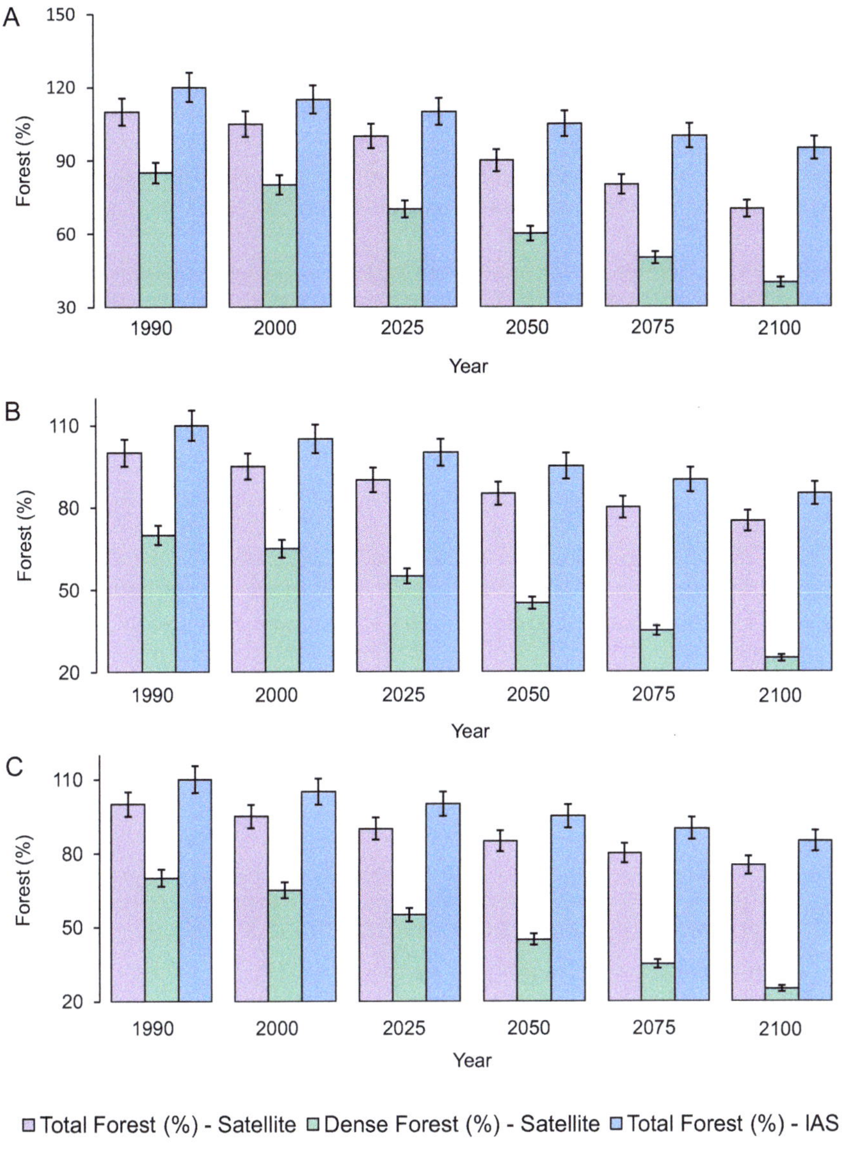

Fig. 3 Percentage of forest area remaining over time in the Indian Himalaya (**a**), Western Himalaya (**b**), and Eastern Himalaya (**c**) from 1970 to 2100. Forest area trends include total forest (combined dense and open canopy) and dense forest cover derived from satellite and IAS datasets. Projections from 2025 to 2100 are based on observed deforestation rates from earlier decades. The Indian Himalaya represents overall data, whereas the Western Himalaya includes regions like Jammu & Kashmir, Himachal Pradesh, and Uttarakhand. The Eastern Himalaya focuses on states such as Arunachal Pradesh and Sikkim

Brahmaputra, and Yamuna originate in the Himalaya. Because of these characteristics, Himalaya is a popular destination for both domestic and international travelers (Badola et al., 2018).

In the above sections, though the greatest attractions of the Himalaya has been emphasized, but Mount Everest remains what it once was: the ultimate frontier for adventurers. Tourism in the Himalayas can have both positive (e.g., economic development, conservation awareness, financial support for conservation, and promotion of local culture) and negative impacts (e.g., deforestation and fragmentation of habitats, soil erosion, compaction, loss of topsoil, water, air and noise pollution, and resource depletion) on biogeography, with the negative impacts often outweighing the positive due to inadequate infrastructure and unsustainable practices. Improperly disposed plastic and other pollutants, left by climbers on Mount Everest has become a major issue, producing pollution and spreading disease on the world's highest peak. Because it is difficult to avoid, people should assume greater responsibility. The pathogens can easily transmit to the drinking water of the individuals present, causing diseases that will travel down the mountain with them (Mahar et al., 2024).

Community cleanup projects in India are now underway. The nonprofit waste warriors are attempting to change all of that. They target a zonal area and then conduct a cleanup. There are also school-based programs that transmit messages and raise awareness about health and sanitation. Establishing buffer zones around protected areas, stronger environmental regulations and enforcement, ecotourism practices can help minimize this serious concern in sensitive ecosystems. According to the report from the Water and Sanitation Program at the World Bank, insufficient sanitation had cost India over $54 billion, or 6.4% of its GDP, in 2006 (Badola et al., 2018).

4.3 Impact of Developmental Activities and Natural Processes on Biogeography of Himalaya

While the ongoing construction of roads and houses exposes the mountains to disaster hazards, this fragile Himalayan region has also been the site of multiple dams and hydroelectric projects (Verma et al., 2021). Although government officials deny any link between hydroelectric projects and the deteriorating situation in Joshimath, several experts and citizens believe otherwise. The calamities in Joshimath highlighted the dangers of large-scale dams in these hilly areas. Awasthi et al. (2024) revealed that the line of sight (LOS) land deformation velocity for the Joshimath region, calculated for the year 2022–2023 using a PSInSAR-based approach, varies from −89.326 to +94.46 mm/year. In 2013, the chair of a Supreme Court committee concluded that, the presence of dams exacerbates the impact of flooding (Verma et al., 2021). Without the dam, the river would simply continue to flow. It would move the solid stones, occasionally breaking it, but it would continue to move. But once a dam would over there, it needs to put up a lot of energy to break through that dam barrier, and in doing so, it unleashes a lot of energy that is quite damaging (Awasthi et al., 2024). Experts argue that huge constructions frequently exacerbate disasters in Uttarakhand, the state with the most dams. However, hydropower, which is considered a clean energy source, account for around 15% of India's power generation capacity by 2023, and dams are also regarded as a reliable addition to the energy basket. Because solar and wind energy are inherently variable, the storage water project provides food, water, and energy security (Kumar & Majid, 2020). Agricultural laborers are among those most affected people by land subsidence at Joshimath, a town in Uttarakhand with a history of the phenomenon. The issue in the town worsened in early 2023 (Awasthi et al., 2024). In addition to land subsidence in various villages, the hilly regions of Uttarakhand and other Himalayan states experience frequent calamities such as floods, cloud bursts, and landslides.

5 Conservation Strategies for the Himalaya

5.1 Cross-Border Collaborations

For years, the local communities have exchanged information and wisdom across borders. Now, environmental specialists are banding together to build one of history's largest conservation efforts: the Great People's Forest of the Eastern Himalaya, a network of groups from the highlands to mangroves. Inspired by India's historic G20 leadership, initiatives have been taken to plant 1 billion trees for both nature and humans. The moto of the initiative is if people will take care of the forest, the forest will take care of its inhabitants (Rasolofoson et al., 2015). Working with governments and corporations, this movement demonstrates the actual potential of forests—the Great People's Forests of the Eastern Himalaya, for the people of South Asia and the world. Examples of some other cross-border collaborations include the Himalayan Peace Park Proposal, Kangchenjunga Landscape Conservation, regional symposiums and meetings, and indicator frameworks (Premlata et al., 2024).

5.2 Biodiversity Conservation

Several biodiversity conservation strategies have been employed till date worldwide to protect nature and benefit humanity on the globe, from the world's tallest mountains to the world's largest mangrove forest and the world's most densely populated biodiversity hotspot. Across Bangladesh, Bhutan, India, and Nepal, Conservation International began working with the Balipara Foundation of Assam and other partners in 2023 on an ambition to end deforestation in the region. Conservation International has formed a collaboration with the inhabitants of the Eastern Himalaya to introduce mountains to mangroves. These multi-pronged conservation strategies safeguard critical habitats, protect endangered species, implement monitoring and research, and restore degraded ecosystems (Rasolofoson et al., 2015).

5.3 India's Strategy to Sustain the Himalayan Ecosystem

To protect against these changes, India has established the *National Action Plan on Climate Change* (NAP), which is an internal strategy for implementing environmentally sustainable development utilizing domestic resources (Haigh et al., 1995). The NAP funds efforts that achieve development goals while simultaneously contributing to the effective and efficient management of climate change. In addition, the National Mission for Sustaining the Himalayan Ecosystem intends to establish appropriate management and policy strategies for sustaining and safeguarding the Himalayan ecosystem, as well as to engage with Indian Himalayan state governments (Rasolofoson et al., 2015). The mission aims to develop management techniques for conserving and protecting Himalayan glaciers and mountain ecosystems by strengthening Himalayan ecosystem monitoring, establishing observational and monitoring networks, promoting community-based management, and enhancing regional collaboration (Byravan & Rajan, 2013). Furthermore, raising environmental awareness among the local populace is an important step in developing long-term solutions for the vulnerable Himalayan ecosystem (Haigh et al., 1995).

5.4 Best Practices for Tree Plantation

Trees are currently being planted throughout the region following best practices like choosing native species, selecting the right sapling size, planting depth, and spacing, long-term management and regular monitoring. Environmental professionals, who are part of the "Mountains to Mangroves" effort are underway to begin the first phase of their work, which will outcome 60,000 hectares of forests and mangroves to be protected and restored in the Eastern Himalayas, including 20 million trees planted. This is the first step in a quest to plant 1 billion new trees, protect and restore 1 million hectares of forest area.

5.5 Regional Success Stories from Nepal

According to the recent reports, progress for Himalayan conservation is already being adopted in the region as Nepal has demonstrated to the world how government-owned forests can be turned over to local people to manage, as well as established a benefit-sharing structure to ensure that $REDD^+$ (Reducing Emissions from

Deforestation and Forest Degradation) works at the grassroots level. In this effort, Nepal and other countries in the Hindukush Himalaya collaborate to guarantee that community-based REDD$^+$ programs benefit local populations while also reaching national and global REDD$^+$ performance goals (Muttaqin et al., 2019). To do so, REDD$^+$ initiatives consider the interests and stakes of people who rely on forest resources, such as indigenous peoples, impoverished and marginalized groups, and women, develop robust MRV systems, including a National Timber Resource Database, access Results-Based Payments (FCPF and the LEAF Coalition). REDD$^+$ provides a unique opportunity for governments, local people, and other stakeholders to reduce emissions from mountain forests at a minimal cost. Some other examples of such key initiatives include the World Wildlife Fund (WWF) Nepal, and Scaling conservation of Himalayan plants and fungi through sustainable trade (Wahyudi et al., 2024).

5.6 Other Strategies for Conservation

(i) Engaging Local Communities: The local community should become involved in the conservation projects. Local groups working to preserve the Himalayan region should be encouraged to develop an effective plan for assessing the region's carrying capacity (Sarin, 2001). The responsible authorities would also incorporate local communities in the policy framework for policy formulation. Academic and non-academic institutions should define what should happen in a region, and involving the government in linking to their knowledge would increase cooperation (Wu et al., 2020). (ii) Focus on Sustainable Growth and: The government should focus on the carrying capacity for the sustainable populations of the various Himalayan nations. Instead of focusing on the demographic model, the authority would consider each Himalayan state's carrying capacity. The carrying capacity and development should be deferred until the real capacity is determined. (iii) Enhancing Forest Cover: As part of the National Action Plan on Climate Change, the Indian Himalayan Mission aims to sustain the Himalayan ecosystem. The region is also being worked on through projects such as the SECURE Himalaya Project (funded by the GEF and UNDP) and the Indian Himalaya Climate Adaptation Programme (collaboration between the Swiss Agency for Development and Cooperation and the Indian Department of Science & Technology), which attempt to help the Himalayan region adapt successfully to the worst effects of climate change (Muttaqin et al., 2019). These projects may help the conservation of Himalayan region which is severely affected by climate change, with glaciers melting and glacial lakes flooding, resulting in catastrophic catastrophes like flash floods (Premlata et al., 2024).

6 Current Limitations in Forest Conservation and Management

One of the most significant difficulties to improving forest conservation is the current perspective on forest resources. Cut timber and fuel wood are valued higher than standing forests and this economic perspective is what drives deforestation and forest degradation. Local forest-dependent communities play an important role in forest protection. Their lives are inextricably linked to forest ecosystems, and they benefit the most from healthy forests. Other significant limitations in forest conservation include shifting cultivation (Jhum), habitat fragmentation, shifting vegetation patterns, changes in water cycles, agrochemical pollution, poverty and dependence on forest resources, and lack of alternative livelihoods (Wu et al., 2020). Recognizing these, there has been a worldwide resurgence of interest in the forestry organizations. The United Nations Convention on Climate Change now allows developing countries to request funding for performance-based forest management. There is a global framework that provides financial incentives to improve forest management, assisting developing countries in breaking free from deforestation and forest degradation trends (MacDicken et al., 2015). The $REDD^+$ mechanism provides incentives for developing nations to reduce emissions from deforestation and forest degradation. Being a voluntary framework, $REDD^+$ allows participating nations to select whether or not to use funding to improve forest management. However, woods cannot be seen solely as a carbon sink. To be successful, $REDD^+$ must have to propose sensitively. $REDD^+$ has the ability to help with adaptation, poverty reduction, biodiversity conservation, and the preservation of essential ecosystem services (Byravan & Rajan, 2013).

7 Future Directions

(i) Tree Planting as a Catalyst for Rural Development: The Starkey reforestation project supports tree planting as an important component of rural development in some of the world's poorest regions (Muttaqin et al., 2019). The trust's tree planting not only benefits local people who rely on forest resources, but it also helps to offset carbon dioxide emissions (Muttaqin et al., 2019). (ii) Empowering Forest Survival: In some of the woodlands where the villagers have worked, fodder output has increased tenfold to twelvefold. As a result, villagers can make money by selling their excess feed to other communities. Villages that were once disdainful of the conservation project are now requesting assistance and starting their own reforestation programs. (iii) Collaborative Efforts with Local Development Groups: The Bartoy Trust is attempting to rectify the disaster for the sake of people and the environment. They collaborate with a local development group to share knowledge and support projects (Bhatt et al., 2024). Field workers supported by the funding assist farmers in managing the trees and thereby preserving their traditional way of life, which is well adapted to the highlands. (iv) Ensuring Responsible Growth in the Fragile Himalayan Region: To mitigate these effects, the region should be designed to adapt to climate change. In 2020, the central government established rules on the carrying capacity of the Indian Himalayan area. The central government has especially requested that states provide assessments on carrying capacity, notably in the 13 Himalayan states. (Baraj et al., 2024). The path forward should prioritize sustainable growth for future generations. Locals should not pursue unprecedented levels of development, which will harm the entire Himalayan ecology, for example, developed locations, such as tourism destinations, which are suffering unsustainable expansion. Sustainable population growth, biodiversity conservation, and responsible natural resource management should be prioritized. Species in the region which are endangered, and unique to the region, such as particular plants, must be protected (De Jong, 2023). It have to be kept in mind that the region is home to approximately 1,000 unique plant species, all of which should be managed responsibly (Khadka et al., 2024). The findings of Sreevidhya and Balamurugan (2025) showed that biogas reduces the amount of time spend by women in firewood gathering by 15 h per months, decreases the annual greenhouse gas emissions by 1938.6 kg CO_2e through firewood usage reduction by 5% in high-altitude community of Nala, Rudraprayag, Uttarakhand.

8 Conclusive Remarks

This chapter concludes that for maintaining the equilibrium within the ecosystems of Himalaya, multidimensional approaches considering various interconnected factors within the environment and their impact on human well-being have to be attained. For instance, the integration of sustainable forest management plans, community engagement programs, and adaptable techniques has the ability to address challenges locally as well as globally. This chapter emphasizes why it is important to incorporate local knowledge, build community resilience, and advocate eco-friendly alternatives to fight against deforestation. Hence, climate change exhibit severe influences, such as glaciers melt, causing flooding, to compete with those issues, the region's carrying capacity has to be examined properly. At the same time, as it is critical to safeguard Himalaya's ecological health, particularly given the recent environmental catastrophe, people have to consider promoting sustainable development to avoid the worst effects of climate change such as Himalayan glaciers retreat. Subsequently, if these glaciers continue to melt at their current rate, all 13 Himalayan states would be flooded, natural calamities, including cyclones, landslides, and flash floods may become more frequent, perhaps resulting in agricultural losses. Likewise, the practice of sustainable water management system meeting current water needs without compromising future water needs may make the region adaptive to these repercussions.

Declaration Conflict of Interest: The author declares no competing interests.

References

Ahmed, R., Rawat, M., Wani, G. F., Ahmad, S. T., Ahmed, P., Jain, S. K., Meraj, G., Mir, R. A., Rather, A. F., & Farooq, M. (2022). Glacial lake outburst flood hazard and risk assessment of Gangabal Lake in the Upper Jhelum Basin of Kashmir Himalaya using geospatial technology and hydrodynamic modeling. *Remote Sensing, 14*(23), 5957. https://doi.org/10.3390/rs14235957

Arora, M., Dhiman, V., & Sharma, R. (2023). Exploring the dimensions of spirituality, wellness and value creation amidst Himalayan regions promoting entrepreneurship and sustainability. *Journal of Tourismology, 0*(0), 0–0. https://doi.org/10.26650/jot.2023.9.2.1327877.

Awasthi, S., Jain, K., Sahoo, S., Kumar, R., Goswami, A., Joshi, G. C., Kulkarni, A. V., & Srivastava, D. C. (2024). Analyzing Joshimath's sinking: Causes, consequences, and future prospects with remote sensing techniques. *Scientific Reports, 14*(1), 10876. https://doi.org/10.1038/s41598-024-60276-3.

Badola, R., Hussain, S. A., Dobriyal, P., Manral, U., Barthwal, S., Rastogi, A., & Gill, A. K. (2018). Institutional arrangements for managing tourism in the Indian Himalayan protected areas. *Tourism Management, 66*, 1–12. https://doi.org/10.1016/j.tourman.2017.10.020.

Baraj, B., Mishra, M., Sudarsan, D., da Silva, R. M., & Santos, C. A. G. (2024). Climate change and resilience, adaptation, and sustainability of agriculture in India: A bibliometric review. *Heliyon, 10*(8), Article e29586. https://doi.org/10.1016/j.heliyon.2024.e29586.

Baral Gautam, Y., Pelkonen, P., & Halder, P. (2013). Perceptions of bioenergy among Nepalese foresters–survey results and policy implications. *Renewable Energy, 57*, 533–538. https://doi.org/10.1016/j.renene.2013.02.017.

Basnet, D., Kandel, P., Chettri, N., Yang, Y., Lodhi, M. S., Htun, N. Z., Uddin, K., & Sharma, E. (2019). Biodiversity research trends and gaps from the confluence of three global biodiversity hotspots in the Far-Eastern Himalaya. *International Journal of Ecology, 2019*, 1–14. https://doi.org/10.1155/2019/1323419.

Bateman, I. J., Anderson, K., Argles, A., Belcher, C., Betts, R. A., Binner, A., et al. (2023). A review of planting principles to identify the right place for the right tree for 'net zero plus' woodlands: Applying a place based natural capital framework for sustainable, efficient and equitable decisions. *People and Nature, 5*(2), 271–301. https://doi.org/10.1002/pan3.10331.

Bhambri, R., Sain, K., Chand, P., Srivastava, D., Tiwari, S. K., & Yadav, J. S. (2023). Frontal changes of Gangotri glacier, Garhwal Himalaya, between 1935 and 2022. *Journal of the Geological Society of India, 99*(2), 169–172. https://doi.org/10.1007/s12594-023-2283-3

Bhatt, H., Pant Jugran, H., & Pandey, R. (2024). Cultural ecosystem services nexus with Socio-Cultural attributes and traditional ecological knowledge for managing community forests of Indian Western Himalaya. *Ecological Indicators, 166*, Article 112379. https://doi.org/10.1016/j.ecolind.2024.112379.

Byers, A. C., Chand, M. B., Lala, J., Shrestha, M., Byers, E. A., & Watanabe, T. (2020). Reconstructing the history of glacial lake outburst floods (GLOF) in the Kanchenjunga conservation area, East Nepal: An interdisciplinary approach. *Sustainability, 12*(13), 5407. https://doi.org/10.3390/su12135407

Byravan, S., & Rajan, S. C. (2013). An evaluation of India's national action plan on climate change. *SSRN Electronic Journal*. https://doi.org/10.2139/ssrn.2195819.

Dandabathula, G., Chintala, S. R., Ghosh, S., Balakrishnan, P., & Jha, C. S. (2021). Exploring the nexus between Indian forestry and the sustainable development goals. *Regional Sustainability, 2*(4), 308–323. https://doi.org/10.1016/j.regsus.2022.01.002.

De Jong, W. (2023). Forest transition and ecosystem services in Himalaya. In *Climate change in the Himalayas* (pp. 47–57). Elsevier. https://doi.org/10.1016/B978-0-443-19415-3.00012-8.

Dimri, A. P., Bookhagen, B., Stoffel, M., & Yasunari, T. (Eds.). (2020). *Himalayan weather and climate and their impact on the environment*. Springer International Publishing. https://doi.org/10.1007/978-3-030-29684-1.

Haigh, M. J., Rawat, J. S., Rawat, M. S., Bartarya, S. K., & Rai, S. P. (1995). Interactions between forest and landslide activity along new highways in the Kumaun Himalaya. *Forest Ecology and Management, 78*(1–3), 173–189. https://doi.org/10.1016/0378-1127(95)03584-5.
Haq, S. M., Amjad, M. S., Waheed, M., Bussmann, R. W., & Proćków, J. (2022). The floristic quality assessment index as ecological health indicator for forest vegetation: A case study from Zabarwan Mountain Range, Himalayas. *Ecological Indicators, 145*, Article 109670. https://doi.org/10.1016/j.ecolind.2022.109670.
Heinrich, M., Jiang, H., Scotti, F., Booker, A., Walt, H., Weckerle, C., & Maake, C. (2021). Medicinal plants from the Himalayan region for potential novel antimicrobial and anti-inflammatory skin treatments. *Journal of Pharmacy and Pharmacology, 73*(7), 956–967. https://doi.org/10.1093/jpp/rgab039.
Isyaku, U. (2021). What motivates communities to participate in forest conservation? A study of REDD+ pilot sites in Cross River, Nigeria. *Forest Policy and Economics, 133*, Article 102598. https://doi.org/10.1016/j.forpol.2021.102598.
Jameel, M. K., Mustafa, M. A., Ahmed, H. S., Mohammed, A., Jassim Ghazy, H., Shakir, M. N., et al. (2024). Biogas: Production, properties, applications, economic and challenges: A review. *Results in Chemistry, 7*, 101549. https://doi.org/10.1016/j.rechem.2024.101549.
Johnson, R. M., Pandey, B. W., Chand, K., Davies, C. L., Edwards, D., Edwards, E., Jeffers, J., King, K., Kuniyal, J. C., Mishra, H., Phillips, V., Roy, N., Seviour, J., Sharma, D. D., Sharma, P., Singh, H., & Singh, R. B. (2025). 'HiFlo-DAT': A flood hazard event-disaster database for the Kullu District, Himachal Pradesh, Indian Himalaya. *International Journal of Disaster Risk Reduction, 120*, 105336. https://doi.org/10.1016/j.ijdrr.2025.105336
Joshi, R., Satyal, P., & Setzer, W. (2016). Himalayan aromatic medicinal plants: A review of their ethnopharmacology, volatile phytochemistry, and biological activities. *Medicines, 3*(1), 6. https://doi.org/10.3390/medicines3010006.
Kabeyi, M. J. B., & Olanrewaju, O. A. (2022). Biogas production and applications in the sustainable energy transition. *Journal of Energy, 2022*, 1–43. https://doi.org/10.1155/2022/8750221
Kattel, G. R. (2022). Climate warming in the Himalayas threatens biodiversity, ecosystem functioning and ecosystem services in the 21st century: Is there a better solution? *Biodiversity and Conservation, 31*(8–9), 2017–2044. https://doi.org/10.1007/s10531-022-02417-6.
Khadka, D., Wu, B., Bhatta, S., Paudel, H. R., Fu, K., Cui, D., & Shi, S. (2024). Endemic Prunus species in the high Himalayas, Eastern Nepal, seeking conservation. *Heliyon, 10*(20), Article e38168. https://doi.org/10.1016/j.heliyon.2024.e38168.
Khan, A. A., Pant, N. C., Sarkar, A., Tandon, S. K., & Thamban, M. (2017). The Himalayan cryosphere: A critical assessment and evaluation of glacial melt fraction in the Bhagirathi basin. *Geoscience Frontiers, 8*(1), 107–115. https://doi.org/10.1016/j.gsf.2015.12.009.
Kumar, J, C. R., & Majid, M. A. (2020). Renewable energy for sustainable development in India: current status, future prospects, challenges, employment, and investment opportunities. *Energy, Sustainability and Society, 10*(1), 2. https://doi.org/10.1186/s13705-019-0232-1.
Lee, E., Carrivick, J. L., Quincey, D. J., Cook, S. J., James, W. H. M., & Brown, L. E. (2021). Accelerated mass loss of Himalayan glaciers since the Little Ice Age. *Scientific Reports, 11*(1), 24284. https://doi.org/10.1038/s41598-021-03805-8.
MacDicken, K. G., Sola, P., Hall, J. E., Sabogal, C., Tadoum, M., & de Wasseige, C. (2015). Global progress toward sustainable forest management. *Forest Ecology and Management, 352*, 47–56. https://doi.org/10.1016/j.foreco.2015.02.005.
Mahar, N., Dobriyal, P., Badola, R., & Hussain, S. A. (2024). Tourism on the roof of the world: Socio-ecological impacts of tourism on the Indian Trans-Himalaya. *Land Use Policy, 138*, Article 107031. https://doi.org/10.1016/j.landusepol.2023.107031.
Mukherji, A., Sinisalo, A., Nüsser, M., Garrard, R., & Eriksson, M. (2019). Contributions of the cryosphere to mountain communities in the Hindu Kush Himalaya: A review. *Regional Environmental Change, 19*(5), 1311–1326. https://doi.org/10.1007/s10113-019-01484-w.
Muttaqin, M. Z., Alviya, I., Lugina, M., Hamdani, F. A. U., & Indartik. (2019). Developing community-based forest ecosystem service management to reduce emissions from deforestation

and forest degradation. *Forest Policy and Economics, 108*, 101938. https://doi.org/10.1016/j.forpol.2019.05.024.
Naz, A., Amjad, M. S., Umair, M., Altaf, M., & Ni, J. (2022). Medicinal plants used as therapeutic medicine in the Himalayan region of Azad Jammu and Kashmir. *South African Journal of Botany, 150*, 194–216. https://doi.org/10.1016/j.sajb.2022.07.007.
O'Neill, A. R. (2019). Evaluating high-altitude Ramsar wetlands in the Eastern Himalayas. *Global Ecology and Conservation, 20*, Article e00715. https://doi.org/10.1016/j.gecco.2019.e00715.
Poudel, U., Gouli, M. R., Hu, K., Khadka, N., Regmi, R. K., & Thapa, B. R. (2025). Multi breach GLOF hazard and exposure analysis of Birendra Lake in the Manaslu Region of Nepal. *Natural Hazards Research.* https://doi.org/10.1016/j.nhres.2025.03.007
Premlata Kumar, R., Hajam, Y. A., & Giri, A. (2024). Threats, challenges, and conservation strategies of Himalayan faunal biodiversity. In *Role of science and technology for sustainable future* (pp. 321–344). Springer Nature Singapore. https://doi.org/10.1007/978-981-97-0710-2_19.
Raman, R., Das, P., Mandal, S., Vijayan, V., AR, A., & Nedungadi, P. (2024). The impact of Gen Z's pro-environmental behavior on sustainable development goals through tree planting. *Sustainable Futures, 8*, 100251. https://doi.org/10.1016/j.sftr.2024.100251.
Rasolofoson, R. A., Ferraro, P. J., Jenkins, C. N., & Jones, J. P. G. (2015). Effectiveness of community forest management at reducing deforestation in Madagascar. *Biological Conservation, 184*, 271–277. https://doi.org/10.1016/j.biocon.2015.01.027.
Rasul, G. (2014). Food, water, and energy security in South Asia: A nexus perspective from the Hindu Kush Himalayan region☆. *Environmental Science & Policy, 39*, 35–48. https://doi.org/10.1016/j.envsci.2014.01.010.
Sarin, M. (2001). Empowerment and disempowerment of forest women in Uttarakhand, India. *Gender, Technology and Development, 5*(3), 341–364. https://doi.org/10.1177/097185240100500301.
Sattar, A., Haritashya, U. K., Kargel, J. S., Leonard, G. J., Shugar, D. H., & Chase, D. V. (2021). Modeling lake outburst and downstream hazard assessment of the Lower Barun Glacial Lake, Nepal Himalaya. *Journal of Hydrology, 598*, Article 126208. https://doi.org/10.1016/j.jhydrol.2021.126208.
Sharma, A., Sharma, N., Gupta, S., & Gupta, M. (2023). Emergence and evolution of a new glacial lake during last climate normal by using remote sensing and GIS in Sikkim, India. *Remote Sensing Applications: Society and Environment, 32*, 101047. https://doi.org/10.1016/j.rsase.2023.101047
Sreevidhya, C., & Balamurugan, S. (2025). An assessment of biogas adoption potential: A pathway to sustainable development in Nala, Rudraprayag, Uttarakhand, India. *Sustainable Futures, 10*, 100856. https://doi.org/10.1016/j.sftr.2025.100856
Verma, A., Schmidt-Vogt, D., De Alban, J. D. T., Lim, C. L., & Webb, E. L. (2021). Drivers and mechanisms of forest change in the Himalayas. *Global Environmental Change, 68*, Article 102244. https://doi.org/10.1016/j.gloenvcha.2021.102244.
Wahyudi, R., Marjaka, W., Silangen, C., Fajar, M., Dharmawan, I. W. S., & Mariamah. (2024). Effectiveness, efficiency, and equity in jurisdictional REDD+ benefit distribution mechanisms: Insights from Jambi province, Indonesia. *Trees, Forests and People, 18*, 100726. https://doi.org/10.1016/j.tfp.2024.100726.
Wani, S. A., Mugal, M. A., Dar, F. A., Sudhakar Reddy, C., Rashid, I., & Khuroo, A. A. (2023). Biodiversity data synthesis on trees of Indian Himalayan Region: Policy and management implications. *Global Ecology and Conservation, 48*, Article e02698. https://doi.org/10.1016/j.gecco.2023.e02698.
Wu, Y., Xie, L., Yuan, Z., Jiang, S., Liu, W., & Sheng, H. (2020). Investigating public biodiversity conservation awareness based on the propagation of wildlife-related incidents on the Sina Weibo social media platform. *Environmental Research Letters, 15*(9), Article 094082. https://doi.org/10.1088/1748-9326/ab9ed1.
Xu, C., & Yang, L. (2024). Evaluation of land resources carrying capacity based on entropy weight and cloud similarity. *Scientific Reports, 14*(1). https://doi.org/10.1038/s41598-024-59692-2

Himalayan Alpine Meadows: Functions, Threats, and Goals to Achieve Sustainability

Shreya Basak and Arindam Sarkar

Abstract The Himalayan alpine meadows are not only high-altitude grasslands but also crucial ecosystems situated at elevations ranging from 3,000 to 5,000 m. These areas experience low temperatures and receive precipitation in the form of rainfall. Alpine grasslands in different parts of the Himalayas also receive snowfall during the monsoon and cold waves from western disturbances. This unique ecosystem is not only home to a rich biodiversity but also a testament to the adaptability of life in the face of environmental extremes, such as cold waves, strong winds, limited sunlight, and a short growing season. Despite the importance of this ecosystem, there is a lack of scientific documentation about the Himalayan alpine meadows, which hinders further exploration and understanding. The objectives of the present study are to examine the mechanisms, factors of growth, ecosystem, and services of specific areas in the Himalayan region. The study aims to identify major threats and urgent suggestions for the future sustainability of the Himalayan biodiversity hotspot. It has been conducted by mapping, quantifying, and interpreting satellite data using geospatial software. Case-specific ground truth verification is also done when necessary. A study reveals climate change accelerates physical and mechanical changes in Himalayan alpine meadows. Human activities in this region significantly alter the blooming season in Himalayan alpine meadows, which typically occurs in the summer. Climate change consequences are rapidly enhancing the potential for soil erosion and land degradation. The conservation and restoration of alpine meadows are primary concerns in maintaining biodiversity and ecological balance in the world's most significant carbon sink. Reducing environmental pressure on these meadows is the primary concern of scientists in achieving ecological sustainability.

Keywords Grassland · Cold climate · Change · Detection · Modification · Management

S. Basak (✉)
Department of Geography, University of Calcutta, Kolkata, India
e-mail: shreyabasak0607@gmail.com

A. Sarkar
Department of Geography, Purash Kanpur Haridas Nandi College, University of Calcutta, Kolkata, India

© The Author(s), under exclusive license to Springer Nature Switzerland AG 2025

S. Ghosh (ed.), *The Himalaya Dilemma*, Springer Natural Hazards,
https://doi.org/10.1007/978-3-031-95083-4_26

1 Introduction

'Alpine' originated from the Latin word 'albus,' meaning white or snow-covered. This term refers to 'Alpine,' which denotes a highly mountainous region that extends beyond the treeline and is characterized by low-growing herbs and scattered shrubs (Billings, 1973; Rau, 1975). This alpine meadow is located in the northern part of the Himalayas. The distribution of alpine meadows generally depends on the geographical location, including the origin area's topography, slope, distance from the equator, climate, seasons, such as summer and winter precipitation, direction of sunlight, and other factors. Alpine meadows generally grow at a temperature of 10 °C during the summer (Rawat, 2007a, b). This type of meadow is primarily dominated by other mosses, herbaceous trees, and small shrubs, all of which are located at high altitudes and latitudes in the northern part of the Himalayas (Mark & Bliss, 1970).

The alpine meadow spans four biogeographic areas in northwestern India: The Ladakh Mountain region, the northwestern part of the Himalayas, the Western Himalayas, Changthang, and the Kibber Plateaus (Rodgers & Panwar, 1988). Alpine zones cover mainly 33% of the geographical area of the Himalayan region, including 25.88% of vegetated areas and 7.12% of perceptual snow areas (Lal et al., 1987). The alpine meadows originated on gentle, smooth slopes ranging from 3500 to 4500 m (Kala, 1997). The alpine meadows are also very popular for their bright colors, an adaptation that maximizes entomophily at low temperatures (Billings, 1973; Mani, 1978). Extremely low temperatures disrupt metabolic processes, but rarely, some indigenous plants thrive in the alpine region. The low temperature, an indirect effect, and the availability of nutrients, a direct effect, could impact the growth of alpine vegetation (Korner & Larcher, 1988).

HAM is an ecologically sensitive and biologically rich ecosystem of the globe that is crucial for hydrological regulation, biodiversity conservation, and local livelihoods. There has been limited research on evaluating ecosystems, their ecological functions, ecosystem services, and the status of the system. Few research results have been found on the long-term effect of climate change, glacial retreat, and vegetation change on meadow dynamics across HAM. Significant Gaps in policy frameworks for the conservation of HAM have also been established. Based on the identified research gap, the following research questions have been developed to address aspects and functions of the ecosystem, dangers and degradation mechanisms in the HAM, and conservation objectives aimed at achieving sustainability through the involvement of the local community. The primary purpose of this research is to assess the functions, current status, threats, and sustainable management objectives of the HAM. The research on Himalayan alpine meadows attempts to present an integrated scientific review of their ecological services, emerging threats, and sustainable management policies. It is interdisciplinary research that entails the integration of ecological, socio-economic, and policy aspects. The study will present in-depth knowledge regarding the functions, threats, and conservation targets of Himalayan alpine meadows, providing helpful information on sustainable management practices and policy interventions to ensure long-term ecological conservation.

2 The Area of the Study: Great Himalayan Alpine Meadows

Himalayan alpine meadows are high-altitude grassland ecosystems situated in the alpine zone of the Himalayan Mountain range, typically at elevations ranging from 3,000 to 5,500 m. These meadows are characterized by low temperatures, short growing seasons, and plant species that are specifically adapted to extreme climatic conditions (Körner and Larcher, 1988). They are primarily situated in the Western Himalayas, particularly in Jammu and Kashmir, Himachal Pradesh, and Uttarakhand, with notable locations such as the Valley of Flowers and Roop Kund Meadows (Rai et al., 2003). Additionally, these meadows extend into the Central Himalayas, encompassing regions like Nepal and Sikkim, as well as parts of the Eastern Himalayas, including Arunachal Pradesh, Bhutan, and Tibet. The Valley of Flowers in Uttarakhand is a UNESCO World Heritage Site renowned for its diverse array of alpine flora. Tsomgo Lake in Sikkim, sitting at 3,750 m above mean sea level (MSL), is surrounded by alpine meadows, as are the areas with shrubs and rocky outcrops in Tawang, Arunachal Pradesh. Himalayan alpine meadows are characterized by rolling hills, plateau-like surfaces, and scattered rocky outcrops. They also influence the glacial valleys of the Himalayas, which are characterized by shallow, rocky, and poorly developed soils. In summer, temperatures in this area range from 5 to 15 °C, and the growing season is notably short. In winter, temperatures can plummet to −25 °C, accompanied by heavy snowfall. Annual precipitation in these meadows can vary between 800 and 2,500 mm, depending on the region. Strong katabatic winds emanating from glaciers have a significant impact on the local microclimate in Himalayan alpine meadows.

3 Data and Methodology

This study uses a comprehensive data collection approach that incorporates both primary and secondary sources. Extensive field surveys have been conducted multiple times in the Sandakphu and Phalut regions to collect data on vegetation types and patterns, as well as soil characteristics, and to assess grazing pressure in the area. Interact with local villagers and forest officials to assess traditional management practices, perceptions of meadow degradation, and the dependency on livelihoods. GPS has been used in the field to identify and locate various aspects related to amateur radio, also known as HAM. Remote sensing data has been used for mapping and interpretation purposes. A review of scientific literature, government reports, and conservation policies on HAM ecosystems was conducted in the preliminary stage of the research. Meteorological data have been collected from the Indian Meteorological Department and WorldClim to interpret temperature, precipitation, and snowfall patterns over the area during the last few decades. Data on species distribution and conservation status were published by the Wildlife Institute of India (WII), the Forest Survey of India (FSI), and the International Union for Conservation of Nature (IUCN) and analyzed in the study.

Intensive Surveys of local pastoral communities have been done in the field regarding their dependence on alpine meadows. Documentation of traditional grazing and resource management practices has been thoroughly conducted. An estimation of vulnerability to meadow degradation has also been performed by gathering information on income dependency, resource availability, and adaptive capacity among the residents of the HAM. An analysis of existing meadow conservation initiatives has been conducted in the field. The effectiveness of conservation strategies has also been evaluated using secondary data and expert interviews. The methodology of the study focused on the identification of ecological functions and services, such as grazing and biodiversity support; mapping the spatial distribution of threats, including grazing, tourism, and encroachment; and developing conservation and restoration strategies tailored to meadows.

4 Result and Discussion

4.1 Factors of Growth Himalayan Alpine Meadow (HAM)

Himalayan Alpine Meadows occur in high-altitude areas, predominantly at elevations of 3000–4500 m. These meadows exist in various climatic zones having dissimilar features. It is not easy for the meadows to survive in low-oxygen and extreme conditions. In the Himalayas, grasslands are typically found in cooler temperatures. Throughout the growing season, the whole meadow zone is in bloom. During the day, temperature ranges vary slightly, but at night, the temperature goes below the freezing point. In very high topography, alpine meadows are impacted because of a deficiency of uncertain climatic occurrences. On steep slopes, soil is unable to develop well because in flat regions, water can be retained in the soil particles, which is beneficial for plant growth. On stony ground, grassland develops in a suitable manner since there are fewer nutrients available. Steepness in slope is unable to assist in the creation of humification. However, such vegetation types respond to these challenging conditions by establishing deep roots, enabling them to utilize the optimal amount of nutrients. Water is provided to meadows through the melting of glaciers, which is conducive to vegetation growth. Alpine meadows rely on the timing and availability of water. During the daytime, alpine meadows receive ample solar radiation, which facilitates the establishment of plants, such as leaves and cuticles. Winds in the high-elevational grassland zone are powerful. It is involved in both the process of evaporation and plant growth.

4.2 Mechanics, Dynamics Equilibrium and Bio-Diversity Hotspot

4.2.1 Types of Himalayan Alpine Meadow (HAM)

Alpine meadows are found in various altitude areas in the Himalayas, characterized by topography, temperature, water availability, rainfall patterns, and soil composition.

4.2.2 Herbaceous Meadows

This meadow community grows in lower altitudinal zones, likely at elevations between 3,000 and 4,500 m. These meadow categories require a gentle slope, a valley, high water, and adequate soil nutrient availability (Saxena et al., 2002). These alpine communities are also known as alpine arid pastures (Rawat, 2007a, b). These meadows are characterized by a dominant vegetation of Potentilla-Geranium, Festuca-Stipa, and Carex (Rawat, 2007a, b). The herbaceous alpine communities are adapted to a cold climate where many hardy grasses and herbs are located, such as Sipa, Melica, Sedum, Draba, Saussurea, *and Christolea* (Rawat, 2007a, b). These meadows are cultivated in large numbers when water availability and soil nutrients are abundant during the growing season (Jain & Sinha, 2004).

4.2.3 Alpine Shrub and Scrub Meadows

The shrub and scrub meadows community is found in the greater Himalayas, which are located above the herbaceous meadows. It generally grows between 4,000 and 5,000 m in altitude, where the slope is very high and ridges are exposed (Khanal et al., 2012). In the western and northwestern parts of the Himalayas, these communities are characterized by the presence of Betulautilis-Rhododendron campanulatum, *Abies Pindrow*, and *Quercus Semecarpifolia* (Rawat, 2007a, b). These meadow communities provide shelter to many species in the Himalayan region. The alpine shrub clumps in frost and snowbed regions, and among them, many plants have deep roots that collect moisture and nutrients from the rocky soil.

4.3 Functions of Himalayan Alpine Meadows (HAM)

4.3.1 Ecological Functions

Biodiversity Hotspot

Alpine Meadows is a natural herbaceous plant community located or originating beyond the natural forest and scrub vegetation limit (Lal et al., 1987). This meadow formation covers approximately 171.646 km^2, or 25% of the Indian Himalayan area (Lal et al., 1987).

Floral diversity of HAM

Some grasses and plant vegetation in the western Himalayas are classified as Gramineae or Poaceae (Uniyal et al., 1994). *Carex* and *Festuca* are also included in this category (Table 1). There are 73 species of pervasive grass category flora located at elevations of 350–1800 m; 51 species of flora category flora located at elevations of 2200–2500 and 2550–3000 m; and 62 species originated above 3000 m (Songh et al., 1980). Over half of Uttar Pradesh's Himalayan region is covered by only 18% of the 450 species (Uniyal et al., 1994). Himalayan Alpine Meadow (HAM) is also known as *Bugyal* in the Garhwal and Kumaon regions, *Dhar* or *Thanch* in Himachal Pradesh, and *Marg* in Jammu and Kashmir (Rawat, 2007a, b).

Wildlife

The alpine zone is the habitat of various wild species or faunal communities, and these species are all characterized by adapting to extreme cold climates, high altitudes, and seasonal changes.

Mammalian Fauna

Wild goats, sheep, rodents, and other animals inhabit the grassland zone. Various kinds of mammalian species in habitat in alpine zone *Includes Goral (Nemobaedus Goral); Himalayan Tahr (Hemitragus Jemlabicus); Himalayan Ibex (Capra Ibex Sibirica); Blue Sheep (Pseudois Hayaur); Tibetan Antelope (Pseudois Hodgsoni); Tibetan Gazelle (Procapra Picticaudata), Tibetan Argali or Hayan (Ovis Ammon Hodgsoni); Tibetan Wild Ass (Equus Kiang); Ladakh Urial or Shapu (Ovis Vignei Vignei); Snow Leopard (Uncia Uncia); Tibetan Wolf (Canis Lipus Chanko).* These are all grazing mammals that inhabit high-altitude grasslands and scrublands. The Goral mammal species mainly inhabit the western Himalayas, and it has been found that this species category feeds almost entirely on grasses (92.2% in the cold season and 98.3% in the warm season) (Mishra & Johnsingh, 1996).

Birds

Table 1 Zone of floristic composition and species dominance as identified by Tsuchida and Numata (1983)

Zone	Altitude (m)	Flora category
I	Less than 1100	*Cynodon Dactylon, Cbrysopogan Aciculatus, Desmodium Frifiorum*
II	1100–2600	*Paspalum Scorbiculatum, Pycreus Sanguinolentus, Eimbristylis Spp., Setaria Spp.*
III	2600–3800	Carex spp., poa spp.
IV	Greater than 3800	*Carex Species, Calamogrostis, Festuca* and *Agrostis*

In the Himalayan alpine zone, bird species are the most diverse, highly threatened, and vulnerable compared to other forested habitats. This natural evidence is found in the western Himalayas. Nearly 405 bird species are found in Jammu and Kashmir, 375 in Himachal Pradesh, and 642 in Arunachal Pradesh, all of which are located in highly forested areas (Datta et al., 1998). However, Meadows supports some types of highly vulnerable and threatened bird species such as- *Snow Cocks (Tetraogallus Tibetanus and T. Himalayanus); Cheer Pheasant (Catreus Wallichii); Snow Partridge (Lerwa Lerwa); Tibetan Sandgrouse (Syrraptes Tibetanus); Chukar Partridge (Alectoris Chukar); Mountain Quail (Ophrysia Super Ciliosa)* (Ali and Ripley, 1983).

Reptiles/Amphibians

Some cold-blooded reptile species that live in the alpine zone include the Himalayan Rock Agama (Laudakia tuberculata), a kind of Himalayan lizard; toads and Frogs (Nanorana Spp.), a kind of glacial stream species.

Insects/Pollinators

Some rare insects in the meadow region play a significant role in the alpine ecosystem. For example, the Himalayan Honeybee (Apis Laboriosa) is renowned for its cliff-nesting habits and is the largest honeybee in the world; Alpine Butterflies (Parnassius) are high-altitude insects; and Bumblebees (Bombus spp.) play a significant role in pollination in Himalayan Alpine Meadows (HAM).

Medicinal Plant

Medicinal plants are a significant component of healthcare in the Himalayan region (Hamilton, 2004). The Himalayas are recognized as global biodiversity hotspots and are renowned for their rich cultural heritage (Rawal et al., 2013). A considerable number of medicinal and aromatic plants (MAPs) have been identified in the alpine zone of northwestern India (Dhar & Kachroo, 1983; Rawat, 1984; Kala, 1997; Singh and Rawat, 1999; Samant et al., 1998). Among the Himalayan states, many local communities depend on these medicinal or herbal plants, which are essential for their consumption and economic extraction. Fourteen types of MAPs are recognized globally by IUCN as threatened, and 40 types of MAPs are identified as regional in Uttarakhand (Chauhan, 1999; Ved et al., 2003). Most medicinal plants are grown in the alpine zone and have rhizomatous or tuberous roots (Rawat, 2007a, b). Excavating breakable alpine soil for collecting herbal plants, followed by subsequent flattening and livestock grazing, spread unwanted plants and caused soil erosion. Some medicinal plants grow at high altitudes in the western Himalayan alpine zone. HAM is associated with several medical plants (Table 2) and has significant market value.

4.3.2 Carbon Sequestration

Vegetation stocks carbon in the western Himalayas, which is ecologically significant in terms of local, regional, and global carbon reserves. Alpine Meadows are mainly

Table 2 Location and uses of medicinal plant species in Himalayan Alpine Meadows

Medicinal plant's name	Altitude (m)	Uses
Dactylorhiza Hatagirea	2500–4500	As Vrishya, Ayurveda helps heal burns, ulcers, and wounds. It is anti-inflammatory, treats diarrhea, dysentery, and indigestion, and is used post-illness recovery to restore body strength. Additionally, it is helpful in managing bone fractures and joint pain
Picrorhiza Kurrooa	3000–5000	Treat liver disorders, jaundice, hepatitis, cirrhosis, acidity, chronic cough, and bronchitis; also treat malaria and reduce fever
Aconitum Heterophyllum	2500–4000	It treats diarrhea, vomiting, and dysentery, reduces fever and respiratory congestion, and acts as a natural analgesic, relieving colic
Jurinea Macrocephala	2500–4500	Treating wounds, skin infections, inflammations, reducing swelling, stomach ulcers, constipation, and using mild laxatives and digestive tonics

rich in grasses, mosses, and herbaceous plants. These plants absorb CO_2, the main gas, if the carbon level increases through photosynthesis. Such plants—including leaves and roots—store carbon in both the living and dead parts above the ground or soil surface, as well as in the roots or other plant parts below the soil surface. However, due to cold weather, the decomposition process of plants is very slow, resulting in the long-term storage of carbon and organic matter in soil particles. Such deep-rooted perennial plants increase carbon storage in soil particles below the surface by transforming organic carbon. Some Alpine Meadows, grown in permafrost layers, store carbon underground due to their higher elevation. Low temperatures lead to reduced microbial activity, thereby preventing the release of CO_2. Alpine Meadows are grown in thick soil layers made with humus and peat, which are rich in organic carbon. In the alpine zone, dead plant parts decompose slowly on the soil surface, resulting in long-term carbon storage within soil particles. These soil types convert carbon in the atmosphere as CO_2 into stable organic compounds. However, human interventions in land management and natural ecosystem practices increase greenhouse gas emissions in the atmosphere. The Himalayan Alpine Meadows (HAM) is the most significant carbon sink in the world. Sustainable management is crucial for providing livelihood opportunities, mitigating the vulnerability of natural resources, and enhancing carbon sequestration capabilities.

4.3.3 Conservation of Soil

Soil conservation is critical in alpine regions due to the terrain landscape. These vulnerable ecological functions are interrupted due to extreme erosion by wind, water, and gravity. However, soil conservation is essential to maintain the fertility and stability of land in more degrading conditions. The Alpine Meadows have dense roots, which leads to soil stability and prevents soil erosion. The deep roots of plants in alpine zones bind soil particles, preventing them from being displaced. They also help by holding the upper layers of soil. At high altitudes, the soil is typically shallow, rocky, fragile, and prone to erosion, a natural phenomenon. The precipitation rate is very high at high altitudes, differing from rain or snow, which leads to increased runoff rates. However, in alpine regions where vegetation is abundant at high altitudes, it helps reduce the erosive process, absorb soil moisture, and allow water infiltration through the in-ground plant roots. Due to the lack of natural barriers and open landscapes, the wind is strongly blown in alpine regions, and the soil is massively eroded. To protect the organic matter in surface soil from winds, it is necessary to have sufficient grasses and shrubs spread evenly over the whole landscape.

4.3.4 Regulation of Himalayan Microclimate

Himalayan Alpine Meadows significantly regulate microclimate by maintaining temperature and humidity levels, which also help support the surrounding ecosystem (Bency et al., 2022). Alpine zones are located in elevation areas with a high solar radiation rate. However, due to dense vegetation cover, solar radiation is absorbed during the daytime, and at night, heat is released gradually. These processes prevent extreme temperature fluctuations in meadow areas and help establish a stable microclimate, which is necessary for the habitation of various flora and fauna species and their adaptation practices.

Humidity helps regulate the absorption of soil moisture and evapotranspiration in the Himalayan alpine meadows. Due to the presence of a large number of plants, vegetation releases vast amounts of water vapor into the atmosphere, which helps increase the local humidity level and control the dryness of the local atmosphere, influencing the formation of precipitation processes. The balanced humidity maintains the sustainability of biodiversity in the Himalayan Alpine Meadows (HAM), including the conservation of rare plant and animal species by creating a comfortable living environment for pollinators. This stable environment contributes to the preservation of the Himalayan ecosystem by reducing the threat level to plant and animal species.

4.3.5 Hydrological Functions

Water Retention and Regulation

Himalayan Alpine Meadows (HAM) plays a crucial role in water retention and regulation. These alpine meadows act as a natural sponge. Due to the dense vegetation and high organic matter content in the soil surface of the alpine zone, it has a significant capacity for absorbing rain and snowmelt water. These waters are stored in soil particles, which prevents the rate of runoff and reduces the likelihood of flash floods. The grassland vegetation allows the release of stored moisture slowly from the soil over time. Gradual water ensures a steady water supply to rivers or streams throughout all seasons, even during periods of low precipitation. Many rivers originate from the Himalayas, and their water flow regulation depends on the Himalayan Alpine Meadows (HAM). In lower-altitude or foothill regions of the Himalayas, irrigation water, drinking water supply, and water for hydropower generation are sourced from the Himalayan Alpine Meadow (HAM) zone. This zone maintains the soil structure by bearing the supplied water, preventing extreme erosion rates, and reducing landslide risks, which is natural in hilly terrain regions.

Natural Regulation of Glacier

The Himalayan Alpine Meadow (HAM) regulates the temperature and humidity in the low-elevation landscape surrounding Himalayan glaciers. It can control microclimatic conditions by absorbing local temperature and gradually releasing solar radiation. This helps minimize extreme temperature variations, which directly impact glacier melting. During the day, the Himalayan Alpine Meadow (HAM) absorbs solar radiation, which helps reduce the high rate of glacier melting. Nevertheless, at night, HAM helps release stored heat and prevent a significant temperature drop, which can cause ice cracks or accelerate glacier melting. HAM works as a wind buffer by reducing the rapid rate of glacier melting. Harsh wind flow removes excessive snow cover from the land surface in high-altitude regions, but Ham helps by preventing wind speeds and preserving the ice mass.

Preventing flash flood

HAM, which includes the flora community, plays a significant role in Flash flood mitigation by controlling the rate of surface runoff. The diverse vegetation, including grasses and herbs, enhances soil stability, reduces the erosion rate, and regulates the rate of glacier melting, a crucial factor in flash floods in high-elevation regions. HAM acts as a natural barrier for flood mitigation, decreasing the rate of flash flood risk.

4.3.6 Socio-Economic Functions

Pastoralism

Himalayan alpine meadows are associated with *Pastoral* communities. This community traditionally practices pastoralism, utilizing grazing grounds for livestock such as Himalayan yaks and sheep in high-altitude alpine zones. Traditional *pastoralist* practices rely on seasonal grazing, where livestock animals are moved to rich grazing lands for nourishment (Rawat and Uniyal, 1993). In the alpine meadows, the history of livestock grazing has been ongoing for several centuries, primarily in the northwestern Himalayas (Tucker, 1986). Recently, 1.2 million livestock grazing farms have been supported in these regions, which include Himalayan sheep, goats, cattle, yaks, horses, and donkeys (Kawosa, 1988). Pastoralist communities practice a migratory livestock grazing pattern, where they move livestock to higher altitudes during the summer and to lower valleys in alpine meadows during winter. These communities depend on livestock products, including dairy, wool, meat, and trade. All these products, such as cheese from dairy and yak butter, are essential for local economic growth. The Agro-Pastoralists in the western and central parts of the Himalayas keep more cattle because those areas are easily accessible for free grazing, and they are unable to dispose of some religious sentiments (Sundriyal, 1995). According to some authors, livestock grazing is crucial for maintaining species diversity in meadow biodiversity (Naithani et al., 1992; Negi et al., 1993; Saberwal, 1996), as it helps prevent shrub encroachment while promoting the growth of native grasses. According to Rawat and Rodgers (1988), moderate grazing levels may enhance the diversity of herbaceous species in the alpine zone. However, these grazing animals sometimes affect plant communities by defoliating plants, removing and redistributing nutrients, and impacting soil when livestock repeatedly walk over grassland or vegetation, resulting in soil compaction, damage to vegetation, and an increase in soil erosion rate.

Medicinal Plant Harvesting

Many commercial alpine medicinal plants are harvested and collected from the Himalayan region by local communities, which helps increase their household income. A recent study reported the annual trade amounts of thousands of tonnes of roots, tubers, and leaves of medicinal plants, valued at millions of US dollars (Olsen and Larsen, 2003). All these plants are harvested for traditional, Ayurvedic, and pharmaceutical uses, with their primary market in India, and are traded at both the national and local levels in Nepal. In high altitudes, these valuable herbal medicinal plants have been used in local healing practices, including Picrorhiza Kurrooa, Yarsagumba (Ophiocordyceps sinensis), Jurinea Macrocephala, Kutki (Picrorhiza Kurrooa), Jatamansi (Nardostachys Jatamansi), and Atis (Aconitum Heterophyllum), among others. The trade of these medicinal plants helps increase their income, supports traditional healing practices, and boosts the local economy through the export of herbs. Some rare and threatened medicinal plants are found in the Valley of Flowers National Park, particularly in the grazing and unprotected alpine meadow zone (Kala et al., 1998).

Tourism and Recreation

The Himalayas are considered a haven for every traveller to breathe. Recently, this type of tourism has gained significant popularity in the foothills of the Trans-Himalayan region. Ecotourism refers to travel with a minimal impact on ecologically sensitive areas. It ensures the benefits of nature and local cultures and supports the local communities. Mountain areas are rich in natural resources, including wildlife and biodiversity, which are the primary attractions for tourists. This tourism is an alternative source of economic activity for people living around the place. These practices boost their income level. Currently, ecotourism is recognized as a key sector that helps link development and strategy in sustainable tourism, including alternative or village tourism.

Tourists visit here for mountaineering, trekking, and river rafting. For tourists, many homestays, hotels, motels, dhabas, and stalls are established along the roadsides. These all help increase the local people's employment opportunities and improve their living standards. These are all enormous sources of national income (Sati & Kumar, 2004). Primarily, people are employed as guides and porters in tourist spots. Due to the increase in tourism, there has been a rise in demand for handicrafts made by local artisans. All these reasons are related to economic development. Tourism helps encourage the conservation of nature due to its economic value. Tourism activity promotes cultural exchange between locals and visitors. The visitors experience the local Himalayan traditional lifestyle practices. The government's increased focus on tourism primarily centers on the development or improvement of transportation and communication, healthcare facilities, and skills development programs in education to encourage local people to practice eco-friendly tourism and its benefits for their socio-economic development and growth.

4.3.7 Cultural Function

Sacred Place

Himalayan alpine meadows are not only renowned for their beautiful ecology, but they are also celebrated for their rich cultural heritage, diverse spiritual beliefs, and diverse religious practices. Local communities often believe in the benefits of spiritual retreats. They are linked with Hinduism and Buddhism. Some religious places are famous which are including under the Himalayan Alpine Meadows such as Valley of Flowers which is located in Uttarakhand, the local people believe that Lord Hanuman found the Sanjeevani herbal plant from this valley; Roopkund, which is present in Uttarakhand; Kailash Man Sarovar it is located is Tibet- India birder, famous for Lord Shiva, believed by Hindu, Buddhist, and Jaina. In the northeastern part, the Dzukou Valley, located on the Nagaland-Manipur border, is also renowned for its Naga tribes. They believe that some ancestral spirits are present. Based on these religious and sacred beliefs, the local community is committed to conserving this grassland from human activities.

Rich in cultural satisfactions

Alpine Meadows are rich in cultural traditions and concepts, including the celebration of traditional festivals and rituals, such as the Hemis festival in Ladakh by the Buddhist community and the Nanda Devi Raj Jat Yatra in Uttarakhand. Some traditional songs, including folk songs, are created based on the relationship between nature and humans. They believe that meadows are gifts from nature. In ancient times, Monks meditated in dense meadow zones to enhance their inner peace.

Myths and Religious Practices

The grassland zone is enriched with mythological histories, which are passed down from generation to generation among the local villagers. These mythological stories influence their religious practices and customs. From Mahabharata, we know the journey of the Pandavas. They trek through many meadows, which are known as the way of heaven. In the Hindu religion, we are familiar with another story of Lord Shiva and Parvati. Ancient legends associate the alpine meadow with annual pilgrimages, such as the one to Manimahesh Lake. This lake is located in Himachal Pradesh. Some indigenous tribal communities, such as the Lepchas and Monpas, believe that powerful spirits live in meadows. In some communities, local people believe livestock is their god. Because, according to them, their crops and herbs will grow in response to their blessings. These types of mythologies promote environmental conservation by preventing deforestation and overexploitation of natural resources.

4.3.8 Water Buffer of Climate Change

Issues related to climate change in Himalayan Alpine Meadows (HAM)

Climate change presents issues in the Himalayan Alpine Meadows. Due to global warming, temperature rises, precipitation patterns are changed, and uncertain rainfall rates increase. The rising temperature is melting the glacier cover of this region, which directly impacts the water regulation system. Sequentially, the change in precipitation patterns leads to different hazard conditions, such as landslides, floods, and droughts. These uncertain condition impacts the ecological balance. In the grassland zone, vegetation is shifting due to temperature increase; many species are changing their habitat because they cannot adapt to this climate change. The whole biodiversity faces a high risk of ecosystem losses. Due to changes in precipitation, the likelihood of landslides, which can lead to soil erosion, increases. Rapid climate change is not only a future threat but also a reality of the present scenario. These grassland zones experience glacier melting, which affects the natural storage of water, fresh water supply to the human population, and wildlife in this region.

Crisis of ecosystem and natural regulators

The extreme climatic events lead to the diversity of native plants, mosses, and grasses. These all regulate soil composition, carbon sequestration, and other factors. Sometimes, a temperature rise can be adopted by some species. This extreme climatic event impacts groundwater infiltration, regulates soil nutrients, and reduces the ecological species. Due to excessive rainfall, soil erosion occurs in high-elevation areas, leading to the loss of vegetation in meadows. Without natural vegetation, the ecosystem cannot survive during flooding and periods of water unavailability.

Natural buffer of climate variations

In the Himalayan region, alpine meadows create a natural buffer zone against climate change. These grasslands function as a tanker of water by absorbing excessive water due to heavy rainfall. These zones help prevent floods and other hazardous conditions. The highly vegetational grassland provides a cooling environment through releasing extreme temperatures. This natural buffer is also disrupted due to intensive precipitation and rising temperatures. These buffer zones protect the diversity by absorbing unpredictable water availability. Global warming leads to the shifting of plant and animal species, which is also known as Phenology. Meadow's region is enriched with flowers and fruit during the growing season. However, the rising temperature makes disturbance between the flowers and pollinators.

5 Threats to the Himalayan Alpine Meadows

5.1 *Anthropogenic Threats*

Due to anthropogenic activity, the Himalayan alpine meadows face ecosystem threats that disrupt biodiversity, livelihood, and the natural environment.

5.1.1 Hosting of High-Profile Wedding Partly

In the Himalayan alpine meadows, high-profile weddings pose a significant environmental threat to fragile ecosystems. Human activity contributes to the destruction of soil and vegetation, and it also affects local communities. The considerable pressure from guests, stages, and flooring compacts the fine alpine soil, reducing its water absorption potential, which is essential for plant life. Overcrowding contributes to the local environment becoming more polluted due to excessive amounts of decorative materials, food packets, food waste, detergents, and other biodegradable materials. The crowd can disturb wildlife with the loud party music, which is also one reason for their habitat changes. As a result, the whole biodiversity is in a vulnerable condition. This type of human activity also disrupts the local community's livelihoods by affecting their pastoral activities. Hosting a high-profile party can lead to a loss of cultural and spiritual significance in the Himalayan alpine region.

5.1.2 Garbage from Tourists

The Himalayas are very popular with tourists for trekking and camping. However, this generates plastic bottles, wrappers, and packets in the grassland, which do not decompose in cold environments. They also dispose of glass or metal products, which create threats to wild animals and livestock. Their leftover waste disrupts the natural ecosystem, including the food chain and food web in the grassland area. Sometimes, visitors dispose of chemical waste, such as toiletries, in mountainous streams, which can affect the local people's drinking water and harm wild animals in this region. These issues collectively lead to health problems for the local wildlife and community. The habitat spoils the natural beauty of the mountain region and increases the pollution rate in the Himalayas.

5.1.3 Cattle Grazing or Intensive Livestock Grazing

Intensive livestock grazing in Himalayan alpine meadows by sheep, yaks, and goats poses a significant threat to biodiversity and contributes to environmental degradation. Unregulated grazing reduces vegetation cover on the land's surface, exposing the soil and making it susceptible to erosion by wind and water. This high erosion rate reduces water infiltration, which affects plant regrowth. Continuous grazing practices degrade native meadows and herbal plants, reducing plant diversity and negatively impacting wild habitats and pollinators. Overgrazing exacerbates land desertification, resulting in barrenness, and reduces the availability of nutrients in vegetation, which is the primary food source for livestock. Due to this malnourishment of livestock, they produce low-quality milk and wool, which negatively impacts their economy.

The wild populations of the Himalayan Musk deer, Ibex, Hangul, and markhor play a crucial role in maintaining ecological balance in the Himalayan ecosystem. These wild ungulates affect water regulation and soil retention. Nevertheless, sometimes, when their population declines, the number of livestock animals tends to increase. This wild ungulate poses a threat to the grassland zone. The overgrazing of wild ungulates reduces the abundance of many herbs, shrubs, and grasses, which take longer to grow due to harsh climatic conditions. The grazing of wild ungulates affects vegetation cover and makes the soil vulnerable to erosion by decreasing the number of nutrient particles.

5.1.4 Unmonitored Tourism

Mass tourism in the Himalayan Alpine Meadows creates the environment, ecology, biodiversity, and socio-cultural aspects. The fragile ecosystem is located in a high grassland zone, characterized by a unique flora and fauna. Unmonitored tourism activities increase the footfall rate due to adventurous activities like trekking and camping, resulting in excessive land degradation. Non-biodegradable products, such as plastic waste, wrappers, and food packets, are often left behind by tourists, which creates soil and water pollution in nature. Tourists create noise that disrupts the natural habitat of wild species, such as snow leopards and Himalayan musk deer. Due to anthropogenic activities, a reduction in vegetation cover affects water retention and increases the risk of landslides.

5.1.5 Poor Waste Management

Nowadays, Himalayan grasslands are widespread due to tourism, trekking, and pilgrimage activities. However, some waste management issues have long-term impacts on the regional ecology and environment. Tourists often dump waste on the lakeshore, near streams, or at glaciers, polluting the local community's drinking water sources and harming wildlife. This is also harmful to the aquatic ecosystem. This polluted water spreads various diseases among the local communities and is also toxic to animals. Due to the lack of proper waste management, soil stability is very weak, and as a result, the rate of landslide risk is very high in Monsson. The decomposition of plastic materials releases a significant amount of carbon into the environment, contributing to glacier melting and global warming.

5.1.6 Overexploitation of Medicinal Herbs for Commercial Purposes

Based on Kala's (2005) opinion, 138 herbal healers have likely documented indigenous uses. These threatened medicinal plants inhabit the buffer zone villages of the Indian Himalayas, where 83 Tibetan healers and 55 Ayurvedic healers live. Due to the rising global demand, medicinal plants from Himalayan alpine meadows are being overexploited for commercial purposes. In the Indian Himalayas, 32 medicinal herbal plants are threatened (Kala, 2005). Due to the increase in global demand, Himalayan medicinal plants are being harvested excessively in unregulated ways, resulting in their rapid decline in status. Some of the plants take many years to regenerate. Overexploitation of these types of plants reduces soil fertility, which threatens the regrowth of plants. Many medicinal plants contribute to the natural ecosystem by supporting local pollinators and insects; however, continuous exploitation disrupts this ecosystem chain.

The most threatened medicinal plants are in the Valley of Flowers, a protected area. These threatened species are used to heal 45 ailments, including *Arnebia Euchroma* (Royle ex Benth), *Johnston*, and *Ephedra Gerardiana Wall. Ex Stapf*, (Kala, 2005). The trade in medicinal plants is a significant source of income for local communities. The collection and over-harvesting affect the resources hub and pose a threat to their future earnings. In the era of globalization, the rate of illegal trade, activity, and smuggling of medicinal plants is increasing due to their high commercial value, making it very challenging to implement effective measures. This overexploitation leads to the degradation of local biodiversity, soil quality, and communities, as well as threats to wildlife, ultimately damaging the ecosystem. The value of medicinal plants in the fragile alpine meadows also disappears.

5.1.7 Deforestation in the Lower Elevation

In lower elevations, alpine vegetation rapidly responds to the undulating topography, wide variations in altitude, soil components, temperature, rainfall, and biodiversity. In this elevation, plants help stabilize the soil. Nevertheless, deforestation is a significant issue for the conservation of vegetation. The world has undergone rapid transformation due to industrialization and urbanization, resulting in a high rate of deforestation in lower elevations. Plants' roots bind soil particles. However, deforestation increases the rate of soil erosion and surface runoff and controls water regulation, availability, and moisture absorption and release. In lower elevations, it also helps in groundwater infiltration. However, due to deforestation, the groundwater recharge rate also decreases, which affects the growth of the alpine meadows. Deforestation contributes to regional warming by altering climate patterns and reducing carbon sequestration, resulting in increased temperatures in that region, which in turn affects habitat availability.

5.1.8 Decline in Traditional Pastoralism

Pastoralism is a cultural tradition in the Himalayan region. Indigenous communities traditionally practice pastoralism. Grazing maintains the alpine grassland, which helps to control shrub expansion. However, when pastoral declines, shrubs expand into the grassland area at lower altitudes, affecting meadow habitats. Pastoralism is a cultural tradition. Pastoral decline encompasses sustainable grazing practices, the utilization of medicinal plants, and practical landscape management processes. Due to pastoralism, land use changes in higher-altitude areas, such as tourism, road construction, expansion, and numerous hydropower and hydroelectric projects, pose direct threats to alpine meadows. The decline of pastoralism practices disrupts ecological balance in alpine grassland.

5.1.9 Land-Use Changes

Human encroachment to the construction of settlement

Himalayas are a major tourist attraction. It attracts millions of visitors every year. For this tourism expansion, many infrastructure projects have been developed at high altitudes. Many habitats, including those of plant and animal species, are destroyed for the construction of settlements, lodges, hotels, and roads. Excessive anthropogenic pressure increases the pressure on natural resources by extracting firewood, water, fodder, fiber, and medicinal herbs. Human activity, including the disposal of waste such as plastics, metals, and glass fragments, harms plants and wildlife. Irregulating tourism threatens soil compaction, wild ungulates, and waste dispersion.

Construction of road and high-rise building

Good communication is required for economic growth. Road construction has a direct impact on alpine meadows, disrupting the habitats of wildlife, including the Himalayan tahr, musk deer, and blue sheep. Many trees are cut down for road and building construction, which reduces soil stability and increases the risk of landslides in alpine meadows. Vehicles also contribute to air pollution, which is harming the fragile alpine flora. Pollutants from roads increase the carbon footprint in these regions, contributing to climate change through glacial melting and rising temperatures.

Large-scale infrastructural development by altering the natural hydrological pattern of streams

Some large-scale infrastructure developments, such as hydropower plants, dams, and embankments, directly changed the hydrological patterns of mountainous streams and rivers in the Himalayas. These types of developmental projects alter the natural flow of rivers. For this reason, the water supply is sometimes reduced, which affects the distribution of alpine meadows. Natural habitats are also affected by changes in stream flow, which impact aquatic ecosystems. Some wild animals also suffer from changes in hydrological patterns. Alterations in flow patterns are the primary cause of flash floods and landslides in alpine meadows, posing a significant threat to local villagers.

5.2 *Climate Change*

5.2.1 Retreat of Glacier

In the Himalayan region, glacier retreat poses a significant threat to the ecosystem in alpine meadows. Many floras and faunas exist in glaciated areas due to climatic stability. However, in the era of globalization, due to climate change, glaciers are melting at a rapid rate. As a result, water shortages affect soil moisture, a crucial factor for plant growth. Various endemic flora and fauna species have habitats in this cold climatic condition. However, glacier retreat disrupts their habitats, resulting in a huge biodiversity loss. Glacier retreat exposes the upper layer of rocks and soil, increasing the erosion rate, landslides, and flash floods in alpine grassland. The local communities in the meadow's region depend on the grassland ecosystem, which is affected by glacier retreat. This threatens their livelihoods and reduces the availability of fodder.

5.2.2 Thaw of Permafrost

The permafrost thaw poses a threat to alpine meadows by impacting hydrology, carbon sequestration, ecosystems, soil stability, and biodiversity. Thawing permafrost releases carbon in the form of CO_2 and CH_4. As a result, the climate changes, and temperatures increase. High temperatures increase surface runoff, disrupting the hydrological balance. Water availability is reduced in dry periods, and water molecules are inter-logged in soils. Permafrost serves as a binding agent for soil particles. However, the thawing process affects the structure and components of soil, making it more susceptible to erosion. This soil erosion threatens biodiversity loss and declines unique grassland flora and fauna species.

5.2.3 Rise of Temperature

Due to global warming, temperatures have increased, posing a threat to the Himalayan alpine grassland, which is home to some of the most fragile ecosystems. These grassland types are located above the treeline in high-altitude areas, characterized by unique biodiversity. With the increase in temperature, treelines are shifting, and as a result, the local species in meadow areas are adapting their habitats. Among them, many endemic species do not exist in hot climatic conditions. As a result, alpine biodiversity is lost. Nowadays, rising temperatures are forcing the migration of ecosystems due to limited space and resources. High temperatures in the meadow region affect livestock grazing by affecting food availability. Rising temperatures interrupt pollinators' life cycles and threaten the whole ecosystem chain. Warmer weather in Himalayan alpine grasslands affects the ecological balance by spreading pests and pathogens.

5.2.4 Torrential Rainfall in Short Periods/Intervals of Time

In the Himalayan region, torrential rainfall is a frequent weather condition that occurs over a short period due to climate change. Due to this significant change in precipitation patterns, the upper layer of thin soil in alpine meadows erodes through downwash at high elevations. The nutrient-rich topsoil is essential for plant growth in meadow zones. This torrential rainfall increases the risk of landslides, which can destroy habitats for flora and fauna in grassland regions. These extreme weather events capture excessive water quickly due to poorly drained soil. This flooding weakens plant roots, reduces biodiversity in the ecosystem, and disturbs pollinators and their breeding and nesting patterns.

5.2.5 Increased Number of Cloudbursts and Flood

Himalayan alpine meadows are situated at high altitudes, where the climate undergoes continuous changes, such as cloud bursts—a type of sudden, extreme rainfall in a specific region. This uncertain climate event leads to flash floods and landslides, which destroy the local habitat ecosystem and the ecological balance in grassland biodiversity. Cloudbursts cause rapid surface runoff, which exfoliates the thin upper layer of topsoil of alpine meadows. This increases erosion, creating deep gully channels and altering the natural landscape. This specific condition led to a flood, which washed away the grass cover on the land, resulting in food shortages for livestock and wild animals. When grazing animals move forcefully to lower elevations due to food shortages, human- wildlife conflict in the region is created. Due to the destruction of grassland regions, alpine wetlands, which are also habitats for migratory birds, are impacted. The local community is dependent on the resources of the meadow area. However, the cloudbursts threaten their livelihoods and economic-related activities, forcing them to migrate and resulting in economic and habitat loss.

5.2.6 Unpredictable Drought

Drought, which depends on snowmelt and monsoonal rainfall, is one serious threat to Himalayan alpine grassland. This prolonged dry period degrades soil by reducing moisture and nutrient availability. This unpredictable condition sometimes increases the chance of wildfires due to lengthy dry conditions. The fires can quickly spread across large meadows, disrupting the ecological balance. Eco-tourism activities in the Himalayan region depend on a healthy green landscape. However, the extension of dry periods impacts it, which is directly involved with economic stability.

5.2.7 Expansion of Non-native Plants or Invasive Species

The growth of non-native species disrupts the fragile local ecosystem. The growth of invasive species is faster than that of native species. Non-native species can proliferate in areas with lower sunlight, water, and soil nutrients. As a result, the local native plant diversity is lost rapidly. These non-native species lead to changes in soil chemistry that are unsuitable for native plants. Invasive plants absorb more water, resulting in dry wetlands and reduced water availability for native plants. Some non-native plants are toxic to domestic livestock, such as yaks, sheep, and goats. These animals struggle to find suitable food, and for this reason, they migrate to lower altitudes.

5.2.8 Loss of Endemic Flora

The Himalayas are home to a diverse array of unique species. Many endemic and rare species of flora and fauna are struggling to survive in extreme climatic conditions. These rare species are threatened by numerous natural, environmental, and anthropogenic factors, leading to the loss of biodiversity and disruption of ecological processes. Due to the first spreading of non-native plants, native plants are affected. Non-native plant species compete with native plants for soil nutrients, solar radiation, and space. For this, it is challenging to survive against the odds of native plants. The rise in temperature due to global warming is another factor contributing to the decline of endemic flora. At high altitudes where extreme cold weather exists, it is difficult for fragile ecosystems to habitat if the temperature rises and the precipitation pattern changes.

6 Achieving Sustainability in the Himalayan Alpine Meadow (HAM)

6.1 Adaptation and Mitigation of Climate Change

6.1.1 Adaptation with the Ecosystem and Ecological Behaviour

The Himalayas are highly affected by global climate change, where glaciers are melting rapidly. The local community uses the grassland zone for grazing and collecting medicinal plants. In the case of the Himalayan Alpine Meadows, climate change also monitored the future conditions of the grassland zone. In the meadow region, people must be concerned about preventing the cutting down of treelines and shrubs of alpine grassland, as these types of grasslands always protect the local habitat and mitigate the risk of landslides due to high elevation. The local community should focus on afforestation, especially of rare and vulnerable plant species, which help stabilize the meadows' biodiversity. The Himalayan region is rich in medicinal plants and fodder grass. In high altitudes, some plants are cultivated for local communities. These should all be limited for sustainable use because excessive collection increases pressure on natural ecosystems.

The Himalayas are popular for being a hub of medicinal plants. These types of herbal plants should be identified first, as they are valuable and must be conserved through planned cultivation, sustainable harvesting, and the prevention of illegal marketing. The regeneration of native plants should focus on preventing the growth of non-native plants, as all local plants are essential for restoring soil fertility and supporting local wildlife. Overgrazing practices have always degraded land and reduced soil fertility. In high altitudes, the terrace farming pattern must be followed to prevent soil erosion, sustain local agricultural processes, and encourage the cultivation of rare or indigenous crops. Pastoralism practices and agroforestry management help prevent the decline in carbon sequestration rates by acting as a carbon sink and restoring greenhouse gas emissions.

6.1.2 Management and Regulation of Himalayan Water Resources

Himalayan Alpine Meadows play a significant role in holding and regulating water resources, acting as a natural sponge in high-elevation areas. However, due to climate change, the glaciers and snow are melting, as well as the seasonal water shortages in this altitudinal region. To prevent this glacier from retreating, artificial glaciers should be created. This scientific activity is very popular in Ladakh and is well known as Ice Stupas. This activity is conducted during winter, when water is stored in a solid form, which is then released gradually in liquid form for livestock. This process is also beneficial for maintaining water availability in grassland zones, which are fragile natural ecosystems that include endangered species such as the Himalayan

Musk Deer and the Himalayan Ibex. This scientific application is designed to help maintain soil moisture and prevent rapid surface runoff.

In the Himalayan region, the primary source of water is springs. However, due to global warming, climate change, and deforestation, the decline of spring discharge is attributed to these causes. To control the condition, various techniques have been adapted, including treatment of the catchment area, construction of check dams to recharge the groundwater level, and increasing the plantation of deep-rooted native plants. These techniques are all beneficial for restoring spring water flows. Spring rejuvenation prevents the alpine meadows from experiencing dry conditions and provides drinking water for local communities, wildlife, and the surrounding area. These techniques play a significant role by maintaining the local economy through attracting ecotourism and sustaining agricultural practices.

6.1.3 Measurement and Monitoring of Climate Resilience

In the Himalayan region, the grassland is a significant biodiversity zone that is particularly vulnerable to climate change, which includes glacier retreat, changes in weather patterns, reduced water retention, and shifting vegetation. To address these climate-related issues, it is necessary to establish an automated weather station that records temperature, humidity, precipitation, wind speed, and wind direction for monitoring and predicting flash floods, landslides, droughts, and excessive snowfall, which can impact local communities and wildlife. The Automated Radar system helps predict changes in weather conditions, such as monsoonal shifts and precipitation intensity. This system provides early warnings for natural hazards, such as flash floods, landslides, and glacier outbursts. The establishment of the ecological observatory for the long term helps by providing data on biodiversity loss, species migration, and how these conditions are related to glacier retreat and vegetation shifting. Satellite-related technologies are applicable for detecting changes in glacier cover zones, ice shrinkage, changes in land patterns, and land degradation. By using satellite data, mapping has been conducted to identify and analyse treeline shifting and grassland degradation caused by global warming and climate change. Machine Learning (ML) and Artificial Intelligence (AI) applications can analyze GLOFs, intensive rainfall, future vegetation shifts, species diversity, and trends in climatic data within the local Himalayan fragile meadow ecosystem.

6.2 Conservation and Management of Himalayan Ecology and Ecosystem

6.2.1 Control Over Grazing

In the Himalayan alpine meadows, overgrazing has a profound impact on land-use management. However, sustainable grazing practices can mitigate the rate of land impact. Implementing rotational livestock grazing in meadow areas across different regions helps regrow grasses and maintain biodiversity. Overgrazing control can reduce the pressure of livestock on specific vegetational regions, allowing native species to regenerate and thrive. The number of grazing livestock and wild animals should be limited according to the capacity level. This maintenance process can protect certain alpine grasses, which are crucial for maintaining ecosystem stability. Overgrazing control increases the practices of home-based fodder cultivation, which benefits fragile alpine ecosystems. Some types of fodder grasses and barley are cultivated for livestock, which can reduce the dependency on grasslands. Home-based fodder cultivation practices reduce the likelihood of long-distance grazing.

6.2.2 Conservation of Biological Diversity

Himalayan Alpine Meadows have a unique biodiversity. Many native and medicinal plants are maintaining the local fragile ecosystem balance. This conservation effort is essential for these native species through seed banking and reforestation, aiming to protect the rare species (Rawat, 1997, 1998). Many plants possess vital medicinal properties that offer economic benefits. Traditional farming practices in high altitudes minimize the chances of land degradation and promote the practice of small kitchen gardens. Sustainable, localized cultivation also reduces the dependency rate on Himalayan alpine meadows and supports local communities in planting crops in their households for food security. Recently, the illegal marketing of medicinal plants has increased based on their value. Unscientific cultivation and uncontrolled collection cannot ensure its long-term availability. For this, it is essential to demarcate specific scientific zones for the cultivation, harvesting, and conservation of this resource. This implication can help control the overcollection and exploitation of medicinal plants, allowing these all-herbal plants to regenerate. Sometimes, herbal gardens are also affected by tourists. Generating conservation projects can raise awareness about biodiversity conservation among local communities and tourists.

6.2.3 Erosion Control and Management of Soil Through Bio-Engineering Method

Himalayan Alpine Meadows are a highly soil erosion-prone region due to their high elevation, steep slopes, weak soil particles, and overgrazing practices. For stabilizing soil conditions, bio-engineering techniques are adapted. These techniques are also beneficial for vegetation restoration and enhancing ecosystem resilience through the application of jute geo-textile. This geo-textile application process prevents soil loss and reduces runoff in alpine meadows. This technique serves as a biodegradable cover for the soil, which is beneficial for the growth of native plants. In high-elevation regions, small check dams are sometimes constructed to slow down water flow, which helps prevent the creation of gullies and surface runoff. The construction of dams also helps to increase groundwater infiltration, which is essential for alpine growth and development. Afforestation can help reduce soil erosion by creating a natural barrier of vegetation. It also impacts land degradation by binding soil and absorbing excess moisture, thereby reducing the risk of landslides. Plant parts, such as roots, leaves, and branches, as well as cow dung, are incorporated into the soil, which increases soil fertility and promotes biodiversity. The deep ploughing method increases the erosion rate in hilly regions. However, the light ploughing method helps to preserve the soil nutrients and reduce water runoff on the surface.

6.2.4 Carbon Sequestration Programs

In the Himalayan region, carbon is stored in soil particles. Alpine meadows are grown in humus and peat soils, which are rich in organic carbon. Native plant species are supported by enhancing to store carbon in alpine zones. These species all absorb CO_2 through afforestation and grassland restoration in their deep roots. Overgrazing increases the likelihood of carbon emissions from the soil. For this, it is essential to maintain carbon sinks in meadows in the long term, ensuring the region can avoid extreme climatic conditions and provide a habitat for natural fauna and flora species.

6.3 Community-Based Management Strategies

6.3.1 Diversification of Sustainable Livelihoods

Himalayan Alpine Meadows are rich with fragile ecosystems. To protect and conserve this ecosystem, some essential sustainable conservation strategies are needed, which also benefit the local communities economically. To manage land degradation, the community should encourage high-altitude crop cultivation strategies, thereby reducing the impact of overgrazing pressure on the land surface. The shift in their livelihood from grazing to sustainable agriculture resulted in a decrease in the soil erosion rate and the likelihood of landslide hazards. They must adopt medicinal plant cultivation strategies to develop their economy in both local and global markets. In the Himalayan region, livestock animals provide a range of economic products. They can continue wool processing, dairy product production, and handmade craft activities to develop their financial structure.

6.3.2 Promotion of Eco-Tourism and Homestay Instead of Construction of High-Rise Hotels

The Himalayas are among the most popular tourist destinations. The presence of natural environment attracts more tourists here. For this, we need more ecotourism strategies based on natural systems, such as bird sanctuaries, wildlife safaris, and glacier trekking, among others. However, sometimes, the wildlife is also harmed by the tourists. They can pollute the natural environment by disposal of non-biodegradable products. To prevent this situation, we need to establish eco-tourism circuits in alpine meadows, thereby preventing habitat destruction. This approach would enable every tourist to support the habitat restoration of wild animals, insects, and pollinators and participate in conservation projects. For the development of ecotourism, they built some homestays, which play a vital role in cultural mixing with local communities. The culture of homestay preserves local traditional knowledge, offers organic food, and promotes reduced deforestation. These are all increases in large hotel construction. The ecotourism practices benefited the local communities economically by providing them employment in many tourist-related activities.

6.3.3 Community Entrepreneurship to Preserve Himalayan Alpine Meadow (HAM)

In the Himalayas, several strategies have been adopted to preserve alpine meadows, including encouraging the local community to engage in the use, processing, and trade of natural products in both global and regional markets, such as herbal plants, organic food, wool, and dairy products. There must be developed some eco-tourism buildings where mainly local communities are involved. They manage the tourism sites, caring for the plant species and creating handmade crafts collaboratively. Wool processing is very famous in the Himalayas. To incorporate these activities within economic boundaries, specific strategies must be adopted that encourage local women to weave wool, package organic herbs, and create various craft products, among others.

6.3.4 Incorporation of Indigenous Knowledge of Pastoral and Agrarian Communities in Sustainable Meadows Management

Pastoralism in the Himalayan grassland zone helps to practices a sustainable grazing process. This traditional transhumance practices reduce the overexploitation of plants, grasses, and shrubs in pasturelands due to overgrazing. These indigenous traditional techniques provide knowledge of soil conservation and its importance, including the significance of soil nutrients for cultivation, the impact of overgrazing on these nutrients, and methods for conserving them through pastoralism practices, such as creating water channels and applying organic manure. This traditional knowledge helps secure the entire biodiversity's habitat, reducing the threat to fragile ecosystems.

6.3.5 Participatory Management of Forest

Local communities should follow the afforestation process and lead many restoration programs for alpine meadows to ensure a long-term conservation method. Communities must follow Joint Forest Management (JFM). It is a collaboration approach. Here, the communities work collaborates with some government agencies to manage, conserve, and restore the Himalayan alpine meadows and the local ecosystems. Similarly, these agencies provide them with some knowledge about sustainable grazing and pastureland management practices, the benefits of afforestation, and the conservation of medicinal plants, among other topics.

6.3.6 Engage Local Communities in Conservation Through Incentives

Engage local communities in conservation through incentives in JFM to ensure they are economically benefited from participating in this management programme. From this incentive's conservation strategies, local communities earn money by guiding tourists on treks and collecting entry fees. This process encouraged local communities to conserve the meadows, protect wild animals, and preserve biodiversity. These incentives for conservation practices of nature and species provide job opportunities for local communities and generate revenue, reducing the dependency rate in grazing. They also offer knowledge about sustainable resource use, encourage improvements in biodiversity and ecosystem value, promote conservation, and foster stewardship in meadows.

6.3.7 Capacity Building Training

Capacity-building training is crucial for engaging local communities in sustainable conservation practices. In Himalayan alpine meadows, their livelihood directly depends on sustainable conservation practices. In these training, mainly focused on their skill development about eco-tourism management; sustainable agricultural practices; provide training on organic farming; medicinal plant cultivation; seasonal livestock grazing; local product development which are produces from livestock; teach them about climate resilience, soil erosion controlling method; agroforestry; afforestation; biodiversity conservation; forest restoration etc. These training programmes also discussed how they economically benefited their community through employment. The capacity-building training helps preserve the fragile alpine meadows while improving their livelihoods by developing eco-friendly skills and knowledge.

6.3.8 Involve Local Community in Policy Framing

Involving local communities such as villagers, pastoralists, and indigenous groups in policy framing to provide knowledge about resource benefits, management, and conservation. Initially, they need to establish a local council for conservation, where they can participate in afforestation and the restoration of meadows. Provides them with knowledge about the responsibility of ecosystem management. Villagers must be assured about policy alignment in cases involving their traditional practices. Involve local communities by conducting meetings, programs, and workshops on grazing limitations, tourist regulations, biodiversity conservation, waste management, and other relevant topics. By following policy-framing practices, we can mitigate the conflict between governmental conservation laws and traditional practices of local communities, promote ecological balance, and achieve other objectives.

6.4 Governance and Policy Implementation

6.4.1 Strict Implementation of Conservation Policies

The government should strictly implement policies related to environmental conservation and the protection of ecosystems from species loss. Baumgartner and Hartmann (2001) state that funding and time constraints can hinder the implementation of a plant biodiversity programme. The programs are designed based on existing laws, limited ecological knowledge, and consideration of biodiversity as a pool of resources (Baumgartner & Hartmann, 2001). The government should continue to strictly enforce penalties for environmental violations, including the improper disposal of waste, habitat destruction, and deforestation. To prevent the destruction of biodiversity through these activities, utilize advanced technologies such as satellite imagery and drone applications to identify and monitor illegal activities. The involvement of local communities and raising awareness in the conservation of alpine meadows helps enhance the possibility of long-term sustainability (Matthew, 2016). Policy implications for conserving the fragile ecosystem include ensuring the participation of local people in decision-making. These policies are essential for implementation among various environmental NGOs, governmental agencies, scientists, researchers, and local indigenous communities.

6.4.2 Proper and Scientific Management of Protected Areas

In the Himalayan alpine meadows, numerous flora and fauna communities are present in varying numbers. This grassland has a unique ecosystem. To protect them adequately, adopt some scientific management strategies, such as creating a buffer zone to protect vulnerable species in a high-value conservation zone, which helps reduce human impact. This also involves identifying and regulating endangered species, as well as researching and collecting information on species behaviour, biosphere functions, and environmental change. In scientific management includes adaptation management strategies for conserve species and reduce the challenges; involving the local communities to create nature base eco- tourism and grazing control; reduce the changes of carbon footprint by managing land degradation; targeting local communities to continue awareness programme; monitored climatic trends and mitigate its impacts on habitat restoration and ecosystem relocation.

6.4.3 Cross-Border Cooperation

The Himalayas comprise a single ecosystem shared by four significant countries: India, Nepal, Bhutan, and China. Himalayan Alpine Meadows are located in the regions of these four countries, where some endangered species face numerous environmental challenges and threats, including overgrazing, soil erosion, deforestation, glacier retreat, climate change, and habitat loss. Based on these challenges, cross-border cooperation is very crucial among the countries. These four countries must collaborate on research to maintain biodiversity and ecological balance and reduce the threat to ecosystems in alpine grasslands. These joint research and environmental monitoring help by sharing data about the species and the process of their conservation techniques.

Every country has its conservation laws and policies, but conflict issues can affect cross-border cooperation in habitat conservation, overgrazing regulation, deforestation management, and wildlife protection. However, it ensures that their collaborative conservation efforts align with all harmonized policies. Countries must share a monitoring platform to track changes in vegetation, the number of species, and habitats. This information enables a quick response to threats and conservation actions based on recent data. Cross-border cooperation among these countries is highly significant for providing benefits through biodiversity enhancement, livelihood improvement, and the application of better strategies, which creates a safeguard for the Himalayan alpine meadows and their ecosystems.

6.4.4 Climate-Resilient Framework, Policymaking, and Implementation

In Himalayan alpine meadows, various endemic flora and fauna species are present in threatened conditions due to temperature rise, changes in precipitation patterns, and the retreat of glaciers. To prevent these challenging conditions, it is essential to implement a climate resilience framework, ensuring that native species are not lost due to land degradation or deforestation. These uncertain changes in climatic conditions must be mitigated by implementing sustainable land use patterns, such as rotational grazing, to prevent habitat degradation due to overgrazing.

Some essential policies are needed to address all challenges that affect local biodiversity, encouraging local communities through the application of community participation policies. This will prevent unmonitored tourism and unregulated livelihoods from disrupting the ecological balance. Arranging various management programs based on sustainable resource utilization practices for local villagers is also crucial. These all are possible if there is a strong bond between government bodies and local communities. As a result, climatic impacts can be mitigated and addressed so that they do not affect the fragile local ecosystem.

6.5 *Scientific and Technological Innovations Drive R&D Approaches*

6.5.1 Use of Geospatial Technologies

Himalayan Alpine Meadows have a unique ecosystem. However, most of these species are highly endangered due to human activities. The application of some scientific and technological innovations can protect all these vulnerable species and also ensure sustainable management. The use of a Geographical Information System (GIS), Remote Sensing (RS), and a Global Positioning System (GPS) is vital for creating maps of the habitat locations of these species and for monitoring alpine meadows. These technologies are applied to detail the chances of land, the rate of deforestation, and how land degradation is related to habitat loss. The application of all these geospatial technologies provides data about temperature increases, soil erosion, and changes in precipitation patterns. Based on these data, GIS applications help create management strategy models to protect the future scenario of grassland areas.

6.5.2 Promote Scientific Studies on Soil Health Monitoring and Hydrological Restoration, as Well as the Management of HAM Ecosystems

The promotion of geospatial technologies provides data about soil health and hydrological functions in the Himalayan Alpine Meadows. Scientific research helps to identify the soil nutrient composition. Soil condition has been monitored for its importance on the growth of plant species. Hydrological restoration is crucial for the development of alpine meadows. From the Himalayas, water comes in the form of fresh water, which is used to produce drinking water for humans and wildlife. By applying scientific principles, water restoration can effectively control soil erosion and protect vegetation from degradation, ultimately stabilizing a fragile ecosystem.

6.5.3 Adequate Funding for R&D and Enhanced Collaboration Opportunities Between Government Agencies, Academic Institutions, NGOs, and Local People in HAM

In the Himalayan Alpine ecosystem, adequate funding is essential for protecting biodiversity, which is closely linked to local regional development. Financial support is necessary for research, monitoring, technological innovations, capacity building, cooperation, and the promotion of long-term development projects. Governmental Agencies must collaborate with the local community for learning and concern about the conservation of the local ecosystem in the Himalayan Alpine Meadows. The involvement of NGOs and policymakers can help manage biodiversity by providing traditional knowledge of sustainable livelihood practices from local communities. Some academic institutions with specialized expertise in resources implement such policies on a large scale and collaborate with local, state, or central government authorities to coordinate sustainable development. This collaboration helps ensure a practical approach to protecting the Meadows and creates a balance in economic growth.

6.5.4 Place Extra Emphasis on Traditional Knowledge Systems (TKS) for Informed Policy-Making and Effective Implementation

The Traditional Knowledge System (TKS) offers a sustainable approach to conserving natural resources in the Himalayan Alpine Meadows. Many indigenous communities have inhabited the Himalayas for centuries. They sustain and develop their traditional knowledge by practicing it, which is related to water retention, land use, and biodiversity. For conservation and application, it is essential to emphasize the Traditional Knowledge System (TKS) in these communities. To enable policy-making and research to incorporate better opportunities within the regional environment. To restore and manage the Himalayan Alpine Meadows requires more advanced scientific technology, research procedures, and geospatial innovations to promote soil composition and monitor hydrological restoration, ensuring the resilience of the Himalayan Alpine Meadows and mitigating its impacts due to extreme climate change.

7 Major Findings

HAM is a biological hotspot and a globally recognized area of interest. It is associated with a diverse colony of endemic and medicinal plant species, including Saussurea obvallata, Aconitum violaceum, and Dactylorhiza hatagirea. HAM plays a significant role in maintaining gene pools of rare and threatened species, many of which hold pharmaceutical and ecological importance. It is also associated with the endangered snow leopard (Panthera uncia), Himalayan musk deer (Moschus chrysogaster), and Himalayan monal (Lophophorus impejanus). HAM acts as a water regulator of the Himalayan landscape, functioning as natural sponges that absorb and gradually release glacial meltwater, thereby sustaining river flow during the dry season. The vegetation cover conserves soil and stabilizes regional slopes, reducing the risks of various natural calamities, such as landslides. Furthermore, HAM contributes to carbon sequestration, which mitigates the impacts of climate change by storing atmospheric carbon in the soil and vegetation. Himalayan alpine meadows are crucial for the livelihoods of pastoral communities, including the Bhotiyas, Gujjars, and Gaddis, who engage in seasonal grazing. HAM holds excellent cultural and spiritual significance, being linked with religious practices, festivals, and local folklore.

The meadows provide fodder resources during transhumance and serve as gathering grounds for medicinal herbs, which contribute to the region's traditional healthcare practices. Seasonal grass generates a substantial amount of organic matter annually, which enriches the soil of HAM (Adhikari et al. 2007, 2010). However, HAM is under threat from overgrazing, immense pressure from international and domestic tourists, climate change- induced temperature rises, the retreat of Himalayan glaciers and ice melting, as well as rapid species invasions. Overgrazing by livestock poses a significant threat to vegetation cover, soil erosion, and biodiversity loss. HAM is also a popular trekking route for adventurers. Trekking activities in the region lead to trampling, soil degradation, and waste accumulation, thereby disrupting fragile alpine ecosystems. Currently, invasive species such as Artemisia vulgaris and Polygonum polystachyum threaten the native flora, altering the biodiversity of meadows. Implementing sustainable management practices, such as rotational grazing and community-based conservation, is crucial to mitigate degradation in the HAM area. Moreover, promoting environmental awareness through the enhancement of ecotourism practices and controlled trekking along various trails should help reduce anthropogenic impact and environmental degradation. Long-term monitoring programs and ecological restoration initiatives are essential for safeguarding these fragile ecosystems.

8 Conclusion

HAM plays a significant role in maintaining ecological stability, regulating hydrological systems, and preserving biodiversity. HAM is threatened by climate change, overgrazing, uncontrolled tourism, and invasive species, which jeopardize its sustainability. Integrated conservation strategies must be adopted to get sustainability. Community participation, eco-friendly tourism practices, and scientific monitoring should be implemented to resolve the issues. Implementing climate-resilient management and supporting sustainable livelihoods for residents are crucial for protecting these vital ecosystems. Future conservation initiatives should concentrate on rehabilitating degraded meadows, safeguarding native species, and enhancing community resilience to ensure the long-term sustainability of Himalayan alpine meadows.

References

Adhikari, B. S., Babu, M. M., Saklani, P. L., & Rawat, G. S. (2007). Distribution, use pattern and prospects for conservation of medicinal shrubs in Uttaranchal State, India. *Journal of Mountain Science, 4*, 155–180.

Adhikari, B. S., Babu, M. M., Saklani, P. L., & Rawat, G. S. (2010). Medicinal plants diversity and their conservation status in Wildlife Institute of India (WII) campus, Dehradun. *Ethnobotanical Leaflets, 2010*(1), 6.

Ali, S., & Ripley, S. D. (1983). *Handbook of the birds of India and Pakistan.* Compact edition. Oxford University Press and BNHS, Mumbai. Ali, S., & Ripley, S. D. (1995). *The pictorial guide to the birds of Indian Sub-continent.* Oxford University Press and BNHS, Mumbai. Baskaran, S. T. (1992). Sighting of Dusky Horned Owl. *Newsletter for Birdwatchers, 32*(9), 10.

Baumgärtner, J., & Hartmann, J. (2001). The design and implementation of sustainable plant diversity conservation program for alpine meadows and pastures. *Journal of Agricultural and Environmental Ethics, 14*, 67–83.

Bency, D., Bency, D., Grießinger, J., Tomar, N., Singh, C., & Bräuning, A. (2022). Tree-ring oxygen isotope variations in subalpine firs from the Western Himalaya capture spring season temperature signals. *Forests, 13*(3), 437.

Billings, W. D. (1973). Arctic and alpine vegetations: Similarities, differences, and susceptibility to disturbance. *BioScience, 23*(12), 697–704.

Chauhan, N. S. (1999). *Medicinal and aromatic plants of Himachal Pradesh.* Indus Publishing.

Datta, A., Singh, P., Athreya, R. M., & Karthikeyan, S. (1998). Birds of Pakhui wildlife sanctuary in Western Arunachal Pradesh, North East India. *Newsletter for Birdwatchers, 38*(6), 91–96.

Dhar, U., & Kachroo, P. (1983). Alpine flora of Kashmir Himalaya. *Brittonia, 35*(4), 379.

Hamilton, A. C. (2004). Medicinal plants, conservation and livelihoods. *Biodiversity & Conservation, 13*, 1477–1517.

Jain, S. K., & Sinha, R. (2004). *Sustainable development and ecological conservation of the Himalayas.* Concept Publishing Company.

Kala, C. P. (1997). Ethnobotany and ethnopharmacology of Indian alpine plants. *International Journal of Sustainable Development & World Ecology, 4*(3), 193–202.

Kala, C. P. (2005). Indigenous uses, population density, and conservation of threatened medicinal plants in protected areas of the Indian Himalayas. *Conservation Biology, 19*(2), 368–378.

Kala, C. P., Rawat, G. S., & Uniyal, V. K. (1998). Ecology and conservation of the Valley of Flowers National Park, Garhwal Himalaya. *Mountain Research and Development, 18*(1), 51–61.

Kawosa, R. (1988). *Livestock grazing practices in the Himalayas: An ecological and socio-economic perspective.* Himalayan Studies Publications.

Khanal, N. R., Watanabe, T., & Kazama, F. (2012). Vegetation dynamics in the Himalayas: Patterns and processes. *Journal of Mountain Science, 9*(5), 677–689. https://doi.org/10.1007/s11629-012-2311-3.

Körner, C., & Larcher, W. (1988). Plant life in cold climates. *Symposia of the society for experimental biology, 42*, 25–57.

Lal, A. K., Sah, M. P., Sharma, B. P., & Pal, D. (1987). On the land use practices in Pauri Garhwal district with emphasis on socio-economic status and degradation of land and ecosystem. *Western Himalaya, Environment, Problem and Development, Gyanodaya Prakashan, Nainital.*

Mani, M. S. (1978). *Ecology and phytogeography of high altitude plants of the Northwest Himalaya* (p. 205).

Mark, A. F., & Bliss, L. C. (1970). The high-alpine vegetation of central Otago, New Zealand. *New Zealand Journal of Botany, 8*(4), 381–451.

Matthew, A. (2016). The implications of policy instability on engineering projects in the renewable energy sector. https://doi.org/10.14264/uql.2017.191.

Mishra, C., & Johnsingh, A. J. T. (1996). On habitat selection by the goral Nemorhaedus goral bedfordi (Bovidae, Artiodactyla). *Journal of Zoology, 240*(3), 573–580.

Naithani, B. D., Nandi, S. K., & Singh, S. P. (1992). Impact of livestock grazing on Alpine Meadows of the Western Himalayas. *Journal of Himalayan Ecology, 5*(2), 87–94.
Negi, G. C. S., Singh, S. P., & Rawat, Y. S. (1993). The influence of grazing on the structure and species composition of high-altitude grasslands in the Himalayas. *Journal of Range Management, 46*(4), 301–306.
Olsen, C. S., & Larsen, H. O. (2003). Alpine medicinal plant trade and its implications for conservation and livelihoods. *Mountain Research and Development, 23*(1), 10–19. https://doi.org/10.1659/0276-4741(2003)023[0010:AMPATI]2.0.CO;2
Rai, I. D., Bharti, R. R., Adhikari, B. S., & Rawat, G. S. (2003). Structure and functioning of timberline vegetation in the western Himalayas: A. *High-altitude rangelands and their interfaces in the Hindu Kush Himalayas, 91.*
Rau, M. A. (1975). *High-altitude flowering plants of the West Himalaya.* Howrah: Botanical Survey of India.
Rawal, R. S., Bhatt, I. D., Sekar, K. C., & Nandi, S. K. (2013). The Himalayan biodiversity: richness, representativeness, uniqueness and life-support values. *GB Pant Institute of Himalayan Environment and Development (GBPIHED), Kosi-Katarmal, Almora, Uttarakhand, India, 84pp.*
Rawat, G. S. (1997). Conservation status of forests and wildlife in the Eastern Ghats, India. *Environmental Conservation, 24*(4), 307–315.
Rawat, G. S. (1998). Temperate and alpine grasslands of the Himalayas: Ecology and conservation. *Parks, 8*(3), 27–36.
Rawat, G. S. (2007b). Pastoral practices, wild mammals, and conservation status of alpine meadows in the Western Himalayas. *Journal-Bombay Natural History Society, 104*(1), 5.
Rawat, G. S., & Rodgers, W. A. (1988). Ecological status and conservation of Himalayan alpine meadows in relation to grazing and wildlife. *Biological Conservation, 44*(1), 69–77. https://doi.org/10.1016/0006-3207(88)90051-5
Rawat, G. S., & Uniyal, V. K. (1993). Pastoralism and plant conservation: The Valley of Flowers Dilemma. *Environmental Conservation, 20*(2), 164–167.
Rawat, G. S. (1984). Studies on the high-altitude flowering plants of Kumaon Himalaya (p. 453). Ph.D. thesis.
Rawat, G. S. (2007). *Alpine vegetation of the western Himalaya: Species diversity, community structure, dynamics and aspects of conservation* (p. 239). Doctoral dissertation, D.Sc Thesis, Kumaun University, Nainital.
Rodgers, W. A., & Panwar, H. S. (1988). *Planning a wildlife-protected area network in India.*
Saberwal, V. K. (1996). Pastoral politics: Gaddi grazing, degradation, and biodiversity conservation in Himachal Pradesh, India. *Conservation Biology, 10*(3), 741–749.
Samant, S. S., Dhar, U., & Palni, L. M. S. (1998). Medicinal plants of Indian Himalaya: Diversity, distribution and potential values. *Nainital: Gyanodaya Prakashan.*
Sati, V. P., & Kumar, K. (2004). Tourism and regional development in the Indian Himalaya: A socio-economic and environmental perspective. *Journal of Mountain Science, 1*(1), 43–54.
Saxena, K. G., Rao, K. S., Sen, K. K., Maikhuri, R. K., & Semwal, R. L. (2002). Integrated natural resource management: approaches and lessons from the Himalaya. *Conservation Ecology, 5*(2).
Singh, S. K., & Rawat, G. S. (1999). Floral diversity and vegetation structure in the Great Himalayan National Park, Western Himalaya. *Wildlife Institute of India, 2*(5), 1–28.
Songh, J., Sharma, S., & Mehta, P. (1980). Distribution and diversity of grass species in the Himalayan region. *Journal of Himalayan Botany, 12*(3), 45–56.
Sundriyal, R. C. (1995). Agro-pastoralism in the Indian Himalaya: Livelihood patterns and environmental concerns. *Mountain Research and Development, 15*(2), 145–156.
Tsuchida, S., & Numata, M. (1983). Zone of floristic composition and species dominance. *Journal of Vegetation Science, 12*(2), 123–134.
Tucker, R. P. (1986). The evolution of transhumant grazing in the Punjab Himalayas. *Mountain Research and Development*, 17–28.
Uniyal, B. P., Singh, D. K., & Chowdhery, H. J. (1994). *Flora of Uttar Pradesh Himalaya: Volume 1.* Botanical Survey of India.

Ved, D. K., Kinhal, G. A., Ravikumar, K., Prabhakaran, V., Ghate, U., Shankar, R. V., & Indresha, J. H. (2003). *Conservation assessment and management prioritization for medicinal plants of Jammu and Kashmir, Himachal Pradesh, and Uttarakhand*. Foundation for Revitalization of Local Health Traditions (FRLHT).

Mapping Rill and Gully Erosion Hotspots in the Lish and Gish Catchment, Darjiling Himalaya Using Remote Sensing and GIS

Amit Kumar Jana and Gouri Sankar Bhunia

Abstract The land stability, agricultural productivity, and watershed management are at risk owing to soil erosion from rill and gully formation. In this study, we used geospatial tools to focus on Mapping Rill and Gully Erosion Hot spots in the Lish and Gish Catchment Using Remote Sensing and GIS, through an investigation of hydrology, relief, and land use to examine possible vulnerability to erosion. The hydrological elements we considered consisted of, rainfall intensity, drainage frequency, drainage density, and stream ordering; analyzed to describe the characteristics of water flow and contribution to erosion. The relief aspects we focused on were, the types of relative relief, dissection index, roughness index, and slope; analyzed to assess the landscape's vulnerability to rill and gully formation. The land use factors included soil type and vegetation characteristics; subsequently incorporated to relate land cover to erosion controls. To summarize the results, the impact of high rainfall intensity, steep slope, high drainage density, and low order streams were significantly contributing to erosion processes, particularly in the central and western parts of the catchment. Areas of gravelly loamy and loamy skeletal soils exhibited high proportions of erosion, although dense vegetation reduces the risk of erosion. The results have corroborated with previous studies, this study emphasizes the importance of sustainable watershed management, soil conservation, and afforestation programs as soil erosion controls on rill and gully. The present study illustrated the benefits of geospatial tools for monitoring, examining, and management of soil erosion and makes the case for proactively researching erosion dynamics as it relates to climatic change.

Keywords Rill erosion · Gully erosion · Remote sensing · GIS · Hydrological factors · Relief factors · Land use · Drainage density · Slope · Soil erosion hotspot · Sustainable watershed management

A. K. Jana (✉) · G. S. Bhunia
Department of Geography and Applied Geography, University of North Bengal, Darjiling, India
e-mail: amitjanah@gmail.com

G. S. Bhunia
e-mail: rsgis_gouri@rediff.com

© The Author(s), under exclusive license to Springer Nature Switzerland AG 2025
S. Ghosh (ed.), *The Himalaya Dilemma*, Springer Natural Hazards,
https://doi.org/10.1007/978-3-031-95083-4_27

1 Introduction

Rill and gully erosion processes are important land degradation issues impacting landscapes globally. These types of soil erosion are caused mainly by intense rain, incorrect land use, and lack of vegetation cover (Poesen et al., 2003). Globally, regions like the Loess Plateau in China, the Ethiopian Highlands, the Great Plains of North America, and parts of Europe and Australia have been greatly affected by rill and gully erosion (Morgan, 2009). In India, rill and gully erosion pose a serious threat, especially in the semi-arid and sub-humid regions. The Chambal Ravines in Madhya Pradesh, Rajasthan, and Uttar Pradesh are prime examples of landscapes impacted by gully erosion that have been studied intensively, where agriculture lands have been significantly degraded by deep and wide gully formation (Pandey et al., 2016). In fact, rill and gully erosion are among the most destructive types of soil erosion that cause major landscape degradation, reduced arable land, and sedimentation in water bodies. Rill erosion, is when shallow linear channels are formed by surface runoff and gully erosion is the next stage when deeper and wider channels make the land unusable (Sharda et al., 2013). We must identify and map areas that experience rill and gully erosion to inform land use, conservation planning, and mitigation strategies.

The Lish and Gish lip catchment, situated in a high-intensity rainfall and rugged terrain region, has experienced considerable land degradation from rill and gully erosion processes. Erosion processes responsible for these features occur from the concentration of surface runoff on the land surface until it displaces and, in some cases, removes topsoil. Erosion of the land surface produces deep incisional channels leading to new or existing formations of permanent networks of gullies (Government of India, 2020). As the area impacted by gullies continues to expand, the region will increasingly become susceptible to issues of agricultural productivity, management of water resources, and performance of infrastructural land stabilization.

Advancements in geospatial technology, including Remote Sensing (RS) and Geographic Information Systems (GIS), have provided efficient tools for identifying and mapping rill and gully erosion-prone areas. Remote sensing data, including optical and radar imagery, has been widely used for detecting erosion features. High-resolution satellite images (e.g., Sentinel-2, Landsat 8, and WorldView) enable precise identification of erosion-prone areas (Borrelli et al., 2014; Vrieling et al., 2007). The integration of high-resolution satellite imagery, digital elevation models (DEM), and hydrological modeling facilitates precise delineation and monitoring of erosion features. Digital terrain analysis using DEMs, such as Shuttle Radar Topography Mission (SRTM) and Advanced Spaceborne Thermal Emission and Reflection Radiometer (ASTER), provides critical insights into slope gradients, drainage patterns, and erosion potential (Hancock et al., 2008). GIS-based spatial modeling, including the use of topographic indices derived from DEMs, helps delineate potential erosion-prone zones (Martínez-Casasnovas et al., 2009). Recent advancements in artificial intelligence (AI) have improved erosion mapping accuracy. Supervised classification techniques, such as Random Forest and Support Vector Machines, have been used to identify erosion hotspots (Rahmati et al., 2017). Deep learning models, including Convolutional Neural Networks (CNNs), have shown promising results in automated detection of gully features from satellite imagery (Huang et al., 2020). Present study aims to review existing methodologies for mapping rill and gully erosion hotspots, analyze their effectiveness, and explore new approaches that leverage modern geospatial and computational tools.

2 Study Area

The Lish and Gish catchment is situated in West Bengal's Darjiling district, located in the Eastern Himalaya (Fig. 1). The terrain is mountainous and generally rugged, with elevations ranging from about 500 m to more than 2,000 m above sea level. The climate is categorized as a temperate subtropical highland climate featuring highly variable yearly rainfall due to the effects of the monsoon ranging from 2,000 mm or more rainfall/year. The geology is considered relatively young, where the various geological formations consist the metamorphic rocks of phyllites and schists, thus making the landscape prone to landslides during heavy rainfall associated with monsoon systems. Soils vary widely within the catchments highly based on elevation, slope steepness, vegetation cover and underlying geology. The Lish and Gish rivers are tributaries of the Teesta River which moves south and originates from a glacier in Sikkim where it drains the region. The catchment area is rich in biodiversity but tend to have forest dominated by *Sal* (*Shorea robusta*), oak, and many orchids. The faunal diversity includes several mammals such as civets, mongooses, and Himalayan black bears. In many instances, the ecological balance has been increasingly disrupted by anthropogenic factors, such as deforestation and habitat disruption produced by increasing development and agricultural activities.

3 Materials and Methods

The data collection procedure for determining susceptibility to rill and gully erosion in the Lish and Gish catchment area involved primary spatial datasets as well as secondary data sources. The combination of all the datasets allowed for analysis that was comprehensive and geospatially assured. The catchment was delineated using Survey of India (SOI) topographic sheets revealing essential hydrological and topographic information. In order to maintain a spatially distributed analysis, the whole study area was divided into 2 × 2 km grid cells. For each grid location, various hydrological factors and relief factors were calculated in a systematic manner. A Shuttle Radar Topography Mission (SRTM) Digital Elevation Model (DEM) was used to extract critical hydrological and relief parameters: elevation, slope, relative relief, drainage density, dissection index, and roughness index.

Information about the different soil types was obtained from the National Bureau of Soil Survey (NBSS), which produced a soil classification map of the region. This

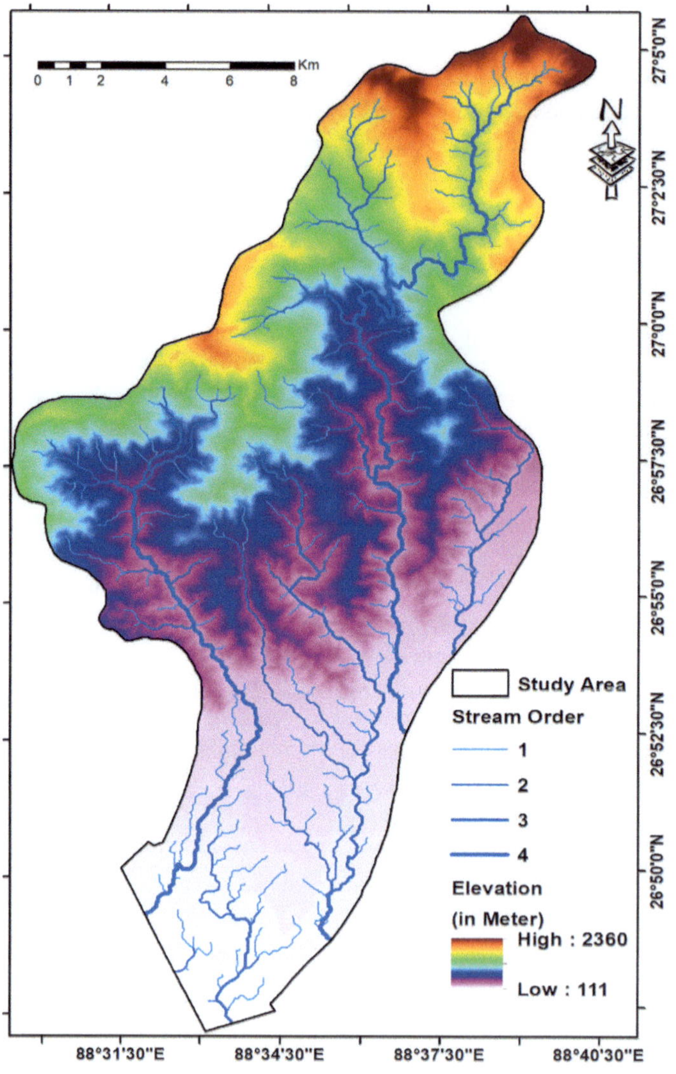

Fig. 1 Location map of the Lish and Gish catchment area

secondary data was important in determining soil erodibility, texture and composition, as all of these aspects are relevant to rill and gully formation. For vegetation analysis, data captured by the Landsat 8 Operational Land Imager (OLI) sensor was used to calculate the Normalized Difference Vegetation Index (NDVI). All spatial datasets (the SOI topographical sheets, terrain maps, derived from DEM, and soil maps) were geo-referenced to ensure spatial integrity and accuracy, and were projected in the Universal Transverse Mercator (UTM) coordinate system using WGS 84 datum.

3.1 Methods

Hydrological factors, i.e., rainfall, drainage frequency, drainage density, and stream ordering, are important in determining the extent of erosion susceptibility of rills and gullies. Each hydrological factor provides important information on the erosion potential of a catchment area. Rainfall is one mechanism relating to surface runoff, and drainage frequency and density are related to the extent of flow from surface water. These parameters are obtained using a broad array of geospatial and statistical techniques including often, hydrological modeling based on GIS.

3.1.1 Hydrological Factors

Drainage Frequency (Streams/km^2)

Drainage frequency (D_f) refers to the number of stream channels per unit area and is a measure of landscape dissection and runoff potential. It is calculated as:

$$D_f = \frac{N}{A}$$

where

N—Total number of stream segments

A—Basin area (km^2).

Higher drainage frequency indicates greater surface runoff and increased erosion potential.

Drainage Density (km/km^2)

Drainage density (D_d) is a key indicator of surface runoff characteristics and the degree of landscape development. It is calculated as:

$$D_d = \frac{L}{A}$$

where,

L—Total length of all streams (km)

A—Basin area (km^2).

A high drainage density implies impermeable surfaces, steep slopes, and a high potential for rill and gully erosion.

Stream Ordering (Strahler's Method)

Stream ordering classifies river networks based on hierarchy. The Strahler method assigns orders to streams as follows:

- First-order streams: Smallest streams with no tributaries.
- Second-order streams: Formed by the junction of two first-order streams.
- Higher-order streams: Formed by the convergence of two streams of the same order.

The equation for stream order is:

$$O = \max(O_1, O_2) + 1$$

where streams of the same order combine to form the next order. Higher stream order signifies a growing hydrological complexity and, therefore, a greater water transport capacity. It systematically estimates rill-gully erosion area spatial influence on proximity to the catchment area applying a multi ring buffer zone.

3.1.2 Relief Factors

Factors such as Relative Relief, Dissection Index, Roughness Index, and Slope are considered relief relation properties that play an important role in determining whether the area is susceptible to rill and gully erosion. These factors affect the surface runoff, soil detachment, and drainage patterns.

Relative Relief

The variation between the highest and lowest elevations in a certain area is known as relative relief. Understanding the roughness and erosion potential of the terrain requires this crucial characteristic.

$$RR = H_{\max} - H_{\min}$$

where:

H_{max} = Maximum elevation (m)

H_{min} = Minimum elevation (m).

More gravitational force acting on soil particles, steeper slopes, and a greater propensity for gully and rill development are all indicators of higher relative relief.

Dissection Index

The degree of terrain dissection brought on by erosion is measured by the Dissection Index (DI). It is a ratio of absolute relief to relative relief that sheds light on the processes of denudation and landscape maturity.

$$DI = \frac{RR}{H_{\max}}$$

where,

RR = Relative relief

H_{max} = Maximum elevation (m).

Values range from 0 (flat terrain) to 1 (highly dissected mountainous regions). A higher DI suggests greater terrain fragmentation and susceptibility to gully formation.

Roughness Index

Roughness Index quantifies surface irregularities and helps assess the difficulty of water flow across a landscape. It is computed as:

$$RI = \frac{Z_{sd}}{Z_{mean}}$$

where

Z_{sd} = Standard deviation of elevation values within a given grid

Z_{mean} = Mean elevation of the grid.

A higher roughness index indicates greater undulations in the terrain, which enhances water turbulence and soil detachment, increasing erosion risk.

Slope

A key component in determining the velocity of water runoff is slope, which is the rate at which elevation changes. GIS slope methods based on DEM data are used to calculate it:

$$S = \tan^{-1}\left(\frac{\Delta H}{D}\right)x\left(\frac{180}{\pi}\right)$$

where,

ΔH = Elevation difference between two points (m)

D = Horizontal distance between points (m).

Higher slopes lead to faster water runoff and increased erosion potential, especially in loosely compacted soils.

3.1.3 Land Use Factors

NDVI

NDVI is a remote sensing-based index that measures vegetation health and density, which directly affects soil erosion susceptibility. Higher vegetation cover reduces erosion by stabilizing soil, while low NDVI values indicate barren or degraded land with a higher risk of rill and gully formation.

$$NDVI = \frac{NIR - Red}{NIR + Red}$$

where:

- NIR (Near-Infrared Band): Reflectance in the near-infrared spectrum (healthy vegetation reflects more in this band)
- RED (Red Band): Reflectance in the red spectrum (healthy vegetation absorbs more in this band).

A lower NDVI value (<0.2) indicates exposed surfaces prone to erosion, whereas higher NDVI values (>0.5) signify dense vegetation, which reduces runoff and erosion.

Soil Type

The soil map was used in the Lish and Gish catchment area obtained from NBSS which provides soil classification and distribution data across the country. The soil map was geo-referenced and registered in the UTM projection system WGS 84 datum reference system to be consistent with other spatial datasets. Then, a digital soil map was prepared based on heads-up digitizing, which is the method of drawing out the soil borders and related features over a scanned map by using the GIS program. After digitizing, the attribute information of the soil map was organized and stored in a geospatial database.

4 Results

4.1 Hydrological Factor

The spatial distribution of rainfall intensity, drainage frequency, drainage density, and stream ordering directly influence the runoff generation, soil erosion potential, and water movement in the catchment. The areal extent and percentage coverage of each hydrological sub-parameter provide insights into the erosion-prone regions, helping to assess watershed stability and potential intervention areas.

4.1.1 Rainfall

Rainfall is one of the main factors of violation of rill and gully formation, because the intensity of precipitation causes an increase of surface air flow and soil detachment. The large area of high to extreme rainfall (~75.9% of the area) implies that intensity of precipitation is a dominating factor controlling the susceptibility of any area to erosion. 49.36 km^2 49.031 (17.05%), where the generation of a moderate output occurs (Fig. 2), High rainfall (3601–4000 mm) area is 78.84 km^2 (27.23%), having a

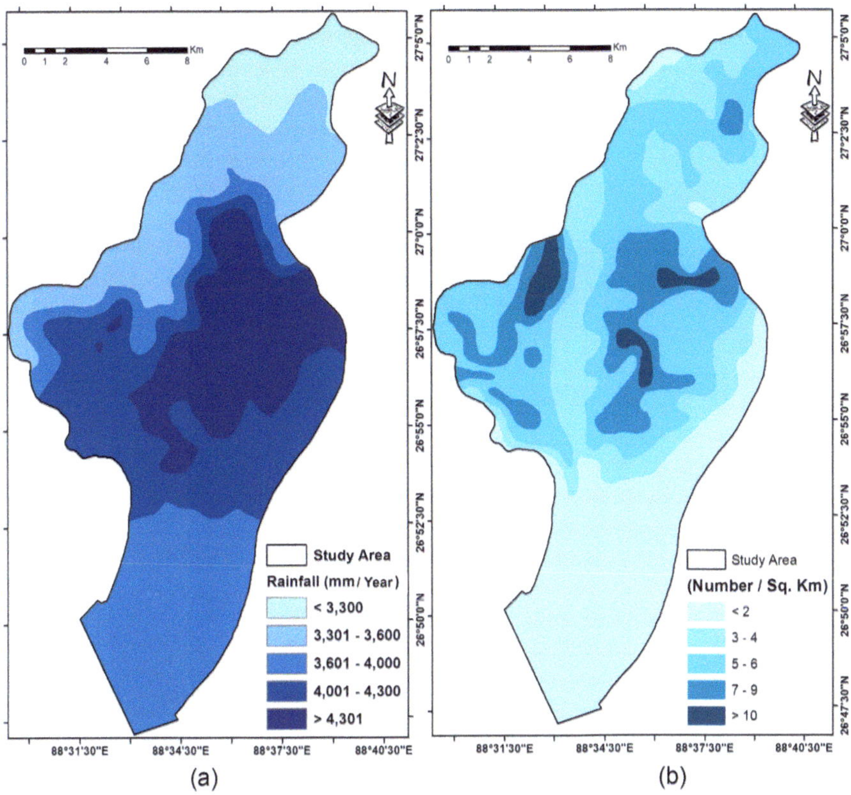

Fig. 2 Geographical variation of **a** rainfall and **b** drainage frequency of the Lish and Gish catchment area

notable impact on hydrological response of the catchment. It accounts for 77.52 sq. as the area with very high rainfall (4001–4300 mm) km (26.78%), promoting rill and gully formation (Table 1). High rainfall intensity is a typical feature in the middle of the catchment. On the contrary, the northern portion of the catchment is characterized by lower rainfall density, lower runoff, and a relatively stable hydrological regime, meaning a lower risk of erosion. The southern domain of the study area shows moderate rainfall density, which provides an average to balanced water potential for vegetation growth and runoff process with moderate erosion capacity.

4.1.2 Drainage Frequency

Drainage frequency, in the context of the number of streams per square kilometer, measures the quantity of surface water channels and their ability to transport the running-off water. A greater frequency of streams is often the cause of a higher erosion potential because of the runoff being concentrated in specific areas. The

Table 1 Areal characteristics of hydrological factors of Lish and Gish catchment area

Hydrological factor	Sub-parameters	Area (in km^2)	Percent
Rainfall (mm)	<3300 (Low)	20.42	7.05
	3301–3600 (Moderate)	49.36	17.05
	3601–4000 (High)	78.84	27.23
	4001–4300 (Very high)	77.52	26.78
	>4301 (Extreme)	63.38	21.89
Drainage frequency (streams/km^2)	<2 (Very Low)	85.32	29.47
	3–4 (Low)	63.53	21.94
	4–5 (Moderate)	26.05	9.00
	6–7 (High)	67.27	23.23
	8–9 (Very high)	39.16	13.53
	>10 (Extreme)	8.19	2.83
Drainage density (km/km^2)	<1.0 (Low)	98.41	33.99
	1.1–2.0 (Moderate)	40.22	13.89
	2.1–3.0 (High)	66.52	22.98
	3.1–4.0 (Very high)	55.30	19.10
	>4.1 (Extreme)	29.07	10.04
Stream ordering	4th Order (Low erosion/stable)	146.13	50.47
	3rd Order (Moderate erosion)	94.36	32.59
	2nd Order (High erosion)	46.03	15.90
	1st Order (Very high erosion)	3.00	1.04

high and extra high drainage frequency categories include about 39.6% of the area, thus pointing to the zones where runoff could be intensified leading to rill and gully development. Very low drainage frequency (<2 streams/km^2) is present in 85.32 km^2 (29.47%), thus indicating relatively non-erosive areas with low surface flow concentration. Low drainage frequency (3–4 streams/km^2) is found in 63.53 km^2 (21.94%), hence showing the regions with moderate stability. Furthermore, moderate drainage frequency (4–5 streams/km^2) covers 26.05 km^2 (9.00%), so it means that the area is prone to some level of erosion. The area with the highest drainage frequency of 6–7 streams/km^2 and equivalent to 67.27 km^2 (23.23%) has the largest volume of water and can thus be an area where the processes of erosion are said to be vigorous. While most runoff is expected to occur in the area with very high drainage frequency (8–9 streams/km^2) which extends over 39.16 km^2 (13.53%) and the area will have the highest erosion risk.

The areas in the drain region show high drainage frequency (small patches of central part of the catchment), signifying the dense network of streams for quick runoff and concentrated flow that are the main drivers of soil erosion and gully formation. On the other hand, the northern portion of the catchment displays a medium drainage frequency, indicating a moderately developed stream network that balances water delivery without intensifying the erosive capacity of hillslope processes. The southern section of the catchment, on the other hand, is classified as having a weaker drainage density, where directional infiltration from a low amount of stream channels leads to slower run-off and a stable hydrological period.

4.1.3 Drainage Density

Drainage density (Fig. 2), which is the total length of all streams in a given area to the area, is a measure of land dissection and potential runoff. More than 50% range of the area is the high to extreme drainage density which shows that the land surface is highly dissection and a significant part of water movement is present leading to erosion. Moderate drainage density (1.0–2.0 km/km^2) accounts for 13.78 km^2 (33.99%), suggesting relatively stable areas with little dissection. Moderate drainage density (1.1–2.0 km/km^2) occupies 40.22 km^2 (13.89%), representing moderate dissection. High drainage density (2.1–3.0 km/km^2) accounts for 66.52 km^2 (22.98%), where erosion processes are active. Very high drainage density (3.1–4.0 km/km^2) extends over 55.30 km^2 (19.10%), suggesting highly dissected terrain.

For the Lish and Gish catchment area, the spatial arrangement of drainage density (Fig. 3) is uneven as this factors into understanding hydrological activity and erosion in the area. The core of the study area captures a zone with high drainage density, which means the stream channel patterns are greatly condensed, which might pose a problem due to high soil erosion or channel digging Deep streambeds soil scouring. This is in contrast to the moderate drainage density in the northern part of the catchment basin, where stream channels are further apart. In this region, runoff is slower and erosive forces are significantly less than the central region. The southern part of the catchment has an extremely low drainage density which indicates fewer streams is present. This suggests that water movement is more diffuse and the rate of surface erosion is much lower.

4.1.4 Stream Ordering

Stream ordering categorizes streams by their hierarchical structure, where streams of higher order are more stable and those of lower order are more susceptible to erosion. The widespread occurrence of the lower order streams (1st and 2nd Order) suggests the presence of actively eroding landscapes—particularly in region of high relief and steep slopes—since in approximately 16.9% of the catchment area was observed to have 1st and 2nd Order streams. 4th Order streams (Low Erosion/Stable zones) are dominant in 146.13 km^2 (50.47% of the area) which indicates mature, well developed drainage networks are present and relatively stable conditions exist. 3rd Order streams (Moderate Erosion zones) are present in 94.36 km^2 (32.59% of the area) which experience some amount of soil erosion and channel instability. 2nd Order streams (High Erosion zones) cover 46.03 km^2 (15.90%) which indicates areas of moderate

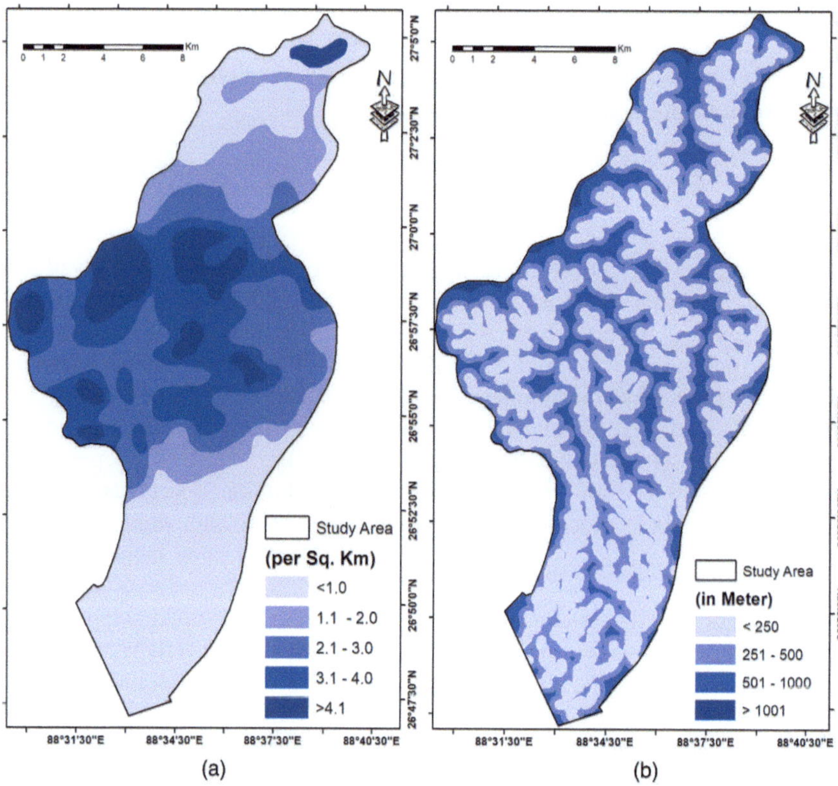

Fig. 3 Geographical variation of **c** drainage density and **d** stream ordering of the Lish and Gish catchment area

to severe instability and actively eroding territory. 1st Order streams (Very High Erosion zones) are limited to 3.00 km^2 (1.04%) where extreme vulnerability to gully erosion is present due to loose soil and weak channel structure.

4.1.5 Relief Factors

The topographical features of a catchment area have a great impact on water runoff, soil erosion, and hydrological response. The Lish and Gish catchment display features of physiographic division and corresponding relief—such as relative relief, dissection index, roughness index, slope change, and relief step—which determine soil erosion and land degradation. The categorization of these factors is helpful in describing the processes of relief formation, the degree of rugged topography, and the degree of erosion vulnerability for efficient utilisation in watershed management and conservation methods.

4.1.6 Relative Relief

Relative relief (Table 2), which measures the difference in elevation between the highest and lowest points within a specific area, is a critical factor in runoff velocity and erosional processes. High to extreme relief existing in around 60 percent of the area suggests a significant portion of the catchment is prone to eroding rapidly, along with rapid water drainage, highlighting the need for stabilization in the region, such as vegetative cover and soil conservation structures. Low relief (<100 m) accounts for 78.26 km^2 (27.03%) which represents slightly declining stable surfaces with gentle contours which are free from surface flow and erosion. Moderate relief (101–200 m) occurs over 37.49 km^2 (12.95%) which has mild slopes that result in moderate surface runoff accompanied by moderate erosion. High relief (201–300 m) accounts for 87.01 km^2 (30.05%) which extends to steeper slopes with increased surface flow, thus higher potential for erosion. Very high relief (301-400 m) spans 58.08 km^2 (20.06%) which depicts areas of marked elevation change with high concentration of surface flow. The lower part or the southern section of the catchment consists of a wide area of low relative relief (<100 m) indicating these regions are characterized by minimal elevation differences leading to gentle slopes with reduced surface runoff. In contrast, the northern region exhibits a medium relative relief, suggesting moderate elevation variations that contribute to a balanced hydrological regime with intermediate runoff and erosion potential. Meanwhile, the central and western parts of the catchment display high relative relief values, reflecting pronounced elevation differences and steeper slopes.

4.1.7 Dissection Index

The dissection index (Fig. 4) is a measure representing the extent terrain is fragmented by vertically dissecting the terrain through fluvial erosion. The moderate to high dissection index in almost 45% of the area suggests that the greater portion of the catchment is being sculpted through fluvial erosion, deepening channels and gullies. Low dissection index (<3.0) covers 159.99 km^2 (55.26%) which shows that these landscapes are relatively stable with little dissection of landforms. Moderate dissection index (3.1–5.0) occupies 115.55 km^2 (39.91%), indicating regions where streams and rivers have partially dissected the landscape. High dissection index (>5.1) covers 13.98 km^2 (4.83%) marking areas of severe fragmentation of land surface and active erosional processes.

The northern and southern section's exhibit lower DI values, suggesting these regions have relatively undissected, smoother landforms with fewer deep channels or incised valleys. In contrast, the central and western parts of the catchment display medium DI values, indicating a moderate level of dissection where fluvial processes have begun to carve the landscape, creating a balance between stable and actively eroding terrains. Notably, within the central part of the catchment, small pockets exhibit high DI values, pointing to localized zones of intense erosion where the landscape has been significantly incised, resulting in a highly dissected terrain.

Table 2 Areal distribution of relief factors of Lish and Gish catchment area

Parameters	Sub-categories	Area (in km^2)	Percent
Relative relief (in Mt)	<100	78.26	27.03
	101–200	37.49	12.95
	201–300	87.01	30.05
	301–400	58.08	20.06
	>401	28.69	9.91
Dissection index	<3.00	159.99	55.26
	3.10–5.0	115.55	39.91
	>5.10	13.98	4.83
Roughness index	<20	83.76	28.93
	21–40	24.77	8.56
	41–60	157.69	54.46
	>61	23.30	8.05
Slope (in degree)	<3.0	74.28	25.65
	3.1–5.0	11.53	3.98
	5.1–10.0	9.57	3.31
	10.1–15.0	27.70	9.57
	15.1–20.0	90.08	31.11
	20.1–25.0	63.64	21.98
	>25.1	12.72	4.39

4.1.8 Roughness Index

In order to evaluate surface irregularity and hydrological flow resistance, the roughness index calculates the elevation variability of the terrain. More than 60% of the region has high to extreme roughness, which indicates that significant areas of the catchment are vulnerable to soil separation and swiftly flowing runoff, requiring erosion control measures. Low roughness (<20) covers 83.76 km^2 (28.93%), indicating comparatively stable and smooth surfaces with uniform runoff distribution. Moderate roughness (21–40) indicates somewhat undulating terrain with moderate surface abnormalities, covering 24.77 km^2 (8.56%). High roughness (41–60) predominates in 157.69 km^2 (54.46%), emphasizing regions with rough and uneven terrain where surface runoff is concentrated, hence raising the risk of erosion. The area covered by extreme roughness (>61) is 23.30 km^2 (8.05%), indicating extremely difficult terrain with significant relief fluctuations that raise the risk of landslides and rapid water drainage.

The Roughness Index (RI) varies by zone in the Lish and Gish catchment basin. Lower RI values in the southern region indicate smoother terrain with less noticeable surface flaws. The north and center, on the other hand, have medium RI values, indicating a moderate degree of surface undulation that has a moderate impact on

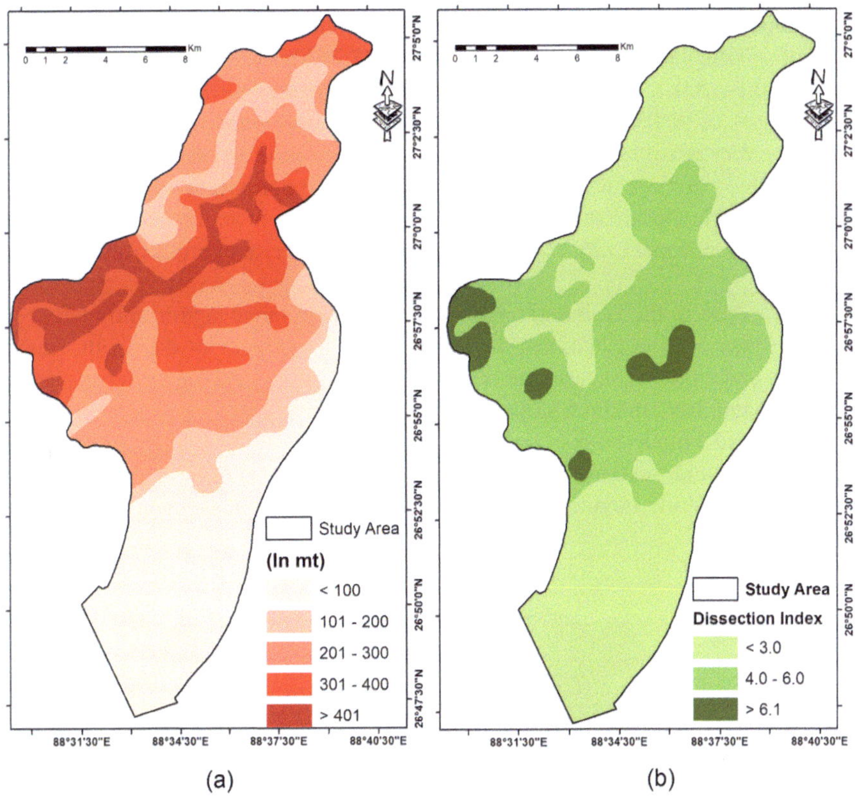

Fig. 4 Spatial variation of **a** relative relief and **b** Dissection index of the Lish and Gish catchment area

erosion and water flow. Furthermore, there are a few isolated sites in the center region where the RI is high, indicating regions with noticeable surface roughness and rocky topography that may give rise to localized erosion hotspots.

4.1.9 Slope

Steeper slopes stimulate quicker surface water movement and higher rates of erosion, making them important factors in determining runoff velocity, soil stability, and erosion potential. The necessity of land stabilizing measures like terracing, afforestation, and check dams to reduce the risk of erosion is highlighted by the fact that over 50% of the watershed has slopes higher than 15°. The 74.28 km^2 (25.65%) with gentle slopes (<3.0°) are stable areas with low erosion hazards, strong water infiltration, and little runoff (Fig. 5). 11.53 km^2 (3.98%) are covered by moderate slopes (3.1–5.0°), with moderate water flow and little erosion. 9.57 km^2 (3.31%) of moderately steep slopes (5.1–10.0°) are where surface runoff starts to increase, raising the possibility of soil displacement. With a total area of 27.70 km^2 (9.57%) covered by steep slopes (10.1–15.0°), surface erosion susceptibility and runoff speed are elevated. The majority of 90.08 km^2 (31.11%) is made up of very steep slopes (15.1–20.0°), which are very erosive and cause gully formation and rapid water movement. 63.64 km^2 (21.98%) of extremely steep slopes (20.1–25.0°) necessitate vigorous conservation efforts due to the severely degraded soil stability. Over 12.72 km^2 (4.39%), there are severe slopes (>25.1°), indicating areas at high risk of erosion and possible landslides.

Slope values in the Lish and Gish catchment basin range significantly between areas. Because of the catchment's comparatively low slopes in the south, runoff velocities and erosion concerns are often lower. The central and northern regions, on the other hand, have medium slopes, with moderate gradients that support balanced water flow and a modest risk of soil erosion. High slopes also characterize the catchment's middle and western regions, increasing the likelihood of erosion and causing rapid surface runoff. Targeted soil conservation and land management methods are

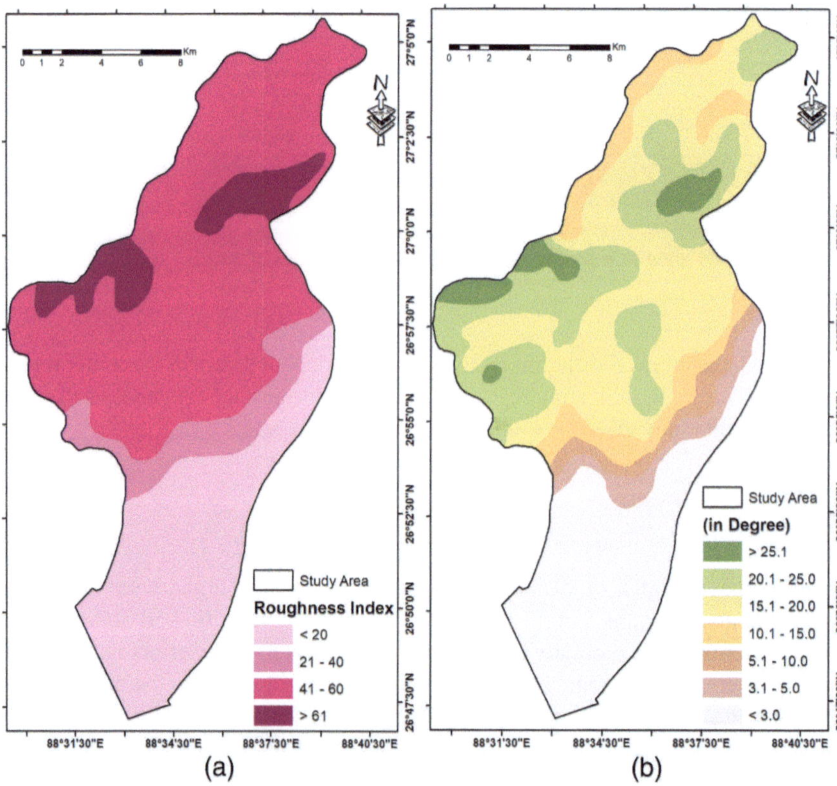

Fig. 5 Spatial variation of **c** roughness index and **d** average slope (in degree) in the Lish and Gish catchment area

informed by this spatial variation in slope, which is essential for comprehending the hydrological dynamics and erosion processes within the catchment.

4.2 Land Use Factor

Soil type and vegetation characteristics (NDVI values) are the main land use elements that are examined; both are important in determining the ecological balance and erosion dynamics of the watershed. Gravelly loamy to loamy skeletal soils make up the majority of the soil composition of the Lish and Gish catchment area, which spans 96.65 km^2 (33.38%) of the total area (Table 3). These soils, which are generally found in hilly areas, have moderate to high permeability, which lowers the likelihood of erosion and surface runoff. Coarse loamy soils (75.11 km^2, 25.94%) are the second most common type of soil; their loose structure makes them susceptible to detachment during periods of heavy rainfall. This transitional composition, which balances erosion resistance and water retention ability, is represented by the gravelly loamy to coarse loamy soil type (79.10 km^2, 27.32%). There are only a few pockets of fine loamy to coarse loamy soils (8.25 km^2, 2.85%) and gravelly loamy soils (8.01 km^2, 2.77%), usually in places where soil erosion is more noticeable because of steep slopes and little vegetation cover.

The distribution of soil types in the Lish and Gish catchment area shows distinct spatial variation. The coarse loamy soil that predominates in the southern portion of the watershed usually provides moderate drainage and stability. The soils in the central and northern regions are primarily categorized as gravelly loamy to loamy

Table 3 Areal characteristics of landuse factors of Lish and Gish catchment area

Landuse factors	Sub-parameters	Area (in km^2)	Percent
Soil type	Gravelly Loamy to Loamy Skeletal	96.65	33.38
	Coarse Loamy	75.11	25.94
	Gravelly Loamy to Coarse Loamy	79.10	27.32
	Gravelly Loamy	8.01	2.77
	Fine Loamy to Coarse Loamy	8.25	2.85
	River	22.40	7.74
Vegetation characteristics (NDVI value)	<0.0	14.47	5.00
	0.01–0.20	33.92	11.71
	0.21–0.35	38.06	13.14
	0.36–0.50	110.55	38.18
	>0.51	92.53	31.96

skeletal, which reflects a balance between structural support and permeability. Indicative of transitional soil textures that might affect erosion and water retention, the western and northern regions also frequently feature a mixture of gravelly loamy to coarse loamy soils. Additionally, fine loamy to coarse loamy soils, which may be more prone to erosion in specific circumstances, are found in small pockets within the watershed. The significance of region-specific land management and conservation measures is highlighted by the regional variation in soil types.

In order to stabilize soil, decrease runoff, and improve infiltration, vegetation is essential. Five levels of vegetation cover are distinguished by the Normalized Difference Vegetation Index (NDVI), which represents the density and condition of the catchment's plant life. 14.47 km^2 (5.00%) of bare land (<0.0 NDVI) are places of extreme degradation, frequently linked to anthropogenic disturbances, landslides, or active erosion. 33.92 km^2 (11.71%) of land is sparsely vegetated (0.01–0.20 NDVI), where soil erosion is probably caused by less plant cover and more surface runoff. Areas with mixed land use, agriculture, and regenerating forests make up 38.06 km^2 (13.14%) of moderately vegetated land (0.21–0.35 NDVI), which offers moderate soil stability. By absorbing rainwater and promoting soil cohesiveness, dense vegetation (0.36–0.50 NDVI), which covers 110.55 km^2 (38.18%), significantly lowers the risk of erosion. 92.53 km^2 (31.96%) are covered by very dense vegetation (>0.51 NDVI), mostly in wooded areas where canopy cover efficiently reduces erosion, controls water balance, and improves soil fertility.

Different patterns for vegetation density can be seen in the Normalized Difference plant Index (NDVI) distribution in the Lish and Gish catchment region. Lower NDVI values in the catchment's lower and southern regions suggest less vegetation and possibly degraded land cover (Fig. 6). On the other hand, the catchment's eastern sector and a few isolated areas in the north have medium NDVI values, which indicate a moderate density of vegetation. High NDVI values, on the other hand, indicate extensive and healthy plant cover in the catchment's middle and western regions, which probably contributes significantly to stabilizing the terrain and lowering soil erosion.

4.3 Rill-Gully Susceptibility Region

The Weighted Overlay Analysis (WOA), a GIS-based multi-criteria decision-making (MCDM) technique, has been used to evaluate the Lish and Gish catchment area's vulnerability to rill and gully erosion. Several predisposing factors are combined in this method, and each is given a rank and weight according to how much of an impact it has on erosion processes. Climate, hydrology, topography, and land-use parameters are among the predisposing elements that have a major influence on the extent and spatial distribution of erosion. These variables are combined into a single erosion susceptibility index by the weighted overlay model, which enables a methodical assessment of erosion-prone areas.

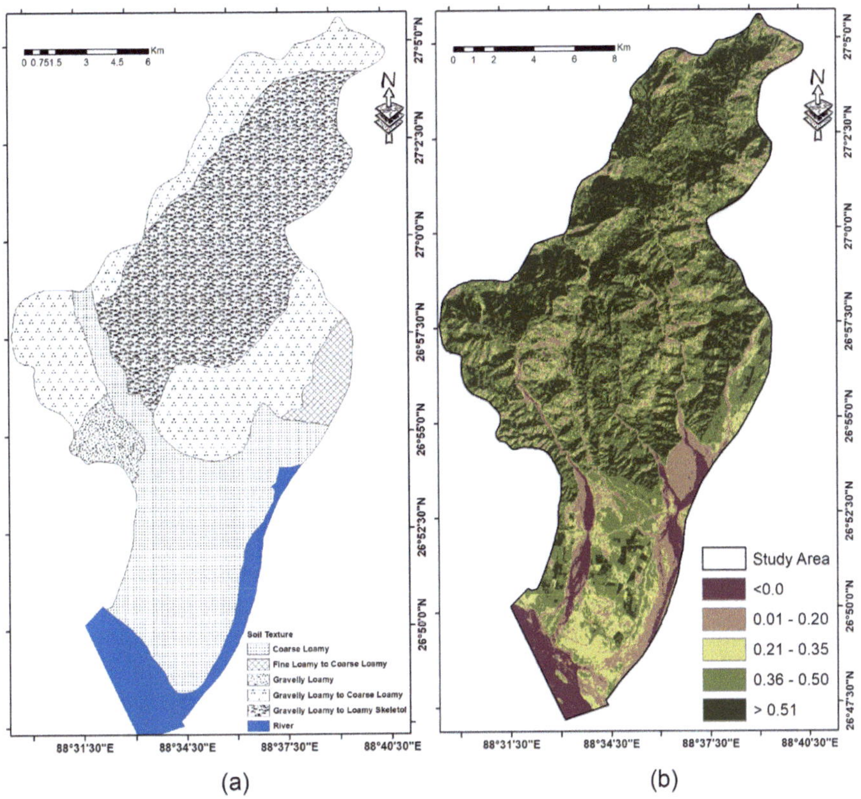

Fig. 6 Geographical variation of **a** soil characteristics and **b** vegetation cover index of the Lish and Gish catchment area

The main cause of soil erosion is rainfall because heavy rains cause surface runoff, separate soil particles, and cause rill formation. Rainfall threshold categories range from low (<3300 mm/year, Rank 1) to extreme (>4301 mm/year, Rank 5). Higher rainfall areas are more vulnerable to gully and rill erosion because of stronger runoff and infiltration restrictions. Drainage frequency is a measure of the density of stream networks and their capacity to move eroded sediments. It is expressed in streams per square kilometer. While a low drainage frequency (<2 streams/km^2, Rank 1) denotes a stable terrain with little vulnerability to erosion, a higher drainage frequency (>10 streams/km^2, Rank 5) indicates a rapidly eroding landscape with more surface runoff and channel incision. The amount of surface runoff concentration is reflected in drainage density, which is the total length of streams per unit area. Low drainage density (<1.0 km/km^2, Rank 1) areas have less channelization, which lowers the danger of erosion. On the other hand, rapid runoff in high-density zones (>4.1 km/km^2, Rank 5) promotes the growth of gullies. One essential geomorphological factor that shows how streams are categorized hierarchically is stream order. While lower-order streams (1st order, Rank 4) have steeper slopes and more sediment transport potential, which encourages the formation of gullies, higher-order streams (4th order, Rank 1) are more stable and less erosive.

Erosion intensity and runoff velocity are strongly influenced by relative relief, which is the vertical elevation variations within a unit region. Low relative relief (<100 m, Rank 1) areas have little erosion, while high relief (>401 m, Rank 5) has more gravitational forces, which causes soil separation and gully deepening. The dissection index measures the extent to which erosional processes have dissected landforms and is a measure of terrain fragmentation. High values (>6.1, Rank 4) indicate intense landscape erosion, which causes gully headward extension, whereas low values (<3, Rank 2) indicate stable landforms. Erosion susceptibility and surface irregularity are reflected in the roughness index. In undulating landscapes, severely uneven terrains (>61, Rank 4) disturb flow patterns and increase erosion, whereas smooth surfaces (<20, Rank 1) avoid excessive runoff. A key factor in soil erosion is slope gradient, which affects sediment movement and runoff velocity. Significant rill and gully expansion is caused by high-energy flows generated by steep slopes (>20%, Rank 5), while flat areas (<5%, Rank 1) permit more infiltration and less erosion.

Permeability, structure, and texture all affect how easily soil erodes. While gravelly loamy soils (Rank 1) provide superior resistance to erosion, coarse loamy soils (Rank 5) are more prone to rill and gully formation. Because of their low cohesiveness and quick detachment when affected by rainfall, soils with a high silt and fine loam content are especially vulnerable. Because it improves infiltration and captures rainwater, vegetation is essential for stabilizing soil and lowering the risk of erosion. Five categories of land cover have been identified by the Normalized Difference Vegetation Index (NDVI): bare land (0–0.2, Rank 4), sparse vegetation (0.21–0.35, Rank 3), moderate cover (0.36–0.50, Rank 2), dense vegetation (>0.51, Rank 1), and degraded land (<0, Rank 5). Due to their low root binding capacity and direct soil exposure to runoff forces, degraded and bare soils are particularly vulnerable to erosion (Table 4).

Five categories have been established to classify the vulnerability of various regions within the catchment area (Table 5 and Fig. 7). These categories are based on land use, soil type, hydrological factors, and topography characteristics. The weighted overlay analysis (WOA) and weighted sum model (WSM) based on GIS have been used to study the geographical distribution of these erosion-prone zones, incorporating important variables like rainfall, slope, relative relief, NDVI, soil type, drainage density, and dissection index.

The Very Low Susceptibility Zone (58.60 km^2, 20.24%) denotes regions where gully and rill formation is not as likely to occur. These areas are generally found in stable soil areas with low drainage density, low slope, and dense plant cover. Significantly less runoff and soil detachment occurs when there is thick forest cover and well-structured soil profiles.

Although the landscape in the Low Susceptibility Zone (56.73 km^2, 19.59%) is somewhat undulating, erosion is managed by stable soil types and a modest vegetation density. However, limited erosion activity is caused by moderate drainage frequency and localized slope fluctuations. In these places, rill initiation can be further

Table 4 Weighted overlay analysis of predisposing factor for Rill and gully erosion susceptibility in Lish and Gish catchment area

Factor	Category	Rank (1 = Low, 5 = High)	Weight (%)
Rainfall (mm/yr)	<3300 (Low)	1	15
	3301–3600 (Moderate)	2	
	3601–4000 (High)	3	
	4001–4300 (Very high)	4	
	>4301 (Extreme)	5	
Drainage frequency (No. of streams/km^2)	<2 (Very low)	1	8
	3–4 (Low)	2	
	4–5 (Moderate)	3	
	6–7 (High)	4	
	8–9 (Very high)		
	>10 (Extreme)	5	
Drainage density (km/km^2)	<1.0 (Low)	1	7
	1.1–2.0 (Moderate)	2	
	2.1–3.0 (High)	3	
	3.1–4.0 (Very high)	4	
	>4.1 (Extreme)	5	
Stream ordering	4th Order (Low erosion/stable)	1	5
	3rd Order (Moderate erosion)	2	
	2nd Order (High erosion)	3	
	1st Order (Very high erosion)	4	
Relative relief (m)	<100 (Low)	1	10
	101–200 (Moderate)	2	
	201–300 (High)	3	
	301–400 (Very high)	4	
	>401 (Extreme)	5	
Dissection index	<3 (Very low)	2	8
	4.0–6.0 (Moderate)	3	
	>6.1 (High)	4	
Roughness index	<20 (Smooth)	1	6
	21–40 (Moderate)	2	
	41–60 (Irregular)	3	
	>61 (Very irregular)	4	
Average slope (%)	<5 (Flat)	1	12
	5.1–10 (Gentle)	2	
	10.1–15 (Moderate)	3	

(continued)

Table 4 (continued)

Factor	Category	Rank (1 = Low, 5 = High)	Weight (%)
	15.1–20 (Steep)	4	
	>20 (Very steep)	5	
Soil type	Gravelly Loamy to Loamy Skeletol (Low erosion potential)	1	9
	Gravel Loamy to Coarse Loamy (Moderate erosion potential)	2	
	Gravel Loamy (High erosion potential)	3	
	Fine Loamy to Corase Loamy (Very high)	4	
	Coarse Loamy (Extreme)	5	
Vegetation cover (NDVI)	>0.51 (Dense)	1	10
	0.36–0.50 (Moderate)	2	
	0.21–0.35 (Sparse)	3	
	0–0.2 (Bare)	4	
	<0 (Degraded land)	5	

Table 5 Areal distribution of rill and gully potential area in Lish and Gish catchment area

Rill-gully potential	Area (in km^2)	Percent
Very low	58.60	20.24
Low	56.73	19.59
Moderate	35.73	12.34
High	81.94	28.30
Very high	56.54	19.53

avoided by using sustainable land management techniques. Higher runoff, moderate slopes, and sporadic plant cover make the Moderate Susceptibility Zone (35.73 km^2, 12.34%) rather vulnerable to gully and rill erosion. The medium erodibility of the soils in these regions is accelerated by the presence of agricultural practices or patches of deforested land. Steep slopes, a high dissection index, and little vegetation cover all contribute to the High Susceptibility Zone (81.94 km^2, 28.30%), which indicates vigorous rill and gully formation. Deforestation, intensive human activity, and inappropriate land-use practices all contribute to soil erosion. High drainage density and steep gradient slopes allow surface runoff to occur quickly, resulting in the creation of distinct gullies and rills.

The most susceptible area to severe rill and gully erosion is the Very High Susceptibility Zone (56.54 km^2, 19.53%), which is defined by steep slopes (>30°), high relative relief, loose soil structure, and little vegetation cover. Gully extension and deep incision are accelerated by heavy rainfall and swift flow. To prevent further

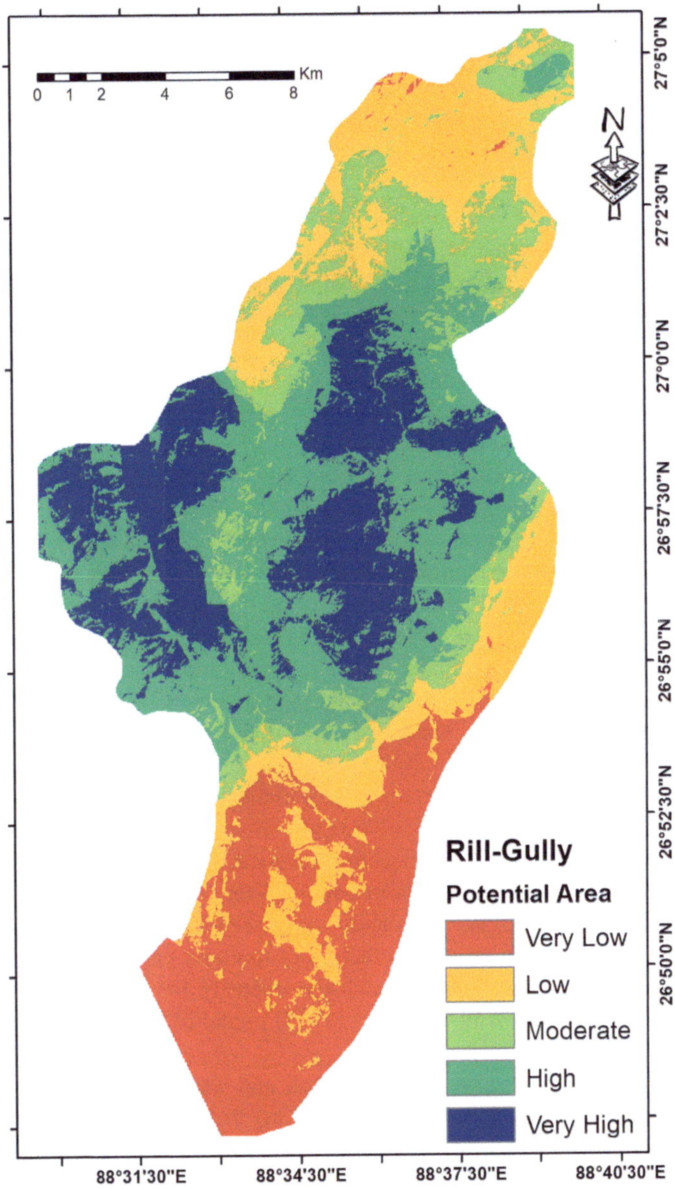

Fig. 7 Geographical variation of weighted overlay analysis of the Lish and Gish catchment area

deterioration, these areas urgently need watershed management and soil protection measures.

5 Discussion

Soil erosion in the form of rill and gully development is a major environmental challenge, leading to land degradation, sediment deposition, and loss of agricultural productivity. The identification of erosion hotspots is essential for effective land management and conservation planning. In the Lish and Gish catchment, remote sensing (RS) and geographic information systems (GIS) were utilized to assess erosion susceptibility based on hydrological, relief, and land use factors.

Soil erosion is mainly driven by rainfall since it influences surface runoff and sediment transport. The middle part of the catchment area receives very high rainfall (>3600 mm), which consequently increases runoff and subsequent gully formation. The southern and northern parts, where moderate to low rainfall occurs, are relatively less erodible. This finding is conformable with previous works (Wischmeier & Smith, 1978), which indicated that higher intensity of rainfall is directly proportional to the severity of erosion. It has been revealed that a place with a high drainage frequency (>6 streams/km^2) and wide drainage density (>4 km/km^2) mostly towards the central and western sides yields much rill and gully type of erosion. Such places have high concentration of runoff which accelerates the process of erosion. The northern and lower southern parts, with medium to low drainage frequency and lower drainage density, are less susceptible by virtue of their relative position in the basin. This has been corroborated by Poesen et al. (2003) who noted that higher densities lead to more surface water movement and thus more erosion. The 1st and 2nd order streams being dominant in central and western parts indicate that active erosion is taking place because there is high runoff velocity and unregulated flow. The higher-order streams (3rd and 4th order) in the northern and southern parts can be construed as offering better stability for channels and less intensity for erosion processes; this has considerable support from Strahler's (1957) model of stream hierarchy related to potential for erosion.

Erosion intensity is observed in the central and western regions and is highly correlated with high relative relaxation (>300 m) that increases gravity force and accelerates runoff and soil shifts. Moderate RR values (100°3) in the northern region indicate low sensitivity, while the southern part of the low RR (<100 m) remains relatively stable. These results are correct in the study of Sharda et al. (2013) found that higher RR values were strongly associated with heavy gully erosion. DI values are low in the north and south, with fewer average off-road sections, and medium sized DI values dominate the central and western areas. High DI bags in the central region show active gully development. Similarly, in contrast to high RI values in small pockets in the central catchment area, low RI values in the south are implied by landlords that contribute to local erosion. These observations are consistent with the study by Morgan (2009) and combined index values with higher roughness and greater sensitivity to erosion. The best climbing (>15°) concentrates in the central and western areas, accelerating drainage and strengthening the rill and gully formations. Medium slopes in the north and central areas (3°10°) exhibit moderate erosion activity, while lower slopes (<3°) exhibit stable conditions in the south. These results are in line with the USLE model of the universal soil loss equation (Wischmeier & Smith, 1978) that emphasizes the slope as the main erosion coefficient.

Soil composition has a major effect on erosion. This study shows that coarse loam soils in the south are relatively resistant to erosion, while the gravel clay blocks are highly sensitive to loamy beds in the central and northern skeletal beds. The western and northern regions have gravel loams and some loamy soils that are too coarse, but small pouches in the catchment contain fine loamy clay soils. A study by ICAR-NBSS & LUP (2019) confirms that skeletal and gravel floors are susceptible to engulfing erosion due to loose structure and poor moisture retention. Vegetation acts as a natural barrier to erosion. NDVI analysis shows low values in the southern region. This indicates a high risk of sparse vegetation and erosion. Medium NDVI values in the eastern and northern catchment areas indicate moderate vegetation ranges, whereas high NDVI values in the central and western areas indicate Dichsen vegetation, which reduces the risk of erosion. These results were shown in Sharda et al. (2013) highlights the role of vegetation in soil surface stabilization and reduces drainage-induced erosion.

6 Conclusion and Way Forward

Using remote sensing and GIS, studies on mapping of rill and gully erosion hotspots in the Lish and Gish catchments provided important insights into erosion areas by integrating hydrology, fatigue and land use factors. The results show that high intensity, high strength, high drainage frequency, density and low power rainfall in the central and western parts of the catchment contribute to heavy rill and gully erosion. Similarly, further soil erosion occurs due to increased drainage rates due to steep slopes, high relative mitigation, and increased indicators of dissection and roughness. The role of soil type and vegetation characteristics is also clear: gravel and clay skeletal soils are susceptible to erosion, and areas with dense vegetation have a lower risk of erosion. These results are consistent with previous testing and increase the need for target erosion control strategies.

In the future, sustainable practices in surface catchments should be prioritized to mitigate soil erosion in identified hot spots. Afforestation and Agroforestry initiatives can improve vegetation coverage and reduce surface runoff and soil shifts. Implementing soil protection measures such as contours, dams, and nutrient filter strips can stabilize highly erosible areas. Additionally, the integration of advanced geospatial models and machine learning technologies can improve erosion prediction and actual time monitoring, and support proactive decision-making. Further research should also focus on the impact of climate change on erosion dynamics for the development of adaptive management strategies. Combining remote sensations allows political decision makers and stakeholders to ensure GIS and field intervention in sustainable land use plans, with the long-term impact of lease and rill and gully erosion in the Lish and Gish catchment.

References

Borrelli, P., et al. (2014). Use of remote sensing and GIS for mapping soil erosion risk. *Geoderma, 223*, 145–156.
Government of India. (2020). *Watershed development guidelines*. Ministry of Rural Development.
Hancock, G. R., et al. (2008). Gully erosion prediction using GIS and digital elevation models. *Earth Surface Processes and Landforms, 33*(2), 217–234.
Huang, X., et al. (2020). Deep learning for gully erosion detection using high-resolution satellite imagery. *Remote Sensing, 12*(10), 1652.
Martínez-Casasnovas, J. A., et al. (2009). Topographic threshold conditions for gully initiation in agricultural fields. *Earth Surface Processes and Landforms, 34*(5), 677–686.
Morgan, R. P. C. (2005). *Soil erosion and conservation* (3rd ed.). Oxford: Blackwell Publishing.
Morgan, R. P. C. (2009). *Soil erosion and conservation*. Wiley & Sons.
Nyssen, J., et al. (2006). Gully erosion in the Ethiopian highlands: Impact and control measures. *CATENA, 65*(2), 131–141.
Nyssen, J., Poesen, J., Moeyersons, J., Luyten, E., Veyret-Picot, M., Deckers, J., & Haile, M. (2006). Spatial and temporal variability of rainfall and implications on runoff and soil loss in Northern Ethiopia. *Catena, 70*(3), 293–305. https://doi.org/10.1016/j.catena.2006.01.003
Pandey, A., Himanshu, S. K., Mishra, S. K., & Singh, V. P. (2016). Physically-based soil erosion and sediment yield models revisited. *CATENA, 147*, 595–620.
Poesen, J., et al. (2003). Gully erosion and environmental change: Importance and research needs. *CATENA, 50*(2–4), 91–133.
Rahmati, O., et al. (2017). Application of machine learning models for mapping gully erosion susceptibility. *Geomorphology, 298*, 118–137.
Sharda, V. N., Dogra, P., & Prakash, C. (2013). Assessment of global soil erosion trends and mitigation measures: A review. *Indian Journal of Soil Conservation, 41*(1), 1–14.
Strahler, A. N. (1957). Quantitative analysis of watershed geomorphology. *EOS, Transactions American Geophysical Union, 38*(6), 913–920.
Valentin, C., Poesen, J., & Li, Y. (2005). Gully erosion: Impacts, factors and control. *Catena, 63*(2–3), 132–153. https://doi.org/10.1016/j.catena.2005.06.001
Vrieling, A., et al. (2007). Mapping gully erosion using high-resolution imagery and digital terrain analysis. *Land Degradation & Development, 18*(2), 159–173.
Wischmeier, W. H., & Smith, D. D. (1978). *Predicting rainfall Erosion losses—a guide to conservation planning*. USDA Agriculture Handbook.